AF743732

BD 034051127 4

Continuous Bivariate Distributions, Emphasising Applications

Continuous Bivariate Distributions, Emphasising Applications

T. P. Hutchinson
Department of Civil Engineering,
University of Adelaide,
South Australia

and

C. D. Lai
Department of Mathematics and Statistics,
Massey University, Palmerston North,
New Zealand

Rumsby Scientific Publishing

P.O. Box 76, Rundle Mall, Adelaide, South Australia 5000

© Copyright 1990 T.P.Hutchinson.
First published January 1990.
Available from: Rumsby Scientific Publishing, P.O. Box 76, Rundle Mall, Adelaide, South Australia 5000.
ISBN 0 7316 7206 2.

UNIVERSITY OF BRADFORD
WITHDRAWN AND DISCARDED
FEB 1994
0340511274 L519.237 HUT
LIBRARY

Acknowledgements

The authors thank those who have helped their researches and writing, whether in person or through correspondence. T.P.H. is grateful to the Department of Mathematics and Statistics, Massey University, for hospitality which enabled him to collaborate with C.D.L., and to Mrs. Julie Freund for assistance with the indexing. C.D.L. wishes to express his appreciation of the encouragement and support of his wife and children, and also acknowledges the many helpful comments he has had from Dr. Terry Moore.

List of Chapters

I	**Background**	**1**
1	Introduction	3
2	Univariate Distributions	9
3	Overview of Methods of Constructing Bivariate Distributions	31
II	**Families of Continuous Bivariate Distributions**	**45**
4	The Bivariate Normal Distribution	47
5	Distributions Expressed as Copulas	76
6	Simple Forms for the Bivariate Density Function	96
7	Further Distributions Related to the Bivariate Normal	108
8	Bivariate Distributions Constructed by the Conditional Approach	127
9	Bivariate Distributions Arising in Reliability Contexts	138
10	Miscellaneous Bivariate Distributions	165
III	**Mostly Theory**	**175**
11	Measures of Correlation	177
12	Concepts of Stochastic Dependence	193
13	Rank Correlation Properties	202
14	Diagonal Expansion	211
15	Variables in Common	218
16	Overview of Generating Random Variates	225
17	Reliability Models Leading to Bivariate Distributions	235
18	Reliability Classes: Concepts of Ageing	243
IV	**Mostly Applications**	**251**
19	Applications of the Bivariate Normal Distribution	253
20	Construction of the Bivariate Pareto Distribution by Compounding, with Applications	268
21	Joint Distributions of Injury Severity: A Bivariate Normal Model	282
22	Bivariate Generalisations of Probit Analysis	292
23	Bivariate Models in Signal Detection Theory	300
24	Examples of Data	311
	References	**361**
	Index	**405**

Contents

I Background 1

1 Introduction 3
1.1 Opening Statement 3
1.2 Contents of Part I 4
1.3 Description of the Structure of this Book 4
1.4 What this Book Emphasises, What it Omits 4
1.5 Explanation of Conventions, Notations, and Abbreviations . . 5
1.5.1 Abbreviations 6
1.5.2 Conventions 6
1.5.3 Symbols 6
1.5.4 Some Functions 7
1.6 The Future of the Subject 7

2 Univariate Distributions 9
2.1 Introduction 9
2.2 Notation and Definitions 10
2.2.1 Notation 10
2.2.2 Explanations 10
2.2.2.1 Moment Generating Function 10
2.2.2.2 Characteristic Function 11
2.2.2.3 Cumulant Generating Function 11
2.2.2.4 Interrelationships of Moments and Cumulants 11
2.3 Generalities About Univariate Shape 11
2.3.1 Location and Scale 11
2.3.2 Skewness and Kurtosis 12
2.3.3 Tail Behaviour 12
2.3.4 Some Multi-Parameter Systems of Univariate Distributions 12
2.3.5 Reliability Classes 12
2.4 Normal Distribution, and Transformations 13
2.4.1 Normal Distribution 13
2.4.2 Lognormal Distribution 13
2.4.3 Johnson's System 13
2.4.4 Box-Cox Power Transformations to Normality 14
2.4.5 The g and h Family 14
2.4.6 Efron (1982a) 14
2.4.7 Distributions of a Ratio 14
2.4.8 Compound Normal Distributions 14
2.5 Beta Distributions 14
2.5.1 The First Kind 14
2.5.2 Uniform Distribution 15

	2.5.3	Inverted Beta Distribution	16
2.6	Exponential, Gamma, Weibull, and Stacy Distributions		16
	2.6.1	Exponential Distribution	16
	2.6.2	Gamma Distribution	16
	2.6.3	Chi-Squared and Chi Distributions	17
	2.6.4	Weibull Distribution	17
	2.6.5	Stacy Distribution	17
	2.6.6	Comments on Skew Distributions	18
	2.6.7	Compound Exponential Distributions	18
2.7	Logistic, Laplace, and Cauchy Distributions		18
	2.7.1	Logistic Distribution	18
	2.7.2	Laplace Distribution	18
	2.7.3	The Generalised Error Distribution	19
	2.7.4	Cauchy Distribution	19
2.8	Extreme-Value Distributions		19
	2.8.1	Type I	19
	2.8.2	Types II and III	19
2.9	Pareto Distribution		19
2.10	The Pearson System		20
2.11	The Burr System		20
2.12	t- and F-Distributions		21
	2.12.1	t-Distribution	21
	2.12.2	F-Distribution	21
2.13	Noncentral Distributions		21
2.14	Some Lesser-Known Distributions		22
	2.14.1	Inverse Gaussian Distribution	22
	2.14.2	Meixner Hypergeometric Distribution	22
	2.14.3	Hyperbolic Distributions	22
	2.14.4	Stable Distributions	22
2.15	What's in the Encyclopaedia?		22
	2.15.1	Named Univariate Distributions	23
	2.15.2	Theory and Concepts	25
	2.15.3	Specialised Areas of Application	29
2.16	What's in Patil et al's Bibliography?		30
3	**Overview of Methods of Constructing Bivariate Distributions**		**31**
3.1	Introduction		31
	3.1.1	Contents of this Chapter	31
	3.1.2	The Fréchet Bounds	31
	3.1.3	Transformations	32
	3.1.4	Computer Algebra	32
3.2	The Marginal Transformation Method		33
	3.2.1	General Description	33
	3.2.2	Johnson's Translation Method	33
	3.2.3	Uniform Representation	34
	3.2.4	Some Properties Unaffected by Transformation	34
	3.2.5	Example	35
3.3	Some Inelegant Methods		35
3.4	Mixing and Compounding		35
	3.4.1	Mixing	35
	3.4.2	Compounding	36
3.5	Regression and Conditional Distribution		37
3.6	Trivariate Reduction		37

3.7 Weighted Linear Combination 38
3.8 Limits of Discrete Distributions 38
3.8.1 The Bivariate Bernoulli Distribution 38
3.8.2 A Bivariate Exponential Distribution 38
3.8.3 A Bivariate Gamma Distribution 39
3.9 Differential Equation Methods 39
3.9.1 Pearson's System 39
3.9.1.1 Derivation of Equation 39
3.9.1.2 Case 1 40
3.9.1.3 Case 2 40
3.9.1.4 Case 3 40
3.9.1.5 Remarks 41
3.9.2 The Fokker-Planck Equation 41
3.9.3 The Ali-Mikhail-Haq Distribution 41
3.10 Diagonal Expansion 41
3.11 Bivariate Edgeworth Expansion 41
3.11.1 Univariate Edgeworth, Gram-Charlier, and Cornish-Fisher Expansions 41
3.11.2 Bivariate Edgeworth Series 42
3.11.3 An Application: Wind Velocity at the Ocean Surface 43
3.11.4 A Second Application: Statistical Spectroscopy . . . 43

II Families of Continuous Bivariate Distributions 45

4 The Bivariate Normal Distribution 47
4.1 Organisation of Part II 47
4.2 Introduction to Chapter 4 47
4.3 Basic Formulae 49
4.3.1 Notation 49
4.3.2 Support 49
4.3.3 Formula for p.d.f. 49
4.3.4 Formula for (Cumulative) Distribution Function . . . 49
4.3.5 Univariate Properties 49
4.3.6 Correlation Coefficients 49
4.3.7 Conditional Properties 49
4.3.8 Moments and Absolute Moments 49
4.4 Methods of Derivation 50
4.4.1 Differential Equation Method 50
4.4.2 Compounding 50
4.4.3 Trivariate Reduction 50
4.4.4 Bivariate Central Limit Theorem 50
4.5 Characterisations 51
4.6 Relation to Other Distributions 51
4.7 References to Illustrations 51
4.8 References to the Generation of Random Variates 52
4.9 Other Interesting Properties 52
4.10 Notes on Some More Specialised Fields 53
4.11 Applications 53
4.12 Computation of Bivariate Normal Integrals 53
4.12.1 The Short Answer 53
4.12.2 Algorithms—Rectangles 54
4.12.3 Algorithms—Triangles 55

4.12.4 Algorithms—Owen's T Function 55
4.12.5 Algorithms—Wedge-Shaped Domain 56
4.12.6 Algorithms—Arbitrary Polygons 57
4.12.7 Tables 57
4.12.8 Computer Programs 57
4.12.9 Literature Reviews 58
4.13 Testing for Bivariate Normality 58
4.13.1 How Might Bivariate Normality Fail? 58
4.13.2 Outliers 58
4.13.3 Graphical Checks 59
4.13.3.1 Univariate Plotting 59
4.13.3.2 Scatterplots 59
4.13.3.3 Radii and Angles 60
4.13.3.4 Projection Pursuit 60
4.13.3.5 The Kernel Method 60
4.13.3.6 Haar Distributions 61
4.13.4 Formal Tests—Univariate Normality 61
4.13.4.1 Why Test? 61
4.13.4.2 Chi-Squared Test 61
4.13.4.3 Moment Tests 62
4.13.4.4 Tests Based on the Empirical Distribution Function 62
4.13.4.5 Probability Plots 62
4.13.4.6 The CPIT Approach 62
4.13.5 Formal Tests—Bivariate Normality 62
4.13.5.1 Chi-Squared Test 62
4.13.5.2 Tests Based on the Empirical Distribution Function 63
4.13.5.3 Tests Based on the Empirical Characteristic Function 63
4.13.5.4 Malkovich and Afifi (1973) 63
4.13.5.5 Cox and Small (1978) 63
4.13.5.6 Hawkins (1981) 63
4.13.5.7 Bera and John (1983) 63
4.13.5.8 Bivariate Skewness and Kurtosis 64
4.13.5.9 Best and Rayner's Comparisons 64
4.13.5.10 Asymptotically χ^2_2? 65
4.13.6 Tests of Bivariate Normality After Transformations . 65
4.13.7 Some Comments and Suggestions 66
4.14 Univariate Transformations 67
4.14.1 The Bivariate Lognormal Distribution 67
4.14.2 Johnson's System 68
4.14.2.1 Derivation 68
4.14.2.2 Formula for p.d.f. 68
4.14.2.3 Univariate Properties 68
4.14.2.4 Conditional Properties 68
4.14.2.5 References to Illustrations 68
4.14.2.6 Applications 68
4.14.3 The Uniform Representation 69
4.14.4 The g and h Transformation 69
4.14.5 Effect of Transformations on Correlation 69
4.15 The Logistic-Normal Distribution 70
4.16 Truncated Bivariate Normal Distributions 71

4.16.1 Properties 71
4.16.2 Application to Selection Procedures 71
4.17 Bivariate Normal Mixtures 72
4.17.1 Construction 72
4.17.2 References to Illustrations 73
4.17.3 Applications 73
4.17.4 Generalisation and Compounding 73
4.17.5 Properties of a Special Case 73
4.17.6 Sharpening a Scatterplot 74
4.17.7 Digression Analysis 74
4.18 Some Distributions Constructed for Pedagogic Purposes 74
4.18.1 Simple Examples with Normal Marginals 74
4.18.2 Normal Marginals with Linear Regressions 75
4.18.3 Linear Combinations of Normal Marginals 75
4.18.4 Uncorrelated Non-Normal Bivariate Distributions with Normal Marginals 75
4.19 Cross-References to Other Distributions with Normal Marginals 75

5 Distributions Expressed as Copulas **76**

5.1 Introduction 76
5.1.1 Outline of this Chapter 76
5.1.2 Three Explanations 77
5.1.3 Five Sources of Illustrations 77
5.1.4 Three Sources of Algorithms for Random Variate Generation 78
5.2 Farlie-Gumbel-Morgenstern and Related Distributions 78
5.2.1 Support 78
5.2.2 Formula for (Cumulative) Distribution Function . . . 78
5.2.3 Formula for p.d.f. 78
5.2.4 Univariate Properties 78
5.2.5 Correlation Coefficients 78
5.2.6 Conditional Properties 78
5.2.7 References to Illustrations 79
5.2.8 References to the Generation of Random Variates . . 79
5.2.9 Further Results 79
5.2.10 Remarks 79
5.2.11 Applications 80
5.2.12 Univariate Transformations 80
5.2.13 Iterated F-G-M Distributions 80
5.2.14 A Switched-Source Model 81
5.2.15 Ordinal Contingency Tables 81
5.2.16 Other Related Distributions 81
5.3 Ali-Mikhail-Haq Distribution 82
5.3.1 Support 82
5.3.2 Formula for (Cumulative) Distribution Function . . . 82
5.3.3 Formula for p.d.f. 82
5.3.4 Univariate Properties 82
5.3.5 Correlation Coefficients 82
5.3.6 Derivation 82
5.3.7 References to Illustrations 82
5.3.8 References to the Generation of Random Variates . . 83
5.3.9 Remarks 83
5.4 Frank's Distribution 83

5.4.1 Support . . . 83
5.4.2 Formula for (Cumulative) Distribution Function . . . 83
5.4.3 Formula for p.d.f. . . . 83
5.4.4 Univariate Properties . . . 83
5.4.5 Correlation Coefficients . . . 83
5.4.6 Derivation . . . 84
5.4.7 References to Illustrations . . . 84
5.4.8 References to the Generation of Random Variates . . . 84
5.4.9 Remarks . . . 84
5.5 Distribution of Cuadras and Augé . . . 84
5.6 Gumbel-Hougaard Copula . . . 84
5.6.1 Support . . . 84
5.6.2 Formula for (Cumulative) Distribution Function . . . 84
5.6.3 Univariate Properties . . . 84
5.6.4 Correlation Coefficients . . . 85
5.6.5 Derivation . . . 85
5.6.6 References to Illustrations . . . 85
5.6.7 Remarks . . . 85
5.6.8 Fields of Application . . . 85
5.6.9 Transformations of the Marginals . . . 86
5.6.9.1 Exponential . . . 86
5.6.9.2 Weibull . . . 86
5.6.9.3 Logistic . . . 86
5.6.9.4 Extreme-Value . . . 86
5.7 Plackett's Distribution . . . 86
5.7.1 Support . . . 86
5.7.2 Formula for (Cumulative) Distribution Function . . . 86
5.7.3 Formula for p.d.f. . . . 86
5.7.4 Univariate Properties . . . 87
5.7.5 Correlation Coefficients . . . 87
5.7.6 Conditional Properties . . . 87
5.7.7 References to Illustrations . . . 87
5.7.8 References to the Generation of Random Variates . . . 87
5.7.9 Remarks . . . 87
5.7.10 Fields of Application . . . 87
5.8 Bivariate Pareto Distribution . . . 88
5.8.1 Support . . . 88
5.8.2 Formula for (Cumulative) Distribution Function . . . 88
5.8.3 Formula for p.d.f. . . . 88
5.8.4 Univariate Properties . . . 88
5.8.5 Correlation Coefficients . . . 88
5.8.6 Conditional Properties . . . 88
5.8.7 Derivation . . . 88
5.8.8 References to Illustrations . . . 89
5.8.9 References to the Generation of Random Variates . . . 89
5.8.10 Remarks . . . 89
5.8.11 Applications . . . 89
5.8.12 Univariate Transformations . . . 89
5.8.13 The Durling-Pareto Distribution . . . 89
5.8.13.1 Formula for (Cumulative) Distribution Function . . . 89
5.8.13.2 Derivation . . . 90
5.8.13.3 Remarks . . . 90

5.8.13.4 The Special Case $c = 1$ 90
5.8.13.5 A Related Distribution 90
5.8.14 Pareto Conditionals 90
5.8.15 Generalisation 90
5.9 Pareto Copula 91
5.9.1 Support 91
5.9.2 Formula for (Cumulative) Distribution Function . . . 91
5.9.3 References to Illustrations 91
5.9.4 References to the Generation of Random Variates . . 91
5.9.5 Remarks 91
5.9.6 Fields of Application 91
5.10 Gumbel's Bivariate Exponential Distribution, Type I 92
5.10.1 Support 92
5.10.2 Formula for (Cumulative) Distribution Function . . . 92
5.10.3 Formula for p.d.f. 92
5.10.4 Univariate Properties 93
5.10.5 Correlation Coefficients 93
5.10.6 Conditional Properties 93
5.10.7 References to Illustrations 93
5.10.8 References to the Generation of Random Variates . . 93
5.10.9 Remarks 93
5.10.10 An Application 94
5.11 Gumbel-Barnett Copula 94
5.11.1 Reason for the Name 94
5.11.2 Support 94
5.11.3 Formula for (Cumulative) Distribution Function . . . 94
5.11.4 Formula for p.d.f. 94
5.11.5 References to Illustrations 94
5.11.6 References to the Generation of Random Variates . . 94
5.11.7 Remark 95

6 Simple Forms for the Bivariate Density Function 96
6.1 Introduction 96
6.2 Bivariate t-Distribution 97
6.2.1 Support 97
6.2.2 Formula for p.d.f. 97
6.2.3 Univariate Properties 97
6.2.4 Correlation Coefficients 97
6.2.5 Moments 97
6.2.6 Conditional Properties 97
6.2.7 Derivation 97
6.2.8 References to Illustrations 98
6.2.9 References to the Generation of Random Variates . . 98
6.2.10 Remarks 98
6.2.11 Fields of Application 98
6.2.12 References to Tables or Algorithms 98
6.2.13 Bivariate Cauchy Distribution 99
6.2.14 Generalisations 99
6.3 Bivariate F-Distribution 99
6.3.1 Support 99
6.3.2 Formula for p.d.f. 99
6.3.3 Formula for (Cumulative) Distribution Function . . . 100
6.3.4 Univariate Properties 100

6.3.5 Correlation Coefficients 100
6.3.6 Moments 100
6.3.7 Conditional Properties 100
6.3.8 Methods of Derivation 100
6.3.9 Relation to Other Distributions 100
6.3.10 Fields of Application 101
6.3.11 References to Tables or Algorithms 101
6.4 Bivariate Pearson Type II Distribution 101
6.4.1 Support 101
6.4.2 Formula for p.d.f. 101
6.4.3 Univariate Properties 101
6.4.4 Correlation Coefficients 101
6.4.5 Conditional Properties 101
6.4.6 Relation to Other Distributions 102
6.4.7 References to Illustrations 102
6.4.8 References to the Generation of Random Variates . . 102
6.4.9 Remarks 102
6.4.10 References to Tables or Algorithms 102
6.5 Bivariate Beta Distribution 102
6.5.1 Support 102
6.5.2 Formula for p.d.f. 102
6.5.3 Univariate Properties 102
6.5.4 Correlation Coefficients 102
6.5.5 Moments 103
6.5.6 Conditional Properties 103
6.5.7 Methods of Derivation 103
6.5.8 Relation to Other Distributions 103
6.5.9 References to Illustrations 103
6.5.10 References to the Generation of Random Variates . . 103
6.5.11 Remarks 103
6.5.12 Fields of Application 104
6.5.13 References to Tables or Algorithms 104
6.5.14 Generalisations 104
6.5.15 Bivariate Inverted Beta Distribution 105
6.5.15.1 Support 105
6.5.15.2 Formula for p.d.f. 105
6.5.15.3 Formula for (Cumulative) Distribution Function 105
6.5.15.4 Generalisation 105
6.5.15.5 Further Information 105
6.6 Bivariate Liouville-Dirichlet Distribution 106
6.7 Bivariate Logistic Distributions 106
6.8 Bivariate Burr Distributions 106
6.9 Rhodes' Distribution 107
6.9.1 Support 107
6.9.2 Formula for p.d.f. 107
6.9.3 Derivation 107
6.9.4 Remarks 107

7 Further Distributions Related to the Bivariate Normal **108**
7.1 Introduction 108
7.2 A Class of Elliptically-Contoured Bivariate Densities 108
7.2.1 The Growing Importance of this Class 108
7.2.2 Formula for p.d.f. 109
7.2.3 Alternative Definition 109
7.2.4 Well-Known Members of this Family 109
7.2.5 Formula for (Cumulative) Distribution Function . . . 109
7.2.6 Characteristic Function 109
7.2.7 Conditional Properties 109
7.2.8 Derivation 110
7.2.9 References to the Generation of Random Variates . . 110
7.2.10 Remarks 110
7.2.11 Fields of Application 110
7.2.12 Elliptical Compound Bivariate Normal Distributions 111
7.3 Nachtsheim and Johnson (1988) 111
7.4 Bivariate Exponential Power Distributions 111
7.4.1 Support 111
7.4.2 Formula for p.d.f. 112
7.4.3 Univariate Properties 112
7.4.4 Further Properties 112
7.4.5 Remark 112
7.5 Kibble's Bivariate Gamma Distribution 112
7.5.1 Support 112
7.5.2 Formula for p.d.f. 112
7.5.3 Formula for (Cumulative) Distribution Function . . . 113
7.5.4 Univariate Properties 113
7.5.5 Correlation Coefficients 113
7.5.6 Moment Generating Function 113
7.5.7 Conditional Properties 113
7.5.8 Derivation 114
7.5.9 Relation to Other Distributions 114
7.5.10 Generalisations 114
7.5.11 References to Illustrations 114
7.5.12 Remarks 114
7.5.13 Fields of Application 115
7.5.14 References to Tables or Algorithms 115
7.5.15 Transformations of the Marginals 115
7.6 Noncentral Bivariate Chi-Squared Distribution 115
7.7 Jensen's Bivariate Gamma Distribution 116
7.7.1 Support 116
7.7.2 Formula for p.d.f. 116
7.7.3 Univariate Properties 116
7.7.4 Correlation Coefficients 116
7.7.5 Characteristic Function 117
7.7.6 Derivation 117
7.7.7 References to Illustrations 117
7.7.8 Remarks 117
7.7.9 Fields of Application 117
7.7.10 References to Tables or Algorithms 117
7.7.11 Gunst and Webster (1973) and Related Distributions 118
7.7.12 Interrelationships 119
7.8 Jensen's Bivariate F-Distribution 119

7.8.1 Support 119
7.8.2 Formula for p.d.f. 119
7.8.3 Formula for (Cumulative) Distribution Function . . . 119
7.8.4 Univariate Properties 120
7.8.5 Derivation 120
7.8.6 Relation to Other Distributions 120
7.8.7 Field of Application 120
7.8.8 References to Tables or Algorithms 120
7.9 The Bivariate Inverse Gaussian Distribution 120
7.9.1 Support 120
7.9.2 Formula for p.d.f. 121
7.9.3 Univariate Properties 121
7.9.4 Correlation Coefficients 121
7.9.5 Conditional Properties 121
7.9.6 Derivation 121
7.9.7 References to Illustrations 121
7.9.8 Remarks 121
7.10 The Bivariate Hyperbolic Distribution 122
7.10.1 Support 122
7.10.2 Formula for p.d.f. 122
7.10.3 Univariate Properties 122
7.10.4 Derivation 122
7.10.5 References to Illustrations 123
7.10.6 Remarks 123
7.10.7 Fields of Application 123
7.11 Bivariate Symmetric Stable Distributions 123
7.11.1 Explanations 123
7.11.2 Characteristic Function 124
7.11.3 Probability Densities 124
7.11.4 Generalisation 124
7.11.5 Association Parameter 124
7.11.5.1 Press (1972a) 124
7.11.5.2 de Silva (1978) 125
7.11.6 Remarks 125
7.11.7 Application 125

8 Bivariate Distributions Constructed by the Conditional Approach 127
8.1 Introduction 127
8.1.1 Contents of this Chapter 127
8.1.2 Approximating d.f.'s Using the Conditional Approach 127
8.1.3 Compatibility of Conditional Densities 127
8.1.4 The Marginal and Conditional Distributions of the Same Variate 127
8.2 Exponential Conditionals 128
8.2.1 Support 128
8.2.2 Formula for p.d.f. 128
8.2.3 Formula for (Cumulative) Distribution Function . . . 128
8.2.4 Univariate Properties 128
8.2.5 Correlation Coefficients 128
8.2.6 Relation to Other Distributions 128
8.2.7 Remarks 128
8.2.8 Fields of Application 129

8.3 Pareto Conditionals . . . 129
8.3.1 Support . . . 129
8.3.2 Formula for p.d.f. . . . 129
8.3.3 Univariate Properties . . . 129
8.3.4 Remarks . . . 129
8.4 Beta Conditionals . . . 129
8.4.1 Support . . . 129
8.4.2 Formula for p.d.f. . . . 129
8.4.3 Univariate Properties . . . 129
8.4.4 Remarks . . . 129
8.4.5 Another Distribution . . . 130
8.5 Normal Conditionals . . . 130
8.6 McKay's Bivariate Gamma Distribution . . . 130
8.6.1 Support . . . 130
8.6.2 Formula for p.d.f. . . . 130
8.6.3 Formula for (Cumulative) Distribution Function . . . 130
8.6.4 Univariate Properties . . . 130
8.6.5 Correlation Coefficients . . . 130
8.6.6 Conditional Properties . . . 130
8.6.7 Methods of Derivation . . . 130
8.6.8 References to Illustrations . . . 131
8.6.9 Remarks . . . 131
8.6.10 Dussauchoy and Berland's Bivariate Gamma Distribution . . . 131
8.6.11 Some Variants of this Distribution . . . 131
8.7 Dubey's Distribution . . . 132
8.8 Blumen and Ypelaar's Distribution . . . 132
8.8.1 Support . . . 132
8.8.2 Formula for p.d.f. . . . 132
8.8.3 Univariate Properties . . . 132
8.8.4 Conditional Properties . . . 132
8.8.5 Remark . . . 132
8.9 Exponential Dispersion Models . . . 132
8.10 Four Densities Given by Barndorff-Nielsen and Blæsild . . . 132
8.11 Continuous Bivariate Densities with a Discontinuous Marginal Density . . . 133
8.12 The Conditional Approach in Modelling . . . 133
8.12.1 Beta-Stacy Distribution . . . 133
8.12.2 Sample Skewness and Kurtosis . . . 133
8.12.3 Business Risk Analysis . . . 134
8.12.4 Intercropping . . . 134
8.12.5 Winds and Waves, Rain and Floods . . . 135
8.12.5.1 Height and Period of Waves of the Sea . . . 135
8.12.5.2 Wind Speeds . . . 135
8.12.5.3 Wind Speed and Wave Height . . . 136
8.12.5.4 Storm Surge and Wave Height . . . 136
8.12.5.5 Floods . . . 136
8.12.5.6 Streamflow and Rain . . . 136
8.12.5.7 Rain . . . 136

9 Bivariate Distributions Arising in Reliability Contexts **138**
9.1 Introduction 138
9.2 Marshall and Olkin's Bivariate Exponential Distribution 139
9.2.1 Support 139
9.2.2 Formula for (Cumulative) Distribution Function . . . 139
9.2.3 Formula for p.d.f. 139
9.2.4 Univariate Properties 139
9.2.5 Correlation Coefficients 139
9.2.6 Derivation 140
9.2.7 Remarks 140
9.2.8 Fields of Application 141
9.2.9 Transformation to Uniform Marginals 141
9.2.10 Transformation to Weibull Marginals 141
9.2.11 Transformation to Extreme-Value Marginals 141
9.2.12 Transformation of Marginals—The Approach of Muliere and Scarsini 141
9.2.13 Generalisations 142
9.3 Freund's Bivariate Distribution 142
9.3.1 Support 142
9.3.2 Formula for p.d.f. 143
9.3.3 Formula for (Cumulative) Distribution Function . . . 143
9.3.4 Univariate Properties 143
9.3.5 Correlation Coefficients 143
9.3.6 Conditional Properties 143
9.3.7 Moment Generating Function 143
9.3.8 Derivation 143
9.3.9 References to Illustrations 144
9.3.10 Remarks 144
9.3.11 Fields of Application 144
9.3.12 Transformation of the Marginals 144
9.3.13 Compounding 144
9.3.14 Becker and Roux's Generalisation 144
9.3.15 Bhattacharya and Holla's Generalisations 145
9.4 Hashino and Sugi's Distribution 145
9.4.1 Support 145
9.4.2 Formula for p.d.f. 145
9.4.3 Remarks 145
9.4.4 An Application 146
9.5 ACBVE of Block and Basu 146
9.5.1 Support 146
9.5.2 Formula for p.d.f. 146
9.5.3 Formula for (Cumulative) Distribution Function . . . 146
9.5.4 Univariate Properties 146
9.5.5 Correlation Coefficients 146
9.5.6 Moment Generating Function 147
9.5.7 Derivation 147
9.5.8 Remarks 147
9.5.9 Applications 147
9.6 Sarkar's Distribution 147
9.6.1 Support 147
9.6.2 Formula for p.d.f. 148
9.6.3 Formula for (Cumulative) Distribution Function . . . 148
9.6.4 Univariate Properties 148

9.6.5 Correlation Coefficient 148
9.6.6 Derivation 148
9.6.7 Relation to Marshall and Olkin's Distribution 148
9.7 Comparison of Four Distributions 148
9.8 The Generalisation by Friday and Patil 149
9.9 The Bivariate Distribution of Tosch and Holmes 149
9.10 The Method of Lawrance and Lewis 150
9.10.1 General Form 150
9.10.2 Model EP1 150
9.10.3 Model EP3 150
9.10.4 Model EP5 150
9.10.5 Models with Negative Correlation 151
9.10.6 Models with Uniform Marginals 151
9.10.7 Mixture Models 151
9.10.8 Models with Line Singularities 151
9.11 Raftery's Scheme 151
9.11.1 General Form 151
9.11.2 First Special Case 151
9.11.3 Second Special Case 152
9.11.4 Formula for p.d.f. 152
9.11.5 Derivation 152
9.11.6 References to Illustrations 152
9.11.7 Remarks 152
9.11.8 Applications 152
9.12 Downton's Bivariate Exponential Distribution 152
9.12.1 Support 152
9.12.2 Formula for p.d.f. 152
9.12.3 Formula for (Cumulative) Distribution Function . . . 153
9.12.4 Univariate Properties 153
9.12.5 Correlation Coefficients 153
9.12.6 Conditional Properties 153
9.12.7 Moment Generating Function 153
9.12.8 Regression 153
9.12.9 Derivation 153
9.12.10 References to Illustrations 154
9.12.11 Random Variate Generation 154
9.12.12 Remarks 154
9.12.13 Fields of Application 154
9.12.14 References to Tables or Algorithms 155
9.12.15 Weibull Marginals 155
9.12.16 A Bivariate Laplace Distribution 155
9.13 Bivariate Phase-Type Distributions 156
9.14 Cowan's Bivariate Exponential Distribution 156
9.14.1 Support 156
9.14.2 Formula for (Cumulative) Distribution Function . . . 156
9.14.3 Formula for p.d.f. 157
9.14.4 Univariate Properties 157
9.14.5 Correlation Coefficients 157
9.14.6 Conditional Properties 157
9.14.7 Derivation 157
9.14.8 References to Illustrations 157
9.14.9 Remarks 157
9.14.10 Transformation of the Marginals 157

9.15 Other Distributions with Exponential Marginals 157
9.16 Bivariate Weibull Distributions 158
9.17 Bivariate Extreme-Value Distributions 158
9.17.1 Preliminaries 158
9.17.2 Support 159
9.17.3 Formula for (Cumulative) Distribution Function . . . 159
9.17.4 Properties of the Dependence Function A 159
9.17.5 Differentiable Models 160
9.17.5.1 Example 1 160
9.17.5.2 Example 2 160
9.17.6 Non-Differentiable Models 160
9.17.6.1 Example 3 160
9.17.6.2 Example 4 160
9.17.6.3 Example 5 160
9.17.7 Tawn's Extension of Differentiable Models 160
9.17.7.1 Background 160
9.17.7.2 Example 6 161
9.17.7.3 Example 7 161
9.17.8 Correlations 161
9.17.9 Methods of Derivation 162
9.17.10 References to Illustrations 162
9.17.11 Applications 162
9.18 Competing Risks 163
9.18.1 Format of Data 163
9.18.2 Methods of Analysis 163
9.18.3 Unusual Applications 164

10 Miscellaneous Bivariate Distributions **165**
10.1 Contents of this Chapter 165
10.2 Multiplicative Trivariate Reduction 165
10.2.1 Introductory Explanation 165
10.2.2 Bryson and Johnson (1982) 166
10.2.3 Gokhale (1973) 166
10.2.4 Ulrich (1984) 166
10.3 Cherian's Bivariate Gamma Distribution 167
10.3.1 Formula for p.d.f. 167
10.3.2 Univariate Properties 167
10.3.3 Correlation Coefficients 167
10.3.4 Moment Generating Function 167
10.3.5 Conditional Properties 167
10.3.6 Derivation 167
10.3.7 References to the Generation of Random Variates . . 167
10.3.8 Remarks 167
10.3.9 Application 168
10.4 Weighted Linear Combination 168
10.4.1 Derivation 168
10.4.2 Formula for p.d.f. 168
10.4.3 Correlation Coefficients 168
10.4.4 Remark 168
10.5 Khintchine Mixture 168
10.5.1 Derivation 168
10.5.2 Exponential Marginals 169
10.5.3 Normal Marginals 169

10.5.4 References to the Generation of Random Variates . . 169
10.6 Fox's H-Function as a p.d.f. 169
10.7 Kimeldorf and Sampson's Bivariate Uniform Distribution . . . 170
10.7.1 Support 170
10.7.2 Formula for p.d.f. 170
10.7.3 Formula for (Cumulative) Distribution Function . . . 170
10.7.4 Univariate Properties 170
10.7.5 Correlation Coefficients 170
10.8 Densities for Wave Height and Period 170
10.9 A Distribution Having a Singularity Along $Y = X$ 171
10.9.1 The Distribution 171
10.9.2 Cohen's Kappa 171
10.9.3 Criticisms of Cohen's Kappa 172
10.9.4 Cohen's Normalisation of Kappa 172
10.10 The ARMA Context 173

III Mostly Theory **175**

11 Measures of Correlation **177**
11.1 Organisation of Part III 177
11.2 Introduction 177
11.3 Total Dependence 178
11.3.1 Functions 178
11.3.2 Mutual Complete Dependence 178
11.3.3 Monotone Dependence 179
11.3.4 Functional and Implicit Dependence 179
11.3.5 Overview 179
11.4 Conditions on a Measure of Dependence 179
11.5 Pearson's Product-Moment Correlation Coefficient 181
11.5.1 Definition 181
11.5.2 Interpretation of Correlation 182
11.5.3 Correlation Ratio 183
11.5.4 Chebyshev's Inequality 183
11.6 Maximal Correlation (Sup Correlation) 183
11.7 Monotone Correlation 184
11.7.1 Definition and Properties 184
11.7.2 Concordant and Discordant Monotone Correlations . 185
11.8 Rank Correlations 185
11.8.1 Introductory Explanation 185
11.8.2 Kendall's Tau 185
11.8.3 Grade Correlation 186
11.8.4 Remarks 186
11.9 Measure of Schweizer and Wolff, and Related Measures 187
11.10 Matrix of Correlations 188
11.11 Tetrachoric and Polychoric Correlation 188
11.12 Compatibility with Perfect Rank Ordering 189
11.13 Conclusions on Measures of Dependence 189
11.14 Local Dependence Function 190
11.14.1 Definition of Local Dependence 190
11.14.2 Properties of the Local Dependence Function 190
11.15 Regional Dependence 191
11.15.1 Preliminaries 191

11.15.2 Quasi-Independence and Quasi-Independent Projection 191
11.15.3 A Measure of Regional Dependence 192

12 Concepts of Stochastic Dependence **193**
12.1 Introduction 193
12.2 Important Conditions Describing Positive Dependence 193
12.2.1 Four Basic Conditions 193
12.2.2 The Relative Stringency of the Conditions 194
12.3 Further Discussion of Concepts of Positive Dependence 195
12.3.1 PQD 195
12.3.2 PQDE 196
12.3.3 Association 196
12.3.4 LTD and RTI 197
12.3.5 PRD and DTP 197
12.3.6 Interrelationships 198
12.3.7 More Concepts 198
12.4 Negative Dependence 198
12.4.1 Concepts 198
12.4.2 Neutrality 199
12.5 Monotonic Quadrant Dependence Function 199
12.5.1 Explanation 199
12.5.2 Interpretation of $\mu_{Y,X}$ 200
12.5.3 Properties of $\mu_{Y,X}$ 200
12.5.4 Remarks 200
12.6 Orderings of Positive Dependence 201

13 Rank Correlation Properties **202**
13.1 Kendall's Tau and Spearman's Rho 202
13.2 Farlie-Gumbel-Morgenstern and Related Distributions 203
13.2.1 Farlie-Gumbel-Morgenstern 203
13.2.2 Iterated F-G-M Distribution 203
13.2.3 A More General Distribution 204
13.3 Ali-Mikhail-Haq Family 204
13.4 Frank's System 204
13.5 Bivariate Pareto Distribution 205
13.6 Marshall and Olkin's Distribution 205
13.7 Kimeldorf and Sampson's Distribution 206
13.8 Weighted Linear Combination 206
13.9 Trivariate Reduction 206
13.10 A Distribution Uniform on Part of the Unit Square 207
13.11 A Distribution Suggested by R B Nelsen 207
13.12 Singly-Truncated Bivariate Normal Distribution 207
13.13 Discussion 207

14 Diagonal Expansion **211**
14.1 Introduction 211
14.2 A General Expansion of a Bivariate Distribution 211
14.3 ϕ^2-Bounded Bivariate Distributions 212
14.4 Biorthogonal Property and Diagonal Expansion 212
14.5 Conditions for Polynomial Biorthogonal Property 212
14.6 Canonical Sequence and Moment Sequence 213
14.6.1 Normal Marginals 213
14.6.2 Other Marginals 214

14.6.3 The Case of Identical F and G 215
14.6.4 The Case of Different F and G 216
14.7 Remarks 217
14.8 Diagonal Expansion and Construction of Markov Processes . . 217

15 Variables in Common **218**
15.1 Introduction 218
15.2 General Description 219
15.3 Additive Model 219
15.3.1 Background 219
15.3.2 Meixner Classes 220
15.3.3 Correlation 220
15.3.4 Symmetric Stable Distributions 220
15.3.5 Bivariate Triangular Distribution 221
15.3.6 Summing Several i.i.d. Variables 221
15.4 Bivariate Distributions Having Common Denominator 221
15.4.1 Explanation 221
15.4.2 Compounding 222
15.4.3 Example 1: Bivariate Cauchy Distribution 222
15.4.4 Example 2: Bivariate t-Distribution 222
15.4.5 Example 3: Bivariate F-Distribution 222
15.4.6 Example 4: Jensen's Bivariate F-Distribution 222
15.4.7 Example 5: Bivariate Pareto Distribution 222
15.5 The Transformation Being the Minimum 223
15.6 Other Forms of the Variables in Common Technique 223
15.6.1 Bivariate Chi-Squared Distribution 223
15.6.2 Bivariate Beta Distribution 224
15.6.3 Bivariate Z-Distribution 224
15.6.4 Cross-References 224

16 Overview of Generating Random Variates **225**
16.1 Introduction 225
16.2 The Univariate Case—The Major Approaches 225
16.2.1 Contents of this Section 225
16.2.2 Inverse Probability Integral Transform 226
16.2.3 Composition 226
16.2.4 Acceptance/Rejection 227
16.2.5 Ratio of Uniform Deviates 227
16.2.6 Transformations 227
16.3 The Univariate Case—Some Specific Distributions 227
16.3.1 The Normal Distribution 227
16.3.2 The Gamma Distribution 228
16.3.3 The Beta Distribution 228
16.3.4 The t-Distribution 229
16.3.5 Other Distributions 229
16.3.6 IMSL and NAG 229
16.3.7 Microcomputers and Random Variate Generation . . 230
16.4 The Bivariate Case—General Methods 230
16.4.1 The Setting 230
16.4.2 The Conditional Distribution Method 230
16.4.3 The Transformation Method 231
16.4.4 Methods Reflecting the Distribution's Construction . 231
16.5 The Bivariate Case—Some Specific Distributions 231

16.5.1 Bivariate Normal Distribution 231
16.5.2 Elliptically-Contoured Bivariate Distributions 232
16.5.3 The Gumbel Type I Bivariate Exponential Distribution 232
16.5.4 Generation of Bivariate Distributions with Given Marginals 233
16.5.4.1 Background 233
16.5.4.2 Weighted Linear Combination and Trivariate Reduction 233
16.5.4.3 Schmeiser and Lal's Methods 233
16.5.4.4 Cubic Transformation of Normals 234

17 Reliability Models Leading to Bivariate Distributions 235
17.1 Introduction 235
17.2 Two Bivariate Exponential Distributions 235
17.2.1 Marshall and Olkin's Shock Model 235
17.2.1.1 The Distribution 235
17.2.1.2 Fatal Shocks 236
17.2.1.3 Non-Fatal Shocks 236
17.2.1.4 A Random Sums Model 236
17.2.2 The Lack of Memory Property of Marshall and Olkin's Distribution 236
17.2.3 Downton's Successive Damage Model 237
17.3 Bivariate Exponentials and a Geometric Compounding Scheme 238
17.3.1 Background 238
17.3.2 Probability Generating Function 238
17.3.3 Bivariate Geometric Distribution 238
17.3.4 Bivariate Geometric Distribution Arising from a Shock Model 238
17.3.5 Infinitely Divisible Bivariate Exponential Distributions 240
17.3.6 Bivariate Cumulative Damage Models 240
17.3.6.1 The Univariate Case 240
17.3.6.2 The Bivariate Case 241
17.4 Freund's Bivariate Model 241
17.5 Bivariate Distributions Arising from Components Sharing a Common Environment 241
17.6 Results on $\Pr(X > Y)$: A Stress-Strength Model 242
17.6.1 The Basic Idea 242
17.6.2 Marshall and Olkin's Model 242
17.6.3 Downton's Model 242
17.6.4 Independent Truncated Bivariate Normal Model . . . 242

18 Reliability Classes: Concepts of Ageing 243
18.1 Introduction 243
18.2 Reliability Classes—Univariate Ideas 243
18.2.1 Introducing the Jargon 243
18.2.2 Definitions 244
18.2.3 Interrelationships 245
18.2.4 Properties of IFR and DFR 245
18.2.5 Applications 246
18.2.6 Ordering 246
18.2.7 Further Classes 246
18.3 Bivariate Reliability Classes 247
18.3.1 Alternative Requirements 247
18.3.2 Bivariate IFR 247

18.3.3 Bivariate IFRA 248
18.3.4 Bivariate NBU 249
18.3.5 Bivariate NBUE, DMRL, and HNBUE 250
18.4 Concluding Remarks 250

IV Mostly Applications **251**

19 Applications of the Bivariate Normal Distribution **253**
19.1 Organisation of Part IV 253
19.2 Introduction to Chapter 19 253
19.3 Biology 254
19.3.1 Size and Shape of Animals 254
19.3.2 The Dimensions of Trees 254
19.3.3 The Home Ranges of Animals 254
19.3.4 Probit Analysis 255
19.4 Medicine 255
19.4.1 Epidemiology 255
19.4.2 Blood Pressure Measurements 255
19.4.3 Mixtures—Graphical Methods 255
19.4.4 Mixtures—Normals and Haemophilia Carriers 256
19.4.5 Pharmacokinetics 256
19.4.6 Length and Diameter of Airborne Fibres 256
19.4.7 Leukemia 256
19.4.8 Severity of Injury in Road Accidents 256
19.4.9 Evaluation of X-Ray Images 256
19.4.10 Remark 256
19.5 Genetics: The Multifactorial Model of Disease Transmission . . 256
19.5.1 General Description 256
19.5.2 Correlated Traits 257
19.5.3 Mixtures—Twins of Unknown Zygosity 257
19.5.3.1 The Problem 257
19.5.3.2 The Model 258
19.5.3.3 The Data 259
19.5.3.4 Some Notes on Fitting the Model to the Data 259
19.5.3.5 Results 259
19.5.4 Mixtures—A Paper by Tan 260
19.5.5 A Veterinary Example 260
19.6 Psychology 261
19.6.1 Mental Abilities 261
19.6.2 Replicated Paired Comparison Tests 261
19.6.3 Signal Detection Theory 261
19.7 Air, Earth, Water, and Climate 261
19.7.1 Wind Speed 261
19.7.2 Meteorological Conditions at Two Times 262
19.7.3 Rainfall 262
19.7.4 Rainfall Duration and Amount 262
19.7.5 Cloud-Seeding Experiments 262
19.7.6 Errors—Tropical Cyclone Forecasts 262
19.7.7 River Flows and Rain 262
19.7.8 River Floods 263
19.7.9 Tides at Two Times 263

19.7.10 Height and Period of Waves 263
19.7.11 Rock Discontinuities 263
19.7.12 Earthquake Motion 263
19.7.13 Slope Stability 263
19.8 Physical and Technological Sciences 263
19.8.1 Measuring Instrument Errors and a Calibration System 263
19.8.2 Structural Reliability 264
19.8.3 Strength of Lumber 264
19.8.4 Destructive Testing 265
19.8.5 Thermodynamics 265
19.8.6 Astronomy 265
19.8.7 Radar 265
19.8.8 Optics 265
19.8.9 Shooting at a Target 265
19.8.10 Forces in Wood Grinding 266
19.9 Sociology and Economics 266
19.9.1 Household Size and Income 266
19.9.2 Commodities Trading 266
19.9.3 Blackjack Gambling 266
19.9.4 Policy Analysis 266
19.9.5 Sample Selection Equations 267

20 Construction of the Bivariate Pareto Distribution by Compounding, with Applications 268
20.1 Description of the Method 268
20.2 Ordinal-Scale Assessments by Two Observers 269
20.2.1 The Topic, in a Medical Context 269
20.2.2 Application of the Model 270
20.2.3 Results 270
20.2.4 Comments 271
20.3 Injury Severity in Road Accidents 271
20.3.1 The Topic and the Model 271
20.3.2 Results 272
20.3.3 A Further Model and Results 272
20.4 The Multifactorial Model of Disease Transmission 273
20.4.1 The Topic, and the Relevance of the Pareto Distribution 273
20.4.2 Data and Results 274
20.4.3 Comments on Estimated Correlations 274
20.4.4 A Trivariate Approximation 275
20.5 Replicated Paired Comparison Tests 275
20.5.1 The Topic 275
20.5.2 The Model 275
20.5.3 Results and Comments 276
20.6 Compounding of an F-G-M Distribution, and Application to Trace Metal Concentration 278
20.6.1 Derivation of the Distribution 278
20.6.2 Description of Some Properties 278
20.6.3 Application 278
20.7 Generalising the Bivariate Pareto Distribution 279
20.7.1 Introduction 279
20.7.2 Distribution GG1 280
20.7.3 Distributions GGD and SGD 280

20.7.4 Properties 281
20.7.5 An Alternative 281
20.7.6 A Bayesian Application 281

21 Joint Distributions of Injury Severity: A Bivariate Normal Model **282**
21.1 Introduction 282
21.2 A Variables-in-Common Approach 282
21.3 Results 284
21.3.1 First Model 284
21.3.2 Second Model 284
21.4 Comments on the Fit 287
21.5 Practical Conclusions 288
21.6 Discussion of the Realism of the Model 288
21.7 Two Ideas for the Future 289
21.7.1 Semiparametrics? 289
21.7.2 Equations Adapted from Econometrics? 290
21.8 Appendix. Injury Severities of Two Occupants in Single-Vehicle Crashes 291
21.8.1 The Data 291
21.8.2 The Statistical Model 291
21.8.3 Results 291

22 Bivariate Generalisations of Probit Analysis **292**
22.1 Outline of Probit Analysis 292
22.2 Two Stimulants 292
22.2.1 Two Stimulants in the Conventional Setting 292
22.2.2 Econometrics: Two Utilities 294
22.2.3 Gap Acceptance by Drivers 294
22.3 One Stimulus, Three Alternative Outcomes 295
22.3.1 Three Ordered Responses 295
22.3.2 Aitchison and Bennett's Idea 295
22.3.3 The Multinomial Logit Model 295
22.4 Bivariate Response 296
22.4.1 Medical Applications 296
22.4.2 Econometric and Sociological Applications 297
22.4.2.1 Morimune 297
22.4.2.2 Amemiya 297
22.4.2.3 Sample Selection Equations 297
22.4.2.4 Three Other Studies 298
22.4.2.5 The Cost of Partial Observability 298

23 Bivariate Models in Signal Detection Theory **300**
23.1 Signal Detection Theory in Psychophysics 300
23.1.1 Background 300
23.1.2 Mathematical Formulation 301
23.1.3 The ROC Curve 301
23.1.4 Alternative Shapes of ROC 302
23.1.5 Other Experimental Situations 302
23.1.5.1 Rating 302
23.1.5.2 Forced-Choice 303
23.1.5.3 Second-Choice 303
23.1.5.4 Same-Different 303

23.1.6 Remark 303
23.2 The Effect of Signal Probability on the Slope of the ROC . . . 303
23.3 Evaluation of X-ray Images 305
23.4 Information Retrieval Systems 306
23.5 Performance in Two-Alternative Choice Tests 307
23.6 The ABX Experiment 309
23.7 Graham et al 309
23.8 Overview 310

24 Examples of Data **311**
24.1 Introduction 311
24.1.1 Purpose of the Chapter 311
24.1.2 Reader, Beware! 311
24.1.2.1 Some Quotations 311
24.1.2.2 A Response 312
24.1.2.3 Alternatives to Analysis as a Distribution 312
24.1.2.4 Conclusion 312
24.1.3 Organisation of the Material 312
24.1.4 Data in Other Chapters 313
24.2 Reliability 313
24.3 Extremes 315
24.4 Two Measurements of the Same Thing 317
24.4.1 Data Presented as Exact Values 317
24.4.2 Data Presented as Frequency Tables 317
24.4.3 References to Scatterplots 324
24.4.4 Cross-References 325
24.5 Compositions (Totalling 100%) 325
24.5.1 Aitchison's Book and Datasets 325
24.5.2 Further References 326
24.6 A Few Discrete Examples 327
24.7 Shapes of Living Things 333
24.7.1 Data Presented as Exact Values 333
24.7.2 Data Presented as Frequency Tables 335
24.7.3 References to Scatterplots 337
24.8 Biological Sciences 337
24.8.1 Data Presented as Exact Values 337
24.8.2 Data Presented as Frequency Tables 338
24.8.3 References to Scatterplots 338
24.8.4 Cross-References 340
24.9 Medicine and Human Biology 340
24.9.1 Data Presented as Exact Values 340
24.9.2 Data Presented as Frequency Tables 341
24.9.3 References to Scatterplots 343
24.9.4 Cross-References 344
24.10 Mental Abilities 344
24.10.1 Data Presented as Exact Values 344
24.10.2 Data Presented as Frequency Tables 345
24.10.3 Further References 346
24.11 Sociology and Economics 346
24.11.1 Data Presented as Exact Values 346
24.11.2 Data Presented as Frequency Tables 347
24.11.3 Cross-References 351
24.12 Air, Earth, Water, and Climate 351

24.12.1 Data Presented as Exact Values 351
24.12.2 Data Presented as Frequency Tables 352
24.12.3 References to Scatterplots 356
24.12.4 Cross-References 358
24.13 Physical and Technological Sciences 358
24.13.1 Data Presented as Exact Values 358
24.13.2 Data Presented as Frequency Tables 359
24.13.3 Cross-References 359
24.14 Miscellaneous 360
24.14.1 Data Presented as Exact Values 360
24.14.2 References to Scatterplots 360

References **361**

Index **405**

Part I

Background

Chapter 1

Introduction

1.1 Opening Statement

Whether one approaches statistics in a data-driven fashion, or by constructing a mathematical model of reality, distributions are central to the subject—in the first case, because of the desire to describe the data parsimoniously, in the second case, to describe the random disturbance terms in the equations.

When we have only a single variable, our concern is naturally with univariate distributions. When we have lots of them, our interest is with multivariate distributions. In between lies the topic of bivariate distributions, the simplest case that incorporates interdependencies of different variables, the subject of this book.

We have found the books by Mardia (1970a) and Johnson and Kotz (1972) of inestimable value over the years. So far as we can judge, many others have, too. Some excuse is therefore needed for offering the present work to the reader:

- An important change has occurred over the last 10 or 20 years in the way continuous bivariate distributions are viewed. It is now widely appreciated that monotonic transformations applied to the X and Y axes can make the marginals take any form that is required. The specifically *bi*variate aspects of the subject enter when we have decided on some standard form for the marginals (e.g., uniform, normal, or exponential).

- Neither Mardia nor Johnson and Kotz are strong on applications. In Chapters 20–23, we describe at length some applications of continuous bivariate distributions that we are familiar with, in Chapter 19 we refer to many applications of the bivariate normal distribution, and throughout the book we give short accounts of the ways in which various other distributions have been used. Chapter 24 is a collection of empirical bivariate datasets.

- And the third excuse, perhaps a mundane one, is that much research on continuous bivariate distributions has been conducted over the past 20 years that has been published only in journal articles or conference proceedings, and it is time it was brought together in book form.

This is primarily a specialist monograph for the postgraduate and practising statistician. However, by including so many descriptions of how particular bivariate distributions have been used in this area of the technologies or that area of the social sciences, we feel this book will benefit the student of the wider field of statistical modelling, as well as of the specific field of statistical distributions.

Don't believe all you see in print. Parents tell their children not to credit all they read in newspapers. Well, the same goes for specialist monographs. Trying to be helpful to the reader, we have at times ventured into areas we understand only imperfectly.[1] No doubt, therefore, there are errors to be found herein. And no book can hope to be entirely free of misprints. We will appreciate news of mistakes found, and other suggestions for improving and updating our work.

[1] Beginning at page 1 and continuing to page 412, some critics will say.

1.2 Contents of Part I

The remaining Sections of this Chapter are as follows:

1.3 Description of the structure of this book.

1.4 What this book emphasises, what it omits.

1.5 Explanation of conventions, notations, and abbreviations.

1.6 The future of the subject.

There are two other Chapters in Part I—on continuous univariate distributions, and an overview of methods of constructing continuous bivariate distributions.

1.3 Description of the Structure of this Book

There are four Parts to this book:

I. Preliminaries (Chapters 1–3).

II. Families of continuous bivariate distributions (Chapters 4–10).

III. Mostly theory (Chapters 11–18).

IV. Mostly applications (Chapters 19–24).

The heart of the book is Part II, the account of the various families of continuous bivariate distributions. There is no single best way of classifying these, any more than there is of classifying univariate distributions. We hope our arrangement of them into seven Chapters will be found comprehensible. It is described in Section 4.1; suffice it to say here that we start with the bivariate normal distribution (Chapter 4), and then go on to less familiar ones, finally resorting to "Miscellaneous" in Chapter 10.

As already mentioned, we prepare the way with Part I, on univariate distributions and methods of constructing distributions.

Part III gives an account of a number of abstract issues regarding the construction and properties of continuous bivariate distributions. Ways in which two variables can be dependent on one another are dealt with in Chapters 11–13. Then come Chapters on diagonal expansion, variables-in-common, and methods of generating random deviates. Finally, there are two Chapters on how bivariate distributions arise in reliability models and on the concepts of ageing that are met in that context.

In Part IV, we turn to applications. Chapter 19 is a series of brief accounts of applications of the bivariate normal distribution. (Included in the descriptions of the various distributions of Chapters 5–10 are summaries of the applications made of them. To include the applications of the bivariate normal in Chapter 4 would have made it over-long, however.) Chapters 20–23 are on the method of compounding, the variables-in-common model as applied in a road accident data analysis setting, bivariate versions of probit analysis, and bivariate models in signal detection theory. It is becoming something of a trend to conclude a book with a collection of relevant datasets; it is a trend which we welcome and are happy to join, with our Chapter 24 having many tables of bivariate data, either listed as (x, y) pairs or given as two-way tables of frequencies.

1.4 What this Book Emphasises, What it Omits

As already mentioned, current practice is that one often studies a bivariate distribution in three parts—the marginal distribution of X, the marginal distribution of Y, and the association structure that connects X and Y. The first two fall into the province of univariate distributions, and it is only the third that is distinctively bivariate. One often, therefore, adopts some standard form for the marginals of X and Y—uniform, normal, and exponential are popular choices—and asks what are the characteristics of a bivariate distribution when its marginals have been transformed into the standard form. This separation of marginal distributions from bivariate structure has been a productive approach over the last 20 years or so, and we give a good deal of emphasis to it.

Another strong point of this book, we feel, is the large number of models for constructing bivariate distributions that we discuss, and the uses that these have found, whether in the statistical or the applied literature.

On the other hand, the book is about statistical distributions, their formulae, and their properties, so it does not give great weight to strongly data-oriented methods that try to directly read the message some given empirical data is telling us. Nevertheless, some attention is given to this—see particularly Section 4.13, and the data analyses reported in Chapters 20 and 21 may be helpful in this context, too.

Characterisations of distributions—that is, descriptions of properties from which may be deduced a distribution (as represented by, for instance, a formula for its p.d.f. or d.f.)—are mentioned from time to time, but we have not attempted a systematic survey. Papers on characterisations tend to be spattered with THEOREMS, LEMMAS, and COROLLARIES.[2] In fact, we have rather neglected most aspects of the subject where heavy mathematical development is required.

There is little specifically on time series models—it seems that the normal distribution is overwhelmingly the preferred one here, though we come across one or two uses of other bivariate distributions in this context, which are mentioned where appropriate.

The omission that will raise most eyebrows, we imagine, is any serious account of methods of estimating and testing the parameters of the distributions. The chief reason for not doing anything much on this was that we didn't feel like it. But we can clothe the decision with some sort of rational justification—that parameter estimation methods are well understood in the statistical community, that there are excellent accounts of the subject elsewhere, that computing is so cheap that one need not bother with refinements, that errors in model specification and arising from extrapolation are likely to be far larger than those arising in the estimation process, and so on.

Neither in Part IV nor elsewhere do we give much attention to the "application" of any bivariate distribution in robustness studies. (For instance, suppose a statistical test can be shown theoretically to be valid on the assumption of an underlying bivariate normal distribution. Theoretical results may not be available other than for this case. One then selects some distributions that are sufficiently different from the bivariate normal, and determines the performance of the statistic under study using random samples generated from these distributions.) No, the applications which we describe are generally much more closely tied in to the real world than are robustness studies. We have two further reasons for not listing robustness studies among the applications which we consider:

- The basic idea of such studies is very well known in principle.
- Anything other than the basic idea is almost totally unknown (by us, perhaps by everyone). Specifically, what are the distributions that ought to be included in a robustness study?[3]

We do, however, devote a Chapter to the generation of random variates from specified distributions, which is basic whenever simulation is used to study robustness.

1.5 Explanation of Conventions, Notations, and Abbreviations

Note (i) this is not a complete list, only a collection of those we have most commonly used, and (ii) we have at no stage fully rationalised our notations, with the result that the same symbol can mean different things in different parts of the book. For instance, R usually means the set of real numbers, but it might mean radius or it might denote the correlation coefficient. We do hope, however, that we have not left in any ambiguity that might actually be confusing.

[2] We are never quite sure if, behind all this impenetrable finery, the emperor may not have holes in his underwear.

[3] Actually, having articulated that question, we find we can surmise an answer. Assuming that it is known what marginal distributions are to be considered, and thus the question only refers to the bivariate structure, we suggest including (i) the bivariate distribution consisting of a mixture of perfect correlation and independence, and (ii) the Kimeldorf-Sampson distribution of Section 10.7. (It is conjectured in Section 13.13 that these distributions respectively provide the minimum and maximum values of the grade correlation, for a given value of Kendall's tau, that are attainable by sensible bivariate distributions.)

1.5.1 Abbreviations

c.f.—characteristic function.

d.f.—(cumulative) distribution function, or degrees of freedom.

i.i.d.—independent identically distributed.

m.g.f.—moment generating function.

p.d.f.—probability density function.

p.g.f.—probability generating function.

r.v.—random variable.

1.5.2 Conventions

Capital letters are used to stand for the names of variables, and lower case to represent their values.

We have not usually been too concerned about whether an interval is specified as open or closed, or whether relations $<$ or $\leq$ should be specified. We think the few places where care is necessary will be readily apparent. Similarly, we have usually preferred to say "increasing" where "nondecreasing" would be more strictly correct, and "decreasing" rather than "nonincreasing".

Location and scale parameters have often been omitted from formulae for probability distributions. They can readily be introduced into the d.f. by replacing x with $\frac{x-\alpha}{\beta}$. As to how they are introduced into the p.d.f., see Section 3.1.3.

The usual uses of brackets (whether round, square, or curly) are to group a mathematical expression or to enclose the argument of a function. Occasionally, they have more meaning, as when $[0,1]$ refers to the range $0 < x < 1$ whilst $(0,1)$ refers to the range $0 \leq x \leq 1$, $[x]$ means the integer part of x, or curly brackets refer to sets (thus, $\{x : c\}$ means the set of all x such that condition c is satisfied). At one or two points where there might otherwise be confusion, a function $\xi(x)$ is instead written as ξ_x.

It is common to say that (X, Y) has a bivariate ∇ distribution if both marginals are univariate ∇, whatever the form of the dependence between the two variates.[4] The chief exception to this is that "bivariate normal" refers to *the* bivariate normal distribution, the one that is the subject of our Chapter 4.

Variable or variate? Not 1% of the statistical community knows precisely how these terms should be used, and we do not claim to be in that select band. Roughly, "variate" is used when emphasis is on the name of a quantity, and "variable" when emphasis is on the value it takes.

1.5.3 Symbols

Though they may perhaps be used for something else occasionally, the usual meanings of $F(x)$, $G(y)$, and $H(x,y)$ are the d.f.'s of X, Y, and (X,Y), respectively, and the corresponding p.d.f.'s are denoted $f(x)$, $g(y)$, and $h(x,y)$.

φ is used for the characteristic function, M for the moment generating function, and μ'_{ij} and μ_{ij} for moments about zero and about the mean, respectively.

Pr denotes probability, E expectation, s.d. standard deviation, var variance, cov covariance, and corr correlation.

[4] This is the third definition in Seshadri and Patil (1963, p. 220); their first definition of a bivariate ∇ distribution is much stronger, namely, that both marginals and both sets of conditionals are univariate ∇.

R, R^2, and R^n often mean the sets of real numbers, of pairs of real numbers, and of n-dimensional real numbers, respectively, but they sometimes take other meanings.

The meaning of a bar above a symbol should be clear from the context—e.g., $\bar{H}$ denotes the bivariate survival function, i.e., the upper right volume under the probability density surface, and $\bar{x}$ the sample mean.

Vectors and matrices are given boldface symbols, such as $\mathbf{x}$ or $\mathbf{\Sigma}$. A prime ($'$) denotes transpose. (A vector is usually thought of as being a column vector, and its transpose $\mathbf{x}'$ as being a row vector.)

$\sim$ usually means "is distributed as", as in $X \sim U(0,1)$ meaning that X has the uniform distribution over $[0,1]$, or $\mathbf{Z} \sim N(\boldsymbol{\mu}, \mathbf{\Sigma})$ meaning that $\mathbf{Z}$ has a normal distribution with mean $\boldsymbol{\mu}$ and covariance matrix $\mathbf{\Sigma}$.

$\propto$ means "is proportional to".

$\binom{n}{r}$ means the number of ways of choosing r objects from n, i.e., $\frac{n!}{r!(n-r)!}$.

Set relational symbols have their usual meanings. The most common one is $\in$, meaning "in".

1.5.4 Some Functions

log means $\log_e$.

$I_{\{c\}}$ is the indicator function, i.e., is 1 if condition c is satisfied, and is 0 if not.

δ_{ij} is usually the Kronecker delta, i.e., is 1 if $i = j$ and is 0 otherwise.

sgn takes the value -1 if its argument is negative, 0 if its argument is 0, and $+1$ if its argument is positive.

ϕ and Φ are usually the p.d.f. and d.f. of the standardised univariate normal distribution.

ψ and Ψ are usually the p.d.f. and d.f. of the standardised bivariate normal distribution.

$\Gamma(\alpha)$, $\Gamma_x(\alpha)$, $B(\alpha, \beta)$, $B_x(\alpha, \beta)$ are complete and incomplete gamma functions, and complete and incomplete beta functions. See footnote 2 in Chapter 2.

I_ν and K_ν are used for modified Bessel functions.

${}_2F_1$ is Gauss' hypergeometric function, and ${}_3F_2$ is a generalised hypergeometric function. See footnote 3 in Chapter 13.

Other occasionally-used special functions (Appell's and Saran's hypergeometric functions, Debye functions, the exponential integral, the dilogarithm) are also explained when introduced. The reader should take note that nomenclature and notation for some of these functions are not fully standardised. A recent and well-received book on special functions is by Spanier and Oldham (1987).

1.6 The Future of the Subject

As the statistician "certainly cannot express the univariate frequency distribution by less than four momental constants, so he cannot expect to be let off a bivariate distribution with less than fifteen" (Pearson, 1925). It is not now—if it ever was—popular to describe an empirical dataset with the fourteen moments of up to the fourth order. (Pearson's fifteenth constant is the total frequency.) Curiously, though, we can foresee that the current preoccupation with one-parameter families of bivariate distributions will soon result in a revival of interest in multi-parameter systems. The reason for this is the separation of marginal distributions from bivariate structure, and thinking about $4 + 4 + 3$ parameters is less formidable than thinking about all 11 at once. One might agree with Pearson that the marginal distributions need four parameters each, in order to get right the skewness and the heaviness of the tails. As to the bivariate structure, the illustrations in Barnett (1980) and Johnson (1987) bring home the variety of bivariate shapes that are possible even when the marginals are fixed (e.g., are uniform), and that one has been very lucky if one's data is adequately described by a one-parameter family. That single parameter will usually reflect, in the case of uniform marginals, the spread of the distribution along the line $Y = 1 - X$ (the bigger the spread, the less the association between X and Y). One can easily imagine needing at least two more parameters, to reflect the peakedness along the lines $Y = 1 - X$ and $Y = X$. Now we are up to a total of eleven parameters, not so far from Pearson's fourteen, and we haven't thought about asymmetry yet.

Although we think that viewing bivariate distributions in this way, as marginals plus association structure, will be productive for the next 20 or 30 years, we should also point to its inherently unsatisfactory nature—it is difficult to imagine that the causative mechanism that generates a bivariate distribution actually acts in this way. We can hope there will be growing success in understanding the physical mechanisms that generate bivariate distributions in specific applications areas; but these will, we imagine, usually produce x and y simultaneously, or will produce the bivariate distribution as a marginal $f(x)$ and a set of conditionals $g(y|x)$. We suspect Hougaard (1987) had much the same point in mind when he said of the Farlie-Gumbel-Morgenstern distribution, "It is not clear whether there is any probability mechanism which generates distributions of this type."

Taking this point a little further, if we look in Chapter 5 at the derivations that underlie distributions in uniform-marginals form (i.e., what we shall call the copula), we fail to find any unity:

Distribution	Section	Method of derivation
Pareto	5.8–5.9	Compounding
Gumbel-Hougaard	5.6	Compounding
Plackett	5.7	Constant global cross ratio
Ali-Mikhail-Haq	5.3	Differential equation
Frank	5.4	Associativity of H and $x + y - H$
Cuadras-Augé	5.5	Marginal transformation of Marshall-Olkin
Farlie-Gumbel-Morgenstern	5.2	Compact formula
Gumbel-Barnett	5.10–5.11	Compact formula

A major open question is whether there exists any physical mechanism that generates a particular copula, which can be combined with mechanisms that generate the marginal distributions, resulting in the construction of all features of the distribution. (One common idea in practice is that of compounding plus loss of information through only observing X and Y on an ordinal scale, but obtaining a copula due to limitations on the measurement process is rather different from inherently generating a copula.)

We would also like to call attention to two particular technical areas of statistical research that we see as impacting on bivariate distributions in the future—computer algebra, and semiparametric analysis.

- The first may lead to a revolution in the questions we ask: at present, we often (i) write down equations that model the process of interest, (ii) deduce a bivariate distribution, and (iii) fit the distribution to data. We see the second of these as an interesting, important, and enlightening activity. But why? If computers were good enough at algebra, and quick enough with numerical calculations—and both of these conditions may soon be true—we might be able to estimate the parameters of the model and assess its goodness of fit directly from the equations that define it, without needing the intermediate step of deriving the p.d.f. or d.f. of the distribution. For a little more on this, see Section 3.1.4.

- The idea behind semiparametrics is that there are some features of a model which we regard as its heart, and we really want to know if they are healthy or not; and there are others that are peripheral, we are not interested in their details, we do not want to make specific assumptions about them just in order to get explicit algebraic or numerical results, and we would prefer our conclusions about the heart to be robust with respect to the choices of these peripheral assumptions. For a few more words on this, see Section 21.7.1.

Chapter 2

Univariate Distributions

2.1 Introduction

Most of this Chapter is given over to a consideration of the major continuous univariate distributions:

2.4 Normal distribution, and transformations.

2.5 Beta distributions.

2.6 Exponential, gamma, Weibull, and Stacy distributions.

2.7 Logistic, Laplace, and Cauchy distributions.

2.8 Extreme value distributions.

2.9 Pareto distribution.

2.10 The Pearson system.

2.11 The Burr system.

2.12 t- and F-distributions.

2.13 Noncentral distributions.

2.14 Some lesser-known distributions (inverse Gaussian, Meixner hypergeometric, hyperbolic, and stable distributions).

But before we commence that, we first have Sections (2.2 and 2.3) on notation and definitions and on generalities about univariate shape; and the Chapter concludes (Sections 2.15 and 2.16) with a guide to how to find out more about univariate distributions in the recently-published *Encyclopedia of Statistical Sciences*, and a description of the bibliography by Patil, Boswell, and Ratnaparkhi (1984).

As well as the *Encyclopedia*, there are several other excellent works useful as regards continuous univariate distributions. Johnson and Kotz (1970a,b) have devoted two volumes of their four-volume set to continuous univariate distributions. Ord's (1972) monograph is deservedly well-known, also. Hastings and Peacock (1975) is a convenient handbook that shows graphs of densities and variate relationships for several distributions. Another useful compendium is by Patel et al (1976); Chapters 3 and 4 of Manoukian (1986) give many distributions and relations between them. Extensive collections of illustrations of p.d.f.'s may be found in Hirano et al (1983) (105 graphs, each with typically about 5 curves shown, grouped in 25 families of distributions), and in Patil, Boswell, and Ratnaparkhi (1984) (see Section 2.16 below). As to bibliographies, that by Patil and colleagues has already been mentioned, and the earlier one by Haight (1961) is noteworthy, also.

2.2 Notation and Definitions

2.2.1 Notation

In the univariate case, the (cumulative) distribution function and the probability density function will be denoted $F(x)$ and $f(x)$, respectively. The following is a list of terms and symbols:

Term	Symbol	Brief explanation
Moment generating function	$M(t)$	$E[e^{tX}]$
Characteristic function	$\varphi(t)$	$E[e^{itX}]$
Cumulant generating function	$K(t)$	$\log\varphi(t)$
rth moment (about the origin)	μ'_r	$E[X^r]$
rth central moment	μ_r	$E[(X-\mu)^r], \mu = \mu'_1$
rth cumulant	κ_r	The coefficient of $(it)^r/r!$ in the expansion of $K(t)$
Variance	σ^2	μ_2
Coefficient of skewness	$\alpha_3 = \sqrt{\beta_1}$	μ_3/σ^3
Coefficient of kurtosis	$\alpha_4 = \beta_2$	μ_4/σ^4
Coefficient of variation		σ/μ
Survival function	$\bar{F}(x)$	$1 - F(x)$
Hazard function	$r(x)$	$f(x)/[1-F(x)]$
Sample mean	$\bar{X}$	$\sum_{i=1}^n X_i/n$
Sample variance	s^2	$\sum_{i=1}^n (X_i-\bar{X})^2/(n-1)$

(In the bivariate context, μ_1 and μ_2 will often be used for the means of the two variables. There is unlikely to be any confusion over this.)

2.2.2 Explanations

2.2.2.1 Moment Generating Function

Let X be a r.v. with d.f. F and density f. Let

$$M(t) = E(e^{tX}) = \int_{-\infty}^{\infty} e^{tx} f(x)\,dx. \tag{2.1}$$

We say that $M(t)$ is the m.g.f. of F if the integral is convergent for all values of t belonging to an interval which contains the origin. The existence of the m.g.f. is not assured for all distributions; if it does exist, it will uniquely determine the distribution. When it exists, it may be written as

$$M(t) = \sum_{j=0}^{\infty} \mu'_j \frac{t^j}{j!}. \tag{2.2}$$

(This implies μ'_j is $M^{(j)}(0)$, i.e., the jth derivative of M, evaluated at 0.) Note that it is possible to have μ'_j existing for all j and yet $M(t)$ not existing.

Let X_1 and X_2 be two independent r.v.'s, with m.g.f.'s $M_1(t)$ and $M_2(t)$, respectively. It is easy to show that the m.g.f. of $X_1 + X_2$ is $M_1(t)M_2(t)$. Hence the m.g.f. is a convenient tool for the study of the sum of independent r.v.'s.

For univariate distributions, the existence (finiteness) of a moment of some particular order implies the existence of all moments of lower order.[1]

[1]Is the same true for bivariate distributions? No. What we can say is that if moments of orders (κ,ℓ), (k,λ), and (k,ℓ) exist, then so do all moments of orders (m,n), where $\kappa \le m \le k$ and $\lambda \le n \le \ell$ (van der Vaart, 1973).

2.2.2.2 Characteristic Function

The c.f. φ of a d.f. F is a complex-valued function defined by

$$\begin{aligned}\varphi(t) &= E(e^{itX}) && (2.3)\\ &= \int_{-\infty}^{\infty} e^{itx} f(x)\,dx && (2.4)\\ &= \int_{-\infty}^{\infty} \cos tx\; f(x)\,dx + i\int_{-\infty}^{\infty} \sin tx\; f(x)\,dx && (2.5)\end{aligned}$$

(where $i = \sqrt{-1}$), for all real t.

The c.f. uniquely determines the distribution. It satisfies (i) $\varphi(0) = 1$, (ii) $|\varphi(t)| \leq 1$ for all real t, and (iii) $\varphi(-t) = \overline{\varphi(t)}$, where the bar denotes the complex conjugate.

Unlike the m.g.f., $\varphi(t)$ exists for all distributions.

Suppose that F has finite moments μ'_j up to order n, then $\varphi^{(j)}(0) = i^j\mu'_j$ (for $1 \leq k \leq n$), where $\varphi^{(j)}$ is the jth derivative of φ.

The c.f. can be inverted to give the p.d.f.:

$$f(x) = \frac{1}{2\pi}\int_{-\infty}^{\infty} e^{-itx}\varphi(t)\,dt. \qquad (2.6)$$

If X_1 and X_2 are independent r.v.'s with c.f.'s φ_1 and φ_2, respectively, the c.f. of the sum is the product of the c.f.'s, $\varphi_1(t)\varphi_2(t)$.

An overview of the characteristic function is given by Laha (1982). The books by Lukacs (1970, 1983) are key works.

2.2.2.3 Cumulant Generating Function

Let $K(t) = \log\varphi(t)$. Then $K(t)$ is known as the cumulant generating function. Assuming again that the first n moments of F exist,

$$K(t) = \sum_{j=1}^{n} \frac{\kappa_j}{j!}(it)^j + o(|t|^n)$$

as $t \to 0$. The coefficients κ_j in this expansion are called the cumulants (or semi-invariants) of F. Clearly,

$$\kappa_j = i^{-j}K^{(j)}(0),$$

where $K^{(j)}(0)$ is the jth derivative of K, evaluated at 0.

The normal distribution has the unique characterising property that all its cumulants of order 3 and higher are zero.

2.2.2.4 Interrelationships of Moments and Cumulants

The relations between the lower moments about the origin μ'_j, central moments μ_j, and cumulants κ_j are as follows:

$$\begin{aligned}&\kappa_1 = \mu'_1 = \mu \text{ (the mean)} && \mu'_1 = \kappa_1\\ &\kappa_2 = \mu'_2 - {\mu'_1}^2 = \sigma^2 \text{ (the variance)} && \mu'_2 = \kappa_2 + \kappa_1^2\\ &\kappa_3 = \mu'_3 - 3\mu'_1\mu'_2 + 2{\mu'_1}^3 = \mu_3 && \mu'_3 = \kappa_3 + 3\kappa_2\kappa_1 + \kappa_1^3\\ &\kappa_4 = \mu'_4 - 3{\mu'_2}^2 - 4\mu'_1\mu'_3 + 12{\mu'_1}^2\mu_2 - 6{\mu'_1}^4 && \mu'_4 = \kappa_4 + 3\kappa_2^2 + 4\kappa_1\kappa_2 + 6\kappa_1^2\kappa_2 + \kappa_1^4.\end{aligned}$$

2.3 Generalities About Univariate Shape

2.3.1 Location and Scale

If $F(x)$ is a cumulative distribution of a variate X, we may introduce a location parameter a and a scale parameter b into the formula by writing $F\left(\frac{x-a}{b}\right)$. These parameters a and b are often the mean and standard distribution, respectively, but they need not be—(i) these quantities may not be finite (in such a case, we might set a = median and b = semi-interquartile range), and (ii) it may be more convenient for distributions whose p.d.f. is zero for $X < x_0$ to set $a = x_0$ instead of a = mean.

2.3.2 Skewness and Kurtosis

The most common measure of skewness is the normalised third central moment,

$$\alpha_3 = \sqrt{\beta_1} = \frac{\mu_3}{\mu_2^{3/2}}. \tag{2.7}$$

For symmetric p.d.f.'s such as the normal, this is zero.

The normalised fourth central moment,

$$\alpha_4 = \beta_2 = \frac{\mu_4}{\mu_2^2}, \tag{2.8}$$

is the usual measure of kurtosis. The normal distribution has $\beta_2 = 3$, so sometimes $\gamma_2 = \beta_2 - 3$ is referred to as the "excess of kurtosis". There is some controversy as to what kurtosis actually means, but a distribution with $\beta_2 < 3$ ("platykurtic") usually is less sharply peaked in the centre and has thinner tails than the normal distribution having the same standard deviation, whereas a distribution with $\beta_2 > 3$ ("leptokurtic") usually is more sharply peaked in the centre and has fatter tails than the normal distribution having the same standard deviation.

For all distributions, $\beta_2 \geq \beta_1 + 1$.

The shape of a distribution is not completely determined by the values of β_1 and β_2. Nevertheless, these two quantities are a help when we have decided on a particular family of distributions (e.g., the Pearson, or the Johnson), because we can plot them on a chart marked with what regions of (β_1, β_2) correspond to which member of the family, and hence make the choice of member.

2.3.3 Tail Behaviour

In considering this subject, we are not interested in tail behaviour as affected by the standard deviation or other measure of scale—we assume such effects have been discarded by some process of standardisation. Even when this has been done, it is still possible to classify distributions as short-, medium-, or long-tailed. See Parzen (1979) and Schuster (1984).

2.3.4 Some Multi-Parameter Systems of Univariate Distributions

Among systems of univariate distributions having several parameters—typically, four, so that skewness and tail-heaviness can be got right when fitting to emirical data—are Pearson's, the transformed normal system of Johnson, the transformed logistic system of Tadikamalla and Johnson, the generalised lambda, and Tukey's g and h. Mendoza and Iglewicz (1983) used these five to fit twelve of the symmetric distributions commonly used in simulation studies, and compared them in terms of ease of fitting and of goodness of fit at selected percentiles. Pearson et al (1979) compared the percentage points of distributions chosen from the Pearson, Johnson, and Burr systems.

2.3.5 Reliability Classes

Patel (1973) classifies 15 continuous distributions as to whether they have the increasing (or decreasing) failure rate property, and whether they have the increasing (or decreasing) failure rate on average property. He also states whether they belong to the exponential family.

A table of formulae including the reliability function ($\bar{F}$), the hazard function, and the mean residual life function is given by Sheikh et al (1987); the distributions included are the normal, gamma, and Weibull, and also their reciprocals.

2.4 Normal Distribution, and Transformations

2.4.1 Normal Distribution

The normal (Gaussian) distribution is symmetric about μ, and assigns positive probability density to all of the real line:

$$f = \frac{1}{\sqrt{2\pi}\sigma} \exp[-(x-\mu)^2/2\sigma^2]. \tag{2.9}$$

For the unit normal (the standard form), the density is denoted ϕ; the argument is often written as z, rather than x:

$$\phi(z) = \frac{1}{\sqrt{2\pi}} \exp(-z^2/2). \tag{2.10}$$

$F(x)$: the symbol Φ is used. There is no elementary expression.
Properties:

$$\begin{aligned} E(X) &= \mu & (2.11)\\ \text{var}(X) &= \sigma^2 & (2.12)\\ \kappa_r &= 0, \text{ for } r > 2. \end{aligned}$$

The mode and the median, as well as the mean, are μ.

2.4.2 Lognormal Distribution

For $x > 0$:

$$f = (x\sqrt{2\pi}\sigma)^{-1} \exp[-(\log x - \xi)^2/2\sigma^2]. \tag{2.13}$$

Letting Φ be the standardised normal distribution function,

$$F = \Phi\left(\frac{\log x - \xi}{\sigma}\right). \tag{2.14}$$

Properties:

$$\begin{aligned} \mu &= \exp(\xi + \tfrac{1}{2}\sigma^2) & (2.15)\\ \text{var}(X) &= e^{2\xi}e^{\sigma^2}/(e^{\sigma^2}-1). & (2.16) \end{aligned}$$

If $\log X$ has a normal distribution, then X is said to have a lognormal distribution.

2.4.3 Johnson's System

Johnson's (1949) system of distributions is obtained by starting with a unit normal variate Z and applying one of several simple transformations to it:

$$Z = \gamma + \delta T(Y), \tag{2.17}$$

where

- $T(Y) = \log Y$ gives the lognormal family, denoted S_L.
- $T(Y) = \sinh^{-1} Y$ gives the S_U system with unbounded range, $-\infty < Y < \infty$.
- $T(Y) = \log[Y/(1-Y)]$ gives the S_B family with bounded range, $0 < Y < 1$.
- The normal distribution may be considered within this family (take $T(Y) = Y$), and be denoted S_N.

Making one of the above choices determines the shape of the distribution. Location and scale parameters may be introduced by setting $Y = (X-a)/b$.

Detailed discussions may be found in Johnson and Kotz (1970a, Section 4.3) and Bowman and Shenton (1983). DeBrota et al (1988) have provided software to help in the choice of an appropriate member of this system (e.g., by fitting to data).

2.4.4 Box-Cox Power Transformations to Normality

If X is not normally distributed, a power function of it might be. The Box-Cox transformation is

$$\left.\begin{array}{ll}(X^\lambda - 1)/\lambda & \text{for } \lambda \neq 0 \\ \log X & \text{for } \lambda = 0\end{array}\right\} \tag{2.18}$$

2.4.5 The g and h Family

This family is obtained by starting with a standard normal variate Z and applying the transformation

$$T_{g,h}(Z) = \frac{e^{gZ} - 1}{g} \exp(hZ^2/2), \tag{2.19}$$

where g and h are constants, the former controlling asymmetry or skewness and the latter elongation or extent to which the tails are stretched relative to the normal.

When $g = 0$, a symmetric distribution is obtained from $Z \exp(hZ^2/2)$. When $h = 0$, the lognormal distribution is obtained.

2.4.6 Efron (1982a)

Efron (1982a) considered the question of whether there is a single transformation $Y = a(X)$ such that Y has nearly a normal distribution, when the distribution of X comes from some one-parameter family of distributions. Efron developed a theory to answer this question without considering the form of a; and, in those cases where the answer is positive, he gave formulae for calculating a.

2.4.7 Distributions of a Ratio

Rogers and Tukey (1972) have studied distributions obtained from Z/v, where Z has a normal distribution, and v is a positive r.v. that is independent of Z. Among special cases of this are:

- The normal distribution itself (the denominator being constant),
- t-distributions (the denominator being the square root of a chi-squared variate), including the special case of the Cauchy distribution (the denominator being half-normal),
- The so-called contaminated distributions (the denominator taking only two values), and
- The slash distribution (the denominator being uniformly distributed).

2.4.8 Compound Normal Distributions

Starting from a normal distribution of X, denoted in the usual way, $N(\mu, \sigma^2)$, we may now suppose that μ or σ^2 are themselves random variables.

- If μ has a normal distribution, $N(\xi, \sigma_\mu^2)$, then the distribution of X will also be normal, $N(\xi, \sigma^2 + \sigma_\mu^2)$.
- If $X \sim N(\mu + \beta U, \sigma^2 U)$, with U being a random variable, the resulting distribution of X is called a normal variance-mean mixture, see Barndorff-Nielsen et al (1982). If $\beta = 0$, it is a normal variance mixture.

2.5 Beta Distributions

2.5.1 The First Kind

For $0 \leq x \leq 1$:

$$f = \frac{1}{B(p,q)} x^{p-1}(1-x)^{q-1}, \tag{2.20}$$

where p and q are shape parameters and $B(p,q)$ is the beta function.[2]

The d.f. cannot be expressed in a closed form other than as an incomplete beta function.

Notation: we use Beta(α,β) to denote the beta distribution with shape parameters α and β.

Properties:

$$\mu'_r = \frac{B(p+r,q)}{B(p,q)} \tag{2.21}$$

$$\mu = \frac{p}{p+q} \tag{2.22}$$

$$\sigma^2 = pq(p+q)^{-2}(p+q+1)^{-1}. \tag{2.23}$$

For $p > 1, q > 1$, the mode is at $(p-1)/(p+q+2)$.

2.5.2 Uniform Distribution

A special case of the beta distribution is the uniform distribution over the range $0 < x < 1$. The following equations are written in terms of the more general case, $a < x < b$:

$$f = \frac{1}{b-a} \tag{2.24}$$

$$\tag{2.25}$$

$$F = \frac{x-a}{b-a}. \tag{2.26}$$

(Outside this range, f is 0 and F is either 0 or 1.)

Properties:

$$\mu_r = \left\{ \begin{array}{ll} 0 & \text{for } r \text{ odd} \\ (\frac{b-a}{2})^r/(r+1) & \text{for } r \text{ even} \end{array} \right\} \tag{2.27}$$

$$\mu = (b-a)/2 \tag{2.28}$$

$$\sigma^2 = (b-a)^2/12. \tag{2.29}$$

One reason why this distribution is so important is its role in generating random variates—let U be uniformly distributed over $[0,1]$; then $F^{-1}(U)$ has distribution F. See Section 16.2.2.

---

[2]The beta function with arguments α and β is defined by

$$B(\alpha,\beta) = \int_0^1 t^{\alpha-1}(1-t)^{\beta-1}\,dt$$

$(\alpha > 0, \beta > 0)$. The incomplete beta function is defined by

$$B_x(\alpha,\beta) = \int_0^x t^{\alpha-1}(1-t)^{\beta-1}\,dt.$$

We shall see that the beta function is related to the gamma function. With argument α, this is defined by

$$\Gamma(\alpha) = \int_0^\infty t^{\alpha-1}e^{-t}\,dt.$$

It satisfies the recurrence relation

$$\Gamma(\alpha+1) = \alpha\Gamma(\alpha).$$

Also, $\Gamma(\frac{1}{2}) = \sqrt{\pi}$, $\Gamma(1) = 1$, and if α is an integer, $\Gamma(\alpha+1) = \alpha!$. The incomplete gamma function is defined as

$$\Gamma_x(\alpha) = \int_0^x t^{\alpha-1}e^{-t}\,dt.$$

Recent methods for computing Γ_x are those of DiDonato and Morris (1986) and Shea (1988).

The beta and gamma functions are connected by the following equation:

$$B(\alpha,\beta) = \Gamma(\alpha)\Gamma(\beta)/\Gamma(\alpha+\beta).$$

2.5.3 Inverted Beta Distribution

This is also known as the beta distribution of the second kind, as the F, and as the Pearson type VI distribution. Its p.d.f. is:

$$f = \frac{1}{B(\alpha,\beta)}\frac{x^{\alpha-1}}{(1+x)^{\alpha+\beta}} \tag{2.30}$$

$(x > 0)$. α and β are shape parameters.

F can be expressed in terms of an incomplete beta function.

Properties:

$$\mu'_r = B(\alpha + r, \beta + 2 - r). \tag{2.31}$$

This is a transformation of the beta distribution. Suppose X has a beta distribution. Then $X/(1-X)$ is distributed as (2.30). A convenient summary of the interrelationships between the beta, inverted beta, gamma, t-, F-, and Cauchy distributions is at p. 430 of Devroye (1986).

When we take the logarithmic transformation of an inverted beta variate, the resulting distribution is sometimes termed the Z-distribution. For $-\infty < x < \infty$, $\lambda_1 > 0$, $\lambda_2 > 0$,

$$f = \frac{1}{B(\lambda_1,\lambda_2)}\frac{e^{-\lambda_2 x}}{(1+e^{-x})^{\lambda_1+\lambda_2}}. \tag{2.32}$$

Its properties include $M(t) = \frac{\Gamma(\lambda_1+t)\Gamma(\lambda_2-t)}{\Gamma(\lambda_1)\Gamma(\lambda_2)}$, and $\kappa_r = \psi^{(r-1)}(\lambda_1) + (-1)^r\psi^{(r-1)}(\lambda_2)$, where $\psi^{(r)}(t) = \frac{d^r[\Gamma'(t)/\Gamma(t)]}{dt^r}$. When $\lambda_1 + \lambda_2 = 1$, this becomes an example of the Meixner hypergeometric distribution (Section 2.14.2).

2.6 Exponential, Gamma, Weibull, and Stacy Distributions

2.6.1 Exponential Distribution

For $x \geq 0$ and scale parameter $\lambda > 0$:

$$f = \lambda e^{-\lambda x} \tag{2.33}$$
$$F = 1 - e^{-\lambda x}. \tag{2.34}$$

Properties:

$$\mu'_r = r!/\lambda^r \tag{2.35}$$
$$\mu = 1/\lambda \tag{2.36}$$
$$\text{median} = \log 2/\lambda \tag{2.37}$$
$$\text{mode} = 0 \tag{2.38}$$
$$\sigma^2 = 1/\lambda^2. \tag{2.39}$$

This distribution is characterised by the "lack of memory" property, i.e.,

$$\Pr(X \leq x + y | X > y) = \Pr(X \leq x). \tag{2.40}$$

2.6.2 Gamma Distribution

For $x > 0$, $\alpha > 0$, $\beta > 0$:

$$f = \frac{x^{\alpha-1}\exp(-x/\beta)}{\beta^\alpha\Gamma(\alpha)}, \tag{2.41}$$

where $\Gamma(\alpha)$ is the gamma function, defined in a footnote earlier.

To write an expression for F, it is necessary to use the incomplete gamma function:

$$F = \Gamma_{x/\beta}(\alpha)/\Gamma(\alpha). \tag{2.42}$$

Properties:

$$\mu'_r = \beta^r \prod_{i=0}^{r-1}(\alpha + i) \tag{2.43}$$

$$\mu = \alpha\beta \tag{2.44}$$

$$\sigma^2 = \alpha\beta^2. \tag{2.45}$$

For $\alpha \geq 1$, the mode is at $\beta(\alpha - 1)$.

Notation: We use Gamma(α, β) to denote the gamma distribution with shape parameter α and scaling parameter β.

The Erlang distribution is simply a gamma distribution with α being a positive integer.

2.6.3 Chi-Squared and Chi Distributions

The chi-squared distribution is the gamma distribution written in a slightly different way (and often thought of in different contexts). ν, effectively a shape parameter, is referred to as the number of degrees of freedom of the distribution. For $x > 0$:

$$f = \frac{x^{(\nu-2)/2}\exp(-x/2)}{2^{\nu/2}\Gamma(\nu/2)}. \tag{2.46}$$

The chi-squared variate may be obtained as the sum of ν squared independent standard normal variates.

As to the chi distribution, $\chi_\nu = \sqrt{\chi^2_\nu}$ has the density function

$$f = [2^{(\nu-2)/2}\Gamma(\nu/2)]^{-1}x^{\nu-1}\exp(-x^2/2), \tag{2.47}$$

and moments

$$\mu'_r = 2^{r/2}\Gamma[(\nu + r)/2]/\Gamma(\nu/2). \tag{2.48}$$

The case $\nu = 2$ is sometimes called the Rayleigh distribution.

2.6.4 Weibull Distribution

For positive x, α (a shape parameter), and λ (a scale parameter):

$$f = \alpha\lambda(\lambda x)^{\alpha-1}e^{-(\lambda x)^\alpha} \tag{2.49}$$

$$F = 1 - e^{-(\lambda x)^\alpha}. \tag{2.50}$$

Properties:

$$\mu'_r = \lambda^{-r}\Gamma[(\alpha + r)/\alpha] \tag{2.51}$$

$$\mu = \lambda^{-1}\Gamma[(\alpha + 1)/\alpha] \tag{2.52}$$

$$\sigma^2 = \lambda^{-2}\left\{\Gamma\left(\tfrac{\alpha+2}{\alpha}\right) - \left[\Gamma\left(\tfrac{\alpha+1}{\alpha}\right)\right]^2\right\}. \tag{2.53}$$

2.6.5 Stacy Distribution

Seeing the formulae (2.41) and (2.49) for p.d.f.'s, it takes no great leap of the intellect to imagine this one:

$$f = \frac{cx^{c\alpha-1}\exp[-(x/\beta)^c]}{\beta^{c\alpha}\Gamma(\alpha)}. \tag{2.54}$$

This is generally called the Stacy distribution, after Stacy (1962), but it dates back at least as far as Knibbs (1911). In the study of hydrology, at least in the U.S.S.R., it is known as the Kritsky-Menkel distribution (Sokolov et al, 1976, Section 2.3.3.1).

Properties:

$$\mu'_r = \beta^r\Gamma(\alpha + r/c)/\Gamma(\alpha). \tag{2.55}$$

When $\beta = 1$, the cumulant generating function is

$$K(t) = \log\Gamma(\alpha + t/i) - \log\Gamma(\alpha). \tag{2.56}$$

2.6.6 Comments on Skew Distributions

At a gross level, the shapes of the gamma, Weibull, and lognormal distributions are similar. If the starting point is a free parameter (so that the p.d.f. is nonzero for $X > a$, instead of $X > 0$), they all have three parameters. In this form, methods of estimating the parameters have been compared by Kappenman (1985).

2.6.7 Compound Exponential Distributions

Because of its lack-of-memory property, the exponential distribution is often considered to be the embodiment of true randomness. However, in the life-testing context, it can easily be imagined that the specimens on test differ in their quality, and hence the lifetimes do not have an exponential distribution. This is the compounding model, i.e., the exponential distribution's parameter λ itself is a random variable with some distribution.

If λ has a gamma distribution, the result is a Pareto distribution (Section 2.9 below).

Bhattacharya and Kumar (1986) considered the case of $1/\lambda$ having an inverse Gaussian distribution. They obtain a p.d.f. that involves a modified Bessel function of the third kind; the distribution has a decreasing failure rate. Earlier, Bhattacharya and Holla (1965) and Bhattacharya (1966) had considered $1/\lambda$ having various more elementary distributions.

2.7 Logistic, Laplace, and Cauchy Distributions

These are grouped together because they are symmetric and assign probability density to all of the real line, i.e., their support is from $-\infty$ to ∞, and thus may be seen as competitors of the normal distribution.

2.7.1 Logistic Distribution

For $-\infty < x < \infty$, and scale parameter $\beta > 0$:

$$f = \beta^{-1}\{\exp[-(x-\alpha)/\beta]\}\{1+\exp[-(x-\alpha)/\beta]\}^{-2} \quad (2.57)$$

$$F = \{1+\exp[-(x-\alpha)/\beta]\}^{-1} \quad (2.58)$$

$$= \tfrac{1}{2}\left[1+\tanh\left(\tfrac{x-\alpha}{2\beta}\right)\right]. \quad (2.59)$$

The mean, median, and mode are all α. The variance is $\beta^2\pi^2/3$.

Johnson's system of transformations (see Section 2.4.3) can be applied to a logistic variate, instead of starting with a normal variate (Tadikamalla and Johnson, 1982).

Tukey's lambda distribution may be regarded as a generalisation of the logistic. For this, instead of there being a nice equation for F in terms of x, there is an equation for x in terms of F:

$$x = [F^{\lambda} - (1-F)^{\lambda}]/\lambda. \quad (2.60)$$

On letting $\lambda \to 0$, we find $F = (1+e^{-x})^{-1}$. An extended family may be written as

$$x = \lambda_1 + [F^{\lambda_3} - (1-F)^{\lambda_4}]/\lambda_2. \quad (2.61)$$

2.7.2 Laplace Distribution

This is also known as the double exponential distribution. For $-\infty < x < \infty$ and $\phi > 0$:

$$f = \tfrac{1}{2}\phi^{-1}\exp(-|x-\theta|/\phi) \quad (2.62)$$

$$F = \left\{\begin{array}{ll} \tfrac{1}{2}\exp[-(\theta-x)/\phi] & \text{for } x \le \theta \\ 1-\tfrac{1}{2}\exp[-(x-\theta)/\phi] & \text{for } x \ge \theta \end{array}\right\} \quad (2.63)$$

The mean, median, and mode are all θ. The variance is $2\phi^2$.

Johnson's system of transformations (see Section 2.4.3) can be applied to a Laplace variate, instead of starting with a normal variate (Johnson, 1954).

2.7.3 The Generalised Error Distribution

To subsume the normal and Laplace distributions within one family, we may write:

$$f = \left[2^{(\delta+2)/2}\Gamma\left(\tfrac{\delta}{2}+1\right)\right]^{-1}\phi^{-1}\exp\left(-\frac{1}{2}\left|\frac{x-\theta}{\phi}\right|^{2/\delta}\right). \tag{2.64}$$

2.7.4 Cauchy Distribution

For $-\infty < x < \infty$, and scale parameter $\lambda > 0$:

$$f = (\pi\lambda)^{-1}\{1+[(x-\theta)/\lambda]^2\}^{-1} \tag{2.65}$$
$$F = \tfrac{1}{2} + \pi^{-1}\tan^{-1}[(x-\theta)/\lambda]. \tag{2.66}$$

The moments do not exist. However, θ and λ are location and scale parameters, respectively, and may be regarded as being analogous to mean and standard deviation.

Both median and mode are at θ.

This distribution, like the normal, is "stable", meaning that the distribution of the sample mean is of the same form as the parent distribution. In contrast to the normal distribution, the distribution of the sample mean has the *same* scale parameter as the parent distribution.

2.8 Extreme-Value Distributions

2.8.1 Type I

This is also known as Gumbel's extreme-value distribution. The support of X is the whole real line, from $-\infty$ to ∞. In largest extreme-value form,

$$F = \exp[-\exp(-x)] \tag{2.67}$$
$$f = \exp(-x)\exp[-\exp(-x)]. \tag{2.68}$$

In smallest extreme-value form,

$$F = 1-\exp(-\exp x) \tag{2.69}$$
$$f = \exp x\exp(-\exp x). \tag{2.70}$$

2.8.2 Types II and III

These include the Weibull distribution and ones closely similar to it.

Type II largest—for $x \geq 0$, $F = \exp(-x^{-\alpha})$.

Type III largest—for $x < 0$, $F = \exp(-|x|^{\alpha})$.

Type II smallest—alias Weibull—for $x \geq 0$, $F = 1-\exp(-x^{\alpha})$.

Type III smallest—for $x < 0$, $F = 1-\exp(-|x|^{-\alpha})$.

2.9 Pareto Distribution

For $x \geq k$ and $a > 0$:

$$f = \frac{ak^a}{x^{a+1}} \tag{2.71}$$
$$F = 1-\left(\frac{k}{x}\right)^a. \tag{2.72}$$

Properties:

$$\begin{aligned}\mu'_r &= \frac{ak^r}{a-r} &(2.73)\\ \mu &= ak(a-1) &(2.74)\\ \sigma^2 &= ak^2(a-1)^{-2}(a-2)^{-1}, \text{ if } a>2. &(2.75)\end{aligned}$$

This is sometimes referred as the "Pareto distribution of the first kind".

Another form of this distribution, known as "Pareto distribution of the second kind" (sometimes called the Lomax distribution) is given by, for $x \geq 0$,

$$\begin{aligned}F &= 1 - c^a/(x+c)^a &(2.76)\\ f &= ac^a/(x+c)^{a+1}. &(2.77)\end{aligned}$$

See Johnson and Kotz (1970a, p. 234) for details.

There is now a monograph devoted to Pareto distributions—Arnold (1983).

2.10 The Pearson System

All members of Karl Pearson's system of continuous densities satisfy the differential equation

$$\frac{df}{dx} = \frac{(x-a)\,f(x)}{b_0 + b_1 x + b_2 x^2}. \tag{2.78}$$

For $b_1 = b_2 = 0$, the density f is normal. There are twelve other types, many of which are better-known under other names, as shown in the following table.

Common name	Type	Density	Support
Beta (shifted)	I	$(1+x)^{m_1}(1-x)^{m_2}$	-1 to 1
	II	$(1-x^2)^m$	-1 to 1
Gamma	III	$x^m e^{-x}$	0 to ∞
	IV	$(1+x^2)^{-m}\exp(-v\tan^{-1}x)$	$-\infty$ to ∞
Reciprocal of gamma	V	$x^{-m}\exp(-x^{-1})$	0 to ∞
Inverted beta (F)	VI	$x^{m_2}(1+x)^{-m_1}$	0 to ∞
t	VII	$(1+x^2)^{-m}$	$-\infty$ to ∞
	VIII	$(1+x)^{-m}$	0 to 1
	IX	$(1+x)^m$	0 to 1
Exponential	X	e^{-x}	0 to ∞
Pareto	XI	x^{-m}	1 to ∞
	XII	$[(1+x)/(1-x)]^m$	-1 to 1

2.11 The Burr System

There are twelve types of Burr distribution. The two most important are as below. In both cases, the parameters c and k are positive; as usual, location and scale parameters can be introduced if required.

Type XII: for $x > 0$,

$$F = 1 - (1 + x^c)^{-k}. \tag{2.79}$$

Type III: for $x > 0$,

$$F = (1 + x^{-c})^{-k}. \tag{2.80}$$

2.12 t- and F-Distributions

The common feature of these distributions is that they are used chiefly not in describing the variability of some directly-observed quantity like length or time, but are obtained as the theoretical distribution of some calculated statistic.

2.12.1 t-Distribution

With ν being the degrees of freedom (effectively a shape parameter), for $x > 0$:

$$f = \frac{1}{\sqrt{\nu}B(\frac{1}{2}, \frac{\nu}{2})} \left(1 + \frac{x^2}{\nu}\right)^{-(\nu+1)/2} . \tag{2.81}$$

There is no convenient expression for F, except for the cases $\nu = 1, 2, 3$.

The mean is 0. The variance is $\nu/(\nu - 2)$, for $\nu > 2$.

The ratio $Z \Big/ \sqrt{X/\nu}$ has a t-distribution, if Z has a standard normal distribution, X has a chi-squared distribution with ν degrees of freedom, and Z and X are independent.

2.12.2 F-Distribution

This distribution is effectively the inverted beta (Section 2.5.3), written in a slightly different way. The pair ν_1 and ν_2, effectively two shape parameters, is referred to as the degrees of freedom of the distribution.

For $x > 0$:

$$f = \frac{\Gamma[(\nu_1 + \nu_2)/2]}{\Gamma(\nu_1/2)\Gamma(\nu_2/2)} \left(\frac{\nu_1}{\nu_2}\right)^{\nu_1/2} x^{(\nu_1-2)/2} \left(1 + \frac{\nu_1}{\nu_2}x\right)^{-(\nu_1+\nu_2)/2} . \tag{2.82}$$

$F(x)$ cannot be expressed in an elementary form.

For $\nu_2 > 2$, the mean is $\nu_2/(\nu_2 - 2)$. For $\nu_2 > 4$, the variance is $2\nu_2^2(\nu_1 + \nu_2 - 2)/[\nu_1(\nu_2 - 2)^2(\nu_2 - 4)]$. For $\nu_1 > 1$, the mode is $\nu_2(\nu_1 - 2)/[\nu_1(\nu_2 + 2)]$.

The ratio $(X_1/\nu_1)/(X_2/\nu_2)$ has a F-distribution, if X_1 and X_2 are independent chi-squared variates with ν_1 and ν_2 degrees of freedom, respectively. (The chi-squared—i.e., the gamma—is not the only distribution for which this is true, see Section 9.14 of Springer, 1979.)

2.13 Noncentral Distributions

The noncentral chi-squared variate, with ν degrees of freedom and noncentrality parameter λ, arises as $\sum_{i=1}^{\nu}(Z_i + a_i)^2$, where the Z's are independent standard normal variates and $\lambda = \sum_{i=1}^{\nu} a_i^2$.

The noncentral F-variate is obtained from the ratio of a noncentral chi-squared variate to an independent chi-squared variate, $\frac{\nu_2 \sum_{i=1}^{\nu_1}(Z_i+a_i)^2}{\nu_1 \sum_{i=\nu_1+1}^{\nu_1+\nu_2} Z_i^2}$. The doubly noncentral F-variate is similarly obtained from the ratio of two independent noncentral chi-squared variates.

The noncentral t-variate, with ν degrees of freedom and noncentrality parameter δ, arises as $(Z + \delta)/\sqrt{X/\nu}$, where Z is a standard normal variate and X is an independent chi-squared variate with ν degrees of freedom. The doubly noncentral t-variate is similarly obtained if X has a noncentral chi-squared distribution.

The noncentral beta variate is obtained as $X/(X+Y)$, where Y and X are independent chi-squared and noncentral chi-squared variates, respectively. If they are both noncentral chi-squared variates, $X/(X+Y)$ has the doubly noncentral beta distribution.

These distributions do not have elementary expressions for their p.d.f.'s or d.f.'s.

2.14 Some Lesser-Known Distributions

2.14.1 Inverse Gaussian Distribution

This is also known as the Wald distribution. For $x > 0$ and $\phi > 0$:

$$f = \sqrt{\frac{\phi}{2\pi}} e^{\phi} x^{-3/2} \exp\left[-\tfrac{1}{2}\phi(x + x^{-1})\right] \tag{2.83}$$

$$F = \Phi\left[(x-1)\sqrt{\phi/x}\right] + e^{2\phi}\Phi\left[-(x+1)\sqrt{\phi/x}\right]. \tag{2.84}$$

Properties: $\mu = 1$, and $\sigma^2 = \phi^{-1}$. When the mean μ is other than 1, the Wald distribution is generally known as the inverse Gaussian (because of the inverse relationship between the cumulant generating function of this distribution and that of the normal (Gaussian) distribution). In this case, the density is

$$f = \sqrt{\frac{\phi}{2\pi}} x^{-3/2} \exp\left[-\frac{\phi(x-\mu)^2}{2\mu^2 x}\right], \tag{2.85}$$

and the variance is now $\phi^{-1}\mu^3$.

2.14.2 Meixner Hypergeometric Distribution

$$f = [\pi\Gamma(a)]^{-1} 2^{a-2} \left|\Gamma\left(\frac{a}{2} + \frac{ix}{2}\right)\right|^2 e^{\gamma x}(\cos\gamma)^a \tag{2.86}$$

(in which $|\gamma| < \frac{\pi}{2}$, and $a > 0$). This is called the generalised hyperbolic secant distribution if $\gamma = 0$ (Harkness and Harkness, 1968). If, in addition, $a = 1$, it is known as the hyperbolic secant distribution.

$F(x)$ can be expressed through the incomplete beta function.

Properties:

$$\mu = a\tan\gamma \tag{2.87}$$

$$\sigma^2 = a[1 + (\tan\gamma)^2]. \tag{2.88}$$

2.14.3 Hyperbolic Distributions

The logarithm of the p.d.f. is a hyperbola. Omitting location and scale parameters,

$$f \propto \exp\left[-\zeta\left(\sqrt{(1+\eta^2)(1+x^2)} - \eta x\right)\right]. \tag{2.89}$$

2.14.4 Stable Distributions

If X's are i.i.d. r.v.'s and there exist constants $a_n > 0$ and b_n such that $a_n^{-1}\sum_{i=1}^{n} X_i - b_n$ has the same distribution as the X's, then this distribution is said to be "stable". $a_n = n^{1/\alpha}$, where α is known as the characteristic exponent ($0 < \alpha \leq 2$); $\alpha = 2$ for the normal distribution and $\alpha = 1$ for the Cauchy. In addition to α and scaling and centring constants, a skew parameter β is involved. The expression for the characteristic function is reasonably simple, but not that for the p.d.f. (except for certain very special cases). The chief area of application is in modelling certain economic phenomena that are claimed to have heavy-tailed distributions.

2.15 What's in the Encyclopaedia?

The encyclopaedia that we refer to is the *Encyclopedia of Statistical Sciences*.[3] We anticipate that for many years this extraordinary and admirable work will be the reference of first resort for a great many

[3] Published in 9 volumes over the years 1982–1988. (A supplementary volume is anticipated.) Publisher: Wiley, New York. Editors-in-Chief: S Kotz and N L Johnson.

topics of interest to statisticians. We give here a brief guide to those of the substantial entries concerning univariate distributions which are of most relevance to us.

Our lists are three in number:

- Named univariate distributions.
- Theory and concepts.
- Specialised areas of application.

After the name of the entry, we give its author, the year of publication, and a few words of explanation. (Those entries without authors named are by the editors of the encyclopaedia.)

2.15.1 Named Univariate Distributions

1. Bessel function distributions. 1982. (The density resulting from linear functions of independent chi-squared variables with common degrees of freedom involves a modified Bessel function of either the first or second kind.)

2. Burr distributions. R N Rodriguez. 1982. (The account of the univariate system is some four pages in length. Some densities are sketched, and a moment ratio diagram given.)

3a. Chi-square distribution. H O Lancaster. 1982. (About two pages in length. Oriented towards the distribution's use in goodness-of-fit testing.)

3b. Doubly noncentral distributions. 1982. (Brief explanation.)

4. Exponential distribution. J Galambos. 1982. (Definitions and history; basic properties; statistical inference.)

5. Exponential families. O Barndorff-Nielsen. 1982. (These have probability (point or density) functions of the form $b(\theta)a(x)\exp[q(\theta)t(x)]$, θ being a parameter; x and θ may be multidimensional. Continuous examples include the normal, gamma, and beta distributions. Headings in this article: first properties; regular exponential families; curved exponential families; incompletely observed exponential situations; concluding remarks.)

6. Extreme-value distributions. N R Mann and N D Singpurwalla. 1982. (Preliminaries; distribution of the extreme values; estimation.)

7. Frequency curves, systems of. R N Rodriguez. 1983. (Historical background; theoretical genesis of the commonly used systems; summary of the main functional forms; techniques for fitting; technical problems subsequent to fitting; special applications. Most discussion is of the Pearson system, series expansions, the Burr system, the symmetric Tukey lambda distributions, and Johnson's translation system.)

8. Gamma distribution. L Bain. 1983. (A six-page account that includes characterisations, applications, estimation, and generalisations.)

9. g-and-h-distributions. D C Hoaglin. 1983. ("In fitting a theoretical distribution to data, it is often advantageous to use quantiles rather than moments or the method of maximum likelihood, especially if one is concerned that a few observations may not fit in well with the rest. The family of g-and-h-distributions, introduced by Tukey," is defined in terms of quantiles relative to the standard normal distribution.)

10. Gumbel distribution. J Tiago de Oliveira. 1983. (Basic results; important characteristics; estimation of parameters; other statistical problems; tables and random numbers; extensions; applications.)

11. H-function distribution. M D Springer. 1983. (This refers to p.d.f.'s that are proportional to the H-function, a rather general transcendental function that includes many of the special functions of applied mathematics as special cases. The range of the variate is from 0 to ∞, and among the well-known distributions that are included in this class are the half-normal, beta, gamma, half-Cauchy, half-t, F, and Weibull. Products, quotients, and rational powers (but, in general, not sums and differences) of independent H-function r.v.'s are themselves H-function r.v.'s.)

12. Hyperbolic distributions. O Barndorff-Nielsen and P Blæsild. 1983. (For these, the logarithm of the p.d.f. is a hyperbola. In addition to location and scale parameters, there are two others. After an introduction, this article discusses the univariate case, the multivariate case, and related distributions.)

13. Inverse distributions. J L Folks. 1983. (One use of this term is to describe the distributions of X and Y when $Y = 1/X$, or, more generally, when any one-to-one transformation connects X and Y. Beta distributions of the first and second kinds provide an example. See also 2.15.2—16 below.)

14. Inverse Gaussian distribution. J L Folks. 1983. (A two-page account covering history, formulae, properties, estimation, and applications.)

15. Inverted beta distribution. M V Ratnaparkhi. 1983. (Six types of application; definition and structure; generalised inverted beta distribution.)

16. Johnson's system of distributions. K O Bowman and L R Shenton. 1983. (A twelve-page account: introduction, including graphs of the density; the S_U system; the S_B system; examples; general remarks.)

17a. Laplace distribution. J K Ord. 1983. (Introduction; properties; estimation; related distributions; bivariate distributions.)

17b. Lambda distributions. 1983. (The expression for X in terms of F is $X = [F^\lambda + (1-F)^\lambda]/\lambda$.)

18. Linear exponential family. G P Patil. 1985. (This refers to p.d.f.'s that are proportional to $a(x)\exp(\omega x)$; x and ω may be multidimensional.)

19. Linear hazard rate distribution. C B Read. 1985. (As its name suggests, for this distribution, $f/(1-F)$ is a linear function of x.)

20a. Logistic distribution. H J Malik. 1985. (Introduction and basic formulae; order statistics; estimation; multivariate logistic distributions; tables.)

20b. Log-Laplace distribution. 1985. (An expression for the moments is included.)

21. Lognormal distribution. C E Antle. 1985. (Important properties; estimation of parameters.)

22. MacDonald distribution. 1985. (The p.d.f. involves a modified Bessel function of the third kind.)

23. Mixture distributions. B S Everitt. 1985. (Definitions; applications; commonly-occurring mixture distributions; identifiability; estimation; detecting mixture distributions.)

24. Modified normal distributions. 1985. (These are certain compound distributions derived from the normal.)

25. Natural exponential families. C N Morris. 1985. (These include the normal and gamma, but not the lognormal or beta, distributions.)

26. Noncentral chi-square distribution. M Tiku. 1985. (Definition; m.g.f.; reproductivity; decomposition; probability integral; approximations; applications.)

27. Noncentral F-distribution. M Tiku. 1985. (Definition; moments; probability integral; approximations; applications.)

28. Noncentral t-distribution. D B Owen. 1985. (Definition; representations; applications; tables.)

29. Normal distribution. C B Read. 1985. (Introduction; historical background; tabulation and computation; large-sample role; choice of a standard normal density; transformations to normality—approximations; role in sampling theory.)

30. Omega distribution. P W Mielke. 1985. (In addition to location and scale, this distribution has one further parameter. The Laplace, logistic, and uniform distributions are included as special cases.)

31. Pareto distribution. B C Arnold. 1985. (Introduction and history; the classical Pareto distribution and related variants; distributional properties; inference; related distributions.)

32. Partially truncated distributions. 1985. (As distinct from full truncation, this refers to a proportion of values which fall within a certain interval being discarded, rather than all of them.)

33. Pearson system of distributions. J K Ord. 1985. (Introduction; the distributions; fitting; the discrete system; related computer programs.)

34. Perks' distributions. 1985. (These distributions generalise the logistic: with both m_1 and m_2 usually being small, $f = [\sum_{n=0}^{m_1} a_{1n}\exp(-n\theta x)]/[\sum_{n=0}^{m_2} a_{2n}\exp(-n\theta x)]$.)

35. Phase type distributions. M Shaked and J G Shanthikumar. 1985. (Introduction; univariate PH distributions; renewal processes of phase type; multivariate PH distributions; applications of PH distributions.)

36. Planck distribution. 1986. (In standard form, the p.d.f. is proportional to $x^\gamma/(e^x - 1)$.)

37. Poly-t distributions. 1986. (The p.d.f.'s are proportional to products of densities of Pearson type VII distributions.)

38. Quartic exponential distribution. 1986. ($f \propto \exp(\text{quartic in } x)$.)

39. Randomized gamma distribution. K Hirano. 1986. (This is a kind of compound gamma distribution.)

40. Rayleigh distribution. K Hirano. 1986. (Here, this term is used as another name for the chi distribution, $f \propto x^{n-1}\exp(-x^2/2)$.)

41. Sargan distributions. 1988. (These are generalisations of the Laplace distribution, having $f \propto e^{-|x|}(1+\sum_{i=1}^{n}\alpha_i|x|^i)$.)

42. Skew-normal distributions. 1988. ($f \propto \phi\Phi$, ϕ and Φ being the standard normal density and distribution functions.)

43. Slash distribution. K Kafadar. 1988. (This arises from the ratio of a normal r.v. to an independent uniform r.v. Its density is $[\phi(0)-\phi(x)]/x^2$, where ϕ is the standard normal density.)

44. Stable distributions. D Monrad and W Stout. 1988. (Introductory explanation; the characteristic function; examples of symmetric stable distributions are the normal and the Cauchy; skew stable distributions also occur; the central limit problem; modelling electrical fields induced by charged particles in random motion; modelling economic phenomena; criticisms of the empirically-based arguments in the econometric applications.)

45. Subexponential distributions. E Willekens. 1988. (These are a class of long-tailed distributions.)

46. T_1- and T_2-classes of distributions. L Bondesson. 1988. (T_1-distributions were introduced by Thorin in 1977 under the name "generalised Γ-convolutions". An extension, the T_2-distributions, was introduced by Bondesson in 1981 to obtain a class of infinitely divisible distributions containing all mixtures of exponential distributions.)

47. t distribution. C B Read. 1988. (A two-page account, useful for the references to algorithms and tables.)

48. Trapezoidal distributions. 1988. (The name reflects the shape of the p.d.f.)

49. Uniform (or rectangular) distributions. C B Read. 1988. (Discusses the discrete and continuous univariate cases, including applications, properties, and characterisations, inference, related distributions, and the bivariate case.)

50. Wakeby distributions. 1988. (This is a five-parameter class said to be popular in hydrology. It is defined in terms of the inverse of the d.f., $x(F)$, by $x(F) = \xi + \theta\beta^{-1}[1-(1-F)^\beta] - \gamma\delta^{-1}[1-(1-F)^\delta]$.)

51. Weibull distribution. C E Antle and L J Bain. 1988. (Definitions and historical remarks; relation to extreme-value distributions; properties; statistical inference for the two-parameter Weibull; statistical inference for the three-parameter Weibull; competing models.)

52. Weibull-exponential distribution. 1988. (Below a change point, this is an exponential distribution; above, a Weibull hazard rate is superposed.)

53. Weighted distributions. G P Patil, C R Rao, and M Zelen. (This topic concerns the situation in which observations of X do not enter the record with equal probability, but with probability dependent upon x. An important special case is when this probability is proportional to x—i.e., size-based sampling. Introduction; univariate weighted distributions; weight functions and properties; bivariate weighted distributions; posterior and weighted distributions; weighted distributions in stochastic population dynamics; applications.)

54. Whittaker-type distributions. 1988. (The p.d.f. of a stable distribution with characteristic exponent $\frac{2}{3}$ can be written explicitly in terms of a Whittaker function.)

55. Zero degrees of freedom. A F Siegel. 1988. (The noncentral chi-squared distribution with zero degrees of freedom is a mixture of a finite probability at $X = 0$ and a continuous distribution over $X > 0$. Properties of t- and F-distributions with zero degrees of freedom follow naturally.)

2.15.2 Theory and Concepts

1. Approximations to distributions. K O Bowman and L R Shenton. 1982. (Historical developments; test statistics in distribution; the Pearson system; translation systems; series developments; multivariate densities; discrete distributions.)

2. Characteristic functions. R G Laha. 1982. (Some elementary properties; basic theorems; some criteria for a complex-valued function to be a characteristic function; infinitely divisible distributions; general decomposition theorems; analytic characteristic functions; some generalisations.)

3. Characterizations of distributions. J Galambos. 1982. (Introduction; characterisations through limit theorems; normal distribution; exponential distribution; transformation to exponentiality; gamma distributions.)

4. Cornish-Fisher and Edgeworth expansions. E D McCune and H L Gray. 1982. (The subject is the representation of one distribution function in terms of a second, and of a quantile of one distribution in terms of the corresponding quantile of a second. This second distribution is usually the normal. The article is divided into: the expansions; applications of the Edgeworth expansion; generalisations and modifications of Edgeworth expansions; multivariate Edgeworth expansions; applications of Cornish-Fisher expansions; generalisations and modifications of the Cornish-Fisher expansions; concluding remarks.)

5. Departures from normality, tests for. R B D'Agostino. 1982. (First tests for departures from normality; later developments; power studies.)

6. Domain of attraction. W Stout. 1982. ("Recall that a random phenomenon is often modelled by a normal random variable, the justification being that the phenomenon results from the summation of many independent and roughly identically distributed random quantities. However, this argument is invalid if we are unwilling to assume that the summed random quantities have finite variances. Because of the theory of domains of attraction, one is forced in the infinite variances case to consider that a random phenomenon may be modelled by a nonnormal stable random variable.")

7. EDF statistics. M A Stephens. 1982. (Definitions of the major measures of discrepancy between distribution functions; distributions; goodness of fit and power, first with the distribution being completely specified, then with unknown location or scale parameters; miscellaneous topics. There are separate encyclopaedia entries on such cases as the Anderson-Darling test and Kolmogorov-Smirnov statistics.)

8. Extremal processes. S I Resnick. 1982. (This is a tool for the study of the stochastic behaviour of maxima and records.)

9. Generating functions. W L Smith. 1983. (Includes a page on moment generating functions.)

10. Generation of random variables. D C Hoaglin. 1983. (Uniform random numbers; quality of congruential generators; nonuniform continuous distributions; nonuniform discrete distributions; literature.)

11. Geometry in statistics: Convexity. M Shaked. 1983. (Includes mention of densities whose logarithm is concave, such as the normal, gamma, beta, and Weibull.)

12. Hazard rate and other classifications of distributions. J K Patel. 1983. (Explains concepts like increasing hazard rate, increasing hazard rate on the average, new better than used, new better than used in expectation, and decreasing mean residual life.)

13. Heavy-tailed distributions. M C Bryson. 1983. (Discusses various notions of what is meant by "heavy-tailed", e.g., infinite variance.)

14. Infinite divisibility. F W Steutel. 1983. (X is infinitely divisible if, for any n, there exist i.i.d. r.v.'s W_i such that X has the same shape of distribution as $\sum_{i=1}^{n} W_i$. Many well-known distributions are infinitely divisible, including the normal, lognormal, t, F, and chi-squared.)

15. Integral transforms. M D Springer. 1983. (An example is the characteristic function, as it is in fact a Fourier integral transform. The Laplace or bilateral Laplace transforms may sometimes be useful instead. The Mellin integral transform is a natural tool for deriving the distribution of products, quotients, and rational powers of continuous r.v.'s, and a modified version is applicable even to r.v.'s that can take negative as well as positive values.)

16. Inverse distributions. J L Folks. 1983. (One use of this term is to describe the distributions of X and Y when there is an inverse relationship between certain functions, such as their cumulant generating functions. The normal and inverse Gaussian distributions are inverses in this sense. See also 2.15.1—13 above.)

17. Laws of error I: Development of the concept. C Eisenhart. 1983. (A historical account: introduction and summary; Simpson; Lagrange; Lambert; Bernoulli; Laplace.)

18. Laws of error II: The Gaussian distribution. C Eisenhart. 1983. (A historical account: Gauss's derivation; criticism of Gauss's derivation; Laplace provides support for the law; Adrain's two derivations; further "proofs" of Gauss's law of error; empirical support.)

19. Laws of error III: Later (non-Gaussian) distributions. C Eisenhart. 1983. (A historical account: discordant observations and mixtures of Gaussian laws of error; failure of the "arithmetic mean rule"; corrections for skewness or kurtosis; Pearson curves; concluding remarks.)

20. *L* class laws. B M Schreiber. 1983. (These distributions are a subclass of the infinitely divisible ones, and include are the stable ones.)

21. Limit theorem, central. C C Heyde. 1983. (This refers to convergence in distribution to the normal law for a sum of an increasing number of random variables. In this article, five theorems are stated and discussed.)

22. Link tests. D Pregibon. 1985. (A model frequently used in dose-response experiments is $g(\mu)$ = linear function of explanatory variables, where μ is the mean of the dependent variable and g is the inverse of some univariate distribution function. This article discusses testing the adequacy of a particular hypothesised link function, g.)

23. Log-concave and log-convex distributions. F W Steutel. 1985. (These terms refer to densities whose logarithm is a concave or convex function, respectively.)

24. Lorenz curve. C Dagum. 1985. (This is the first moment distribution function of a probability distribution. In the economic context, it is a plot of the cumulative proportion of income received against the cumulative proportion of people receiving it, they having been listed in order from least to greatest income.)

25. Luce's choice axiom and generalizations. D J Strauss. 1985. (This refers to the probabilities of a certain choice when made from a full set of alternatives and when made from a subset. The connexion with univariate distributions is through random utility models.)

26. Mean, median, mode, and skewness. H L MacGillivray. 1985. (Introduction; measures of location; the ordering of mean, median, and mode; skewness and measures thereof.)

27. Model construction: Selection of distributions. C P Quesenberry. 1985. (This article considers the problem of, given a sample, selecting which of a collection of families of distribution functions is the best-fitting family: invariant selection rules; maximum likelihood selection rules; other techniques.)

28. Moment problem. V K Rohatgi. 1985. (This is the problem of characterising a distribution from its set of moments.)

29. Moment-ratio diagrams. 1985. (These are used to show the combinations of β_1 ($= \mu_3^2/\mu_2^3$) and β_2 ($= \mu_4/\mu_2^2$) that are possible for a class of distributions. The traditional form has β_1 on the horizontal axis and β_2 plotted downwards on the vertical axis. Two other forms: $\sqrt{\beta_1}$ vs. β_2; and Craig's suggestion, β_1 vs. $(2\beta_2 - 3\beta_1 - 6)/(\beta_2 + 3)$.)

30. Normal extremes. J P Cohen. 1985. (Concerned with the distribution of the largest observation in a sample from a normal distribution, and its limiting behaviour for large sample sizes.)

31. Ordering distributions by dispersion. M Shaked. 1985. (Intuitive notions; orderings based on functions of $X - \mu_X$ and $Y - \mu_Y$; ordering based on the quantile functions; applications.)

32. Ordering of distributions, partial. H Oja. 1985. (Introduction; partial ordering of univariate distributions—location, scale, other; partial ordering of multivariate distributions.)

33. Ordering, starshaped. R L Dykstra. 1985. (This refers to functions ϕ such that $\phi(\alpha x) \leq \alpha\phi(x)$ for x being positive and α being between 0 and 1. There are applications in reliability theory.)

34. Order statistics. H A David. 1985. (Motivation from applications; distribution theory and estimation; asymptotic theory; further applications.)

35. Orthogonal expansions. R C Griffiths. 1985. (Univariate; bivariate; multivariate.)

36. Outlier-prone distribution. R F Green. 1985. (Definition of Neyman and Scott. Examples: the normal distribution is outlier-resistant, the Cauchy is outlier-prone.)

37. Percentiles, estimation of. J P Keating and R C Tripathi. 1985. (Definition; location and scale parameter distributions; universal techniques, parametric and nonparametric.)

38. Pólya type 2 frequency (PF_2) distributions. B R Rao. 1986. (This concept is of importance in reliability theory, being closely connected with the increasing failure rate property. Normal, gamma, Laplace, and some Weibull distributions are among those which are PF_2.)

39. Probability integral transformations. C P Quesenberry. 1986. (This refers to the well-known result that if X has a continuous distribution function $F(x)$, then the random variable $U = F(x)$ has a uniform distribution over 0 to 1. This article deals with multivariate and conditional transformations as well as the univariate.)

40. Probability plotting. G L Sievers. 1986. (Cumulative distribution function plots; quantile and percent plots; optimal plotting positions; gamma and half-normal plots; two-sample problems; plotting for

symmetry.)

41. Quantile estimation. C E Davis and S M Steinberg. 1986. (Distribution-free methods and distribution-specific ones are both dealt with.)

42. Quantiles. R L Eubank. 1986. (Introduction, including a table of the distribution, quantile, density, and density-quantile functions of the normal, lognormal, exponential, Weibull, extreme value, logistic, Pareto, Cauchy, uniform, reciprocal of a uniform, and Laplace laws; asymptotic properties of, and nonparametric inference for, sample quantiles; parameter estimation; descriptive statistics and exploratory data analysis; conditional quantiles.)

43. Radial error. V Chew. 1986. (One may derive from the distribution of X and Y, the distribution of the radial distance $\sqrt{X^2+Y^2}$.)

44. Radon transformation. 1986. (One may derive from the distribution of X and Y, the distribution of the projection of (X, Y) on to the line $Y = X \tan\theta$.)

45. Random utility models. D J Strauss. 1986. (The idea is that a subject's utility for a stimulus is a random quantity, and that he or she chooses according to which utility variable is, on that realisation of the set of utilities, the largest. The distributions of the utilities may, for example, be normal, with common standard deviation but different means.)

46. Relevation. L A Baxter. 1988. (This is a concept used in reliability: the relevation of two lifetime distributions F and G is $1 - F(x) + [1 - G(x)] \int_0^x f'/(1-G)$.)

47. Reliability, nonparametric methods in. J F Lawless. 1988. (Includes a discussion of nonparametric estimation of a lifetime distribution.)

48. Reliability, probabilistic. A D Basu. 1988. (Introduction; parametric models—exponential, Weibull, and gamma distributions; ageing concepts, including increasing failure rate, increasing failure rate in average, new better than used, decreasing mean residual life, new better than used in expectation, and harmonic new better than used in expectation; reliability of complex systems.)

49. Reproductive models. O E Barndorff-Nielsen and P Blæsild. 1988. (This entry is one of the encyclopaedia's less comprehensible ones, but the concept seems to be something to do with the distribution of a sample mean belonging to the same family as the distribution of the individual observations.)

50. Saddle point approximations. G H Weiss. 1988. ("The saddle point method, or steepest descent method, is a technique for asymptotic evaluation of [certain] integrals... [that] frequently appear in probability theory, the most common source being the inversion formula for the distribution of a sum of identically distributed random variables... [It] is especially useful when an accurate representation is desired for the tail of a distribution.")

51. Self-reciprocal distributions. 1988. (The reciprocal distribution of one with characteristic function $\varphi(t)$ is one with p.d.f. proportional to $\varphi(x)$; if the reciprocal distribution is the same as the original one, it is called self-reciprocal.)

52. Semi-parametric models. D Oakes. 1988. (These include both unknown parameters and unknown functions; interest will usually centre on the parameters, the unknown functions being a sort of nuisance parameter; the motivation is to avoid restrictive assumptions about secondary aspects of a problem while preserving a tight formulation for the features of primary concern. Such models are certainly of much broader application than only to univariate distributions, but the estimation of a mean when the distributional form is not known provides one example.)

53. Shock models. W S Griffith. 1988. (Includes discussion of the various stochastic ageing processes mentioned under "Reliability, probabilistic", above.)

54. Shot-noise processes and distributions. L Bondesson. 1988. (Many statistical distributions appear as marginal distributions for shot-noise processes, a class of stochastic processes arising in various applied fields such as hydrology, insurance mathematics, and queueing theory.)

55. Spacings. C B Read. 1988. (Spacings are differences between observations from a distribution, when ordered from smallest to largest. This article considers uniform spacings, spacings constructed from exponential variables, tests, and mth order spacings.)

56. Statistical catastrophe theory. L Cobb. 1988. (Introduction; time-series analysis; exponential families; response surface analysis; equivalence of densities.)

57. Stochastic ordering. W Whitt. 1988. (Introduction; stochastic orderings on general spaces; coupling—constructing ordered random elements; establishing stochastic order; stochastic ordering and

conditioning; applications.)

58. Tail ordering. 1988. (F_2 is said to be tail-ordered with respect to F_1 if $F_1^{-1}[F_2(x)] - x$ is a nondecreasing function of x. There are connexions with concepts like increasing failure rate.)

59. Tail probabilities. B M Hill. 1988. (Discusses (i) inference about the parameters of the tails of a distribution when the overall form of the distribution is unknown, and (ii) an aspect of Bayesian inference.)

60. Total positivity. J S Kim and F Proschan. 1988. (Introduction; definition and basic properties; applications to statistical decision theory; applications in probability and stochastic processes; applications in inventory problems; applications in reliability and life testing; multivariate total positivity.)

61. Total time on test transform. B Bergman and B Klefsjö. 1988. (This is an idea mostly used in the analysis of failure data, though there is a close relation to the Lorenz curve of economics.)

62. Uniformity, tests of. M A Stephens. 1988. (Introduction; tests of uniformity; transformations from uniforms to uniforms; tests arising from tests of exponentiality; practical application of tests.)

63. Unimodality. 1988. (Explanation; consequences for moments; the discrete case.)

64. Utility theory. P C Fishburn. 1988. (This account does not include probabilistic utility, for which, see 2.15.2—25 and 2.15.2—45. Introduction; ordered preferences; expected utility; risk attitudes; subjective expected utility; new directions.)

65. Van Zwet tail ordering. 1988. (According to this, F_1 has lighter tails than F_2 if $F_2^{-1}[F_1(x)]$ is convex for $x \geq 0$.)

66. Variance dilation. M Shaked. 1988. (This is a notion of "more variable".)

67. Variance, upper bounds. J W Seaman and P L Odell. 1988. (Basic inequalities; variance inequalities and dispersion orderings; inequalities for special applications.)

2.15.3 Specialised Areas of Application

1. Actuarial statistics—Life. E A Lew. 1982. (Includes some formulae for the "force of mortality", $f/(1-F)$.)

2. Fatigue models. S C Saunders. 1983. (This refers to failure of materials, principally metals, after a period of sustaining a load of fluctuating magnitude.)

3. Income distribution models. C Dagum. 1983. (Pareto's model; generating systems, including Pearson's, D'Addario's, and the generalised logistic-Burr; properties; application.)

4. Life testing. N R Mann and N D Singpurwalla. 1983. (Introduction; censoring patterns; accelerated life testing; sequential life tests.)

5. Particle-size statistics. A S Goldman. 1985. (Applications areas; the lognormal and hyperbolic distributions; shape and its definition; practical problems with the data; packing density.)

6. Phase problem of X-ray crystallography, probabilistic methods in the. H Hauptman. 1985. (The phase problem; the normalised structure factors; the structure invariants; the fundamental principle of direct methods; the neighborhood principle; the neighborhoods of the structure invariants; conditional probability distributions of the structure invariants; summary.)

7. Risk theory. H L Seal. 1988. (Mentions Pareto and lognormal distributions as descriptions of insurance claim sizes.)

8. Survival analysis. P K Andersen and M Væth. 1988. (Survival data; nonparametric survival models; parametric survival models; inference for parametric models; regression models; some extensions; an example.)

9. Survival analysis, grouped data. J A Deddens and G G Koch. 1988. (Introduction; data structure; statistical methods; examples.)

10. Thurstone's theory of comparative judgment. D B MacKay. 1988. (This is a psychophysical idea: subjects are asked to express a dominance judgment about two stimuli; it is envisaged the outcome is determined by a comparison of two r.v.'s; often, these are assumed to be normally-distributed, though other distributional assumptions are sometimes made.)

2.16 What's in Patil et al's Bibliography?

The bibliography that we refer to is the *Dictionary and Classified Bibliography of Statistical Distributions in Scientific Work, Volume 2, Continuous Univariate Models* (Patil, Boswell, and Ratnaparkhi, 1984).

The main contents of this are as follows.

- Pages 11–159 list the basic properties of 151 distributions.
- There are 15 charts showing transformations that interrelate the various distributions.
- There are 97 graphs of selected p.d.f.'s; each graph typically has about 3 curves on it, corresponding to different values of a parameter of the distribution.
- A bibliography of over 5000 entries, arranged alphabetically by author, with an index that uses the name of the distribution and the type of statistical inference or application.
- An addendum to the bibliography, covering the period 1980–1983, and including over 300 entries.

Chapter 3

Overview of Methods of Constructing Bivariate Distributions

3.1 Introduction

3.1.1 Contents of this Chapter

This Chapter describes a number of methods that have been used with success in obtaining bivariate distributions:

3.2 The marginal transformation method.

3.3 Some inelegant methods.

3.4 Mixing and compounding.

3.5 Regression and conditional distribution.

3.6 Trivariate reduction.

3.7 Weighted linear combination.

3.8 Limits of discrete distributions.

3.9 Differential equation methods.

3.10 Diagonal expansion.

3.11 Bivariate Edgeworth expansion.

Before starting on the main business of the Chapter, we have a few words to say about the Fréchet bounds, about transformations, and about computer algebra.

3.1.2 The Fréchet Bounds

For given marginals F and G, what limits must a bivariate d.f. H satisfy, as a result of its p.d.f. (or probability function) needing to be non-negative everywhere? Hoeffding (1940) and Fréchet (1951) answered this as follows:

$$H^-(x,y) \leq H(x,y) \leq H^+(x,y), \tag{3.1}$$

where

$$H^+(x,y) = \min[F(x), G(y)] \tag{3.2}$$

and

$$H^-(x,y) = \max[F(x) + G(y) - 1, 0]. \tag{3.3}$$

It is easy to verify that the Fréchet bounds H^+ and H^- are themselves d.f.'s, and that they have maximum and minimum correlations for the given marginals. Also, H^+ concentrates all the probability on the increasing curve $F(x) = G(y)$, and H^- concentrates all the probability on the decreasing curve $F(x) + G(y) = 1$. For any F and G, $[F^{-1}(U), G^{-1}(U)]$ has d.f. H^+ and $[F^{-1}(U), G^{-1}(1-U)]$ has d.f. H^- (where U denotes a variate that is uniformly distributed between 0 and 1). For proofs and a discussion on H^+ and H^-, see, for example, Whitt (1976), who also proved that convolution of identical bivariate distributions results in an increase of the upper Fréchet bound and a decrease of the lower.

In order to have notation for the independent case, we further define

$$H^0(x,y) = F(x)G(y). \tag{3.4}$$

Devroye (1986, p. 581) uses the adjective "comprehensive" for any family of distributions that includes H^+, H^0, and H^-.

What if the distribution is restricted to the region $X \leq Y$? W Smith (1983) has shown that in this case, the bounds on H become:

$$G(y) - \max\{0, \min[G(y) - G(x), F(y) - F(x)]\} \leq H(x,y) \leq \min[F(x), G(y)]. \tag{3.5}$$

(The lower bound is not necessarily a distribution function. A sufficient condition for it to be one is that there exists an x_0 such that $g(x) > f(x)$ for $x > x_0$, and $g(x) < f(x)$ for $x < x_0$.)

Regarding Fréchet bounds for *multi*variate distributions, see Kwerel (1983).

3.1.3 Transformations

Suppose we have a density $h(x,y)$, and we create new variates $A(X,Y)$ and $B(X,Y)$. What is the joint density of A and B? To answer this, we first need to obtain X and Y in terms of A and B. Letting, as usual, the values of variates X, Y, A, and B be x, y, a, and b, the density of (A,B) is then $h[x(a,b), y(a,b)]\,|J|$, where J is the "Jacobian", given by

$$J = \begin{vmatrix} \frac{\partial x}{\partial a} & \frac{\partial x}{\partial b} \\ \frac{\partial y}{\partial a} & \frac{\partial y}{\partial b} \end{vmatrix} = \frac{\partial x}{\partial a}\frac{\partial y}{\partial b} - \frac{\partial y}{\partial a}\frac{\partial x}{\partial b} = \frac{1}{\frac{\partial a}{\partial x}\frac{\partial b}{\partial y} - \frac{\partial b}{\partial x}\frac{\partial a}{\partial y}}. \tag{3.6}$$

For a fuller explanation, which includes pictures of a rectangle being transformed into a distorted rectangle, see Blake (1979, Section 7.2). Transformations often met with include $X+Y$, $X-Y$, XY, X/Y, $X/(X+Y)$, $\sqrt{X^2+Y^2}$, and $\tan^{-1}(Y/X)$. See, for example, Blake (1979, Section 7.2).

Now consider the special case of transforming the marginals. Suppose X and Y are each uniformly distributed between 0 and 1, and we transform the marginals so that they become F and G (with densities f and g). In this case, $A \equiv F^{-1}$ and $B \equiv G^{-1}$, so that $X \equiv F$ and $Y \equiv G$. So the density of (A,B) will be $h[F(a), G(b)]\frac{\partial F}{\partial a}\frac{\partial G}{\partial b} = h[F(a), G(b)]f(a)g(b)$.

Physicists have apparently found it helpful to put the conditions that a p.d.f. has to satisfy (non-negative and integrates to 1), together with what happens under transformation of the marginals, into the following form. Bivariate densities having $f(x)$ and $g(y)$ as their marginal densities and $F(x)$ and $G(y)$ as their marginal d.f.'s must be of the form $h = fg[1 + a(F,G)]$, where $a(u,v)$ is any function on the unit square that is bounded below by -1 and which satisfies $\int_0^1 a(u,v)\,du = \int_0^1 a(u,v)\,dv = 0$. See Finch and Groblicki (1984), and also Cohen and Zaparovanny (1980).

3.1.4 Computer Algebra

In some areas of statistics, the results we are able to obtain are determined by what we are clever enough to manipulate mathematically, rather than by anything of real conceptual depth. For instance, suppose we have a bivariate survival function $\bar{H}_1(x,y)$ and a bivariate p.d.f. $h_2(\theta_1, \theta_2)$. Then we can obtain another bivariate survival function by compounding: $\int_0^\infty \int_0^\infty \bar{H}_1(\theta_1 x, \theta_2 y) h_2(\theta_1, \theta_2)\, d\theta_1\, d\theta_2$ (as in Sections 20.3.3 and 20.7). Our results are determined by our ingenuity in choosing $\bar{H}_1$ and h_2 such that the double integration is tractable.

The increasing sophistication and widening availability of packages for computerised algebraic manipulation, such as MACSYMA[1] and REDUCE,[2] gives hope that this limitation may diminish in importance in the next few years. See, for example, Steele (1985) and Bryan-Jones (1987) for more on this.[3] Actually, there have been attempts to apply computer algebra to statistical problems for a number of years—for example, Dunn and Houston (1975) reported development of a program to decode the probability generating function of a discrete distribution. Perhaps the most important type of symbolic manipulation is integration: a paper by Davenport (1985) reviews this (though not specifically with application to statistics).

Indeed, we can ask why should we want an explicit expression for $\int\int \bar{H}_1 h_2$. One could say that this itself contains all the modelling information—and we should be looking to fit this directly to data.

- One can imagine the interfacing of computer algebra packages with those for model-fitting, so that for given $\bar{H}_1$ and h_2, the algebra part solves the double integral and passes the result to the model-fitting part.

- Or, number-crunching becoming as fast as it is, the double integral could be evaluated numerically whenever required by the model-fitting package.

The above discussion has been couched in terms of the compounding method for constructing distributions, but applies equally to other methods, too.

3.2 The Marginal Transformation Method

3.2.1 General Description

The idea here, usually attributed to Nataf (1962), is to start with any bivariate distribution, and apply monotonic transformations to X and Y, in order to arrive at a bivariate distribution with marginals that are different from the original ones. In the univariate context, familiar examples are (i) transforming the normal distribution so that it becomes lognormal, and (ii) transforming the exponential distribution so that it becomes Weibull.

The emphasis may take either of two forms:

- Start with the bivariate normal distribution. Accept its description of how X and Y are interconnected as satisfactory, but suppose normal marginals are unsatisfactory for the purpose at hand. Transform the marginals so that they become what is required.

- Start with an empirical or unfamiliar bivariate distribution. In order to compare its contours or other properties with the bivariate normal, free from the influence of the forms of the marginals, transform its marginals to be normal.

Other distributions are sometimes used as standards—uniform and exponential marginals are two examples. In particular, we will meet in Chapter 5 a number of distributions written as their uniform representations, from which it is easy to transform to any other marginals required.

The algebra of this method has been given in Section 3.1.3.

3.2.2 Johnson's Translation Method

Johnson (1949) considered starting with the bivariate normal distribution, and transforming X and/or Y so that the marginals

- Remained the same,

[1] See, for example, Rand (1984).

[2] See, for example, Rayna (1987). At pp. 29–31, there is a brief description of symbolic integration.

[3] We may note that computer algebra is not the only possible approach to computerised manipulation of statistical distributions. Groner and Cook (1986) describe a method in which a distribution is represented by a frequency table, rather than by an algebraic expression. Their paper is on univariate distributions, but they foreshadow another on correlated variates and multivariate distributions.

- Became lognormal,
- Became logit-normal, or
- Became $\sinh^{-1}$-normal.

This is traditionally referred to as a "translation" method, though we think "transformation" would be a better term.

Subscripts N, L, B, and U are used for the four alternatives. There being four choices for X and four for Y, sixteen possible bivariate distributions result. They are listed in Table 2 (p. 17) of Johnson and Kotz (1972). For example, h_{NN} is the bivariate normal density whereas h_{LL} is the bivariate lognormal density function. See Section 4.14.2 for more on Johnson's translation method.

It is well-known that Pearson's product-moment correlation is not affected by linear transformations of X and Y. But what of applying non-linear transformations? The answer is that if we start with the bivariate normal distribution and apply do this, the correlation becomes smaller (in absolute magnitude). See Section 4.14.5 for more on this.

3.2.3 Uniform Representation

Important advantages of considering distributions after their marginals have been made uniform are:

- Independence does not usually have a clear geometric meaning, in that the graph of the joint p.d.f. of X and Y provides us no insight as to whether or not X and Y are independent. However, independence takes on a geometric meaning for variates U and V with uniform marginals—they are independent if and only if their joint p.d.f. is constant. Any variation in value of p.d.f. is indicative of dependence between U and V.

- The uniform representation—*copula* is an alternative term—is the natural framework in which to discuss nonparametric measures of correlation, such as Kendall's τ and the grade correlation ρ_g.

Chapter 5 presents distributions that are expressed as copulas.

3.2.4 Some Properties Unaffected by Transformation

For any family H_θ $(-1 \leq \theta \leq 1)$ of d.f.'s having absolutely continuous marginals F and G, consider the five conditions listed below.

1. The upper Fréchet bound corresponds to $\theta = 1$, i.e., $H_1(x, y) = \min[F(x), G(y)]$.

2. At $\theta = 0$, X and Y are independent, i.e., $H_0(x, y) = F(x)G(y)$.

3. The lower Fréchet bound corresponds to $\theta = -1$, i.e., $H_{-1}(x, y) = \max[F(x) + G(y) - 1, 0]$.

4. For fixed x, y, H_θ is continuous in θ in $[-1, 1]$.

5. For fixed θ in $(-1, 1)$, H_θ is an absolutely continuous d.f.

The following result was given by Kimeldorf and Sampson (1975a). Let $H = \{H_\theta : -1 \leq \theta \leq 1\}$ be a family of d.f.'s with fixed marginals F_1, G_1, and satisfying any subset of (1)—(5). Let F_2 and G_2 be continuous d.f.'s. Then

$$J = \{J_\theta(x, y) = H_\theta[F_1^{-1}F_2(x), G_1^{-1}G_2(y)], -1 \leq \theta \leq 1\} \tag{3.7}$$

is a family of d.f.'s with fixed marginals F_2 and G_2, which satisfies the same subset of (1)–(5) as does H_θ.

3.2.5 Example

Suppose we pick one of the distributions whose d.f. is simple in form and whose marginals are uniform—for example, Frank's (Section 5.4):

$$H_\alpha = \log_\alpha \left[1 + \frac{(\alpha^x - 1)(\alpha^y - 1)}{\alpha - 1}\right] \tag{3.8}$$

($0 < \alpha$, $\alpha \neq 1$). Then if we want a distribution with marginals F and G, we can write:

$$J_\alpha(x, y) = \log_\alpha \left[1 + \frac{(\alpha^{F(x)} - 1)(\alpha^{G(y)} - 1)}{\alpha - 1}\right]. \tag{3.9}$$

3.3 Some Inelegant Methods

The transformation method of Section 3.2 is pretty trivial—all that is done is to take one distribution, and stretch or compress in the X and/or the Y direction. Other methods that may be thought to be so trivial as to be inelegant include the following:

- Let the formula for h take one form for some region of the (X, Y) plane, and another form for the remaining region. (A particular example occurs when the p.d.f. of a unimodal distribution is reduced to c within the contour $h = c$, and then h is rescaled so that it becomes a p.d.f. again.)
- Take an existing distribution and truncate it, singly or doubly, in one or both the variates.
- Take an existing distribution and extend its region of support by reflecting the p.d.f. into the previously empty area.
- Take an existing p.d.f. $h(x, y)$ and multiply it by some function $a(x, y)$. Provided the volume under the surface remains finite, the result can be treated as proportional to a probability density. A special case of this method is where $a(x, y)$ is a risk function, so that the densities in the surviving and non-surviving (or, more generally, selected and non-selected) populations are $(1 - a)h$ and ah.

3.4 Mixing and Compounding

3.4.1 Mixing

One of easier ways to generate bivariate distributions is to use the method of mixing together two distributions. Suppose H_1 and H_2 are two bivariate distribution functions, then

$$H(x, y) = \theta H_1(x, y) + (1 - \theta)H_2(x, y) \tag{3.10}$$

($0 \leq \theta \leq 1$) is a new bivariate distribution. Examples are readily found in the Fréchet bounds:

- Fréchet (1951) himself suggested a one-parameter family of bivariate distributions which attained the Fréchet bounds at the limits of the parameter θ:

  $$H(x, y) = \theta H^-(x, y) + (1 - \theta)H^+(x, y) \tag{3.11}$$

  ($0 < \theta < 1$). However, this family does not include H^0.

- A second simple example of a one-parameter family with a meaningful θ that includes H^+ and H^- is that given in Mardia (1970a, p. 33):

  $$H_\theta(x, y) = \tfrac{1}{2}\theta^2(1 + \theta)H^+(x, y) + (1 - \theta^2)H^0(x, y) + \tfrac{1}{2}\theta^2(1 - \theta)H^-(x, y) \tag{3.12}$$

  ($-1 \leq \theta \leq 1$). This family does include H^0.

Kimeldorf and Sampson (1975a) generalised the idea to get

$$L_\theta = t(\theta)H_\theta + [1 - t(\theta)]K_\theta \tag{3.13}$$

$(-1 \leq \theta \leq 1)$, where $\{H_\theta\}$ and $\{K_\theta\}$ are two families of d.f.'s having the same marginals and satisfying the conditions (1)–(5) given in Section 3.2.4. Here t is a continuous function mapping $[-1, 1]$ into $[0, 1]$. This generalisation allows us to generate a wide range of bivariate distributions, though its usefulness is questionable.

For mixtures of two bivariate normal distributions, see Section 4.17 and Johnson (1987, pp. 55–62).

The above concept of mixture can be extended to three components—an example is (3.10) above, though usually two of the proportions will be free parameters, not merely one—or more. A mixture of infinitely many components is called compounding, see Section 3.4.2 below.

For more details and accounts of applications of mixture distributions, see Everitt (1985) and McLachlan and Basford (1988).

3.4.2 Compounding

The idea of generating distributions by compounding[4] has a long history, particularly in the univariate context. Motivation is often from survival time applications in biological or engineering sciences, and this applies to the bivariate case as much as to the univariate. Let X and Y be two random variables with parameters θ_1 and θ_2, respectively. For a given value of (θ_1, θ_2), X and Y are assumed to be independent. The basic idea of compounding is to say that θ_1 and θ_2 are themselves random variables, not constants, and the observed distribution of X and Y results from integrating over the (unobserved) distribution of θ_1 and θ_2. It is usual to assume that θ_1 and θ_2 are identically equal—thus only a single integration is performed—but sometimes they are assumed to be merely correlated, resulting in an integration with respect to their bivariate distribution being necessary. If θ_1 and θ_2 are independent, independence of X and Y results. A well-known example of constructing a distribution by compounding is that of the Pareto distribution, from assigning a gamma distribution to the scale parameter of an exponential distribution—see Section 20.1.

It is noteworthy that if θ_1 and θ_2 are identical, playing the role of a scale parameter of F and G, then compounding is equivalent to the method of trivariate reduction—see Chapter 15, especially Section 15.4.2.

In Chapter 20 the method of compounding and its applications will be discussed in detail.

To conclude this Section, we shall now give an example which is due to Gaver (1970). This would not fit well in Chapter 20; it does serve to illustrate that the joint moment generating function of the compound distribution is obtained by summing (or integrating) over the distribution of θ (the common value of θ_1 and θ_2).

Let X and Y have the same gamma distribution with the shape parameter $\theta + k$ (θ is an integer, $k > 0$ need not be an integer). For a given value of θ, X and Y are independent with moment generating functions $(1-s)^{-(\theta+k)}$ and $(1-t)^{-(\theta+k)}$, respectively. Assuming that θ has a negative binomial distribution with the probability generating function given by

$$G_k(z) = \sum_{n=0}^{\infty} b_n(k)z^n = \left(\frac{\alpha}{1 + \alpha - z}\right)^k \tag{3.14}$$

(where $b_n(k)$ is the probability that θ takes the value n, and k and $\alpha > 0$ are the two parameters of the negative binomial distribution), then the joint moment generating function of X and Y is

$$\begin{aligned}
M(s,t) &= E(e^{sX+tY}) \\
&= \sum_{n=0}^{\infty} E(e^{sX+tY}|\theta = n)\Pr(\theta = n) \\
&= \sum_{n=0}^{\infty} b_n(k)[(1-s)(1-t)]^{-n}[(1-s)(1-t)]^{-k} \\
&= G_k\{[(1-s)(1-t)]^{-1}\}[(1-s)(1-t)]^{-k} \\
&= \left(1 - \frac{\alpha+1}{\alpha}s - \frac{\alpha+1}{\alpha}t + \frac{\alpha+1}{\alpha}st\right)^{-k},
\end{aligned} \tag{3.15}$$

[4]Some authors refer to compounding as mixing.

Table 3.1: Some bivariate distributions derived from regression models.

$u(x)$	$v(x)$	Type of bivariate distribution	Sections
linear	constant	normal	Chapter 4
linear	linear	beta, Pareto, F	6.5, 5.8, 6.3
constant	linear	McKay	8.6
linear	parabolic	t, Cauchy, Pearson type II	6.2, 6.2.13, 6.4
r.h.*	r.h.*	Conditional distributions are gamma, $h \propto (x+b_1)^{\gamma_1}(y+b_2)^{\gamma_2}\exp[\gamma(x+c_1)(y+c_2)]$	See Mardia (1970a, p. 47)

* r.h. denotes rectangular hyperbola, i.e., of the form $1/(x+a)$.

which is the moment generating function of Kibble's bivariate gamma distribution (see Section 7.5) with correlation coefficient $(1+\alpha)^{-1}$.

3.5 Regression and Conditional Distribution

Most families of distributions considered so far have had their marginals specified. Narumi (1923a,b) took a different approach. His attack on the problem of creating bivariate distributions was by specifying the regression and scedastic (conditional standard deviation) curves. An account of his work is given in Chapter 6 of Mardia (1970a).

Consider bivariate distributions with p.d.f.'s $h(x,y) = f(x)g(y|x)$ of the form

$$f(x)\,g\left(\frac{y-u(x)}{v(x)}\right), \tag{3.16}$$

where u and v are, respectively, the regression and scedastic curves of Y on X. The conditional density $f(x|y)$ also can be written in the form as above. Some bivariate distributions that can be written in this form are summarised in Table 3.1.

An example of the conditional distribution method is given by Mihram and Hultquist (1967), who constructed a bivariate warning-time/failure-time distribution which is generally known as the beta-Stacy bivariate distribution (Johnson et al, 1984). The authors assumed that the first variable Y has Stacy's (1962) distribution (see Section 2.6.5). The conditional distribution of X given $Y=y$ is then specified as beta with parameters p and q. The resulting joint distribution of (X,Y) has density (8.11).

Chapter 8 is relevant to this method, and in Section 8.1.3 is a condition due to Abrahams and Thomas (1984) on the compatibility of the two conditional distributions, $f(x|y)$ and $g(y|x)$.

3.6 Trivariate Reduction

The trivariate reduction or variables-in-common technique is in many contexts an intuitively appealing method for deriving bivariate distributions. Basically, the idea is to create a pair of dependent random variables from three or more random variables. In many cases, these initial random variables are independent, but occasionally they may be dependent—an example of the latter is the construction of a bivariate t-distribution from two variates that have a standardised correlated bivariate normal distribution and one that has a chi-distribution (see Section 6.2.7). An important aspect of this method is that the functions connecting these random variables to the two dependent random variables are generally elementary ones; random realisations of the latter can therefore be generated as easily as of the former.

A broad definition of the variables-in-common technique is as follows. Set

$$\left.\begin{array}{rcl} X &=& T_1(X_1,X_2,X_3) \\ Y &=& T_2(X_1,X_2,X_3) \end{array}\right\} \tag{3.17}$$

where X_1, X_2, X_3 are not necessarily independent or identically distributed.

For other possible definitions and further discussion of this topic, see Chapter 15.

3.7 Weighted Linear Combination

In many simulation applications, it is required to generate dependent pairs of continuous random variables for which there is limited information on the joint distribution. The example that Johnson and Tenenbein (1981) give is that of a portfolio analysis simulation, in which a joint distribution of stock and bond returns may have to be specified. Because of lack of data, it may be difficult to specify completely the joint distribution of stock and bond returns. However, it may be realistic (so Johnson and Tenenbein say) to specify the marginal distributions and some measures of dependence between the random variables.

The weighted linear combination (WLC) technique is as follows. Let

$$\left.\begin{array}{rcl} X & = & U_1 \\ Y & = & cU_1 + (1-c)U_2 \end{array}\right\} \tag{3.18}$$

where U_1 and U_2 are independent and identically distributed with common probability density function f, and c is a constant $(0 \leq c \leq 1)$.

Johnson and Tenenbein (1981) were concerned with using WLC in the case where the specified measure of dependence was Kendall's τ or the grade correlation ρ_g. For some choices of f, they obtained equations connecting c to τ and ρ_g. They handled the problem of getting the marginals right by means of the transformation method discussed in Section 3.2 above. A slightly more general model than WLC in which Y is an arbitrary linear combination of U_1 and U_2 was considered earlier by Jogdeo (1964). A general model in which both X and Y are linear combinations of U_1 and U_2 has been treated in Chapter 5 of Mardia (1970a).

3.8 Limits of Discrete Distributions

3.8.1 The Bivariate Bernoulli Distribution

It is well known that many of the univariate distributions have their genesis in the Bernoulli distribution, and are obtained as sums or limits.[5] Marshall and Olkin (1985b) extended these elementary probability ideas to two dimensions, and obtained a number of bivariate distributions.

A random variable (X, Y) is said to have a bivariate Bernoulli distribution if it has only the four possible values $(1,1)$, $(1,0)$, $(0,1)$, and $(0,0)$, these occurring with probabilities p_{11}, p_{10}, p_{01}, and p_{00}, respectively. We also set $p_{1+} = p_{11} + p_{10} = 1 - p_{0+}$ and $p_{+1} = p_{11} + p_{01} = 1 - p_{+0}$ (notation of Marshall and Olkin).

Many of the bivariate distributions obtained by Marshall and Olkin are discrete; as this book is concerned with continuous distributions, we only mention the construction of a bivariate exponential as the limit of a bivariate geometric distribution, and a bivariate gamma as the limit of a bivariate negative binomial distribution.

3.8.2 A Bivariate Exponential Distribution

If (X_1, Y_1), (X_2, Y_2), ... is a sequence of i.i.d. bivariate Bernoulli variates, and U and V are the number of 0's before the first 1 among the X's and among the Y's, respectively, then U and V each have a geometric distribution, in general not independent. The bivariate distribution function of U and V is given by

$$\Pr(U = u, V = v) = \left\{ \begin{array}{ll} p_{00}^u p_{10} p_{+0}^{v-u-1} p_{+1} & \text{if } 0 \leq u < v \\ & \\ p_{00}^u p_{11} & \text{if } 0 \leq u = v \end{array} \right\} \tag{3.19}$$

and

$$\Pr(U \geq u, V \geq v) = p_{00}^u p_{+0}^{v-u} \text{ for } 0 \leq u \leq v. \tag{3.20}$$

[5] Examples are the binomial, negative binomial, Poisson, and gamma (integer shape parameter) distributions.

Now, obtain a bivariate exponential distribution as the limit of the bivariate geometric distribution (3.20): if independent Bernoulli variates (X_1, Y_1), (X_2, Y_2), ... are observed at times $\frac{1}{n}$, $\frac{2}{n}$, ..., then

$$\Pr(U > t_1, V > t_2) = \left\{ \begin{array}{ll} p_{00}^{[nt_1]} p_{+0}^{[nt_2]-[nt_1]} & \text{if } t_1 < t_2 \\ p_{00}^{[nt_2]} p_{0+}^{[nt_1]-[nt_2]} & \text{if } t_1 > t_2 \end{array} \right\} \tag{3.21}$$

(provided nt_1 and nt_2 are not integers), where $[a]$ denotes the integer part of a. Writing $\lambda_{ij} = np_{ij}$ and passing to the limit, we find

$$\lim_{n\to\infty} \Pr(U > t_1, V > t_2) = \exp[-\lambda_{10}t_1 - \lambda_{01}t_2 - \lambda_{11}\max(t_1, t_2)] \tag{3.22}$$

$(t_1, t_2 \geq 0)$. This is the bivariate exponential distribution of Marshall and Olkin (1967a)—for more details, see Sections 9.2 and 17.2.

3.8.3 A Bivariate Gamma Distribution

If (X_1, Y_1), (X_2, Y_2), ... is a sequence of i.i.d. bivariate Bernoulli variates, and (for positive integers r, s) U and V are the number of 0's before the rth 1 among the X's and before the sth 1 among the Y's, respectively, then U and V each have a negative binomial distribution, in general not independent.

The bivariate negative binomial distribution obtained in this way has untidy expressions for its probability function and for its cumulative distribution function, and we shall not give them here. See Marshall and Olkin (1985b). Proceeding as formerly, if independent Bernoulli variates (X_1, Y_1), (X_2, Y_2), ... are observed at times $\frac{1}{n}$, $\frac{2}{n}$, ..., then on setting $\lambda_{ij} = np_{ij}$, Marshall and Olkin (1985b) showed that

$$\lim_{n\to\infty} \Pr(U > t_1, V > t_2) = \sum \frac{t_1^{a+l-i}\lambda_{11}^i \lambda_{10}^{l-i}\lambda_{01}^{a-i}}{i!(a-i)!(l-i)!} \exp\left[-(\lambda_{11} + \lambda_{10} + \lambda_{01})t_1 \frac{\lambda_{+1}^m}{m!}\right] \exp[-\lambda_{+1}(t_2 - t_1)^m] \tag{3.23}$$

(for $0 \leq t_1 \leq t_2$ and with nt_1 and nt_2 not being integers), where the summation is over those values of a, i, l, m such that $r - 1 \geq l \geq i \geq 0$, $s - 1 \geq a \geq i \geq 0$, and $s - 1 - a \geq m \geq 0$. This distribution has marginals that are gamma distributions with integer shape parameters r and s, respectively.

3.9 Differential Equation Methods

3.9.1 Pearson's System

3.9.1.1 Derivation of Equation

Karl Pearson derived a family of univariate distributions through the differential equation

$$\frac{1}{f}\frac{df}{dx} = \frac{x - a}{b_0 + b_1x + b_2x^2}, \tag{3.24}$$

where a, b_0, b_1 and b_2 are constants. The univariate Pearson family of distributions has been discussed fully in Chapter 12 of Johnson and Kotz (1970a).

Early efforts to generalise this method to two dimensions were unsuccessful, including those of Pearson (1923) himself. But Van Uven (1947a,b, 1948a,b) succeeded. Note that the left hand side of (3.24) is $d(\log f)/dx$. In the case of Pearson, he started with this pair of differential equations:

$$\left.\begin{array}{rcl} \frac{\partial \log h}{\partial x} & = & \frac{C_1}{Q} \\ \frac{\partial \log h}{\partial y} & = & \frac{C_2}{Q} \end{array}\right\} \tag{3.25}$$

where h is the joint p.d.f. of X and Y, C_1 and C_2 are cubic functions of both x and y, and Q is a quadratic function of both x and y. On the other hand, Van Uven started with

$$\left.\begin{array}{rcl} \frac{\partial \log h}{\partial x} & = & \frac{L_1}{Q_1} \\ \frac{\partial \log h}{\partial y} & = & \frac{L_2}{Q_2} \end{array}\right\} \tag{3.26}$$

where h is the joint p.d.f. of X and Y, L_1 and L_2 are linear functions of both x and y, and Q_1 and Q_2 are quadratic (or, possibly, linear) functions of both x and y. On fixing either x or y, it is clear that the conditional distributions of either variable, given the other, satisfying differential equations of the form (3.26), belong to the univariate Pearson family. A detailed discussion of the solutions to the above differential equations is in Mardia (1970a, pp. 5–9). We may condense it as follows.

From (3.26), we obtain by a simple differentiation that

$$\frac{\partial^2 \log h}{\partial x \partial y} = \frac{\partial(\frac{L_1}{Q_1})}{\partial y} = \frac{\partial(\frac{L_2}{Q_2})}{\partial x} \tag{3.27}$$

$$Q_2\frac{\partial L_1}{\partial y} - Q_1\frac{\partial L_2}{\partial x} = \frac{L_1 Q_2}{Q_1}\frac{\partial Q_1}{\partial y} - \frac{L_2 Q_1}{Q_2}\frac{\partial Q_2}{\partial x} \tag{3.28}$$

The nature of the solution of (3.28) depends mainly on the structure of Q_1 and Q_2, and the usefulness of the solution will depend on their having common factors. (If Q_1 and Q_2 do not have a common factor, then X and Y are independent, as shown at p. 8 of Mardia, 1970a.) Important cases are:

1. Q_1 and Q_2 have a common linear factor.

2. Q_1 and Q_2 are identical.

3. Q_2 is a linear factor of Q_1, i.e., $Q_1 = LQ_2$.

3.9.1.2 Case 1

The solution has the form

$$h = k_0(ax+b)^{p_1}(cy+d)^{p_2}(a_1 x + b_1 y + c_1)^{p_3}. \tag{3.29}$$

This family of distributions includes the bivariate beta, Pareto, and F-distributions, and the two given by

$$h = \frac{\Gamma(-p_2)x^{p_1}y^{p_2}(-1-x+y)^{p_3}}{\Gamma(1+p_1)\Gamma(1+p_2)\Gamma(-p_1-p_2-p_3-2)} \tag{3.30}$$

(for $p_1, p_3 > -1$; $(p_1+p_2+p_3) < -3$; $y-1 > x > 0$) and

$$h = \frac{\Gamma(-p_1)x^{p_1}y^{p_2}(-1+x-y)^{p_3}}{\Gamma(1+p_2)\Gamma(1+p_3)\Gamma(-p_1-p_2-p_3-2)} \tag{3.31}$$

(for $p_2, p_3 > -1$; $(p_1+p_2+p_3) < -3$; $x-1 > y > 0$).

The last two equations are effectively equivalent, though they are listed as distinct types (respectively, II$a\gamma$ and II$a\delta$) in Johnson and Kotz (1972, p. 8).

3.9.1.3 Case 2

When $Q_1 = Q_2$, the solution is

$$h = k_0(ax^2 + 2bxy + cy^2 + 2dx + 2ey + c_0)^p. \tag{3.32}$$

Examples are the bivariate Cauchy, t-, and Pearson type II distributions. The bivariate normal distribution is a limit of Case 2, in which $a = c = -(1-\rho^2)/2$, $b = \rho/(1-\rho^2)$, $c_0 = 1$, $d = e = 0$, $k_0 = (2\pi\sqrt{1-\rho^2})^{-1}$, and $p \to \infty$.

3.9.1.4 Case 3

When $Q_1 = LQ_2$, we get

$$h = k_0(ax+b)^p(a_1 x + b_1 y + c_1)^q \exp(-cy). \tag{3.33}$$

This family includes McKay's bivariate gamma distribution (McKay, 1934).

3.9.1.5 Remarks

In fact, Van Uven considered all possible cases, but other solutions do not have both marginals of the same form.

The system of bivariate distributions obtained through (3.26) is generally known as the family of bivariate Pearson distributions. Any member of this family may be called a Pearson type i distribution, $i = \mathrm{I}, \mathrm{II}, \ldots \mathrm{VII}$, if the marginals are type i. For example, the bivariate t-distribution may be called a bivariate Pearson type VII distribution. However, Van Uven's designation is different—see Section 6.1.

In general, the conditional distribution of a member of the family of bivariate Pearson distributions is a univariate Pearson distribution, though it may not have the same form as the marginals (Mardia, 1970a, p. 10).

3.9.2 The Fokker-Planck Equation

Another family of bivariate distributions that was constructed from a differential equation is given in Wong and Thomas (1962) and Wong (1964). The differential equation concerned is the Fokker-Planck equation in diffusion theory

$$\frac{\partial^2}{\partial x^2}[B(x)p] - \frac{\partial}{\partial x}[A(x)p] = \frac{\partial p}{\partial t}, \tag{3.34}$$

where $p = p(x|x_0, t)$, $0 < t < \infty$, and the variables are x and t rather than x and y. $A(x)$ and $B(x)$ are called the "infinitesimal" mean and variance of the underlying Markov process in continuous time with continuous states. The solutions to the above equation are the Markov transitional probability density functions. (Consult a text on stochastic processes, such as Chapter 5 of Cox and Miller (1965), for further information and explanation of the jargon.)

The joint densities $h(x_0, x)$ obtained from the conditional densities p form a family which includes some members of the Pearson system, e.g., the bivariate normal and types I and II, and Kibble's bivariate gamma (to be discussed in Section 7.5). The equilibrium density $f(x) = \lim_{t\to\infty} p(x|x_0, t)$ satisfies the Pearson differential equation (3.24) when $A(x)$ and $B(x)$ are linear and quadratic functions, respectively, and the latter is non-negative.

3.9.3 The Ali-Mikhail-Haq Distribution

See Section 5.3 for this distribution and its derivation from a differential equation.

3.10 Diagonal Expansion

A more detailed discussion of diagonal expansion is given in Chapter 14. Here we briefly outline the concept.

Under the condition of H being ϕ^2-bounded (see Section 14.3), it may be represented as

$$dH(x, y) = dF(x)\, dG(y) \sum_{i=0}^{\infty} \rho_i \xi_i'(x) \eta_i'(y), \tag{3.35}$$

where ξ_i' and η_i' are the canonical variables and the ρ's are the canonical correlations (see Section 14.4 for these terms). Interest is chiefly in the case where ξ_i' and η_i' are polynomials, as it is then possible to characterise $\{\rho_i\}$ for which the above infinite sum is convergent and nonnegative.

3.11 Bivariate Edgeworth Expansion

3.11.1 Univariate Edgeworth, Gram-Charlier, and Cornish-Fisher Expansions

A brief introduction to these named expansions in the univariate case is necessary before proceeding to discuss the bivariate Edgeworth expansion.

Let F be a distribution function with known cumulants κ_i and Φ be the standard normal distribution function. Both Gram-Charlier and Edgeworth expansions are representations of F in terms of Φ and κ_i. They are formally equivalent. However, from a practical point of view, the Edgeworth expansion differs as it is a rearrangement[6] of the Gram-Charlier series, often having better asymptotic properties (but Johnson and Kotz (1970a, p. 19) said it had no general theoretical superiority).

The Cornish-Fisher expansion is the representation of the quantiles of Φ in terms of the quantiles of F, i.e., Φ^{-1} in terms of F^{-1}. (If F^{-1} is expressed in terms of Φ^{-1}, it is called the inverse Cornish-Fisher expansion.) This is done essentially through the Edgeworth expansion.

For further information, see Johnson and Kotz (1970a, pp. 15–22, 33–35) and Kendall and Stuart (1977, pp. 169–179).

3.11.2 Bivariate Edgeworth Series

In Section 3.10 and Chapter 14 we discuss the expansion of a bivariate distribution in terms of orthogonal polynomials and marginal distributions. Here we consider another type of expansion of a bivariate distribution, namely, the bivariate Edgeworth series expansion, which is an extension of the univariate Edgeworth expansion. Briefly speaking, we expand a bivariate density function h in the series of derivatives of the standardised bivariate normal density ϕ such that

$$h(x,y) = \phi(x,y;\rho) + \sum_{m+n\geq 3} (-1)^{m+n} A_{mn} \frac{D_1^m}{m!}\frac{D_2^n}{n!}\phi(x,y;\rho), \tag{3.36}$$

where the A_{mn} may be expressed in terms of the cumulants of X and Y, and $D_1 = \partial/\partial x$, $D_2 = \partial/\partial y$.

Similarly, the joint distribution function is expanded as

$$H(x,y) = \Phi(x,y;\rho) + \sum_{m+n\geq 3} (-1)^{m+n} A_{mn} \frac{D_1^{m-1}}{m!}\frac{D_2^{n-1}}{n!}\phi(x,y;\rho). \tag{3.37}$$

The characteristic function of (3.36) is given by

$$\exp\left[-\tfrac{1}{2}(s^2+2\rho st+t^2) + \sum_{m+n\geq 3} \frac{\kappa_{mn}}{m!n!}(is)^m(it)^n\right], \tag{3.38}$$

where the κ's are the joint cumulants of X and Y.

What all this means is that h is represented as $\phi\,\times$ a polynomial in x and y, i.e., $h = \phi\sum a_{mn}x^m y^n$. The distribution obtained by considering terms up to $m+n=4$ is given in Pearson (1925). This "fifteen constant" bivariate distribution is also known as the type AA distribution.

Chapter 3 of Mardia (1970a) gives a historical account of the bivariate Edgeworth expansion as well as describing how the type AA distribution was fitted to Johannsen's bean data. See also Rodriguez (1983b, pp. 235–239). The type AA distribution was also applied by Mitropol'skii (1966, pp. 67–70) to the diameters and heights of pine trees. The bivariate distribution (BVESD) obtained by having the terms $m+n=3,4,6$ was considered by Gayen (1951), who fitted it to data from five populations. Subrahmaniam and Gajjar (1980) obtained the moments of the sample correlation coefficient of BVESD. For simulation of the bivariate distribution, see Kocherlakota et al (1985).

As in the univariate case, the sum of a finite number of terms of the expansion may give negative frequencies, particularly near the tails. Mardia (1970a, Section 3.5) remarked that the approach of approximating a bivariate distribution by the bivariate Edgeworth expansion is not satisfactory. This is echoed by McCune and Gray (1982) who concluded their survey of the subject by saying "Although the theory of Cornish-Fisher and Edgeworth expansions has had a rich history, it is doubtful that these expansions will play a large role in the future of statistics."

[6] Whilst the sum of a rearranged finite series is unchanged, the sum of a rearranged infinite series may change, if the series is not absolutely convergent.

3.11.3 An Application: Wind Velocity at the Ocean Surface

For this specialised application, we feel obliged to follow quite closely the explanations of Frieden (1983, Section 3.15.9) and Cox and Munk (1954).

Cox and Munk photographed from an aeroplane the sun's glitter pattern on the ocean surface, and translated the statistics of the glitter into the statistics of the slope distribution of the ocean surface; that is, of the joint distribution of wave slope in the direction of the wind (X) and transverse to the wind direction (Y). Conceivably, this could be the basis of a method of measuring the wind velocity at the ocean surface.

"If the sea surface were absolutely calm, a single, mirror-like reflection of the sun would be seen at the horizontal specular point. In the usual case there are thousands of 'dancing' highlights. At each highlight there must be a water facet, possibly quite small, which is so inclined as to reflect an incoming ray from the sun towards the observer. The farther the highlighted facet is from the horizontal specular point, the larger must be this inclination. The width of the glitter pattern is therefore an indication of the maximum slope of the sea surface."—Cox and Munk (1954). In fact, these authors measured the variation in brightness within the glitter pattern, rather than computing maximum slopes from the outer boundaries, and thus obtained more detailed information.

In choosing a functional form for $h(x, y)$ in this case, two factors to consider are the following.

- The p.d.f. of X should be skewed, as waves tend to lean away from the wind, having gentler slopes on the windward side than on the leeward side.

- There should be no such skew for the p.d.f. of Y, since waves transverse to the wind are not directly formed by the wind, but rather by leakage of energy from the longitudinal wave motion.

Consequently, the following form of the two-dimensional Gram-Charlier expansion was fitted to experimental data:

$$h = \phi(x)\phi(y)[1 + \alpha_{12}H_1(x)H_2(y) + \alpha_{30}H_3(x) + \alpha_{04}H_4(y) + \alpha_{22}H_2(x)H_2(y) + \alpha_{40}H_4(x)], \qquad (3.39)$$

where the H_i's are the Hermite polynomials.

3.11.4 A Second Application: Statistical Spectroscopy

As a result of analytical and numerical studies showing that the higher bivariate cumulants of the relevant variables are quite small, Kota (1984) concluded that it was meaningful to employ an expansion around a bivariate normal density—specifically, the bivariate Edgeworth expansion—for a certain bivariate density of importance in statistical spectroscopy. A second paper is that of Kota and Potbhare (1985).

Part II

Families of Continuous Bivariate Distributions

Chapter 4

The Bivariate Normal Distribution

4.1 Organisation of Part II

Part II of this book gives the basic facts about the major families of bivariate distributions. It is organised into seven Chapters. We start, in Chapter 4, with the bivariate normal distribution, though towards the end of the Chapter some distributions are included that are obtained in an elementary way from the bivariate normal, such as by univariate transformation, truncation, or mixture. Chapter 5 is on distributions expressed in the form of their copula, i.e., their uniform representation. We have said already that a major change in viewpoint over recent years has been the separation of bivariate structure from the distributions of the marginals, and one aspect of this has been the study of copulas, the form taken when a distribution is transformed to have uniform marginals. Perhaps the best-known copula is that which we refer to as the Farlie-Gumbel-Morgenstern, and that is the first one encountered in Chapter 5; we have chosen to include in this Chapter one or two distributions, such as the bivariate Pareto, that are more familiar when their marginals are something other than uniform. In Chapter 6, we deal with distributions that have a simple form for the p.d.f. This Chapter includes, but is not restricted to, some of the sampling distributions obtained from the normal, such as the t and the F. Further distributions connected with the bivariate normal are described in Chapter 7. Some of these are extensions of the bivariate t, F, and gamma distributions. Others are families that the bivariate normal can be seen as a special case of, such as elliptically-contoured distributions, or the class of "stable" distributions. It is conceptually easy to construct bivariate distributions by the conditional method, $h(x,y) = g(y|x)f(x)$, and that is the subject of Chapter 8. Chapter 9 is concerned with distributions arising in reliability contexts—many of the distributions there have exponential marginals, or some other close connexion with the univariate exponential distribution. Finally, Chapter 10 is a collection of miscellaneous distributions that did not seem to fit in well anywhere else.

We have already (Section 1.1) mentioned the usefulness of the books by Mardia (1970a) and Johnson and Kotz (1972). An extensive bibliography has been given by Patil, Boswell, Ratnaparkhi, and Roux (1985); the earlier work by Haight (1961) should also be mentioned.

4.2 Introduction to Chapter 4

At kindergarten, if we want to pass our statistics exams, we have to know why the (univariate) normal distribution is important—specifically, that the random variables that occur in many situations are approximately normally distributed, and that it arises in theoretical work as an approximation to the distribution of many statistics, such as sums of independent random variables. More or less the same reasons apply to the bivariate normal distribution. "But the prime stimulus has undoubtedly arisen from the strange tractability of the normal model: a facility of manipulation which is absent when we consider almost any other multivariate data-generating mechanism."—Barnett (1979). And we may also note the following views:

- "In multivariate analysis, the only distribution leading to tractable inference is the multivariate normal"—Mardia (1985).

- "The only type of bivariate distribution with which most of us feel much familiarity (other than the joint distribution of a pair of independent random variables) is the bivariate normal distribution"—Anscombe (1981, p. 305).

On the other hand:

- "But who has ever seen a multivariate normal sample?" asks Barnett (1979) rhetorically, and goes on to choose, without any conscious bias in their selection, three bivariate datasets from the published literature, which all turn out to be grossly non-normal.

- "The only sure defence against a successful disproof of the assumption of multivariate normality is to abstain from collecting, or presenting, too much data!"—wording adapted from Burnaby (1966, p. 109).

Origins of the bivariate normal are found in the first half of the nineteenth century, in the work of Laplace, Plana, Gauss, and Bravais. Seal (1967) and Lancaster (1972) have given accounts of these developments. The latter pointed out that the early authors derived the bivariate normal as the joint distribution of linear forms in independently distributed normal variables, but did not define a coefficient of correlation; the distribution was used as the basis for a theory of measurement error. Francis Galton (b. 1822, d. 1911), in analysing the measurements of the heights of parents and their adult children, studied the structure of a bivariate normal density function. He observed that the marginal distributions of the data were normal, the array (conditional) means fell on nearly a straight line, the array variances were constant, and the contours of equal frequency were ellipses. He was also the first to recognise the need for a measure of correlation in bivariate data. Since his time, the growth in use of the bivariate normal has been such as to produce the comments already quoted.

The structure of this Chapter is as follows:

4.3 Basic formulae.

4.4 Methods of derivation.

4.5 Characterisations.

4.6 Relation to other distributions.

4.7 References to illustrations.

4.8 References to the generation of random variates.

4.9 Other interesting properties.

4.10 Notes on some more specialised fields.

4.11 Applications.

4.12 Computation of bivariate normal integrals.

4.13 Testing for bivariate normality.

4.14 Univariate transformations.

4.15 The logistic-normal distribution.

4.16 Truncated bivariate normal distributions.

4.17 Bivariate normal mixtures.

4.18 Some distributions constructed for pedagogic purposes.

4.19 Cross-references to other distributions with normal marginals.

For further information, see Johnson and Kotz (1972, Chapter 36), Kendall and Stuart (1977, Chapter 15; 1979, Chapters 18 and 26), and Patel and Read (1982, Chapter 10).

4.3 Basic Formulae

4.3.1 Notation

In this Chapter, we let ϕ and Φ be the p.d.f. and (cumulative) d.f. of the standardised univariate normal distribution, and, similarly, ψ and Ψ be the p.d.f. and d.f. of the standardised bivariate normal distribution. The upper right volume under the probability density surface is denoted $L(x, y; \rho)$.

When we need notation for the means and variances, we use $E(X) = \mu_1$, $E(Y) = \mu_2$, $\text{var}(X) = \sigma_1^2$, and $\text{var}(Y) = \sigma_2^2$.

4.3.2 Support

The support (see Section 5.1.2 for explanation of this jargon) is the whole plane.

4.3.3 Formula for p.d.f.

$$\psi = \left(2\pi\sqrt{1-\rho^2}\right)^{-1} \exp\left[-\tfrac{1}{2(1-\rho^2)}(x^2 - 2\rho xy + y^2)\right] \tag{4.1}$$

Location and scale parameters can be introduced, as usual, by replacing x and y by $(x - \mu_1)/\sigma_1$ and $(y - \mu_2)/\sigma_2$, respectively. The contours are elliptical.

4.3.4 Formula for (Cumulative) Distribution Function

An expression in terms of elementary functions does not exist. See Section 4.12 for methods of computation.

4.3.5 Univariate Properties

Normal marginals.

4.3.6 Correlation Coefficients

ρ in equation (4.1) is Pearson's product-moment correlation. Kendall's τ and the grade correlation ρ_g may readily be expressed in terms of ρ—see Sections 11.8 and 13.1.

4.3.7 Conditional Properties

The conditional distributions are also normal; the regression is linear and the conditional variance is constant:

$$\begin{aligned} E(Y|X = x) &= \mu_2 + \rho\sigma_2(x - \mu_1)/\sigma_1 && (4.2) \\ \text{var}(Y|X = x) &= \sigma_2^2(1 - \rho^2), && (4.3) \end{aligned}$$

in which location and scale parameters have been included.

4.3.8 Moments and Absolute Moments

Assuming the distribution is in the standardised form, we have:

$$\begin{aligned} M(s,t) &= E(e^{sX+tY}) \\ &= \exp[\tfrac{1}{2}(s^2 + 2st\rho + t^2)] && (4.4) \\ \mu_{mn} &= (m + n - 1)\rho\mu_{m-1\,n-1} + (m-1)(n-1)(1-\rho^2)\mu_{m-2\,n-2}, && (4.5) \end{aligned}$$

where μ_{mn} is the central product-moment, $E[(X-\mu_1)^m(Y-\mu_2)^n]$. It is convenient to write μ_{mn} in different forms according to whether m and n are both even, both odd, or one is even and the other odd:

$$\mu_{2m\,2n} = \frac{(2m)!(2n)!}{2^{m+n}} \sum_{j=0}^{\min(m,n)} \frac{(2\rho)^{2j}}{(m-j)!(n-j)!(2j)!} \tag{4.6}$$

$$\mu_{2m+1\,2n+1} = \frac{(2m+1)!(2n+1)!}{2^{m+n}}\rho \sum_{j=0}^{\min(m,n)} \frac{(2\rho)^{2j}}{(m-j)!(n-j)!(2j+1)!} \tag{4.7}$$

$$\mu_{2m\,2n+1} = 0. \tag{4.8}$$

Pearson and Young (1918) tabulated the values of μ_{mn} for m and n up to 10.

Let ν_{mn} denote the joint absolute moment $E(|x^m y^n|)$, and set $\tau = \sqrt{1-\rho^2}$. Then

$$\begin{array}{ll}
\nu_{11} = 2(\tau + \rho\sin^{-1}\rho)/\pi & \nu_{22} = 1 + 2\rho^2 \\
\nu_{12} = (1+\rho^2)\sqrt{2/\pi} & \nu_{23} = 2(1+3\rho^2)\sqrt{2/\pi} \\
\nu_{13} = 2[\tau(2+\rho^2) + 3\rho\sin^{-1}\rho]/\pi & \nu_{24} = 3(1+4\rho^2) \\
\nu_{14} = (3+6\rho^2-\rho^4)\sqrt{2/\pi} & \nu_{33} = 2[(4+11\rho^2)\tau + 3\rho(3+2\rho^2)\sin^{-1}\rho]/\pi.
\end{array}$$

For further formulae, up to $m+n \le 12$, see Nabeya (1951). Generally,

$$\nu_{mn} = \pi^{-1}2^{(m+n)/2}(1-\rho^2)^{m+n+1} \sum_{k=0}^{\infty} \Gamma\left(\tfrac{m+1}{2}+k\right)\Gamma\left(\tfrac{n+1}{2}+k\right)\frac{(2\rho)^{2k}}{(2k)!}, \tag{4.9}$$

which can alternatively be expressed in terms of Gauss' hypergeometric function ${}_2F_1$.

More extensive collections of formulae can be found in, for example, Johnson and Kotz (1972, pp. 91–93), Patel and Read (1982, Section 10.4), and Kendall and Stuart (1977, paragraphs 3.27–3.29).

4.4 Methods of Derivation

The bivariate normal distribution can be derived in many ways. We give here four of them.

4.4.1 Differential Equation Method

The bivariate normal distribution may be obtained by solving a pair of differential equations, $\frac{\partial \log h}{\partial x} = \frac{L_1}{Q}$ and $\frac{\partial \log h}{\partial y} = \frac{L_2}{Q}$, where h is the joint p.d.f. of x and y, L_1 and L_2 are linear functions of both x and y, and Q is a quadratic function of x and y. (See Section 3.9.1.3.)

4.4.2 Compounding

If X and Y have independent univariate normal distributions, each with mean μ and standard deviation 1, then the joint distribution of X and Y is circular normal centred on (μ, μ). Now, if μ itself has a normal distribution with mean 0 and standard deviation σ, then (X, Y) is bivariate normal with mean at $(0,0)$, variances of $1+\sigma^2$, and correlation coefficient $\sigma/\sqrt{1+\sigma^2}$.

4.4.3 Trivariate Reduction

Let X_i $(i = 1,2,3)$ be three independent univariate normal random variables. Then $X = X_1 + X_3$ and $Y = X_2 + X_3$ have a bivariate normal distribution. This is termed trivariate reduction—see Section 3.6 and Chapter 15 for more on this.

4.4.4 Bivariate Central Limit Theorem

Let $(X_1, X_2), \ldots (X_n, Y_n)$ be i.i.d. random vectors with finite second moments and correlation coefficient ρ. Then, as $n \to \infty$, the sampling distribution of $(\bar{X}, \bar{Y})$ tends to a bivariate normal distribution whose correlation coefficient is ρ, the same as in the parent distribution.

4.5 Characterisations

- All cumulants and cross-cumulants of order higher than 2 are zero.

- For any constants a and b, not both zero, $aX + bY$ has a normal distribution (Johnson and Kotz, 1972, p. 59).

- Suppose X and Y have a bivariate exponential-type distribution.[1] Then (X, Y) is bivariate normal if and only if (i) $E(X|Y = y)$ and $E(Y|X = x)$ are both linear, and (ii) $X + Y$ is normal (Johnson and Kotz, 1972, p. 86).

- Brucker (1979) showed that (X, Y) has a bivariate normal distribution if and only if the conditional distribution of each component, given the value of the other component, is normal, with linear regression and constant variance. Fraser and Streit (1980) gave a modification of Brucker's conditions.

- If $X - aY$ and Y are independent, and $Y - bX$ and X are independent, for all a, b such that $ab \neq 0$ or 1, then (X, Y) has a normal distribution (Rao, 1975, pp. 1–13).

- The sample mean vector $(\bar{X}, \bar{Y})$ of a random sample from a bivariate normal population and the elements (S_1^2, S_2^2, R) which determine the sample variance-covariance matrix, where $R = \sum(X_i - \bar{X})(Y_i - \bar{Y}) \Big/ \sqrt{\sum(X_i - \bar{X})^2 \sum(Y_i - \bar{Y})^2}$, are independent (Kendall and Stuart, 1977, p. 413).

- Let X, Y, U_1, and U_2 be random variables, and a and b be constants, such that (i) $Z_1 = X + aY + U_1$ and (Y, U_1, U_2) are independent, and (ii) $Z_2 = bX + Y + U_2$ and (X, U_1, U_2) are independent. Then (Z_1, Z_2) has a bivariate normal distribution if $a \neq 0$, $b \neq 0$; further, $(Z_1, Z_2$ and (U_1, U_2) are independent (Khatri and Rao, 1976, pp. 83–84).

- More advanced forms of characterisation can be found in Johnson and Kotz (1972, pp. 59–62) and Mathai and Pederzoli (1977, Chapter 10).

4.6 Relation to Other Distributions

- Let $H(x, y)$ be a ϕ^2-bounded[2] bivariate distribution, with standard normal marginals; then the density function can be written as a mixture (finite or infinite) of bivariate normal densities:

$$h = \int_{-1}^{1} \psi(x, y; \rho)\, d\mu(\rho), \tag{4.10}$$

  where $\psi(x, y; \rho)$ is the standardised bivariate normal density, and $\mu(.)$ is a distribution function over $[-1, 1]$.

- Kibble's bivariate gamma distribution may be obtained from the bivariate normal. See Section 7.5.8.

- Suppose (X_1, X_2) has the standardised bivariate normal distribution, and X_0, independent of X_1 and X_2, has a chi-squared distribution with ν degrees of freedom. Then $X = X_1/\sqrt{X_0/\nu}$ and $Y = X_2/\sqrt{X_0/\nu}$ have a bivariate t-distribution with ν degrees of freedom. See Section 6.2 for this distribution.

4.7 References to Illustrations

Practically every introductory statistics textbook has some sort of illustration of the bivariate normal distribution. The following examples may be noted.

[1] For this class of distributions, the density is of the form $h = a(x, y) \exp[x\theta_1 + y\theta_2 - q(\theta_1, \theta_2)]$, where θ_1 and θ_2 are two parameters, and $a(x, y) \geq 0$ is a function of x and y (Bildikar and Patil, 1968).

[2] See Section 14.3 for explanation.

- Contours of density: Johnson and Kotz (1972, pp. 88–90), Johnson (1987, pp. 51–53).
- Plot of density surface: Rodriguez (1982b), Johnson and Kotz (1972, p. 89–90), and Johnson (1987, pp. 51–54).
- Contour plot of the uniform representation of the bivariate normal: Barnett (1980).
- Contour and three-dimensional plots after the marginals have been transformed to be exponential: Johnson et al (1981).
- Zelen and Severo (1960) have plotted graphs of $L(h, 0; \rho)$ for various ranges of h and ρ.

4.8 References to the Generation of Random Variates

See Section 16.5.1.

4.9 Other Interesting Properties

- X and $Y - E(Y|X = x)$ are independent.
- $\rho = 0$ if and only X and Y are independent. ρ describes the strength of linear relationship between X and Y.
- Let (X_1, Y_1) and (X_2, Y_2) be two standard bivariate normal distributions, with correlation coefficients ρ_1 and ρ_2, respectively. If $\rho_1 \geq \rho_2$, then $\Pr(X_1 > x, Y_1 > y) \geq \Pr(X_2 > x, Y_2 > y)$, which is known as Slepian's inequality (Gupta, 1963a, p. 805). Alternatively, $\Pr(X_1 < x, Y_1 < y) \leq \Pr(X_2 < x, Y_2 < y)$. We then say that (X_1, Y_1) has larger quadrant dependence than (X_2, Y_2). In other words, the bivariate normal distribution with fixed marginals is ordered by quadrant dependence (Section 12.3.1) through the correlation ρ.
- By letting $\rho_2 = 0$ in the above inequality, we have $\Pr(X \leq x, Y \leq y) \geq \Pr(X \leq x)\Pr(Y \leq y)$ if $\rho \geq 0$, for all x, y; and there is a similar inequality if $\rho \leq 0$. Moreover, $\Pr(|X| \leq x, |Y| \leq y) \geq \Pr(|X| \leq x)\Pr(|Y| \leq y)$, where $x \geq 0$, $y \geq 0$. For these inequalities, see Tong (1980, pp. 8–15), who considered them for multivariate, rather than bivariate, distributions, so that the nature of the variance-covariance matrix played an important part in the discussion.
- $X + Y$ has a univariate normal distribution; more generally, so does $aX + bY$.
- The magnitude of the vector sum—i.e., $\sqrt{X^2 + Y^2}$—has a Rayleigh distribution if X and Y are i.i.d. normal variates with zero means. For the case where X and Y have the general bivariate normal distribution, Chou and Corotis (1983) have given some results. See also Section 19.7.1.
- $\psi(x, y)$ can be expanded diagonally in terms of ρ and the Hermite polynomials, i.e.

$$\psi = \phi(x)\phi(y)\sum_{j=0}^{\infty} \rho^j H_j(x)H_j(y) \tag{4.11}$$

(Cramér, 1946, p. 133). Here, the H_j are normalised so that the integral $\int \phi(x)H_i(x)H_j(x)$ is 1 if $i = j$ and is 0 if $i \neq j$; also, $H_0 = 1$. Since

$$-\frac{d}{dx}[H_{j-1}(x)\phi(x)] = \sqrt{j}H_j(x)\phi(x) \tag{4.12}$$

(Kendall and Stuart, 1979, p. 326), it follows that $\Psi(x, y)$ also has a diagonal expansion in terms of ρ and the Hermite polynomials,

$$\Psi = \phi(x)\phi(y)\sum_{j=1}^{\infty} \frac{\rho^j}{j} H_{j-1}(x)H_{j-1}(y) + \Phi(x)\Phi(y). \tag{4.13}$$

- $T = (1-\rho^2)^{-1}(X^2 - 2\rho XY + Y^2)$ has an exponential distribution with mean 2. Hence the integral of (4.1) over the interior of the ellipse $x^2 - 2\rho xy + y^2 = k$ is $\Pr[T \leq k(1-\rho^2)] = 1 - \exp[-\frac{1}{2}k(1-\rho^2)]$ (Johnson and Kotz, 1972, p. 16).

- Any bivariate distribution obtained from the bivariate normal by separate transformation of X and Y has a correlation that in absolute value cannot exceed $|\rho|$ (e.g., Kendall and Stuart, 1979, p. 600). See Section 4.14.5.

- If Ψ_1, Ψ_2, Ψ_3, and Ψ_4 are four bivariate normal distribution functions such that $\Psi_1\Psi_2 \equiv \Psi_3\Psi_4$, then Ψ_3, Ψ_4, are the same two functions as Ψ_1, Ψ_2 (Anderson and Ghurye, 1978).

- If (X, Y) has a standardised bivariate normal distribution, the ratio X/Y has a Cauchy distribution with p.d.f. $\pi^{-1}(1-\rho^2)^{1/2}/(1-2\rho u + u^2)$. (The bivariate normal is not the only distribution for which this is true, see Section 9.14 of Springer, 1979.) Hinkley (1969) approximated the cumulative distribution function of X/Y when (X, Y) is not standardised. For later results, see Springer (1979, Section 4.8.3), Aroian (1986), and references cited there. For the product XY, see Craig (1936), Haldane (1942), Aroian (1947, 1978), and Springer (1979, Section 4.8.3).

- For multidimensional central limit theorems, see Heyde (1985). (The subject is apparently uninteresting when thinking of ordinary numbers, because so much of the unidimensional theory carries over directly.)

4.10 Notes on Some More Specialised Fields

- Formulae relating to the application of this distribution in the "competing risks" context—for which, see Section 9.18—are to be found in David and Moeschberger (1978, Chapter 4).

- The quantising of a two-dimensional random variable is of interest, so it seems, to electrical engineers. We quote from Bucklew and Gallagher (1978): "Consider a two dimensional random variable **X** whose bivariate density is circularly symmetric and we desire to represent this quantity by a finite set of values. One possible representation of **X** leads to a Cartesian co-ordinate system expression wherein we individually quantize the two rectangular components of the random variable. Another common representation leads to a polar co-ordinate representation where we quantize the magnitude and phase angle of **X**... we obtain a simple criterion by which to determine whether polar format or rectangular format gives a smaller mean square quantization error."

- The following description of planar random movement is adapted from van Zyl (1987). A particle, starting from the origin, jumps a random length U on the plane. All directions for the jump are equally likely. Thus, in an obvious notation, after n such jumps the particle is at coordinates $(\sum_{i=1}^{n} u_i \cos\theta_i, \sum_{i=1}^{n} u_i \sin\theta_i)$, where θ is uniformly distributed over the range 0 to 2π. van Zyl derives the characteristic function of the distribution of the position, considers the normal approximation, and derives approximate results concerning the distance of the particle from the origin.

4.11 Applications

Discussion of applications is deferred to Part IV of this book. In particular, Chapter 19 collects together brief accounts of over 30 subject areas where the bivariate normal distribution has been used.

4.12 Computation of Bivariate Normal Integrals

4.12.1 The Short Answer

The short answer is as follows.

- As to tables of the bivariate normal integral: National Bureau of Standards (1959) and Japanese Standards Association (1972) have tables of L.

- For a computer program: that of Donnelly (1973) is in a widely-available publication, and is written in FORTRAN. Perhaps Baughman's (1988) program supercedes it, but we do not have experience with that one yet.

The remainder of this Section expands upon this, and is divided into several subsections on algorithms, and then tables, computer programs, and references to reviews of the subject are considered. We consider only L and related quantities, not, for instance, integrals over an offset circle (for which, see Groenewoud et al, 1967, and Patel and Read, 1982, Section 10.3).

In passing, we may note that computation of the *uni*variate normal d.f. is not straightforward when one is interested in the tails of the distribution, as one may be in safety contexts (see Rosenblueth, 1985).

4.12.2 Algorithms—Rectangles

In the early development of the subject, interest centred on the upper right volume under the density surface, what would be called in the reliability context the "survival function". It is conventional to use the symbol L for this:

$$L(h,k;\rho) = \Pr(X > h, Y > k) = \int_h^\infty \int_k^\infty \psi(x,y;\rho)\,dy\,dx. \tag{4.14}$$

A special case is: $L(0,0;\rho) = \frac{1}{4} + \frac{1}{2\pi}\sin^{-1}\rho$. And we also have:

$$L(h,k;\rho) = L(h,0;\rho_{hk}) + L(k,0;\rho_{hk}) - \tfrac{1}{2}(1-\delta_{hk}),$$

where δ_{hk} is 0 or 1 depending on whether h and k have the same sign or not, and $\rho_{hk} = \frac{(\rho h-k)a_h}{\sqrt{h^2-2\rho hk+k^2}}$, with a_h being 1 or -1 depending on whether h is positive or negative.

Relationships involving the distribution function Ψ:

$$\Psi(h,k;\rho) = L(-h,-k;\rho) = L(h,k;\rho) + \Phi(h) + \Phi(k) - 1$$

$$\Psi(-h,k;\rho) = L(h,-k;\rho) = 1 - \Phi(h) - L(h,k;-\rho).$$

Pearson (1901) gave a method for evaluating L as a power series in ρ involving tetrachoric functions. His method provides a good approximation for small $|\rho|$. Computation of L can also be via the functions V and T that will be discussed shortly.

Drezner (1978) presented a simple algorithm for Ψ based on numerical integration of the density function using the Gauss quadrature method.

Divgi (1979a) calculated Ψ by firstly transforming X and Y into two independent standard normal variates, and then approximating $1-\Phi(x)$ by $x\phi(x)\sum_{k=0}^{n} d_{nk}x^k$, where d_{nk} are coefficients given by Divgi.

Bouver and Bargmann (1979) approximated the bivariate normal integral by:

$$(b-a)\sum_{i=1}^{n} \frac{w_i}{2}\frac{e^{-x^2/2}}{\sqrt{2\pi}}\,\Phi\left(\frac{k-x\rho}{\sqrt{1-\rho^2}}\right), \tag{4.15}$$

where $x = a + h(1+c_i)/2$ and $a = -10$ with step size $h = (b-a)/30$. Here w_i and c_i are the weights and abscissas of the Gauss-Legendre numerical integration rule.

We understand that Bjerager and Skov (1982) have given an approximate formula for Ψ.

A simple way of obtaining a close approximation is to write $\Psi(h,k;\rho) = \Pr(X<h)\Pr(Y<k|X<h)$. Now, work out the mean and standard deviation of $Y|X<h$, and to find $\Pr(Y<k|X<h)$, assume it has a normal distribution. The repeated application of this is a familiar strategy to approximate the n-dimensional cumulative normal distribution ($n \geq 3$), since there is a paucity of competing methods. Mee and Owen (1983) investigated its usefulness for the bivariate case. In particular, they gave guidance as to how the calculation should be performed—as $\Psi(h,k;\rho)$, $\Psi(k,h;\rho)$, $\Phi(k) - \Psi(-h,k;-\rho)$, or $\Phi(h) - \Psi(-k,h;\rho)$, depending on the values of h and k.

Foulley and Gianola (1984) approximated L in terms of 10 positive roots of Hermite polynomials of order 20.

Wang (1987) proposed a method for computing the bivariate normal probability integral over any rectangular region, with reasonable accuracy and without the need for numerical integration. By observing that the cross-product ratio (see Section 11.14.1) for infinitesimal rectangular regions of the bivariate normal is constant, the method involves an iterative proportional fitting algorithm for the row and column marginal totals in a two-way contingency table which is constructed from discretised normal probabilities. A table of integrals for $\rho = 1/2$ is given by Wang.

Results in terms of the bivariate (and multivariate) Mills' ratio—that is, ratio of tail volume L to bounding ordinate ψ—have been given by Savage (1962) and Ruben (1964). Savage finds upper and lower inequalities for this ratio, and Ruben an asymptotic expansion.

4.12.3 Algorithms—Triangles

$V(h, k)$ is defined by:

$$V(h,k) = \frac{1}{2\pi}\int_0^h \int_0^{kx/h} \exp[-(x^2+y^2)/2]\, dy\, dx \tag{4.16}$$

$(h, k \geq 0)$. This is the integral of the standard circular normal density over the triangle with vertices $(0,0)$, $(h,0)$, and (h,k). Clearly, $V(0,k) = 0 = V(h,0)$. The following relation holds between L and V:

$$L(h,k;\rho) = 1 - \tfrac{1}{2}[\Phi(h) + \Phi(k)] - \tfrac{1}{2\pi}\cos^{-1}\rho + V\left(h, \frac{k-\rho h}{\sqrt{1-\rho^2}}\right) + V\left(k, \frac{h-\rho k}{\sqrt{1-\rho^2}}\right),$$

and other relations involving V are:

$$V(h,k) = -V(-h,k) = V(-h,-k) = -V(h,-k)$$

$$V(h,k) + V(k,h) = [\Phi(h) - \tfrac{1}{2}][\Phi(h) - 1]$$

$$V(h,\infty) = \tfrac{1}{2}[\Phi(h) - \tfrac{1}{2}]$$

$$V(\infty,k) = 0$$

Nicholson (1943) gave tables of V to 6 decimal places.

4.12.4 Algorithms—Owen's T Function

For computational purposes, it is easier to work with Owen's T function than with V, where

$$\begin{aligned} T(h,\lambda) &= \frac{\pi}{2}\int_0^\lambda (1+x^2)^{-1}\exp[-h^2(1+x^2)/2]\, dx \\ &= (2\pi)^{-1}\tan^{-1}\lambda - V(h,\lambda h) \end{aligned} \tag{4.17}$$

Here $(2\pi)^{-1}\tan^{-1}\lambda$ is the integral of the circular normal density $\psi(x,y;0)$ over the sector in the positive quadrant that is bounded by the lines $y = 0$ and $y = \lambda x$.

We note the following relations, which show that computation of T for $h \geq 0$ and $0 \leq \lambda \leq 1$ is sufficient to obtain T for any other values of the arguments:

$$T(\lambda h, \lambda^{-1}) = \tfrac{1}{2}[\Phi(h) + \Phi(\lambda h)] - \Phi(h)\Phi(\lambda h) - T(h,\lambda)$$

$$T(h,\lambda) = T(-h,\lambda) = -T(h,-\lambda)$$

$$T(h,0) = 0.$$

Ψ can be expressed in terms of Φ and T, as follows:

$$\Psi(h,k;\rho) = \tfrac{1}{2}[\Phi(h) + \Phi(k)] - T\left(h, \frac{k-\rho h}{h\sqrt{1-\rho^2}}\right) - T\left(h, \frac{h-\rho k}{k\sqrt{1-\rho^2}}\right) - b,$$

where $b = 0$ if $hk > 0$, or $hk = 0$ with $h + k \geq 0$, and $b = \frac{1}{2}$, otherwise.

Owen (1980) has published a collection of integral formulae involving the normal distribution. Most of these concern the univariate function, but there are several relating to Ψ and/or T.

Much effort has been devoted to searching for an accurate numerical approximation to T. Owen (1956) showed that

$$T(h, \lambda) = (2\pi)^{-1} \left(\tan^{-1} \lambda - \sum_{j=0}^{\infty} c_j \lambda^{2j+1} \right), \tag{4.18}$$

with

$$c_j = \frac{(-1)^j}{2j+1} \left[1 - e^{-h^2/2} \sum_{i=0}^{j} \frac{(\frac{1}{2}h^2)^i}{i!} \right].$$

For small values of h and λ, convergence is rapid, and the formula is a useful one for computing T. Amos (1969) has given instructive comparisons of computer times to calculate Ψ using various formulae; he recommended (4.18) as being generally preferable.

Borth (1973) agreed with the use of (4.18) as an approximation to T with a desired accuracy of 10^{-7} if $h \leq 1.6$ or $\lambda \leq 0.3$. For faster convergence with higher values of h or λ, he gave the following modification: approximate $(1 + x^2)^{-1}$ over the range -1 to 1 by a polynomial of degree $2m$, $\sum_{k=0}^{m} a_{2k} x^{2k}$. Then T may be approximated by

$$(2\pi)^{-1} \exp(-h^2/2) \sum_{k=0}^{m} a_{2k} I_{2k}(w)(h/\sqrt{2})^{-(2k+1)}, \tag{4.19}$$

where $w = h\lambda/\sqrt{2}$ and I is obtained by the iterative relation

$$I_{2k}(w) = \tfrac{1}{2}[(2k-1)I_{2k-2}(w) - w^{2k-1} \exp(-w^2)]$$

with $I_0(w) = \sqrt{\pi} \left[\Phi(w\sqrt{2}) - \frac{1}{2} \right]$. Borth recommended this modification if $h > 1.6$ and $\lambda > 0.3$, noting that if $h > 5.2$, then $T < 10^{-7}$, and that the required accuracy is attained if $m = 6$. This use of Owen's algorithm with Borth's modification combines speed of computation with accuracy.

Sowden and Ashford (1969) suggested a composite method of computing L, incorporating Owen's algorithm (4.18).

For small h and λ, Daley (1974) claimed that Simpson's numerical integration rule to evaluate (4.17) yields a better result than the power series expansion (4.18).

Young and Minder (1974) also gave an algorithm for calculating T over all values of h and λ. (There are corrections and improvements to this by various authors.)

4.12.5 Algorithms—Wedge-Shaped Domain

Grauslund and Lind (1986) considered the integral of the standard circular normal density over a wedge-shaped domain,

$$I(h, k) = \int\int_D \phi(x)\phi(y)\, dx\, dy, \tag{4.20}$$

where the integral is taken over the region D defined by $y \geq k$, $x \geq hy/k$.

I can be reduced to a single integral:

$$I(h, k) = k\phi(k) \int_h^{\infty} \frac{\phi(x)}{k^2 + x^2}\, dx. \tag{4.21}$$

The following identities hold: $I(h, k) = \Phi(-k) - I(-h, k) = \frac{1}{2} - I(h, -k) = \Phi(-h)\Phi(-k) - I(k, h)$. The integral may be evaluated by numerical integration. Alternatively, Grauslund and Lind obtained a simple approximation, which they claimed to be suitable for many technical applications: introduce the function

$$I_1(h, k) = \frac{k}{2h} \left[\Phi \left(-\sqrt{(h^2 + k^2)/2} \right) \right]^2$$

which is, when $h > k$, a first approximation and a lower bound of I. In order to obtain greater accuracy, write I in the form

$$I(h,k) = c(h,k)I_1(h,k)$$

where c is a correction factor function. Grauslund and Lind give a table of c in terms of h and k, plus two approximations:

- $c = 1.053$ if $2 \leq h \leq 8$ and $0 \leq k \leq 8$.
- $c = a_1 + a_2k + a_3h^2 + a_4hk + a_5h^3 + a_6h^2k + a_7hk^2$, the values of the a's being given in their article.

Gideon and Gurland (1978) consider essentially the same function, though in different notation. They approximate it by $d(r,\theta)\frac{1}{2}[1-\Phi(r)]$, where $r = \sqrt{h^2+k^2}$, $\theta = \tan^{-1}\frac{k}{h}$, and $d(r,\theta)$ is approximated by $b_0\theta + (b_1 + b_2r)r\theta + (b_3 + b_4r)r\theta^3 + (b_5 + b_6r)r\theta^5$, the values of the b's being given in their article (being different for different ranges of r).

4.12.6 Algorithms Arbitrary Polygons

Cadwell (1951) was possibly the first to consider the bivariate normal integral over an arbitrary polygon. His procedure was to transform X and Y into independent standard normal variates, by a rotation of axes followed by a change of scale, and then make use of the V function discussed earlier. His work was developed more formally in National Bureau of Standards (1959). See also Johnson and Kotz (1972, pp. 99–100). We note that a linear transformation of the type mentioned above will leave unaltered the property of the polygon as to whether it is convex, simple, or self-intersecting.

By making use of these linear transformations, DiDonato et al (1980) and DiDonato and Hageman (1982) developed an efficient programmable procedure to calculate bivariate normal integrals over arbitrary polygons.

4.12.7 Tables

We now summarise the major sets of tables relevant to the bivariate normal integral. For more details, see National Bureau of Standards (1959) and Greenwood and Hartley (1962, pp. 119–122).

Pearson (1931)	$L(h,k;\rho)$
Nicholson (1943)	$V(h,k)$
National Bureau of Standards (1959)	$L(h,k;\rho)$, $V(h,\lambda h)$, and $V(\lambda h,h)$
Japanese Standards Association (1972)	$L(h,k;\rho)$ and $V(h,\lambda h)$
Owen (1956, 1962)	$T(h,\lambda)$
Smirnov and Bol'shev (1962)	$T(h,\lambda)$ and $T(h,1)$

4.12.8 Computer Programs

We have come across the following.

Donnelly (1973)	$L(h,k;\rho)$	FORTRAN code listed
Young and Minder (1974) (corrections and remarks by other authors)	$T(h,\lambda)$	FORTRAN code listed
Divgi (1979a)	$L(h,k;\rho)$	
Bouver and Bargmann (1979)	$\Psi(h,k)$	FORTRAN code listed
DiDonato and Hageman (1982)	Integral over polygon	
Baughman (1988)	$L(h,k;\rho)$	FORTRAN code listed

We understand that the IMSL package includes a routine for evaluating the bivariate normal integral, and so do NWA STATPAK, which is microcomputer-oriented (and for which we rely upon Siegel, 1985, p. 224, for the information), and STATLIB (Brelsford and Relles, 1981, p. 370). STATLIB's method is unsophisticated—Simpson's-rule integration of the univariate cumulative normal.

4.12.9 Literature Reviews

Extensive reviews of the subject may be found in National Bureau of Standards (1959), Gupta (1963a,b), Johnson and Kotz (1972, Chapter 36), Martynov (1981), and Patel and Read (1982, Chapter 10).

4.13 Testing for Bivariate Normality

Some of the discussion in this Section is of general issues of discerning shape in empirical bivariate data, but is included here as the bivariate normal is so often the benchmark.

4.13.1 How Might Bivariate Normality Fail?

As was pointed out earlier, the bivariate normal has been used extensively in empirical research. The question arises as to how we can know if the two random variables have a joint distribution that is bivariate normal. Among reviews of assessing normality with bivariate (or higher-dimensioned) data, we call attention to Kowalski (1970), Andrews et al (1971), Gnanadesikan (1977), Mardia (1980), Small (1985), Csörgő (1986), and Looney (1986). There are many possible ways of departing from the bivariate normal; as a first step, we may classify them as:

- Failure of the marginals to be normal,
- Normal marginals, but failure to be bivariate normal, or
- Failure to be bivariate normal after univariate transformations have made the marginals normal.

Broadly speaking, there are two methods of checking bivariate normality—graphical procedures, and formal tests of significance. The structure of the main part of this Section is as follows:

- Graphical checks,
- Formal tests—univariate normality,
- Formal tests—bivariate normality,
- Tests of bivariate normality after transformations,
- Some comments and suggestions.

We first, however, make some remarks about outliers.

4.13.2 Outliers

Failure to be bivariate normal may apparently be due to "outliers"—one observation, or a few, that seems to be separate from the others.

- The concept of an outlier is by no means as straightforward as it is for a univariate sample, and we refer to Barnett and Lewis (1984, Chapter 9) for further discussion of this. Other accounts are those of Hawkins (1980, Chapter 8) and Barnett (1983b).
- One important idea is to represent a multivariate observation $\mathbf{x}$ by some distance measure $(\mathbf{x}-\mathbf{x}_0)'\boldsymbol{\Omega}^{-1}(\mathbf{x}-\mathbf{x}_0)$, where $\mathbf{x}_0$ is a measure of location and $\boldsymbol{\Omega}$ is a measure of scatter—$\mathbf{x}_0$ might be the population mean (if known), or the sample mean, and $\boldsymbol{\Omega}$ might be the population or sample variance-covariance matrix. Then the observations can be ordered according to this measure.

- For a transformation approach to handling outliers, see Barnett (1983a) and Barnett and Lewis (1984, Section 9.3.4). The idea involves obtaining standard normal variates U_1 and U_2 by means of the equations $F(x) = \Phi(u_1)$, $G(y|X = x) = \Phi(u_2)$.

- For outlier detection when the "linear structural model" applies—i.e., there are unobserved variates Z_1 and Z_2 connected by $Z_2 = \alpha + \beta Z_1$, with observed variates $X = Z_1 + \varepsilon_1$ and $Y = Z_2 + \varepsilon_2$—see Barnett (1985b).

- For the "influence function" approach to outlier detection, see Chernick (1983). The idea is to identify which observations have the biggest effect on some statistic of interest, such as the mean, or the correlation. Chernick reports an application to monthly consumption/generation data for power plants.

- Bacon-Shone and Fung (1987) have proposed a graphical method which they claim is good in detecting multiple outliers.

- For multivariate "trimming" (i.e., removal of extreme values), see Ruppert (1988).

- Building on the univariate ideas of Green (1976), Mathar (1985) discusses the classification of bivariate and multivariate distributions as "outlier-resistant" or "outlier-prone": a multivariate distribution is absolutely outlier-resistant if, with increasing sample size, the difference between the largest and second-largest distances from the origin converges to zero in probability; it is relatively outlier-resistant if the corresponding ratio converges to one in probability. However, with these definitions, the outlier-behaviour of a multivariate distribution is determined by its marginals, the dependence structure does not affect it. Consequently, we don't like the definitions much—they can hardly be said to be of multidimensional relevance if the marginals contain all the information.

- As an example of an applied paper, we draw attention to Clark et al (1987), whose variates were diastolic and systolic blood pressure; a feature of their investigation was how the identification of observations as outliers or not changed as various covariates (e.g., nature of activity when blood pressure was measured) were taken into consideration.

4.13.3 Graphical Checks

4.13.3.1 Univariate Plotting

To check univariate normality, arrange the observations in order of size, calculate suitable plotting positions, and plot on the special graph paper that is available for this purpose (or convert the plotting positions to equivalent normal deviates and use ordinary graph paper). A straight line indicates normality. An excellent account of this kind of technique is by D'Agostino (1986a). See also Harter (1984), Sievers (1986), and (for censored data) Michael and Schucany (1986). The plotting position $i/(n+1)$ is often used for the ith observation in an ordered sample of size n, but the best choice is a matter of some controversy—in the applied literature (e.g., Cunnane, 1978) as well as in the statistical.

4.13.3.2 Scatterplots

A well-known method that can be used to check bivariate normality is to draw a scatter diagram. If the sample observations do come from a bivariate normal distribution, the points will roughly lie within an elliptical region with a heavier concentration near the middle, and with a gradually decreasing concentration away from the middle. A scatterplot may give indications of non-normality or reveal outliers that, if included in the analysis, may give rise to spurious indications of departure from normality or perhaps conceal real departure from normality. For a listing of a program written in APL that creates a scatterplot and superimposes contours of the bivariate normal p.d.f., see Bouver and Bargmann (1981). For the "sharpening" of a scatterplot to reveal its structure more clearly, see Section 4.17.6. For a review of convex hulls[3] and other methods of "peeling" bivariate data (with rectangles, ellipses, etc.), see Green (1981, 1985).

[3] The convex hull of a set of points is the smallest convex set containing all of them.

For many ideas about elaborating scatterplots to bring out their meaning more clearly, see Chambers et al (1983, especially Chapter 4).

In many situations, however, a scatterplot will be inconclusive and a formal test of goodness of fit may be required. Note also that a necessary condition for a bivariate normal is that the conditional means are linear and the conditional variances are constant. Therefore a plot of these statistics can be helpful in assessing bivariate normality.

4.13.3.3 Radii and Angles

Another approach to assessing bivariate normality, which is based on radii and angles, is discussed by Gnanadesikan (1977, Chapter 5). The rationale for this method is as follows. Let $(X_1, X_2)'$ denote the bivariate normal vector with variance-covariance matrix $\mathbf{\Sigma}$. First, transform the original variates X_1 and X_2 to independent[4] standard normal variates X and Y using

$$\begin{pmatrix} X \\ Y \end{pmatrix} = \mathbf{\Sigma}^{-1/2} \begin{pmatrix} X_1 - \mu_1 \\ X_2 - \mu_2 \end{pmatrix}. \tag{4.22}$$

Second, transform (X, Y) to polar coordinates (R, Θ). Then, under the hypothesis of bivariate normality, R^2 has a χ^2_2 distribution (i.e., exponential with mean 2), and Θ has a uniform distribution over the range 0 to 2π. These consequences may be tested graphically—by plotting sample quantiles of R^2 versus quantiles of the exponential distribution with mean 2, and plotting sample quantiles of the angle Θ versus those of the uniform distribution. For illustration, see Gnanadesikan (1977, Exhibits 28i,j, 29d,e). If bivariate normality holds, the two plots should be approximately linear. However, if $\boldsymbol{\mu}' = (\mu_1, \mu_2)$ and $\mathbf{\Sigma}$ are estimated, the distributional properties of R and Θ are only approximate. For $n \geq 25$, the approximation is good. The radii-and-angles approach, though informal, is an informative graphical aid. However, as in the case of scatterplots, the test may be inconclusive, particularly with small samples.

4.13.3.4 Projection Pursuit

Though it is aimed at the multivariate situation rather than only two dimensions, the method of projection pursuit (Friedman and Tukey, 1974; Friedman and Stuetzle, 1982; Tukey and Tukey, 1981b) should be mentioned. The strategy is to "pursue the projection"—i.e., find the vector—that most clearly shows up the non-normality of the data. Implementation on interactive computer graphics systems in envisaged. At each step, an augmenting function is estimated as the ratio of the data to the model when projected on to a certain vector. The final model is the product of the initial model (e.g., multivariate normal) and a series of augmenting functions.

Somewhat in the same style is an idea described by Gnanadesikan (1977, pp. 142–143), in which univariate Box-Cox transformations—see Section 4.13.6 below—are repeatedly applied in different directions.

4.13.3.5 The Kernel Method

Silverman (1986, Chapter 4) argues persuasively that both two-dimensional histograms and scatterplots are poor aids to grasping the structure of bivariate data, and proposes the "kernel" method as an improvement. He describes this method of estimating a density as being a sum of bumps centred at the observations.

- Choose a kernel function $K(.)$ and a window width w.
- The density at a point $\mathbf{x}$ is then estimated to be

$$\hat{h}(\mathbf{x}) = \frac{1}{nw^2} \sum_{i=1}^{n} K\left(\frac{\mathbf{x} - \mathbf{x}_i}{w}\right), \tag{4.23}$$

  where $\mathbf{x}_1, \mathbf{x}_2, \ldots \mathbf{x}_n$ are the n data points in our sample.

[4] There are infinitely many ways to transform a bivariate normal vector into two independent normal variates by decomposing $\mathbf{\Sigma}$ into products of two matrices. In addition to $\mathbf{\Sigma} = \mathbf{\Sigma}^{1/2}\mathbf{\Sigma}^{1/2}$ as represented by (4.22), the Choleski decomposition $\mathbf{\Sigma} = \mathbf{L}\mathbf{L}'$, where $\mathbf{L}$ is a lower triangular matrix, is popular.

- $K(.)$ is usually a radially-symmetric unimodal p.d.f.—for instance, the bivariate normal.
- The data should be re-scaled to avoid extreme differences of spread in the various coordinate directions (or else the single smoothing parameter w replaced by a vector).
- Appropriate computer graphics are then used to display $\hat{h}(\mathbf{x})$ as a surface or contour plot.

Although one may fear that different $K(.)$ and w may give different results, Silverman does provide extensive discussion about these choices. He gives an example (also in Silverman, 1981) of 100 data points drawn from a bivariate normal mixture, smoothed using window widths 1.2, 2.2, and 2.8; the first appears undersmoothed, the last oversmoothed. And let us remember that the classes from which a histogram is constructed are arbitrary, too, and that the scales chosen for a scatterplot affect the subjective impression it gives.

Other accounts of this are by Everitt and Hand (1981, Section 5.3) and Chambers et al (1983, Section 4.10).

Tanabe et al (1988) have published FORTRAN subroutines for computing bivariate (and univariate) density estimates, using a Bayesian nonparametric method.

4.13.3.6 Haar Distributions

We are hardly any the wiser about what these are after reading the article by Wachter (1983), but we did notice the following paragraph: "On the whole, Haar measures have gained prominence in statistics with the realisation that many consequences of multivariate normal assumptions do not depend on normality itself but only on rotational symmetries. For graphically based data analysis, symmetry assumptions are often preferable to parametric distributional assumptions such as normality. Thus, curiously enough, data analytic emphasis in multivariate statistics has promoted ties with the highly mathematical theory of Haar distributions."

4.13.4 Formal Tests—Univariate Normality

4.13.4.1 Why Test?

As mentioned earlier, graphical checks of bivariate normality may be inconclusive and hence a formal test of significance will perhaps give a more objective guide as to the suitability of the bivariate normal distribution. Again as mentioned earlier, there are many different ways in which an empirical distribution may deviate from bivariate normality. This suggests that equally many different techniques may be needed to spot such deviation.[5] According to Small (1985), "there is no single best method, and choice should be guided by what departure might be expected a priori or would have the most serious consequences".

We deal here with tests on the marginals, to see if they are normal (remember that marginal normality is a necessary, though not sufficient, condition for bivariate normality). Fuller accounts are by D'Agostino (1982, 1986b), Koziol (1986), and (for censored data) Michael and Schucany (1986).

4.13.4.2 Chi-Squared Test

Group the observations into a number of ranges of the variate. Determine the expected numbers that would fall into these groups if the normal distribution was valid. Compute one of the statistics that asymptotically has a chi-squared distribution, such as $\sum(O-E)^2/E$.

The advantage of this is its ease and elementary character. A minor technical disadvantage is that one never knows precisely how to carry out the grouping—having too few groups loses much information, whereas having too many groups means there are few observations per group, and hence the chi-squared approximation to the test statistic is of dubious validity. But the major disadvantage is the loss of information concerning the ordering of the groups, and the consequent loss of power. As an example, suppose the pattern of residuals were $++---+++---++$. If X^2 just failed to indicate statistical significance, we would nevertheless suspect the empirical distribution differed from the normal in kurtosis.

[5] But there is a danger here—if lots of different tests are conducted, a "significant" result is quite likely to come about, purely by chance.

For more on this, see Moore (1986).

4.13.4.3 Moment Tests

The skewness ($\sqrt{b_1}$) and kurtosis (b_2) statistics are defined by $\sqrt{b_1} = m_3/m_2^{3/2}$ and $b_2 = m_4/m_2^2$, where m_i is the ith sample moment about the mean.

$\sqrt{b_1}/\sqrt{6/n}$ and $(b_2-3)/\sqrt{24/n}$ are both asymptotically normal. Consequently, D'Agostino and Pearson (1973) suggested adding together the squares of the standardised variates corresponding to these sample statistics, and treating the result as a χ_2^2 variate (i.e., exponential with mean 2), this test being sensitive to departures from normality in both skewness and kurtosis, though naturally not as powerful with regard to either as a specific test for that feature would be. And, of course, it is not sensitive to departures from normality other than those which are reflected by the skewness and kurtosis statistics.

Tests based on $\sqrt{b_1}$ and b_2 are reviewed by Bowman and Shenton (1986) and D'Agostino (1986b).

4.13.4.4 Tests Based on the Empirical Distribution Function

The best-known of these is the Kolmogorov-Smirnov test. Using the notations F_n for the empirical d.f. based on a sample size of n, and F for the hypothesised distribution, the test involves calculating

$$D_n = \sup_{-\infty<x<\infty} |F_n(x) - F(x)|. \tag{4.24}$$

The Cramér-von Mises test involves calculating

$$W_n^2 = \int_{-\infty}^{\infty} [F_n(x) - F(x)]^2 A[F(x)]\, dF(x), \tag{4.25}$$

where A is a non-negative weight function. There are other alternatives, too.

There are many difficulties with such tests. One is their frequent absence from elementary textbooks and computer packages. Another is that textbook writers, when they do include them, rarely convince the reader that they understand the subject; perhaps it is really the case that no-one understands their properties when the parameters of the distribution have been estimated from the sample, and the sample size is small to moderate. Another is the labour involved. (We wrote the last few sentences before seeing the chapter by Stephens (1986a). This seems to be very thorough and helpful.)

4.13.4.5 Probability Plots

Tests have been proposed (e.g., by Shapiro and Wilk, 1965) that are based upon how far a probability plot of the type described in Section 4.13.3.1 is from a straight line, i.e., upon its correlation coefficient. For a review of this approach, see Stephens (1986b).

4.13.4.6 The CPIT Approach

CPIT stands for conditional probability integral transformation. See Quesenberry (1986a,b) for this method.

4.13.5 Formal Tests—Bivariate Normality

4.13.5.1 Chi-Squared Test

It is easy, in principle, to compare observed and expected numbers in a discretised bivariate distribution by means of Pearson's X^2 or similar statistic, but the disadvantages mentioned in Section 4.13.4.2 remain.

4.13.5.2 Tests Based on the Empirical Distribution Function

Such tests have been proposed, but they do not seem to have received wide acceptance yet. One such is that using a statistic of Cramér-von Mises type, described by Pettitt (1979), who also discusses (at pp. 707–708) the kinds of departure from normality that the test is and is not sensitive to. A related approach is via Rosenblatt's (1952) multivariate probability integral transformation or the conditional probability integral transformation (see Quesenberry, 1986a, and references therein).

4.13.5.3 Tests Based on the Empirical Characteristic Function

Tests of this type are put forward by Csörgő (1984, 1986) and Baringhaus and Henze (1988). The former also mentions he has an analogous test for Marshall and Olkin's (1967a) distribution. See also Csörgő (1989).

4.13.5.4 Malkovich and Afifi (1973)

These authors generalised the univariate skewness and kurtosis statistics, and the W statistics proposed by Shapiro and Wilk (1965), to the bivariate case by using Roy's (1953) union-intersection principle (for which, see Arnold, 1988). They made use of the property of the bivariate normal distribution that any linear combination of X and Y is univariate normal. Formally, they defined bivariate skewness as $\max_{\mathbf{c}}[\beta_1(\mathbf{c})]$, where

$$\beta_1(\mathbf{c}) = \frac{\{E[c_1(X-\mu_1)+c_2(Y-\mu_2)]^3\}^2}{[\text{var}(c_1X+c_2Y)]^3},$$

and bivariate kurtosis as $\max_{\mathbf{c}}\{[\beta_2(\mathbf{c})-3]^2\}$, where

$$\beta_2(\mathbf{c}) = \frac{E[c_1(X-\mu_1)+c_2(Y-\mu_2)]^4}{[\text{var}(c_1X+c_2Y)]^2},$$

for some vector $\mathbf{c}' = (c_1, c_2)$. Using Roy's principle, one retains the null hypothesis of bivariate normality if $\max_{\mathbf{c}}[b_1(\mathbf{c})] \leq k_{b_1}$ and $\max_{\mathbf{c}}\{[b_2(\mathbf{c}) - k]^2\} \leq k_{b_2}$, where $b_1(\mathbf{c})$ and $b_2(\mathbf{c})$ are the sample counterparts of $\beta_1(\mathbf{c})$ and $\beta_2(\mathbf{c})$, with k_{b_1}, k_{b_2}, and k being constants, such that $k \longrightarrow 3$ as the sample size becomes infinitely large.

According to Bera and John (1983), these tests are conceptually simple but computationally burdensome.

4.13.5.5 Cox and Small (1978)

These authors based their method on the extent of nonlinearity of the regression line. Specifically, it involves the coefficients of the quadratic terms when Y is regressed on X and X^2, and X is regressed on Y and Y^2. A statistic that is asymptotically χ^2_2 (i.e., exponential with mean 2) can be calculated. A disadvantage of this procedure is that the bivariate normal is not the only distribution having normal marginals and linear regressions, there are many others. (Any mixture of two bivariate normal distributions having the same means and standard deviations provides an example.)

4.13.5.6 Hawkins (1981)

In this paper, a procedure was proposed that can be used to test for normality and homoscedasticity simultaneously. Considerable use is made of it in the book by McLachlan and Basford (1988).

4.13.5.7 Bera and John (1983)

These authors considered the bivariate Pearson family of distributions (which includes the bivariate normal). They used Rao's (1948b) score principle to develop four tests for bivariate normality. Each of the test statistics was shown to have an asymptotic chi-squared distribution. They also compared the powers of their tests with those of Mardia's (1970b) tests (to be discussed shortly).

4.13.5.8 Bivariate Skewness and Kurtosis

Many test statistics involve sample product moments, and are asymptotically distributed as chi-squared variates.

Let $(x_1, y_1), (x_2, y_2), \ldots (x_n, y_n)$ be n pairs of independent observations from a bivariate population. Define

$$\begin{pmatrix} z_{1i} \\ z_{2i} \end{pmatrix} = \begin{pmatrix} \hat{\sigma}_1^2 & r\hat{\sigma}_1\hat{\sigma}_2 \\ r\hat{\sigma}_1\hat{\sigma}_2 & \hat{\sigma}_2^2 \end{pmatrix}^{-1/2} \begin{pmatrix} x_i - \bar{x} \\ y_i - \bar{y} \end{pmatrix} \tag{4.26}$$

for $i = 1, 2, \ldots n$, where $\bar{x}$ and $\bar{y}$ are the sample means, the $\hat{\sigma}^2$'s are the maximum likelihood estimates of the variances, i.e., they are the sample variances with n used as the divisor, and r is the sample correlation coefficient. Next, let us write

$$m_{ij} = \frac{1}{n}\sum_{k=1}^{n} z_{1k}^i z_{2k}^j. \tag{4.27}$$

We may now define test statistics in terms of sample product moments. Mardia's (1970b) tests are based on sample measures of bivariate skewness and kurtosis, defined by, respectively,

$$b_{1,2} = m_{30}^2 + m_{03}^2 + 3(m_{21}^2 + m_{12}^2) \tag{4.28}$$

and

$$b_{2,2} = \frac{1}{n}\sum_{j=1}^{n}(z_{1j}^2 + z_{2j}^2)^2 = m_{40} + m_{04} + 2m_{22}. \tag{4.29}$$

Univariate skewness and kurtosis are functions of the third and fourth central moments, respectively. Here, $b_{1,2}$ is a function of product moments of order 3, and $b_{2,2}$ is a function of product moments of order 4, this indicating their appropriateness as bivariate skewness and kurtosis measures. But note that, in each case, the first two terms are functions of univariate statistics—so these formulae do not give measures free of these. Tests derived from these measures are large sample tests (Mardia, 1974); indeed, Reyment (1971) confirmed that Mardia's procedures do not stabilise until a large number of observations have been included in the calculations. Mardia (1974) gives tables of the critical values of $b_{1,2}$ and $b_{2,2}$ for several common values of n and levels of significance.

Mardia has formulated several other test statistics; among them are S_W^2 and C_W^2, which combine $b_{1,2}$ and $b_{2,2}$—see Mardia and Foster (1983, pp. 212–213) and Mardia (1985).

Monte Carlo comparisons of the behaviour of measures introduced by Mardia, Malkovich and Afifi, and others, in circumstances where the distribution is a mixture of two bivariate normals, were made by Isogai (1983a,b), who then attempted to clarify the meanings of these statistics as diagnostic tools for measuring non-normality. Other comparisons of several proposals include those by Ulrich (1984) and Booker et al (1984).

Schwager (1985) discusses notions of multivariate skewness and kurtosis proposed by Mardia, Malkovich and Afifi, Isogai, and others.

According to the assessment of Looney (1986), $b_{1,2}$ and $b_{2,2}$ are the most thoroughly developed tests for bivariate normality. In particular, there is a published algorithm (Mardia and Zemroch, 1975). But they appear to be less powerful than that of Bera and John (1983).

4.13.5.9 Best and Rayner's Comparisons

Koziol (1986, 1987) discussed certain statistics in the style of Neyman's "smooth" tests. Best and Rayner (1988) give Koziol's formulae adapted for the bivariate case. The first of these is simply $nb_{1,2}/6$. The second is defined by

$$\hat{U}_4^2 = n\left[\frac{(m_{22}-1)^2}{4} + \frac{m_{31}^2 + m_{13}^2}{6} + \frac{(m_{04}-3)^2 + (m_{40}-3)^2}{24}\right]. \tag{4.30}$$

Best and Rayner (1988) compared the (approximate) powers of the test statistics for $\hat{U}_3^2$, $\hat{U}_4^2$, $(\hat{U}_3^2 + \hat{U}_4^2)$, $b_{2,2}$, and S_W^2 with the level of significance set at 5%, under the alternatives

- X and Y being independent lognormal variables,
- X and Y being independent uniform variables,
- X and Y being independent t_4 variables (i.e., having the t-distribution with 4 degrees of freedom), and
- (X, Y) having various bivariate normal mixture distributions.

Their conclusion was that no single statistic dominates, though $(\hat{U}_3^2 + \hat{U}_4^2)$ usually does better than S_W^2, a statistic favoured by Mardia.

4.13.5.10 Asymptotically χ_2^2?

For a number of proposals, it happens that the test statistic is asymptotically distributed as χ_2^2. For the two tests they investigated, Mason and Young (1985) found that this approximation can be too conservative, and lead to inappropriate rejection of normality, when the population parameters in the formulae are replaced by their sample estimates.

4.13.6 Tests of Bivariate Normality After Transformations

A popular approach[6] to the understanding of bivariate distributions with non-normal marginals is to (i) transform the marginals to normality, (ii) check that the bivariate distribution is roughly bivariate normal in appearance, and then (iii) proceed with the analysis under the assumption of bivariate normality. Separate statements can then be made about the univariate transformations that were necessary and the conclusions drawn from the bivariate transformed observations. This procedure, we feel, has much to recommend it; a technical disadvantage is that the properties of tests for bivariate normality are even less well understood when applied to transformed data than when applied to raw observations, we believe. Occasionally, a more fundamental objection arises: that is when we have an explicit model for how the distribution is constructed, albeit with some uncertainties, as when we are assuming a trivariate reduction model but do not know the forms of the component distributions; in such a case, the bivariate distribution is intimately tied up with its marginals, and a procedure that separates the treatment of the individual variables from the treatment of their association may be thought undesirable.

If we follow this stategy, we have to decide how to transform the marginals to normality:

- Do we enforce exact normality, by calculating $\Phi^{-1}[F(x)]$ for each observation x? If so, then a question will arise as to how to interpret this in the context of a sample, i.e., whether the best estimate of F is $i/(n+1)$ when x is the ith smallest observation in the sample of size n, or something else.
- Alternatively, do we insist on some easily-comprehended transformation, such as the logarithm, or a power function? If so, how much effort should we put into the search for the best transformation? Should we just try one or two of the best-known ones? Should we consider a whole parametric family, allowing the data to determine the parameter that gives the best fit?

As to easily-comprehended transformations, we note:

- Probably the most popular single choice is the logarithm.
- Johnson's (1949) system of bivariate distributions consists of the following transformations applied to the marginals of a bivariate normal: logit, $\sinh^{-1}$, log, and none. If choosing from this set, we can thus work with the original observations and one of Johnson's distributions, if we prefer doing that to working with transformed observations and the bivariate normal. Further details of this system are given in Section 4.14.2.

[6] See, especially, Kowalski (1970). The major Sections of that paper were: tests for univariate normality; tests for bivariate normality; the coordinate transformation to normality and its estimation; summary of results of tests for bivariate normality; application to correlation theory.

- The most popular parametric family is that of Box and Cox (1964); see Box and Tiao (1973, Chapter 10) and, for computer programs, Howarth and Earle (1979) and Liem (1980) and references therein. Andrews et al (1971) extended this to bivariate distributions. The family is that of power transformations, with the logarithm as a special case. The next paragraph gives more details.

The Box-Cox transformation of (x, y) to (z_1, z_2) is as follows:

$$z_1 = \left\{ \begin{array}{ll} (x^{\lambda_1} - 1)/\lambda_1 & \text{for } \lambda_1 \neq 0 \\ \log x & \text{for } \lambda_1 = 0 \end{array} \right\} \tag{4.31}$$

$$z_2 = \left\{ \begin{array}{ll} (y^{\lambda_2} - 1)/\lambda_2 & \text{for } \lambda_2 \neq 0 \\ \log y & \text{for } \lambda_2 = 0 \end{array} \right\} \tag{4.32}$$

One might choose the λ's so as to make the z's as (univariate) normal as possible—the method of Box and Cox would be applied to each variable separately. But perhaps bivariate normality is not thereby optimised. A likelihood approach to achieving joint normality is as follows. First, express the bivariate density function of (X, Y) in terms of a bivariate normal density with mean $\boldsymbol{\mu}$ and variance-covariance matrix $\boldsymbol{\Sigma}$. Next, find the log likelihood function of $\boldsymbol{\mu}$, $\boldsymbol{\Sigma}$, and $\boldsymbol{\lambda}$, where $\boldsymbol{\lambda}' = (\lambda_1, \lambda_2)$. By keeping λ_1 and λ_2 fixed temporarily, we can find the maximum likelihood estimates $\hat{\boldsymbol{\mu}}$ and $\hat{\boldsymbol{\Sigma}}$, and the maximised log likelihood function is now

$$L^*(\lambda_1, \lambda_2) = -\frac{n}{2} \log |\hat{\Sigma}| + (\lambda_1 - 1) \sum_{i=1}^{n} \log x_i + (\lambda_2 - 1) \sum_{i=1}^{n} \log y_i. \tag{4.33}$$

Also, maximum likelihood estimates $\hat{\lambda}_1$ and $\hat{\lambda}_2$ may be obtained by numerically maximising (4.33). Andrews et al (1971) showed that $2[L^*(\hat{\lambda}_1, \hat{\lambda}_2) - L^*(1, 1)]$ has asymptotically a χ_2^2 distribution (i.e., exponential with mean 2). Rejection of the null hypothesis $\boldsymbol{\lambda}' = (1, 1)$ implies non-normality of the untransformed data.

4.13.7 Some Comments and Suggestions

This subject is in a rather unsatisfactory state. The various pieces of knowledge do not seem to be well integrated. There seem to be gaps between them. There seems to be rather little experience with such tests as have been put forward. Hence there is lack of knowledge of their properties—behaviour with samples of small to moderate size, and with observations that are merely crudely grouped, are two areas we have in mind. And even if we knew how to measure non-normality, would we also know how much non-normality is important?

But perhaps we ought not to be surprised—the most that anyone does in the univariate context is to plot the observations on normal graph paper. Sometimes the result is roughly a straight line, sometimes it isn't. Very often there is a slight consistent curve throughout the range, or a straight line in the middle is spoilt by apparent nonlinearities at the ends. In practice, noone goes on to perform a Kolmogorov-Smirnov test, or knows what to do with the answer if they do. Is it any wonder that the bivariate context is worse?

One thing that sometimes helps is thought. Stimulated by contact with data and by knowledge of the various types of departure from bivariate normality that have been discussed theoretically, the human brain can sometimes reach understanding of the message in the data.

If data and theory are not sufficient in abstract, then the context may serve to focus ideas. In particular:

- Does the mechanism generating the data direct attention to a particular class of distributions, such as those constructed by compounding, or from a univariate distribution of X and a set of conditional distributions of Y given X?

- What use is the result going to be put to? For instance, is the correlation coefficient important? Is the survival function $\Pr(X > x, Y > y)$ important?

Many univariate distributions form a hierarchy—for example, the exponential is a special case of the gamma, which is a special case of the Stacy, which in turn can have a shift parameter included, and so on. This provides a natural environment for testing goodness-of-fit: fit a three- or four-parameter distribution, and test whether the parameter values are consistent with some special case that is a one- or two-parameter distribution. Such a procedure is much less common with bivariate distributions, because such hierarchies are not so well-known. Nevertheless, it is a desirable one, if practicable.

If more specific procedures do not come to mind, we suggest that implied above. In summary:

- Consider the marginals. Ask what shape they have. Answer this by various forms of probability plotting, calculation of moments, and comparison of the goodness-of-fit of members of a hierarchy of distributions.
- If the marginals appear to be non-normal, transform them to normality.
- Does bivariate normality hold? Answer this by the various graphical procedures and calculations (e.g., of moments) discussed above.

We note that if a scatter diagram prepared after transformation to marginal normality still fails to be bivariate normal, then a comparison with Figure 1 of Johnson et al (1984, p. 242) may provide some insight into the underlying bivariate distribution.

Other choices of the standard form for the marginals might be equally or more suitable than the normal. The uniform is the obvious competitor—in this case, Mardia (1970a, p. 81) suggests focussing attention on the regression and scedastic curves. We are unaware of anyone yet preparing the diagrams that Mardia suggests.

4.14 Univariate Transformations

4.14.1 The Bivariate Lognormal Distribution

If $\log X$ and $\log Y$ have a bivariate normal distribution with means μ_1 and μ_2, variances σ_1^2 and σ_2^2, and correlation ρ, then:

$$E(Y|X=x) = x^{\rho\sigma_2/\sigma_1} \exp\left[-\tfrac{1}{2}(1-\rho^2)\sigma_2^2 + \mu_2 - \rho\sigma_2\mu_1/\sigma_1\right] \tag{4.34}$$

$$\text{var}(Y|X=x) = \omega'(\omega'-1)x^{2\rho\sigma_2/\sigma_1} \exp[2(\mu_2 - \rho\sigma_2\mu_1/\sigma_1)], \tag{4.35}$$

where $\omega' = \exp[(1-\rho^2)\sigma_2^2]$ (Johnson and Kotz, 1972, p. 19).

The joint moments are:

$$\mu'_{ij} = E(X^iY^j) = \exp[i\mu_1 + j\mu_2 + \tfrac{1}{2}(i^2\sigma_1^2 + 2ij\rho\sigma_1\sigma_2 + j^2\sigma_2^2)]. \tag{4.36}$$

In particular,

$$\text{cov}(X,Y) = [\exp(\rho\sigma_1\sigma_2) - 1]\exp[\mu_1 + \mu_2 + (\sigma_1^2 + \sigma_2^2)/2] \tag{4.37}$$

and

$$\text{corr}(X,Y) = \frac{\exp(\rho\sigma_1\sigma_2) - 1}{\sqrt{[\exp(\sigma_1^2) - 1][\exp(\sigma_2^2) - 1]}}. \tag{4.38}$$

(For the meaningfulness, or otherwise, of the correlation coefficient in this case, see Section 4.14.5.)

Thomopoulos and Longinow (1984) have concisely listed the basic properties of the bivariate lognormal distribution. The application they envisage for it is to structural reliability analyses in which load and resistance are correlated. (See also Section 19.8.2 below.)

For applications, see Sections 19.3.1, 19.4.3, 19.4.4, 19.4.5, 19.4.6, 19.7.4, 19.7.5, 19.7.8, 19.7.10, 19.7.12, 19.8.4, 19.9.1, and 19.9.2.

4.14.2 Johnson's System

4.14.2.1 Derivation

Bivariate distributions with specified marginals may be obtained from the bivariate normal, by stretching and compressing the X and Y axes as required. N L Johnson (1949) constructed what has become a well-known system of distributions in this manner. The bivariate distributions are denoted S_{IJ}, in which one variable has an S_I distribution and the other has an S_J distribution, where I and J can be B, U, L, or N (standing for bounded, unbounded, lognormal, and normal). Thus the variables

$$Z_1 = \gamma_1 + \delta_1 a_I\left(\tfrac{X-\xi_1}{\lambda_1}\right) \qquad (4.39)$$

$$Z_2 = \gamma_2 + \delta_2 a_J\left(\tfrac{Y-\xi_2}{\lambda_2}\right) \qquad (4.40)$$

(where $a_B(y) = \log[y/(1-y)]$, $a_U(y) = \sinh^{-1} y$, $a_L(y) = \log y$, and $a_N(y) = y$) are standardised (unit) normal variables, with correlation coefficient ρ.

Chapter 5 of M E Johnson's (1987) book is given over to this system, so we can be brief. The great advantage is that the simplicity of derivation makes variate generation for simulation studies equally simple. See also Rodriguez (1983b, pp. 239–240).

For the case S_{NL}, see Yuan (1933), Crofts (1969), Crofts and Owen (1972), Suzuki (1983), and Suzuki et al (1984).

4.14.2.2 Formula for p.d.f.

The joint distribution of X and Y, so defined, has nine parameters: γ_1, γ_2, δ_1, δ_2, ξ_1, ξ_2, λ_1, λ_2, and ρ. The standard form of the distribution is obtained by taking $\xi_1 = \xi_2 = 0$, $\lambda_1 = \lambda_2 = 1$. The density is

$$h = \delta_1\delta_2\psi[\gamma_1 + \delta_1 f_I(x), \gamma_2 + \delta_2 f_J(y); \rho]. \qquad (4.41)$$

4.14.2.3 Univariate Properties

X and Y have S_I and S_J distributions, respectively.

4.14.2.4 Conditional Properties

The conditional distribution of Y, given $X = x$, is of the same system (S_J) as Y, but with γ_2, δ_2 replaced by $(1-\rho^2)^{-1/2}\{\gamma_2 - \rho[\gamma_1 + \delta_1 f_I(x)]\}$, $(1-\rho^2)^{-1/2}\delta_2$ respectively (Johnson and Kotz, 1972, pp. 15–17).

The regressions and conditional variances are given in Mardia (1970a, pp. 25–26), and a table of median regressions in Johnson and Kotz (1972, p. 17) and Rodriguez (1983b).

4.14.2.5 References to Illustrations

Johnson (1987, p. 63) remarks that "The only difficulty in employing the system in simulation work is to specify appropriate parameter combinations to meet the needs of particular applications." To assist, he gives numerous contour plots for the S_{LL}, S_{UU}, and S_{BB} cases. (In number, 24, 40, and 84 contour plots, plus 5 density surface plots of S_{BB} distributions.) These could be equally useful as an aid to distribution selection, when wondering whether any are suitable to fit to an empirical dataset.

DeBrota et al (1988) describe software for fitting the univariate Johnson system to data, and to assist in subjective visual choice of an appropriate member of the system in the absence of data. They mention at the end that they are developing corresponding multivariate software.

4.14.2.6 Applications

See Sections 19.3.2 and 19.9.4.

4.14.3 The Uniform Representation

An explicit expression for this is in, for example, Barnett (1980) and Johnson et al (1981). Barnett illustrates contours of the p.d.f. for the case $\rho = 0.8$.

4.14.4 The g and h Transformation

We mentioned Tukey's g and h family of univariate distributions in Section 2.4.5. Johnson (1987, pp. 205–206) suggests applying this transformation to the marginals of a bivariate normal distribution.

4.14.5 Effect of Transformations on Correlation

It is well known that if we start with a bivariate normal distribution, and apply any nonlinear transformations to the marginals, Pearson's product-moment correlation coefficient is smaller (in absolute magnitude) in the resulting distribution than in the original bivariate normal one. (Rank correlation coefficients are, of course, unaltered, provided the transformations are monotonic.)

An extensive quantitative study of the effect of marginal transformations on the correlation coefficient was reported by Der Kiureghian and Liu (1986). Their dependent variable was the ratio ρ/ρ_t, where $\rho =$ correlation coefficient in the normal distribution, and $\rho_t =$ correlation coefficient in the distribution after transformation of the marginals. They gave a series of empirically-derived formulae for calculating this ratio, based upon ρ_t and δ_t, where δ_t is the coefficient of variation in the transformed distribution. (Notice they were envisaging ρ_t being known and ρ being wanted, not vice versa.) The marginal distributions which they considered were the following: uniform, shifted exponential, shifted Rayleigh, type I largest value, type I smallest value, lognormal, gamma, type II largest value, type III smallest value.

- The simplest formulae were $\frac{\rho}{\rho_t} =$ constant, for one marginal being normal and the other being one of the first five in the list above.

- At the other extreme, the most complicated formula was $\frac{\rho}{\rho_t} =$ a 19-term polynomial in ρ, δ_{t_1}, and δ_{t_2}, for both marginals being type II largest value.

Bhatt and Dave (1964) give an expression for the correlation between the variates that result when two standard normal variates with correlation ρ are subjected to arbitrary polynomial transformations. The expression is in terms of ρ and the coefficients when the transformations are written in terms of the Hermite polynomials. In the special case when both transformations are quadratic, $a_0 + a_1x + a_2x^2$ and $b_0 + b_1y + b_2y^2$, the correlation becomes

$$\frac{a_1b_1\rho + 2a_2b_2\rho^2}{\sqrt{(a_1^2 + 2a_2^2)(b_1^2 + 2b_2^2)}}. \tag{4.42}$$

For the case where both transformations are cubic, see also Vale and Maurelli (1983).

As to the lognormal distribution, on setting $\rho = -1$ and $\rho = 1$ in (4.38), we find

$$\frac{\exp(-\sigma_1\sigma_2) - 1}{\sqrt{[\exp(\sigma_1^2) - 1][\exp(\sigma_2^2) - 1]}} \leq \operatorname{corr}(X, Y) \leq \frac{\exp(\sigma_1\sigma_2) - 1}{\sqrt{[\exp(\sigma_1^2) - 1][\exp(\sigma_2^2) - 1]}}. \tag{4.43}$$

Adapting slightly from Romano and Siegel (1986, Section 4.22), "This has some striking implications. If, for example, we restrict ourselves to the family of distributions with $\sigma_1 = 1$ and $\sigma_2 = 4$ but we allow any values for the means and the correlation between $\log X$ and $\log Y$, then the correlation between X and Y is constrained to lie in the interval from -0.000251 to 0.01372! Such a result raises a serious question in practice about how to interpret the correlation between lognormal random variables. Clearly, small correlations may be very misleading because a correlation of 0.01372 indicates, in fact, that X and Y are perfectly functionally (but nonlinearly) related."

The general shape of the univariate gamma distribution makes it a competitor of the lognormal for fitting to data. See Moran (1967) for the range of correlations possible in a bivariate distribution with gamma marginals having specified shape parameters.

We have already seen, equation (4.38), what the correlation in a normal distribution becomes when the variates are exponentiated; Bhatt and Dave (1965) have given some results for the correlation between $\sum_{i=0}^{n} a_i \exp(\alpha_i x)$ and $\sum_{i=0}^{n} b_i \exp(\beta_i y)$ (with $\alpha_0 = \beta_0 = 0$), (X, Y) having a standard bivariate normal distribution with correlation ρ. Special cases mentioned include:

- $a_0 + a \cosh \alpha x$, $b_0 + b \cosh \beta y$, for which the correlation is $\sinh^2(\rho\alpha\beta/2)/[\sinh(\alpha^2/2)\sinh(\beta^2/2)]$, and
- $a_0 + a \sinh \alpha x$, $b_0 + b \sinh \beta y$, which is Johnson's S_U distribution; in this case, the correlation is found to be $\sinh(\rho\alpha\beta)/\sqrt{\sinh(\alpha^2)\sinh(\beta^2)}$. For the general case, the distribution of (X, Y) not being standardised, see equation (5.2) of Johnson (1987).

Lindqvist (1976) argued that correlation coefficients—in particular, when they are used as input to a factor analysis—should not be based upon variates that are skewed. He gave a listing of a computer program that chooses and applies a transformation of the form $\log(X + \text{constant})$ if the skew of the raw data is unacceptably large. Factor analyses of 13 constituents of 566 rock specimens were performed on such transformed data, and on data for which all variates had simply been log-transformed, and results of the former found preferable.

On the other hand, McDonald (1960) argued that the change in r when making transformations even of bivariate data that is grossly non-normal, such as is encountered in hydrology, is usually of little practical importance. McDonald's evidence was from precipitation data from Arizona. (However, of the 14 correlations investigated, in half of them taking the log transformation changed r by at least .05. As one usually wants to be confident about the first decimal place, perhaps it would be wise to go to the trouble of choosing the right transformation, despite what McDonald says.[7])

4.15 The Logistic-Normal Distribution

Suppose that variates Z_i have a multivariate normal distribution. Then $Y_i = \exp(Z_i)$ have a multivariate lognormal distribution. Further, $X_i = Y_i/(1 + \sum Y_i) = \exp(Z_i)/[1 + \sum \exp(Z_i)]$ have a multivariate logistic-normal distribution[8] (Aitchison and Shen, 1980; Aitchison, 1982). Johnson and Kotz (1972, p. 20) attribute this idea to R L Obenchain.

Since the Y's are positive, the X's are restricted to the range 0 to 1. This distribution is, therefore, especially suited to the description of compositional data—that is, the proportions of different components that make up some assemblage (e.g., different oxides in a rock). It has more parameters than the multivariate beta (Dirichlet) distribution, and so is more flexible. For fitting it to data, see Bacon-Shone (1985).

The total $\sum X_i$ is less than 1, since it is $\sum Y_i/(1 + \sum Y_i)$, and $1 - \sum X_i$ is the proportion of the $(n+1)$th component in the total when the X's refer to the first n components.

With the proposal of this flexible distribution, and the publication of the book by Aitchison (1986), this whole field has received new impetus and is likely to see productive interaction of data with theory over the next few years. See also Sections 12.4.2 and 24.5. Aitchison (1985) proposed a distribution that subsumes both the (multivariate) beta and logistic-normal as special cases.

---

[7]If one is solely interested in a measure of correlation, not in the marginal distributions, one might calculate a rank correlation from the data, and then convert it to an equivalent ρ by equation (13.4) or (13.5), as appropriate.

[8]Here is a comparison with the S_{BB} distribution:

	X	Y
Johnson's S_{BB} (marginal transformations)	$\exp(Z_1)/[1 + \exp(Z_1)]$	$\exp(Z_2)/[1 + \exp(Z_2)]$
Logistic-normal	$\exp(Z_1)/[1 + \exp(Z_1) + \exp(Z_2)]$	$\exp(Z_2)/[1 + \exp(Z_1) + \exp(Z_2)]$

4.16 Truncated Bivariate Normal Distributions

4.16.1 Properties

The most common form of truncation of a standardised bivariate normal distribution is single truncation, from above or below, with respect to one of the variables. We shall consider the case where $X > h$. Thus the support is $h < X < \infty$, $-\infty < Y < \infty$, and the p.d.f. is $\psi(x, y; \rho)/\Phi(-h)$.

The marginal density of the truncated variable X is obviously $\phi(x)/\Phi(-h)$. The marginal density of Y is $\frac{\phi(y)}{\Phi(-h)}\Phi\left(\frac{-h+\rho y}{\sqrt{1-\rho^2}}\right)$ (Chou and Owen, 1984, p. 2538).

Let E_T and var_T be used to denote the mean and variance after truncation. Also, let $q(h)$ be the hazard rate (failure rate) $\phi(h)/\Phi(-h)$, i.e., the inverse of Mills' ratio. Then (Rao et al, 1968, pp. 434–435):

$$E_T(X) = q(h) \tag{4.44}$$

$$E_T(Y) = \rho q(h) \tag{4.45}$$

$$\mathrm{var}_T(X) = 1 - q(h)[q(h) - h] \tag{4.46}$$

$$\mathrm{var}_T(Y) = 1 - \rho^2 q(h)[q(h) - h]. \tag{4.47}$$

Pearson's product-moment correlation is

$$\rho_T = \rho\sqrt{\mathrm{var}_T(X)/\mathrm{var}_T(Y)} = \frac{\rho}{\sqrt{\rho^2 + \frac{1-\rho^2}{\mathrm{var}_T(X)}}}. \tag{4.48}$$

Since $\mathrm{var}_T(X) \le \mathrm{var}_T(Y)$, it follows that $|\rho_T| \le |\rho|$.

The conditional distribution of Y, given $X = x$, is normal with mean ρx and standard deviation $(1 - \rho^2)^{-1/2}$, i.e., a single truncation does not affect the regression. However, the regression of X on Y is given by:

$$E_T(X|Y = y) = \rho y + \sqrt{1 - \rho^2}\, q\left(\frac{h - \rho y}{\sqrt{1 - \rho^2}}\right) \tag{4.49}$$

(Johnson and Kotz, 1972, p. 113).

The moment generating function is

$$M(s, t) = \frac{\Phi(s + \rho t - h)}{\Phi(-h)} \exp[(s^2 + 2\rho st + t^2)/2]. \tag{4.50}$$

For further results, including truncation of both variables, see Johnson and Kotz (1972, Section 36.7) and references cited therein. Regier and Hamdan (1971) and Gajjar and Subrahmaniam (1978) have given a number of results, both algebraic and numerical, for the case of single truncation in both variables. Kovner and Patil (1973) obtained expressions for the moments up to order 4 when both variables are doubly truncated. For some formulae relating to the truncated bivariate lognormal distribution, see Lien (1985). Judging by the title, the paper by Shah and Parikh (1964) is relevant, too.

Nath (1972) derived the moments of a linearly truncated bivariate normal distribution such that the support is of the form $w_1X + w_2Y \ge a$.

Brunden (1978) discussed the probability contours and a goodness-of-fit test for the singly-truncated bivariate normal distribution.

4.16.2 Application to Selection Procedures

The context envisaged is that of the quality of performance of a manufactured item, or perhaps of an employee. The items that are put into service, or the employees who are hired, are those that score above some threshold level on a screening test. Some measure of performance in service becomes available at some later date; there is substantial, but less than perfect, correlation between scores on the two tests. (Academic readers will immediately think of students' performance at school and performance at university as an example.)

If it is assumed that the joint distribution of performances in the unselected population is bivariate normal, then the relevant distribution for items in service is that whose properties are given above. Equation (4.48), for example, tells us what the correlation will be.[9]

This area is extensively discussed in National Bureau of Standards (1959, Section 2.6). More recent work includes the following:

- Problem: knowing the marginal properties and correlation, determine k from known values of h and p, where $\Pr(Y > k|X > h) = p$. Chou and Owen (1984) obtained an approximation using the method of Cornish-Fisher expansion. As this involved the bivariate cumulants κ_{ij}, Chou and Owen gave a method for calculating these: $\kappa_{ij} = \rho^j \kappa_{i+j\,0}$, where $\kappa_{\ell 0}$ is given in a table of theirs for $\ell = 1$ to 8 and $h = -3.0(0.2)3.0$. See also Odeh and Owen (1980).

- Problem: knowing the marginal properties and correlation, determine h from known values of k, ζ, ℓ, and m, such that we are assured (with degree of confidence ζ) that the number of units satisfying $Y \le k$ is at least ℓ in the group of m units satisfying $X \le h$ (there being as many units available for screening as are necessary). This may be thought of as the problem facing a supplier who wants to reject as few of his items as possible, subject to being reasonably confident that the proportion of sub-standard items is sufficiently low. Owen et al (1981) obtained results. See also Madsen (1982).

- Problem: what if there is an upper limit of acceptability for Y, as well as a lower one? See Li and Owen (1979).

- Problem: what if the mean and standard deviation of X are not known in advance, but have to be estimated from a preliminary sample? See Owen and Haas (1978) and Odeh and Owen (1980) for this.

- Davis and Jalkanen (1988) gave a practical example of reduced correlation in a truncated sample. The subject was amounts of gold and silver in samples from drill holes from a goldfield. For the whole sample, the correlation between these quantities was 0.61, but for the 28% of samples that contained the most gold—and are thus of most interest—the correlation was only 0.26.

- A related idea in genetics is described in Section 19.5.2.

- See Section 19.6.1 for the mental abilities context.

Another account, which includes further references to work by D B Owen and colleagues, is that of Owen (1988). For a selection problem in which the requirement of bivariate normality is dropped, see Section 15.2 below.

4.17 Bivariate Normal Mixtures

4.17.1 Construction

Suppose we mix together two bivariate normal distributions. The density is then

$$h = p\psi_1 + (1-p)\psi_2, \tag{4.51}$$

and we shall denote the means μ_{Xi} and μ_{Yi}, the standard deviations σ_{Xi} and σ_{Yi}, and the correlation coefficients ρ_i. Johnson (1987, pp. 56–61) has provided 60 contour plots, with different mixing proportions, means, standard deviations, and correlations, to indicate the range of appearance of bivariate normal mixtures. The covariance is

$$p\rho_1\sigma_{X1}\sigma_{Y1} + (1-p)\rho_2\sigma_{X2}\sigma_{Y2} + p(1-p)(\mu_{X1}-\mu_{X2})(\mu_{Y1}-\mu_{Y2}) \tag{4.52}$$

[9] But if the distribution is not bivariate normal, then it is possible for the correlation to be increased rather than decreased: suppose $Y = a + bX + \varepsilon$. Then the correlation between X and Y is $b^2/(b^2 + \sigma_\varepsilon^2/\sigma_X^2)$. Consequently, if truncation of X increases the variance of X rather than decreasing it, the correlation is increased. This happens for distributions of non-negative r.v.'s that have nonzero density at the origin and a coefficient of variation greater than 1; this class includes decreasing hazard rate distributions. These points were made by Mullooly (1988).

(cf. Johnson, 1987, p. 57).

When p is close to 0 or 1, a bivariate normal mixture can be considered as a single bivariate normal that has been "contaminated" by admixture with a small proportion of another one.

The subject of cluster analysis may be viewed as an attempt to fit a mixture of normal distributions to a dataset. Usually, there are many more variables than two, so we shall not discuss this other than to mention the paper by Wolfe (1970), which explicitly treats the subject in this manner.

Tarter and Silvers (1975) described an interactive computer graphical method for decomposing mixtures consisting of two or more bivariate normal components. (See also Titterington et al, 1985, pp. 142–145.) Though largely on the univariate case, the book by Everitt and Hand (1981) deals with the multivariate case to some extent (at pp. 26, 35, 43–44, 47–48, 115–117, and 123–124). It is especially useful as regards methods of parameter estimation. Much the same remarks could be made about the book by Titterington et al (1985). The Appendix to McLachlan and Basford (1988) contains FORTRAN programs for fitting mixtures of multivariate normal distributions; there is much other material of interest in that book, especially regarding testing for multivariate normality and identification of outliers.

4.17.2 References to Illustrations

Johnson (1987, pp. 56–61) has given 60 contour plots: in all of them, $\mu_{X1} = \mu_{Y1} = 0$, $\sigma_{X1} = \sigma_{Y1} = \sigma_{X2} = \sigma_{Y2} = 1$; all combinations ($2 \times 5 \times 6$) are shown of (i) $p = 0.5$ or 0.9, (ii) $\rho_1 = -0.9, -0.5, 0.0, 0.5$, or 0.9, and (iii) $(\rho_2, \mu_{X2}, \mu_{Y2}) = (0, 0.5, 0.5), (0, 1.0, 1.0), (0, 1.5, 1.5), (0, 2.0, 2.0), (0.5, 1.0, 1.0)$, or $(0.9, 1.0, 1.0)$.

There is an illustration of a three-component density in Everitt (1985).

4.17.3 Applications

See Sections 19.3.1, 19.4.3, 19.4.4, 19.5.2, 19.5.3, 19.5.4, 19.7.6, and 19.8.6.

4.17.4 Generalisation and Compounding

One may generalise equation (4.51) to a finite mixture of the form $h(x, y) = \sum_{i=1}^{n} p_i \psi_i$, in which $\sum p = 1$. A mixture of infinitely many bivariate normals with the same mean leads to a class of elliptical compound bivariate normal distributions (Section 7.2.12).

4.17.5 Properties of a Special Case

We shall now assume the two component distributions are equal in their vectors of means and standard deviations. We shall take the means to be 0 and the standard deviations to be 1.

- The correlation is $p\rho_1 + (1-p)\rho_2$.
- The conditional distribution of Y given $X = x$ is the mixture of $N(\rho_1 x, 1 - \rho_1^2)$ and $N(\rho_2 x, 1 - \rho_2^2)$ in the proportions $p : 1 - p$.
- The regression of Y on X is linear, $E(Y|X = x) = [p\rho_1 + (1-p)\rho_2]x$.
- The conditional variance is a quadratic function of x:

$$\text{var}(Y|X = x) = 1 - [p\rho_1^2 + (1-p)\rho_2^2] + \{p\rho_1^2 + (1-p)\rho_2^2 - [p\rho_1 + (1-p)\rho_2]^2\}x^2.$$

- For a more detailed discussion of these conditional properties, see Kowalski (1973).
- The special case having $p = 0.5$ and $\rho_1 = -\rho_2$ has been considered by several authors, Lancaster (1959) and Sarmanov (1966), for example. It is an example of a bivariate distribution with normal marginals whose variates are uncorrelated yet dependent; its canonical correlation coefficient (see Section 14.5) has the property $c_n = \rho^n$ when n is even, 0 when n is odd (here, $\rho = \rho_1 = -\rho_2$). Another example of a bivariate distribution with dependent normal marginals having zero correlation is given by Ruymgaart (1973).

4.17.6 Sharpening a Scatterplot

"Sharpening" a scatterplot aims to reveal its structure more clearly by increasing the impact of points that are typical, at the expense of atypical points. Green (1988) presents an example of 150 points generated from a mixture of two normal distributions, in which the two clusters show up more clearly when the points in regions of low estimated probability density are plotted with smaller symbols. A similar example is shown by Tukey and Tukey (1981b). See also Chambers et al (1983, especially Section 4.10).

4.17.7 Digression Analysis

Digression analysis places emphasis on the data, rather than the distribution, and it regresses Y on X, rather than treating them symmetrically. Instead of fitting one regression line to empirical points (x_i, y_i), it fits two lines, with each point being supposed to be associated with the line that it is nearest to. This is a natural, though not entirely appropriate, thing to do if it is assumed the points are from two populations, mixed together. Thus what is done (when the regressions are straight lines) is to minimise

$$\sum \min\{[y-(\alpha_1+\alpha_2 x)]^2, [y-(\alpha_3+\alpha_4 x)]^2\} \tag{4.53}$$

with respect to the α's. For more on this, see Mustonen (1982).

"Switching regressions" is another phrase used for much the same thing (e.g., Quandt and Ramsey, 1978). Regression methods do not necessarily rely upon any bivariate distribution $H(X,Y)$, of course, the conditional distribution of Y given X being emphasised, instead.

4.18 Some Distributions Constructed for Pedagogic Purposes

Various distributions have been constructed to show that normal marginals are necessary but not sufficient for the joint distribution to be *the* bivariate normal (in the sense of (4.1)). Actually, we need do no more than start with any bivariate distribution with continuous marginals, and transform the marginals to be normal. Other distributions illustrate that linearity of the regression is not a sufficient condition, either. In this Section, we collect some classroom examples that are intended to dispel possible misconceptions.

4.18.1 Simple Examples with Normal Marginals

A list of such examples was presented in Kowalski (1973). Here we shall illustrate how some such examples can be obtained by manipulating (mixing and shifting probability masses) the bivariate normal distribution.

Example 1 (Lancaster, 1959):

$$h = [\psi(x,y;\rho) + \psi(x,y;-\rho)]/2. \tag{4.54}$$

A special case of this is given in Van Yzeren (1972), $\rho = \frac{1}{2}$. The following corresponds to $\rho = 1$: let X be a normal variate, and let Y be X or $-X$ with equal probability (Broffitt, 1986). More generally, $\alpha\psi(x,y;\rho_1) + (1-\alpha)\psi(x,y;\rho_2)$ also has normal marginals.

Example 2 (Anderson, 1958, p. 37): let (X,Y) have a standardised uncorrelated bivariate normal distribution. Draw identical circles in each of the four quadrants of the plane, each having the same position with respect to the origin. With the circles being numbered clockwise around the origin, transfer the probability mass in circle 1 to 2, and transfer the probability mass in circle 3 to 4. The resulting distribution still possesses normal marginals.

Example 3 (Flury, 1986): let X and Y be independent normal variates. Divide the plane into octants bordered by $x = 0$, $y = 0$, $x + y = 0$, and $x - y = 0$. Shade alternate octants. Transfer all the probability masses in the blank areas to the next shaded area. The resulting distribution has normal marginals. This can easily be extended to X and Y being correlated—the probability masses in the blank areas are now transferred by reflecting them about their boundary line $x = 0$ or $y = 0$.

See also Romano and Siegel (1986, Section 2.10).

4.18.2 Normal Marginals with Linear Regressions

Linearity of the regressions, even in association with marginal normality, is not a sufficient condition for bivariate normality. It is easy to show that $\alpha\psi_1 + (1-\alpha)\psi_2$ (from Example 1) has linear regressions (Kowalski, 1973).

Example 4 (Ruymgaart, 1973):

$$h = \psi(x, y; 0) + \lambda u(x)u(y), \tag{4.55}$$

where $u(t) = \sin|t|$ for $-2\pi \leq t \leq 2\pi$, and is 0, otherwise, and λ is chosen to prevent h assuming negative values. In this example, the regressions are linear (specifically, flat). Furthermore, X and Y are uncorrelated. Nevertheless, h is still not the bivariate normal density (4.1).

4.18.3 Linear Combinations of Normal Marginals

It is well-known (Section 4.5) that (X, Y) has a normal distribution if and only if all linear combinations of X and Y are normal. Melnick and Tenenbein (1982) gave an example in which n linear combinations of two normal marginals are normal, for a large integer n, yet the distribution is not bivariate normal.

4.18.4 Uncorrelated Non-Normal Bivariate Distributions with Normal Marginals

The variates of Examples 1 and 4 are uncorrelated. Another example is the following.

Example 5 (Melnick and Tenenbein, 1982): let X have a standard normal distribution, and Y be defined as:

$$Y = \left\{ \begin{array}{ll} X & \text{if } |X| \leq 1.54 \\ -X & \text{if } |X| > 1.54 \end{array} \right\} \tag{4.56}$$

Then X and Y are uncorrelated. Here, 1.54 (correct to three significant digits) is the solution to the integral equation $\Phi(c) = .75 + c\phi(c)$.

4.19 Cross-References to Other Distributions with Normal Marginals

Mention is made at several other points, especially in Chapter 5, to distributions that have normal marginals.

Copula	Section
Farlie-Gumbel-Morgenstern	5.2.5, 5.2.7, 5.2.12
Ali-Mikhail-Haq	5.3.7
Plackett	5.7.5. 5.7.7, 5.7.9
Pareto	5.9.3
Gumbel-Barnett	5.11.5
Khintchine mixture	10.5.3

In addition, Section 8.5 deals with distributions for which both sets of conditional densities are normal.

Chapter 5

Distributions Expressed as Copulas

5.1 Introduction

5.1.1 Outline of this Chapter

The feature common to the distributions in this Chapter is that $H(x,y)$ is a simple function of $F(x)$ and $G(y)$. Thus the emphasis is on the uniform case $F(x) = x$, $G(y) = y$ (for $0 \leq x, y \leq 1$), and the corresponding forms for other marginals may readily be obtained by substituting F for x and G for y. (The density becomes $f(x)g(y)h(F,G)$.)

When the marginals are uniform, independence of X and Y implies a flat p.d.f.—*any* deviation from this indicates some form of dependence.

Bivariate distributions discussed elsewhere in this book are sometimes thought of in terms of their copulas. Indeed, the distribution of Kimeldorf and Sampson (1975a) that is discussed in Section 10.7 already has uniform marginals and needs no transformations to bring it to that form. The reason this and others have not been included in this Chapter is that the formula for the copula is not simple.

The great majority of the copulas described below have a single parameter that reflects the strength of mutual dependence between X and Y. To emphasise its role, we could have used the same symbol in all these cases. We have not done this, as for some distributions it is usual to find α used, others θ, yet others c.

The Sections of this Chapter are as follows:

5.2 Farlie-Gumbel-Morgenstern and related distributions.

5.3 Ali-Mikhail-Haq distribution.

5.4 Frank's distribution.

5.5 Distribution of Cuadras and Augé.

5.6 Gumbel-Hougaard copula.

5.7 Plackett's distribution.

5.8 Bivariate Pareto distribution.

5.9 Pareto copula.

5.10 Gumbel's bivariate exponential distribution, type I.

5.11 Gumbel-Barnett copula.

5.1.2 Three Explanations

Now let us briefly explain some jargon:

Support. This is the region of values of X and Y over which the p.d.f. is nonzero. For brevity, we refer to the three most common regions of support as: *the unit square* (meaning $0 \leq x, y \leq 1$), *the positive quadrant* (meaning $x, y \geq 0$), and *the whole plane* (meaning any real values of x and y).

Copula. This is the form the distribution takes when the marginals have been transformed to be uniform (over the range 0 to 1). Thus the support of a copula is necessarily the unit square, $0 \leq x, y \leq 1$. More or less equivalent terms are "uniform representation" and "dependence function". (The latter term is popular in connexion with extreme-value distributions. See, for example, Galambos (1978, p. 250) and Deheuvels (1980, 1982).) The dictionary definition of copula is "something that connects", and the word is used here to indicate that it is what interconnects the marginals to produce the joint distribution.

If one replaces H by $\bar{H}$, x by $1-x$, and y by $1-y$ in the formula, one is effectively moving the origin of the coordinate system from $(0,0)$ to $(1,1)$, and measuring the variates in the reverse direction. Although this is such a trivial procedure, the two distributions are perhaps best regarded as distinct, as the results of fitting them to data are different (unless there is symmetry). We shall see an example of this in Section 20.4, when the Pareto copula is fitted to some data on psychiatric disturbances in people and their relatives.

Archimedean copula. If it happens that $a(H) = a(x) + a(y)$ for some function a, the copula is said to be Archimedean (Genest and MacKay, 1986a). In several equations below, the relevant H is put into this form.

5.1.3 Five Sources of Illustrations

In a number of Sections below, we will cite references that contain illustrations of distributions. We now outline the five most important of these.

Conway (1981). Conway's graphs were of bivariate distribution contours. That is, for uniform marginals $F(x) = x$ and $G(y) = y$, y as a function of x was plotted such that a contour of the (cumulative) distribution was the result, i.e., $H(x, y) = c$, a constant. The paper (i) gave such contours for various c for three reference distributions (upper and lower Fréchet bounds, and independence), (ii) gave the $c = 0.2$ contour for distributions having various strengths of correlation drawn from the Farlie-Gumbel-Morgenstern, Ali-Mikhail-Haq, Plackett, Marshall-Olkin, and Gumbel-Hougaard families, and (iii) presented some geometric interpretations of properties of bivariate distributions.

Barnett (1980). The contours in this paper were of probability density. The distributions were again transformed to have uniform marginals; the bivariate normal, Farlie-Gumbel-Morgenstern, Plackett, Cauchy, and Gumbel-Barnett were the ones included.

Johnson et al (1981). This showed both contour and three-dimensional plots of the p.d.f.'s of a number of distributions after their marginals had been transformed to be either normal or exponential. The well-known distributions included were the Farlie-Gumbel-Morgenstern, Plackett, Cauchy, and Gumbel's type I exponential, plus the bivariate normal transformed to exponential marginals. However, the main purpose of Johnson et al was to give similar plots for distributions obtained by a trivariate reduction technique (see Section 10.2.2) and by Khintchine mixture (see Section 10.5).

Johnson et al (1984). In this, there are 18 small contour plots of the p.d.f.'s of distributions after their marginals had been transformed to be normal. The well-known distributions included were the normal, Farlie-Gumbel-Morgenstern, Ali-Mikhail-Haq, Plackett, Gumbel's type I exponential, and Pareto.

Johnson (1987). Chapters 9–10 of this presented contour and three-dimensional plots of the p.d.f.'s of the following distributions: Farlie-Gumbel-Morgenstern (uniform, normal, and exponential marginals), Ali-Mikhail-Haq (normal marginals), Plackett (contour plots only; uniform, normal, and exponential marginals), Gumbel's type I exponential (uniform, normal, and exponential marginals), Pareto (uniform and normal marginals; contour plots only for exponential marginals), and Cook and Johnson's generalised Pareto that we will consider in Section 20.6 (contour plots only; uniform, normal, and exponential marginals; plus one three-dimensional plot of normal marginals).

Thinking of contours of p.d.f., the subject of unimodality (or otherwise) of multivariate distributions comes to mind. This topic is too advanced mathematically for us, and we refer the reader to Dharmadhikari and Joag-Dev (1985).

5.1.4 Three Sources of Algorithms for Random Variate Generation

Our main discussion of random variate generation is in Chapter 16. We do, however, cite useful sources of algorithms in various Sections below. Two of the most important are Johnson et al (1984) and Johnson (1987), just mentioned in Section 5.1.3. A third is the book by Devroye (1986), though bivariate distributions play a comparatively minor role there.

5.2 Farlie-Gumbel-Morgenstern and Related Distributions

5.2.1 Support

The unit square.

5.2.2 Formula for (Cumulative) Distribution Function

$$H = xy[1 + \alpha(1 - x)(1 - y)] \tag{5.1}$$

$(-1 \leq \alpha \leq 1)$.

5.2.3 Formula for p.d.f.

$$h = 1 + \alpha(1 - 2x)(1 - 2y) \tag{5.2}$$

5.2.4 Univariate Properties

Uniform marginals.

5.2.5 Correlation Coefficients

See Chapter 13. Remember that for any distribution whose marginals are uniform, Pearson's product-moment correlation is the same as the grade correlation.

As to the product-moment correlation after the marginals have been transformed to something other than uniform, Gumbel (1960b) and Schucany et al (1978) have (i) shown that it cannot exceed $\frac{1}{3}$, and (ii) calculated what it is for some well-known distributions—for example, $\frac{\alpha}{\pi}$ for normal marginals, and $\frac{\alpha}{4}$ for exponential.

5.2.6 Conditional Properties

The regression $E(Y|X = x)$ is linear in x.

5.2.7 References to Illustrations

Conway (1981); Barnett (1980).

After the marginals have been transformed to be normal: Johnson et al (1981, 1984); Cook and Johnson (1986); Johnson (1987, Section 10.1).

After the marginals have been transformed to be exponential: Johnson et al (1981); Johnson (1987, Section 10.1).

After the marginals have been transformed to be Burr type III: Rodriguez (1980).

5.2.8 References to the Generation of Random Variates

Devroye (1986, pp. 580–581); Johnson et al (1984, p. 235); Johnson (1987, pp. 184–185). (In all three cases, the conditional method is used.)

5.2.9 Further Results

- Among the results given by Mikhail, Chasnov, and Wooldridge (1987) are the regression curves when the marginals are exponential. (See also the papers cited in Section 5.2.12.)

- Mukherjee and Sasmal (1977) have worked out the properties of a two-component system whose components' lifetimes have the F-G-M distribution with standard exponential marginals. The properties are such things as the densities, means, m.g.f.'s, and tail probabilities of $\min(X,Y)$, $\max(X,Y)$, and $X+Y$, these being of relevance to series, parallel, and standby systems, respectively. Mukherjee and Sasmal compared some of the results (densities and means of $\min(X,Y)$ and $\max(X,Y)$) with those for the Downton (1970) and Marshall and Olkin (1967a) distributions.

- Tolley and Norman (1979) obtained some results relevant to epidemiological applications, the marginals being exponential. See Section 9.15.

- Lingappaiah (1984) was also concerned with properties of the F-G-M distribution relevant to the reliability context, but with gamma marginals.

- Building upon a paper by Phillips (1981), Kotz and Johnson (1984) considered a model in which components 1 and 2 were subject to "revealed" and "unrevealed" faults, respectively, with (Y,Z) having an F-G-M distribution, where Y = time between unrevealed faults and Z = time from an unrevealed fault to a revealed fault.

- In the sample selection context—see Section 19.9.5 for the meaning of this—Ray et al (1980) give results for the disturbances having logistic marginals, with the copula being either the F-G-M or the Pareto.

5.2.10 Remarks

- Eyraud (1936) is generally credited with being the first to suggest a distribution of this type, but "Farlie-Gumbel-Morgenstern" seems to be the most common name at the moment, after Morgenstern (1956), Farlie (1960), and Gumbel (1960a,b, 1961).

- This copula is not Archimedean (Genest and MacKay, 1986a).

- The p.d.f. is symmetric in the point $(\frac{1}{2},\frac{1}{2})$, i.e., is the same at $(1-x,1-y)$ as at (x,y), and so a distinct distribution is not obtained when $\bar{H}$, $1-x$, and $1-y$ are substituted for H, x, and y.

- Another account of this distribution is by Conway (1983).

5.2.11 Applications

- Cook and Johnson (1986) included this distribution (with lognormal marginals) as one they fitted to datasets on the joint occurrence of certain trace elements in water. (See also Section 20.6.)

- As mentioned in Section 19.4.1, Halperin et al (1979) used this distribution, with exponential marginals, as a starting point when considering how a population p.d.f. $h(x, y)$ is altered in the surviving and non-surviving groups by a risk function $a(x, y)$. (X and Y were blood pressure and cigarette smoking in the study by Halperin et al.)

- Durling (1974) utilised this distribution with logistic marginals in reanalysing seven previously-published datasets on the effects of mixtures of poisons. (See also Section 22.2.1.)

- Chinchilli and Breen (1985) used a 6-variate version of this distribution with logistic marginals to analyse multivariate binary response data arising in toxicological experiments—specifically, tumour incidence at six different organ sites, mice having been exposed to one of five dosages of a possible carcinogen (data from Brown and Fears, 1981).

- Thinking now of "lifetimes" in the component reliability context, Teichmann (1986) used this distribution for (U_1, U_2), with U_i being a measure of association between an external factor and the failure of unit i—specifically, it was the ratio of how much the external factor increases the probability of failure by to how much an always-fatal factor would increase the probability of failure by.

5.2.12 Univariate Transformations

The case of exponential marginals was considered by Gumbel (1960a,b), that of normal marginals by Gumbel (1958, 1960b), that of logistic marginals by Gumbel (1961, Section 6), that of Burr type III marginals by Rodriguez (1980), that of gamma marginals by D'Este (1981), that of Pareto marginals by Arnold (1983, Section 6.2.5), who cites Conway (1979), that of "inverse Rayleigh" marginals (i.e., $F = \exp(-\theta/x^2)$) by Mukherjee and Saran (1984), and that of Burr type XII marginals by Bagchi and Samanta (1985).

5.2.13 Iterated F-G-M Distributions

For the singly-iterated case, the distribution function H and p.d.f. h are respectively:

$$\begin{aligned} H &= xy[1 + \alpha(1-x)(1-y) + \beta xy(1-x)(1-y)] && (5.3)\\ h &= 1 + \alpha(1-2x)(1-2y) + \beta xy(2-3x)(2-3y), && (5.4) \end{aligned}$$

where the valid combinations of α and β are $-1 \le \alpha \le 1$, $-1-\alpha \le \beta \le \left(3 - \alpha + \sqrt{9 - 6\alpha - 3\alpha^2}\right)/2$.

This distribution is obtained (Johnson and Kotz, 1977; Kotz and Johnson, 1977) by realising that (5.1) may alternatively be written in terms of the survival function $\bar{H}$ as

$$\bar{H} = (1-x)(1-y)(1+\alpha xy). \qquad (5.5)$$

Now, replace the independence survival function term $(1-x)(1-y)$ in (5.1) by this survival function of an F-G-M distribution, having a possibly different association parameter, β/α (say) instead of α. The result is (5.3). This process could be repeated.

Johnson and Kotz (1983) have obtained some results for the iterated F-G-M distribution in the relevation context.[1]

[1] "Relevation" is the name given to the distribution of failure time of a replacement from ageing stock: it is envisaged that when a component with survival distribution function (s.d.f.) $\bar{F}_1$ fails, it is replaced by a component that is not new, but is of the same age (as if it had been manufactured or installed at the same time) and which has s.d.f. $\bar{F}_2$. Then the s.d.f. of the time from the start until both components have failed is $\bar{F}_1(x) + \bar{F}_2(x)\int_0^x f_1(t)/\bar{F}_2(t)\,dt$. This expression is called the relevation of $\bar{F}_1$ with $\bar{F}_2$.

If the life lengths are not independent, the expression becomes $\bar{F}_1(x) + \int_0^x \frac{S(x|t)}{S(t|t)} f_1(t)\,dt$, where $S(y|x)$ is the probability that the second component's lifetime exceeds y given that the first component's lifetime equals x.

See also Johnson and Kotz (1981) and Baxter (1988).

The first iteration increases the maximum attainable correlation to over 0.4, see also Section 13.2.2. However, very little further increase of the maximum correlation is achievable with further iterations (Kotz and Johnson, 1977).

Lin (1987) has suggested another way of iterating the F-G-M distribution: start with (5.5), and replace xy by (5.1). After substituting for $\bar{H}$ in terms of H, the result is

$$H = xy[1 + \alpha(1-x)(1-y) + \beta(1-x)^2(1-y)^2] \tag{5.6}$$

at the first step. This is a special case, $\alpha_2 = 0$, of distribution (15) of Kotz and Johnson (1977).

5.2.14 A Switched-Source Model

For general marginals, the density is $f(x)g(y)\{1 + \alpha[1 - 2F(x)][1 - 2G(y)]\}$. The density

$$a(x)a(y)[1 + \alpha b(x)b(y)] \tag{5.7}$$

arises from a mixture model governed by a Markov process: imagine a source producing observations from a density f_1, another source producing observations from a density f_2, a switch connecting one or other of these sources to the output, a Markov process governing the operation of the switch, and X and Y being observations at two points in time (Willett and Thomas, 1985, 1987).

5.2.15 Ordinal Contingency Tables

The nonidentical marginals case of (5.2) is $a(x)c(y)[1 + \alpha b(x)d(y)]$. This looks very much like the "rank-2 canonical correlation model" to describe structure in ordinal contingency tables (Gilula, 1984; Gilula et al, 1988; see also Goodman, 1986).

Now, instead of generalising (5.1) and comparing it with contingency table models, we shall explicitly write (5.1) in contingency table form, and see what restrictions on the parameters of a contingency table model are effectively being imposed. The probability within a rectangle $x_0 < X < x_1$, $y_0 < Y < y_1$, is $H_{11} - H_{01} - H_{10} + H_{00}$ (in what we hope is an obvious notation), which equals

$$\begin{aligned}&(x_1 - x_0)(y_1 - y_0) + \alpha[x_1(1-x_1) - x_0(1-x_0)][y_1(1-y_1) - y_0(1-y_0)]\\ &= (x_1 - x_0)(y_1 - y_0)[1 + \alpha(1 - x_1 - x_0)(1 - y_1 - y_0)].\end{aligned}$$

Comparing this with equation (2.2) of Goodman (1986), we see that $(1 - x_1 - x_0)$ and $(1 - y_1 - y_0)$ play the role of row scores and column scores—in effect, Goodman's model U.

5.2.16 Other Related Distributions

- Farlie (1960) introduced the more general expression $H(x,y) = F(x)G(y)\{1 + \alpha A[F(x)]B[G(y)]\}$.
- Rodriguez (1980, p. 48), in the context of Burr type III marginals, made passing reference to $H = FG[1 + \alpha(1 - F^a)(1 - G^b)]$.
- Regarding Cook and Johnson's (1986) paper on a compound F-G-M distribution, see Section 20.6.
- Regarding a distribution obtained by Khintchine mixture using the F-G-M distribution as the bivariate uniform that is one of the starting points, see Section 10.5.3 and Johnson (1987, pp. 157–159).
- Cambanis (1977) briefly mentions $H = xy[1 + \beta(1-x) + \beta(1-y) + \alpha(1-x)(1-y)]$, which arises as the conditional distribution in a multivariate F-G-M distribution.
- The following distribution was denoted u_8 in Kimeldorf and Sampson (1975b):

$$H = xy[1 + \alpha(1-x)(1-y) + \beta(1-x^2)(1-y^2)] \tag{5.8}$$
$$h = 1 + \alpha(1-2x)(1-2y) + \beta(1-3x^2)(1-3y^2). \tag{5.9}$$

For correlation coefficients, see Chapter 13.

- Sarmanov (1974) makes brief mention of the following distribution, obtained from the first two Legendre polynomials:

$$H = xy[1 + 3\lambda(1-x)(1-y) + 5\lambda^2(1-x)(1-2x)(1-y)(1-2y)] \quad (5.10)$$
$$h = 1 + 3\lambda(1-2x)(1-2y) + \tfrac{5}{4}\lambda^2[3(1-2x)^2-1][3(1-2y)^2-1]. \quad (5.11)$$

The correlation is λ.

5.3 Ali-Mikhail-Haq Distribution

5.3.1 Support

The unit square.

5.3.2 Formula for (Cumulative) Distribution Function

$$H = \frac{xy}{1-\alpha(1-x)(1-y)} \quad (5.12)$$

$(-1 \le \alpha \le 1)$.

5.3.3 Formula for p.d.f.

$$h = \frac{1-\alpha+2\alpha\frac{xy}{1-\alpha(1-x)(1-y)}}{[1-\alpha(1-x)(1-y)]^2} \quad (5.13)$$

5.3.4 Univariate Properties

Uniform marginals.

5.3.5 Correlation Coefficients

See Chapter 13. Remember that for any distribution whose marginals are uniform, Pearson's product-moment correlation is the same as the grade correlation.

The product-moment correlation is restricted to the range -0.271 to 0.478. As to this after the marginals have been transformed to be something other than uniform, the range is -0.227 to 0.290 for exponential marginals, and approximately -0.300 to 0.600 for normal marginals (Johnson, 1987, pp. 202–203, crediting these results to Conway, 1979).

5.3.6 Derivation

This distribution was introduced by Ali et al (1978). They obtained it by solving the differential equation

$$\frac{\partial^2 C(x,y)}{\partial A(x)\partial B(y)} = 1-\alpha, \quad (5.14)$$

where $A(x)$ and $B(y)$ are the marginal odds functions corresponding to $F(x)$ and $G(y)$, i.e., $A = (1-F)/F$ and $B = (1-G)/G$ (so that $F = 1/(1+A)$ and $G = 1/(1+B)$), and C is a bivariate generalisation, $C = (1-H)/H$.

5.3.7 References to Illustrations

Conway (1981).

After transformation to normal marginals: Johnson (1987, Section 10.4); Johnson et al (1984).

5.3.8 References to the Generation of Random Variates

Johnson (1987, pp. 201–202); Johnson et al (1984, pp. 236–237).

5.3.9 Remarks

- This distribution is an example of an Archimedean copula (Genest and MacKay, 1986):

$$\log\left[\frac{1+\alpha(H-1)}{H}\right] = \log\left[\frac{1+\alpha(x-1)}{x}\right] + \log\left[\frac{1+\alpha(y-1)}{y}\right].$$

- The distribution function may be written as

$$H = xy[1+\alpha(1-x)(1-y)] + \sum_{i=2}^{\infty} \alpha^i (1-x)^i (1-y)^i, \tag{5.15}$$

  and the first term is the Farlie-Gumbel-Morgenstern distribution discussed in the previous Section. (Johnson, 1987, p. 203, credits this observation to Conway, 1979.)

- Ali et al (1978) showed that the distribution is positive quadrant dependent, left tail decreasing, and positive regression dependent. (See Chapter 12 for these concepts.)

- Mikhail, Falwell, Bogue, and Weaver (1987) give some further results, including the (mean) regression curves when the marginals are logistic. They also correct errors in the calculations of the median regression by Ali et al (1978).

5.4 Frank's Distribution

5.4.1 Support

The unit square.

5.4.2 Formula for (Cumulative) Distribution Function

$$H = \log_\alpha\left[1 + \frac{(\alpha^x-1)(\alpha^y-1)}{\alpha-1}\right] \tag{5.16}$$

($\alpha > 0$, $\alpha \neq 1$).

5.4.3 Formula for p.d.f.

$$h = \frac{(\alpha-1)\log\alpha\,\alpha^{x+y}}{[\alpha-1+(\alpha^x-1)(\alpha^y-1)]^2} \tag{5.17}$$

5.4.4 Univariate Properties

Uniform marginals.

5.4.5 Correlation Coefficients

See Chapter 13. Remember that for any distribution whose marginals are uniform, Pearson's product-moment correlation is the same as the grade correlation.

Positive association corresponds to $0 < \alpha < 1$, independence to the limit as $\alpha \to 1$, and negative association to $\alpha > 1$.

Nelsen (1986) gives an expression for Blomqvist's medial correlation coefficient.[2]

[2] This is $\Pr[(X-\tilde{x})(Y-\tilde{y})>0] - \Pr[(X-\tilde{x})(Y-\tilde{y})<0]$, $\tilde{x}$ and $\tilde{y}$ being the medians of X and Y. It may also be written as $4H(\frac{1}{2},\frac{1}{2})-1$.

5.4.6 Derivation

This is the distribution such that both H and $\hat{H} = x + y - H$ are associative, using this to mean that $H[H(x,y),z] = H[x,H(y,z)]$ and similarly for $\hat{H}$ (Frank, 1979). In a typescript version of the paper by Nelsen (1986), it is said that there does not appear to be a probabilistic interpretation of this associativity; we have not found one, either.

Expressed in other words, this is the only Archimedean copula for which $H(x,y) = \bar{H}(1-x, 1-y)$.

5.4.7 References to Illustrations

Genest (1987) illustrates the density when $\ln\alpha = -10$; in a typescript version of the paper by Nelsen, there are contours of the cumulative distribution function, like Conway (1981) gave for other distributions.

5.4.8 References to the Generation of Random Variates

Nelsen (1986); Genest (1987).

5.4.9 Remarks

- This distribution is an example of an Archimedean copula (Genest and MacKay, 1986a):

$$\log\left(\frac{1-\alpha^H}{1-\alpha}\right) = \log\left(\frac{1-\alpha^x}{1-\alpha}\right) + \log\left(\frac{1-\alpha^y}{1-\alpha}\right)$$

- The p.d.f. is symmetric in the point $(\frac{1}{2}, \frac{1}{2})$, i.e., is the same at $(1-x, 1-y)$ as at (x,y), and so a distinct distribution is not obtained when $\bar{H}$, $1-x$, and $1-y$ are substituted for H, x, and y.

- When $0 < \alpha < 1$, this distribution is positive likelihood ratio dependent (see Section 12.3.5 for this concept) (Genest, 1987).

- This distribution has the "monotone regression dependence" property (Bilodeau, 1989).

5.5 Distribution of Cuadras and Augé

For this distribution, put forward by Cuadras and Augé (1981), $H = xy[\max(x,y)]^{-c}$, with c being between 0 and 1. It is usually met in the form with exponential marginals, i.e., the distribution of Marshall and Olkin. See Section 9.2.9. Nelsen (1989) considers a generalisation of this, $H = \min(xy^{1-\beta}, x^{1-\alpha}y)$.

5.6 Gumbel-Hougaard Copula

5.6.1 Support

The unit square.

5.6.2 Formula for (Cumulative) Distribution Function

$$(-\log H)^\alpha = (-\log x)^\alpha + (-\log y)^\alpha \tag{5.18}$$

$(\alpha \geq 1)$.

5.6.3 Univariate Properties

Uniform marginals.

5.6.4 Correlation Coefficients

Kendall's τ is $(\alpha - 1)/\alpha$ (Genest and MacKay, 1986a).

The correlation between $\log X$ and $\log Y$ is $1 - \alpha^2$.

5.6.5 Derivation

Perhaps surprisingly, this distribution can be obtained in a straightforward manner by compounding (Hougaard, 1986, 1988?). Let the hazard functions be $\theta\lambda(x)$ and $\theta\lambda(y)$. Remember that hazard function $= \frac{\text{p.d.f.}}{\text{survival probability}}$; hence the survival probabilities are $e^{-\theta\Lambda(x)}$ and $e^{-\theta\Lambda(y)}$, where $\Lambda(t) = \int_0^t \lambda(u)\,du$, and the joint survival probability $\bar{H}$ is $e^{-\theta[\Lambda(x)+\Lambda(y)]}$. Now average this over all possible values of θ. Let us suppose that θ has a positive stable distribution, with parameter α. This means that it is given by the Laplace transform $E(e^{-s\theta}) = e^{-s^\alpha}$. Thus the expectation of $e^{-\theta[\Lambda(x)+\lambda(y)]}$ is $e^{-[\Lambda(x)+\Lambda(y)]^\alpha}$. Finally, we might suppose that $\lambda(u)$ is of the Weibull form, $\varepsilon\gamma u^{\gamma-1}$, in which case $\Lambda(t) = \varepsilon t^\gamma$, so that

$$\bar{H} = \exp[-(\varepsilon x^\gamma + \varepsilon y^\gamma)^\alpha]. \tag{5.19}$$

If we replace H, x, and y in (5.18) by $\bar{H}$, $1-x$, and $1-y$, we get $(-\log \bar{H})^\alpha = [-\log(1-x)]^\alpha + [-\log(1-y)]^\alpha$, which is (5.19) after specialising to $\alpha = 1/\gamma$ and transforming the marginals $\varepsilon^{1/\gamma}x \to -\log(1-x)$, and similarly for y.

The Pareto distribution is obtained in a similar manner, but with θ having a gamma distribution. Hougaard (1986, p. 676) mentions the possibility of using a distribution that subsumes both the gamma and the positive stable distributions, in order to arrive at a bivariate distribution that subsumes both the Gumbel-Hougaard and Pareto copulas.

Independently, Crowder (1989) had the same idea, but added a new wrinkle to it. His distribution, in bivariate form, is

$$\bar{H} = \exp[\kappa^\alpha - (\kappa + \varepsilon x^\gamma + \varepsilon y^\gamma)^\alpha], \tag{5.20}$$

where we see an extra parameter κ has been introduced. (He also allows the ε's and γ's to be different for X and Y.) An interpretation of κ is in terms of selecting such that $Z > z_0$ from a population having trivariate survival distribution $\exp[-(\varepsilon x^\gamma + \varepsilon y^\gamma + \varepsilon z^\gamma)^\alpha]$. Crowder goes on to discuss some dependence and association properties, hazard functions and failure rates, the marginal distributions, the density function, the distribution of minima, and the fitting to data.

5.6.6 References to Illustrations

Conway (1981).

Crowder (1989) gives nine contour plots of densities from the family (5.20). (The marginals were not made uniform.)

5.6.7 Remarks

We give this copula the name we have because of its appearance in papers by Gumbel (1960a, 1961), and Hougaard's (1986, 1988?) providing a derivation for it.

Clearly, from the form of (5.18), it is an example of an Archimedean copula (Genest and MacKay, 1986a).

5.6.8 Fields of Application

- Gumbel and Mustafi (1967) fitted this distribution, in extreme-value form, to data on the sizes of the annual floods (1918–1950) of the Fox River (Wisconsin) at two points. (See Sections 9.17.11 and 24.3—2 below.)

- Hougaard (1986) used a trivariate version of this distribution to analyse data on tumour appearance in rats—50 litters, each of one drug-treated and two control animals.

- Hougaard (1988?) analysed the insulation failure data given in Section 24.2—2 below using a trivariate form of the Weibull version of this distribution.

- Crowder (1989) fitted (5.20) to data on the sensitivity of rats to tactile stimulation, having or not having received an analgesic drug. (See Section 24.8.1—7 below.)

5.6.9 Transformations of the Marginals

5.6.9.1 Exponential

Gumbel (1960a) briefly mentioned this distribution in exponential form. In this case, the upper right volume under the probability density surface is:

$$\bar{H} = \exp[-(\lambda_1 x^\beta + \lambda_2 y^\beta)^{1/\beta}]. \tag{5.21}$$

5.6.9.2 Weibull

See L Lee (1979), who gives a scheme for generating random realisations from this distribution.

5.6.9.3 Logistic

This distribution is equation (6.2) of Gumbel (1961), who briefly considered the case of logistic marginals.

5.6.9.4 Extreme-Value

Gumbel and Mustafi (1967) considered the case of marginals having extreme-value distributions, i.e., $H = \exp[-(e^{-\beta x} + e^{-\beta y})^{1/\beta}]$. In this case, the product-moment correlation coefficient is $1 - \beta^{-2}$.

5.7 Plackett's Distribution

5.7.1 Support

The unit square.

5.7.2 Formula for (Cumulative) Distribution Function

Showing the derivation and interpretation of the distribution:

$$\frac{H(1 - x - y + H)}{(x - H)(y - H)} = \psi. \tag{5.22}$$

That is, if the support is divided into four regions by dichotomising X and Y,

$$\frac{\text{Probability in the lower left region} \times \text{Probability in the upper right region}}{\text{Probability in the upper left region} \times \text{Probability in the lower right region}} = \text{a constant},$$

independent of where the variates are dichotomised.

Expressed alternatively,

$$H = \frac{[1 + (\psi - 1)(x + y)] - \sqrt{[1 + (\psi - 1)(x + y)]^2 - 4\psi(\psi - 1)xy}}{2(\psi - 1)}. \tag{5.23}$$

The other root is not a proper distribution function, not falling within the Fréchet bounds.

5.7.3 Formula for p.d.f.

$$h = \frac{\psi[(\psi - 1)(x + y - 2xy) + 1]}{\{[1 + (\psi - 1)(x + y)]^2 - 4\psi(\psi - 1)xy\}^{3/2}} \tag{5.24}$$

5.7.4 Univariate Properties

Uniform marginals.

5.7.5 Correlation Coefficients

Pearson's product-moment and the grade correlation are:

$$\frac{\psi+1}{\psi-1}-\frac{2\psi}{(\psi-1)^2}\log\psi. \tag{5.25}$$

For the product-moment correlation when the marginals are normal, see Mardia (1967).

Kendall's τ does not seem to be known as a function of ψ.

5.7.6 Conditional Properties

The regression of Y on X is linear.

After the marginals have been transformed to be normal: the conditional densities are skew, and the regression of Y on X is non-linear (Pearson, 1913a).

5.7.7 References to Illustrations

Barnett (1980); Conway (1981); Morgan (1984, Figure 5.17); Johnson (1987, Section 10.2).

After the marginals have been transformed to be either normal or exponential: Johnson et al (1981); Johnson (1987, Section 10.2).

Contours of probability density after the marginals have been transformed to be normal: Pearson (1913a), who also shows the p.d.f. surface; Anscombe (1981, pp. 308, 310); Johnson et al (1984).

5.7.8 References to the Generation of Random Variates

Devroye (1986, pp. 588–589); Johnson (1987, p. 193); Johnson et al (1984, p. 236).

5.7.9 Remarks

- Interest in this distribution was stimulated by the papers of Plackett (1965) and Mardia (1967), but in fact it can be traced in the contingency table literature back to the days of Yule and Karl Pearson (see Goodman, 1981a).

- As compared with the bivariate normal distribution, the outer contours of p.d.f. of this distribution with normal marginals are more nearly circular—their ellipticity is less than that of the inner ones (Pearson, 1913a; Anscombe, 1981, pp. 306–310).

- For low correlation, this distribution is equivalent to the F-G-M, in the sense that if we set $\psi = 1-\alpha$ in (5.22), expand in terms of α, and then let α be small so that we can neglect α^2 and higher terms, we arrive at (5.1).

- Arnold (1983, Section 6.2.5) makes brief mention of the Pareto-marginals version of this distribution, citing Conway (1979).

- Another account of this distribution is by Conway (1986).

5.7.10 Fields of Application

- This distribution has received considerable attention in the contingency table literature, where it is known as the constant global cross ratio model. Suppose one has a square table of frequencies, the categories of the dimensions being ordinal. Then if the model of independence fails, and a degree of positive (or negative) association is evident, one model that has a single degree of freedom to describe the association is the bivariate normal. But this is inconvenient to handle computationally with most of the present-day packages for modelling tables of frequencies. Another model having a single association model is Plackett's distribution. This is much easier computationally. Recent work in this vein includes the following: Mardia (1970a, Example 8.1), Wahrendorf (1980), Anscombe (1981, Chapter 12), Goodman (1981a), Dale (1983, 1984, 1985, 1986).

- In the context of bivariate probit models, Amemiya (1985, p. 319) mentions that Lee (1982) applied Plackett's distribution with logistic marginals to the data of Ashford and Sowden (1970) and Morimune (1979) (see Section 22.4).
- Mardia (1970c) fitted the S_U-marginals version of this distribution to Johannsen's bean data (see Section 24.7.2—1).

5.8 Bivariate Pareto Distribution

We depart from our usual pattern by describing this distribution, with Pareto marginals, before its copula.

5.8.1 Support

The positive quadrant.

5.8.2 Formula for (Cumulative) Distribution Function

The upper right volume under the p.d.f. is:

$$\bar{H} = (1 + x + y)^{-c} \tag{5.26}$$

$(c > 0)$.

5.8.3 Formula for p.d.f.

$$h = c(c+1)(1 + x + y)^{-(c+2)} \tag{5.27}$$

5.8.4 Univariate Properties

Pareto marginals.

5.8.5 Correlation Coefficients

Pearson's product-moment correlation is $1/c$ (for $c > 2$). See Section 13.5 for the rank correlations.

5.8.6 Conditional Properties

The regression of Y on X is linear, $E(Y|X = x) = (x + 1)/c$, and the conditional variance is quadratic, $\mathrm{var}(Y|X = x) = \frac{c+1}{(c-1)c^2}(x + 1)^2$ for $c > 1$.

5.8.7 Derivation

By the method of compounding (Section 20.1): start with X and Y having independent exponential distributions with the same scale parameter, and then take the scale parameter to have a gamma distribution.

More generally, start with $\Pr(X > x) = [1 - C(x)]^\theta$ and $\Pr(Y > y) = [1 - D(y)]^\theta$, where C and D are distribution functions, take θ to have a gamma distribution, and find the only difference from (5.26) is that monotonic transformations have been applied to X and Y.

If compounding of the scale parameter is applied to an F-G-M distribution that has exponential marginals, instead of to the independence distribution with exponential marginals, a distribution proposed and used by Cook and Johnson (1986) is the result—see Section 20.6.

5.8.8 References to Illustrations

See Section 5.9.3.

5.8.9 References to the Generation of Random Variates

See Section 5.9.4.

5.8.10 Remarks

- Barnett (1979, 1983b) has considered testing for the presence of an outlier in a dataset assumed to come from this distribution. (See also Barnett and Lewis, 1984, Section 9.3.3.) An alternative proposal is in Barnett (1983a), and involves transformation to independent normal variates.

- The bivariate failure rate is decreasing (Nayak, 1987).

- The product moments are $E(X^rY^s) = \Gamma(c-r-s)\Gamma(r+1)\Gamma(s+1)/\Gamma(c)$ if $r+s < c$, and ∞, otherwise (Nayak, 1987).

- Mardia (1962) wrote this distribution in the form $h \propto (bx+ay-ab)^{-(c+2)}$, with $x > a > 0$, $y > b > 0$. For the distribution of XY and X/Y in this case, see Malik and Trudel (1985).

5.8.11 Applications

See Section 5.9.6.

5.8.12 Univariate Transformations

In the Takahasi-Burr (or bivariate Burr type XII) distribution, x and y in the distribution function are replaced by powers of themselves (Takahasi, 1965). Further results, oriented towards the repeated measurements experimental paradigm, for this case are given by Crowder (1985). For generation of random variates following the method of the distribution's derivation (scale mixture), see Devroye (1986, pp. 557–558). Arnold (1983, p. 249) calls this a type IV Pareto distribution.

Rodriguez (1980) has discussed the bivariate Burr type III distribution, i.e., $H = (1+x^{-a}+y^{-b})^{-c}$. In that report, there is a derivation (by compounding an extreme-value distribution with a gamma), algebraic expressions for the conditional density, conditional distribution, conditional moments, and correlation, and a number of illustrations of probability density surfaces.

Satterthwaite and Hutchinson (1978) replaced x and y in the distribution function by e^{-x} and e^{-y}. Gumbel (1961) had previously done this in the special case $c = 1$, thus getting a distribution whose marginals are logistic; however, it lacks an association parameter.

Cook and Johnson (1981) and Johnson (1987, Chapter 9) have treated this copula (whether in Pareto, Takahasi (1965), or Satterthwaite and Hutchinson (1978) form) systematically, including providing several plots of densities. Cook and Johnson (1986) and Johnson (1987, Section 9.2) generalise the distribution further—see our Section 20.6.

5.8.13 The Durling-Pareto Distribution

5.8.13.1 Formula for (Cumulative) Distribution Function

We use the above name for the distribution for which the upper right volume under the probability density surface is

$$\bar{H} = (1+x+y+kxy)^{-c} \tag{5.28}$$

$(0 < c;\ 0 \leq k \leq c+1)$.

5.8.13.2 Derivation

Hutchinson (1979b, 1981b) showed that, like (5.26), the Durling-Pareto distribution may be derived by a compounding model—see Section 20.3.3. See also Arnold (1983, Section 6.2.4).

5.8.13.3 Remarks

- Durling (1975) actually proposed the extra term in the context of the Takahasi-Burr distribution (see Section 5.8.12 above), rather than the simpler Pareto. Some properties of Durling's distribution were given by Bagchi and Samanta (1985).
- Durling gives the (product-moment) correlation coefficient for the general case in which x and y are each raised to some power, and the expression may be simplified considerably for (5.28).
- We will see an application of this distribution, and of the special case with $c = 1$, in Section 20.3.3, when the severities of injury to vehicle drivers in head-on collisions between two vehicles of equal mass are considered.

5.8.13.4 The Special Case $c = 1$

Suppose we have a number of 2×2 contingency tables, each of which corresponds to some particular x and some particular y, and we want to fit the distribution $\bar{H} = (1 + \alpha x + \beta y + k\alpha\beta xy)^{-1}$ to them. Notice that unknown scale parameters have now been included. This special case of (5.28) with $c = 1$ is very convenient in these circumstances, because we would write (in an obvious notation) $p_{11} = (1 + \alpha x + \beta y + k\alpha\beta xy)^{-1}$, $p_{10} + p_{11} = (1 + \alpha x)^{-1}$, and $p_{01} + p_{11} = (1 + \beta y)^{-1}$. Then we could estimate α and β from the marginals,

$$\begin{aligned} \frac{1 - (p_{10} + p_{11})}{p_{10} + p_{11}} &= \alpha x \\ \frac{1 - (p_{01} + p_{11})}{p_{01} + p_{11}} &= \beta y, \end{aligned}$$

and k is found from

$$\frac{\frac{1-p_{11}}{p_{11}} - \frac{1-(p_{10}+p_{11})}{p_{10}+p_{11}} - \frac{1-(p_{01}+p_{11})}{p_{01}+p_{11}}}{\left[\frac{1-(p_{10}+p_{11})}{p_{10}+p_{11}}\right]\left[\frac{1-(p_{01}+p_{11})}{p_{01}+p_{11}}\right]} = k. \tag{5.29}$$

For applications of this in transformed form, see our Sections 22.4.2.1 and 22.4.2.2, based upon Morimune (1979) and Amemiya (1975).

5.8.13.5 A Related Distribution

Rodriguez (1980) introduced a similar term into the bivariate Burr type III distribution, thus getting $H = (1 + x^{-a} + y^{-b} + kx^{-a}y^{-b})^{-c}$. He included a number of plots of probability density surfaces of this distribution in the report. This distribution (with location and scale parameters included) was used by Rodriguez and Taniguchi (1980) to describe the joint distribution of customers' and expert raters' assessments of octane number requirements of cars; see also Section 24.4.1—1 below.

5.8.14 Pareto Conditionals

See Section 8.3.

5.8.15 Generalisation

See Section 20.7 for distributions which generalise the bivariate Pareto while retaining the general method of derivation, compounding.

5.9 Pareto Copula

5.9.1 Support

The unit square.

5.9.2 Formula for (Cumulative) Distribution Function

$$H = (x^{-1/c} + y^{-1/c} - 1)^{-c} \tag{5.30}$$

($c > 0$). This is not symmetric about $y = 1 - x$, and so the distribution of $(1 - X, 1 - Y)$ is different from that of (X, Y). The form

$$\bar{H} = [(1 - x)^{-1/c} + (1 - y)^{-1/c} - 1]^{-c} \tag{5.31}$$

for the upper right volume under a p.d.f. has received considerable attention lately—Oakes (1982, 1986), Cox and Oakes (1984, Section 10.3).

5.9.3 References to Illustrations

With Pareto marginals: Lindley and Singpurwalla (1986).

After the marginals have been transformed to be normal: Cook and Johnson (1981, 1986); M E Johnson (1986); Johnson et al (1984).

Uniform, normal, or exponential marginals: Johnson (1987, Chapter 9).

Contours of the density of $H = (1 + e^{-x} + e^{-y})^{-c}$: Satterthwaite and Hutchinson (1978).

An illustration of a different kind is a plot of points generated from the distribution: for the Pareto marginals case (equation (5.27)) with $c = 2.5$, Barnett (1979) gives an example.

5.9.4 References to the Generation of Random Variates

Devroye (1986, pp. 604–605); Johnson et al (1984, pp. 237–238).

5.9.5 Remarks

- As already mentioned in Section 5.8.12, Johnson (1987, Section 9.1) gives an account of this distribution that pays more explicit attention to the marginals than we have done. Johnson refers to this as the Burr-Pareto-logistic family. (For random variate generation, see Devroye, 1986, pp. 600–603.)
- This distribution is an example of an Archimedean copula (Genest and MacKay, 1986a):

$$H^{-1/c} - 1 = x^{-1/c} - 1 + y^{-1/c} - 1.$$

- Ray et al (1980) give results relevant to the sample selection context. (See Sections 19.9.5 and 22.4.2.3.)
- This distribution has the "monotone regression dependence" property (Bilodeau, 1989).

5.9.6 Fields of Application

- We will outline in Chapter 20 several applications described by Hutchinson (1979b):
  - Ordinal-scale assessments by two observers.
  - Injury severity in road accidents.
  - The multifactorial model of disease transmission.
  - Replicated paired comparison tests.

- Cook and Johnson (1981, 1986) fitted this distribution, and others, with lognormal marginals to data on the joint distribution of certain trace elements (e.g., caesium and scandium) in water. (See also Section 20.6.)

- Concerning association in bivariate life tables, Clayton (1978) deduced that a bivariate survivor function must be of the form $\bar{H} = [1 + a(x) + b(y)]^{-c}$ if:

$$h\bar{H} = c\int_x^{\infty} h(u, y)\,du \int_y^{\infty} h(x, v)\,dv. \tag{5.32}$$

  Clayton's language is of deaths of fathers and sons from some chronic disease, with association stemming from common environmental or genetic influences. The above equation arises as follows:

  - Consider the ratio of (i) the age-specific death rate for sons given that the father died at age y, to (ii) the age-specific death rate for sons given that the father survived beyond age y. This ratio is assumed to be independent of the son's age.
  - As a symmetric form of association is being considered, an analogous assumption holds for the ratio of fathers' age-specific death rates.
  - The proportionality property $\frac{h}{\bar{H}} = c\frac{\partial}{\partial x}(-\log \bar{H})\frac{\partial}{\partial y}(-\log \bar{H})$ then holds. (The left hand side of this equation is the bivariate failure rate, and the right hand side is $c \times$ the product of the hazard function for sons of fathers who survive until y and the hazard function for fathers of sons who survive until x.)

  See also Oakes (1982, 1986) and Clayton and Cuzick (1985a,b).

- See Section 9.18.2 for a note on how Klein and Moeschberger (1988) have used this form of association in the "competing risks" context.

- The bivariate Burr distribution, both with and without the extra association term introduced by Durling (1975), was used by Durling (1974) in reanalysing seven previously-published datasets on the effects of mixtures of poisons. (See also Section 22.2.1.)

- The Takahasi-Burr distribution (see Section 5.8.12 above), in its quadrivariate form, was applied by Crowder (1985) in a repeated measurements context—specifically, response times of rats to painful stimuli, at four intervals after receiving a dose of an analgesic drug.

5.10 Gumbel's Bivariate Exponential Distribution, Type I

We depart from our usual pattern by describing this distribution, with exponential marginals, before its copula.

5.10.1 Support

The positive quadrant.

5.10.2 Formula for (Cumulative) Distribution Function

$$H = 1 - e^{-x} - e^{-y} + e^{-(x+y+\theta xy)} \tag{5.33}$$

$(0 \leq \theta \leq 1)$.

5.10.3 Formula for p.d.f.

$$h = e^{-(x+y+\theta xy)}[(1 + \theta x)(1 + \theta y) - \theta] \tag{5.34}$$

5.10.4 Univariate Properties

Exponential marginals.

5.10.5 Correlation Coefficients

$$\rho = -1 + \int_0^\infty \frac{e^{-y}}{1+\theta y}\,dy. \tag{5.35}$$

Gumbel (1960a) plots ρ as a function of θ. A compact expression may be obtained in terms of the exponential integral (but care is always necessary with this function, as nomenclature and notation are not fully standardised). For $\theta = 0$, X and Y are independent and $\rho = 0$. As θ increases, ρ becomes increasingly negative, reaching -0.404 at $\theta = 1$. Thus this distribution is unusual in being oriented towards negative correlation. (Of course, positive correlation could be obtained by changing X to $-X$ or Y to $-Y$.)

Barnett (1985a) has considered methods of estimating θ—maximum likelihood and from the product-moment correlation.

5.10.6 Conditional Properties

Gumbel (1960a) gives the following results:

$$\begin{aligned} g(y|x) &= e^{-y(1+\theta x)}[(1+\theta x)(1+\theta y)-\theta] && (5.36)\\ E(Y|X=x) &= (1+\theta+\theta x)(1+\theta x)^{-2} && (5.37)\\ \operatorname{var}(Y|X=x) &= \frac{(1+\theta+\theta x)^2 - 2\theta^2}{(1+\theta x)^4}. && (5.38) \end{aligned}$$

5.10.7 References to Illustrations

See Section 5.11.5.

5.10.8 References to the Generation of Random Variates

See Section 5.11.6.

5.10.9 Remarks

- This distribution is characterised by

$$\left.\begin{aligned} E(X-x|X>x \text{ and } Y>y) &= E(X|Y>y)\\ E(Y-y|X>x \text{ and } Y>y) &= E(Y|X>x) \end{aligned}\right\} \tag{5.39}$$

  which is a form of the lack of memory property (K R M Nair and N U Nair, 1988; see also N U Nair and V K R Nair, 1988).

- Barnett (1979, 1983b) and Barnett and Lewis (1984, Section 9.3.2) have considered testing for the presence of an outlier in a dataset assumed to come from this distribution. An alternative proposal is in Barnett (1983a), and involves transformation to independent normal variates (see Section 4.13.2).

- In the structural reliability context, Der Kiureghian and Liu (1986) utilised this distribution (with $\theta = 1$) in the course of demonstrating a procedure to approximate multivariate integrals by transforming the marginals to normality and assuming multivariate normality. See also Grigoriu (1983, Example 2).

5.10.10 An Application

In describing this, let us quote the opening words of the paper by Moore and Clarke (1981): "The rainfall runoff models referred to in the title of this paper are (1) those that attempt to describe explicitly both the storage of precipitated water within a river basin and the translation or routing of water that is in temporary storage to the basin outfall and (2) those in which the parameters of the model are estimated from existing records of mean areal rainfall, Penman potential evaporation E_T, or some similar measure of evaporative demand, and streamflow." On pp. 1373–1374 of the paper is a Section titled "A bivariate exponential storage-translation model". This introduces distribution (5.33), the justification being that it has exponential marginals and that the correlation is negative ("a basin with thin soils in the higher altitude areas that are farthest from the basin outfall is likely to have s and t negatively correlated"). The variables s and t are respectively the depth of a (hypothesised) storage element and the time taken for runoff to reach the catchment outfall.

Moore and Clarke did not present in detail the results of using (5.33), saying "Application of storage-translation models using the more complex distribution functions... did not lead to any appreciable improvement in model performance... One exception... gives a correlation of −0.37 between s and t."

5.11 Gumbel-Barnett Copula

5.11.1 Reason for the Name

Gumbel (1960a,b) suggested the exponential-marginals form of this copula (see Section 5.10); Barnett (1980) included the uniform-marginals case among the distributions he considered.

5.11.2 Support

The unit square.

5.11.3 Formula for (Cumulative) Distribution Function

The upper right volume under the probability density surface is:

$$\bar{H} = (1-x)(1-y)e^{-\theta \log(1-x)\log(1-y)}. \tag{5.40}$$

5.11.4 Formula for p.d.f.

$$h = \{-\theta + [1-\theta\log(1-x)][1-\theta\log(1-y)]\}e^{-\theta\log(1-x)\log(1-y)} \tag{5.41}$$

5.11.5 References to Illustrations

In the form with uniform marginals: Barnett (1980) (the Correction noted in our list of References refers to this illustration), Johnson (1987, Section 10.3).

After transformation to normal marginals: Johnson et al (1981, 1984), Johnson (1987, Section 10.3).

After transformation to exponential marginals: Johnson (1987, Section 10.3).

An illustration of a different kind is a plot of points generated from the distribution: for the exponential-marginals case (equation (5.34)), Barnett (1979) gives an example with $\theta = 1$, and Barnett (1983) an example with $\theta = 0.5$.

5.11.6 References to the Generation of Random Variates

Johnson (1987, p. 198); Johnson et al (1984, p.237).

Elderton and Johnson (1969, p. 138), Johnson and Kotz (1972, Table 1 in Chapter 34), and Rodriguez (1983b) give versions of the above table in which expressions for the p.d.f.'s, the supports, and restrictions on the parameter values are included.

6.2 Bivariate t-Distribution

6.2.1 Support

The whole plane.

6.2.2 Formula for p.d.f.

$$h = \frac{1}{2\pi\sqrt{1-\rho^2}}\left[1+\frac{1}{\nu(1-\rho^2)}(x^2-2\rho xy+y^2)\right]^{-(\nu+2)/2} \tag{6.1}$$

$(\nu > 0;\ -1 < \rho < 1)$.

6.2.3 Univariate Properties

t-distribution marginals, having ν degrees of freedom.

6.2.4 Correlation Coefficients

For $\nu > 2$, Pearson's product-moment correlation is ρ. For $0 < \nu \leq 2$, ρ represents the gradient of the major axis of the elliptical contours.

This distribution is an example of zero correlation not necessarily implying independence. (See also Sections 6.2.7 and 6.2.10.)

6.2.5 Moments

From the method of derivation, the product-moments are easily found:

$$\mu'_{rs} = E(X^rY^s) = \nu^{(r+s)/2}E(X_1^rX_2^s)E(S^{-(r+s)}), \tag{6.2}$$

where $E(X_1^rX_2^s)$ is simply the product moment of the bivariate normal, and

$$E(S^{-(r+s)}) = 2^{-(r+s)/2}\Gamma\left(\tfrac{\nu-r-s}{2}\right)/\Gamma(\nu/2). \tag{6.3}$$

If X_1 and X_2 are independent, i.e., $\rho = 0$, then μ'_{rs} is zero unless both r and s are even, in which case it is

$$\nu^{(r+s)/2}\frac{[1.3.5\ldots(2r-1)][1.3.5\ldots(2s-1)]}{(\nu-2)(\nu-4)\ldots(\nu-r-s)} \tag{6.4}$$

(Johnson and Kotz, 1972, pp. 135–136).

The characteristic function of this distribution is given in Sutradhar (1986).

6.2.6 Conditional Properties

When $X = x$, the following linear transformation of Y, $U = \left[\frac{\nu(\nu+1)}{\nu+x^2}\right]^{1/2}\frac{Y-\rho x}{\sqrt{1-\rho^2}}$, has a t-distribution with $\nu+1$ degrees of freedom. The regression is linear, $E(Y|X=x) = \rho x$, and the conditional variance is quadratic, $\frac{\nu}{\nu-1}(1-\rho^2)(1+x^2/\nu)$ (Mardia, 1970a, p. 92).

6.2.7 Derivation

By a form of trivariate reduction: let (X_1, X_2) have the standardised bivariate normal distribution, with correlation coefficient ρ, and S, independent of X_1 and X_2, be distributed as χ_ν (i.e., the square root of a χ^2_ν variate). Then $X = X_1\sqrt{\nu}/S$ and $Y = X_2\sqrt{\nu}/S$ have the bivariate t-distribution given above.

6.2.8 References to Illustrations

Contours of p.d.f.: Devlin et al (1976).

Illustration of p.d.f. surface: Johnson (1987, pp. 119–122, 124).

6.2.9 References to the Generation of Random Variates

Johnson (1987, pp. 120–121); Johnson et al (1984, p. 235); Vàduva (1985).

6.2.10 Remarks

- This is also known as the Pearson type VII distribution, though the p.d.f. of the latter usually appears in the following form:

$$h = \frac{-m\sqrt{(1-\rho^2)}}{\pi k^m}(k + x^2 - 2xy\rho + y^2)^{m-1} \tag{6.5}$$

  $(m < 0;\ -1 < \rho < 1;\ k > 0)$.

- For the special case $\rho = 0$, $\nu = 1$, the bivariate Cauchy distribution, see below, Section 6.2.13.
- For $\rho = 0$, X^2 and Y^2 have a bivariate F-distribution (see Section 6.3).
- As $\nu \to \infty$, so this distribution becomes bivariate normal.
- The contours of probability density are ellipses. (See Section 7.2 for more on elliptical distributions.)
- The quantity $(X^2 - 2\rho XY + Y^2)/[2(1-\rho^2)]$ has an F-distribution with $(2, \nu)$ degrees of freedom (Johnson et al, 1984).
- For this distribution, zero correlation does not imply independence of X and Y. This is because they are dependent through (i) X_1 and X_2 having a bivariate normal distribution with correlation ρ, and (ii) there being a denominator variable in common, S. Hence, even when $\rho = 0$, the dependence from the second source remains. In fact, apart from the bivariate normal, all the elliptically-contoured bivariate distributions defined in Section 7.2 have this property.
- For the distributions of XY and X/Y, see Malik and Trudel (1985) and Wilcox (1985).
- For probability inequalities applicable to the multivariate t-distribution, see Tong (1980, Section 3.1).

6.2.11 Fields of Application

- While this distribution is not often used to fit data, tables of its percentage points are required in the applications of multiple comparison procedures, ranking and selection procedures, and estimation of rank parameters. For a more detailed discussion, see Dunnett and Sobel (1954), Gupta (1963a), Johnson and Kotz (1972, pp. 145–146), and Chen (1979).
- Pearson (1924) fitted the distribution to two sets of data from some 30 years earlier on the joint distribution of the numbers of cards of a given suit that two players of whist hold in their hands.
- Econometricians make extensive use of systems of linear simultaneous equations. In estimating the parameters appearing in these equations, it is usual to assume the stochastic terms—the disturbances—to have a multivariate normal distribution. Concerned with the possibility that the actual distribution has thicker tails than the normal, and hence that too great a weight is given to outliers by conventional methods of estimation, Prucha and Kelejian (1984) proposed alternative methods that are better for multivariate t and other thick-tailed distributions.

6.2.12 References to Tables or Algorithms

Johnson and Kotz (1972, pp. 137–140) list many references to tables. More recent tables include those of Chen (1979), Gupta et al (1985), Wilcox (1986), and Bechhofer and Dunnett (1987).

6.2.13 Bivariate Cauchy Distribution

This distribution, a special case of the bivariate t ($\rho = 0$, $\nu = 1$), is of only limited interest as it has no correlation parameter.

The support is the whole plane.

The p.d.f. is

$$h = \tfrac{1}{2\pi}(1 + x^2 + y^2)^{-3/2}. \tag{6.6}$$

(Location and scale factors can readily be introduced if required.)

The cumulative distribution function is

$$H = \frac{1}{4} + \frac{1}{2\pi}\left(\tan^{-1} x + \tan^{-1} y + \tan^{-1} \frac{xy}{\sqrt{1 + x^2 + y^2}}\right). \tag{6.7}$$

The marginals are Cauchy; hence their means and standard deviations are undefined (and other measures of location and spread have to be used).

The conditional p.d.f. of Y, given $X = x$, is $\frac{1}{2}(1 + x^2)/(1 + x^2 + y^2)^{3/2}$. Hence $Y \Big/ \sqrt{\frac{1}{2}(1 + x^2)}$, conditional on $X = x$, has a t-distribution with two degrees of freedom. The distribution of any linear combination of X and Y is Cauchy (Ferguson, 1962).

For contours of probability densities, see Devlin et al (1976). For contours (and also the p.d.f.) after transformation to uniform marginals, see Barnett (1980). For contours after transformation to normal marginals, see Johnson et al (1984).

For generation of random variates, see Devroye (1986, p. 555) and Johnson et al (1984).

6.2.14 Generalisations

- Johnson and Kotz (1972, Chapter 37) consider the derivation of more general forms of bivariate t-distribution, as

$$\left.\begin{aligned} X &= (X_1 + \delta_1)\sqrt{\nu_1}/S_1 \\ Y &= (X_2 + \delta_2)\sqrt{\nu_2}/S_2 \end{aligned}\right\} \tag{6.8}$$

  where the δ's are non-centrality parameters, the X's have some sort of joint normal distribution, and the S's have some sort of joint chi distribution. See also Ramig and Nelson (1980).

- Poly (or multiple) t-distributions are those whose densities are given by the product of two or more terms like the right hand side of (6.1) (Editors of the ESS, 1986b; Press, 1972b, Section 6.2.5).

6.3 Bivariate F-Distribution

6.3.1 Support

The positive quadrant.

6.3.2 Formula for p.d.f.

$$h = Kx^{(\nu_1-2)/2}y^{(\nu_2-2)/2}\left(1 + \frac{\nu_1 x + \nu_2 y}{\nu_0}\right)^{-(\nu_0+\nu_1+\nu_2)/2}, \tag{6.9}$$

where the ν's are positive and are referred to as the "degrees of freedom", and the constant K is

$$\Gamma\left(\tfrac{\nu_0+\nu_1+\nu_2}{2}\right)\nu_0^{-(\nu_0+\nu_1+\nu_2)/2}\frac{\nu_0^{\nu_0/2}\nu_1^{\nu_1/2}\nu_2^{\nu_2/2}}{\Gamma(\nu_0/2)\Gamma(\nu_1/2)\Gamma(\nu_2/2)}.$$

6.3.3 Formula for (Cumulative) Distribution Function

There is no elementary form for $H(x,y)$, but it may be written in terms of F_2, Appell's hypergeometric function of two variables—see Amos and Bulgren (1972), Hutchinson (1979b, 1981b), and Sections 6.5.15.3 and 20.7.2.

6.3.4 Univariate Properties

The marginals of X and Y are F-distributions with (ν_1, ν_0) and (ν_2, ν_0) degrees of freedom, respectively.

6.3.5 Correlation Coefficients

Pearson's product-moment correlation is $\sqrt{\frac{\nu_1\nu_2}{(\nu_0+\nu_1-2)(\nu_0+\nu_2-2)}}$, for $\nu_0 > 4$.

6.3.6 Moments

The product moments are

$$E(X^r Y^s) = \frac{\Gamma(\frac{1}{2}\nu_0 - r - s)\Gamma(\frac{1}{2}\nu_1 + r)\Gamma(\frac{1}{2}\nu_2 + s)}{\Gamma(\nu_0/2)\Gamma(\nu_1/2)\Gamma(\nu_2/2)(\nu_1/\nu_0)^r(\nu_2/\nu_0)^s}$$

if $r + s < \nu_0/2$, being undefined otherwise (Nayak, 1987).

6.3.7 Conditional Properties

$(\nu_0 + \nu_1)Y/(\nu_0 + \nu_1 x)$, conditional on $X = x$, has an F-distribution with degrees of freedom $(\nu_2, \nu_0 + \nu_1)$. The regression is linear, $E(Y|X = x) = (\nu_0 + \nu_1 x)/(\nu_0 + \nu_1 - 2)$ (for $\nu_0 > 2$)—Mardia (1970a, p. 93), Nayak (1987).

6.3.8 Methods of Derivation

This distribution may be obtained by transforming the bivariate t-distribution discussed earlier. More precisely, if (X,Y) is the bivariate t variate with ν_0 degrees of freedom and $\rho = 0$, then (X^2, Y^2) has a bivariate F-distribution with degrees of freedom ν_0, 1, and 1. However, this method does not lead to a bivariate F-distribution with other values of ν_1 and ν_2.

Alternatively, consider a trivariate reduction technique, with $X = \frac{X_1/\nu_1}{X_0/\nu_0}$, $Y = \frac{X_2/\nu_2}{X_0/\nu_0}$, where X_0, X_1, and X_2 have independent chi-squared distributions with degrees of freedom ν_0, ν_1, and ν_2, respectively. Then X and Y have a bivariate F-distribution with degrees of freedom ν_0, ν_1, and ν_2. The distribution may also be obtained by the method of compounding (equivalent to the method of trivariate reduction in some situations).

6.3.9 Relation to Other Distributions

- It is related to the bivariate t-distribution as indicated in Section 6.3.8.
- The bivariate inverted beta distribution (see Section 6.5.15) is essentially the bivariate F-distribution, written in a slightly different form.
- For the distribution of XY and X/Y, see Malik and Trudel (1985).
- A non-central generalisation of this distribution has been given by Feingold and Korsog (1986). This is obtained by letting $X = \frac{X_1/\nu_1}{X_0/\nu_0}$, $Y = \frac{X_2/\nu_2}{X_0/\nu_0}$, where X_1 and X_2 now have non-central chi-squared distributions (for which, see Section 2.13).
- Another generalisation is Krishnaiah's (1964, 1965) bivariate F-distribution, obtained by the trivariate reduction method just mentioned, but with X_1 and X_2 now being correlated (central) chi-squared variates; specifically, their joint distribution is Kibble's bivariate gamma (see Section 7.5).

- A generalisation of Krishnaiah's bivariate F-distribution is Jensen's bivariate F, which is obtained through two quadratic forms from a multivariate normal distribution and a chi-squared distribution. See Section 7.8 for details.

- Adegboye and Gupta (1981) studied a bivariate F-distribution arising from $X = \frac{X_1/\nu_1}{X_0/\nu_0}$, $Y = \frac{X_2/\nu_2}{X_1/\nu_1}$, where X_0, X_1, and X_2 have independent chi-squared distributions with degrees of freedom ν_0, ν_1, and ν_2, respectively.

6.3.10 Fields of Application

The distribution is rarely used to fit data. However, tables of its percentage points are required in analysis of variance and experimental design in general (Johnson and Kotz, 1972, pp. 240–241). This distribution is closely related to the bivariate beta distribution, and the applications of the latter to compositional data (see Sections 6.5.12 and 24.5) are sometimes expressed in such a way that the bivariate F is the one applicable (Ratnaparkhi, 1983b).

6.3.11 References to Tables or Algorithms

Amos and Bulgren (1972) recognised that the cumulative distribution can be expressed in terms of Appell's F_2 function. Tiao and Guttman (1965) expressed the integral in terms of a finite sum of incomplete beta functions.

Hewett and Bulgren (1971) have shown that if ν_1 and ν_2 are equal, then for any a and b such that $0 \leq a \leq b \leq \infty$,

$$\Pr(a < X \leq b, a < Y \leq b) \geq \Pr(a < X \leq b)\Pr(a < Y \leq b) \tag{6.10}$$

Numerical studies by them show that the right-hand side of (6.10) is quite a good approximation to the left. Accuracy increases as ν_0 increases, but decreases as ν_1 and ν_2 increase.

Hamdy et al (1988) gave an algorithm to compute the lower and upper percentage points of $\min(X, Y)$. See also the references given there.

6.4 Bivariate Pearson Type II Distribution

6.4.1 Support

Within the ellipse $x^2 - 2\rho xy + y^2 = 1 - \rho^2$ (which itself lies within the unit square).

6.4.2 Formula for p.d.f.

$$h = \frac{(\nu+1)}{\pi\sqrt{1-\rho^2}}\left[1 - \frac{x^2 - 2\rho xy + y^2}{1-\rho^2}\right]^{\nu} \tag{6.11}$$

($\nu > 1$; $-1 < \rho < 1$).

6.4.3 Univariate Properties

Pearson type II marginals, i.e., $f(x) = (1-x^2)^{\nu+\frac{1}{2}}/B(\frac{1}{2}, \nu+\frac{3}{2})$ and similarly for $g(y)$.

6.4.4 Correlation Coefficients

ρ in equation (6.11) is Pearson's product-moment correlation.

6.4.5 Conditional Properties

The conditional distribution of one variable given the other is also Pearson type II.

6.4.6 Relation to Other Distributions

Let $U = (aX - bY)^2$ and $V = (aY - bX)^2$, where $a = \frac{\sqrt{1+\rho}+\sqrt{1-\rho}}{2\sqrt{1-\rho^2}}$ and $b = \frac{\sqrt{1+\rho}-\sqrt{1-\rho}}{2\sqrt{1-\rho^2}}$. Then U and V have a bivariate beta distribution whose density is $\frac{n+1}{\pi}\frac{(1-u-v)^n}{\sqrt{uv}}$.

6.4.7 References to Illustrations

Illustration of p.d.f. surface: M E Johnson (1986; 1987, pp. 111–117, 123).

6.4.8 References to the Generation of Random Variates

Johnson (1987, pp. 115–116, 123); Johnson et al (1984, p. 235).

6.4.9 Remarks

- This is type IIIaβ in van Uven's classification.
- Along with the bivariate normal and t distributions, this distribution is a well-known member of the class of elliptically-contoured distributions. These will be discussed in Section 7.2.
- The quantity $(X^2 - 2\rho XY + Y^2)/(1-\rho^2)$ has a Beta$(1, n+1)$ distribution (see, for example, Johnson et al, 1984).
- The cumulative distribution has a diagonal expansion in terms of orthogonal (Gegenbauer) polynomials (McFadden, 1966). See Chapter 14 regarding diagonal expansion and its applications.

6.4.10 References to Tables or Algorithms

An algorithm for computing the bivariate probability integral could be developed using the results of Parrish and Bargmann (1981).

Joshi and Lalitha (1985) developed a recurrence formula for the evaluation of $\bar{H}$.

6.5 Bivariate Beta Distribution

6.5.1 Support

That part of the unit square such that $x + y \leq 1$.

6.5.2 Formula for p.d.f.

$$h = \frac{\Gamma(\theta_1+\theta_2+\theta_3)}{\Gamma(\theta_1)\Gamma(\theta_2)\Gamma(\theta_3)} x^{\theta_1-1} y^{\theta_2-1} (1-x-y)^{\theta_3-1} \tag{6.12}$$

6.5.3 Univariate Properties

Beta marginals: $X \sim$ Beta$(\theta_1, \theta_2+\theta_3)$, $Y \sim$ Beta$(\theta_2, \theta_1+\theta_3)$.

6.5.4 Correlation Coefficients

Pearson's product-moment correlation is $-\sqrt{\frac{\theta_1\theta_2}{(\theta_1+\theta_3)(\theta_2+\theta_3)}}$. Thus, as might be expected from its support and its application to joint distributions of proportions, this distribution is unusual in being oriented towards negative correlation—to get positive correlation, we would have to change X to $-X$ or Y to $-Y$.

6.5.5 Moments

The product moments are:

$$\mu'_{rs} = \frac{\Gamma(\theta_1 + r)\Gamma(\theta_2 + s)\Gamma(\theta_1 + \theta_2 + \theta_3)}{\Gamma(\theta_1 + \theta_2 + \theta_3 + r + s)\Gamma(\theta_1)\Gamma(\theta_2)} \tag{6.13}$$

(Wilks, 1963, p. 179).

6.5.6 Conditional Properties

$Y/(1-x)$ conditional on $X = x$ has a Beta(θ_2, θ_3) distribution.

6.5.7 Methods of Derivation

This distribution may be obtained by a form of trivariate reduction: if $X_i \sim$ Gamma$(\theta_i, 1)$, then $X_1/(X_1 + X_2 + X_3)$ and $X_2/(X_1 + X_2 + X_3)$ jointly have a bivariate beta distribution (see, for example, Wilks (1963, p. 179)). Also, if $X_i \sim$ Beta$(\theta_i, 1)$, then $X_1/(X_1 + X_2 + X_3)$ and $X_2/(X_1 + X_2 + X_3)$, conditional on $X_1 + X_2 + X_3 \leq 1$, again have a bivariate beta distribution (Loukas, 1984).

6.5.8 Relation to Other Distributions

- This distribution is a bivariate Pearson type I distribution; it is also often referred to as a bivariate Dirichlet distribution.
- For the relation between this distribution and the bivariate Pearson type II, see Section 6.4.6 above.
- Beta conditionals: see Section 8.4.

6.5.9 References to Illustrations

Hoyer and Mayer (1976); Kellogg and Barnes (1989).

6.5.10 References to the Generation of Random Variates

Because of the method of derivation, generation of random variates is straightforward—Devroye (1986, pp. 593–596), see also Macomber and Myers (1978) and Văduva (1985).

6.5.11 Remarks

- The beta distribution variates are "neutral", in the following sense: X and $Y/(1-X)$ are independent and, the distribution being symmetric in x and y, so are Y and $X/(1-Y)$. For more on this, see Section 12.4.2.
- $H(x, y)$ has a diagonal expansion in terms of orthogonal (shifted Jacobi) polynomials (Lee, 1971). See Chapter 14 regarding diagonal expansion.
- If (i) h takes the product form $a_1(x)a_2(y)a_3(1-x-y)$, (ii) at least one of the a_i is a power function, and (iii) the regressions $E(Y|X)$ and $E(X|Y)$ are both linear, then h is the bivariate beta (Rao and Sinha, 1988).
- $X + Y$ has a beta distribution with parameters $\theta_1 + \theta_2$ and θ_3. Also, $X + Y$ is independent of X/Y, which has an inverted beta distribution.
- There is an introduction to the multivariate case by the Editors of the ESS (1982b).

6.5.12 Fields of Application

- This distribution chiefly arises in the context of a trivariate distribution of three quantities that must sum to 1—as, for example, the probabilities of events, or the proportions of substances in a mixture, that are mutually exclusive and exhaustive. When considering just two of these quantities, a bivariate beta distribution may be a natural assumption to adopt. See Section 24.5 for more on this.

- Chatfield (1975) gave a particularly clear example of the just-mentioned general context. The subject is the joint distribution of brand shares, that is, the proportions of brands $1, 2, \ldots n$ of some consumer product that are bought by customers. (The bivariate distribution (6.12) will derive from $n = 3$.) Chatfield says that the following two conditions are approximately correct in most product fields:

  - A consumer's rates of buying different brands are independent.
  - A consumer's brand-shares are independent of his or her total rate of buying.

  The joint distribution of brand shares must then follow the multivariate beta distribution, because of the following characterisation theorem. Suppose Y_i to be independent positive r.v.'s, and set $T = \sum Y_i$ and $X_i = Y_i/T$; then if each X_i is independent of T, then the joint distribution of the X's is multivariate beta.

  See Goodhardt et al (1984) for a later and more comprehensive account of work in this field.

- Hoyer and Mayer (1976) have used this distribution in modelling the proportions of the electorate who vote for the candidates in a two-candidate election (these two proportions adding to less than 1 because of abstentions). They say that this distribution "is sufficiently versatile to model many natural phenomena; yet it demonstrates a degree of simplicity such that a candidate who is reasonably adept at estimating probabilities could easily use our model to make a fairly accurate estimate of the actual joint distribution of proportions of his and his opponent's vote for a fixed set of political strategies."

- A-Grivas and Asaoka (1982) used a bivariate beta distribution (presumably this one) to describe the joint distribution of two soil strength parameters.

- In Bayesian statistics, the beta distribution is popular as a prior because it is conjugate with respect to the binomial distribution, i.e., the posterior distribution is also beta. Similarly, the multivariate beta and the multinomial distributions go together in the same way. An example of such analysis is by Apostolakis and Moieni (1987). These authors considered a system of three identical components, subject to shocks that knock out 0, 1, 2, or 3 of them, in the style of Marshall and Olkin's model (for which, see Section 9.2). Apostolakis and Moieni supposed that the state of knowledge regarding the vector of probabilities (p_0, p_1, p_2, p_3) could be described as a multivariate beta distribution.

6.5.13 References to Tables or Algorithms

For algorithms evaluating the cumulative distribution function, see Parrish and Bargmann (1981), who used this distribution as an illustration of their general technique for evaluation of cumulative bivariate probabilities.

Nabar (1988) has derived upper and lower bounds for H.

6.5.14 Generalisations

- See Connor and Mosimann (1969) and Lochner (1975) for the density

$$h = [B(\alpha_1, \beta_1)B(\alpha_2, \beta_2)]^{-1}x^{\alpha_1-1}y^{\alpha_2-1}(1-x)^{\beta_1-(\alpha_2+\beta_2)}(1-x-y)^{\beta_2-1} \tag{6.14}$$

  (with support $x \geq 0$, $y \geq 0$, $x + y \leq 1$).

- The bivariate Tukey lambda distribution, briefly considered by Johnson and Kotz (1973a), is:

$$\left.\begin{array}{rcl} X &=& [U^{\lambda}-(1-U)^{\lambda}]/\lambda \\ Y &=& [V^{\mu}-(1-V)^{\mu}]/\mu \end{array}\right\} \tag{6.15}$$

  where (U, V) has a bivariate beta distribution. The result is a mess ("mathematically not very elegant"—Johnson and Kotz; "almost intractable"—James, 1975).

- If (U, V) again has a bivariate beta distribution, a distribution of (X, Y) is defined implicitly by $U = \sqrt{XY}$, $V = \sqrt{(1-X)(1-Y)}$; this is briefly mentioned by Mardia (1970a, p. 88).

- Ulrich (1984) has proposed a "bivariate beta mixture" distribution, which he used in a robustness study. Within each rectangle that the unit square is divided into, the p.d.f. is proportional to the product of a beta distribution of X and a beta distribution of Y; the constants of proportionality are different for the different rectangles.

- Attributing the idea to L J Savage, Dickey (1983) gave some attention to the distribution obtained by (first) scaling, and (second) renormalising to sum to unity, i.e.,

$$\left.\begin{array}{rcl} X &=& aU/(aU+bV) \\ Y &=& bV/(aU+bV) \end{array}\right\} \tag{6.16}$$

  with (U, V) having a bivariate beta distribution.

- For another generalisation, see Nagarsenker (1970).

6.5.15 Bivariate Inverted Beta Distribution

6.5.15.1 Support

The positive quadrant.

6.5.15.2 Formula for p.d.f.

$$h = \frac{\Gamma(\alpha_1+\alpha_2+\alpha_3)}{\Gamma(\alpha_1)\Gamma(\alpha_2)\Gamma(\alpha_3)} \frac{x^{\alpha_1-1}y^{\alpha_2-1}}{(1+x+y)^{\alpha_1+\alpha_2+\alpha_3}} \tag{6.17}$$

6.5.15.3 Formula for (Cumulative) Distribution Function

$$H = \frac{\Gamma(\alpha_1+\alpha_2+\alpha_3)}{\Gamma(\alpha_1+1)\Gamma(\alpha_2+1)\Gamma(\alpha_3)} x^{\alpha_1}y^{\alpha_2} F_2(\alpha_1+\alpha_2+\alpha_3; \alpha_1, \alpha_2; \alpha_1+1, \alpha_2+1; -x, -y) \tag{6.18}$$

$$= \frac{\Gamma(\alpha_1+\alpha_2+\alpha_3)}{\Gamma(\alpha_1+1)\Gamma(\alpha_2+1)\Gamma(\alpha_3)} \frac{x^{\alpha_1}y^{\alpha_2}}{(1+x+y)^{\alpha_1+\alpha_2+\alpha_3}} F_2\left(\alpha_1+\alpha_2+\alpha_3; 1, 1; \alpha_1+1, \alpha_2+1; \tfrac{x}{1+x+y}, \tfrac{y}{1+x+y}\right), \tag{6.19}$$

where F_2 is Appell's hypergeometric function of two variables (see footnote 6 in Chapter 20 for definition).

6.5.15.4 Generalisation

For $h \propto \frac{x^{\alpha_1-1}y^{\alpha_2-1}}{(1+x^{\beta_1}+y^{\beta_2})^{(\alpha_1/\beta_1)+(\alpha_2/\beta_2)+\alpha_3}}$, see Nagarsenker (1970).

6.5.15.5 Further Information

Comparing equations (6.9) and (6.17), we see this is effectively the bivariate F-distribution of Section 6.3, so see that Section for further information. Another account is by Ratnaparkhi (1983b).

6.6 Bivariate Liouville-Dirichlet Distribution

Ratnaparkhi (1985) uses this term for any distribution whose p.d.f. is proportional to

$$\xi_{x+y}\, x^{\alpha_1-1}y^{\alpha_2-1}, \tag{6.20}$$

where ξ_u is a nonnegative function; the support is $x > 0$, $y > 0$, $a < x+y < b$. Ratnaparkhi gives the examples summarised below.

ξ_u	a	b	Resulting bivariate distribution
$(1-u)^{\alpha_3-1}$	0	1	Beta
$(1+u)^{-\alpha_1-\alpha_2-\alpha_3}$	0	∞	Inverted beta (F)
$u^{\alpha_3}e^{-u}$	0	∞	Gamma, $h \propto (x+y)^{\alpha_3}x^{\alpha_1-1}y^{\alpha_2-1}e^{-(x+y)}$
$(-\log u)^{\alpha_3-1}$	0	1	"Unit-gamma-type", $h \propto x^{\alpha_1-1}y^{\alpha_2-1}[-\log(x+y)]^{\alpha_3-1}$

A much longer treatment of these distributions is by Gupta and Richards (1987): there are theoretical results concerning stochastic representations, transformation properties, complete neutrality, marginal and conditional distributions, regression functions, total positivity and reverse rule properties, and characterisations. Gupta and Richards attribute the third example above to Marshall and Olkin (1979, p. 308), and they also draw attention to the paper by Sivazlian (1981a). Further references are Sivazlian (1981b) and Lingappaiah (1983)—and these include the generalised case where the argument of ξ is $x^{\beta_1}+y^{\beta_2}$, instead of $x+y$.

For generation of random variates, see Devroye (1986, pp. 596–599).

6.7 Bivariate Logistic Distributions

This name is used for bivariate distributions with logistic marginals. We know of no such distributions that are worth discussing here—all are copulas met in Chapter 5 whose marginals have been transformed to be logistic.

The paper that is commonly referenced on this subject is Gumbel (1961). He proposed three distribution with logistic marginals:

- $H = (1+e^{-x}+e^{-y})^{-1}$. This has no association parameter. Satterthwaite and Hutchinson (1978) considered $H = (1+e^{-x}+e^{-y})^{-c}$, but, of course, this is only a marginal transformation of the bivariate Pareto distribution. See Section 5.8.12.
- The Farlie-Gumbel-Morgenstern distribution with logistic marginals. See Section 5.2.
- The copula $(-\log H)^{\alpha} = (-\log x)^{\alpha} + (-\log y)^{\alpha}$, transformed to logistic marginals. See Section 5.6.

6.8 Bivariate Burr Distributions

Bivariate distributions with Burr type III or type XII marginals have received some attention in the literature. Two main methods have been used for their construction:

- The Farlie-Gumbel-Morgenstern method of Section 5.2.
- Compounding, either as a straightforward generalisation of the construction of the Pareto distribution (abbreviated as P in the table below), or the bivariate method which Hutchinson (1979b, 1981b) showed underlies the Durling-Burr distribution (abbreviated as D).

The table following shows where more details of these may be found. Another account of these distributions is by Rodriguez (1983b, pp. 241–244).

Marginals	Construction	Section	References
XII	Compounding (P)	5.8.11	Takahasi (1965), Crowder (1985)
XII	Compounding (D)	5.8.12.3	Durling (1975), Bagchi and Samanta (1985)
XII	F-G-M	5.2.11	Bagchi and Samanta (1985)
III	Compounding (P)	5.8.11	Rodriguez (1980)
III	Compounding (D)	5.8.12.5	Rodriguez (1980), Rodriguez and Taniguchi (1980)
III	Compounding*		Rodriguez (1980)
III	F-G-M	5.2.11	Rodriguez (1980)
III	F-G-M, extended	5.2.15	Rodriguez (1980)

* $\int_0^\infty \min[1,(x/\lambda)^c]\,dF(\lambda)$, where $F(\lambda) = (1-k)(1+\lambda^{-c})^{-k} + k(1+\lambda^{-c})^{-k-1}$. Rodriguez (1980, p. 39) makes only passing mention of this.

6.9 Rhodes' Distribution

6.9.1 Support

The region where $1 - \frac{x}{a} + \frac{y}{b} > 0$ and $1 + \frac{x}{a'} - \frac{y}{b'} > 0$.

6.9.2 Formula for p.d.f.

$$h \propto \left(1 - \frac{x}{a} + \frac{y}{b}\right)^p \left(1 + \frac{x}{a'} - \frac{y}{b'}\right)^{p'} e^{-\ell x - my} \tag{6.21}$$

6.9.3 Derivation

Start with two independent variables having (not necessarily identical) gamma distributions. Let X be some linear combination of them, and Y be some other linear combination. The result is that (X, Y) has Rhodes' distribution.

6.9.4 Remarks

For the properties of this distribution, see Mardia (1970a, p. 40 and pp. 94–95), Johnson and Kotz (1972, pp. 4–5), and, for more detail, Rhodes (1923). Rhodes fitted this distribution to barometric heights observed at Southampton and Laudale (Pearson and Lee, 1897).

Chapter 7

Further Distributions Related to the Bivariate Normal

7.1 Introduction

Towards the end of Chapter 4, we met a number of bivariate distributions obtained from the bivariate normal in some elementary fashion, such as by transformation or truncation. Some distributions in Chapter 6 are also connected with the bivariate normal, especially those with t-, gamma, or F-distributions for marginals. This present Chapter is devoted to extensions of these, and other distributions derived from or generalising the bivariate normal.

The Sections are as follows:

7.2 A class of elliptically-contoured bivariate densities.

7.3 Nachtsheim and Johnson (1988).

7.4 Bivariate exponential power distributions.

7.5 Kibble's bivariate gamma distribution.

7.6 Noncentral bivariate chi-squared distribution.

7.7 Jensen's bivariate gamma distribution.

7.8 Jensen's bivariate F-distribution.

7.9 The bivariate inverse Gaussian distribution.

7.10 The bivariate hyperbolic distribution.

7.11 Bivariate symmetric stable distributions.

7.2 A Class of Elliptically-Contoured Bivariate Densities

7.2.1 The Growing Importance of this Class

This Section is given over to a consideration of a class of distributions whose contours of probability density are ellipses. The expression (7.1) below does not include all bivariate distributions having elliptical density contours; for the ones considered here, the ellipses are concentric with constant eccentricity. Indeed, we shall meet later in this Chapter a distribution, the bivariate hyperbolic, that has elliptical contours yet is not within the class described now.

Johnson (1987, pp. 106–107) says that this class has received considerable attention recently in the statistical literature, in both theoretical and applied contexts. Interest particularly arises because many theoretical results for the bivariate normal distribution also hold in this broader class. For example, let

(X, Y) be an uncorrelated pair from this class. Then X^2/Y^2 has the usual F-distribution, and $X^2/(X^2 + Y^2)$ has the beta distribution, Beta$(\frac{1}{2}, \frac{1}{2})$ (Kelker, 1970). On the application side, members of the class were used to describe the second-order moments of the transformation of a random signal by an instantaneous linear device (McGraw and Wagner, 1968). And van Praag and Wesselman (1989) have shown that many procedures for multivariate analysis in the normal case can be adapted to the elliptical case with the aid of the estimated kurtosis. See also Bentler and Berkane (1985), who go so far as to say "It is becoming apparent that [elliptical] theory has the potential to displace multinormal theory in a variety of applications such as linear structural modelling" (which includes factor analysis and simultaneous equations models). For a review and bibliography of these distributions, see Chmielewski (1981).

7.2.2 Formula for p.d.f.

First, recall some conventions mentioned in Section 1.5.3: boldface symbols will be used for vectors and matrices; $N(\boldsymbol{\mu}, \boldsymbol{\Sigma})$ denotes the bivariate normal variate having mean $\boldsymbol{\mu}$ and covariance matrix $\boldsymbol{\Sigma}$; the transpose of matrix $\mathbf{A}$ is denoted $\mathbf{A}'$.

If the probability density h is a function only of a positive definite[1] quadratic form

$$(\mathbf{x} - \boldsymbol{\mu})'\boldsymbol{\Sigma}^{-1}(\mathbf{x} - \boldsymbol{\mu}), \tag{7.1}$$

then its contours are ellipses; here, $\mathbf{x}' = (x, y)$, $\boldsymbol{\mu}' = (\mu_1, \mu_2)$, and $\boldsymbol{\Sigma}$ is a nonsingular scaling matrix, which is determined only up to a multiplicative constant. Its role is like that of the covariance matrix and indeed, when the latter exists, $\boldsymbol{\Sigma}$ must be proportional to it (see Devlin et al, 1976, for details). If $\boldsymbol{\mu} = \mathbf{0}$ and $\boldsymbol{\Sigma} = \mathbf{I}$ (the identity matrix), the distribution is called spherical.

7.2.3 Alternative Definition

If $\mathbf{X}$ has an elliptically-contoured bivariate distribution of the type considered here, it can be written as

$$\mathbf{X} = R\mathbf{L}\mathbf{U}^{(2)} + \boldsymbol{\mu}, \tag{7.2}$$

where R^2 has the distribution of $(\mathbf{X} - \boldsymbol{\mu})'\boldsymbol{\Sigma}^{-1}(\mathbf{X} - \boldsymbol{\mu})$, $\mathbf{X}' = (X, Y)$, $\boldsymbol{\Sigma} = \mathbf{L}\mathbf{L}'$ (i.e., $\mathbf{L}$ is the lower triangular matrix of the Choleski decomposition of $\boldsymbol{\Sigma}$), and $\mathbf{U}^{(2)}$ is uniformly distributed on the circumference of a unit circle (Cambanis et al, 1981; Johnson et al, 1984). Equation (7.2) may serve as an alternative definition of an elliptically-contoured bivariate distribution.

7.2.4 Well-Known Members of this Family

The family includes the bivariate normal, t (Pearson type VII), Cauchy, and Pearson type II distributions.

7.2.5 Formula for (Cumulative) Distribution Function

Naturally, since the only thing members of this class have in common is the ellipticity of their contours, there is no single closed form for the distribution function.

7.2.6 Characteristic Function

The characteristic function φ depends only on a quadratic form $\mathbf{t}'\boldsymbol{\Sigma}\mathbf{t}$ (assuming $\boldsymbol{\mu} = \mathbf{0}$)—see Kelker (1970) and Johnson (1987, p. 107).

7.2.7 Conditional Properties

The regression of Y on X is linear. The conditional variance, $\text{var}(Y|X = x)$, is independent of x if and only if X and Y have the bivariate normal distribution; more generally,

$$\text{var}(Y|X = x) = a(x)\sigma_2^2(1 - \rho^2) \tag{7.3}$$

for some function a, where σ_2^2 is the variance of Y and ρ is the correlation between X and Y.

[1] A $n \times n$ symmetric matrix $\mathbf{A}$ is positive definite if $\mathbf{x}'\mathbf{A}\mathbf{x} > 0$ for every nonzero $\mathbf{x}$ in R^n.

7.2.8 Derivation

The quadratic form (7.1) may, in a sense, be regarded as a derivation as it stands.

The subclass of "elliptical compound bivariate normal" distributions can be derived easily—see Section 7.2.12.

7.2.9 References to the Generation of Random Variates

See Section 16.5.2, and Johnson et al (1984, pp. 232–235).

7.2.10 Remarks

- All $\alpha X + \beta Y$ with the same variance (if it exists) have the same distribution.
- X and Y are independent if and only if $\mathbf{\Sigma}$ is diagonal and X and Y have a bivariate normal distribution.
- Writing $\mathbf{\Sigma}$ as (c_{ij}), the correlation matrix (assuming it is defined) is given by $\mathbf{\Sigma}/\sqrt{c_{11}c_{22}}$.
- If $\mathbf{X}_1$ and $\mathbf{X}_2$ are members of this class having the same $\mathbf{\Sigma}$, and are independent, then $\mathbf{X}_1 + \mathbf{X}_2$ is also a member of this class, and has the same $\mathbf{\Sigma}$.
- $h(x, y)$ can be represented as
$$\int_0^\infty \psi(\mathbf{x}; v)\frac{dW}{dv}\,dv, \tag{7.4}$$
where $\frac{dW}{dv}$ is a weighting function ($\int \frac{dW}{dv}\,dv = 1$) which may assume negative values, and ψ is the density function corresponding to the bivariate normal distribution $N(\boldsymbol{\mu}, v^{-2}\mathbf{\Sigma})$. For an interpretation of (7.4), see Section 7.2.12.
- Slepian's inequality for the bivariate normal distribution (see Section 4.9) also holds for this wider class (see, for example, Gordon, 1987). Further probability inequalities applicable to this class are given by Tong (1980, Section 4.3).
- When the variances of X and Y are equal, consider the probability over the region $\frac{x^2}{a^2} + a^2y^2 \leq 1$ (which is an ellipse of area π). Shaked and Tong (1988, pp. 338–339) showed that the maximum probability content is obtained when $a = 1$. Also, when the area of a rectangle is fixed, the maximum probability content is obtained when the rectangle becomes a square.
- For properties concerning the moments, see Berkane and Bentler (1986a,b, 1987).
- Suppose a distribution has p.d.f. that is constant within the ellipse
$$\frac{1}{1-\rho^2}\left[\frac{(x-\mu_1)^2}{\sigma_1^2} - \frac{2\rho(x-\mu_1)(y-\mu_2)}{\sigma_1\sigma_2} + \frac{(y-\mu_2)^2}{\sigma_2^2}\right] = 4 \tag{7.5}$$
and is zero outside. This distribution has means μ_1 and μ_2, standard deviations σ_1 and σ_2, and correlation ρ. This ellipse can, of course, be constructed for any distribution with these moments, in which case it is known as the ellipse of concentration. See Editors of the ESS (1982d). (For generation of random variates from such a distribution, see Devroye, 1986, p. 567.)

(Devlin et al, 1976, give the properties above for which no other reference is noted.)

7.2.11 Fields of Application

Other members of the class are used as alternatives to the normal when studying the robustness of statistical tests. For instance, Devlin et al (1976) used them to obtain samples containing outliers, and then compared the robustness of two estimators of the correlation, namely, the product-moment correlation r and the quadrant correlation r_q; the latter is defined by $r_q = \sin(\pi q/2)$, where $q = (n_1 + n_3 - n_2 - n_4)/n$, n_i being the number of observations in the ith quadrant using the coordinate-wise medians as the origin, and $n = \sum n_i$. They found that r is not robust, whereas r_q is quite robust.

7.2.12 Elliptical Compound Bivariate Normal Distributions

If W in (7.4) is the cumulative distribution of a positive random variable V, then we may say h is an elliptical compound bivariate normal distribution.[2] Obviously, the bivariate normal is itself a member of this class (take V to be a positive constant); other members are longer-tailed (Devlin et al, 1976, p. 369).

It follows from (7.4) that a special property is that $\mathbf{X} = V^{-1}\mathbf{Z} + \boldsymbol{\mu}$, where $\mathbf{Z} \sim N(\mathbf{0}, \Sigma)$ and V and $\mathbf{Z}$ are independent. The question arises as to whether every compound bivariate normal distribution is elliptically contoured. The answer is no, as will be seen from the following:

- If $\mathbf{Z} \sim N(\mathbf{0}, \Sigma)$ and V is independent of $\mathbf{Z}$, then it is clear that the density of $V^{-1}\mathbf{Z}$ is a function only of a positive definite quadratic form, and hence it is elliptically contoured. For example, let $V = S/\sqrt{\nu}$, where S has a chi-distribution with ν degrees of freedom; then $(S/\sqrt{\nu})^{-1}\mathbf{Z}$ has a bivariate t-distribution with ν degrees of freedom.

- Now, if $\mathbf{Z} \sim N(\boldsymbol{\mu}, \Sigma)$, with $\boldsymbol{\mu} \neq \mathbf{0}$, then

$$(S/\sqrt{\nu})^{-1}\mathbf{Z} = (S/\sqrt{\nu})^{-1}(\mathbf{Z} - \boldsymbol{\mu}) + \boldsymbol{\mu}(S/\sqrt{\nu})^{-1},$$

  which does not have a bivariate t-distribution, and it is not elliptically contoured since the second term destroys the symmetry.

7.3 Nachtsheim and Johnson (1988)

The bivariate normal distribution may be written as $R\mathbf{L}\mathbf{B}^{(2)}$, where R is a χ_2-variate, $\mathbf{L}$ is the Choleski factorisation of the scaling matrix, $\mathbf{B}^{(2)}$ is a variate having the uniform distribution on the circle, and R and $\mathbf{B}^{(2)}$ are independent. Elliptical distributions are the same, except R has some other distribution. Nachtsheim and Johnson (1988) take up the idea of R's distribution remaining χ_2, but letting $\mathbf{B}^{(2)}$'s distribution on the circle be something other than uniform. (Their account was not restricted to the bivariate case.) Johnson (1987, Chapter 7) presents much the same material, and includes 40 plots of p.d.f. contours. A brief summary is in M E Johnson (1986).

Thus we let $a(\theta, r)$ be a density function expressed in terms of angle and radius, so that

$$h(x, y) = (x^2 + y^2)^{-1/2}\, a\left(\tan^{-1}\frac{y}{x}, \sqrt{x^2 + y^2}\right). \tag{7.6}$$

If R is a χ_2-variate and independent of Θ,

$$h(x, y) = \exp\left[-\tfrac{1}{2}(x^2 + y^2)\right]\, b\left(\tan^{-1}\frac{y}{x}\right), \tag{7.7}$$

where b is the p.d.f. of the angle Θ.

Nachtsheim and Johnson investigate six choices of $b(\theta)$—cardiod, triangular, von Mises, wrapped Cauchy, wrapped normal, and offset normal—and the resulting bivariate p.d.f.'s. They present contour plots of the resulting densities h, and algorithms for generating random variates from the densities b.

7.4 Bivariate Exponential Power Distributions

7.4.1 Support

The whole plane.

[2] Some authors refer to it as a normal variance mixture.

7.4.2 Formula for p.d.f.

This is cumbersome, and the reader is referred to the following places in the paper by Taguchi (1978):

- Equation (3.39), which gives the distribution "in the narrow sense", with one parameter for the shape of the marginals, and one for the association between the variates.
- Equation (3.1), which gives the distribution "in the wider sense", with an additional parameter that introduces asymmetry.
- Theorem 3.7, which represents the wider distribution as a mixture of distributions of the narrower type.
- Theorem 3.3, which gives a method of generalising the wider distribution to the case of marginals with different shapes.

7.4.3 Univariate Properties

Exponential power marginals.

7.4.4 Further Properties

See Taguchi (1978) for more details. (The name "generalised error distribution" is used in that paper.)

7.4.5 Remark

This is suggested as a generalisation of the bivariate normal just as the exponential power density, proportional to $\exp(-|x|^\alpha)$, is suggested as a generalisation of the univariate normal. And, clearly, just as we can have the univariate Laplace distribution as a special case, so a form of bivariate Laplace distribution is a special case of the bivariate exponential power distribution.

7.5 Kibble's Bivariate Gamma Distribution

7.5.1 Support

The positive quadrant.

7.5.2 Formula for p.d.f.

$$h = f_\alpha(x) f_\alpha(y) \frac{\Gamma(\alpha)}{1-\rho} \exp\left[-\frac{\rho(x+y)}{1-\rho}\right] (xy\rho)^{-(\alpha-1)/2} I_{\alpha-1}\left(\frac{2\sqrt{xy\rho}}{1-\rho}\right) \tag{7.8}$$

$(0 \le \rho < 1)$, where $f_\alpha(t) = t^{\alpha-1}e^{-t}/\Gamma(\alpha)$ and I_ν is the modified Bessel function of the first kind and order ν. The probability density function may also be expressed in terms of Laguerre polynomials[3] $L_j^{(\alpha-1)}$:

$$h = f_\alpha(x) f_\alpha(y) \sum_{j=0}^{\infty} \rho^j L_j^{(\alpha-1)}(x) L_j^{(\alpha-1)}(y) \frac{\Gamma(\alpha)\Gamma(j+1)}{\Gamma(j+\alpha)}. \tag{7.9}$$

[3] Laguerre polynomials $L_n^{(\alpha)}$ are defined as

$$L_n^{(\alpha)}(x) = \sum_{j=0}^{n} \binom{n+\alpha}{n-j} \frac{(-x)^j}{j!}.$$

These are the orthogonal polynomials that are defined on the gamma density function.

An alternative expression of the probability density function, obtained by Krishnaiah (1963) (see also Krishnaiah, 1983), is as follows:

$$h = \frac{(1-\rho)^{-\alpha}}{\Gamma(\alpha)} \sum_{j=0}^{\infty} a_j (xy)^{\alpha+j-1} \exp\left(-\frac{x+y}{1-\rho}\right), \tag{7.10}$$

where $a_j = \frac{1}{\Gamma(\alpha+j)j!} \frac{\rho^j}{(1-\rho)^{2j}}$.

7.5.3 Formula for (Cumulative) Distribution Function

Expressed as an infinite series in Laguerre polynomials,

$$H = F_\alpha(x)F_\alpha(y) + \alpha \sum_{j=0}^{\infty} \frac{\rho^{j+1}}{(j+1)} \frac{\Gamma(\alpha+1)\Gamma(j+1)}{\Gamma(j+\alpha+1)} L_j^{(\alpha)}(x) L_j^{(\alpha)}(y) f_{\alpha+1}(x) f_{\alpha+1}(y), \tag{7.11}$$

where $F_\alpha(t) = \int_0^t f_\alpha(u)\,du$. See Lai and Moore (1984) for details.

Alternatively,

$$H = \frac{(1-\rho)^{\alpha}}{\Gamma(\alpha)} \sum_{j=0}^{\infty} c_j F_{\alpha+j-1}[(1-\rho)^{-1}x] F_{\alpha+j-1}[(1-\rho)^{-1}y], \tag{7.12}$$

where $c_j = \frac{\Gamma(\alpha+j)\rho^j}{j!}$ (see Johnson and Kotz, 1972, p. 221).

7.5.4 Univariate Properties

Gamma marginals, with shape parameter α.

7.5.5 Correlation Coefficients

ρ in equation (7.8) is Pearson's product-moment correlation.

7.5.6 Moment Generating Function

The moment generating function is

$$M(s,t) = [(1-s)(1-t) - \rho st]^{-\alpha} \tag{7.13}$$

($\rho > 0$). Thus the moments μ'_{rs} can be obtained easily.

The m.g.f. was first given by Wicksell (1933), but the explicit form of the density (7.8) is due to Kibble (1941). Some authors refer to the distribution as the Kibble-Wicksell bivariate gamma distribution. Vere-Jones (1967) has shown that it is infinitely divisible.[4]

7.5.7 Conditional Properties

The regression is linear:

$$E(Y|X=x) = \rho(x-\alpha) + \alpha. \tag{7.14}$$

The conditional variance is also linear:

$$\text{var}(Y|X=x) = (1-\rho)[2\rho x + \alpha(1-\rho)] \tag{7.15}$$

(Mardia, 1970a, p. 88).

[4]A bivariate distribution with characteristic function φ is said to be infinitely divisible if $\varphi^{1/n}$ is also a characteristic function, for every positive integer n. In terms of r.v.'s, this means that for each $n \geq 1$, the r.v. with characteristic function φ can be written as $\mathbf{X} = \sum_{j=1}^{n} \mathbf{X}_{nj}$, where $\mathbf{X}_{nj}$ $(1 \leq j \leq n)$ are independent and identically distributed with characteristic function $\varphi^{1/n}$.

7.5.8 Derivation

In the univariate situation, the derivation of the chi-squared distribution as the sum of squared normal variates is familiar. Now, let $(X_1, Y_1), \ldots (X_n, Y_n)$ be a random sample of size n from the bivariate normal population with mean $\mathbf{0}$ and variance-covariance matrix $\boldsymbol{\Sigma} = \begin{pmatrix} \sigma_1^2 & \rho_0\sigma_1\sigma_2 \\ \rho_0\sigma_1\sigma_2 & \sigma_2^2 \end{pmatrix}$. Define $X = \frac{1}{2\sigma_1^2}\sum_{i=1}^n X_i^2$ and $Y = \frac{1}{2\sigma_2^2}\sum_{i=1}^n Y_i^2$. Then after replacing $n/2$ by α in the density function, the distribution of (X, Y) is Kibble's bivariate gamma with $\rho = \rho_0^2$. For generalisations to higher dimensions, see Krishnamoorthy and Parthasarathy (1951) and Krishnaiah and Rao (1961).

Clearly, random variate generation is easy when 2α is a fairly small integer.

7.5.9 Relation to Other Distributions

- Downton's bivariate exponential is a special case of this distribution: see below, Section 9.12.
- According to Khan and Jain (1978), the quantity

$$\frac{u}{u + ax + by} f(x, y; u + ax + by) \tag{7.16}$$

  is a p.d.f. of interest in the theory of the emptiness of reservoirs, u being the initial content of the reservoir and $f(x, y; t)$ being the p.d.f. that amounts ax and by from two sources flow into the reservoir during time t. Khan and Jain consider (7.16) when f is Kibble's distribution: they write out an expression for the p.d.f., and obtain the low moments. (See also Jain and Khan, 1979, pp. 166–167.)

7.5.10 Generalisations

- In Jensen's bivariate gamma distribution, (i) the shape parameters of the marginals are different and they are integers or half-integers, and (ii) the bivariate normal distributions used for the derivation have different correlation coefficients. For this and further generalisations, see Section 7.7.
- Malik and Trudel (1983) rewrote (7.9) as

$$h = (1-\rho)^{\alpha} \sum_{j=0}^{\infty} \frac{\Gamma(\alpha + j)\rho^j (xy)^{\alpha+j-1}}{\Gamma(\alpha) j! [\Gamma(\alpha + j)(1-\rho)^{\alpha+j}]^2} \exp\left(-\frac{x+y}{1-\rho}\right). \tag{7.17}$$

  Then they generalised the above density in the following way:

$$h = (1-\rho)^{(\alpha_1+\alpha_2)/2} \sum_{j=0}^{\infty} \frac{\Gamma\left(\frac{\alpha_1+\alpha_2}{2} + j\right) \rho^j x^{\alpha_1+j-1} y^{\alpha_2+j-1}}{\Gamma\left(\frac{\alpha_1+\alpha_2}{2}\right) j! \Gamma(\alpha_1 + j)\Gamma(\alpha_2 + j)(1-\rho)^{\alpha_1+\alpha_2+2j}} \exp\left(-\frac{x+y}{1-\rho}\right). \tag{7.18}$$

  The marginals are not gamma distributions unless $\alpha_1 = \alpha_2$.

7.5.11 References to Illustrations

Surfaces and contours of probability density, for Kibble's distribution and that given by (7.32) below: Smith et al (1982).

Contours of p.d.f. of Kibble's distribution ($\rho = 0.5$, $\alpha = 1$; and $\rho = 0.5$, $\alpha = 2$): Izawa (1965).

7.5.12 Remarks

- It can be proved easily—see Jensen (1969)—that

$$\Pr(c_1 \le X \le c_2,\ c_1 \le Y \le c_2) \ge \Pr(c_1 \le X \le c_2)\Pr(c_1 \le Y \le c_2). \tag{7.19}$$

  Jensen called this positive dependence, but we will use this term in another way in Chapter 12.

  Also, $\Pr(X \le x,\ Y \le y) \ge \Pr(X \le x)\Pr(Y \le y)$, i.e., X and Y are positive quadrant dependent (see Section 12.3.1 for this term).

- Izawa (1965) gave formulae for the density and moments of the sum, product, and ratio of X and Y.
- For results on the location of the mode, see Brewer et al (1987).
- A brief account of this distribution, in the context of others with gamma maginals, is given by Krishnaiah (1985).

7.5.13 Fields of Application

- Electric counter system. Lampard (1968) built up this distribution in the conditional manner, $h = f(x)g(y|x)$; his context was a system of two reversible counters (i.e., an input can either increase or decrease the cumulative count), with two Poisson inputs (an increase process and a decrease process). Output events occur when either of the cumulative counts decreases to zero. The sequence of time intervals between output events forms a Markov chain, and the joint distribution of successive intervals is of Kibble's form. Lampard also gave an interpretation of the same process in terms of a queueing system.
- Hydrology. Phatarford (1976) used this distribution as a model to describe summer and winter streamflows.
- Rain. As the gamma distribution is a popular univariate choice for the description of amount of rainfall, Izawa (1965) proposed Kibble's distribution to describe the joint distribution of rainfall at two nearby rain gauges.
- Wind gusts. Smith and Adelfang (1981) reported analyses of wind gust data using this distribution. The two variates were magnitude and length of the gust.

7.5.14 References to Tables or Algorithms

For α an integer or half-integer, Gunst and Webster (1973) have a table of upper 5% critical points, and Krishnaiah (1980) gives an algorithm to compute the probability integral.

For arbitrary α, an algorithm to compute the probability integral is given in Lai and Moore (1984).

7.5.15 Transformations of the Marginals

- The joint distribution of $\sqrt{X}$ and $\sqrt{Y}$ is a bivariate chi-distribution (also known as a bivariate Rayleigh distribution). This has been studied by Krishnaiah et al (1963).
- Izawa (1965) has given some attention to distributions for which certain transformations of the variates—square root, cube root, or logarithm—have Kibble's distribution.
- By transforming the marginals to be Pareto in form, Mardia (1962) obtained what he called a type 2 bivariate Pareto distribution.

7.6 Noncentral Bivariate Chi-Squared Distribution

Let (X_i, Y_i), $i = 1, 2, \ldots n$ be n independent random vectors having bivariate normal distributions with means (μ_i, μ_i), identical variances (of both X and Y) σ^2, and identical covariances $\rho_0\sigma^2$.

Further, let $X = \sum_{i=1}^n X_i^2/\sigma^2$ and $Y = \sum_{i=1}^n Y_i^2/\sigma^2$. Then X and Y both have a noncentral chi-squared distribution. Krishnan (1976) showed that their joint distribution has density

$$h = \frac{k}{4}\exp\left[-\frac{x+y}{2(1-\rho)}\right]\sum_{i=0}^{\infty} D_i I_{f_i}\left(\frac{\sqrt{\rho xy}}{1-\rho}\right) I_{f_i}\left(\frac{\sqrt{\lambda x}}{1+\sqrt{\rho}}\right) I_{f_i}\left(\frac{\sqrt{\lambda y}}{1+\sqrt{\rho}}\right), \tag{7.20}$$

where $\lambda = \sum_{i=1}^{n} \mu_i^2$ is the noncentrality parameter, $\rho = \rho_0^2$, I_{f_i} is the modified Bessel function of the first kind and order $f_i = \frac{n}{2} + i - 1$, and k and D_i are given by:

$$\begin{aligned} k &= \exp\left(-\frac{\lambda}{1+\sqrt{\rho}}\right)\left[\frac{2(1+\sqrt{\rho})^2}{\lambda\sqrt{\rho}}\right]^{\frac{n}{2}-1} \Big/ (1-\rho) \\ D_i &= \binom{n+i-3}{n-3}\left(\frac{n}{2}+i-1\right)\Gamma\left(\frac{n}{2}-1\right). \end{aligned}$$

Krishnan also showed that the m.g.f. is given by

$$M(s,t) = [1-2(s+t)+4st(1-\rho)]^{-n/2}\exp\left\{\frac{\lambda[s+t-4st(1-\sqrt{\rho})]}{1-2(s+t)+4st(1-\rho)}\right\}. \tag{7.21}$$

When $\lambda = 0$, we get Kibble's distribution.

7.7 Jensen's Bivariate Gamma Distribution

7.7.1 Support

The positive quadrant.

7.7.2 Formula for p.d.f.

For this generalisation of Kibble's distribution (Jensen, 1970), the p.d.f. has a diagonal expansion in terms of Laguerre polynomials:

$$h = f_{a/2}(x) f_{b/2}(y) \sum_{k=0}^{\infty} \frac{G_k(\boldsymbol{\delta})(k!)^2\Gamma(a/2)\Gamma(b/2)}{\Gamma(k+a/2)\Gamma(k+b/2)} L_k^{(\frac{a}{2}-1)}(x) L_k^{(\frac{b}{2}-1)}(y), \tag{7.22}$$

where a and b are positive integers such that $a \leq b$, f_α is the standard gamma density as before, and

$$\begin{aligned} G_k(\boldsymbol{\delta}) &= G_k(\delta_1, \delta_2, \ldots \delta_a) \\ &= \sum_{j_1, j_2, \ldots j_a} c_{1j_1} c_{2j_2} \cdots c_{aj_a}, \end{aligned} \tag{7.23}$$

where the sum is taken over all integer partitions[5] of k in the second subscript of c, and

$$c_{mj_m} = \frac{\delta_m^{j_m}\Gamma(j_m + \frac{1}{2})}{\Gamma(j_m+1)\Gamma(\frac{1}{2})}.$$

The density function for the equicorrelated case (i.e., all the δ's are equal) with $a = b$ is discussed in Section 7.5; for the case where $a \neq b$, see Krishnamoorthy and Parthasarathy (1951).

7.7.3 Univariate Properties

The marginals are again gamma distributions, but this time with different shape parameters ($a/2$ and $b/2$).

7.7.4 Correlation Coefficients

Pearson's product-moment correlation is

$$\rho = \frac{\rho_1^2 + \rho_2^2 + \ldots \rho_a^2}{\sqrt{ab}}, \tag{7.24}$$

where $\rho_j^2 = \delta_j > 0$, and ρ_j is the correlation coefficient of the bivariate normal distribution which is involved in the derivation of this distribution, Section 7.7.6.

[5] An integer partition of k with a groups is a vector $(j_1, j_2, \ldots j_a)$ such that $\sum_{i=1}^{a} j_i = k$, $0 \leq j_i \leq k$. Each distinct vector is a distinct partition. For example, if $a = k = 2$, then all possible partitions are: $(0,2)$, $(2,0)$, and $(1,1)$.

7.7.5 Characteristic Function

The characteristic function is:

$$\varphi(s,t) = (1-it)^{-(b-a)/2}\prod_{j=1}^{a}[(1-is)(1-it)+st\rho_j^2]^{-1/2}. \tag{7.25}$$

In the equicorrelated case $\rho_1^2 = \rho_2^2 = \ldots \rho_a^2 = \eta$, this becomes

$$\varphi(s,t) = (1-it)^{-(b-a)/2}[(1-is)(1-it)+st\eta]^{-a/2} \tag{7.26}$$

and the correlation is $\eta\sqrt{a/b}$.

7.7.6 Derivation

The distribution may be derived as follows. Let $\mathbf{Z}$ be a normal random vector with $a+b$ components, having zero means and general positive definite variance-covariance matrix $\boldsymbol{\Sigma}$, partitioned as $\mathbf{Z}' = (\mathbf{Z}_1', \mathbf{Z}_2')$, $\boldsymbol{\Sigma} = \begin{pmatrix} \boldsymbol{\Sigma}_{11} & \boldsymbol{\Sigma}_{12} \\ \boldsymbol{\Sigma}_{21} & \boldsymbol{\Sigma}_{22} \end{pmatrix}$, where $\mathbf{Z}_1$ and $\mathbf{Z}_2$ are $(a \times 1)$ and $(b \times 1)$ normal vectors, with $a \le b$, respectively. Here $\boldsymbol{\Sigma}_{11}$ and $\boldsymbol{\Sigma}_{22}$ are identity matrices, and $\boldsymbol{\Sigma}_{12} = \boldsymbol{\Sigma}_{21}' = (\mathbf{D}\,\mathbf{0})$, where $\mathbf{D}$ has the ρ's down the diagonal and zeros elsewhere. Then the quadratic forms $Q_1 = \frac{1}{2}\mathbf{Z}_1'\boldsymbol{\Sigma}_{11}^{-1}\mathbf{Z}_1$ and $Q_2 = \frac{1}{2}\mathbf{Z}_2'\boldsymbol{\Sigma}_{22}^{-1}\mathbf{Z}_2$ have this bivariate gamma distribution.

7.7.7 References to Illustrations

Smith et al (1982); Tubbs (1983b).

7.7.8 Remarks

Jensen (1970) showed that his bivariate gamma distribution can also be expanded diagonally in terms of orthogonal polynomials:

$$h = f_{a/2}(x)f_{b/2}(y)\sum_{k=0}^{\infty} M_k x^{(k)} y^{(k)}, \tag{7.27}$$

where $x^{(k)}$ and $y^{(k)}$ are the normalised[6] Laguerre polynomials, and the normalising constants are

$$M_k = \frac{k!\sqrt{\Gamma(a/2)\Gamma(b/2)}}{\sqrt{\Gamma(\frac{a}{2}+k)\Gamma(\frac{b}{2}+k)}}G_k(\delta).$$

7.7.9 Fields of Application

Modelling of wind gusts—Smith et al (1982), Tubbs (1983b). See Sections 24.12.2—1 and 24.12.3—1 below. An advantage of this distribution is that the shape parameters of the marginals can be unequal.

7.7.10 References to Tables or Algorithms

Table of upper 5% critical points: Gunst and Webster (1973).

Algorithm to calculate the probability integral: Smith et al (1982).

[6] Any orthogonal function or polynomial with respect to a weight function f can be normalised to give $\int \theta_i(x)\theta_j(x)f(x)\,dx = \delta_{ij}$, where δ_{ij} is 1 if $i=j$ and is 0 otherwise. The Laguerre polynomials $L_k^{(\alpha-1)}$ were defined in footnote 3. The normalised Laguerre polynomials are $x^{(k)} = L_k^{(\alpha-1)}\Big/\sqrt{\binom{k+\alpha-1}{k}}$.

7.7.11 Gunst and Webster (1973) and Related Distributions

Gunst and Webster (1973) considered Jensen's bivariate gamma distribution in the case where all of the ρ_i^2 are zero or η. Let m be the number of non-zero ρ_i^2.

- Case (3) of Gunst and Webster (1973). Set $a = m+n$, $b = m+p$, with the m non-zero $\rho_i^2 = \eta$. Then the density is given by

$$h = \frac{(1-\eta)^{-m/2}}{\Gamma(m/2)\Gamma(n/2)\Gamma(p/2)} \sum_{j=0}^{\infty}\sum_{k=0}^{\infty}\sum_{\ell=0}^{\infty} \alpha_{jk\ell} \frac{\eta^{j+k+\ell}}{(1-\eta)^{2j+k+\ell}} x^{N_1+N_2+1} y^{N_1+N_3+1} \exp\left(-\frac{x+y}{2(1-\eta)}\right), \tag{7.28}$$

where $N_1 = \frac{m}{2}+j-1$, $N_2 = \frac{n}{2}+k-1$, $N_3 = \frac{p}{2}+\ell-1$, and

$$\alpha_{jk\ell} = \frac{2^{-(2N_1+N_2+N_3-4)}}{j!k!\ell!} \frac{\Gamma(N_1+1)\Gamma(N_2+1)\Gamma(N_3+1)}{\Gamma(N_1+N_2+2)\Gamma(N_1+N_3+2)}.$$

The correlation coefficient is $\eta m/\sqrt{ab}$.

As to the case where m, n, and p are not necessarily integers, Krishnaiah and Rao (1961) and Krishnaiah (1983) rewrote the m.g.f. in (7.13) as

$$M(s,t) = (1-s)^{-\alpha}(1-t)^{-\alpha}\{1-\rho st[(1-s)(1-t)]^{-1}\}^{-\alpha}.$$

Then the first two α's were replaced by α_1 and α_2, with $\alpha_i \geq \alpha > 0$, to give

$$M(s,t) = (1-s)^{-\alpha_1}(1-t)^{-\alpha_2}\{1-\rho st[(1-s)(1-t)]^{-1}\}^{-\alpha}. \tag{7.29}$$

(It is clear from here that the marginals have shape parameters α_1 and α_2.) The above m.g.f. was inverted to obtain the density

$$h = f_{\alpha_1}(x) f_{\alpha_2}(y) \sum_{j=0}^{\infty} \rho^j j! \frac{\Gamma(\alpha+j)}{\Gamma(\alpha)} \frac{\Gamma(\alpha_1)}{\Gamma(\alpha_1+j)} \frac{\Gamma(\alpha_2)}{\Gamma(\alpha_2+j)} L_j^{(\alpha_1-1)}(x) L_j^{(\alpha_2-1)}(y), \tag{7.30}$$

which is an alternative expression for the density function given in (7.28). (Note that the Laguerre polynomial $L_j(x,\alpha)$ defined in Krishnaiah (1983) is $j!L_j^{\alpha-1}(x)$.) Sarmanov (1974) also constructed the same bivariate gamma distribution.

- Case (2) of Gunst and Webster (1973). Here, $a = m$, $b = m+p$. This is the equicorrelated case of Jensen's bivariate gamma, i.e., all the δ's are equal. The density is:

$$h = \frac{(1-\eta)^{-m/2}}{\Gamma(m/2)\Gamma(p/2)} \sum_{j=0}^{\infty}\sum_{k=0}^{\infty} \alpha_{jk} \frac{\eta^{j+k}}{(1-\eta)^{2j+k}} x^{N_1} y^{N_1+N_2+1} \exp\left(-\frac{x+y}{2(1-\eta)}\right), \tag{7.31}$$

where $N_1 = \frac{m}{2}+j-1$, $N_2 = \frac{p}{2}+k-1$, and $\alpha_{jk} = \frac{2^{-(2N_1+N_2+3)}\Gamma(N_2+1)}{j!k!\Gamma(N_1+N_2+2)}$. The correlation coefficient is $\eta\sqrt{a/b}$.

As to the case where m and p are not necessarily integers, Smith et al (1982) replaced $a/2$ and $b/2$ by γ_1 and γ_2, respectively. They showed that the probability density function can be written as

$$h = \frac{x^{\gamma_1-1}y^{\gamma_2-1}\exp[(x+y)/(1-\eta)]}{(1-\eta)^{\gamma_1}\Gamma(\gamma_1)\Gamma(\gamma_2-\gamma_1)} \sum_{k=0}^{\infty} a_k I_{\gamma_2+k-1}\left(\frac{2\sqrt{\eta xy}}{1-\eta}\right), \tag{7.32}$$

where $a_k = \frac{(\eta y)^k \Gamma(\gamma_2-\gamma_1+k)(1-\eta)^{\gamma_2-1}}{k!(\eta xy)^{(\gamma_2+k-1)/2}}$, and η is a dependency parameter satisfying $0 < \eta < 1$ and $\eta = \rho(\gamma_2/\gamma_1)^{1/2}$, in which ρ is the correlation between X and Y. See Brewer et al (1987) and Smith et al (1982) for further details. (The expression for a_k given in those papers is incorrect.)

Brewer et al (1987) gave some results concerning the location of the mode of distributions (7.8) and (7.32). See Tubbs and Smith (1985) and Tubbs (1983a) for the distribution of the ratio X/Y.

7.7.12 Interrelationships

Interrelationships between the distributions of Kibble (1941), Jensen (1970), Gunst and Webster (1973), Smith et al (1982), Krishnaiah (1983), and Malik and Trudel (1983) are as follows.

Krishnaiah (a)	$\xrightarrow{(b)}$	Smith et al	$\xrightarrow{(c)}$	Kibble	$\xleftarrow{(c)}$	Malik and Trudel (d)
$\downarrow$		$\downarrow$				
Gunst and Webster Case 3 (a)	$\xrightarrow{(b)}$	Gunst and Webster Case 2	$\xleftarrow{(e)}$	Jensen (f)		

Notes:

- The arrows downwards mean that the α_i are restricted to be integers or half-integers.

(a) Parameter α, no greater than α_1 or α_2, is present.

(b) Parameter α is dropped.

(c) α_1 and α_2 are set equal.

(d) The marginals are not gamma distributions.

(e) The correlations are set equal.

(f) $\alpha_1 \leq \alpha_2$, the α_i being integers or half-integers. $\rho_1, \rho_2, \ldots \rho_{2\alpha_1}$ are non-zero but may be different.

7.8 Jensen's Bivariate F-Distribution

7.8.1 Support

The positive quadrant.

7.8.2 Formula for p.d.f.

$$h = \sum_{k=0}^{\infty} G_k(\boldsymbol{\delta}) \sum_{m=0}^{k} \sum_{n=0}^{k} d_{mn}(k) \frac{x^{(r+2m-2)/2} y^{(s+2n-2)/2}}{(1 + \frac{r}{\nu}x + \frac{s}{\nu}y)^{(r+s+2m+2n+\nu)/2}}, \tag{7.33}$$

in which $\nu, r \leq s$, are degrees of freedom of three chi-squared variates, the definition of $d_{mn}(k)$ is

$$d_{mn}(k) = (-1)^{m+n} \binom{k}{m} \binom{k}{n} \frac{r^{(r+2m)/2} s^{(s+2n)/2}\, \Gamma(\frac{r+s+2m+2n+\nu}{2})}{\nu^{(r+s+2m+2n)/2}\, \Gamma(\frac{r+2m}{2}) \Gamma(\frac{s+2n}{2}) \Gamma(\frac{\nu}{2})},$$

and $G_k(\boldsymbol{\delta})$ is as in (7.23).

Recall that in Section 7.7.6 we partitioned a normal random vector $\mathbf{Z}' = (\mathbf{Z}_1', \mathbf{Z}_2')$, where $\mathbf{Z}_1' = (Z_{11}, \ldots Z_{1r})$ and $\mathbf{Z}_2' = (Z_{21}, \ldots Z_{2r}, \ldots Z_{2s})$ such that the pairs $(Z_{11}, Z_{21}), \ldots (Z_{1r}, Z_{2r})$ are independent standardised bivariate normal vectors with correlations $\rho_1, \ldots \rho_r$, respectively. The summation in $G_k(\boldsymbol{\delta})$ is taken over all non-negative integer partitions of k.

7.8.3 Formula for (Cumulative) Distribution Function

$$H = \sum_{k=0}^{\infty} G_k(\boldsymbol{\delta}) \sum_{m=0}^{k} \sum_{n=0}^{k} d_{mn}(k) I_2(\boldsymbol{\alpha}, \boldsymbol{\beta}, c) \tag{7.34}$$

where $\boldsymbol{\alpha}' = (\alpha_1, \alpha_2) = (\frac{r}{2} + m, \frac{s}{2} + n)$, $\boldsymbol{\beta}' = (\beta_1, \beta_2) = (\frac{rx}{\nu}, \frac{sy}{\nu})$, $c = \nu/2$, and

$$I_2(\boldsymbol{\alpha}, \boldsymbol{\beta}, c) = \int_0^{\infty} \frac{z^{c-1} e^{-z}}{\Gamma(c)} \Gamma_{\beta_1 z}(\alpha_1) \Gamma_{\beta_2 z}(\alpha_2)\, dz$$

(Recall from footnote 2 of Chapter 2 that $\Gamma_x(\alpha)$ is the incomplete gamma function.) I_2 can be expressed as an infinite sum involving incomplete beta functions. See McAllister et al (1981) for details.

7.8.4 Univariate Properties

The marginals are F-distributions, that of X having r and ν degrees of freedom, that of Y having s and ν degrees of freedom.

7.8.5 Derivation

Suppose Q_1 and Q_2 jointly follow Jensen's bivariate gamma distribution, and have degrees of freedom r and s, respectively. Further, let V have a gamma distribution with ν degrees of freedom, independent of (Q_1, Q_2). Then $X = \frac{\nu Q_1}{rV}$ and $Y = \frac{\nu Q_2}{sV}$ follow Jensen's bivariate F-distribution (Jensen, 1970).

7.8.6 Relation to Other Distributions

When $r = s$ and Q_1 and Q_2 are independent (i.e., $\rho_1 = \rho_2 = \ldots \rho_r = 0$), this reduces to the bivariate F-distribution discussed in Section 6.3.

For $r = s$ and equal ρ's, it reduces to the bivariate version of the multivariate F-distribution given by Krishnaiah (1964, 1965). Briefly, let X_1 and X_2 have Kibble's distribution with degrees of freedom. If V, independent of X_1 and X_2, is a gamma variate with ν degrees of freedom, then $X = \frac{\nu X_1}{rV}$ and $Y = \frac{\nu X_2}{rV}$ have Krishnaiah's F-distribution.

7.8.7 Field of Application

We do not know of this distribution being fitted to data. It arises in analysis of variance and certain life testing experiments. In particular, it is the joint distribution of the F-statistics corresponding to the two factors for a two-way ANOVA without interaction (McAllister and Holland, 1980). McAllister (1983) applied the distribution to the example of Searle (1971, p. 262) concerning the effects on cooking speed of pan type and brand of stove, and determined the exact experiment-wise error rate (the probability of at least one incorrect rejection of the two null hypotheses associated the two marginals) for the data. For other applications, see Krishnaiah (1965) and Jensen (1970).

7.8.8 References to Tables or Algorithms

Much effort has been devoted to evaluate the joint cumulative probability integrals. Most of the algorithms involve sequences of incomplete beta functions. (These can be easily generated as they satisfy certain computationally-simple three-term recurrence relations.)

- For the case of $r = s$ and equal ρ's, see Krishnaiah and Armitage (1965), Amos and Bulgren (1972), and Schuurmann et al (1975). The last of these contains a table of percentage points.

- For the general case, see McAllister et al (1981) and McAllister (1983). One portion of the algorithm of McAllister et al (1981) involves the calculation of coefficients $G_k(\delta)$ that require summing over all nonnegative integer partitions of $0, 1, 2, \ldots N$ of size r. R E Johnson (1986) presents an alternative method, using successive convolution, for generating these coefficients, and this significantly reduces computation time. McAllister (1983) presents a detailed description of a system of subroutines for efficient computation of the probability integral of this distribution.

7.9 The Bivariate Inverse Gaussian Distribution

7.9.1 Support

The positive quadrant.

7.9.2 Formula for p.d.f.

$$\begin{aligned} h \;=\; & \tfrac{1}{4\pi}\sqrt{\tfrac{\lambda_1\lambda_2}{x^3y^3(1-\nu^2)}}\Big\{\exp\Big[\tfrac{-1}{2(1-\rho^2)}\Big(\tfrac{\lambda_1}{\mu_1^2}\tfrac{(x-\mu_1)^2}{x} - \tfrac{2\nu}{\mu_1\mu_2}\sqrt{\tfrac{\lambda_1\lambda_2}{xy}}(x-\mu_1)(y-\mu_2) + \tfrac{\lambda_2}{\mu_2^2}\tfrac{(y-\mu_2)^2}{y}\Big)\Big] + \\ & \exp\Big[\tfrac{-1}{2(1-\nu^2)}\Big(\tfrac{\lambda_1}{\mu_1^2}\tfrac{(x-\mu_1)^2}{x} + \tfrac{2\nu}{\mu_1\mu_2}\sqrt{\tfrac{\lambda_1\lambda_2}{xy}}(x-\mu_1)(y-\mu_2) + \tfrac{\lambda_2}{\mu_2^2}\tfrac{(y-\mu_2)^2}{y}\Big)\Big]\Big\}, \end{aligned} \tag{7.35}$$

where the μ's and the λ's are positive, and $-1 < \nu < 1$.

7.9.3 Univariate Properties

Inverse Gaussian (Wald) marginals, with means μ_i and variances μ_i^3/λ_i.

7.9.4 Correlation Coefficients

This distribution is unusual in that Pearson's product-moment correlation coefficient is always zero. But X and Y are independent if and only if the parameter ν is zero. ν is the correlation coefficient of the standardised bivariate normal distribution.

7.9.5 Conditional Properties

The regression is constant, $E(Y|X=x) = \mu_2$.

The conditional variance is not constant, and takes large values at extreme values of x:

$$\operatorname{var}(Y|X=x) = \frac{\mu_2^3}{\lambda_2}\left[(1-\nu^2) + \frac{\nu^2\lambda_1(x-\mu_1)^2}{\mu_1^2 x}\right]. \tag{7.36}$$

7.9.6 Derivation

Let (Z_1, Z_2) be have the standardised bivariate normal distribution with correlation ν. Define $U = Z_1^2$ and $V = Z_2^2$. (Clearly, U and V have the chi-squared distribution with one degree of freedom.) Then (U, V) has Kibble's bivariate gamma distribution (see Section 7.5) with $\alpha = \frac{1}{2}$. Consider the two-to-one transformations $U = (X-\mu_1)^2/(\mu_1^2 X)$ and $V = (Y-\mu_2)^2/(\mu_2^2 Y)$; then, on solving for X and Y, their joint density is (7.35).

7.9.7 References to Illustrations

Surfaces and contour plots of the p.d.f.: Kocherlakota (1986).

7.9.8 Remarks

- For further results, including joint moments and an infinite series representation of the distribution function, see Kocherlakota (1986).
- Wasan and Buckholtz (1973) derived a partial differential equation which, when solved under suitable boundary conditions, leads to a density of a bivariate inverse Gaussian process; they gave two examples, one with independent and one with dependent variables. For the latter,

$$h = \frac{s(t-s)}{2\pi\sqrt{(x-y)^3y^3}}\exp\left\{-\frac{[x-y-(t-s)]^2}{2(x-y)} - \frac{(y-s)^2}{2y}\right\} \tag{7.37}$$

for $x > y > 0$, with $t > s > 0$. See also Wasan (1968). The above (7.37) is the joint density of (X, Y), where $X = Z + Y$ and Z and Y are independent inverse Gaussian variates.

- Another bivariate inverse Gaussian distribution is described by Al-Hussaini and Abd-el-Hakim (1981). For this, the support is naturally the positive quadrant again, and the p.d.f. is

$$h = f(x)g(y)[1 + \rho\Lambda(x, y)], \tag{7.38}$$

where the parameter ρ equals the product-moment correlation coefficient, f and g are univariate inverse Gaussian densities with parameters (λ_1, μ_1) and (λ_2, μ_2), respectively, and

$$\Lambda(x, y) = 8\sqrt{\frac{\lambda_1\lambda_2}{\mu_1^3\mu_2^3}}(x - \mu_1)(y - \mu_2)\exp\left\{-\left[\frac{\lambda_1(x-\mu_1)^2}{2\mu_1^2 x} + \frac{\lambda_2(y-\mu_2)^2}{2\mu_2^2 y}\right]\right\}. \tag{7.39}$$

For further details, see the paper cited. Kocherlakota (1986) argues that (7.35) is a more natural extension of the univariate distribution than this.

- An abstract by Banerjee (1977) mentions a bivariate inverse Gaussian distribution for which the regression takes the form $E(Y|X = x) = \alpha - \beta/x$.

- Iyengar and Patwardhan (1988) have recently reviewed the inverse Gaussian distribution; one section of their paper is on the bivariate case; they regard the proposal of Al-Hussaini and Abd-el-Hakim as rather artificial, and do not mention that of Kocherlakota. See also Iyengar (1985).

7.10 The Bivariate Hyperbolic Distribution

7.10.1 Support

The whole plane.

7.10.2 Formula for p.d.f.

It is convenient to write this in multidimensional form,

$$h = \frac{\kappa^3 e^{\delta\kappa}}{2\pi\alpha(1+\delta\kappa)}\exp\{-\alpha[\delta^2 + (\mathbf{x}-\boldsymbol{\mu})'\boldsymbol{\Delta}^{-1}(\mathbf{x}-\boldsymbol{\mu})]^{1/2} + \boldsymbol{\beta}'(\mathbf{x}-\boldsymbol{\mu})\}, \tag{7.40}$$

where δ is a scalar parameter, $\boldsymbol{\mu}$ and $\boldsymbol{\beta}$ are vector parameters, $\boldsymbol{\Delta}$ is a symmetric, positive definite, matrix parameter such that $|\boldsymbol{\Delta}| = 1$, and $\kappa = \alpha^2 - \boldsymbol{\beta}'\boldsymbol{\Delta}\boldsymbol{\beta}$. In the bivariate case,

$$h = \frac{\kappa^3 e^{\delta\kappa}}{2\pi\alpha(1+\delta\kappa)}\exp\{-\alpha[\delta^2 + (x-\mu_1)^2\delta_{22} - 2(x-\mu_1)(y-\mu_2)\delta_{12} + (y-\mu_2)^2\delta_{11}]^{1/2} + \beta_1(x-\mu_1) + \beta_2(y-\mu_2)\}, \tag{7.41}$$

where $\boldsymbol{\Delta} = \begin{pmatrix} \delta_{11} & \delta_{12} \\ \delta_{21} & \delta_{22} \end{pmatrix}$, and $\boldsymbol{\beta}' = (\beta_1, \beta_2)$.

The graph of the log-density is a hyperboloid, hence the name of this distribution.

7.10.3 Univariate Properties

The marginals are hyperbolic distributions; the p.d.f.'s take the same form as (7.40), except that the vectors and the matrix become scalars. See Section 2.14.3.

The mean and variance of the distribution are $\boldsymbol{\mu} + b_{\kappa\delta}\boldsymbol{\Delta\beta}$ and $c_{\kappa\delta}(\boldsymbol{\Delta\beta})'(\boldsymbol{\Delta\beta}) + b_{\kappa\delta}\boldsymbol{\Delta}$, respectively, where $b_{\kappa\delta} = (\delta^2\kappa^2 + 3\delta\kappa + 3)\kappa^{-2}(1+\delta\kappa)^{-1}$ and $c_{\kappa\delta} = (\delta^3\kappa^3 + 6\delta^2\kappa^2 + 12\delta\kappa + 6)\kappa^{-4}(1+\delta\kappa)^{-2}$.

7.10.4 Derivation

Suppose we have a bivariate normal distribution $N(\boldsymbol{\xi}, \boldsymbol{\Sigma})$ such that the mean $\boldsymbol{\xi}$ and the covariance matrix $\boldsymbol{\Sigma} = \sigma^2\boldsymbol{\Delta}$ are related by $\boldsymbol{\xi} = \boldsymbol{\mu} + \sigma^2\boldsymbol{\beta}$. Here $\boldsymbol{\Delta}$ is a symmetric, positive definite, matrix parameter such that the determinant $|\Delta| = 1$. Suppose σ^2 has the generalised inverse Gaussian distribution with p.d.f. given by

$$\frac{(\kappa/\delta)^\lambda}{2K_\lambda(\delta\kappa)} x^{1/2} \exp[-(\delta^2 x^{-1} + \kappa^2 x)/2] \tag{7.42}$$

for $x > 0$, with $\lambda = 3/2$ and K_λ being the modified Bessel function of the third kind[7] with index λ. The bivariate hyperbolic distribution is obtained by compounding the bivariate normal with the generalised inverse Gaussian. In other words, the bivariate hyperbolic distribution is a compound distribution, with the bivariate normal being the compounded and the generalised inverse Gaussian being the compounding distribution.

7.10.5 References to Illustrations

Contours of probability density: Blæsild and Jensen (1981); Blæsild (1981).

7.10.6 Remarks

- The key references for this distribution are Barndorff-Nielsen (1977, 1978) and Blæsild (1981). Another account is at pp. 704–705 of Barndorff-Nielsen and Blæsild (1983a).
- Barndorff-Nielsen proposed using this distribution to represent a two-dimensional Brownian motion with drift $\boldsymbol{\beta}$, started at $\boldsymbol{\mu}$, and observed at a random time σ^2.
- The contours of probability density are ellipses. But they are not concentric like those of the distributions of Section 7.2.
- X and Y cannot be independent (Blæsild, 1981, p. 254). But if δ_{12} and one of the β's are zero, then X and Y are uncorrelated.
- The bivariate hyperbolic distribution, a normal variance-mean mixture, has longer tails than the bivariate normal. (Recall from Section 7.2.12 that a compound bivariate normal, which is a normal variance-mixture with $\boldsymbol{\beta} = \mathbf{0}$, also has longer tails.)

7.10.7 Fields of Application

Blæsild (1981) fitted this distribution (with one axis log-transformed) to Johannsen's data on length and width of beans (see 24.7.2—1). There are some further remarks in Blæsild and Jensen (1981).

The distribution in three dimensions arises theoretically in statistical physics (see Blæsild and Jensen, 1981).

7.11 Bivariate Symmetric Stable Distributions

7.11.1 Explanations

A univariate d.f. F is "stable" if, for every c_1, c_2, and positive b_1, b_2, there exist c and (positive) b such that:

$$F\left(\frac{x-c_1}{b_1}\right) * F\left(\frac{x-c_2}{b_2}\right) = F\left(\frac{x-c}{b}\right), \tag{7.43}$$

where $*$ denotes convolution.[8]

By analogy with the univariate case, a bivariate distribution H is said to be stable if, for every $b_1 > 0$, $b_2 > 0$, and real $\mathbf{c}_1$, $\mathbf{c}_2$, there exist $b > 0$ and real $\mathbf{c}$ such that:

$$H\left(\frac{\mathbf{x}-\mathbf{c}_1}{b_1}\right) * H\left(\frac{\mathbf{x}-\mathbf{c}_2}{b_2}\right) = H\left(\frac{\mathbf{x}-\mathbf{c}}{b}\right). \tag{7.44}$$

$\mathbf{X}$ is said to be symmetric about $\mathbf{a}$ if $\mathbf{X}-\mathbf{a}$ and $-(\mathbf{X}-\mathbf{a})$ have the same distribution.

[7] It is related to the I_λ of Section 7.5 by $K_\lambda(z) = \frac{\pi}{2}\frac{I_{-\lambda}(z)-I_\lambda(z)}{\sin\lambda\pi}$.
[8] $F_1 * F_2$ means $\int_{-\infty}^{\infty} F_1(x-t)f_2(t)\,dt$ (and is the same as $F_2 * F_1$, i.e., $\int_{-\infty}^{\infty} F_2(x-t)f_1(t)\,dt$).

7.11.2 Characteristic Function

It has been shown by Press (1972a; 1972b, Chapter 6) that a bivariate stable distribution, symmetric about **a**, of order m, has characteristic function φ such that

$$\log \varphi(\mathbf{t}) = i\mathbf{a}'\mathbf{t} - \frac{1}{2}\sum_{j=1}^{m}(\mathbf{t}'\boldsymbol{\Omega}_j\mathbf{t})^{\alpha/2} \tag{7.45}$$

($0 < \alpha \leq 2$), where $\boldsymbol{\Omega}_j$ is a positive semi-definite[9] matrix. It is assumed that no two of the $\boldsymbol{\Omega}_j$'s are proportional and that $\boldsymbol{\Sigma} = \sum_{j=1}^{m}\boldsymbol{\Omega}_j$ is positive definite. α is called the characteristic exponent.

If $m = 1$, the above distribution is an elliptically-contoured bivariate distribution (Section 7.2). When $\alpha = 1$, this gives the log characteristic function of the bivariate Cauchy distribution discussed in Section 6.2.13. When $\alpha = 2$, it becomes that of the bivariate normal distribution even if $m \neq 1$.

7.11.3 Probability Densities

According to Galambos (1985b), the only multivariate stable densities known in closed form, apart from the multivariate normal, are certain Cauchy distributions.

7.11.4 Generalisation

de Silva (1978, 1979) has generalised (7.45) to a class of symmetric bivariate stable distributions such that

$$\log \varphi(t) = i\mathbf{a}'\mathbf{t} - \gamma(\mathbf{t}), \tag{7.46}$$

where

$$\gamma(\mathbf{t}) = \sum_{j=1}^{m}\left(\sum_{k=1}^{r}|c_{kj}s + d_{kj}t|^{\beta}\right)^{\alpha/\beta},$$

c_{kj} and d_{kj} being real constants, $\mathbf{t}' = (s, t)$, and $0 < \alpha < \beta \leq 2$. He showed that when $\beta = 2$, (7.46) is equivalent to (7.45).

Let us now consider the special case $\beta = 1$, $m = 1$, $r = 3$. Suppose $c_{11} = \lambda_1 > 0$, $d_{11} = 0$; $c_{21} = d_{21} = \lambda_2 > 0$; and $c_{31} = 0$, $d_{31} = \lambda_3 > 0$; then the characteristic function φ (assuming $\mathbf{a} = \mathbf{0}$) is

$$\varphi(s,t) = \exp(-\lambda_1^{\alpha}|s|^{\alpha} - \lambda_2^{\alpha}|s+t|^{\alpha} - \lambda_3^{\alpha}|t|^{\alpha}). \tag{7.47}$$

This particular class of bivariate symmetric stable distributions is considered by de Silva and Griffiths (1980). They showed that these can be obtained by the trivariate reduction method (see Section 15.3.4).

7.11.5 Association Parameter

In bivariate stable distributions with $\alpha < 2$, all second order moments are infinite, and hence the usual Pearson product-moment correlation is undefined. We will see that the usual definition of the correlation coefficient can be extended as below.

7.11.5.1 Press (1972a)

Press (1972a) defined the association parameter ρ for the distribution in (7.45) as follows. Denote element ij of $\boldsymbol{\Omega}_k$ by $\omega_{ij}(k)$ (where $i, j = 1, 2$ and $k = 1, 2, \ldots m$). Then

$$\rho = \frac{\sum_{k=1}^{m}\omega_{12}(k)}{\left[\sum_{k=1}^{m}\omega_{11}(k)\sum_{k=1}^{m}\omega_{22}(k)\right]^{1/2}}. \tag{7.48}$$

When $\alpha = 2$, then $\boldsymbol{\Sigma} = \sum_{j=1}^{m}\boldsymbol{\Omega}_j$ is the covariance matrix of the bivariate normal distribution, and the association parameter ρ becomes the ordinary correlation coefficient. Press showed that ρ defined above satisfies $-1 \leq \rho \leq 1$, and that if X and Y are independent, then $\rho = 0$.

[9] A $n \times n$ symmetric matrix $\mathbf{A}$ is positive semi-definite if $\mathbf{x}'\mathbf{A}\mathbf{x} \geq 0$ for every non-zero $\mathbf{x}$ in R^n. (Compare footnote 1.)

7.11.5.2 de Silva (1978)

Consider a subclass of de Silva's stable distributions (7.46) such that

$$\log \varphi(s,t) = -\sum_{j=1}^{m}(a_j|s|^{\beta} + b_j|t|^{\beta} + c_j|s+\delta t|^{\beta})^{\alpha/\beta}, \tag{7.49}$$

where a_j, b_j, and c_j are non-negative constants, $\delta = \pm 1$, and $0 < \alpha < \beta \leq 2$. de Silva (1978) defined an association parameter ρ^* in the following manner.

Let $U = \frac{e^{isX} - \varphi_X(s)}{[1+|\varphi_X(s)|^2]^{1/2}}$ and $V = \frac{e^{itY} - \varphi_Y(t)}{[1+|\varphi_Y(t)|^2]^{1/2}}$, where φ_X and φ_Y are the characteristic functions of X and Y. Define

$$\rho^* = \limsup_{s,t\to 0} |E(\bar{U}V)|, \tag{7.50}$$

where $\bar{U}$ denotes the complex conjugate of U, and

$$E(\bar{U}V) = \frac{\varphi(-s,t) - \varphi_X(s)\varphi_Y(t)}{[1-|\varphi_X(s)|^2]^{1/2}[1-|\varphi_Y(t)|^2]^{1/2}}. \tag{7.51}$$

de Silva (1978) has shown that $\rho^* = \sum_{j=1}^{m} B_j \big/ 2D$, where $D^2 = [\sum_{j=1}^{m}(a_j+c_j)^{\alpha/\beta}][\sum_{j=1}^{m}(b_j+c_j)^{\alpha/\beta}]$ and $B_j = (a_j+c_j)^{\alpha/\beta} + (b_j+c_j)^{\alpha/\beta} - (a_j+b_j)^{\alpha/\beta}$.

ρ^* reduces to the ordinary correlation coefficient when this class of distributions reduces to the bivariate normal.

Griffiths (1972) (see also de Silva, 1978) proved that X and Y are independent if and only if $\rho^* = 0$.

For a test of independence for the distribution (7.49), based on the empirical characteristic function, see de Silva and Griffiths (1980).

7.11.6 Remarks

The following points are true of all the bivariate symmetric stable distributions discussed in this Section.

- The marginals are symmetric stable distributions with characteristic exponent α. (But a case of a stable distribution with a vector index—namely, $(\frac{1}{2}, \frac{1}{4})$—arises as a special case of the inverse Gaussian/conditional inverse Gaussian distribution of Section 8.10 (Barndorff-Nielsen, 1983).)

- Every linear combination of X and Y (i.e., $aX + bY$) is symmetric stable (Press, 1972a; de Silva, 1978).

- The distribution is infinitely divisible.

A characterisation has been given by Moothathu (1985).

For another account of this class of distributions, see Galambos (1985b).

7.11.7 Application

Investment economists are concerned with optimal selection of portfolios of securities. A tradition has grown up in the statistical side of this work of using symmetric stable (univariate) distributions, partly because of empirical evidence and partly because of the theoretical properties of these distributions. Press (1972a; 1972b, Chapter 12) considered an investment portfolio containing two assets whose price changes, X and Y, follow the bivariate symmetric stable distribution given in (7.45). Suppose the vector of proportions of resources allocated to the two variable-price assets is $\mathbf{c}' = (c_1, c_2)$, so that the return on this allocation of resources is $Q = c_1X + c_2Y$. From Section 7.11.6, it is clear that Q also is symmetric stable. Taking $E(Q) = \mathbf{c}'\mathbf{a}$, the log characteristic function for return is given by

$$\log \varphi_Q(t) = it(\mathbf{c}'\mathbf{a}) - \frac{1}{2}|t|^{\alpha}\sum_{j=1}^{n}(\mathbf{c}'\mathbf{\Omega}_j\mathbf{c})^{\alpha/2}. \tag{7.52}$$

The "risk" associated with the allocation is taken to be $\frac{1}{2}\sum_{j=1}^{m}(\mathbf{c}'\boldsymbol{\Omega}_j\mathbf{c})$. Define the set of "efficient" portfolios as those for which it is not possible to achieve greater expected return without increasing risk. Press showed that this set can be obtained as the solution to a programming problem with objective function

$$\lambda(\mathbf{c}'\mathbf{a}) - \frac{1}{2}\sum_{1}^{m}(\mathbf{c}'\boldsymbol{\Omega}_j\mathbf{c})^{\alpha/2} \tag{7.53}$$

$(0 < \lambda < \infty)$. That is, for some fixed, preassigned, λ, maximise (7.53) with respect to $\mathbf{c}$, subject to $c_i \geq 0$ and $c_1 + c_2 = 1$. Attitude towards avoiding risk-taking is what determines λ.

See also Rao (1983).

Chapter 8

Bivariate Distributions Constructed by the Conditional Approach

8.1 Introduction

8.1.1 Contents of this Chapter

In Section 3.5, we outlined the construction of a bivariate p.d.f. as the product of a marginal p.d.f. and a conditional p.d.f., $h(x, y) = f(x)g(y|x)$. This is easily understood, and has been a popular approach in the practical literature, especially when Y can be thought of as caused by, or predictable from, X.

We begin this Chapter by considering distributions such that both sets of conditionals are exponential, Pareto, beta, or normal, in Sections 8.2–8.5. Then Sections 8.6–8.11 give brief accounts of particular distributions constructed by the conditional method: McKay's bivariate gamma distribution and variants of it, Dubey's distribution, Blumen and Ypelaar's distribution, exponential dispersion models, four densities given by Barndorff-Nielsen and Blæsild, and continuous bivariate densities with a discontinuous marginal density. Finally, several applications-oriented papers are summarised in Section 8.12; meteorology and hydrology provide several of the examples.

8.1.2 Approximating d.f.'s Using the Conditional Approach

Parrish and Bargmann (1981) have given a general method for evaluating bivariate d.f.'s, which "utilises a factorisation of the joint density function into the product of a marginal density function and an associated conditional density, permitting the expression of the double integral in a form amenable to the use of specialised Gaussian-type quadrature techniques for numerical evaluation of cumulative probabilities." See also Parrish (1981).

8.1.3 Compatibility of Conditional Densities

As a preliminary to Sections 8.2–8.5, we note the following theorem, due to Abrahams and Thomas (1984). Let $f(x|y)$ and $g(y|x)$ be given conditional density functions. If there exists some bivariate density $h(x, y)$ having these conditional densities, they are said to be "compatible". A necessary and sufficient condition for compatibility is that the ratio $f(x|y)/g(y|x)$ be of the form $a(x)/b(y)$, where $a(.)$ and $b(.)$ are non-negative integrable functions such that $\int a(x)\,dx = \int b(y)\,dy$.

Patil (1965) also reports on the determination of the joint distribution by the two sets of conditional distributions.

8.1.4 The Marginal and Conditional Distributions of the Same Variate

A paper that struck us as slightly unusual was that of Seshadri and Patil (1963), which gave some theoretical results on the extent to which knowledge of the marginal distribution of one variate together with knowledge of the conditional distributions of the same variate serve to determine the bivariate distribution.

8.2 Exponential Conditionals

8.2.1 Support

The positive quadrant.

8.2.2 Formula for p.d.f.

$$h = Ke^{-(x+y+cxy)} \tag{8.1}$$

($c \geq 0$), where the constant K is $\frac{c\exp(-1/c)}{-\text{Ei}(1/c)}$, in which Ei(.) is the exponential integral function, defined as $\text{Ei}(u) = -\int_u^\infty v^{-1}e^{-v}\,dv$. (Beware of the lack of standardisation of nomenclature and notation for such functions as this. For computation, see Amos, 1980.)

8.2.3 Formula for (Cumulative) Distribution Function

The upper right volume under the probability density surface may be written in a compact, though not elementary, form:

$$\bar{H} = \frac{\text{Ei}(c^{-1} + x + y + cxy)}{\text{Ei}(c^{-1})}. \tag{8.2}$$

8.2.4 Univariate Properties

The marginals are not exponential in form. Instead,

$$f = K\frac{\exp(-x)}{1 + cx} \tag{8.3}$$

(where K is as in (8.1)), and similarly for $g(y)$.

8.2.5 Correlation Coefficients

Pearson's product-moment correlation is $\frac{c+K-K^2}{K(1+c-K)}$, where K is the same function of c as before. This is zero when $c = 0$, the case of independence, becomes increasingly negative with increasing c until it reaches approx. -0.32 at about $c = 6$, and then gets less negative, tending slowly to zero as $c \to \infty$.

8.2.6 Relation to Other Distributions

For a more general family, see Arnold and Strauss (1987). And for conditions on the sign of correlation obtainable with such generalisations, see Arnold (1987b).

8.2.7 Remarks

- This is the distribution such that both sets of conditional densities ($g(y|X = x)$ and $f(x|Y = y)$) are exponential—Abrahams and Thomas (1984), and (independently) Arnold and Strauss (1988). Consequently, it is easy to write down the regression equation—Inaba and Shirahata (1986).
- With K being as before, the joint moment generating function is:

$$M(s,t) = \frac{K(c)}{(1-s)(1-t)K\left(\frac{c}{(1-s)(1-t)}\right)}. \tag{8.4}$$

- The bivariate failure rate[1] $h/\bar{H}$ is increasing in both x and y, being:

$$(1+cx)(1+cy)K\left(\frac{c}{(1+cx)(1+cy)}\right). \tag{8.5}$$

- We understand Castillo and Galambos (1987b) have considered the case of Weibull conditionals.

[1] See Section 18.3.2 for other definitions than this.

8.2.8 Fields of Application

As often with the exponential distribution, applications in reliability studies are envisaged. Inaba and Shirahata (1986) fitted this distribution to data on white blood counts and survival times of patients who died of acute myelogenous leukemia (Gross and Clark, 1975, Table 3.3), comparing it with the bivariate normal distribution.

8.3 Pareto Conditionals

8.3.1 Support

The positive quadrant.

8.3.2 Formula for p.d.f.

$$h = K(1 + x + y + \lambda xy)^{-(c+2)}, \tag{8.6}$$

where $\lambda \geq 0$ if $c > 0$, or $\lambda > 0$ if $-1 < c \leq 0$, and the constant K is expressible in terms of ${}_2F_1$, Gauss' hypergeometric function (defined in footnote 3 of Chapter 13).

8.3.3 Univariate Properties

The marginals are not Pareto in form. Instead,

$$f = \frac{K}{c+1}(1 + \lambda x)^{-1}(1 + x)^{-(c+1)} \tag{8.7}$$

(where K is as in (8.6)), and similarly for $g(y)$.

8.3.4 Remarks

This is the distribution such that both sets of conditional densities (of Y given that $X = x$, and of X given that $Y = y$) are Pareto. It is due to Arnold (1987a).

8.4 Beta Conditionals

8.4.1 Support

That part of the unit square such that $x + y \leq 1$.

8.4.2 Formula for p.d.f.

$$h \propto x^{\theta_1 - 1} y^{\theta_2 - 1} (1 - x - y)^{\theta_3 - 1} e^{\eta \log x \log y} \tag{8.8}$$

($\theta_1, \theta_2, \theta_3 > 0$, $\eta \leq 0$, except that if $\eta < 0$ and $\theta_3 > 1$, θ_1 and θ_2 can be zero).

8.4.3 Univariate Properties

The marginals are not beta in form (unless $\eta = 0$). The expression is rather complicated (see James, 1975). The beta (Dirichlet) distribution is characterised by being that member of this family such that at least one of the marginals is univariate beta.

8.4.4 Remarks

This is the distribution such that both sets of conditional densities (of Y given that $X = x$, and of X given that $Y = y$) are beta. It is due to James (1975). See also James (1981, pp. 133–134).

8.4.5 Another Distribution

The above distribution interprets the requirement for the conditional distributions to be beta as follows: beta distributions over the range 0 to $1-x$ (for Y) or $1-y$ (for X), but with the parameters being functions of x or y.

Instead, we might interpret it as: beta distributions with some particular constant exponent, but the range being a function of x or y. Abrahams and Thomas (1984) have shown in this case the p.d.f. must either be (6.12) or proportional to $(x+y)^{\theta_1-1}(1-x-y)^{\theta_3-1}$ (the support of which is that part of the unit square such that $x+y \leq 1$, and which has uniform marginals).

8.5 Normal Conditionals

If the conditional distributions are of constant variance: Brucker (1979) showed that if the conditional distribution of each variate, given the value of the other, is normal, with linear regression and constant variance, then (X, Y) has a bivariate normal distribution. Fraser and Streit (1980) showed that bivariate normality is implied by considerably weaker conditions than these.

The general case: is difficult (Abrahams and Thomas, 1984). See also Castillo and Galambos (1987a).

8.6 McKay's Bivariate Gamma Distribution

8.6.1 Support

$y > x > 0$, i.e., a wedge that is half of the positive quadrant.

8.6.2 Formula for p.d.f.

$$h = \frac{a^{p+q}}{\Gamma(p)\Gamma(q)} x^{p-1}(y-x)^{q-1}e^{-ay} \tag{8.9}$$

$(a, p, q > 0)$.

8.6.3 Formula for (Cumulative) Distribution Function

The p.d.f. may be expressed in terms of the transcendental function known as Fox's H-function. Hence, so may the d.f., and Kellogg and Barnes (1989, Section 4.6) have done this.

8.6.4 Univariate Properties

The distributions of X and Y are gamma, with shape parameters p and $p+q$, respectively.

8.6.5 Correlation Coefficients

Pearson's product-moment correlation is $\sqrt{p/(p+q)}$.

8.6.6 Conditional Properties

$Y - x$ conditional on $X = x$ has a gamma distribution with shape parameter q.

X/y conditional on $Y = y$ has a beta distribution with parameters p and q.

8.6.7 Methods of Derivation

- McKay (1934) derived this distribution as follows: let $(X_1, X_2, \ldots X_N)$ be a random sample from a normal population. Suppose s_N^2 is the sample variance, and let s_n^2 be the variance in a sub-sample of size n. Then s_N^2 and s_n^2 jointly have the McKay bivariate gamma distribution.

- As this is a member of Pearson's system of bivariate distributions, it may be derived by the differential equation method (see Section 3.9.1.4).
- By the conditional approach, as a special case of the beta-Stacy distribution considered by Mihram and Hultquist (1967), to be discussed in Section 8.12.1 below.

8.6.8 References to Illustrations

Plots of the probability density surface (three cases, $a = 2.0$, $p = q = 0.5$; $a = p = q = 0.5$; $a = 1.0$, $p = 0.2$, $q = 0.8$): Kellogg and Barnes (1989).

8.6.9 Remarks

- This is also known as the bivariate Pearson type III distribution, though in Van Uven's designation (see Section 6.1), it is the type IVa.
- One of the examples that Parrish and Bargmann (1981) gave for illustration of their method for evaluating d.f.'s was this distribution.
- For an application, see Section 8.12.5.6 below.

8.6.10 Dussauchoy and Berland's Bivariate Gamma Distribution

This reduces to McKay's bivariate gamma distribution when $a_1 = a_2 = \beta = 1$.

The support is the wedge $y > \beta x > 0$, and within this the p.d.f. h is:

$$\frac{\beta a_2^{\ell_2}}{\Gamma(\ell_1)\Gamma(\ell_2-\ell_1)}(\beta x)^{\ell_1-1}\exp(-a_2 x)(y-\beta x)^{\ell_2-\ell_1-1}\exp\left[-\frac{a_2}{\beta}(y-\beta x)\right]{}_1F_1\left[\ell_1, \ell_2-\ell_1; \left(\frac{a_1}{\beta}-a_2\right)(y-\beta x)\right]$$

($\beta \geq 0$; $0 < a_2 \leq \frac{a_1}{\beta}$; $0 < \ell_1 < \ell_2$), where ${}_1F_1$ is the confluent hypergeometric function.

Remarks:

- The marginal distributions of X and Y are gamma, with shape parameters ℓ_1 and ℓ_2, respectively.
- Pearson's product-moment correlation is $\frac{\beta a_2}{a_1}\sqrt{\ell_1/\ell_2}$.
- For plots of the probability density surface (seven cases), see Berland and Dussauchoy (1973).
- The density has been written above in a form that makes clear the independence of X and $Y - \beta X$.
- For more details, see Dussauchoy and Berland (1972), Berland and Dussauchoy (1973), and (for the multivariate case) Dussauchoy and Berland (1975).
- Berland and Dussauchoy (1973) have applied this distribution to the joint distribution of the charge transported by a microdischarge (of electricity between two electrodes) and the interval of time between two of them.

8.6.11 Some Variants of this Distribution

We now summarise some variations on the theme of Y necessarily being positive, and X necessarily being between 0 and y.

Reference	Distribution of Y	Distribution of X, given $Y = y$
McKay (1934)	Gamma	Beta over the range 0 to y
Mihram and Hultquist (1967)	Stacy	Beta over the range 0 tp y
Block and Rao (1973)	Generalised inverted beta*	Beta over the range 0 to y
Ratnaparkhi (1981)†	Stacy, Pareto, or lognormal	Beta or log-gamma over the range 0 to y

* Density $\propto y^{a-1}(1+y^c)^{-k}$.

† In Ratnaparkhi's paper, the roles of X and Y were reversed from those here.

8.7 Dubey's Distribution

Dubey (1970) gave some properties of the distribution constructed by supposing (i) that Y has a gamma distribution, and (ii) conditional on $Y = y$, X also has a gamma distribution, with constant shape parameter and mean inversely proportional to y.

8.8 Blumen and Ypelaar's Distribution

8.8.1 Support

The unit square.

8.8.2 Formula for p.d.f.

$$h = x^\alpha y^{x^\alpha - 1} \tag{8.10}$$

8.8.3 Univariate Properties

X is uniformly distributed over the range 0 to 1; this is not true of Y, however.

8.8.4 Conditional Properties

Conditional on $X = x$, the cumulative distribution of Y is y^{x^α}.

8.8.5 Remark

It seems that the motive of Blumen and Ypelaar (1980) for constructing this distribution was to obtain one that is (i) tractable for studying the properties of Kendall's tau, and (ii) reasonably similar to the bivariate normal (after appropriate transformations of the marginals).

8.9 Exponential Dispersion Models

Jørgensen (1987) studied general properties of the class of exponential dispersion models, which is the multivariate generalisation of the error distribution of generalised linear models. The paper is outside our scope, but we note that its Section 5 concerns combining a conditional and a marginal distribution, both being exponential dispersion models, to obtain a higher-dimensional exponential dispersion model.

We may add that in the Discussion of the paper, Seshadri (1987) reports obtaining a bivariate exponential dispersion model with gamma marginals.

8.10 Four Densities Given by Barndorff-Nielsen and Blæsild

We note here that, in the course of studying reproductive exponential models, Barndorff-Nielsen and Blæsild (1983b) wrote out four examples of bivariate p.d.f.'s constructed by the conditional approach:

Distribution of X	Distribution of Y given $X = x$	Example no.
Exponential	Inverse Gaussian	1.1
Inverse Gaussian	Normal	4.1
Inverse Gaussian	Inverse Gaussian	4.2
Inverse Gaussian	Gamma	4.3

The inverse Gaussian/inverse Gaussian example is also considered at pp. 360–361 of Barndorff-Nielsen (1983), who remarks that a special case of it (with two of the four parameters being zero) can be said to be bivariate stable of index $(\frac{1}{2}, \frac{1}{4})$, as when a sample of size n is taken, the distribution of $(n^{-2}\sum x_i, n^{-4}\sum y_i)$ is the same whatever n is.

8.11 Continuous Bivariate Densities with a Discontinuous Marginal Density

The conditional approach was used by Romano and Siegel (1986, Section 2.15) to construct (for the fun of it!) a continuous bivariate density that has a discontinuous marginal density. They did this by letting X have an exponential distribution, and Y (conditional upon $X = x$) have a normal distribution with mean $1/x$ and constant variance. The p.d.f. is 0 for $x \leq 0$ and is proportional to $\exp\left[-x - \frac{1}{2}(y - x^{-1})^2\right]$ for $x > 0$. Romano and Siegel show that $h(x, y)$ is continuous everywhere in the plane, but the marginal density $f(x)$, with its jump at $x = 0$, is not continuous.

Also, Clarke (1975) says let a joint p.d.f. be proportional to $|x|\exp[-(|x|+x^2y^2/2)]$, which is continuous. The p.d.f. of X turns out to be $e^{-|x|}/2$ if $x \neq 0$, but is 0 if $x = 0$, so is not continuous. This example is also in Székely (1986, pp. 216–217). Clarke constructs an example for which $h(x, y)$ is continuous everywhere but $f(x)$ is nowhere continuous.

8.12 The Conditional Approach in Modelling

8.12.1 Beta-Stacy Distribution

Mihram and Hultquist (1967) discussed the idea of a warning-time variable, X, for Y = the failure time of a component on test, where $0 < X < Y$. A bivariate distribution was proposed, with Y having Stacy's generalised gamma distribution,[2] and X, conditional on $Y = y$, having a beta distribution over the range 0 to y. The p.d.f. is thus given by

$$h = \frac{|c|}{a^{bc}\Gamma(b)B(p,q)} x^{p-1}(y-x)^{q-1}y^{bc-p-q}\exp[-(y/a)^c] \tag{8.11}$$

if $0 < x < y$, and is 0 otherwise.

Pearson's product-moment correlation is

$$\sqrt{\frac{p^2\mathrm{var}(Y)}{(p+q)^2\mathrm{var}(X)}}, \tag{8.12}$$

where $\mathrm{var}(X)$ is related to the moments of Y by $\mathrm{var}(X) = \frac{p(p+1)E(Y^2)}{(p+q)(p+q+1)} - \frac{p^2[E(Y)]^2}{(p+q)^2}$, and the moments of Y are given by $E(Y^r) = a^r\Gamma[(bc+r)/p]/\Gamma(b)$ for $r/c > -b$ and are undefined otherwise.

The generation of random variates from this distribution is straightforward (Johnson et al, 1984, pp. 238–239).

Setting $c = 1$ and $bc = p + q$, we obtain McKay's bivariate gamma distribution, see Section 8.6.

8.12.2 Sample Skewness and Kurtosis

Shenton and Bowman (1977) considered the joint distribution of sample skewness and kurtosis statistics.[3] They said that it is well-known that in sampling from a normal population, the distributions of $\sqrt{b_1}$ and b_2 are individually satisfactorily approximated by Johnson's S_U distribution, but that little consideration had been given to the joint distribution. When they conducted extensive simulations of $(\sqrt{b_1}, b_2)$, they found that the distribution of $\sqrt{b_1}$ is unimodal for small b_2 but becomes bimodal for large b_2—provided

[2] See Section 2.6.5.

[3] If we have a sample $(x_1, x_2, \ldots x_n)$, whose mean is $\bar{x}$, calculate $m_j = \sum_{i=1}^n (x_i - \bar{x})^j/n$ for $j = 2, 3, 4$; then skewness is $\sqrt{b_1} = m_3/m_2^{3/2}$ and kurtosis is $b_2 = m_4/m_2^2$.

n is not too large (as $n \to \infty$, so $\sqrt{b_1}$ becomes unimodal, whatever b_2 might be). Their approach to the bivariate distribution was to use S_U for the marginal distribution of $\sqrt{b_1}$, and a conditional gamma density for b_2 given the value of $\sqrt{b_1}$. That is,

$$h(\sqrt{b_1}, b_2) = w(\sqrt{b_1})g(b_2|\sqrt{b_1}), \tag{8.13}$$

where w is the S_U p.d.f., and the gamma density g is written in terms of $b_2 - 1 - b_1$, since the constraint $b_2 \geq 1 + b_1$ applies to the relative values of b_2 and $\sqrt{b_1}$:

$$g(b_2|\sqrt{b_1}) = \frac{k}{\Gamma(\theta)}[k(b_2 - 1 - b_1)]^{\theta-1} \exp[-k(b_2 - 1 - b_1)], \tag{8.14}$$

in which θ is a quadratic in $\sqrt{b_1}$.

This work is also described in Section 7.7 of Bowman and Shenton (1986).

8.12.3 Business Risk Analysis

We summarise here an essay by Kottas and Lau (1978). The subject is risk analysis in business, by which is meant determining the stochastic characteristics of secondary variables like profit Z from (i) the stochastic characteristics of primary variables like sales volume Q, unit price P, unit variable cost V, and fixed cost F, and (ii) a functional relationship like $Z = Q(P - V) - F$. The starting point of Kottas and Lau is:

- Emphasis has traditionally been on estimating the individual stochastic characteristics of the primary variables, with their interdependencies being neglected.

- Even when some attempt has been made to model the dependencies, this has often been done in an arbitrary and unsatisfactory way, e.g., merely by specifying a correlation coefficient.

Kottas and Lau review the shortcomings of the product-moment correlation as a measurement of dependence, the specifics one is imposing on a model when making a simple and apparently harmless assumption like bivariate normality or log-normality, and the impracticability of obtaining subjective estimates of higher moments if a more general bivariate distribution is permitted.

The alternative which they suggest is what they call a "functional approach", and consists of trying to get the dependencies of $E(Y)$ and Variance(Y) on x correctly specified. In principle, this might be extended to higher conditional moments, but in practice the shape of the conditional distribution of Y is to be assumed independent of x, only the mean and the spread being allowed to change.

To a statistical audience, the points made by Kottas and Lau may seem uncontroversial and hardly worth saying, but it is a well-written paper, and brings home the necessity in model construction to always stay closely in touch with what is practicable.

8.12.4 Intercropping

This refers to growing two crops simultaneously on the same area of land, and harvesting and processing them separately. Mead et al (1986) say "Amid all the other justifications of the practice of intercropping, the benefit of 'stability' is a recurring theme. However, the concept of stability is variously and poorly defined, and the attempts to express stability in quantitative terms have been statistically unconvincing." Their paper is chiefly about refining the notion of stability, which they do in terms of the relative risks associated with intercropping and monocropping systems.

The data analysed by Mead et al consisted of the financial returns obtained by (i) intercropping sorghum with pigeonpea, and (ii) monocropping sorghum, at 51 site-year combinations in India (7 years, 11 areas, many combinations omitted). A scatterplot of the 51 points reveals:

- A strong correlation between the returns of the two cropping systems,

- A higher average return with intercropping, and

- Suggestions in the shape of the scatter that the relationship between the two returns is curvilinear and heteroscedastic.

Mead et al suggested it was appropriate to quantify the relative risks of the two cropping systems by plotting the risks of "failure" under each system against each other, as the definition of failure varies. That is, plot $\Pr(Y < t)$ against $\Pr(X < t)$ for various levels of t. For a dataset with 51 points, that can be done satisfactorily directly from the datapoints.

However, partly to understand their dataset better, and partly to provide an approach which would be more satisfactory for lesser amounts of data (given that it had been validated on large datasets), Mead et al went on to

- Fit a bivariate distribution to their scatter of points, and
- Calculate a smooth risk vs. risk curve from that distribution.

The approach chosen was (i) to fit a normal distribution to the sum of the returns ($S = X + Y$) from the two systems, and (ii) to assume the conditional distribution of the difference in returns between the two systems had a normal distribution also, both the mean of this and the logarithm of its variance having a quadratic dependence on S. Mead et al gave a contour plot of the resulting distribution.

This method of analysis was repeated on four further datasets, which had resulted from intercropping sorghum with various second crops.

8.12.5 Winds and Waves, Rain and Floods

8.12.5.1 Height and Period of Waves of the Sea

A good deal of empirical data has been published on the joint distribution of wave height and period—see item 8 in Section 24.12.2. Haver (1985) approached some data collected off northern Norway from the conditional point of view:

- The distribution of wave height H that was chosen was an unusual one, being log-normal for small H and Weibull for large H.
- Given H, the "spectral peak period" T was assumed to have a log-normal distribution.

Haver did not assume a functional form for the dependence on H of the parameters of the distribution of T; instead, for each of several ranges of values of H, the mean and variance of $\log T$ were estimated. In Haver's Figure 10, expressions are given for how these are related to H. However, since the expressions are quite messy, and the marginal distribution of H is also, an explicit formula for the joint distribution of H and T would be grotesquely cumbersome.

Another study of this type was by Burrows and Salih (1987). These authors took H to have a Weibull distribution, and the conditional distribution of T to be either Weibull or lognormal; it seems the Weibull distribution was used in shifted, i.e., three-parameter, form. They fitted these and other distributions to data from 18 sites around the British Isles.

For data from the North Sea, Krogstad (1985) takes H to have a Weibull distribution and the conditional distribution of T given H to be normal, with constant mean and variance inversely proportional to H.

Myrhaug and Kjeldsen (1984) analysed data from the North Sea with regard to the joint distribution of wave height and several other variables—crest front steepness, vertical asymmetry factor, and total wave steepness, as well as period. They took height to have a Weibull distribution and the conditional distributions of crest front steepness and period also to have Weibull distributions. The conditional distributions of vertical asymmetry factor and total wave steepness, in contrast, were taken to be lognormal.

8.12.5.2 Wind Speeds

It is of interest in the wind energy industry, so Kaminsky and Kirchhoff (1988) tell us, to estimate the energy available from a wind energy conversion system at one height from data collected at a lower height.

It is common practice to assume that wind speed has a Rayleigh distribution (see Sections 2.6.3 and 4.9). In modelling the joint distribution of wind speeds at two heights, Kaminsky and Kirchhoff therefore required the marginal distributions to have this form, at least roughly. They, in fact, considered two alternatives:

- X has a Rayleigh distribution, Y has a Rayleigh distribution with an origin at $Y = x$ and a constant scale factor. Thus $\Pr(Y < X) = 0$ for this model.
- X has a Rayleigh distribution, Y has a normal distribution with mean $a + bx$ and a constant variance.

Kaminsky and Kirchhoff presented an empirical contour plot of the bivariate distribution of wind speeds—at heights of 32 ft. and 447 ft. at a site in Waterford, Connecticut—along with contour plots of the two distributions that had been fitted to the data. The Rayleigh-normal distribution appeared to be a better fit than the Rayleigh-shifted Rayleigh (but it does have two more parameters).

The use of the symbolic algebra software package MACSYMA to obtain expressions for the marginal distributions of Y in the two cases was a further feature of interest in this study.

8.12.5.3 Wind Speed and Wave Height

Liu (1987) was concerned with the joint distribution of wind speed X and wave height Y on the Great Lakes of North America. He took $Y|X$ to have a gamma distribution; apparently, he envisaged the use of (i) empirically-obtained equations connecting the parameters of this to wind speed, together with (ii) a histogram of wind speeds, in order to calculate a joint distribution.

8.12.5.4 Storm Surge and Wave Height

In a study by Vrijling (1987) concerned with the Dutch dykes, a part was played by the joint distribution of the storm surge level of the sea and the significant wave height, it being supposed that:

- The storm surge level X has an extreme-value distribution, $F = \exp[-e^{-(x-a)/b}]$.
- Given X, wave height is normally distributed, with mean dependent on X and constant variance.

8.12.5.5 Floods

Correia (1987) has considered the duration and peak discharge of floods of a river. He supposed that:

- Flood duration is exponentially distributed.
- For a given duration, the peak discharge has a normal distribution, whose mean is linear in duration and whose variance is constant.

8.12.5.6 Streamflow and Rain

Clarke (1979, 1980) has used McKay's distribution (Section 8.6) with[4] X = annual streamflow, and Y = areal precipitation. The justification is apparently that:

- With McKay's distribution, X, Y, and $Y - X$ (= evaporation) all have gamma distributions, this being a popular univariate choice in hydrology.
- That $Y \geq X$ is reasonable on physical grounds (for watertight basins with little overyear storage).

The motivation for Clarke's work was that X is the variable of chief interest, but there are often only few years of data available for it, with the records of Y being more extensive.

8.12.5.7 Rain

- According to Etoh and Murota (1986), a rainstorm can be adequately described by three characteristics: duration X, maximum intensity Y, and total amount Z. Further, it can be assumed that $Z \propto XY/2$. Consequently, two random variables suffice. Etoh and Murota made the following assumptions:

[4] In Clarke's (1980) notation, the symbols X and Y were interchanged from what is given here.

- X has a gamma distribution.
- $Y = \eta X^a$, where η has a gamma distribution and a is constant (and $0 \leq a \leq 1$, reflecting a less-than-proportionate increase of maximum intensity with duration).

Etoh and Murota had some empirical data from Osaka, and also some results published by Córdova and Rodríguez-Iturbe (1985) for Denver (Colorado) and Boconó (Venezuela). They found that the shapes of the univariate distributions and the values of the correlation coefficients could be approximately reproduced by judicious selection of the parameters of their model.

- Sogawa et al (1986) were concerned with (i) the annual rainfall, and (ii) the annual maximum daily rainfall, each at four places in Nagano prefecture, Japan. In both cases they used a quadrivariate conditional maximum entropy distribution, the meaning of which we do not really understand. To those who read Japanese, perhaps the paper of Sogawa et al (1984) is helpful.

- The method adopted by Snyder and Thomas (1987) was not exactly that of conditional distributions, but here is as good a place as any to summarise it. The subject-matter was agriculture-related variates, such as monthly rainfall and monthly average temperature. After univariate transformations, Snyder and Thomas used "a form-free bivariate distribution based on two-dimensional sliding polynomials", which they found to be "necessary to model the bi-modal and heavy-tailed distributions frequently encountered".

Chapter 9

Bivariate Distributions Arising in Reliability Contexts

9.1 Introduction

When we talk of reliability, we have in mind the failure of an item of equipment or the death of a living organism. We especially think of the time elapsing between the equipment being put into service and its failure. In the bivariate or multivariate context, we are concerned with dependencies between two different failure times, such as those of two components of an electrical or mechanical or biological system.

Equipment lifetime is a necessarily positive variate, so the bivariate distributions considered in this Chapter have the positive quadrant as their support. They may be of relevance to other variates that are always positive, such as inter-event times.

Perhaps the best-known distributions for bivariate failure times are those of Marshall and Olkin, Freund, and Downton; see Sections 9.2, 9.3, and 9.12, respectively. But we will shortly see that there are a good many others available, also.

When thinking of the failure time of a single component, the exponential distribution is often the benchmark. As it is characterised by the lack of memory property—i.e., constant failure rate—it is often thought of as being the embodiment of true randomness. Much of this Chapter, therefore, will concentrate on bivariate distributions having exponential marginals. Block (1985) and Barnett (1985a) have provided clear and concise reviews of bivariate distributions with exponential marginals. Another review of this class, which concentrates on aspects of shock models, is by Basu (1988b). Gross and Miller (1981) listed a few, and remarked that "applications of non-normal distributions to biological and medical data are not as common as they should be." Although we will for several distributions of this type give expressions for the correlation coefficient, we point out that Barnett (1985a) has expressed the view that for highly skew bivariate distributions, the ordinary correlation coefficient is not a very useful measure of association. (See also our Section 4.14.5.) We note that the correlation cannot be less than -0.645 for distributions with exponential marginals (Moran, 1967).

The Chapter is arranged as follows:

9.2 Marshall and Olkin's bivariate exponential distribution.

9.3 Freund's bivariate distribution.

9.4 Hashino and Sugi's distribution.

9.5 ACBVE of Block and Basu.

9.6 Sarkar's distribution.

9.7 Comparison of four distributions.

9.8 The generalisation by Friday and Patil.

9.9 The bivariate distribution of Tosch and Holmes.

9.10 The method of Lawrance and Lewis.

9.11 Raftery's scheme.

9.12 Downton's bivariate exponential distribution.

9.13 Bivariate phase-type distributions.

9.14 Cowan's bivariate exponential distribution.

9.15 Other distributions with exponential marginals.

9.16 Bivariate Weibull distributions.

9.17 Bivariate extreme-value distributions.

9.18 Competing risks.

See Chapter 17 for more on models underlying bivariate distributions for failure-time data. See Chapter 18 for more on concepts of ageing, such as IFR (increasing failure rate) and NBU (new better than used).

9.2 Marshall and Olkin's Bivariate Exponential Distribution

9.2.1 Support

The positive quadrant.

9.2.2 Formula for (Cumulative) Distribution Function

The upper right volume under the probability density surface is (Marshall and Olkin, 1967a):

$$\bar{H} = \exp[-\lambda_1 x - \lambda_2 y - \lambda_{12} \max(x, y)], \tag{9.1}$$

where the λ's are positive.

9.2.3 Formula for p.d.f.

This takes slightly different forms according to which of x and y is the bigger:

$$h = \left\{ \begin{array}{ll} \lambda_2(\lambda_1 + \lambda_{12}) \exp[-(\lambda_1 + \lambda_{12})x - \lambda_2 y] & \text{for } x > y \\ \lambda_1(\lambda_2 + \lambda_{12}) \exp[-\lambda_1 x - (\lambda_2 + \lambda_{12})y] & \text{for } x < y \\ \text{Singularity along the diagonal } x = y & \end{array} \right\} \tag{9.2}$$

The amount of probability within the singularity is $\lambda_{12}/(\lambda_1 + \lambda_2 + \lambda_{12})$.

The singularity[1] in this case is due to the possibility of X exactly equalling Y. In a reliability context, this refers to the simultaneous failure of the two components.

9.2.4 Univariate Properties

Exponential marginals.

9.2.5 Correlation Coefficients

Pearson's product-moment correlation coefficient is $\lambda_{12}/(\lambda_1+\lambda_2+\lambda_{12})$. For the rank correlation coefficients, see Chapter 13.

[1] For a bivariate distribution, a singularity is a point with positive probability, or a line such that every segment has positive probability. (We are not concerned with more complicated forms of singularity.)

9.2.6 Derivation

Suppose there is a two-component system subject to shocks, which may knock out the first component, the second component, or both of them. If these shocks result from independent Poisson processes with parameters λ_1, λ_2, and λ_{12}, respectively, Marshall and Olkin's distribution results. Equivalently, $X = \min(Z_1, Z_3)$ and $Y = \min(Z_2, Z_3)$, where the Z's are independent exponential variates. (This is an example of trivariate reduction.)

9.2.7 Remarks

- This distribution was derived by Marshall and Olkin (1967a). It is sometimes denoted BVE.
- It arises if and only if: the marginals have exponential distributions, $\min(X,Y)$ has an exponential distribution, and $X - Y$ and $\min(X,Y)$ are independent. Other characterisations are given by Basu and Block (1975), Samanta (1975), Obretenov (1985), Azlarov and Volodin (1986, Chapter 9), and Roy and Mukherjee (1989).
- It has the lack of memory property, equation (9.9) below.
- $\bar{H}$ can be written as

$$\bar{H} = \frac{\lambda_1 + \lambda_2}{\lambda}\bar{H}_a + \frac{\lambda_{12}}{\lambda}\bar{H}_s, \tag{9.3}$$

where $\lambda = \lambda_1 + \lambda_2 + \lambda_{12}$, and H_s and H_a are the singular and absolutely continuous parts[2] of H,

$$\bar{H}_s = \exp[-\lambda \max(x,y)] \tag{9.4}$$

$$\bar{H}_a = \frac{\lambda}{\lambda_1 + \lambda_2}\exp[-\lambda_1 x - \lambda_2 y - \lambda_{12}\max(x,y)] - \frac{\lambda_{12}}{\lambda_1 + \lambda_2}\exp[-\lambda \max(x,y)]. \tag{9.5}$$

- In the "competing risks" context—for explanation of this, see Section 9.18—this distribution is fully identified, provided it is known which observations correspond to failure from both causes together, as well as which correspond to failure from each cause alone. This is because the distribution arises from three kinds of shock acting independently—and that leading to failure of types 1 and 2 together can simply be treated as a failure of type 3. (See David and Moeschberger, 1978, Section 4.4.)
- In collecting data to which this distribution is to be applied, it may happen that the nature of the second failure is indeterminate, i.e., it is not known whether the second shock would or would not have knocked out both components had both still been functioning. This leads to difficulties in estimating the λ's. See Shamseldin and Press (1984).
- Klein and Moeshberger (1988) made some calculations of the errors resulting from erroneously assuming component lifetimes have independent exponential distributions when in fact they have Marshall and Olkin's distribution. They did this for both series and parallel systems.
- According to Klein and Basu (1985), if interest centres on estimating $\bar{H}$, the matter is not as simple as merely substituting good estimates of the parameters into (9.1), as the result may be biased to an unacceptable degree; Klein and Basu study methods of bias reduction.
- This distribution and the associated shock model quickly received attention in the reliability literature (Harris, 1968). Papers since then include the following. A brief report on a two-component system with Marshall and Olkin distributions of both life and repair times is by Ramanarayanan and Subramanian (1981). Osaki (1980), Sugasaw and Kaji (1981), and Goel et al (1985) have given some results for a two-component system in which the failures follow this model but other distributions (i.e., of inspection, repair, and interinspection times) are arbitrary. Ebrahimi (1987) has given some

[2] That is, respectively referring to $Y = X$ (with finite probability) and to $Y \neq X$ (where the p.d.f. is finite). More formally, a bivariate distribution H is absolutely continuous if the joint density exists almost everywhere.

results for the case where a two-component system is tested at a number of different stress levels, s_j, and failures follow the Marshall and Olkin distribution with each λ proportional to s_j^p. Osaki et al (1989) have given some results for availability measures of systems in which two units are in series, failure of unit 1 shuts off unit 2, but not vice versa, the lifetimes follow Marshall and Olkin's distribution, the units have arbitrary repair-time distributions, and two alternative assumptions are made about the repair discipline.

- Another account of this distribution is by Marshall and Olkin (1985a).

9.2.8 Fields of Application

Widely claimed to be useful in reliability. Certainly the idea of simultaneous failure of two components is far from being merely an academic plaything; Hagen (1980) gives a review in the context of nuclear power, and points out that introducing redundancy into a system reduces random component failure to insignificance, leading to the common-mode/common-cause type being predominant among system failures.

Rai and Van Ryzin (1984) applied this distribution as a tolerance distribution in a quantal response context, to the occurrence of bladder and liver tumours in mice exposed to one of several alternative dosages of a carcinogen. Actually, the distribution was (i) used in the form with Weibull marginals, and (ii) mixed with a finite probability of tumours occurring even at zero dose.

9.2.9 Transformation to Uniform Marginals

Cuadras and Augé (1981) proposed the following distribution, whose support is the unit square:

$$H = \left\{ \begin{array}{ll} x^{1-c}y & \text{for } x \geq y \\ \\ xy^{1-c} & \text{for } x < y \end{array} \right\} \tag{9.6}$$

The corresponding p.d.f. is:

$$h = \left\{ \begin{array}{ll} (1-c)x^{-c} & \text{for } x > y \\ \\ (1-c)y^{-c} & \text{for } x < y \\ \\ \text{Singularity along the diagonal } x = y & \end{array} \right\} \tag{9.7}$$

Cuadras and Augé did not refer to Marshall and Olkin, so it is likely they were unaware that their distribution was a transformation of one already known. Conway (1981) gave an illustration of the Marshall and Olkin distribution after transformation to uniform marginals, so that illustration is of the Cuadras and Augé distribution.

9.2.10 Transformation to Weibull Marginals

As with other distributions having exponential marginals, this one is sometimes generalised by changing them to Weibull. See Marshall and Olkin (1967a), Moeschberger (1974), and L Lee (1979).

9.2.11 Transformation to Extreme-Value Marginals

As will become evident in Section 9.17.6.2, this distribution is sometimes met in the form with extreme-value marginals, as in some of the papers cited in Section 9.17.1.

9.2.12 Transformation of Marginals—The Approach of Muliere and Scarsini

First, consider the univariate case. Muliere and Scarsini (1987) write a general version of the lack of memory property as follows:

$$\bar{F}(s * t) = \bar{F}(s)\bar{F}(t), \tag{9.8}$$

where $*$ is any binary operation that is associative (i.e., such that $(x * y) * z = x * (y * z)$). Examples:

- $*$ being addition leads to the usual lack of memory characterisation of the exponential distribution: if $\bar{F}(s+t) = \bar{F}(s)\bar{F}(t)$, then $\bar{F}(x) = e^{-\lambda x}$.
- If $x * y = (x^\alpha + y^\alpha)^{1/\alpha}$, then the Weibull distribution $\bar{F}(x) = \exp(-\lambda x^\alpha)$ results.
- If $x * y = xy$, then the Pareto distribution $\bar{F}(x) = x^{-\lambda}$ results.

As to the bivariate case, consider first the following version of bivariate lack of memory:

$$\bar{H}(s_1 + t, s_2 + t) = \bar{H}(s_1, s_2)\bar{H}(t,t). \tag{9.9}$$

For more on this, see Section 17.2.2, but if we assume the marginals are exponential, the solution is the Marshall and Olkin distribution. Now consider

$$\bar{H}(s_1 * t, s_2 * t) = \bar{H}(s_1, s_2)\bar{H}(t,t), \tag{9.10}$$

together with (9.8) for each of the marginals. The solution is:

$$\bar{H}(s,t) = \exp\{-\lambda_1 a(s) - \lambda_2 a(t) - \lambda_{12} a[\max(s,t)]\}, \tag{9.11}$$

$a(.)$ being the (strictly increasing) function corresponding to the operation $*$, i.e., $a(x * y) = a(x) + a(y)$. Examples:

- $*$ being addition leads to the Marshall and Olkin distribution.
- If $x * y = (x^\alpha + y^\alpha)^{1/\alpha}$, the Weibull version of the Marshall and Olkin distribution results, $\bar{H} = \exp[-\lambda_1 x^\alpha - \lambda_2 y^\alpha - \lambda_{12}\max(x^\alpha, y^\alpha)]$.
- If $x * y = xy$, then the result is $\bar{H} = x^{-\lambda_1} y^{-\lambda_2}[\max(x,y)]^{-\lambda_{12}}$, the Pareto version of the Marshall and Olkin distribution. (For related work, see Sections 6.2.1 and 6.2.3 of Arnold, 1983.)

9.2.13 Generalisations

Johnson and Kotz (1972, p. 267) credit Saw (1969) with proposing replacing $\max(x,y)$ in (9.1) by some increasing function of $\max(x,y)$. One choice leads to

$$\bar{H} = [1 + \max(x,y)]^{\lambda_{12}} \exp[-\lambda_1 x - \lambda_2 y - \lambda_{12}\max(x,y)]. \tag{9.12}$$

Marshall and Olkin (1967b) consider generalisations of (9.1), including

$$\bar{H} = \exp\{-\lambda_1 x - \lambda_2 y - \lambda_{12}\max[x, y + \min(x,\delta)]\} \tag{9.13}$$

$(\delta \geq 0)$.

Ohi and Nishida (1979), following up an idea of Itoi et al (1976), we understand, considered the case where component i $(i = 1,2)$ needs k_i shocks before it fails. The bivariate life distribution that results is called a bivariate Erlang distribution (BVEr). Ohi and Nishida showed that:

- X and Y are positively regression dependent (see Sections 12.2.1 and 12.3.5 for this concept).
- BVEr is bivariate NBU, but not bivariate IFR (see Sections 18.3.2 and 18.3.4; we are using the third sense of these terms).

9.3 Freund's Bivariate Distribution

9.3.1 Support

The positive quadrant.

9.3.2 Formula for p.d.f.

$$h = \begin{cases} \alpha\beta' \exp[-\beta' y - (\alpha + \beta - \beta')x] & \text{for } 0 \le x < y \\ \alpha'\beta \exp[-\alpha' x - (\alpha + \beta - \alpha')y] & \text{for } 0 \le y \le x \end{cases} \tag{9.14}$$

where the parameters are all positive.

9.3.3 Formula for (Cumulative) Distribution Function

The upper right volume under the probability density surface is:

$$\bar{H} = \begin{cases} \frac{\alpha}{\alpha+\beta-\beta'} \exp[-(\alpha+\beta-\beta')x - \beta' y] + \frac{\beta-\beta'}{\alpha+\beta-\beta'} \exp[-(\alpha+\beta)y] & \text{for } x \le y \\ \frac{\beta}{\alpha+\beta-\alpha'} \exp[-(\alpha+\beta-\alpha')y - \alpha' x] + \frac{\alpha-\alpha'}{\alpha+\beta-\alpha'} \exp[-(\alpha+\beta)x] & \text{for } x \ge y \end{cases} \tag{9.15}$$

9.3.4 Univariate Properties

The marginals are not exponentials, but mixtures of exponentials. Hence (9.14) is often known as Freund's bivariate exponential extension. The formula for $f(x)$ is:

$$f = \frac{(\alpha - \alpha')(\alpha + \beta)}{\alpha + \beta - \alpha'} e^{-(\alpha+\beta)x} + \frac{\alpha'\beta}{\alpha + \beta - \alpha'} e^{-\alpha' x} \tag{9.16}$$

(provided $\alpha + \beta \neq \alpha'$), and, naturally, the expression for $g(y)$ is similar, with β and β' interchanging with α and α' respectively. The special case $\alpha + \beta = \alpha'$ gives $f = (\alpha'\beta x + \alpha)e^{-\alpha' x}$.

The mean and variance of this distribution are $\frac{\alpha'+\beta}{\alpha'(\alpha+\beta)}$ and $\frac{\alpha'^2+2\alpha\beta+\beta^2}{\alpha'^2(\alpha+\beta)^2}$, respectively.

9.3.5 Correlation Coefficients

Pearson's product-moment correlation is

$$\frac{\alpha'\beta' - \alpha\beta}{\sqrt{(\alpha'^2 + 2\alpha\beta + \beta^2)(\beta'^2 + 2\alpha\beta + \alpha^2)}} \tag{9.17}$$

and is restricted to the range $-\frac{1}{3}$ to 1.

9.3.6 Conditional Properties

The conditional densities are of elementary form but quite cumbersome.

9.3.7 Moment Generating Function

The m.g.f. is

$$M(s,t) = (\alpha + \beta - s - t)^{-1} \left[\frac{\alpha'\beta}{\alpha' - s} + \frac{\alpha\beta'}{\beta' - t} \right]. \tag{9.18}$$

9.3.8 Derivation

This distribution was originally derived by Freund (1961) from reliability considerations. Suppose a system has two components A and B whose lifetimes X and Y have exponential densities $\alpha e^{-\alpha x}$ and $\beta e^{-\beta y}$, respectively. Further suppose that the only dependence between X and Y arises from a failure of either component changing the parameter of the life distribution of the other component—when A fails, the parameter for Y becomes β', and when B fails, the parameter for X becomes α'. Then the joint density of X and Y is (9.14). A special case of Freund's bivariate distribution was also derived by Block and Basu (1974), see Section 9.5 below.

9.3.9 References to Illustrations

Conditional density plot: Johnson and Kotz (1972, p. 265).

9.3.10 Remarks

- For tests of symmetry and independence, see O'Neill (1985).
- Parallel systems with repair: there is some interest in the reliability literature in the probability of system failure when two components are in parallel and repair or replacement of a failed component takes a finite time. In this situation, the probability that the working component fails before the failed one is repaired is of importance. Biswas and Nair (1984) have considered this situation when Freund's distribution is applicable. See also Adachi and Kodama (1980) and Goel et al (1984).
- For parallel systems, Klein and Moeshberger (1986) made some calculations of the errors resulting from erroneously assuming component lifetimes have independent exponential distributions when in fact they have Freund's distribution.
- The study by Klein and Basu (1985), referred to in 9.2.7 above, also encompassed bias reduction techniques for $\bar{H}$ when (9.15) holds.
- Besides variants and generalisations of this distribution that are described in Sections 9.3.12–9.3.15 and 9.4 below, we note a complicated generalisation given by Holla and Bhattacharya (1965), in which replacement of failed components is involved.

9.3.11 Fields of Application

This distribution is said to be useful as a reliability model. It was applied to the Barlow and Proschan (1977) data on failures of caterpillar tractors (see Section 24.2—1) by O'Neill (1985).

9.3.12 Transformation of the Marginals

The power-transformed version of Freund's distribution has been considered by Spurrier and Weier (1981), concentrating on the performance of maximum likelihood estimates (which are not in closed form).

Hashino and Sugi's (1984) extension of this distribution was used with power-transformed observations by Hashino (1985). See below, Section 9.4.

9.3.13 Compounding

Roux and Becker (1981) obtained a compound distribution, which they called a bivariate Bessel distribution, by assuming that $\alpha'' = 1/\alpha'$ is exponentially distributed with density $\exp(-\alpha''/\gamma)/\gamma$, and, similarly, $\beta'' = 1/\beta'$ has density $\exp(-\beta''/\delta)/\delta$. The density is given by

$$h = \begin{cases} 2\beta\gamma^{-1}\exp[-(\alpha+\beta)y]K_0\left(2\frac{\sqrt{x-y}}{\gamma}\right) & \text{for } 0 < y < x \\ 2\alpha\delta^{-1}\exp[-(\alpha+\beta)x]K_0\left(2\frac{\sqrt{y-x}}{\delta}\right) & \text{for } 0 < x < y \end{cases} \tag{9.19}$$

where K_0 is the modified Bessel function of the third kind of order zero.

9.3.14 Becker and Roux's Generalisation

Re-state Freund's model as follows. Suppose that the shocks that knock out components A and B, respectively, are governed by Poisson processes:

- For component A, the Poisson process has rate α when component B is functioning, and rate α' after component B has failed.

- For component B, the Poisson process has rate β when component A is functioning, and rate β' after component A has failed.

Becker and Roux (1981) generalised Freund's distribution by supposing that the component did not fail after a single shock, but that it took a and b shocks, respectively, to destroy them. (These numbers a and b are deterministic, not random.)

The resulting density function is

$$h = \left\{ \begin{array}{ll} \frac{\beta'\alpha^a}{\Gamma(a)\Gamma(b)} x^{a-1}[\beta'(y-x)+\beta x]^{b-1}\exp[-\beta' y-(\alpha+\beta-\beta')x] & \text{for } 0<x<y \\ \frac{\alpha'\beta^b}{\Gamma(a)\Gamma(b)} y^{b-1}[\alpha'(x-y)+\alpha y]^{a-1}\exp[-\alpha' x-(\alpha+\beta-\alpha')y] & \text{for } 0<y<x \end{array} \right\} \tag{9.20}$$

See also Steel and le Roux (1987).

Roux and Becker (1981) obtained a compound distribution by assuming that $1/\alpha'$ and $1/\beta'$ have gamma distributions. The expression for the density is cumbersome, and we omit it.

9.3.15 Bhattacharya and Holla's Generalisations

In model I of Bhattacharya and Holla (1963), it is supposed that after one component fails, the distribution of the other's remaining lifetime becomes Weibull, not exponential. The density is given as proportional to $(y-x)^{q-1}\exp[-\delta(y-x)^q-(\alpha+\beta)x]$ for $0<x<y$, with an analogous expression for $0<y<x$.

In model II of the same paper, the distribution of the remaining component's lifetime becomes gamma after the first failure of a component. The density is given as proportional to $(y-x)^{q-1}\exp[-\delta(y-x)-(\alpha+\beta)x]$ for $0<x<y$, with an analogous expression for $0<y<x$.

9.4 Hashino and Sugi's Distribution

9.4.1 Support

The positive quadrant.

9.4.2 Formula for p.d.f.

$$h = \left\{ \begin{array}{ll} \alpha\beta'\exp[-\beta' y-(\alpha+\beta-\beta')x] & \text{for } 0\le x\le y, \text{ with } x\le\gamma \\ \alpha'\beta\exp[-\alpha' x-(\alpha+\beta-\alpha')y] & \text{for } 0\le y\le x, \text{ with } y\le\gamma \\ ab'\exp[-b'(y-\delta)-(a+b-b')(x-\delta)] & \text{for } \gamma\le x\le y \\ a'b\exp[-a'(x-\delta)-(a+b-a')(y-\delta)] & \text{for } \gamma\le y\le x \end{array} \right\} \tag{9.21}$$

where all the parameters are positive. In fact, there are just 6 free parameters, because of continuity conditions at $X=\gamma$ and $Y=\gamma$.

9.4.3 Remarks

An English-language account of this extension of Freund's distribution is given by Hashino (1985), who attributes it to Hashino and Sugi (1984). Hashino includes formulae for the marginal density of Y, the marginal (cumulative) distribution function of Y, the conditional (cumulative) distribution function of X given Y, and the joint (cumulative) distribution function H.

The distribution was not motivated by the reliability context; rather, it was desired to have a tractable bivariate distribution that is somewhat analogous to the univariate piecewise-exponential distribution.

9.4.4 An Application

The Osaka district suffers from typhoons. When these occur, the river, in its tidal reaches, rises for two reasons: the rain which drains into it, and the storm surge that comes in from the sea. The study by Hashino was of the peak rainfall intensity and the maximum storm surge, for 117 typhoons occurring over an 80-year period. In fitting distribution (9.21), X and Y were transformed to X^m/σ_{xm} and Y^m/σ_{ym}, respectively, the σ's being the standard deviations of the transformed observations. Hashino found large differences (a factor of more than 2) between return periods $1/H(x,y)$ calculated using the fitted distribution and $1/[F(x)G(y)]$ calculated assuming independence.

Two minor points: (i) It appears that the typhoons included in the study were restricted to those for which the storm surge exceeded a certain level; Hashino did not discuss whether this truncation of the sample had any effect on the conclusions. (ii) The correlation coefficient given by Hashino (-0.02) is calculated for the distribution applying to large values of X and Y only (i.e., the last two expressions in (9.21)), not for the distribution as a whole.

9.5 ACBVE of Block and Basu

9.5.1 Support

The positive quadrant.

9.5.2 Formula for p.d.f.

$$h = \begin{cases} \frac{\lambda_1\lambda(\lambda_2+\lambda_{12})}{\lambda_1+\lambda_2}\exp[-\lambda_1 x - (\lambda_2+\lambda_{12})y] & \text{if } x < y \\ \frac{\lambda_2\lambda(\lambda_1+\lambda_{12})}{\lambda_1+\lambda_2}\exp[-(\lambda_1+\lambda_{12})x - \lambda_2 y] & \text{if } x > y \end{cases} \tag{9.22}$$

where the λ's are positive, and $\lambda = \lambda_1 + \lambda_2 + \lambda_{12}$.

9.5.3 Formula for (Cumulative) Distribution Function

The upper right volume under the probability density surface is:

$$\bar{H} = \frac{\lambda}{\lambda_1+\lambda_2}\exp[-\lambda_1 x - \lambda_2 y - \lambda_{12}\max(x,y)] - \frac{\lambda_{12}}{\lambda_1+\lambda_2}\exp[-\lambda\max(x,y)]. \tag{9.23}$$

9.5.4 Univariate Properties

The marginals are not exponential, but rather a negative mixture of two exponentials,

$$\bar{F}(x) = \frac{\lambda}{\lambda_1+\lambda_2}\exp[-(\lambda_1+\lambda_{12})x] - \frac{\lambda_{12}}{\lambda_1+\lambda_2}\exp(-\lambda x), \tag{9.24}$$

and a similar equation for $\bar{G}(y)$.

9.5.5 Correlation Coefficients

Pearson's product-moment correlation coefficient is

$$\frac{(\lambda_1+\lambda_2)^2(\lambda_1+\lambda_{12})(\lambda_2+\lambda_{12}) - \lambda^2\lambda_1\lambda_2}{\sqrt{[(\lambda_1+\lambda_2)^2(\lambda_1+\lambda_{12})^2+\lambda_2(\lambda_2+2\lambda_1)\lambda^2][(\lambda_1+\lambda_2)^2(\lambda_2+\lambda_{12})^2+\lambda_1(\lambda_1+2\lambda_2)\lambda^2]}}. \tag{9.25}$$

(We think Block and Basu, 1974, were wrong in the expression they gave.)

9.5.6 Moment Generating Function

The m.g.f. may be obtained from (9.18) by the substitutions given in Section 9.5.7, and is found to be

$$M(s,t) = \frac{1}{\lambda_1+\lambda_2}\frac{\lambda}{\lambda-(s+t)}\left[\frac{\lambda_1(\lambda_2+\lambda_{12})}{\lambda_2+\lambda_{12}-t}+\frac{(\lambda_1+\lambda_{12})\lambda_2}{\lambda_1+\lambda_{12}-s}\right]. \tag{9.26}$$

9.5.7 Derivation

This distribution was derived by Block and Basu (1974) by omitting the singular part of Marshall and Olkin's distribution; they called it the ACBVE distribution—absolutely continuous part of a bivariate exponential distribution. (See also Block, 1975.) Alternatively, it can be derived by Freund's method, with

$$\left.\begin{aligned} \alpha &= \lambda_1+\lambda_{12}[\lambda_1/(\lambda_1+\lambda_2)] \\ \alpha' &= \lambda_1+\lambda_{12} \\ \beta &= \lambda_2+\lambda_{12}[\lambda_2/(\lambda_1+\lambda_2)] \\ \beta' &= \lambda_2+\lambda_{12}. \end{aligned}\right\} \tag{9.27}$$

9.5.8 Remarks

- $\min(X,Y)$ is an exponential variate.
- $X-Y$ and $\min(X,Y)$ are independent.
- The lack of memory property holds.

9.5.9 Applications

Gross and Lam (1981) considered this distribution suitable in such cases as these:

- Lengths of tumour remission, when a patient receives different treatments on two occasions.
- Lengths of time required for analgesics to take effect, when patients with headaches receive different ones on two occasions.

Gross and Lam were concerned chiefly with developing hypothesis tests for equality of the marginal means. They also made the following suggestion for determining whether Block and Basu's distribution is appropriate:

- Test whether $\min(X,Y)$ has an exponential distribution.
- Test whether $X-Y$ and $\min(X,Y)$ are uncorrelated.
- Test whether $X-Y$ has the distribution given by their equation (4.1).

These three properties, except with independence replacing zero correlation in the second of them, together characterise Block and Basu's distribution.

9.6 Sarkar's Distribution

9.6.1 Support

The positive quadrant.

9.6.2 Formula for p.d.f.

$$h = \left\{ \begin{array}{ll} \frac{\lambda_1\lambda}{(\lambda_1+\lambda_2)^2}\exp[-\lambda_1 x - (\lambda_2+\lambda_{12})y] & \\ \times\,[(\lambda_1+\lambda_2)(\lambda_2+\lambda_{12}) - \lambda_2\lambda\exp(-\lambda_1 y)][A(\lambda_1 x)]^{\gamma}[A(\lambda_2 y)]^{-(1+\gamma)} & \text{if } x \le y \\ & \\ \frac{\lambda_2\lambda}{(\lambda_1+\lambda_2)^2}\exp[-(\lambda_1+\lambda_{12})x - \lambda_2 y] & \\ \times\,[(\lambda_1+\lambda_2)(\lambda_1+\lambda_{12}) - \lambda_1\lambda\exp(-\lambda_2 x)][A(\lambda_1 x)]^{-(1+\gamma)}[A(\lambda_2 y)]^{\gamma} & \text{if } x \ge y \end{array} \right\} \quad (9.28)$$

where the λ's are positive, $\lambda = \lambda_1 + \lambda_2 + \lambda_{12}$, $\gamma = \lambda_{12}/(\lambda_1+\lambda_2)$, and $A(z) = 1 - \exp(-z)$.

9.6.3 Formula for (Cumulative) Distribution Function

$$\bar{H} = \left\{ \begin{array}{ll} \exp[-(\lambda_2+\lambda_{12})y]\{1 - [A(\lambda_1 x)]^{1+\gamma}\}[A(\lambda_1 y)]^{-\gamma} & \text{if } x \le y \\ & \\ \exp[-(\lambda_1+\lambda_{12})x]\{1 - [A(\lambda_2 x)]^{-\gamma}[A(\lambda_2 y)]^{1+\gamma}\} & \text{if } x \ge y \end{array} \right\} \quad (9.29)$$

H is absolutely continuous.

9.6.4 Univariate Properties

Exponential marginals.

9.6.5 Correlation Coefficient

An expression for Pearson's correlation coefficient is given by Sarkar (1987), but is cumbersome.

9.6.6 Derivation

This distribution, sometimes denoted ACBVE$_2$, was derived by Sarkar through the following conditions of characterisation:

- The bivariate distribution is absolutely continuous.
- X and Y are exponential variates with parameters $\lambda_1 + \lambda_{12}$ and $\lambda_2 + \lambda_{12}$, respectively.
- $\min(X, Y)$ is exponential with parameter $\lambda = \lambda_1 + \lambda_2 + \lambda_{12}$.
- $\min(X, Y)$ is independent of $g(X, Y)$, for some g of the form $\ell(x) - \ell(y)$, where ℓ is an increasing function.

9.6.7 Relation to Marshall and Olkin's Distribution

This distribution is obtained from Marshall and Olkin's by requiring absolute continuity of the distribution function H, and replacing the condition of independence of $\min(X, Y)$ and $X - Y$ by the modified condition above. Also, it has lost the lack of memory property.

9.7 Comparison of Four Distributions

At this point, we may compare the properties of the Marshall and Olkin, Block and Basu, Sarkar, and Freund distributions, in the following table.

	Marshall and Olkin	Block and Basu	Sarkar	Freund
Exponential marginals	$\surd$		$\surd$	
Absolutely continuous		$\surd$	$\surd$	$\surd$
Bivariate lack of memory, (9.9)	$\surd$	$\surd$		$\surd$
$\min(X, Y)$ is exponential	$\surd$	$\surd$	$\surd$	$\surd$
$\min(X, Y)$ is independent of $X - Y$	$\surd$	$\surd$	modified	$\surd$

9.8 The Generalisation by Friday and Patil

Friday and Patil (1977) proposed a distribution that subsumes both Freund's and Marshall and Olkin's:

$$\bar{H} = \gamma \bar{H}_A + (1 - \gamma)\bar{H}_B, \tag{9.30}$$

where $\bar{H}_A$ is the survival function corresponding to Freund's distribution (9.15), and $\bar{H}_B$ is the singular distribution $\exp[-(\alpha + \beta)\max(x, y)]$. More fully,

$$H = \left\{ \begin{array}{ll} \theta_1 \exp[-(\alpha + \beta - \beta')x - \beta' y] + (1 - \theta_1)\exp[-(\alpha + \beta)y] & \text{for } x \leq y \\ \theta_2 \exp[-\alpha' x - (\alpha + \beta - \alpha')y] + (1 - \theta_2)\exp[-(\alpha + \beta)x] & \text{for } x \geq y \end{array} \right\} \tag{9.31}$$

where $\theta_1 = \gamma\alpha(\alpha + \beta - \beta')^{-1}$, $\theta_2 = \gamma\beta(\alpha + \beta - \alpha')^{-1}$, and $0 \leq \gamma \leq 1$. This distribution is another that has the lack of memory property, (9.10). It is sometimes denoted BEE.

Friday and Patil also showed that only two independent standard exponential variates are needed to generate a pair (X, Y) with their distribution (9.31), and thus the same is true for Freund's and Marshall and Olkin's distributions. They go on to examine the computational efficiency of their scheme. Some further results for what seems to be this model are given by Itoi et al (1978).

The model of Platz (1984) is another that includes both those of Marshall and Olkin and of Freund, and also one-out-of-three and two-out-of-three systems with identical components.

9.9 The Bivariate Distribution of Tosch and Holmes

This model (Tosch and Holmes, 1980) generalises both that of Marshall and Olkin and that of Freund: it permits simultaneous failure of both components; and the residual lifetime of one component is not independent of the status (working or failed) of the other component. Stated formally,

$$\left. \begin{array}{rcl} X & = & \min(U_1, U_2) + U_3 I_{\{U_1 > U_2\}} \\ Y & = & \min(U_1, U_2) + U_4 I_{\{U_1 \leq U_2\}} \end{array} \right\} \tag{9.32}$$

where the U's are non-negative mutually independent r.v.'s, and $I_{\{.\}}$ is the indicator variable, i.e., is 1 if the condition within the curly brackets is true and is zero if it is false.

In words, if component 1 is the first to fail, then its lifetime X is U_1 and the second component's extra lifetime is U_4; conversely, if component 2 is the first to fail, its lifetime is U_2 and the first component's extra lifetime is U_3.

The cumulative distribution that corresponds to (9.32) cannot be easily obtained in general. However, it can be when U_1 and U_2 are exponentials with scale parameters α and β, respectively, and U_3 and U_4 are exponentials apart from a discontinuity at the origin (i.e., $\Pr(U_3 \leq t) = 1 - q + q[1 - \exp(-\alpha' t)]$ and $\Pr(U_4 \leq t) = 1 - q + q[1 - \exp(-\beta' t)]$, with $0 \leq q \leq 1$).

9.10 The Method of Lawrance and Lewis

9.10.1 General Form

As a preliminary, note that if E_1 and E_2 are i.i.d. exponential variates with unit mean, and (independently of E_1 and E_2) I is 0 or 1 with probabilities β and $1-\beta$, respectively, then $\beta E_1 + IE_2$ also is an exponential variate with unit mean.

The general form of this model (Lawrance and Lewis, 1983) is as follows:

$$\left.\begin{array}{rcl} X & = & \beta_1 V_1 E_1 + I_1 E_2 \\ Y & = & I_2 E_1 + \beta_2 V_2 E_2 \end{array}\right\} \qquad (9.33)$$

where E_1 and E_2 are independent and exponentially distributed, V_1 and V_2 are each either 0 or 1 (not necessarily independently of each other), with $\Pr(V_i = 1) = \alpha_i$, and I_1 and I_2 are each either 0 or 1 (not necessarily independently of each other), with $\Pr(I_i = 1) = (1-\beta_i)/[1-(1-\alpha_i)\beta_i]$.

Lawrance and Lewis call (9.33) the EP+ model. They mention particularly three special cases, EP1, EP3, and EP5.

9.10.2 Model EP1

This takes $\alpha_1 = \alpha_2 = 1$, $\beta_1 = \beta_2$ $(= \beta)$, and $I_1 = I_2$ $(= I)$. Thus

$$\left.\begin{array}{rcl} X & = & \beta E_1 + IE_2 \\ Y & = & IE_1 + \beta E_2 \end{array}\right\} \qquad (9.34)$$

with $\Pr(I = 1) = 1 - \beta$.

The p.d.f. is:

$$h = I_{\{\beta y < x < y/\beta\}}(1+\beta)^{-1}\exp[-(x+y)/(1+\beta)] + \beta^{-1}\exp[-(x+y)/\beta], \qquad (9.35)$$

where $I_{\{.\}}$ is the indicator function, as earlier. Lawrance and Lewis illustrate this for $\beta = 0.5$. The product-moment correlation is $3\beta(1-\beta)$, so it is at $\beta = 0.5$ that it reaches its maximum. The grade correlation is $3\beta(1-\beta)(8+7\beta+\beta^2)/[(1+\beta)^2(2+\beta)^2]$.

9.10.3 Model EP3

This takes $\alpha_1 = \alpha_2 = 1$, with (I_1, I_2) having maximum possible dependency. This last statement means that the possible combinations of values occur with the following probabilities:

	$I_2 = 0$	$I_2 = 1$
$I_1 = 0$	$\min(\beta_1, \beta_2)$	$\max(\beta_1 - \beta_2, 0)$
$I_1 = 1$	$\max(\beta_2 - \beta_1, 0)$	$\min(1-\beta_1, 1-\beta_2)$

Then

$$\left.\begin{array}{rcl} X & = & \beta_1 E_1 + I_1 E_2 \\ Y & = & I_2 E_1 + \beta_2 E_2 \end{array}\right\} \qquad (9.36)$$

with probabilities of the various combinations of values of I_1 and I_2 being as above.

Lawrance and Lewis give expressions for both the product-moment correlation and the grade correlation.

9.10.4 Model EP5

This takes $\alpha_1 = \alpha_2$ $(= \alpha)$, $\beta_1 = \beta_2$ $(= \beta)$, $V_1 = V_2$ $(= V)$, and $I_1 = I_2$ $(= I)$. Thus

$$\left.\begin{array}{rcl} X & = & \beta V E_1 + IE_2 \\ Y & = & IE_1 + \beta V E_2 \end{array}\right\} \qquad (9.37)$$

with $\Pr(V = 1) = \alpha$ and $\Pr(I = 1) = (1-\beta)/[1-(1-\alpha)\beta]$.

The product-moment correlation is $3\alpha\beta(1-\alpha\beta)$.

9.10.5 Models with Negative Correlation

Lawrance and Lewis also discuss a number of analogous models that have negative correlation, still with exponential marginals.

9.10.6 Models with Uniform Marginals

Lawrance and Lewis also discuss the above distributions after they have been transformed to have uniform marginals. They include an illustration of EP1 (with $\beta = 0.32$) after such a transformation.

9.10.7 Mixture Models

Models that can exhibit either positive or negative dependence can be obtained easily by mixing one of those having positive correlation with one of those having negative correlation.

9.10.8 Models with Line Singularities

Models that are like that of Marshall and Olkin in that there is a non-zero probability that $Y = X$ may readily be constructed.

Let (X_1, X_2) be a pair of variates with unit-exponential marginals, such as any of those described above. Let E be an independent unit-exponential variate. Let (I_1, I_2) be an indicator pair, possibly completely or partially dependent, with marginal probabilities $\Pr(I_i = 1) = 1 - \beta_i$. Three methods of obtaining a distribution having a line singularity are as follows:

$$\left.\begin{aligned} X &= I_1X_1 + (1 - I_1)E \\ Y &= I_2X_2 + (1 - I_2)E \end{aligned}\right\} \tag{9.38}$$

$$\left.\begin{aligned} X &= I_1X_1 + \beta_1 E \\ Y &= I_2X_2 + \beta_2 E \end{aligned}\right\} \tag{9.39}$$

$$\left.\begin{aligned} X &= \min(X_1, E) \\ Y &= \min(X_2, E) \end{aligned}\right\} \tag{9.40}$$

9.11 Raftery's Scheme

9.11.1 General Form

In its general form, Raftery's (1984, 1985) scheme for obtaining a bivariate distribution with exponential marginals is

$$\left.\begin{aligned} X &= (1 - p_{10} - p_{11})U + I_1W \\ Y &= (1 - p_{01} - p_{11})V + I_2W \end{aligned}\right\} \tag{9.41}$$

where U, V, W are independent and exponentially distributed, and I_1 and I_2 are each either 0 or 1, with probabilities as set out below:

	$I_2 = 0$	$I_2 = 1$
$I_1 = 0$	p_{00}	p_{01}
$I_1 = 1$	p_{10}	p_{11}

Raftery gives the correlation as $2p_{11} - (p_{01} + p_{11})(p_{10} + p_{11})$. There is also an extension of the model to permit negative correlation. Raftery gives particular attention to the two special cases below.

9.11.2 First Special Case

This sets $p_{01} = p_{10} = 0$, so that

$$\left.\begin{aligned} X &= (1 - p_{11})U + IW \\ Y &= (1 - p_{11})V + IW \end{aligned}\right\} \tag{9.42}$$

and the distribution is a mixture of independence and trivariate reduction.

9.11.3 Second Special Case

This sets $p_{01} = 0$, $p_{10} = 1 - p_{11}$, so that

$$\left.\begin{array}{rcl} X & = & W \\ Y & = & (1-p_{11})V + I_2 W \end{array}\right\} \tag{9.43}$$

and the distribution is a mixture of independence and weighted linear combination.

9.11.4 Formula for p.d.f.

This can be obtained in explicit form, but is cumbersome. See Raftery (1984).

9.11.5 Derivation

The distribution arises from a shock model. This refers to a system that has two components, S_1 and S_2, each of which can be functioning normally, unsatisfactory, or failed. The system is subject to three kinds of shock, governed by independent Poisson processes. These kinds of shock cause normal components to become unsatisfactory, an unsatisfactory S_1 to fail, and an unsatisfactory S_2 to fail, respectively.

9.11.6 References to Illustrations

Contours of p.d.f.: Raftery (1984).

9.11.7 Remarks

- Generation of random variates is easy by following the method of construction, (9.41).
- The distribution can attain the Fréchet bounds.
- O'Cinneide and Raftery (1989) have shown that this distribution is an example of a bivariate phase-type distribution. (For this class, see Section 9.13 and Assaf et al, 1984.)

9.11.8 Applications

Raftery (1984) applies a Weibull version of this model to two datasets:

- 249 pairs of successive failure times of a computer (Cox and Lewis, 1966, p. 16)—it was found that $p_{01} = p_{10}$.
- Proportions of the population who were without a car and who were foreign-born, for the 88 unincorporated places with a population greater than 25000 found in the 1960 U.S. census (Tukey, 1977, p. 323)—it was found that $p_{01} = 0$.

9.12 Downton's Bivariate Exponential Distribution

9.12.1 Support

The positive quadrant.

9.12.2 Formula for p.d.f.

$$h = \frac{1}{1-\rho}\exp[-(x+y)/(1-\rho)]\, I_0\left(\frac{2\sqrt{xy\rho}}{1-\rho}\right), \tag{9.44}$$

where I_0 is the modified Bessel function of the first kind of order zero.

9.12.3 Formula for (Cumulative) Distribution Function

Expressed as an infinite series,

$$H = (1 - e^{-x})(1 - e^{-y}) + \sum_{j=0}^{\infty} \frac{\rho^{j+1}}{(j+1)^2} L_j^{(1)}(x) L_j^{(1)}(y) xy e^{-(x+y)}, \tag{9.45}$$

where the $L_j^{(1)}$ are Laguerre polynomials (see footnote 3 in Chapter 7).

9.12.4 Univariate Properties

Exponential marginals.

9.12.5 Correlation Coefficients

ρ in equation (9.44) is Pearson's product-moment correlation. As to estimation of ρ, Al-Saadi and Young (1980) obtained the maximum likelihood estimator, the method of moments estimator, the sample correlation estimator, and two bias reduction estimators. (See also Nagao and Kadoya, 1971.)

9.12.6 Conditional Properties

The regression $E(Y|X = x)$ and the conditional variance are both linear in x (Nagao and Kadoya, 1971).

9.12.7 Moment Generating Function

The m.g.f. is

$$M(s,t) = [(1-s)(1-t) - \rho st]^{-1}. \tag{9.46}$$

9.12.8 Regression

The regression is linear:

$$E(Y|X = x) = 1 + \rho(x - 1). \tag{9.47}$$

9.12.9 Derivation

In the context of reliability studies, Downton (1970) used a successive damage model to derive this distribution. Consider a system of two components, each being subjected to shocks, the intervals between successive ones having exponential distributions. Suppose the numbers of shocks N_1 and N_2 required for the components to fail follow a bivariate geometric distribution with joint probability generating function

$$P(z_1, z_2) = \frac{z_1 z_2}{1 + \alpha + \beta + \gamma - \alpha z_1 - \beta z_2 - \gamma z_1 z_2}. \tag{9.48}$$

Write

$$(X, Y) = \left(\sum_{i=1}^{N_1} X_i, \sum_{i=1}^{N_2} Y_i \right), \tag{9.49}$$

where X_i and Y_i are the inter-shock intervals, mutually independent exponential variates. Then the component lifetimes (X, Y) have the p.d.f. (9.44). See also Section 17.2.3.

Gaver (1972) gives a slightly different motivation for this distribution. He supposes that two types of shocks are occurring to an item of equipment, fatal and non-fatal. Repairs are only made after a fatal shock has occurred; repairs of all the non-fatal defects are also made then. If it is assumed that the two types of shock both follow Poisson processes and that the time for repair is the sum of (the random number) of exponential variates, then time to failure and time of repair have Downton's bivariate exponential distribution. Expressed concisely, the positive correlation arises because the longer the time to failure, the greater the accumulated non-fatal damage.[3]

---

[3]Gaver (1972) describes another model which contrasts with this one because it leads to negative correlation between

9.12.10 References to Illustrations

An example of the surface of probability density: Nagao and Kadoya (1971).

9.12.11 Random Variate Generation

Let

$$\left.\begin{array}{rcl} X_1^* &=& X(1-\rho) \\ X_2^* &=& Y(1-\rho) \end{array}\right\} \tag{9.50}$$

$(0 \leq \rho \leq 1)$, where X and Y are unit exponential variates. The joint characteristic function $\varphi(s,t)$ of X_1^* and X_2^* satisfies the relation

$$\varphi(s,t) = \psi(s)\psi(t)[(1-\rho) + \rho\varphi(s,t)], \tag{9.51}$$

where ψ is the c.f. of the marginals, given by $\psi(t) = [1 - it(1-\rho)]^{-1}$. To avoid the intermediate generation of bivariate normal variates, Paulson (1973) proposed the following method of random variate generation. Suppose $\varphi_n(s,t)$, $n = 1, 2, \ldots$ is a sequence of characteristic functions which satisfies the recurrence relation $\varphi_n(s,t) = \psi(s)\psi(t)[(1-\rho) + \rho\varphi_{n-1}(s,t)]$. This corresponds to the vector-valued r.v. $\mathbf{Y}_n = \mathbf{U}_n + \mathbf{V}_n\mathbf{Y}_{n-1}$, where $\{\mathbf{U}_n\}$ is a sequence of independent bivariate r.v.'s whose components are independent unit exponential variates, $\{\mathbf{V}_n\}$ is a sequence of matrix-valued r.v.'s which take on the values $\begin{pmatrix} 0 & 0 \\ 0 & 0 \end{pmatrix}$ and $\begin{pmatrix} 1 & 0 \\ 0 & 1 \end{pmatrix}$ with probabilities $(1-\rho)$ and ρ, respectively, $\mathbf{V}_n$ and $\mathbf{U}_n$ being mutually independent, and $\mathbf{Y}_0 = \mathbf{0}$, a vector of zeros.

It is clear that $\varphi_n(s,t)$ converges to $\varphi(s,t)$ and hence $\{\mathbf{Y}_n\}$ converges in distribution to $(X_1^*, X_2^*)'$. Hence (X_1^*, X_2^*) can be generated as accurately as desired by choosing an appropriate value of n; the value $n = 10$ is quite satisfactory for ρ between 0 and 0.6. Finally, we set $(X,Y) = ((1-\rho)^{-1}X_1^*, (1-\rho)^{-1}X_2^*)$.

9.12.12 Remarks

- This distribution is a special case of the bivariate exponential distributions of Hawkes (1972), Paulson (1973), and Arnold (1975a)—for the interrelationship between these, see Section 17.3—and of the bivariate gamma distribution of Kibble (1941).
- A detailed study was given by Nagao and Kadoya (1971).
- $X + Y$ is expressible as the sum of two independent exponential variates with parameters $(1+\sqrt{\rho})^{-1}$ and $(1-\sqrt{\rho})^{-1}$, respectively (Lai, 1985).
- Formulae relating to the use of this distribution in the "competing risks" context—for which, see Section 9.18—are to be found in David and Moeschberger (1978, Section 4.2).

9.12.13 Fields of Application

- Queueing systems. Consider a single-server queueing system such that the interarrival time X and the service time Y have exponential distributions, as is a common assumption in this context. If it is desired to introduce positive correlation (arising from cooperative service) into the model, Downton's distribution is a suitable choice (Conolly and Choo, 1979). Langaris (1986) applied it to a queueing system with infinitely many servers.
- As a model for Wold's Markov dependent process—Lai (1978). Let X_i denote the time interval between the ith and $(i+1)$th events. Assuming that $\{X_i\}$ is a Markov chain, then $N(t)$ = number of events that occur in $(0,t]$ is a Wold's point process. Lai (1978) used Downton's bivariate exponential model to describe the joint distribution of the lengths of successive time intervals.

---

exponential variates. Instead of failures occurring because of shocks from outside, they occur due to inbuilt defects. Let it be supposed that when a failure occurs, detailed inspection of the equipment is made, and all the defects discovered and repaired. A short time to failure is likely to have arisen because there were many inbuilt defects, and is thus likely to be associated with a lengthy time of repair. Conversely, a long time to failure probably comes about because there is only one defect, or very few, in which case repair time will be short.

- Hydrology. Nagao and Kadoya (1971) claimed that this distribution can be used for such pairs of hydrological quantities as streamflow at two points on a river, or rainfall at two locations.

- As a model for the intensity and duration of a storm of rainfall. Córdova and Rodríguez-Iturbe (1985) claim that exponential distributions for these variables have been shown to be sufficiently realistic. They find that independence should not be assumed:

  - Firstly, it is empirically not true (correlations of 0.30 and 0.33 being found in datasets from Boconó, Venezuela, and Boston, Massachusetts).
  - Secondly, some important quantities are highly dependent on the correlation: quantities such as the mean and variance of storm depth (product of intensity and duration) and the probability of non-zero storm surface runoff (if the soil is sufficiently dry and the storm is sufficiently small, there is no surface runoff).

- Height and period of water waves.

  - The Rayleigh distribution, a special case of the Weibull, has cumulative distribution $F(x) = 1 - \exp(-x^2)$ and p.d.f. $f(x) = 2x\exp(-x^2)$. It is a common choice to describe both the periods squared and the heights of waves of the sea—especially the latter, partly because there is theoretical support in the case of a narrow band spectrum. Consequently, Battjes (1971) suggested that a bivariate distribution with Rayleigh marginals be used for the joint distribution of these variables, and put forward (9.44), appropriately transformed.
  - Kimura (1981) suggested the Weibull-marginals version of this distribution for the height and period of water waves. Kimura performed experiments in which random waves were generated in a wave tank, and their heights and periods measured and cross-tabulated. (An example is given at Section 24.12.2—8 below.) Although Kimura claimed that this distribution "shows good applicability for the principal part of the joint distribution", he also admitted that it failed at the edges—the main method of comparing theory with data in this paper was by means of the conditional distribution of height given certain values of period, and the case presented showed reasonable agreement at $T = 0.8, 1.0, 1.2$, but poor agreement at $T = 0.6, 1.4$ (where T is the period expressed in units of its root-mean-square).
  - Burrows and Salih (1987) included the Weibull-marginals version of this distribution among those they fitted to data from around the British Isles.

- Height of water waves. Kimura and Seyama (1985) assumed that the Rayleigh-marginals version of this distribution described the joint distribution of successive wave heights. Their concern was the overtopping of a sea wall that may occur when a group of high waves attacks it.

9.12.14 References to Tables or Algorithms

Algorithm for the probability integral: Lai and Moore (1984).

Tables for the conditional distribution function: Nagao and Kadoya (1971).

9.12.15 Weibull Marginals

Kimura (1981) has given some properties of this distribution when the marginals are transformed to be Weibull distributions. An expression for the general mixed moment is included, from which the correlation coefficient is obtained.

For applications, see Section 9.12.13 above.

9.12.16 A Bivariate Laplace Distribution

The difference of i.i.d. exponential variates has a Laplace distribution, $f(x) = e^{-|x|}/2$ (with scale parameter omitted). This property has been used by Ulrich and Chen (1987) to obtain a distribution with Laplace marginals, by setting

$$\left.\begin{array}{rcl} U & = & X_1 - X_2 \\ V & = & Y_1 - Y_2 \end{array}\right\} \tag{9.52}$$

where the (X_i, Y_i) come from distribution (9.44).

The m.g.f. can easily be written down, $[(1-s)(1-t)-\rho st]^{-1}[(1+s)(1+t)-\rho st]^{-1}$, but the p.d.f. which Ulrich and Chen obtain by inverting this is quite hideous, involving a double infinite series.

Since it is easy to generate Downton variates (as $X = W_1^2 + W_2^2$, $Y = Z_1^2 + Z_2^2$, where (W_i, Z_i) has a bivariate normal distribution), it is also easy to generate variates from the Ulrich and Chen distribution, by then using (9.52).

9.13 Bivariate Phase-Type Distributions

The abstract of the paper by Assaf et al (1984) which defines this class of distributions reads as follows: "A (univariate) random variable is said to be of phase type if it can be represented as the time until absorption in a finite state absorbing Markov chain. Univariate phase type random variables are useful because they arise from processes that are often encountered in applications, they have densities that can be written in a closed form, they possess some useful closure properties, and they can approximate any nonnegative random variable. This paper introduces and discusses several extensions to the multivariate case. It shows that the multivariate random variables possess many of the properties of univariate phase type distributions and derives explicit formulas for various probabilistic quantities of interest. Some examples are included."

To amplify a little upon this:

- The context is a Markov chain. Consider two overlapping subsets of the set of states. For each of these subsets, once it is entered, it is never left. The overlap is the absorption state. The joint distribution of the times of entering the two subsets is what has a bivariate phase-type distribution.

- Assaf et al are able to give an explicit expression for $\bar{H}$, albeit in matrix notation because of the generality of this class of distributions. It is possible for there to be a singular component of the density along $Y = X$.

- Assaf et al discuss the multivariate case, as well as the bivariate, and give some closure theorems.

- Assaf et al also consider the random variables in common model, $(X, Y) = (X_1 + X_3, X_2 + X_3)$, where X_1, X_2, X_3 are independent phase-type r.v.'s. Distributions obtained in this way constitute a subset of bivariate phase-type distributions.

- Examples of bivariate phase-type distributions: Marshall and Olkin (1967a) (see Section 9.2 above), Freund (1961) (Section 9.3) and the generalisation of Becker and Roux (1981) (Section 9.3.14), and Raftery (1984, 1985) (Section 9.11).

Another summary of bivariate phase-type distributions is included in Shaked and Shanthikumar (1985).

9.14 Cowan's Bivariate Exponential Distribution

9.14.1 Support

The positive quadrant.

9.14.2 Formula for (Cumulative) Distribution Function

The upper right volume under the probability density surface is

$$\bar{H} = \exp\left[-\frac{1}{2}\left(x + y + \sqrt{(x+y)^2 - 4\eta xy - 4xy}\right)\right] \tag{9.53}$$

$(0 \leq \eta \leq 1)$. Obviously, scaling factors can be introduced if desired.

9.14.3 Formula for p.d.f.

$$h = \frac{1-\eta}{2S^3}\{4\eta xy + S[S(x+y) + x^2 + y^2 + 2\eta xy]\}\exp[-(x+y+S)/2], \tag{9.54}$$

where $S^2 = (x+y)^2 - 4\eta xy$.

9.14.4 Univariate Properties

Exponential marginals.

9.14.5 Correlation Coefficients

Pearson's product-moment correlation coefficient is $-1 + \frac{2}{\eta}\left[1 + \frac{1-\eta}{\eta}\log(1-\eta)\right]$. The grade correlation is $\frac{3}{8+\eta}\left[4 - \eta - \frac{8(1-\eta)}{\xi}\log\frac{(\eta-\xi)(3\eta+\xi)}{(\eta+\xi)(3\eta-\xi)}\right]$, where $\xi = \sqrt{\eta(8+\eta)}$.

9.14.6 Conditional Properties

The conditional mean and standard deviation of Y given that $X = x$ are not of simple forms, but graphs of these functions have been given by Cowan (1987). That paper also includes a graph of $E(Y|X = x)$ when the marginals have been transformed to be uniform.

9.14.7 Derivation

This distribution was introduced by Cowan (1987). He derived it as the joint distribution of distances, in two directions separated by an angle α, to the nearest lines of a Poisson line process in the plane. (The association parameter η is $(1+\cos\alpha)/2$.)

9.14.8 References to Illustrations

Cowan (1987) gives contours of the p.d.f. and of the d.f., for the case $\alpha = \pi/6$.

9.14.9 Remarks

An expression for the joint characteristic function of X and Y was given by Cowan (1987).

The minimum of X and Y is exponentially distributed.

9.14.10 Transformation of the Marginals

The (cumulative) distribution function when the marginals are uniform is:

$$H = \sqrt{xy}\exp\left(-\frac{1}{2}\sqrt{(\log xy)^2 - 4\eta\log x\log y}\right). \tag{9.55}$$

9.15 Other Distributions with Exponential Marginals

Distributions with exponential marginals may, of course, be obtained by starting with any bivariate distribution of familiar form, and transforming the X and Y axes appropriately. In particular, this may be done with any of the copulas given in Chapter 5—in the expression for H, simply replace x by $1 - e^{-x}$ and y by $1 - e^{-y}$. See especially the Sections listed below.

Distribution	Sections
Farlie-Gumbel-Morgenstern	5.2.5, 5.2.7, 5.2.9, 5.2.11, 5.2.12
Ali-Mikhail-Haq	5.3.5
Gumbel-Hougaard	5.6.9.1
Plackett	5.7.7
Pareto	5.9.3
Gumbel's type I	5.10, 5.11.5

We may also note that Tolley and Norman (1979) remarked that it is common when comparing mortality rates in epidemiological studies to employ the "person years at risk" estimator, this being based upon failure times with an underlying exponential distribution; as individuals are also at risk of dying of unrelated causes, Tolley and Norman generalised the estimator to the competing risks context (for which, see Section 9.18). They went on to study its behaviour when failure times from the two causes have the Farlie-Gumbel-Morgenstern distribution with exponential or Weibull marginals.

9.16 Bivariate Weibull Distributions

Because the univariate Weibull distribution is obtained from the univariate exponential by a simple transformation of the variable, bivariate distributions with Weibull marginals can readily be obtained by starting with any of the bivariate distributions having exponential marginals and transforming X and Y appropriately. Examples where this has been done explicitly may be traced through the Index.

There are many types of bivariate Weibull distributions, and they can be put into five classes, as below. In each case, X and Y are individually taken to have Weibull distributions.

- Class C1. X and Y are independent.
- Class C2. $X = \min(X_1, X_2)$, $Y = \min(X_2, X_3)$, where the X_i are independent, not necessarily identically distributed, Weibull variates.
- Class C3. $\min(aX, bY)$ has a Weibull distribution for every $a > 0$ and $b > 0$.
- Class C4. $\min(X, Y)$ has a Weibull distribution.
- Class C5. The class of all bivariate distributions with Weibull marginals.

L Lee (1979) described the above classes, and showed that the inclusions $C1 \subset C2 \subset C3 \subset C4 \subset C5$ are strict. Another comprehensive treatment can be found in Block and Savits (1980b). For a brief overview, see Jensen (1985).

Applications can easily be imagined in any of the fields where bivariate distributions with exponential marginals are used, especially those such as reliability where the univariate Weibull is a popular generalisation of the univariate exponential. The Weibull distribution, and others derived by the extreme-value approach, can be plausibly applied to the strengths of materials. Warren (1979) suggests using bivariate distributions with Weibull marginals for the joint distribution of modulus of elasticity and modulus of rupture for lumber. (For this subject, see also Sections 19.8.3, 19.8.4, and 24.2—4.)

9.17 Bivariate Extreme-Value Distributions

9.17.1 Preliminaries

As mentioned in Section 2.8, the univariate extreme-value distributions consist of types I, II, and III, which can easily be transformed into each other. Type I is often called Gumbel's extreme-value distribution; the other two types are similar to each other, and the Weibull distribution is included within them. In the bivariate context, marginals are of secondary interest to the dependence structure. Tiago de Oliveira (1962/63, 1975a,b, 1980, 1984), Gumbel and Goldstein (1964), Gumbel (1965), Gumbel and Mustafi (1967), and Galambos (1978, Chapter 5, especially Section 5.4) assumed Gumbel marginals, whereas de Haan and Resnick (1977) chose marginals $F = \exp(-x^{-1})$. All three types can be transformed to exponential variates easily, and we will follow Pickands (1981), Deheuvels (1983, 1985), Smith (1985), and Tawn (1988a) in making that choice; the chief reason is the simplicity of the expressions that result.

Besides those already mentioned, other early papers on this subject include those of Sibuya (1960), Gumbel (1962a,b), and Tiago de Oliveira (1961).

9.17.2 Support

The positive quadrant.

9.17.3 Formula for (Cumulative) Distribution Function

The upper right volume under the probability density surface is:

$$\bar{H} = \exp\left[-(x+y)A\left(\tfrac{y}{x+y}\right)\right], \tag{9.56}$$

where

$$A(w) = \int_0^1 \max[(1-w)q, w(1-q)]\frac{dB}{dq}\,dq, \tag{9.57}$$

in which B is a positive increasing function on $[0,1]$. In order to have unit exponential marginals, we need

$$1 = \int_0^1 q\frac{dB}{dq}\,dq = \int_0^1 (1-q)\frac{dB}{dq}\,dq. \tag{9.58}$$

(To deduce this, we successively set $x = 0$ and $y = 0$ in (9.56). We then find that $A(0)$ and $A(1)$ must both be 1, and put these values into (9.57).) It follows from (9.58) that $\frac{1}{2}B$ is the distribution function of a random variable with mean $\frac{1}{2}$. We call A the dependence function of (X, Y), in accordance with the usage of Pickands (1981) and Tawn (1988a). (Do not confuse this with any other meaning of the term.) For accounts of the connexions between various dependence functions, see Deheuvels (1984) and Weissman (1985).

The expression for the p.d.f. is not very interesting.

9.17.4 Properties of the Dependence Function A

1. $A(0) = A(1) = 1$.

2. $\max(w, 1-w) \le A(w) \le 1$ (where $0 \le w \le 1$).

3. $A(w) = 1$ implies that X and Y are independent. $A(w) = \max(w, 1-w)$ implies that X and Y are equal, i.e., $\Pr(X = Y) = 1$.

4. A is convex, i.e., $A[\lambda x + (1-\lambda)y] \le \lambda A(x) + (1-\lambda)A(y)$.

5. If A_i are dependence functions, so is $\sum_{i=1}^n \alpha_i A_i$, where $\alpha_i \ge 0$ and $\sum_{i=1}^n \alpha_i = 1$.

A may or may not be differentiable. In the former case, H has a joint density everywhere; in the latter, H has a singular component, and is not differentiable in a certain region of its support. We shall consider examples of this family of distributions classified as differentiable, non-differentiable, or Tawn's extension of differentiable. Examples 1–4, 6, and 7 below were studied in Tawn (1988a).

9.17.5 Differentiable Models

9.17.5.1 Example 1

The mixed model, also known as Gumbel's type A bivariate extreme-value distribution, sets $A(w) = \theta w^2 - \theta w + 1$ (for $0 \le \theta \le 1$). Hence

$$\bar{H} = \exp\left[-(x+y) + \frac{\theta xy}{x+y}\right]. \tag{9.59}$$

See Section 9.17.8 for correlations, and Gumbel and Mustafi (1967) for further properties.

In the case of marginals with different parameters, Elandt-Johnson (1978) showed that the two crude hazard rates $h_1(x) = \left.\frac{\partial \log \bar{H}}{\partial x}\right|_{y=x}$ and $h_2(y) = \left.\frac{\partial \log \bar{H}}{\partial y}\right|_{x=y}$ are proportional if and only if the marginal hazard rates $f/\bar{F}$ and $g/\bar{G}$ are proportional.

9.17.5.2 Example 2

The logistic model, also known as the type B bivariate extreme-value distribution, sets $A(w) = [(1-w)^r + w^r]^{1/r}$, for $r \ge 1$. Hence

$$\bar{H} = \exp[-(x^r + y^r)^{1/r}]. \tag{9.60}$$

This was the third type of exponential distribution mentioned, albeit only briefly, by Gumbel (1960a). See Section 9.17.8 for correlations, and also Section 5.6.9.1 and Gumbel and Mustafi (1967) for further details.

9.17.6 Non-Differentiable Models

9.17.6.1 Example 3

The biextremal model, also known as the type C bivariate extreme-value distribution, sets $A(w) = \max(w, 1-\theta w)$, for $0 \le \theta \le 1$. Hence

$$\bar{H} = \exp\{-\max[x + (1-\theta)y, y]\}. \tag{9.61}$$

See Section 9.17.8 for correlations.

9.17.6.2 Example 4

Gumbel's model sets $A(w) = \max[1-\theta w, 1-\theta(1-w)]$ (for $0 \le \theta \le 1$). Hence

$$\bar{H} = \exp[-(1-\theta)(x+y) - \theta\max(x,y)]. \tag{9.62}$$

This is effectively the bivariate exponential distribution of Marshall and Olkin (1967a) (see Section 9.2) See Section 9.17.8 for correlations.

9.17.6.3 Example 5

The natural model sets $A(w) = \frac{\beta-1}{\beta-\alpha}\max(1-w, \alpha w) + \frac{1-\alpha}{\beta-\alpha}\max(1-w, \beta w)$, for $0 \le \alpha \le 1 \le \beta < \infty$. Hence

$$\bar{H} = \exp\{-[(\beta-1)\max(x, \alpha y) + (1-\alpha)\max(x, \beta y)]/(\beta-\alpha)\}. \tag{9.63}$$

See Section 9.17.8 for correlations.

9.17.7 Tawn's Extension of Differentiable Models

9.17.7.1 Background

In the dependence functions for the differentiable models, Tawn (1988a) added an extra parameter ϕ to give further flexibility. This gives us two new models, as follows.

9.17.7.2 Example 6

The asymmetric mixed model sets $A(w) = \phi w^3 + \theta w^2 - (\theta + \phi)w + 1$, for $\theta \geq 0$, $\theta + \phi \leq 1$, $\theta + 2\phi \leq 1$, $\theta + 3\phi \geq 0$. Hence

$$\bar{H} = \exp\left[-(x+y) + xy\frac{(\theta+\phi)x + (\theta+2\phi)y}{(x+y)^2}\right]. \tag{9.64}$$

When $\phi = 0$, we get the mixed model, Example 1.

9.17.7.3 Example 7

The asymmetric logistic model sets $A(w) = [\theta^r(1-w)^r + \phi^r w^r]^{1/r} + (\theta - \phi)w + 1 - \theta$, for $0 \leq \theta \leq 1$, $0 \leq \phi \leq 1$, $r \geq 1$. Hence

$$\bar{H} = \exp[-(1-\theta)x - (1-\phi)y - (\theta^r x^r + \phi^r y^r)^{1/r}]. \tag{9.65}$$

When $\theta = \phi = 1$, we get the logistic model, Example 2. When $\theta = 1$, we have the biextremal model, Example 3, and when $\theta = \phi$, we have Gumbel's model, Example 4.

If $r \to \infty$, we get $A(w) = \max[1 - \phi w, 1 - \theta(1-w)]$, a non-differentiable model with $\Pr\left(Y = \frac{\theta}{\phi}X\right) = \frac{\theta\phi}{\theta+\phi-\theta\phi}$.

9.17.8 Correlations

X and Y are positively correlated. In fact, as was pointed out by Tawn (1988a), they also have the right tail increasing (RTI) property. (See Section 12.3.4 for this concept of positive dependence.)

Pearson's product-moment correlation may be written as

$$\rho = \int_0^1 \frac{dw}{[A(w)]^2} - 1 \tag{9.66}$$

(Tawn, 1988a).

For Example 1 (Tawn, 1988a),

$$\rho = \frac{\sin^{-1}\left(\frac{1}{2}\sqrt{\theta}\right) - \frac{1}{2}\sqrt{\theta\left(1 - \frac{1}{4}\theta\right)}\left(1 - \frac{1}{2}\theta\right)}{\sqrt{\theta\left(1 - \frac{1}{4}\theta\right)^3}}. \tag{9.67}$$

For Example 2 (Tawn, 1988a),

$$\rho = \frac{[\Gamma(1/r)]^2}{r\Gamma(2/r)} - 1. \tag{9.68}$$

For Example 3, $\mathrm{corr}(-\log X, -\log Y)$ (i.e., the correlation when the marginals are Gumbel's extreme-value distribution) is

$$-6\pi^{-2}\int_0^\theta (1-t)^{-1}\log t\, dt, \tag{9.69}$$

which may also be written as $6\pi^{-2}\mathrm{diln}(\theta) + 1$, where diln is the dilogarithm.[4]

For Example 4,

$$\rho = \theta/(2-\theta). \tag{9.70}$$

See Tiago de Oliveira (1980, 1984) for what the correlation coefficients are when the marginals are Gumbel extreme-value distributions, in the case of Examples 1–5; and Tiago de Oliveira (1975b) gives the results for the first four.

---

[4]This is the nomenclature and notation of Spanier and Oldham (1987, p. 231). Both are different from the usage in the key book on the subject, Lewin (1981). A FORTRAN algorithm for calculating the dilogarithm in Lewin's sense has been published by Ginsberg and Zaborowski (1975).

As to the grade correlation, for Example 1,

$$\rho_g = 3\left(2 - \frac{\theta}{4}\right)^{-1}\left[1 + 2\left(2\theta - \frac{\theta^2}{4}\right)^{-1/2}\tan^{-1}\left(\frac{\sqrt{2\theta - \frac{\theta^2}{4}}}{2 - \frac{\theta}{2}}\right)\right] - 3. \tag{9.71}$$

(There is a misprint in the formula given by Gumbel and Mustafi (1967, p. 583). However, their Table 3 is correct.)

For Example 4,

$$\rho_g = 3\theta/(2+\theta). \tag{9.72}$$

Tiago de Oliveira (1984) gives expressions for Kendall's tau in the case of Examples 2 and 5.

Tawn (1988a) suggests $2\left[1 - A\left(\frac{1}{2}\right)\right]$ as another measure of dependence which is unaffected by the choice of marginals.

9.17.9 Methods of Derivation

- Bivariate extreme-value distributions arise as the limiting distributions of normalised componentwise maxima. More formally, let (X_i, Y_i), $i = 1, 2, \ldots n$, be i.i.d. random vectors. Then $(\max(X_i), \max(Y_i))$, after being suitably normalised, has a bivariate extreme-value distribution as its limiting distribution.[5]

- (X, Y) has a bivariate extreme-value distribution with unit exponential marginals if and only if the marginals are unit exponentials and, for every $n \geq 1$, $[\bar{H}(x,y)]^n = \bar{H}(nx, ny)$. Pickands (1981) showed that this equation is satisfied if and only if $\bar{H}$ can be written as (9.56). For this reason, the dependence function determines the type of bivariate extreme-value distribution; it also expresses the asymptotic connexion between two maxima.

- Alternatively, (X, Y) has a bivariate extreme-value distribution with unit exponential marginals if and only if $\min(aX, bY)$ is exponential for all $a, b > 0$.

9.17.10 References to Illustrations

Plots of the bivariate density along $y = x$ of the mixed and logistic models, Examples 1 and 2, with their marginals being of extreme-value type I form, are given by Gumbel and Mustafi (1967) and Johnson and Kotz (1972, p. 255).

9.17.11 Applications

- In the form with extreme-value marginals, the mixed model, Example 1, was used by Posner et al (1969) in their analysis of the performance of a spacecraft command receiver. See also Section 24.3—4.

- In the form with extreme-value marginals, the mixed model, Example 1, and the logistic model, Example 2, were both used by Gumbel and Mustafi (1967) to describe the floods of the Fox River at an upstream and a downstream gauging station. They found the latter fitted better. See also Section 24.3—2. See Gumbel and Goldstein (1964) and Section 24.3—1 for floods of the Ocmulgee River. Tiago de Oliveira (1975b, 1980) mentions that an unpublished paper of 1975 by M A Amaral and M I Gomes entitled "The fitting of bivariate extreme models" has reanalysed these and other datasets.

- The logistic model was also used by Hougaard (1986) to analyse data on tumours in rats. See also Section 5.6.8.

---

[5] We note that in many distributions (such as the bivariate normal), $X_{\max}$ and $Y_{\max}$ may be asymptotically independent (as the sample size tends to infinity) even if X and Y are correlated. This is so if $\bar{H}/(1-H)$ tends to 0 as X and Y tend to ∞. See, for example, Gumbel (1962a,b). The result may be due to Geffroy (1958/59).

- The models of Examples 1, 2, 6, and 7 were used by Tawn (1988a) to describe the annual maximum sea levels at Lowestoft and Sheerness. See also Section 24.3—3.

- Smith (1986) and Tawn (1988b) considered the joint distribution of the r largest observations—they had time series data of sea levels, and were concerned with issues like the improvement in prediction resulting from making use of the 5 or 10 largest values per annum rather than only the largest. Smith's data was from Venice, and Tawn's from Lowestoft and Great Yarmouth.

- The "station-year" method for the analysis of rainfall or flood maxima is motivated as follows: one may be interested in events with very long return periods (i.e., well out in the tail of the distribution), much larger than the lengths of the individual rainfall datasets; to make deductions about such rare events, one might wish to combine all datasets from measuring stations in a region to form a single series. The extent to which this is justified depends on the tail of the joint distribution of the rainfall amounts. See Buishand (1984).

  Buishand (1984) considered the ratio $q = \log H(x,x)/\log F(x)$. In the case of independence, this ratio is 2. For annual maximum daily rainfall data from The Netherlands, Buishand plotted q against F, for pairs of stations different distances apart. The ratio q increase with both F and distance, and seems to be tending to 2. For data restricted to the winter season, the results were more complex.

- Lewis (1975) briefly mentions work by himself and S J Daldry on annual maxima of wind and gust.

A salutary quotation (Klemeš, 1987): "The natural frequencies of flood peaks in the historic series are, in fact, almost never analyzed. We do not learn whether there seems to be any pattern such as clustering of high or low peaks, trend, or some other feature, nor any indication of some hydrological, geophysical or other context that could shed light on the historic flood record. What happens is that the actual time sequencing is completely ignored and the flood record is declared purely random. The ostensible reason for this is to 'simplify the mathematical treatment'. This, however, is rather amusing when one sees how the laudable resolve to keep things simple is then hastily abandoned and the use of the most advanced theories is advocated for the treatment of this artificially random sample on the pretext that the 'greatest amount of information' must be extracted."

9.18 Competing Risks

9.18.1 Format of Data

Passing reference has been made several times to the "competing risks" format of data. As its name suggests, this is concerned with an item of equipment which might fail due to either of two causes, or a living organism which might die from either of two causes. (Actually, it is common for one of the types of "failure" to be end-of-study censoring.) The variables of interest are the times to failure X_1 and X_2 from each cause operating individually, but we cannot observe them both, merely whichever was the smaller (and we also know which one it was that was the smaller). Stating this formally, we observe $Z = \min(X_1, X_2)$ and the label I defined by $Z = X_I$.

9.18.2 Methods of Analysis

How much does this limited information tell us about the bivariate distribution of (X_1, X_2)? Additional information is needed:

- The most popular approach is to assume that X_1 and X_2 are independent. It can then be deduced that

$$\bar{F}_1(t) = \exp\left[-\int_0^t \frac{a_1(u)}{P(u)}\,du\right], \tag{9.73}$$

  where $\bar{F}_1(t)$ is the survivor function for cause 1 (i.e., $\Pr(X_1 > t)$), $a_1(t)$ is the joint probability-p.d.f. that failure is of type 1 and occurs at time t, and $P(t)$ is the probability of surviving past time t. Since a_1/P is observable, $\bar{F}_1$ can be plotted. Naturally, there is a similar equation for $\bar{F}_2$.

- A parameterised functional form for $H(X_1, X_2)$ may be assumed—perhaps, the bivariate normal—and the parameters estimated by some such method as maximum likelihood.

- Klein and Moeschberger (1988) suggested one should let H have arbitrary marginals but assume it has one association parameter (called θ). Then if one is prepared to assume a particular value for θ, one can estimate the marginals. In particular, one might be prepared to put upper and lower bounds on θ, thus obtaining upper and lower bounds on $\bar{F}_1$ and $\bar{F}_2$. The form of H utilised by Klein and Moeschberger was the Pareto-like form associated with the names of Clayton and Oakes and described in Section 5.9.6.

- In an engineering context, it might happen that the two types of failure are failures of two different components, and that $\bar{F}_1$ and $\bar{F}_2$ are known, from experience with each of the components separately. In this case, it may be of interest to test whether there is any dependency between X and Y. See Klein (1986), who employed distributions (5.1) and (5.30) in a simulation study of the test he proposed.

For a fuller account, see David and Moeschberger (1978, especially Chapter 4). Chapter 9 of Cox and Oakes (1984) is on this subject.

9.18.3 Unusual Applications

Hutchinson (1983b) has drawn attention to two other contexts in which the competing risks model has been invented independently of the mainstream work:

- Impact tests are carried out to determine the fracture strength of cadaver legs. A blow is struck, the presence or absence of fracture is noted, the force applied is monitored. The severity of the blow is not the same from one test to another, so that even if fracture did not occur the peak force recorded would differ from one test to another. When fracture occurs, the force does not reach the peak which it would have reached in the absence of fracture, and the peak actually reached corresponds to the fracture strength of the leg. Thus what is known is (i) the minimum of the strength and the force available, and (ii) which of these the measurement is. Reference: Searle et al (1978).

- Suppose we wish to know the distribution of speeds which vehicle drivers would adopt (the so-called "free" speed) if they were not influenced by the presence of other vehicles. Branston (1979) devised a means of finding this. Speeds of vehicles on a motorway were measured, each vehicle being classified as unimpeded or impeded. Thus what is known is (i) the minimum of the free speed of the vehicle and the speed available to it, and (ii) which of these the observed speed is.

Further, it is likely that connexion could be made between this field and the various econometric fields that are outlined in Sections 19.9.5 and 22.4.2. We would interpret d and s in equation (19.2) as being the two times to failure, and our observations would consist of $\min(d, s)$ together with knowing which of the pair any particular observation was. The idea that there is a connexion here was prompted by the paper by Butler et al (1989) (see Section 22.4.2.4 for the work by these authors).

Chapter 10

Miscellaneous Bivariate Distributions

10.1 Contents of this Chapter

Sections 10.2–10.5 are concerned with trivariate reduction and similar methods of constructing bivariate distributions:

10.2 Multiplicative trivariate reduction.

10.3 Cherian's bivariate gamma distribution.

10.4 Weighted linear combination.

10.5 Khintchine mixture.

(See Section 15.4.2 regarding the intimate relation between trivariate reduction and compounding.)
The five remaining Sections of the Chapter are as follows:

10.6 Fox's H-function as a p.d.f.

10.7 Kimeldorf and Sampson's bivariate uniform distribution.

10.8 Densities for wave height and period.

10.9 A distribution having a singularity along $Y = X$.

10.10 The ARMA context.

10.2 Multiplicative Trivariate Reduction

10.2.1 Introductory Explanation

In Chapter 15, we shall consider schemes of trivariate reduction or variables-in-common for the construction of bivariate distributions, i.e.,

$$\left.\begin{aligned} X &= T(X_1, X_3) \\ Y &= T(X_2, X_3) \end{aligned}\right\} \qquad (10.1)$$

The X_i's are often i.i.d. r.v.'s, but they need not be. In the present Section, we discuss the case where the transformation T is multiplication.

10.2.2 Bryson and Johnson (1982)

Bryson and Johnson (1982) (and Johnson, 1987, Chapter 8) draw attention to Khintchine's theorem, which says that any continuous random variable X has a single mode at the origin if and only if it can be expressed as a product

$$X = ZU, \tag{10.2}$$

where Z and U are independent continuous variables, U having a uniform distribution on the unit interval. (See, for instance, Feller, 1971, Section V.9.) For given marginal density of X, f, the density f_Z has to be $-zf'(z)$, where f' is the derivative of f.

Bryson and Johnson present a multiplicative version of trivariate reduction,

$$\left.\begin{array}{rcl} X &=& ZU_1 \\ Y &=& ZU_2 \end{array}\right\} \tag{10.3}$$

where (U_1, U_2) has any bivariate distribution that has uniform marginals. Z is referred to as a "generator" variable. They find that the correlation between X and Y is

$$\left[3 - c_X^{-2} + \rho_{(U)}\left(1 + c_X^{-2}\right)\right]/4 \tag{10.4}$$

where c_X is the common coefficient of variation of X and Y, and $\rho_{(U)}$ is the correlation between U_1 and U_2. (A consequence of Khintchine's theorem is $c_X^2 \geq \frac{1}{3}$.) Specific examples are that if X and Y are exponential, the correlation between them is $(1 + \rho_{(U)})/2$, and if X and Y are normal, their correlation is $(3 + \rho_{(U)})/4$.

Bryson and Johnson also consider the scheme

$$\left.\begin{array}{rcl} X &=& Z_1U_1 \\ Y &=& Z_2U_2 \end{array}\right\} \tag{10.5}$$

where (U_1, U_2) is as before, but the Z's are now i.i.d. r.v.'s instead of identical. In this case, the correlation between X and Y is $\rho_{(U)}/(3c_X^2)$ (which is necessarily less than or equal to $\rho_{(U)}$); if they have normal or other symmetric distributions, they are uncorrelated, though they are independent only if the U's are.

Bryson and Johnson go on to discuss what they call Khintchine mixtures—see Section 10.5.

10.2.3 Gokhale (1973)

Gokhale (1973) gave some attention to the scheme of construction

$$\left.\begin{array}{rcl} X &=& ZV_1 \\ Y &=& ZV_2 \end{array}\right\} \tag{10.6}$$

where V_1, V_2, and Z were independent beta variates, whose parameters were either

- Respectively (a, θ), $(a + m, \theta - m)$, and $(a + \theta, b + m - \theta)$, so that X and Y had beta distributions with parameters $(a, b + m)$ and $(a + m, b)$, respectively; or

- Respectively $(a + \Delta, b - \Delta)$, $(a + \Delta, b' - \Delta)$, and (a, Δ), so that X and Y had beta distributions with parameters (a, b) and (a, b'), respectively.

10.2.4 Ulrich (1984)

Ulrich (1984) has considered

$$\left.\begin{array}{rcl} X &=& Z_1V_1 \\ Y &=& Z_2V_2 \end{array}\right\} \tag{10.7}$$

where the Z_i are independent, having gamma distributions (with unit scale parameter and shape parameter $\alpha_i + \phi$), and the V_i, independent of the Z's but possibly not mutually independent, have beta distributions with parameters α_i and ϕ. The scheme of dependence that Ulrich gives most attention to is that of his bivariate beta mixture (described briefly in Section 6.5.14). He calls the resulting distribution of (X,Y) the "bivariate product gamma".

10.3 Cherian's Bivariate Gamma Distribution

10.3.1 Formula for p.d.f.

$$h = \frac{e^{-(x+y)}}{\Gamma(\theta_1)\Gamma(\theta_2)\Gamma(\theta_3)} \int_0^{\min(x,y)} (x-z)^{\theta_1-1}(y-z)^{\theta_2-1} z^{\theta_3-1} e^z \, dz \tag{10.8}$$

10.3.2 Univariate Properties

Gamma marginals, shape parameters $\theta_1 + \theta_3$ and $\theta_2 + \theta_3$.

10.3.3 Correlation Coefficients

Pearson's product-moment correlation is $\frac{\theta_3}{\sqrt{(\theta_1+\theta_3)(\theta_2+\theta_3)}}$.

Dabrowska (1982) has discussed the behaviour of the monotone quadrant dependence function (for which, see Section 12.5)—whether the tendency for small values of Y to associate with small values of X is bigger or smaller than the tendency of big values of Y to associate with big values of X, for example.

10.3.4 Moment Generating Function

The m.g.f. is:

$$M(s,t) = (1-s)^{-\theta_1}(1-t)^{-\theta_2}(1-s-t)^{-\theta_3}. \tag{10.9}$$

10.3.5 Conditional Properties

The conditional distribution of Y given X is the sum of two independent random variables, one distributed as $X \times$ (standard beta variable, parameters θ_3 and θ_1) and the other as a standard gamma variable with parameter θ_2. The regression is linear, $E(Y|X=x) = \frac{\theta_3}{\theta_3+\theta_1}x + \theta_2$, and the conditional variance is quadratic, $\frac{\theta_1\theta_3}{(\theta_1+\theta_3)^2(1+\theta_1+\theta_3)}x^2 + \theta_2$. Reference: Johnson and Kotz (1972, p. 218).

10.3.6 Derivation

By trivariate reduction: let $X_i \sim \text{Gamma}(\theta_i, 1)$ (for $i = 1,2,3$), and let the X_i's be mutually independent. Then $X = X_1 + X_3$, $Y = X_2 + X_3$ have this distribution.

10.3.7 References to the Generation of Random Variates

The trivariate reduction method is a very easy method both to understand and to program for a computer—Devroye (1986, pp. 587–588). Consequently, this distribution could be used in simulation to generate a bivariate gamma population when only the marginals (gamma) and the correlation coefficient are specified—Schmeiser and Lal (1982).

10.3.8 Remarks

- This distribution originated with Cherian (1941), who considered the case in which $\theta_1 = \theta_2$.
- Independently of Cherian, David and Fix (1961) obtained this distribution and derived a number of properties. In particular, they gave explicit expressions for h for five combinations of small integer values of θ_1, θ_2, and θ_3.
- The joint probability density function has a different expression for $x < y$ and $x > y$—Moran (1967).
- The joint density can be expanded in terms of Laguerre polynomials—Eagleson (1964), Mardia (1970a, p. 75).
- Ghirtis (1967) has studied some properties of estimators for parameters of this distribution.

- For the density of X/Y, see Mielke and Flueck (1976) and Lee et al (1979).
- The class of bivariate gamma distributions having diagonal expansions, considered by Griffiths (1969), includes Cherian's, Kibble's, and Jensen's.

10.3.9 Application

Prékopa and Szántai (1978) used a 6-variate gamma distribution constructed in this style, from sums of 20 independent gamma variates, in a study of the streamflows of a river in six months of the year.

10.4 Weighted Linear Combination

10.4.1 Derivation

$$\left.\begin{aligned} X &= U_1 \\ Y &= cU_1 + (1-c)U_2 \end{aligned}\right\} \tag{10.10}$$

$(0 \le c \le 1)$, where the U's are i.i.d. r.v.'s.

10.4.2 Formula for p.d.f.

If the U's have a negative exponential distribution:

$$h = \tfrac{1}{1-c} e^{-x-y+2cx}, \tag{10.11}$$

the support being part of the positive quadrant.

If the U's have a Laplace distribution:

$$h = \tfrac{1}{4(1-c)} e^{(-\,|x| - |y-cx|)/(1-c)}, \tag{10.12}$$

the support being the whole plane.

If the U's have a uniform distribution:

$$h = \tfrac{1}{1-c}, \tag{10.13}$$

the support being part of the unit square.

For further details, see Johnson and Tenenbein (1979, 1981).

10.4.3 Correlation Coefficients

For the grade correlation and Kendall's τ, see Section 13.8 and Johnson and Tenenbein (1979, 1981).

10.4.4 Remark

For given distribution of the U's, these distributions have the "monotone regression dependence" property, i.e., the degree to which they are regression dependent is a monotone function of the parameter indexing the family, c (Bilodeau, 1989).

10.5 Khintchine Mixture

10.5.1 Derivation

Continuing the discussion of bivariate distributions suggested by Bryson and Johnson (1982) and Johnson (1987, Chapter 8) that we started in Section 10.2, let

$$\left.\begin{aligned} X &= Z_1U_1 \\ Y &= Z_2U_2 \end{aligned}\right\} \tag{10.14}$$

where the U's are uniformly distributed on (0,1), and either:

- the U's are independent, and the Z's are either identical (with probability p) or independent (with probability $1-p$), or
- the Z's are independent, and the U's are either identical (with probability q) or independent (with probability $1-q$).

(As before, the Z's are referred to as "generator" variables.)

10.5.2 Exponential Marginals

If X and Y are to have exponential marginals, Bryson and Johnson give these results:

- The case of independent U's and identical Z's gives a p.d.f. of $-\text{Ei}[-\max(x,y)]$, where Ei(.) is the exponential integral (see Section 8.2.2 for the definition of this, which we hope is the same as Bryson and Johnson's).
- The case of independent Z's and identical U's gives a p.d.f. of $\frac{xy}{(x+y)^3}[2+2(x+y)+(x+y)^2]e^{-(x+y)}$.
- In the fully independent case, the p.d.f. is $e^{-(x+y)}$.

The correlation is $p/2$ if the first and third are mixed in proportions $p:1-p$, $q/3$ if the second and third are mixed in proportions $q:1-q$, and $\frac{p}{2}+\frac{q}{3}$ if all three are mixed in proportions $p:q:1-p-q$.

The following five cases are illustrated (contour and three-dimensional plots of the p.d.f.'s) by Johnson et al (1981): independent U's, independent Z's (i.e., $p=q=0$); independent U's, $p=0.6$; independent U's, identical Z's; independent Z's, $q=0.6$; independent Z's, identical U's. (The final one is also shown in Figure 8.2 of Johnson, 1987.)

10.5.3 Normal Marginals

This case is also treated by Bryson and Johnson, but the formulae are more complicated than in the exponential case.

The following six cases are illustrated (contour and three-dimensional plots of the p.d.f.'s) by Johnson et al (1981): $q=0, p=0$; $q=0, p=0.25$; $q=0, p=0.5$; $q=0, p=0.75$; $q=0, p=1$; $q=0.25, p=0.75$. The two cases $p=1$ and $q=1$ are illustrated in Figures 8.3 and 8.4 of Johnson (1987).

Three examples in which the U's have the Farlie-Gumbel-Morgenstern distribution are also illustrated by Bryson and Johnson (1982) and Johnson (1987, Figures 8.5–8.7). The density in this case is given by these authors as

$$h = \frac{\alpha xy}{2\max(|x|,|y|)}\phi[\max(|x|,|y|)] + \frac{1-\alpha xy}{2}\{1-\Phi[\max(|x|,|y|)]\}. \tag{10.15}$$

There are related illustrations in Johnson et al (1984, p. 242) and Johnson (1986).

10.5.4 References to the Generation of Random Variates

Devroye (1986, pp. 603–604); Johnson et al (1984, pp. 239–240).

10.6 Fox's H-Function as a p.d.f.

One of the higher transcendental functions of mathematics is Fox's H-function. In its one-variable form, the H-function distribution subsumes many of the well-known univariate distributions of positive random variables (Cook and Barnes, 1981; Springer, 1983). The bivariate H-function distribution has been discussed by Kellogg and Barnes (1987). Its p.d.f. is proportional to the H-function of two variables.

- Six examples of special cases are the quarter Cauchy, McKay's bivariate gamma, the bivariate beta, and three types of general exponential distribution, p.d.f. $\propto (x^2+y^2)^\beta \exp[-\alpha(x^2+y^2)]$, p.d.f. $\propto \exp(-\alpha x - \beta y/x)$, and p.d.f. $\propto x^c \exp(-\alpha x - \beta xy)$.

- The distribution of $Z = X^p Y^q$, where p and q are rational, is a univariate H-function.

Further results are given by Kellogg and Barnes (1989), including a demonstration that the cumulative distribution is also a Fox's H-function. The paper also includes 17 illustrations of five special classes of the p.d.f.—McKay's bivariate gamma, the bivariate beta, and the three types of "general exponential" distribution mentioned above.

Whether this distribution is too general to be really useful remains to be seen. Our feeling is that there is a lot of potential for distributions involving advanced mathematical functions, once two types of computer software packages are widely available—algebraic manipulation programs that are clever enough to handle these functions (and are user-friendly), and ones for numerically evaluating them; indeed, it is not too much to hope for merging of the algebraic and numerical operations.

10.7 Kimeldorf and Sampson's Bivariate Uniform Distribution

10.7.1 Support

Part of the unit square.

10.7.2 Formula for p.d.f.

Kimeldorf and Sampson (1975a) constructed a distribution with p.d.f. h as follows:

- β, i.e., a constant, on each of $[\beta]$ squares of side $1/\beta$ arranged corner-to-corner up the diagonal from $(0,0)$ to $(1,1)$, where $[\beta]$ is the largest integer not exceeding β,
- $\beta/(\beta - [\beta])$ on one smaller square of side $1 - [\beta]/\beta$ in the top right corner of the unit square (unless β is an integer), and
- zero elsewhere.

10.7.3 Formula for (Cumulative) Distribution Function

An expression for this can be obtained, but is cumbersome.

10.7.4 Univariate Properties

Uniform marginals.

10.7.5 Correlation Coefficients

See Chapter 13. Remember that for any distribution whose marginals are uniform, Pearson's product-moment correlation is the same as the grade correlation.

10.8 Densities for Wave Height and Period

As will be mentioned in Section 24.12.2—8, Goda (1979) published some data on the joint distribution of the height and period of sea waves. In the same paper, two explicit expressions for p.d.f.'s applicable to this pair of variables were given.

The first was that of Longuet-Higgins:

$$h = \frac{\pi x^2}{4\nu} \exp\left\{-\frac{\pi}{4}x^2\left[1 + \frac{(y-1)^2}{\nu^2}\right]\right\}, \tag{10.16}$$

where x and y are normalised height and normalised period, respectively. The distribution is symmetric about $y = 1$. The correlation is necessarily zero.

The second was that of A Cavanié and colleagues:

$$h = \frac{Ax^2}{y^5} \exp\left\{-B\frac{x^2}{y^4}\left[1 + \frac{(y^2 - \alpha^2)^2}{C}\right]\right\}. \tag{10.17}$$

This distribution is able to reflect the substantial (0.7 or 0.8) correlation that is sometimes found empirically.

The paper by Longuet-Higgins (1975) includes an account of the theory on which (10.16) is based, an illustration of the contours, and some comparisons with observations. The conclusions are worth quoting because they give the flavour of the theory, too: "We have compared the theoretical distribution of wave heights and periods, applicable to a narrow Gaussian wave spectrum, with ocean wave data from the Gulf of Mexico and found good agreement. It has to be remembered, however, that in pressure records, much of the high-frequency tail of the spectrum is reduced, and in unfiltered records of surface elevation the narrow band approximation is probably less applicable. The data were also taken in moderately deep water. In shallow water, especially for steep waves, the distribution can again be expected to be less applicable because of the enhanced nonlinearity. Nevertheless, it appears that under the appropriate conditions the theoretical distribution may be of some use."

Recently, Srokosz and Challenor (1987) have compared three theoretical bivariate height-period distributions with data. That of Cavanié and colleagues we have met above. That of G Lindgren is not available in closed analytical form, but requires numerical integration. The third is due to Longuet-Higgins (1983):

$$h = A\left(\frac{x}{y}\right)^2 \exp\left\{-x^2\left[1 + B(1 - y^{-1})^2\right]\right\}. \tag{10.18}$$

The three datasets used by Srokosz and Challenor were notable for the wave conditions remaining approximately stationary for several hours; two were from off the Scilly Isles, one was from a reservoir. The comparisons were by means of contour diagrams. Lindgren's distribution came out best, but was found to be expensive computationally.

Srokosz and Challenor noted that if bivariate distributions of height and period turn out to be bimodal, this might be taken as evidence for the importance of nonlinearities, or it might simply result from a mixture of two distributions.

Perhaps we should finally mention that this appears to be a far from mature subject—it does not seem to be generally agreed, for instance, what is a wave. One definition is based upon zero-upcrossings. Another is local maxima.

10.9 A Distribution Having a Singularity Along $Y = X$

10.9.1 The Distribution

Let us have two variates, X and Y, each uniformly distributed over the range 0 to 1. With probability p, let Y be equal to X; with probability $1 - p$, let Y be independent of X. Then the bivariate p.d.f. h is flat (with a height of $1 - p$), except along $Y = X$, where there is a singularity containing the probability p.

Our construction of this distribution is motivated by the statistic κ/κ_M proposed by Cohen (1960). This will now be explained. The distribution was previously considered by Konijn (1959).

10.9.2 Cohen's Kappa

Cohen's kappa (κ) is a popular statistic for measuring the agreement between two observers in assigning items to one of two categories. We shall argue below that it is a poor choice and its popularity is undeserved, but first we explain how and why it is calculated. Suppose the counts of items in the cells of our 2×2 table (first observer's categorisation $\times$ second observer's) are N_{ij}; the number of items to which the observers gave the same classification is $N_{11} + N_{22}$ ($= a$, say). The number of items which would have received the same classification had the observers' judgments been randomly paired is (in an obvious notation) $\frac{N_{1\cdot}N_{\cdot 1}}{N_{\cdot\cdot}} + \frac{N_{2\cdot}N_{\cdot 2}}{N_{\cdot\cdot}}$ ($= b$, say). Then κ is the ratio of the amount that agreement exceeds the chance level $\left(\frac{a-b}{N_{\cdot\cdot}}\right)$ to the amount by which perfect agreement would exceed the chance level $\left(1 - \frac{b}{N_{\cdot\cdot}}\right)$. That is, $\kappa = \frac{a-b}{N_{\cdot\cdot}-b}$.

Cohen's kappa is usually justified by saying that some level of agreement is to be expected merely by chance, and therefore that $a/N..$ is a poor statistic. True though this is, it is not exactly a rigorous proof of the merits of κ!

10.9.3 Criticisms of Cohen's Kappa

A number of papers have appeared in recent years that have highlighted the deficiencies of κ. The most important criticism, in our view, is that κ mixes together the following two distinct types of disagreement: those occurring because the observers have different propensities to place items in the positive category rather than in the negative category, and those occurring because they rank-order the items differently (or would do so if asked). The first type can be interpreted as different positioning along an unobserved continuum of a threshold separating positive items from negative ones. If one accepts this distinction, it is inherently impossible to realistically represent the two types by a single number.

10.9.4 Cohen's Normalisation of Kappa

Curiously, Cohen's original paper made much the same point as above about the importance of the marginals: "It may occasionally occur that the setting of the interjudge agreement may impose some inherent difference in category widths between judges. For example, Judge A may be a clinical psychologist working with projective techniques who may make the diagnosis of schizophrenia more frequently than Judge B, a psychiatrist working with screening interview material." We part company with Cohen when he goes on to say that distinguishing between the two types of disagreement is usually unimportant.

For given margins, the maximum possible level of agreement occurs when one of the cells is zero. Cohen denotes the corresponding value of kappa κ_M, and goes on to suggest the ratio κ/κ_M for answering "How much of the marginally-permitted agreement is present?" when the margins are appreciably different.

The ratio κ/κ_M can reflect agreement between two dichotomisations, uncontaminated by threshold differences. Specifically, it does so under the following circumstances: in a proportion p of cases, the two judges agree precisely as to where the item falls on the relevant latent scale; in a proportion $1-p$ of cases, the two judges' opinions are quite independent of each other. The distribution described in Section 10.9.1 results when X and Y are taken to have uniform distributions.

To prove the above, observe that κ/κ_M is the ratio of the extent by which agreement exceeds chance to the maximum amount by which agreement can exceed chance; consequently, if perfect agreements (on the latent scale) between the judges, and random pairings of their judgments, are mixed in the proportions $p : 1-p$, then κ/κ_M will be p, independent of where the two judges position their thresholds.[1]

Thus if we want a measure of association that is unaffected by differences between the marginal distributions, κ is incapable of fulfilling this role; κ/κ_M can do so, but a statistic based upon some other distribution, such as tetrachoric correlation based upon the bivariate normal, will usually be preferable, as the distribution of Section 10.9.1 is very quirky, not smooth.

Agresti (1989) has recently had much the same idea, though in the contingency table rather than continuous bivariate distribution context, and he does not introduce κ_M, but uses a model in which κ_M is necessarily 1.

---

[1]If this is not apparent, write

$$\frac{\kappa}{\kappa_M} = \frac{\text{actual} - \text{chance}}{\text{maximum} - \text{chance}}$$
$$= \frac{p\,\text{maximum} + (1-p)\,\text{chance} - \text{chance}}{\text{maximum} - \text{chance}}$$

$= p$, independent of threshold positions. In the above equations, actual = number of agreements when the scale is dichotomised (whether these arise from the perfect agreement on the latent scale or not), chance = number of agreements arising by chance, and maximum = the maximum number of agreements compatible with the dichotomisations.

10.10 The ARMA Context

Bivariate distributions arising from time series models are outside the scope of this book. Nevertheless, passing mention should be made of the recent paper by Damsleth and El-Shaarawi (1989). These authors considered autoregressive moving average (ARMA) models in which the "noise" has a Laplace distribution instead of the usually-assumed normal. Most of their results are for the AR(1) model $X_t = \phi X_{t-1} + \varepsilon_t$, with ε_t having a Laplace distribution. Damsleth and El-Shaarawi obtain an expression (an infinite series) for the p.d.f. of X (notice that this is not a Laplace distribution). They then extend this to the joint distribution of X_t and X_{t-k}, and give six contour plots of the resulting p.d.f. (for $\phi = 0.25$ and 0.90, and $k = 1$, 5, and 10). At least four of these have a quite different appearance to the corresponding normal contours. Damsleth and El-Shaarawi applied both normal and Laplace AR(1) models to a dataset of 147 weekly observations of sulphate concentration in the Turkey Lakes Watershed, Ontario.

Part III

Mostly Theory

Chapter 11

Measures of Correlation

11.1 Organisation of Part III

There are eight Chapters in Part III of this book. They are mostly concerned with theoretical issues arising with bivariate distributions, and are as listed below.

11. Measures of correlation.
12. Concepts of stochastic dependence.
13. Rank correlation proportion.
14. Diagonal expansion.
15. Variables in common.
16. Overview of generating random variates.
17. Reliability models leading to bivariate distributions.
18. Reliability classes: concepts of ageing.

11.2 Introduction

A measure of dependence indicates in some defined sense how closely X and Y are related; one extreme will include the case of complete linear dependence, and the other extreme will be mutual independence. Although it is customary in bivariate data analysis to compute a correlation coefficient of some sort, one number (or index) alone can never reveal the nature of dependence fully; hence a variety of measures is needed.

The division of the Chapter is as follows:

11.3 Total dependence.

11.4 Conditions on a measure of dependence.

11.5 Pearson's product-moment correlation coefficient.

11.6 Maximal correlation (sup correlation).

11.7 Monotone correlation.

11.8 Rank correlations.

11.9 Measure of Schweizer and Wolff, and related measures.

11.10 Matrix of correlations.

11.11 Tetrachoric and polychoric correlation.

11.12 Compatibility with perfect rank ordering.

11.13 Conclusions on measures of dependence.

11.14 Local dependence function.

11.15 Regional dependence.

11.3 Total Dependence

Let us now examine what an antithesis of stochastic independence is, i.e., clarify the concept of "total" dependence.

11.3.1 Functions

Regarding different definitions of total dependence, it is helpful to remind ourselves what a function is.

- By a function b from a set A to another set B we mean a mapping (rule) which assigns to each x in A a unique element $b(x)$ in B. (Because of the uniqueness requirement, $\pm\sqrt{x}$, for instance, is not a function.)

- b is called *one-to-one* if $b(x) = b(y)$ only when $x = y$.

- b is *onto* if $b(A) = B$. That is, for each y in B there exists at least one x in A such that $b(x) = y$.

- A function b that is one-to-one and onto is called a one-to-one correspondence. Such a function has an inverse, which is denoted by b^{-1}.

- b is said to be Borel measurable if for each α the set $\{x : f(x) > \alpha\}$ is a Borel set, which is typically a countable union of open or closed sets or complements of these. (The reader need not get bogged down with this as most functions we come across are Borel measurable.)

In this Chapter, we assume all the functions are Borel measurable and they are onto.

11.3.2 Mutual Complete Dependence

If each of two random variables X and Y can be predicted from the other, then, intuitively, X is a function of Y and Y is a function of X, and X and Y are dependent on each other. In order to define this more formally, we first need the following definition.

Definition: A random variable Y is completely dependent on X if there exists a function b such that

$$\Pr[Y = b(X)] = 1. \tag{11.1}$$

This equation essentially means that $Y = b(X)$, as we are not too concerned with events of zero probability.

Definition: X and Y are *mutually completely dependent* if (11.1) holds for some one-to-one function b (Lancaster, 1963b).

The concept of mutual complete dependence is directly opposite to stochastic independence, in that mutual complete dependence entails complete predictability of either random variable from the other (i.e., X and Y are mutually determined), while stochastic independence entails X and Y being useless in predicting one another.

11.3.3 Monotone Dependence

Clearly, if a sequence $\{(X_n, Y_n)\}$ of pairs of independent random variables converges in distribution[1] to (X, Y), then X and Y must be mutually independent. However, Kimeldorf and Sampson (1978) constructed a sequence of pairs of mutually completely dependent random variables, all having a uniform distribution on [0,1], which converges in distribution to a pair of independent random variables each having a uniform distribution on [0,1]. From this point of view, mutual complete dependence is not perfectly opposite to independence. This defect of mutual complete dependence motivated Kimeldorf and Sampson (1978) to present a new concept of total statistical dependence, called monotone dependence.

Definition: Let X and Y be continuous random variables. Then Y is *monotonely dependent* on X if there exists a strictly monotone[2] function b for which $\Pr[Y = b(X)] = 1$. It is clear that Y is monotonely dependent on X if and only if X is monotonely dependent on Y. We can therefore make the following further definitions.

Definitions: If b in the previous definition is increasing, X and Y are said to be *increasing dependent*; if b is decreasing, X and Y are said to be *decreasing dependent.*

Note: a function g may be one-to-one and yet not be monotone, e.g., $b(x) = x$, $3 - x$, or x, according to whether x is between 0 and 1, between 1 and 2, or between 2 and 3. Hence monotone dependence is stronger than mutual complete dependence.

Kimeldorf and Sampson (1978) showed that a necessary and sufficient condition that X and Y be increasing (decreasing) monotone dependent is that the joint distribution function of (X, Y) is H^+ (H^-), these being the Fréchet bounds (see Section 3.1.2).

11.3.4 Functional and Implicit Dependence

These are weaker definititions of total dependence.

Definition: X and Y are *functionally dependent* if either $X = a(Y)$ or $Y = b(X)$ for some functions a and b. (See Rényi, 1959, and Jogdeo, 1982.) X and Y are functionally dependent if either X is completely dependent on Y or vice versa. An example is $Y = X^2$.

Definition: X and Y are *implicitly dependent* if there exist two functions a and b such that $a(X) = b(Y)$, with $\text{var}[a(X)] > 0$. (See Rényi, 1970, p. 283.) In other words, there may exist no function connecting X and Y and yet they are related. For example, consider the relation $X^2 + Y^2 = 1$: if we set $a(X) = X^2$ and $b(Y) = 1 - Y^2$, then $a(X) = b(Y)$. Nevertheless, $Y = \pm\sqrt{1 - X^2}$ is not a function as it assigns one value of X to two values of Y.

11.3.5 Overview

We may now list different notions of total dependence in decreasing order of strength:

- Linear dependence,
- Monotone dependence,
- Mutual complete dependence,
- Functional dependence,
- Implicit dependence.

11.4 Conditions on a Measure of Dependence

If X and Y are not totally dependent, then it may be helpful to find some quantities that can measure the strength or degree of dependence between them. If such a measure can be expressed as a scalar, it is often more convenient to refer to it as an index. We may ask what conditions ought an index to satisfy,

[1] The sequence $\{F_n\}$ of distribution functions of the random variables $\{X_n\}$ is said to be convergent in distribution to F if $\lim_{n\to\infty} F_n(x) = F(x)$, for every point for which F is continuous.
[2] A function b is strictly monotone increasing if $b(x) < b(y)$ whenever $x < y$.

in order to be useful. Rényi (1959) proposed a set of seven conditions for this purpose,[3] and showed that the maximal correlation (Section 11.6) fulfils all of them. Lancaster (1982b) has modified and enlarged Rényi's set of axioms to nine conditions.

Let $\delta(X, Y)$ denote an index of dependence between X and Y. The following, apart from the last condition, is Lancaster's version of Rényi's conditions. Condition (9) is taken from Schweizer and Wolff (1981) instead of Lancaster (1982b), as the latter is expressed in highly technical language.

1. $\delta(X, Y)$ is defined for any pair of random variables, neither of them being constant with probability 1. This is to avoid trivialities.

2. $\delta(X, Y) = \delta(Y, X)$. But notice that while independence is a symmetric property, total dependence is not, as one variable may be determined by the other, but not vice versa.

3. $0 \leq \delta(X, Y) \leq 1$. Lancaster says that this is an obvious choice, but not everyone will agree.

4. $\delta(X, Y) = 0$ if and only if X and Y are mutually independent. Notice how strong this condition is made by the "only if".

5. If the functions a and b map the spaces of X and Y in a one-to-one manner, respectively, onto themselves, then $\delta(a(X), b(Y)) = \delta(X, Y)$. The condition means that the index remains invariant under one-to-one transformations of the marginal variables.

6. $\delta(X, Y) = 1$ if and only if X and Y are mutually completely dependent.

7. If X and Y are jointly normal, with correlation coefficient ρ, then $\delta(X, Y) = |\rho|$.

8. In any family of distributions defined by a vector parameter $\boldsymbol{\theta}$, $\delta(X, Y)$ must be a function of $\boldsymbol{\theta}$.

9. If (X, Y) and (X_n, Y_n), $n = 1, 2, \ldots$ are pairs of random variables with joint distributions F and F_n respectively, and if $\{F_n\}$ converges in distribution to F, then $\lim_{n\to\infty} \delta(X_n, Y_n) = \delta(X, Y)$.

Another version of Rényi's axioms for a symmetric, nonparametric, measure of dependence is given in Schweizer and Wolff (1981). A similar set of criteria for a good measure of association (dependence) is also given at pp. 206–207 of Gibbons (1971). So-called nonparametric measures like Kendall's and Spearman's rank correlations will be discussed in Section 11.8.

Comments on the above conditions:

- A curious feature of the list is its mixture of the trivial and/or unhelpful with the strong and/or deep. We would say that (1), (3), (7), and (8) fall into the first category (unless there are subtle consequences to them that elude us), whereas (2), (4), (5), and (6) fall into the second. We are unsure about (9).

- Summarising, conditions (2), (5), (4), and (6) say that we are looking for a measure that is symmetric in X and Y, is defined by the ranks of X and the ranks of Y, attains 0 only in the case of independence and attains 1 whenever there is mutual complete dependence.

- Condition (3) is too restrictive for correlations, as the range of these is traditionally from -1 to $+1$

- (6) is stronger than the original condition which says $\delta(X, Y) = 1$ if either $X = a(Y)$ or $Y = b(X$ for some functions a and b, i.e., if X and Y are functionally dependent. Rényi intentionally left out the converse implication, i.e., $\delta(X, Y) = 1$ only if X and Y are functionally dependent, as he felt it to be too restrictive. The strengthening from functional dependence to mutual complete dependence is possibly due to Lancaster himself.

- Condition (7) is not appropriate to rank correlations—it should be replaced by δ being a strictly increasing function of $|\rho|$, as is done by Schweizer and Wolff (1981).

[3]See references cited by Lancaster (1982b) for work prior to Rényi.

- Schweizer and Wolff (1981) claimed that, at least for nonparametric measures, Rényi's original conditions are too strong.

- The chief point about these axioms is not their virtues or demerits, either individually or as a set, but that they make us think about what we mean by dependence and what we want from a measure of it. They provide a yardstick against which to measure the properties of different measures.

11.5 Pearson's Product-Moment Correlation Coefficient

11.5.1 Definition

Pearson's product-moment correlation is a measure of the strength of the linear relationship between two random variables, its definition being

$$\rho(X,Y) = \frac{\text{cov}(X,Y)}{\sqrt{\text{var}(X)\text{var}(Y)}}, \tag{11.2}$$

$\text{cov}(X,Y) = E\{[X - E(X)][Y - E(Y)]\}$ being the covariance of X and Y, and $\text{var}(X)$ and $\text{var}(Y)$ being the variances of X and Y. If either of the variables is constant, the correlation is undefined. If either has infinite variance, it may be possible to extend this definition, as outlined in Section 7.11.5 for bivariate "stable" distributions. From the definition, it is clear that conditions (1) and (2) of the previous Section are satisfied.

It is clear that $|\rho(X,Y)| \leq 1$; equality occurs only if X and Y are linearly dependent;[4] ρ takes the same sign as the slope of the regression line. Suppose the marginals $F(x)$ and $G(y)$ are given. Then ρ can take all values in the range -1 to 1 if and only if constants α and β exist such that $\alpha + \beta X$ has the same distribution as Y, and the distributions are symmetrical about their means (Moran, 1967).

If X and Y are independent, then $\rho(X,Y) = 0$. But zero correlation does not imply independence—hence, condition (4) of the previous Section is not satisfied. (Between uncorrelatedness and independence lies "semi-independence". This means that $E(Y|X) = E(Y)$ and $E(X|Y) = E(X)$. See Jensen, 1988.) As is well known, adding constants to X and Y does not alter $\rho(X,Y)$, and neither does multiplying X and Y by constant factors with the same sign. As $\rho(X,Y)$ may be negative, condition (3) is violated. Further, $\rho(X,Y)$ is not invariant under monotonic transformations of the marginals, so condition (5) is not satisfied. And as $\rho(X,-X) = -1$, the "if" part of condition (6) is not satisfied. Conditions (7) and (8) are obviously satisfied. Condition (9) is satisfied—we can prove this by using the continuity theorem for two-dimensional characteristic functions (Cramér, 1946, p. 102) and the expansions of such characteristic functions in terms of product moments (Bauer, 1972, pp. 264–265).

As to estimating the correlation in a sample of n bivariate observations $(x_1, y_1), \ldots (x_n, y_n)$, the usual formula is

$$r = \frac{\sum_{i=1}^n (x_i - \bar{x})(y_i - \bar{y})}{\sqrt{\sum_{i=1}^n (x_i - \bar{x})^2 \sum_{i=1}^n (y_i - \bar{y})^2}}, \tag{11.3}$$

where $\bar{x}$ and $\bar{y}$ are the respective sample means.

If $(x_1, y_1), \ldots (x_n, y_n)$ are n independent pairs of observations from a bivariate normal distribution, r is a maximum likelihood and approximately unbiased estimator of ρ. A disadvantage of r is that it is very sensitive to contamination of the sample by outliers. Devlin et al (1975) have compared r with various other estimators of ρ with respect to their robustness; see Ruppert (1988) for ideas on multivariate "trimming" (i.e., removal of extreme values).

$\rho(X,Y)$ will be abbreviated to ρ whenever there is no ambiguity; further, the symbols ρ' and ρ^* will be used for other types of correlation.

---

[4] Perhaps we should point out that weird cases like

$$Y = \begin{cases} X & \text{if } X \text{ is irrational} \\ 0 & \text{if } X \text{ is rational} \end{cases}$$

can be written down (Romano and Siegel, 1986, Example 4.27). This has correlation 1.

The distribution of $z = \frac{1}{2}\log\frac{1+r}{1-r} = \tanh^{-1} r$, called Fisher's variance-stabilising transformation of r, approaches normality (as n increases) much faster than that of r, particularly when $\rho \neq 0$. For a fuller discussion, see Rodriguez (1982b). The article by Mudholkar (1983) includes some comments on the behaviour of this transformation when the parent population is non-normal.

11.5.2 Interpretation of Correlation

Rodriguez (1982b) describes the historical development of correlation, and states that although Karl Pearson was aware that high correlation between two variables may be due to a third variable, this was not generally recognised until Yule's (1926) paper. One aspect of the difficulty of interpreting correlation is that it is still all too easy to confuse it with causation.

Rodriguez argues that, in order to interpret a calculated correlation, an accompanying probability model for the chance variation in the data is necessary, the two most common ones being as follows.

- The bivariate normal distribution. In this case, r estimates the parameter ρ; confidence intervals may be constructed for ρ and hypothesis tests carried out.

- The simple regression model, $y_i = \alpha + \beta x_i +$ random error. Here, r^2 represents the proportion of total variability (as measured by the sum of squares) in the y's which can be explained by the linear regression. That is,
$$r^2 = \frac{\sum_{i=1}^{n}(\hat{y}_i - \bar{y})^2}{\sum_{i=1}^{n}(y_i - \bar{y})^2}, \tag{11.4}$$
where $\hat{y}_i$ is the predicted value of y_i, calculated from the estimated regression equation. In the regression context, the x_i are very often chosen, not random, and thus there is no underlying bivariate distribution for r to be an estimate of a parameter in.

Even so, says Elffers (1980), "It can be difficult (i) to decide when a particular value of ρ indicates association strong enough for a given purpose, and (ii) in a given situation, to weigh the losses involved in obtaining more strongly associated variables against the gains." Elffers therefore puts forward functions of the correlation that can be interpreted as probabilities of taking a wrong decision in certain situations.

Though they are elementary, it is perhaps worth emphasising these four points:

- $r = 0.0$ does not mean that there is no relationship between the x's and the y's. A scatterplot might reveal a clear (though nonlinear) relationship.

- And even if the correlation is close to 1, the relationship may be obviously nonlinear, either to the eye when plotted directly, or because a transformation reveals a relationship that is incompatible with linearity. If X is uniformly distribution over the range 8 to 10, and Y is proportional to X^2, then the correlation between X and Y is approximately 0.999 (Blake, 1979, Example 6.18).

- Lots of different-looking sets of points can all produce the same value of r (see Chambers et al, 1983, Section 4.2, for eight scatterplots, all having $r = 0.7$).

- The value of r calculated from a small sample may be totally misleading if not viewed in the context of its likely sampling error.

In view of the above, the computation of r should be accompanied by the use of such devices as scatterplots. When the data are not from a bivariate normal population, r provides only limited information about the observations.[5] Barnett (1985a), citing two scatterplots in Barnett (1979), has expressed the view that for highly skew bivariate distributions, such as those with exponential marginals, the ordinary correlation coefficient is not a very useful measure of association. See Section 4.14.5 for the limitations of correlation when applied to bivariate lognormal variates.

[5] Nevertheless, it is possible to overstate this point. McDonald (1960), for instance, argued that even the grossly non-normal bivariate distributions met with in meteorology and hydrology do not disturb r to a degree that is of practical importance.

11.5.3 Correlation Ratio

The interpretation of r^2 given in Section 11.4.2, which presumes that var$(Y|X)$ is a constant, suggests writing the theoretical correlation as $\rho^2 = 1 - \frac{\text{var}(Y|X)}{\text{var}(Y)}$. More generally (i.e., beyond the context of linear regression), the quantity η, calculated from $\eta^2 = 1 - \frac{E[Y-E(Y|X)]^2}{\text{var}(Y)}$, is termed the correlation ratio of Y on X, and was introduced by Pearson (1905). For further discussion on this, see Chapter 26 of Kendall and Stuart (1979).

11.5.4 Chebyshev's Inequality

For any univariate distribution with zero mean and unit standard deviation, $\Pr(|X| \leq a) \geq 1 - a^{-2}$, for all $a > 0$. For general mean and standard deviation, the left hand side of the inequality becomes $\Pr(|X - \mu| \leq a\sigma)$.

For any bivariate distribution with zero means, unit standard deviations, and correlation ρ,

$$\Pr(|X| \leq a \text{ and } |Y| \leq a) \geq 1 - \left(1 + \sqrt{1-\rho^2}\right) a^{-2}.$$

More generally,

$$\Pr(|X| \leq a_1 \text{ and } |Y| \leq a_2) \geq 1 - \frac{\frac{a_1}{2a_2} + \frac{a_2}{2a_1} + \sqrt{\left(\frac{a_1}{2a_2} + \frac{a_2}{2a_1}\right)^2 - \rho^2}}{a_1 a_2}.$$

See Tong (1980, Section 7.2).

11.6 Maximal Correlation (Sup Correlation)

A frequently quoted measure of dependence between two random variables X and Y is that of maximal correlation, introduced by Gebelein (1941), studied among others by Rényi (1959) and Sarmanov (1962, 1963), and defined by

$$\rho'(X,Y) = \sup \rho[a(X), b(Y)],$$

where the supremum[6] is taken over all Borel-measurable functions a and b, such that var$[a(X)]$ and var$[b(Y)]$ are finite and nonzero, and where ρ represents the ordinary (Pearson's product-moment) correlation coefficient. The maximal correlation is also known as sup correlation. The properties of maximal correlation as a measure of dependence are discussed in Rényi (1959). This measure satisfies the following:

1. $0 \leq \rho'(X,Y) \leq 1$.

2. $\rho'(X,Y) = \rho'(Y,X)$.

3. $\rho'(X,Y) = 0$ if and only if X and Y are independent. (To see this, consider indicator functions of $X \leq \xi$, $Y \leq \eta$, where ξ, η are varied).

4. If X and Y are mutually completely dependent, then $\rho'(X,Y) = 1$, but the converse is not true. (See Lancaster (1963b) for counter examples, and for necessary and sufficient conditions for the complete mutual dependence of random variables). Hence condition (6) of Section 11.4 fails in part.

5. It is obvious from the definition that $|\rho(X,Y)| \leq \rho'(X,Y)$.

6. $|\rho(X,Y)| = \rho'(X,Y)$ if (X,Y) is the bivariate normal random variable. This is because in this particular case, $|\rho[a(X), b(Y)]| \leq |\rho(X,Y)|$, equality holding only when a and b are identity functions (see Kendall and Stuart, 1979, p. 600).

[6]The supremum of a set A is the least upper bound of A. Those unworried by technicalities may think of the supremum as the maximum.

7. Condition (9) of Section 11.4 is not fulfilled—as mentioned at the beginning of Section 11.3.3, Kimeldorf and Sampson (1978) give an example of a sequence of mutually completely dependent r.v.'s $\{(X_n, Y_n)\}$ converging in distribution to a distribution in which X and Y are independent. Clearly, $\rho'(X_n, Y_n) = 1$, but $\rho'(X, Y) = 0$.

Rényi (1970, p. 283) proved that even if X and Y are only implicitly dependent, then $\rho'(X, Y)$ is still equal to 1.

If the bivariate distribution is ϕ^2-bounded (see Section 14.3), then the maximal correlation equals ρ_1, the first canonical correlation coefficient.

This measure has many good properties. However, according to Hall (1970), it has a number of drawbacks too. For instance, it equals 1 too often,[7] and is generally not readily computable.

11.7 Monotone Correlation

11.7.1 Definition and Properties

At the start of Section 11.3.3, we saw that mutual complete dependence is not incompatible with independence. So they can hardly be opposites! Hence Kimeldorf and Sampson (1978) suggested the notion of monotone dependence. X and Y are monotone dependent if there exists a perfect monotone relation between them. If the random variables are not perfectly monotonically related, it may be useful to measure numerically the degree of monotone dependence between them. One such measure, called monotone correlation, can be defined as follows:

$$\rho^*(X, Y) = \sup \rho[a(X), b(Y)], \tag{11.5}$$

where the supremum is taken over all monotone functions a, b, for which $\text{var}[a(X)]$ and $\text{var}[b(Y)]$ are finite and nonzero.

Properties:

1. $0 \le \rho^*(X, Y) \le 1$.

2. $\rho^*(X, Y) = \rho^*(Y, X)$.

3. $\rho^*(X, Y) = 0$ if and only if X and Y are independent.[8]

4. $|\rho(X, Y)| \le \rho^*(X, Y) \le \rho'(X, Y)$, which is obvious.

5. $|\rho(X, Y)| = \rho^*(X, Y) = \rho'(X, Y)$ if (X, Y) has the bivariate normal distribution.

6. If X and Y are monotone dependent, then $\rho^*(X, Y) = 1$, but the converse is not true (see the appropriate example given in Kimeldorf and Sampson, 1978, p. 899).

7. If (V, W) has the same uniform representation as (X, Y), then $\rho^*(X, Y) = \rho^*(V, W)$.

8. $\rho^*(X, Y) = \sup\{|\rho(V, W)| : (V, W)$ having the same uniform representation as $(X, Y)\}$.

9. $\rho_g(X, Y) \le \rho^*(X, Y) \le \rho'(X, Y)$ where ρ_g is the grade correlation, $\rho_g(X, Y) = \rho[G(X), H(Y)]$. (Note that the grade correlation is the ordinary correlation coefficient of the uniform representation.)

10. ρ^* is invariant under all order-preserving or order-reversing transformations of X and Y, hence it satisfies a weaker version of condition (5) of Section 11.4.

For detailed discussion, see Kimeldorf and Sampson (1978).

[7] As an example, let probability mass $1 - \epsilon$ be uniformly distributed on the unit square, with the remaining ϵ distributed over the upper quadrant formed at $(1, 1)$. Then if a and b are chosen to be indicators of $X > 1$ and $Y > 1$, respectively, the maximal correlation is 1, yet X and Y are almost independent if ϵ is small!

[8] Suppose $\rho^*(X, Y) = 0$. For any real t, define $a_t(x)$ to be 1 if $x < t$, and 0, otherwise. We claim that $\rho[a_s(X), a_t(Y)] = 0$. For if not, then either $\rho[a_s(X), a_t(Y)] > 0$ or $\rho[a_s(X), -a_t(Y)] > 0$, which contradicts the hypothesis. Now, $\rho[a_s(X), a_t(Y)] = 0$ implies that $\Pr(X \le s, Y \le t) = \Pr(X \le s)\Pr(Y \le t)$, which implies independence.

11.7.2 Concordant and Discordant Monotone Correlations

The concept of monotone correlation can be refined by measuring separately the strength of relationship between X and Y in a positive direction and the strength of relationship in a negative direction, i.e., the strength of concordancy and of discordancy between X and Y. The following definitions are due to Kimeldorf et al (1982).

Definition: If a and b in (11.5) are both restricted to be increasing (or, equivalently, both to be decreasing), the resulting measure is called the *concordant monotone correlation* (abbreviated as CMC).

Definition: If a and b are both restricted to be increasing, then DMC $= \inf \rho[a(X), b(Y)]$ is called the *discordant monotone correlation.*

Kimeldorf et al claimed that CMC and DMC have natural interpretations as measures of positive and negative association, respectively, for ordinal random variables.

It is easy to show that for any pair of increasing functions a and b, we have

$$\text{DMC} \leq \rho[a(X), b(Y)] \leq \text{CMC}.$$

Suppose it is desired to impose numeric monotone scalings for a pair of psychological tests; if the CMC and DMC are close, then by the above equation it makes little difference which monotone scales are used. If DMC = CMC = 0, then X and Y are independent; however, it is possible for DMC < CMC = 0 and yet X and Y not be independent. Note also that if X and Y are increasing dependent (Section 11.3.3), then CMC = 1; and if X and Y are decreasing dependent, then DMC = 1.

In some situations, X and Y should have the same scaling—for example, scores on a single test before and after treatment. This leads to two further definitions.

Definition: If $a = b$ in (11.5), then the resulting measure is called the *isoconcordant monotone correlation* (ICMC).

Definition: If $a = b$ in the definition of DMC, then the resulting measure is called the *isodiscordant monotone correlation* (IDMC).

Note that isoscaling (i.e., assuming $a = b$) is not appropriate when X and Y have inherently different ranges of values.

Kimeldorf et al (1982) evaluate these measures of association by means of a nonlinear optimisation algorithm.

Kimeldorf et al (1981) describe an interactive FORTRAN program called MONCOR to compute the monotone correlations described here.

11.8 Rank Correlations

11.8.1 Introductory Explanation

Kendall's tau (τ) and the grade correlation (ρ_g) are the best-known rank correlation coefficients. (Spearman's rho is virtually the same as the latter; for explanation of our nomenclature regarding these, see Section 13.1.) Essentially, these are measures of correlation between rankings, rather than between actual values, of X and Y; they are unaffected by any increasing transformation of X and Y, whereas ρ is unaffected only by linear increasing transformations.[9]

11.8.2 Kendall's Tau

Kendall's tau is defined by

$$\tau = 4 \int\int H(x, y)h(x, y)\, dx\, dy - 1, \tag{11.6}$$

where H and h are, as usual, the d.f. and the p.d.f. of (X, Y). Alternatively,

$$\tau = \text{cov}[\text{sgn}(X_2 - X_1), \text{sgn}(Y_2 - Y_1)]. \tag{11.7}$$

[9] Another competitor is Ulam's metric. For this, the "distance" between two rankings is small if they agree on the relative ordering of a large subset of the items. It originated in DNA research: the dissimilarity between two sequences will be quite small whenever they are almost identical except that one is shifted by a few units relative to the other. See Critchlow (1988).

The unbiased estimate of τ (sample size $= n$) is

$$T = \frac{1}{n(n-1)} \sum_i \sum_j A_{ij}, \tag{11.8}$$

where the summation is over i and j such that $1 \leq i < j \leq n$, and $A_{ij} = \text{sgn}(x_j - x_i)\text{sgn}(y_j - y_i)$.

If the distribution of (X, Y) is bivariate normal with correlation ρ, then $\tau = \frac{2}{\pi} \sin^{-1} \rho$.

11.8.3 Grade Correlation

The grade correlation is defined by

$$\rho_g = 12 \int \int F(x)G(y)h(x,y)\, dx\, dy - 3, \tag{11.9}$$

where F and G are, as usual, the marginal d.f.'s. Thus ρ_g equals corr$[F(X), G(Y)]$, Pearson's product-moment correlation of the uniform representation of H. Equivalently,

$$\rho_g = 12 \int \int H(x,y)f(x)g(y)\, dx\, dy - 3, \tag{11.10}$$

where f and g are the marginal p.d.f.'s. A further expression is

$$\rho_g = \text{cov}[\text{sgn}(X_2 - X_1), \text{sgn}(Y_3 - Y_1)]. \tag{11.11}$$

The grade correlation is the population analogue of the well-known Spearman's rank correlation coefficient, which, for a sample of size n, is

$$R = \frac{12}{n(n^2-1)} \sum_i \left(r_i - \frac{n+1}{2}\right)\left(s_i - \frac{n+1}{2}\right), \tag{11.12}$$

where $r_i = \text{rank}(x_i)$ and $s_i = \text{rank}(y_i)$. Another common form for R is

$$R = 1 - \frac{6 \sum_i d_i^2}{n(n^2-1)}, \tag{11.13}$$

where $d_i = r_i - s_i$.

The expectation of R in a sample of size n is

$$E(R) = \frac{(n-2)\rho_g + 3\tau}{n+1}, \tag{11.14}$$

which tends to ρ_g as n tends to infinity.

If the distribution of (X, Y) is bivariate normal with correlation ρ, then $\rho_g = \frac{6}{\pi} \sin^{-1} \frac{\rho}{2}$.

11.8.4 Remarks

- Independence of X and Y implies that $\tau = \rho_g = 0$, but the converse implication does not hold.
- τ and ρ_g are both restricted to the range -1 to 1, attaining these limits for perfect negative and perfect positive relationships, respectively.
- If X and Y are positive quadrant dependent (see Section 12.3.1), then $\tau \geq 0$ and $\rho_g \geq 0$.
- If two distributions H and H' have the same marginals and H is more concordant than H' (i.e., $H \geq H'$, see Section 12.6), then τ and ρ_g are at least as great for H as for H' (Tchen, 1980, corollary 3.2).
- It has already been said that r in (11.3) is very sensitive to outliers; the sample counterparts of τ and ρ_g are less so, but Gideon and Hollister (1987) proposed a statistic that is even more resistant to the influence of outliers.
- For a review that includes rank correlations and some of the other material of this Chapter, see Joag-Dev (1984).

11.9 Measure of Schweizer and Wolff, and Related Measures

Schweizer and Wolff (1976, 1981) proposed the following procedure. Make monotone transformations so that the marginals become uniform. Thus, if F and G are d.f.'s of X and Y, respectively, then $U = F(X)$ and $V = G(Y)$ are uniformly distributed on [0,1], and their joint distribution function, $C(u, v)$, say, reflects the dependence between X and Y. (C is really the uniform representation of the joint distribution H, and is known as the copula—see Section 5.1.2.) Then they expressed the grade correlation as

$$\rho_g(X,Y) = 12\int_0^1\int_0^1 [C(u,v) - uv]\,du\,dv. \tag{11.15}$$

Observing that the integral in (11.15) is simply the signed volume between the surfaces $z = C(u, v)$ and $z = uv$, and that X and Y are independent if and only if $C(u, v) = uv$, they suggested that any suitably normalised measure of distance such as L_p-distance[10] should yield a symmetric, nonparametric, measure of dependence. By considering $p = 1$, $p = 2$, and $p \to \infty$, they obtained three measures of dependence:

$$\sigma(X,Y) = 12\int_0^1\int_0^1 |C(u,v) - uv|\,du\,dv \tag{11.16}$$

$$\gamma(X,Y) = \sqrt{90\int_0^1\int_0^1 [C(u,v) - uv]^2\,du\,dv} \tag{11.17}$$

$$\kappa(X,Y) = 4\sup_{u,v\in[0,1]} |C(u,v) - uv|. \tag{11.18}$$

Equation (11.17) is equivalent to the Cramér-Von Mises index,

$$\int_{-\infty}^{\infty}\int_{-\infty}^{\infty} [F(x,y) - G(x)H(y)]^2\,dG(x)\,dH(y), \tag{11.19}$$

which is also equivalent to the Φ^2 of Hoeffding (1940) (see Lancaster, 1982b). On the other hand, (11.18) is equivalent to the Kolmogorov-Smirnov measure

$$\sup_{x,y} |F(x,y) - G(x)H(y)|.$$

Schweizer and Wolff (1981) showed that, when evaluated according to a suitably modified version of Rényi's conditions, σ possesses many desirable properties, including, in particular, the condition (9) of Section 11.4.

A comparison of σ with ρ_g may be desirable. They respectively measure the volume and the signed volume between the surfaces $C(u, v)$ and uv. Schweizer and Wolff note that:

- $|\rho_g(X,Y)| \leq \sigma(X,Y)$.
- Equality holds for the bivariate normal distribution.
- The difference can be large.

Kotz and Johnson (1977) compared σ to the mean square contingency ϕ^2 (which is another measure of dependence, to be defined in Section 14.3) when (X, Y) has the iterated F-G-M distribution (see Section 5.2.13). They found the expression for σ to be simpler than that for ϕ^2, whence, as they remarked, comparisons of the dependence properties of this distribution are more clear-cut when σ is used.

---

[10] L_p-distance between two functions a and b on R^2 is defined by $\int |a-b|^p\,dx\,dy$. (A modification is needed to accomodate the limit as $p \to \infty$.)

11.10 Matrix of Correlations

In this Section, we shall present a summary of the relevant aspects of the diagonal expansion method discussed in Chapter 14. Definitions will not be repeated here. See also Lancaster (1982b, p. 335).

Let $\{\xi_i\}$ and $\{\eta_j\}$ be complete orthonormal systems on F and G, respectively, with $\xi_0 = 1 = \eta_0$ (that is, $E(\xi_i \xi_j) = \delta_{ij}$, where δ_{ij} is either 1 or 0 depending on whether i and j are equal or not, and similarly for η.) Let $\rho_{ij} = E(\xi_i \eta_j)$ and $\mathbf{R} = (\rho_{ij})$, for all positive integers i and j, is an infinite matrix. For given F and G, $\mathbf{R}$ completely determines H (Lancaster, 1963a), so that $\mathbf{R}$ can be said to be a matrix measure of dependence. In particular, $\mathbf{R} = \mathbf{0}$ if and only if X is independent of Y. $\mathbf{R}$ is orthogonal[11] if and only if X and Y are mutually completely dependent (Lancaster, 1963b). Special interest arises if $\{\xi_i\}$ and $\{\eta_j\}$ possess the biorthogonal property, i.e., $E(\xi_i \eta_j) = \delta_{ij}\rho_{ij}$; in this case $\mathbf{R}$ is obviously diagonal.

The scalar $\phi^2 = \text{tr } \mathbf{RR}' = \sum_{i=1}^{\infty}\sum_{j=1}^{\infty} \rho_{ij}^2$ is an index for measuring dependence of two random variables. The ϕ^2 here is the same ϕ^2 that will be defined in Section 14.3 and was referred to as the mean square contingency in the last Section. ϕ^2 is zero if and only if X and Y are independent. In the case of the bivariate normal distribution, $\phi^2 + 1 = (1 - \rho^2)^{-1}$.

As we have just mentioned, X and Y being mutually completely dependent implies $\mathbf{R}$ is orthogonal, which in turn implies that $\phi^2 = \infty$. However, ϕ^2 can be infinite without having X and Y mutually completely dependent. Consider a monotonic transformation of ϕ^2 defined by $\lambda(X,Y) = \phi^2/(1+\phi^2)$. It is clear from the present discussion that λ does not satisfy the "necessary" part of condition (6). However, Rényi (1959) showed that λ satisfies conditions (2)–(5) and (7). If the joint distribution is either absolutely continuous or discrete, condition (1) will also be satisfied.

11.11 Tetrachoric and Polychoric Correlation

It is common for data to be recorded on an ordinal scale with only a few steps to it. A typical case from the social sciences is where subjects (respondents) are asked to report whether they approve strongly, approve, are neutral towards, disapprove, or disapprove strongly of some proposal.

When analysing this kind of data, a common approach is to assign an integer to each category and proceed in the analysis as if the results were on an interval scale, with convenient distributional properties. Although this approach may work satisfactorily in some cases, in others it may lead to erroneous results (Olsson, 1980).

The polychoric correlation is suggested in the literature as an appropriate measure of correlation for bivariate tables of such data; it is termed the tetrachoric correlation[12] when applied to 2×2 tables. The idea behind it is now described.

Formally, we denote the observed, ordinal, variables by X and Y; let them have r and s distinct categories, respectively. We assume that X and Y have been generated from some unobserved (latent) variables Z_1 and Z_2 which have a bivariate normal distribution. The relation between X and Z_1 may be written as

$$\begin{array}{rcl} X & = & 1 \text{ if } Z_1 < s_1 \\ X & = & 2 \text{ if } s_1 \le Z_1 < s_2 \\ \vdots & \vdots & \vdots \quad \vdots \\ X & = & r \text{ if } s_{r-1} \le Z_1 \end{array}$$

and correspondingly for Y in terms of class limits $t_1, t_2, \ldots t_{s-1}$ of Z_2. The s's and the t's are sometimes referred to as thresholds.

Interest is often primarily in estimating $\rho(Z_1, Z_2)$, the correlation between Z_1 and Z_2. Suppose we want to do this by means of the maximum likelihood method.[13] Given this general aim, the problem may be solved in at least two different ways. One way is to estimate ρ and the thresholds simultaneously.

---

[11] That is, its transpose and its inverse are the same.

[12] There is some divergence in the literature as to whether "tetrachoric correlation" merely refers to the correlation in an assumed underlying normal distribution, or whether it refers to some particular method of estimating that correlation.

[13] In the case of 2×2 tables, the parameters to be fitted (s_1, t_1, ρ) are precisely the right number to produce an exact fit to the data. The only problem is how best to determine, or approximate ρ from the equation (in an obvious notation) $p_{11} = L(s_1, t_1; \rho)$.

Alternatively, the thresholds are first estimated as the inverse of the normal distribution function, evaluated at the cumulative marginal proportions of the contingency table, and the maximum likelihood estimate of ρ is then computed with the thresholds fixed at their estimates. This may be called a two-step procedure. It has the advantage of greater ease of the numerical calculations, although the former is formally more correct. For practical purposes, the results are almost identical (Olsson, 1979). For a generalisation of these methods to three- and higher-dimensional polytomous ordinal variables, see Lee (1985) and Lee and Poon (1987a,b,c). Divgi (1979b) reports on a FORTRAN program for calculating tetrachoric correlation, and offers to provide a listing of it to any interested reader. Martinson and Hamdan (1975) have published a computer program for calculating the polychoric correlation.

Other accounts of this are by Drasgow (1986) and Harris (1988); the latter, especially, has a number of references to methods of approximating the tetrachoric correlation.

11.12 Compatibility with Perfect Rank Ordering

Suppose we have a two-way ordinal contingency table, as in the previous Section, which we imagine to have arisen from grouping two continuous variates. For simplicity, suppose each variate has been reduced to a dichotomy, so that our table is only a 2×2 one. If the frequencies are $\begin{smallmatrix} 0 & 1 \\ 1 & 2 \end{smallmatrix}$, how well are X and Y correlated?

- One approach is to calculate the tetrachoric correlation, implicitly thinking of the bivariate normal distribution, or to estimate the association parameter of some other bivariate distribution.

- There is an alternative, which is especially relevant if X and Y are two different measures of the same thing (e.g., severity of disease as assessed by two doctors). Ask to what extent is the data compatible with perfect agreement between the X-ordering and the Y-ordering. The set of frequencies $\begin{smallmatrix} 0 & 1 \\ 1 & 2 \end{smallmatrix}$ is compatible with perfect agreement between the two orderings, as it could be that if finer discrimination were insisted upon, the table would become $\begin{smallmatrix} 0 & 0 & 0 & 1 \\ 0 & 0 & 1 & 0 \\ 0 & 1 & 0 & 0 \\ 1 & 0 & 0 & 0 \end{smallmatrix}$. (The original table is obtaining by combining the last three rows and combining the last three columns.) In a sense, we are starting with perfect correlation, not zero, as our null hypothesis, and asking to what extent is the data incompatible. We think that a formalisation of this would run somewhat as follows. Calculate

$$\sup \rho_r[a(Z), b(Z)] \tag{11.20}$$

(where ρ_r is a rank correlation coefficient, such as ρ_g or τ), subject to X being a nondecreasing function of a, and Y being a nondecreasing function of b. (The point is that $X \to a$ and $Y \to b$ are one $\to$ many relationships, not functions.)

Having seen the article by Guttman (1986), we see a good deal of common ground between the second alternative above and Guttman's plea that what he calls "weak" coefficients of monotonicity are sometimes more appropriate than "strong" ones.

11.13 Conclusions on Measures of Dependence

There is, we fear, no universal answer to "What is the best measure of dependence?" According to Lancaster (1982b), for some defined classes of distributions, the absolute product-moment correlation $|\rho|$ is the index of choice—for the bivariate normal distribution, it satisfies all the conditions given in Section 11.4 except for (5); for the random elements in common model,[14] it completely determines the joint distribution. In other classes, there may be indices useful for some purposes—the user needs to think about what purposes have priority. There is inevitably some loss of information in condensing the matrix of correlations to a

[14] See Section 15.3.6.

single index. The absence of an always-best measure should not surprise us, if we reflect on the persistence in the literature of two competing measures of rank correlation, Kendall's and Spearman's.

11.14 Local Dependence Function

11.14.1 Definition of Local Dependence

The ideas of local dependence and regional dependence, to be discussed in this and the following Sections, do not follow in the same line of development as the above, but nevertheless fall within the scope of this Chapter. These concepts were introduced by Holland and Wang (1987a,b).

The local dependence for two continuous random variables X and Y is motivated from the contingency table for two discrete random variables. Consider an $r \times s$ contingency table with cell proportions p_{ij}. For any two pairs of indices (i, j) and (k, l), the cross-product ratio is

$$\alpha_{ij,kl} = (p_{ij}p_{kl})/(p_{il}p_{kj}), \tag{11.21}$$

for $1 \leq i, k \leq (r-1)$ and $1 \leq j, l \leq (s-1)$. Yule and Kendall (1937, Section 5.15) and Goodman (1969) suggested consideration of the following set of local cross-product ratios:

$$\alpha_{ij} = (p_{ij}p_{i+1\,j+1})/(p_{i\,j+1}p_{i+1\,j}), \tag{11.22}$$

for $1 \leq i \leq (r-1)$ and $1 \leq j \leq (s-1)$. Also, let $\gamma_{ij} = \log \alpha_{ij}$. Both α_{ij} and γ_{ij} measure the association in the 2×2 subtables formed at the intersection of pairs of adjacent rows and columns. They are invariant under multiplications of rows and columns.

Now let us go back to the continuous case. Recall the support for X and Y is the region of non-zero p.d.f., which is here denoted $R(h) = \{(x, y) : h(x, y) > 0\}$. Suppose $R(h)$ has been partitioned by a very fine rectangular grid. The probability content of the rectangle containing the point (x, y) with sides dx and dy is approximately $h(x, y)\,dx\,dy$. This probability may be viewed as one cell probability of a large two-way table. For this two-way table, the cross-product ratio in (11.21) may be expressed as

$$\alpha(x, y; u, v) = \frac{h(x, y)h(u, v)}{h(x, v)h(u, y)}, \tag{11.23}$$

for $x < u$ and $y < v$, assuming that the four points are in $R(h)$. The function in (11.23) will be called the *cross-product ratio* function, and its logarithm

$$\theta(x, y; u, v) = \log \alpha(x, y; u, v) \tag{11.24}$$

will be called the *global dependence* function.

To obtain an analogy to γ_{ij}, we consider the global dependence function evaluated at (x, y) and at $(x + dx, y + dy)$. To express this as a rate, we form $\theta(x, y; x + dx, y + dy)/(dx\,dy)$ and let $dx \to 0$, $dy \to 0$. The result is

$$\gamma(x, y) = \lim_{dx,dy\to 0} \frac{\theta(x, y; x + dx, y + dy)}{dx\,dy} = \frac{\partial^2}{\partial x \partial y} \log h(x, y), \tag{11.25}$$

assuming that h is mixed differentiable, i.e., the partial derivative of the second order exists.

11.14.2 Properties of the Local Dependence Function

We shall assume that the regional support $R(h)$ is a rectangle; R^2 may also be regarded as a rectangle for this purpose. (If $R(h)$ is not a rectangle, then the shape of $R(h)$ can introduce dependence between X and Y of a different nature from the local dependence—we take up this issue in Section 11.15.)

The following is a list of the properties of the local dependence function $\gamma(x, y)$.

- $-\infty < \gamma(x, y) < \infty$.

- $\gamma(x, y) = 0$ for all $(x, y) \in R(h)$ if and only if X and Y are independent. $\gamma(x, y)$ reveals more information about the dependence than some other indices; recall, for example, that the product-moment correlation ρ may be zero without X and Y being independent.

- $\gamma(x, y)$ is "margin-free", in the sense that a bivariate density generated from $h(x, y)$ via multiplication by a function of x only and by a function of y only, thus changing the marginals, has an unchanged $\gamma(x, y)$.

- The local dependence functions of the two conditional densities are the same as that of the joint density.

- Holland and Wang (1987b) mentioned that when $\gamma(x, y)$ is constant, any monotone function of that constant will be a "good" measure of association. But when $\gamma(x, y)$ changes sign in $R(h)$, most measures of association will be inadequate or even misleading.

11.15 Regional Dependence

11.15.1 Preliminaries

In this Section, we shall discuss the notion of regional dependence expounded by Holland and Wang (1987a).

In addition to the notation of the previous Section, we will write $R(f) = \{x : f(x) > 0\}$ and $R(g) = \{y : g(y) > 0\}$ for the supports of the marginals. We assume that $R(g)$ is an open[15] convex[16] set of the plane, $R(f)$ and $R(g)$ are open intervals, and that f, g, and h are continuous in their respective regions of support.

Clearly, $R(h)$ is contained in the Cartesian product[17] of $R(f)$ and $R(g)$, denoted $R(f) \times R(g)$. If $R(h)$ is not *equal* to $R(f) \times R(g)$, then there exists a point (x_0, y_0) in $R(f) \times R(g)$ which is not in $R(h)$, at which $h(x_0, y_0) = 0$. Yet $f(x_0)g(y_0) > 0$. So $h(x_0, y_0) \neq f(x_0)g(y_0)$. Therefore, X and Y cannot be independent if their region of support is not a rectangle. This situation is parallel to the effects caused by structural zeros in a two-way contingency table.[18] In this Section, we are concerned with the type of statistical dependence which is "caused" by the region of support.

11.15.2 Quasi-Independence and Quasi-Independent Projection

Let us define the x-section, $R_y(x)$, and the y-section, $R_x(y)$, of $R(h)$ by $R_y(x) = \{x : h(x, y) > 0\}$ and $R_x(y) = \{y : h(x, y) > 0\}$. Clearly, $R_y(x) \subseteq R(f)$ and $R_x(y) \subseteq R(g)$. The following definition of quasi-independence is analogous to quasi-independence in a two-way contingency table.

Definition: X and Y, having density function $h(x, y)$, are said to be quasi-independent if there exist positive functions $f_1(x)$ and $g_1(y)$ such that $h(x, y) = f_1(x)g_1(y)$ for all $(x, y) \in R(h)$.

If $R(h) = R(f) \times R(g)$, quasi-independence implies ordinary independence. If $R(h)$ is a proper subset of $R(f) \times R(g)$, then, as we have seen, X and Y cannot be independent.

Definition: A positive density function $P_h(x, y)$ on $R(h)$ is the quasi-independent projection of $h(x, y)$ on $R(h)$ if there exist positive functions $a(x)$ and $b(y)$ such that the following three equations hold:

$$P_h(x, y) = a(x)b(y) \text{ for all } (x, y) \in R(h) \quad (11.26)$$

$$\int_{R_x(y)} a(x)b(y)\, dy = g(x) \text{ for all } x \in R(f) \quad (11.27)$$

$$\int_{R_y(x)} a(x)b(y)\, dx = h(y) \text{ for all } y \in R(g). \quad (11.28)$$

[15] A subset S of the plane is called open if for every $\mathbf{x} = (x_1, x_2)$ in S, there is a $\delta > 0$ such that each $\mathbf{y} = (y_1, y_2)$ within a distance δ of $\mathbf{x}$ belongs to S.

[16] A set A is convex if x and y being in A implies that $(1 - t)x + ty$ is also in A, for t between 0 and 1.

[17] The Cartesian product of two sets A and B is the set of all ordered pairs (x, y), where $x \in A$ and $y \in B$.

[18] "Structural zeros" refers to one or more of the theoretical frequencies in a contingency table being zero. (This is quite distinct from frequencies happening by chance to be zero in a particular sample.)

The quasi-independent projection of h is a joint density which has the same marginals as those of X and Y, and has the functional form of the product of two independent distributions. The explicit form of $P_h(x,y)$ can be obtained by solving the two integral equations (11.27) and (11.28).

Holland and Wang (1987a) have shown that if $R(f) = (a,b)$ and $R(g) = (c,d)$ are both finite intervals then the quasi-independent projection $P_h(x,y)$ exists uniquely over $R(h)$.

11.15.3 A Measure of Regional Dependence

The regional dependence measure $M(X,Y) = M(f,g,R(h))$ is defined by

$$M(X,Y) = 1 - \frac{1}{c}, \tag{11.29}$$

where

$$c = \int_{R(f)\times R(g)} P_h(x,y)\,dx\,dy. \tag{11.30}$$

For discrete random variables, $M(X,Y)$ can be computed from incomplete two-way contingency tables. The following properties of the measure M can be found in Holland and Wang (1987b).

- $0 \le M(X,Y) \le 1$.
- X and Y being independent implies $M(X,Y) = 0$. But the converse is not true.
- If X and Y are monotonely dependent (Section 11.3.3), then $M(X,Y) = 1$.
- If $h(x,y)$ is a constant throughout $R(h)$, then

$$M(X,Y) = 1 - \frac{\text{Area of } R(h)}{\text{Area of } [R(f) \times R(g)]}.$$

- For fixed $R(f)$ and $R(g)$, let h_1 and h_2 be two constant densities defined inside $R(f) \times R(g)$, such that the marginal densities are all positive in $R(f)$ and $R(g)$. Then $R(h_1) \subseteq R(h_2)$ implies that $M(X_1,Y_1) \ge M(X_2,Y_2)$, where (X_1,X_2) and (X_2,Y_2) have joint densities h_1 and h_2, respectively.
- $M(X,Y)$ is invariant under smooth monotonic transformations of the marginals. (Of course, it changes sign if the transformation of one marginal is increasing and the other is decreasing.)

In common with maximal correlation and monotone correlation, $M(X,Y)$ may not be easy to calculate, in particular, when $R(f)$ and $R(g)$ are not finite intervals.

Chapter 12

Concepts of Stochastic Dependence

12.1 Introduction

PQD, PRD, LTD, RTI—you may have seen abbreviations like these in recent years. We shall try to explain them and their interrelationships in this Chapter. They refer to the form of dependence between two variables. (We shall not be concerned with the body of work relating to a stochastic process evolving over time—Markov chains, martingales, etc.)

Positive dependence means that large values of Y tend to accompany large values of X, and similarly for small values. Discussion of concepts of dependence involves refining, by means of definitions and deductions, this basic idea. The Sections of this Chapter are as follows:

12.2 Important conditions describing positive dependence. (We state some definitions, and examine their relative stringency.)

12.3 Further discussion of concepts of positive dependence. (Takes the ideas of Section 12.2 further. Also includes ones not mentioned there, and interrelationships.)

12.4 Negative dependence. (There are negative analogues of most of the positive dependence concepts. In addition, the idea of "neutrality" is discussed.)

12.5 Monotonic quadrant dependence function. (This is closely connected to the notion of "positive quadrant dependent in expectation".)

12.6 Orderings of positive dependence. (Given any notion of dependence, such as PQD, can we say that one distribution possesses it to a greater degree than does another?)

12.2 Important Conditions Describing Positive Dependence

12.2.1 Four Basic Conditions

Concepts of stochastic dependence for a bivariate distribution play an important part in statistics. For each concept, it is often convenient to refer to the bivariate distributions that satisfy it as a family, a group. We are mainly concerned with positive dependence, though negative dependence will also be discussed briefly.

Various notions of dependence are motivated from applications in statistical reliability (see Barlow and Proschan, 1975, 1981). The baseline, or starting point, of the reliability analysis of systems is independence of the lifetimes of the components. As we shall see especially in Chapters 17 and 18, it is often more realistic to assume some form of positive dependence among components.

Around about 1970, several works discussed different notions of positive dependence between two random variables, and derived some interrelationships among them—for example, Lehmann (1966), Esary et al (1967), Esary and Proschan (1972), R Harris (1970), and Brindley and Thompson (1972), among others. Yanagimoto (1972) unified some of these notions by introducing a family of concepts of positive dependence. Some further notions of positive dependence were introduced by Shaked (1977, 1979, 1982).

Chapter 5 of Tong (1980) covers much the same material as the present Chapter, and a briefer account is by Joag-Dev (1984).

For some concepts of multivariate dependence, see Block and Ting (1981).

Jogdeo (1982) lists the following four basic conditions describing positive dependence; these are in increasing order of stringency. (See also Jogdeo, 1975.)

1. Positive correlation, $\mathrm{cov}(X,Y) \geq 0$.

2. For every pair of increasing functions a and b, defined on R,

$$\mathrm{cov}[a(X), b(Y)] \geq 0. \tag{12.1}$$

   Lehmann (1966) showed this condition is equivalent to $\Pr(X \geq x, Y \geq y) \geq \Pr(X \geq x)\Pr(Y \geq y)$, or equivalently $\Pr(X \leq x, Y \leq y) \geq \Pr(X \leq x)\Pr(Y \leq y)$. We say that (X,Y) is positive quadrant dependent (PQD) if and only if these inequalities hold.

3. Esary et al (1967) introduced the following condition, termed "association". For every pair of functions a and b, defined on R^2, which are increasing in each of the arguments (separately),

$$\mathrm{cov}[a(X,Y), b(X,Y)] \geq 0. \tag{12.2}$$

4. Either for every y, $\Pr(Y > y|X = x)$ is increasing in x, or, for every x, $\Pr(X > x|Y = y)$ is increasing in y. Y is said to be positively regression dependent[1] on X (written as PRD$(Y|X)$) in the former case, and X to be positively regression dependent on Y (PRD$(X|Y)$) in the latter. We will simply write PRD if there is no ambiguity.

12.2.2 The Relative Stringency of the Conditions

We shall now show that these conditions are arranged in increasing order of stringency. Jogdeo (1982) notes that (3) $\Rightarrow$ (2) $\Rightarrow$ (1) is obvious, and states the following identity, which helps in showing that (4) $\Rightarrow$ (3) and that (2) is the same as PQD:

$$x - y = \int_{-\infty}^{\infty} (I_{\{y \leq u\}} - I_{\{x \leq u\}})\, du, \tag{12.3}$$

in which $I_{\{.\}}$ is the indicator function, as in Section 1.5.4. Jogdeo states the following consequences:

1. If (X,Y) is a pair of random variables, then

$$E(X - Y) = \int_{-\infty}^{\infty} [G(u) - F(u)]\, du, \tag{12.4}$$

   where F and G are the d.f.'s of X and Y, respectively. Thus, if

$$\Pr(X > u) \geq \Pr(Y > u) \tag{12.5}$$

   for every u, then $E(X) \geq E(Y)$. Equality holds only if F and G are identical. If (12.5) holds, so is $\Pr[a(X) > v] \geq \Pr[a(Y) > v]$, for every v, and any increasing function a. It follows that

$$E[a(X)] \geq E[a(Y)]. \tag{12.6}$$

[1]While most authors use this terminology, others say that Y is stochastically increasing in X if this condition holds (see, for example, Barlow and Proschan, 1981, p. 142). The word "regression" often implies the taking of an expectation. It follows from consequence (2) in Section 12.2.2 below that

$$E(Y|X=x) = -\int_{-\infty}^{0} \Pr(Y \leq y|X=x)\, dy + \int_{0}^{\infty} \Pr(Y > y|X=x)\, dy$$

Clearly, condition (4) implies that $E(Y|X=x)$ is increasing. But the converse does not hold.

2. We now write X as the difference between its positive part[2] X^+ and negative part X^-. Then, from (12.4) or integration by parts, $E(X) = E(X^+ - X^-) = -\int_{-\infty}^0 F(x)\,dx + \int_0^\infty [1 - F(x)]\,dx$.

3. See Lehmann (1966) or Shea (1983) for a proof of the bivariate version of this, known as Hoeffding's (1940) lemma:
$$\operatorname{cov}(X,Y) = \int_{-\infty}^{\infty}\int_{-\infty}^{\infty} [H(x,y) - F(x)G(y)]\,dx\,dy. \tag{12.7}$$

We see from (12.7) that (X,Y) being PQD implies $\operatorname{cov}(X,Y) \geq 0$, equality holding only if X and Y are independent. Furthermore, if a and b are two increasing real functions, then (X,Y) being PQD implies $(a(X), b(Y))$ is PQD, and thus $\operatorname{cov}[a(X), b(Y)] \geq 0$. Suppose, now, $\operatorname{cov}[a(X), b(Y)] \geq 0$ for all increasing functions a and b. Set $a(X) = I_{\{X \geq x\}}$ and $b(Y) = I_{\{Y \geq y\}}$. Now, $\operatorname{cov}[a(X), b(Y)] = \Pr(X \geq x, Y \geq y) - \Pr(X \geq x)\Pr(Y \geq y) \geq 0$, which means (X,Y) is PQD. Therefore, $\operatorname{cov}[a(X), b(Y)] \geq 0$ for all increasing functions a and b and PQD are equivalent. A particular case is that $\operatorname{cov}[a(X), b(X)] \geq 0$ if a and b are increasing functions.

To see that (4) implies (3), consider the following identity:[3]
$$\operatorname{cov}(X,Y) = \operatorname{cov}[E(X|\mathbf{Z}), E(Y|\mathbf{Z})] + E\{\operatorname{cov}[(X,Y)|\mathbf{Z}]\} \tag{12.8}$$
(in which we have taken expectation over an arbitrary random vector $\mathbf{Z}$). Now, with a and b again being increasing functions,
$$\operatorname{cov}[a(X,Y), b(X,Y)] = \operatorname{cov}\{E[a(X,Y)|X], E[b(X,Y)|X]\} + E\{\operatorname{cov}[a(X,Y), b(X,Y)|X]\}. \tag{12.9}$$
If (4) holds, the expected values in the first term on the right hand side of (12.9) are increasing[4] in X; taken with the result that $\operatorname{cov}[a(X), b(X)] \geq 0$ that we established earlier, this means that this term is nonnegative. Also, a and b being monotone means that the conditional covariance in the second term is nonnegative—and so must be its expected value.

12.3 Further Discussion of Concepts of Positive Dependence

12.3.1 PQD

The concept of PQD is useful in reliability applications, see Barlow and Proschan (1981) and Lai (1986). Consider a system of two components which are arranged in series. By assuming that the two components are independent when they are in fact positively quadrant dependent, we will underestimate the system reliability.

---

[2] The positive part of X is defined as
$$X^+ = \begin{cases} X & \text{if } X \geq 0 \\ 0 & \text{otherwise,} \end{cases}$$
and the negative part of X is defined as
$$X^- = \begin{cases} 0 & \text{if } X \geq 0 \\ -X & \text{otherwise.} \end{cases}$$

[3] The right hand side of the equation which follows is
$$\begin{aligned} & E[E(X|\mathbf{Z})E(Y|\mathbf{Z})] - E[E(X|\mathbf{Z})]E[E(Y|\mathbf{Z})] + E[E(XY|\mathbf{Z}) - E(X|\mathbf{Z})E(Y|\mathbf{Z})] \\ = \;& E[E(X|\mathbf{Z})E(Y|\mathbf{Z})] - E(X)E(Y) + E(XY) - E[E(X|\mathbf{Z})E(Y|\mathbf{Z})] \\ = \;& \operatorname{cov}(X,Y). \end{aligned}$$

[4] $\Pr(Y > y|X = x)$ increasing in x implies that $\Pr[a(X,Y) > a(x,y)|X = x]$ increases in x for every increasing function a defined on R^2. By using the equation in footnote 1, we have
$$E[a(X,Y)|X = x] = -\int_{-\infty}^{0} \Pr[a(X,Y) \leq a(x,y)|X = x]\,dy + \int_0^\infty \Pr[a(X,Y) > a(x,y)|X = x]\,dy,$$
which is therefore increasing in x.

Before presenting further applications of PQD, we need to state a result due to Lehmann (1966): let r and s be a pair of real functions defined on R^n which are monotone in each of their n arguments. The functions r and s are said to be concordant in the ith argument if the directions of the monotonicity for the ith argument are the same (i.e., both functions are simultaneously either increasing or decreasing in the ith argument while the others are kept fixed), and discordant if the directions are opposite. Let $(X_i, Y_i), i = 1, 2, \dots n$ be n independent pairs each satisfying PQD. Suppose r and s are concordant in each of these arguments, then

$$\text{cov}[r(X_1, \dots X_n), s(Y_1, \dots Y_n)] \geq 0. \tag{12.10}$$

The result has the following applications (see also Jogdeo, 1982):

1. Let $r(X_1, X_2) = \text{sgn}(X_2 - X_1)$ and $s(Y_1, Y_2) = \text{sgn}(Y_2 - Y_1)$. Then $\tau = \text{cov}[\text{sgn}(X_2 - X_1), \text{sgn}(Y_2 - Y_1)]$, where τ is Kendall's tau. From (12.10), the condition PQD implies $\tau \geq 0$.

2. Grade correlation $\rho_g = \text{cov}[\text{sgn}(X_2 - X_1), \text{sgn}(Y_3 - Y_1)]$. On letting $r(X_1, X_2, X_3) = X_2 - X_1$ and $s(Y_1, Y_2, Y_3) = Y_3 - Y_1$, we see $\rho_g \geq 0$ under condition PQD.

3. Blomqvist (1950) proposed $(2p_n - 1)$ as a measure of dependence, p_n being the proportion of pairs (X_i, Y_i) that fall in either the positive or the negative quadrants formed by the lines $X = \tilde{x}$, $Y = \tilde{y}$, where $\tilde{x}$ and $\tilde{y}$ are the medians of X and Y. The expectation of this,

$$E(2p_n - 1) = 2[\text{cov}(I_{\{X_i \geq \tilde{x}\}}, I_{\{Y_i \geq \tilde{y}\}}) + \text{cov}(I_{\{X_i \leq \tilde{x}\}}, I_{\{Y_i \leq \tilde{y}\}})], \tag{12.11}$$

is ≥ 0 under PQD.

The class of all PQD distributions with fixed marginals is shown by Bhaskara et al (1987) to be convex. That is, if H_1 and H_2 are both PQD, then so is $\lambda H_1 + (1 - \lambda) H_2$, for $0 \leq \lambda \leq 1$.

12.3.2 PQDE

For any real number x, let Y_x be the r.v. with d.f. $\Pr(Y \leq y | X > x)$. It may easily be verified that $\Pr(Y \leq y | X > x) \leq \Pr(Y \leq y)$ implies

$$E(Y_x) \geq E(Y). \tag{12.12}$$

We then say that Y is positive quadrant dependent on X in expectation (PQDE). Similarly, we say there is negative quadrant dependence in expectation if

$$E(Y_x) \leq E(Y). \tag{12.13}$$

It is clear that PQD $\Rightarrow$ PQDE. Also, we have

$$\begin{aligned} \text{cov}(X, Y) &= \int \int [\bar{H}(x, y) - \bar{F}(x) \bar{G}(y)] \, dx \, dy & (12.14) \\ &= \int \bar{F}(x) \left\{ \int [\Pr(Y > y | X > x) - \bar{G}(y)] \, dy \right\} dx & (12.15) \\ &= \int \bar{F}(x) [E(Y_x) - E(Y)] \, dx, & (12.16) \end{aligned}$$

which is ≥ 0 if X and Y are PQDE. Hence PQDE $\Rightarrow$ cov ≥ 0.

12.3.3 Association

The concept of "association" (i.e., condition (3) in Section 12.2.1) is very useful in reliability, particularly in the context of multivariate (as distinct from merely bivariate) dependence. Jogdeo (1982) defines an n-variate random vector $\mathbf{X} = (X_1, \dots X_n)$ or its distribution to be *associated* if for every pair of increasing real functions a and b, defined on R^n, $\text{cov}[a(\mathbf{X}), b(\mathbf{X})] \geq 0$.

The property of association has a number of consequences; see the list given by Jogdeo (1982). Some of them seem pretty trivial, at least for the bivariate case. We merely note that (i) increasing (or decreasing) functions of associated random variables are associated, and (ii) if $(Y_1, \dots Y_n)$ is associated also, and the X's and the Y's are positive, then $(X_1Y_1, \dots X_nY_n)$ is associated.

For detailed discussion on association and its reliability applications, see Barlow and Proschan (1981). For the relation of association to multivariate total positivity, see Kim and Proschan (1988). As to the association of chi-squared, t-, and F-distributions, see Abdel-Hameed and Sampson (1978).

We shall now introduce some additional concepts of positive dependence, some of which are stronger than those met so far.

12.3.4 LTD and RTI

The following definition is due to Lehmann (1966). We say that Y is left tail decreasing in X (written as LTD($Y|X$) or simply LTD when there is no ambiguity) if

$$\Pr(Y \leq y|X \leq x) \text{ is decreasing in } x \text{ for all } y, \tag{12.17}$$

and Y is right tail increasing in X (RTI($Y|X$) or simply RTI) if

$$\Pr(Y > y|X > x) \text{ is increasing in } x \text{ for all } y. \tag{12.18}$$

Lehmann (1966) showed that PRD $\Rightarrow$ LTD $\Rightarrow$ PQD. We have already seen PRD $\Rightarrow$ ASSOCIATION $\Rightarrow$ PQD.

Esary and Proschan (1972) studied the relationship of LTD and RTI to association, and hence provided a fuller link among the various notions we have considered so far.

12.3.5 PRD and DTP

As for the concept of PRD, Shaked (1977) showed that PRD is equivalent to

$$R(y|X = x) \text{ is decreasing in } x \text{ for all } y \geq 0, \tag{12.19}$$

where R is the conditional hazard function[5] defined as

$$R(y|X \in A) = -\log \Pr(Y > y|X \in A). \tag{12.20}$$

It is clear that Y is RTI if and only if $R(y|X > x)$ is decreasing in x for all $y \geq 0$.

The concept of positive regression dependence can be strengthened further. We say that X and Y are *positive likelihood ratio dependent* (or TP_2, meaning *totally positive of order two*) if

$$h(x_1, y_1)h(x_2, y_2) \geq h(x_1, y_2)h(x_2, y_1) \tag{12.21}$$

whenever $x_1 < x_2$ and $y_1 < y_2$. Lemma 2.2 of Abdel-Hameed and Sampson (1978) gives a sufficient condition for $h(x, y)$ to have this property.[6]

As pointed out in Shaked (1977), this notion of dependence is stronger than any notion we have yet discussed. He also used the classical results of the theory of total positivity[7] to construct a family of concepts of positive dependence called *dependent by total positivity of order two with degree* (m, n) and denoted by DTP(m, n).

Shaked showed that DTP(0,0) is equivalent to positive likelihood ratio dependent, and that DTP(1,1) is equivalent to right corner set increasing (RCSI), a notion introduced by R Harris (1970) meaning that $\Pr(X > x_1, Y > y_1|X > x_2, Y > y_2)$ is increasing in x_2 and y_2 for all x_1 and y_1. By choosing $x_1 = -\infty$, $y_2 = -\infty$, we see that RCSI($Y|X$) implies RTI($Y|X$).

For the multivariate case of DTP, see Lee (1985a).

---

[5]The hazard function is the cumulative hazard rate.

[6]It says let $h(x, y) = k_1(x)k_2(y)k_3(xy)$ for $x \geq 0$, $y \geq 0$, and $h(x, y) = 0$, otherwise; assume $k_1 \geq 0$, $k_2 \geq 0$, $k_3 \geq 0$; if k_3 is nondecreasing and $\log k_3$ is convex, then h is TP_2.

A consequence of this lemma is that the bivariate absolute normal distribution (that is, the distribution of $(|X|, |Y|)$ where (X, Y) has the bivariate normal distribution with zero mean) is TP_2.

[7]A function $K(x, y)$ is said to be totally positive of order n if the determinant of every matrix formed by (k_{ij}), where $k_{ij} = K(x_i, y_j)$, is positive for all $x_1 < x_2 < \dots < x_r$, $y_1 < y_2 < \dots < y_r$, for $r = 1, 2, \dots n$. A function which is totally positive of all finite orders is said simply to be totally positive. See Kim and Proschan (1988).

12.3.6 Interrelationships

In fact, Shaked (1977) showed that positive likelihood ratio dependent implies RCSI. We may summarise the chain of relations in the following (in which Y is conditional on X whenever there is a conditioning):

$$\begin{array}{ccccccccccc} \text{DTP}(1,1) & \Leftrightarrow & \text{RCSI} & \Rightarrow & \text{RTI} & \Rightarrow & \text{ASSOC.} & \Rightarrow & \text{PQD} & \Rightarrow & \text{PQDE} \\ \Uparrow & & \Uparrow & & \Uparrow & & \Uparrow & & & & \Downarrow \\ \text{DTP}(0,0) & \Leftrightarrow & \text{TP}_2 & \Rightarrow & \text{PRD} & \Rightarrow & \text{LTD} & & & & \text{cov} \geq 0 \end{array}$$

Another account of most of these interrelationships is by Ohi and Nishida (1978).

12.3.7 More Concepts

Shaked (1979) introduced further ideas of positive dependence, applicable to interchangeable bivariate random vectors—i.e., random vectors with permutation-invariant distributions. Their names are:

- Diagonal square dependent (DSD),
- Generalised diagonal square dependent (GDSD),
- Positively dependent by mixture (PDM),
- Positively dependent by expansion (PDE),
- Positive definite dependent (PDD).

Explanations:

- DSD means that $\Pr(X \in I \text{ and } Y \in I) \geq \Pr(X \in I)\Pr(Y \in I)$ for every interval I.
- PDM distributions are the joint distributions of $a(X_1, W)$ and $a(X_2, W)$ for some i.i.d. r.v.'s X_1 and X_2, a random vector W which is independent of the X's, and a function a.
- For explanation of the others, see Shaked (1979).

Interrelationships:

$$\begin{array}{ccccccc} & & \text{PDE} & & & & \\ & & \Downarrow & & & & \\ \text{PDM} & \Rightarrow & \text{PDD} & \Rightarrow & \text{GDSD} & \Rightarrow & \text{DSD} \\ & & \Downarrow & & & & \\ & & \text{cov}(X,Y) \geq 0 & & & & \end{array}$$

These concepts seem to be unconnected to those previously discussed: that is, the weak concept PQD is not implied by PDM and PDE, nor does the strong concept TP_2 imply DSD.

12.4 Negative Dependence

12.4.1 Concepts

Some concepts of negative dependence were first introduced by Lehmann (1966), and further developed by others. All of these can be obtained by negative analogues of positive dependence, e.g., when the inequality signs in (1), (2), and (4) (refer to Section 12.2.1) are reversed, we obtain negative dependence. For example, the negative analogue of PQD is negative quadrant dependent (NQD), and there are concepts of NRD, RCSD, and RTD. However, association has no negative analogue—the definition refers to *every* pair of functions a and b, and the choice $a = b$ will necessarily lead to $\text{cov}[a(X,Y), a(X,Y)] \geq 0$. A chain of implications is given below:

$$\text{RCSD} \Rightarrow \text{RTD} \Rightarrow \text{NQD} \Rightarrow \text{cov} \leq 0$$

It is not yet established if NRD $\Rightarrow$ NQD.

Block et al (1982a,b, 1985), Ebrahimi and Ghosh (1981, 1982), Karlin and Rinott (1980), and Lee (1985b) introduced and studied some concepts of multivariate negative dependence. See also the references contained in Block et al (1985).

Lehmann (1966) defined the concept of negative likelihood ratio dependence. This was called reverse regular of order two (RR_2) by Karlin and Rinott (1980) and Block et al (1982a). The latter showed that under a condition which essentially requires the sum of three independent r.v.'s to be fixed, two of them satisfy the RR_2 condition. They also showed that $RR_2 \Rightarrow$ NQD.

12.4.2 Neutrality

We should point out one important context where negative dependence is more natural than positive—that is, when concerned with three proportions or probabilities, X_1, X_2, and X_3, which sum to one, and we focus attention on only two of them. Then (X_1, X_2) is distributed over a triangle. The percentage composition of different minerals in rocks is an example, the percentage of household expenditure spent on different groups of commodities is another. See Section 24.5 for further examples.

The two variables are often taken to have a bivariate beta distribution (Section 6.5). The idea of neutrality was introduced by Connor and Mosimann (1969): X_1 and X_2 are said to be *neutral* if X_i and $X_j/(1-X_i)$ are independent $(i \neq j)$. It is well-known that if X_1 and X_2 have a bivariate beta distribution, then they are neutral (Section 6.5.11), and the converse is also true (Fabius, 1973). It was pointed out by Lehmann (1966) that the bivariate beta distribution is RR_2; hence, it is also NQD. Negative covariance is also noted in Section 6.5.4.

A recent and thorough account—though one which we find hard going—is that by Mosimann (1988). (See also Mosimann, 1975b, and Mosimann and Malley, 1981.) We note, also, that frequently variables that sum to 1 are obtained by dividing more primitive variables by their total, as in $X = X_1/(X_1 + X_2 + X_3)$, $Y = X_2/(X_1 + X_2 + X_3)$, and that spurious correlation may arise through the division by the same quantity—see also footnote 2 in Chapter 15, Pendleton (1986), and Prather (1988).

12.5 Monotonic Quadrant Dependence Function

12.5.1 Explanation

PQDE is based upon comparing $E(Y_x)$ with $E(Y)$. Kowalczyk and Pleszczyńska (1977) introduced the "monotonic quadrant dependence function" to quantify the difference between these two expectations.

Let B be the set of all bivariate r.v.'s with finite marginal means, and let x_p and y_p denote the pth quantiles of X and Y, respectively $(0 < p < 1)$. For each $(X, Y) \in B$ we may define a difference function

$$L_{Y,X}(p) = E(Y|X > x_p) - E(Y). \tag{12.22}$$

We may now define a function which can be used as a measure of strength of monotonic quadrant dependence between X and Y, as follows.

Definitions:

$$\mu^+_{Y,X}(p) = L_{Y,X}(p)/[E(Y|Y > y_p) - E(Y)] \tag{12.23}$$

$$\mu^-_{Y,X}(p) = L_{Y,X}(p)/[E(Y) - E(Y|Y < y_{1-p})] \tag{12.24}$$

$$\mu_{Y,X}(p) = \left\{ \begin{array}{ll} \mu^+_{Y,X} & \text{if } L_{Y,X}(p) \geq 0 \\ \mu^-_{Y,X} & \text{if } L_{Y,X}(p) \leq 0 \end{array} \right\} \tag{12.25}$$

The function $\mu_{Y,X}$ will be called the monotonic quadrant dependence function. Described in words, it compares the improvement in prediction of Y from knowing that X is big to the improvement in prediction of Y from knowing that Y is big.

12.5.2 Interpretation of $\mu_{Y,X}$

From the definitions above, we see that $\mu_{Y,X}(p)$ is a suitably normalised expected value of Y under the condition that X exceeds its pth quantile. It is a measure of strength of monotonic quadrant dependence between X and Y in the following sense: let (X, Y) and (X', Y') be two pairs of r.v.'s from B, having identical marginal distributions; then the positive quadrant dependence between X and Y is said to be stronger than between X' and Y' if $\mu_{Y,X}(p) \geq \mu_{Y',X'}(p)$, for all p between 0 and 1. (This is because $E(Y|X > x_p) > E(Y'|X' > x_p)$ is equivalent to $\mu_{Y,X}(p) > \mu_{Y',X'}(p)$.) The PQD is strongest when $\mu_{Y,X}(p) = 1$ and is weakest when $\mu_{Y,X}(p) = -1$. (If instead of B we consider only distributions for which $E(Y_x) \geq E(Y)$, then PQD is weakest when $\mu_{Y,X}(p) = 0$.)

12.5.3 Properties of $\mu_{Y,X}$

- $-1 \leq \mu_{Y,X}(p) \leq 1$.
- $\mu_{Y,X}(p) = 1 \Leftrightarrow \Pr(X < x_p \text{ and } Y > y_p) = \Pr(X > x_p \text{ and } Y < y_p) = 0$.
- $\mu_{Y,X}(p) = -1 \Leftrightarrow \Pr(X < x_p \text{ and } Y < y_{1-p}) = \Pr(X > x_p \text{ and } Y > y_{1-p}) = 0$.
- Let k and ℓ be two functions such that $F(a) < F(b) \Rightarrow k(a) < k(b)$ and $\ell(a) > \ell(b)$. Then, for any real a and b ($a \neq 0$),

$$\mu_{aY+b,\, k(X)}(p) = (\text{sgn } a)\mu_{Y,X}(p) \qquad (12.26)$$
$$\mu_{aY+b,\, \ell(X)}(p) = (-\text{sgn } a)\mu_{Y,X}(1-p). \qquad (12.27)$$

- $\mu_{Y,X}(p) = 0$ if and only if $E(Y|X) = E(Y)$ almost everywhere (i.e., the probability that they are unequal is 0).
- $\mu_{Y,X}(p)$ is $\mu^+_{Y,X}(p)$ if X and Y are PQDE, and is $\mu^-_{Y,X}$ if X and Y are NQDE.
- If X and Y are PQD, then $\mu_{X,Y} \geq 0$ and $\mu_{Y,X} \geq 0$.
- If X and Y are either PQD or NQD, then $\mu_{Y,X}(p) = 0$ if and only if X and Y are independent.
- If the distributions of (X, Y) and (X', Y') are both in B, and they have the same marginals, then $\mu_{Y,X} = \mu_{Y',X'}$ if and only if $E(Y|X)$ and $E(Y'|X')$ have the same distribution.

12.5.4 Remarks

- $\mu_{Y,X}$ is a function of p, and thus takes on different values for different p.
- $\mu_{Y,X}$ is not symmetric in X and Y: it is more similar to a prediction-improvement index than to a conventional measure of correlation.
- $\mu_{Y,X}$ is invariant under increasing transformations of X and linear increasing transformations of Y. (Recall that the product-moment correlation is invariant under linear increasing transformations of X and Y, and that rank correlation measures are invariant under increasing transformations of X and Y.)
- For sample counterparts of $\mu_{Y,X}$, see Kowalczyk (1977); Kowalczyk and Ledwina (1982) discussed the grade monotone dependence function $\mu_{G(Y),F(X)}$; for some comments aimed at aiding interpretation, see Kowalczyk (1982).

12.6 Orderings of Positive Dependence

Throughout this Section, we let H and H' denote the bivariate distribution functions of (X,Y) and (X',Y'), respectively, having common marginal distributions. We shall now introduce some (partial) orderings which compare the strength of positive dependence of (X,Y) with that of (X',Y').

- H is said to be more PQDE than H', written $H \overset{e}{\geq} H'$, if $E(Y|X > x) \geq E(Y'|X' > x)$ (Kowalczyk and Pleszczyńska, 1977).
- H is said to be more concordant than H' (Cambanis et al, 1976; Tchen, 1980), or more quadrant dependent than H' (Yanagimoto and Okamoto, 1969), written $H \overset{c}{\geq} H'$, if $H(x,y) \geq H'(x,y)$.
- H is said to be more regression dependent than H', written $H \overset{r}{\geq} H'$, if $\Pr(Y \leq y|X = x) \geq \Pr(Y' \leq y'|X' = x)$ implies $\Pr(Y \leq y|X = x') \geq \Pr(Y' \leq y'|X = x')$, for any $x' > x$ (Yanagimoto and Okamoto, 1969).[8]
- H is said to be more associated than H', written $H \overset{a}{\geq} H'$, if there exist functions u and v which map $R(f) \times R(g)$ on to $R(f)$ and $R(g)$, respectively,[9] such that

$$\left.\begin{array}{r} x_1 \leq x_2 \\ y_1 \leq y_2 \end{array}\right\} \Rightarrow \left\{\begin{array}{l} u(x_1,y_1) \leq u(x_2,y_2) \\ v(x_1,y_1) \leq v(x_2,y_2) \end{array}\right.$$

$$\left.\begin{array}{r} u(x_1,y_1) < u(x_2,y_2) \\ v(x_1,y_1) > v(x_2,y_2) \end{array}\right\} \Rightarrow \left\{\begin{array}{l} x_1 < x_2 \\ y_1 > y_2 \end{array}\right.$$

$$(X,Y) \sim (u(X',Y'),v(X',Y'))$$

(Schriever, 1987a,b). In the special case $u(x,y) = x$, H is more regression dependent than H' (as defined above).

The following chain of implications holds: $\overset{r}{\geq} \Rightarrow \overset{a}{\geq} \Rightarrow \overset{c}{\geq} \Rightarrow \overset{e}{\geq}$ (Yanagimoto and Okamoto, 1969; Schriever, 1987b).

Considering the special case $H' = FG$ (i.e., X' and Y' are independent), the following implications hold:

$H \overset{a}{\geq} FG \Rightarrow X$ and Y are associated.

$H \overset{r}{\geq} FG \Rightarrow X$ and Y are PRD.

$H \overset{c}{\geq} FG \Rightarrow X$ and Y are PQD.

$H \overset{e}{\geq} FG \Rightarrow X$ and Y are PQDE.

Kimeldorf and Sampson (1987) have given a list of ten conditions that define a "positive dependence ordering", denoted $\ll$. Of these, the most interesting are (P1) $H \ll H' \Rightarrow H(x,y) \leq H'(x,y)$ for all x,y, and (P6) $(X,Y) \ll (X',Y') \Rightarrow (a(X),Y) \leq (a(X'),Y')$ for all increasing functions a, in an obvious notation. (H and H' are assumed to have the same marginals.)

[8] Bilodeau (1989) uses "monotone regression dependent" to mean that the members of a family of distributions that is indexed by a parameter α are put into order with respect to their degree of regression dependence by the value of α. He gives a necessary and sufficient condition for one copula to be more regression dependent than another.

[9] $R(f)$ and $R(g)$ denote the supports of the marginals, and $R(f) \times R(g)$ is their Cartesian product. See Section 11.15, especially footnote 17 there.

Chapter 13

Rank Correlation Properties

13.1 Kendall's Tau and Spearman's Rho

The best-known rank correlation coefficients are those bearing the names of Kendall and of Spearman. We are not concerned here with formulae for computing them for practical samples, but with the population equivalents. The general properties of these coefficients have been introduced, in the company of other measures of correlation, in Section 11.8. In this Chapter, we ask what formulae connect them to the association parameter of various bivariate distributions.

Kendall's tau is given by

$$\tau = 4\int\int H(x,y)h(x,y)\,dx\,dy - 1, \tag{13.1}$$

where H is the cumulative distribution function and h is the probability density function.

Terminology and notation regarding Spearman's rho varies somewhat between authors. Ours is as follows. The population analogue of Spearman's rho is the grade correlation, given by

$$\rho_g = 12\int\int F(x)G(y)h(x,y)\,dx\,dy - 3. \tag{13.2}$$

The relation between Spearman's rho and the grade correlation is as follows: the expectation of Spearman's rho in samples of size n is

$$\frac{(n-2)\rho_g + 3\tau}{n+1}. \tag{13.3}$$

Thus we use "grade correlation" for the population parameter, reserving "Spearman's rho" for the sample statistic.

For a bivariate normal distribution, correlation coefficient ρ, it is well known that

$$\tau = \tfrac{2}{\pi}\sin^{-1}\rho \tag{13.4}$$

$$\rho_g = \tfrac{6}{\pi}\sin^{-1}\tfrac{\rho}{2}. \tag{13.5}$$

Hence the following relation between τ and ρ_g holds in the bivariate normal distribution:

$$\rho_g = \tfrac{6}{\pi}\sin^{-1}\left(\tfrac{1}{2}\sin\tfrac{\pi\tau}{2}\right). \tag{13.6}$$

We may now ask what is the relation between τ and ρ_g for other bivariate distributions, and can this relation be used to determine what is the bivariate shape of an empirical distribution? (By "bivariate shape", we mean the shape remaining once univariate shape has been discarded by ranking.) In this Chapter, we will give results for many of the bivariate distributions that have been described in Chapters 5–10. Since rank correlations are unaffected by marginal transformations, they are determined by the uniform representation, or copula, of the distribution, and it is natural to pay greatest attention to those distributions already written in this form, as are those in Chapter 5.

Bounds for ρ_g in terms of τ were established by Durbin and Stuart (1951). For $\tau > 0$, these are:

$$\frac{3\tau - 1}{2} \leq \rho_g \leq \frac{1 + 2\tau - \tau^2}{2}. \tag{13.7}$$

These bounds are remarkably wide: for instance, when $\tau = 0$, ρ_g can range between -0.5 and $+0.5$. Daniels (1950) comments that the assumption that τ and ρ_g describe more or less the same aspect of a bivariate population of ranks may be far from true, and suggests circumstances in which the message conveyed by the two coefficients is quite different. ("The worst discrepancy... occurs when the individuals fall into two groups of about equal size, within which corresponding pairs of ranks are nearly all concordant, but between which they are nearly all discordant"—Daniels, 1950, p. 190.) But Fieller et al (1957) do not think this would happen very often, saying that although after transforming the margins to normality, the resulting bivariate distribution will not necessarily be bivariate normal, "we think it likely that in practical situations it would not differ greatly from this form", adding "This is a field in which further investigaton would be of considerable interest".

We have noted elsewhere (Sections 1.6 and 3.1.3) our belief that computer algebra holds great promise for statistical research. Our impression of the field is that the integrations required in this Chapter are of the right level of difficulty to provide a good test of symbolic manipulation software. We shall meet some examples that are pretty simple, others that are elementary but lengthy and cumbersome, others whose results involve special functions such as the dilogarithm, Debye functions, or a hypergeometric function, yet others which are complicated by a singularity, or whose result involves "the largest integer not exceeding", or whose result is of different algebraic form in different ranges of the parameter. We think these problems are not so trivial that all relevant packages would handle them without difficulty, nor so difficult that it is unreasonable to expect solution.

13.2 Farlie-Gumbel-Morgenstern and Related Distributions

13.2.1 Farlie-Gumbel-Morgenstern

We recall from Section 5.2 that this system of bivariate distributions is

$$H(x, y) = F(x)G(y)\{1 + \alpha[1 - F(x)][1 - G(y)]\}. \tag{13.8}$$

Since rank correlation coefficients are unaffected by monotonic transformations of the axes, they are independent of the choice of $F(x)$ and $G(y)$. They turn out to be $\tau = 2\alpha/9$, $\rho_g = \alpha/3$; hence $\rho_g = 3\tau/2$ (Schucany, Parr, and Boyer, 1978). Farlie (1960) showed that this equation also holds if $H(x, y) = F(x)G(y)\{1 + \alpha A[F(x)]B[G(y)]\}$.

13.2.2 Iterated F-G-M Distribution

A generalisation is the (singly-) iterated F-G-M distribution:[1]

$$H = FG + \alpha FG(1 - F)(1 - G) + \beta(FG)^2(1 - F)(1 - G). \tag{13.9}$$

The grade correlation is $\frac{\alpha}{3} + \frac{\beta}{12}$ (equation (8) of Kotz and Johnson, 1977; Huang and Kotz, 1984). Kendall's tau is $\frac{2\alpha}{9} + \frac{\beta}{18} + \frac{\alpha\beta}{450}$ and hence

$$\rho_g = \frac{3\tau}{2} - \frac{\alpha\beta}{300}. \tag{13.10}$$

The same results hold for the distribution of Lin (1987, equation (8)).

[1] See Section 5.2.13 for the valid combinations of α and β.

13.2.3 A More General Distribution

For the distribution

$$H = FG + \alpha FG(1-F)(1-G) + \beta FG(1-F)(1-G)(\gamma+F)(\gamma+G), \tag{13.11}$$

it may be shown that $\tau = \frac{2\alpha}{9} + \frac{\beta(1+2\gamma)^2}{18} + \frac{\alpha\beta}{450}$ and $\rho_g = \frac{\alpha}{3} + \frac{\beta(1+2\gamma)^2}{12}$. Hence the above relation (13.10) between τ and ρ_g still holds.

Distribution (13.11) subsumes (13.9) (set $\gamma = 0$), distribution u_8 of Kimeldorf and Sampson (1975b) (set $\gamma = 1$), and the distribution (5.10) due to Sarmanov (1974) (set $\gamma = -\frac{1}{2}$).

13.3 Ali-Mikhail-Haq Family

As stated in Section 5.3, this general method for constructing bivariate distributions was introduced by Ali, Mikhail, and Haq (1978):

$$H(x,y) = \frac{F(x)G(y)}{1-\alpha[1-F(x)][1-G(y)]}. \tag{13.12}$$

To obtain τ, the algebra is straightforward, though messy, with result:

$$\tau = \frac{3\alpha-2}{3\alpha} - \frac{2(1-\alpha)^2}{3\alpha^2}\log(1-\alpha) \tag{13.13}$$

(Genest and MacKay, 1986a, p. 155).

To obtain ρ_g, the second integration requires finding $\int_0^1 (1-u)^{-1}\log(1-\alpha+\alpha u)\,du$. By substituting $x = \alpha(1-u)$, this becomes $\int_0^\alpha x^{-1}\log(1-x)\,dx$, which is $\operatorname{diln}(1-\alpha)$, diln being the dilogarithm.[2]

The final expression for ρ_g is

$$\rho_g = -\frac{12(1+\alpha)}{\alpha^2}\operatorname{diln}(1-\alpha) - \frac{3(12+\alpha)}{\alpha} - \frac{24(1-\alpha)}{\alpha^2}\log(1-\alpha). \tag{13.14}$$

13.4 Frank's System

Frank's system of bivariate distributions is

$$H(x,y) = \log_\alpha\left\{1 + \frac{[\alpha^{F(x)}-1][\alpha^{G(y)}-1]}{\alpha-1}\right\} \tag{13.15}$$

(see Section 5.4).

Nelsen (1986) and Genest (1987) have shown that

$$\begin{aligned} \tau &= 1 + 4[D_1(\alpha^*)-1]/\alpha^* & (13.16)\\ \rho_g &= 1 + 12[D_2(\alpha^*)-D_1(\alpha^*)]/\alpha^*, & (13.17) \end{aligned}$$

where $\alpha^* = -\log(\alpha)$, and D_1 and D_2 are Debye functions defined by

$$D_k(\beta) = \frac{k}{\beta^k}\int_0^\beta \frac{t^k}{e^t-1}\,dt. \tag{13.18}$$

Nelsen (1986) has plotted τ and ρ_g as functions of α^*.

[2]See footnote 4 in Chapter 9.

13.5 Bivariate Pareto Distribution

This has upper right volume $\bar{H} = (1 + x + y)^{-c}$ (see Section 5.8).

Straightforward algebra shows that $\tau = \frac{1}{1+2c}$. (An equivalent formula is given by Genest and MacKay (1986a, p. 154; 1986b).)

The grade correlation is more difficult. The first integration requires the use of equation 3.197.1 of Gradshteyn and Ryzhik (1965) to show

$$\int_0^\infty (1+x)^{-c}(1+y+x)^{-(c+2)}\,dx = \frac{1}{2c+1}(1+y)^{-(c+2)}\,{}_2F_1\left(c+2,1;2c+2;\frac{y}{1+y}\right),$$

where ${}_2F_1$ is Gauss' hypergeometric function; and the second integration requires Gradshteyn and Ryzhik's equation 7.512.5 to find

$$\int_0^1 (1-v)^{2c}{}_2F_1(c+2,1;2c+2;v)\,dv = \frac{1}{2c+1}{}_3F_2(c+2,1,1;2c+2,2c+2;1),$$

where ${}_3F_2$ is a generalised hypergeometric function.[3]

The final result is

$$\rho_g = \frac{12c(1+c)}{(1+2c)^2}\,{}_3F_2(c+2,1,1;2c+2,2c+2;1) - 3.$$

We have been unable to simplify this.[4] No doubt equation (2.4) of Cook and Johnson (1981), involving an infinite sum of ratios of gamma functions, is equivalent. Oakes (1982, equation (6)) also ran aground at the same point.

13.6 Marshall and Olkin's Distribution

Cuadras and Augé (1981) introduced this distribution on the unit square:

$$H = \left\{ \begin{array}{ll} x^{1-c}y & \text{for } x \geq y \\ xy^{1-c} & \text{for } x < y \end{array} \right\} \qquad (13.19)$$

A slight complicating factor with this is that the p.d.f. has a singularity along $y = x$. Cuadras and Augé determined that Pearson's correlation is $3c/(4-c)$. The marginals are uniform, and so ρ_g is the same.

It may also be shown that $\tau = c/(2-c)$. Hence $\rho_g = 3\tau/(2+\tau)$.

We understand $\rho_g = 3\tau/(2+\tau)$ holds also for the asymmetric case $H = \min(xy^{1-\beta}, x^{1-\alpha}y)$—Nelsen (1989).

This distribution is better known in the form with exponential marginals—Marshall and Olkin (1967a), discussed in Section 9.2.

[3]The infinite series representations of ${}_2F_1$ and ${}_3F_2$ are

$$\begin{aligned} {}_2F_1(\alpha,\beta;\gamma;u) &= \sum_{i=0}^{\infty} \frac{(\alpha,i)(\beta,i)}{(\gamma,i)}\frac{u^i}{i!} \\ {}_3F_2(\alpha,\beta,\gamma;\delta,\epsilon;u) &= \sum_{i=0}^{\infty} \frac{(\alpha,i)(\beta,i)(\gamma,i)}{(\delta,i)(\epsilon,i)}\frac{u^i}{i!} \end{aligned}$$

In these equations, (ν,n) is the Appell symbol, $(\nu,n) = \Gamma(\nu+n)/\Gamma(\nu)$, which is $\nu(\nu+1)(\nu+2)\ldots(\nu+n-1)$ for integer n. (The Pochhammer symbol ν_n differs only typographically.)

Conde and Kalla (1979?) have prepared a table of the values of ${}_2F_1$. The reviewer for *Mathematics of Computation* (Y L L, at vol. 34 (1980), p. 330) felt that the table fills a noted gap, but that most people will prefer to program an algorithm and generate values as required.

[4]Wimp (1981) gives an algorithm for calculating ${}_3F_2(\alpha,\beta,\gamma;\delta,\epsilon;1)$. Wimp says that "if ${}_3F_2(1)$ could be determined as a closed-form expression, e.g., as a ratio of finite products of Gamma functions (as can the function ${}_2F_1(1)$), much of applied mathematics could be substantially simplified."

13.7 Kimeldorf and Sampson's Distribution

Kimeldorf and Sampson (1975a) studied a bivariate distribution on the unit square, with uniform marginals and p.d.f. as follows:

- β on each of $[\beta]$ squares of side $1/\beta$ arranged corner-to-corner up the diagonal from (0,0) towards (1,1), $[\beta]$ being the largest integer not exceeding β,
- $\frac{\beta}{\beta-[\beta]}$ on one smaller square of side $1-[\beta]/\beta$ in the top right corner of the unit square (unless β is an integer),
- and 0 elsewhere.

For this distribution,

$$\rho_g = \frac{[\beta]}{\beta}\frac{3\beta^2 - 3\beta[\beta] + [\beta]^2 - 1}{\beta^2} \tag{13.20}$$

(Johnson and Tenenbein, 1979). Professor R B Nelsen of Lewis and Clark College has shown that

$$\tau = \frac{[\beta]}{\beta}\frac{2\beta - [\beta] - 1}{\beta}.$$

Hence if $1 \le \beta < 2$, $\rho_g = 3\tau/2$; and if β is an integer, $\rho_g = 2\tau - \tau^2$.

13.8 Weighted Linear Combination

The weighted linear combination procedure (see Sections 3.7 and 10.4) generates random variables by setting $X = X_1$ and $Y = cX_1 + (1-c)X_2$, where X_1 and X_2 are i.i.d. r.v.'s with specified p.d.f.

Johnson and Tenenbein (1979, 1981) give τ and ρ_g for three choices of the distribution of X_1 and X_2, as follows.

Exponential: $\tau = c$, $\rho_g = c(3-2c)/(2-c)$, and hence $\rho_g = \tau(3-2\tau)/(2-\tau)$.

Laplace: $\tau = c(3+3c-2c^2)/4$, $\rho_g = c(9-18c^2+14c^3-3c^4)/[2(2-c)^2]$.

Uniform: τ is $\frac{4c-5c^2}{6(1-c)^2}$ for $0 \le c \le 0.5$, and is $\frac{11c^2-6c+1}{6c^2}$ for $0.5 \le c \le 1$. As for ρ_g, it is $\frac{c(10-13c)}{10(1-c)^2}$ for $0 \le c \le 0.5$, and $\frac{3c^3+16c^2-11c+2}{10c^3}$ for $0.5 \le c \le 1$.

13.9 Trivariate Reduction

Johnson and Tenenbein (1979) give τ and ρ_g for random variables generated by setting $X = X_1 + cX_3$ and $Y = X_2 + cX_3$, where X_1, X_2, and X_3 are i.i.d. r.v.'s with specified p.d.f. Johnson and Tenenbein (1979) give τ and ρ_g for three choices of the distribution of the X_i's, as follows.

Exponential: $\tau = \frac{2c^2}{(1+c)(1+2c)}$, $\rho_g = \frac{c^2(2c^2+9c+6)}{(1+c)^2(1+2c)(2+c)}$.

Laplace: $\tau = \frac{c^2(32c^5+125c^4+161c^3+90c^2+22c+2)}{2(1+c)^3(1+2c)^4}$, $\rho_g = \frac{c^2(16c^7+152c^6+588c^5+1122c^4+1104c3+555c^2+132c+12)}{2(1+c)^4(1+2c)^3(2+c)^2}$.

Uniform:

$$\tau = \begin{cases} \frac{c^2(c^2-6c+10)}{15} & \text{for } 0 \le c \le 1 \\ \frac{15c^2-14c+4}{15c^2} & \text{for } 1 \le c \end{cases}$$

$$\rho_g = \begin{cases} \frac{c^2(19c^2-126c+210)}{210} & \text{for } 0 \le c \le 1 \\ \frac{c^7-14c^6+84c^5-280c^4+770c^3-672c^2+238c-24}{210c^3} & \text{for } 1 \le c \le 2 \\ \frac{105c^3-105c+52}{105c^3} & \text{for } 2 \le c \end{cases}$$

Cherian's bivariate gamma distribution is obtained by a form of trivariate reduction, starting with univariate gamma distributions—see Section 10.3. Values of τ and ρ_g can be found for three combinations of (small integer) values of the shape parameters of the X's from Table III of David and Fix (1961).

13.10 A Distribution Uniform on Part of the Unit Square

The form of the expressions for τ and ρ_g makes it tempting to consider distributions on the unit square whose p.d.f. is constant within some region and zero elsewhere. For instance, if the p.d.f. is constant for $x^{1/c} \leq y \leq x^c$ and is 0 elsewhere, it may be shown that

$$\tau = \frac{2c}{1+c} \tag{13.21}$$

$$\rho_g = \frac{c(12c^4+61c^3+97c^2+61c+12)}{(2+c)^2(1+2c)^2(1+c+c^2)}, \tag{13.22}$$

and hence

$$\rho_g = \frac{\tau(2-\tau)(-\tau^4+4\tau^3-56\tau^2+104\tau+192)}{(4-\tau)^2(2+\tau)^2(4-2\tau+\tau^2)}. \tag{13.23}$$

13.11 A Distribution Suggested by R B Nelsen

Professor R B Nelsen suggests this distribution on the unit square:

$$H = \min[x, y, (xy)^{(2-c)/2}]. \tag{13.24}$$

It is similar to that of Section 13.10 in that the p.d.f. is concentrated between $y = x^{(2-c)/c}$ and $y = x^{c/(2-c)}$, and is similar to that of Section 13.6 in that the p.d.f. has singularities. (These occur on the bounding curves.)

Nelsen finds that $\tau = c$ and $\rho_g = 3c(8-5c)/(4-c)^2$.

Nelsen has also obtained results for:

- The asymmetric case of (13.24), $H = \min(x, y, x^{1-\alpha}y^{1-\beta})$ (Nelsen, 1989),
- Several distributions for which the probability mass is concentrated on the two curves $y = a(x)$ and $y = a^{-1}(x)$, an idea which he takes from Sherwood and Taylor (1988),
- The example of Raftery's (1984, 1985) scheme that we have called the first special case in our Section 9.11.2, for which the same relation between τ and ρ_g holds as for (13.24), and
- The EP1 model of Lawrance and Lewis (1983) (see our Section 9.10.2).

13.12 Singly-Truncated Bivariate Normal Distribution

Aitkin and Hume (1965) have considered τ and ρ_g in the singly truncated bivariate normal distribution. Their results are in terms of certain integrals which must be evaluated numerically, though some tables are also included in their paper.

13.13 Discussion

For a given value of τ, how much do distributions differ in their value of ρ_g? Table 13.1 shows that although ρ_g could theoretically take on a very wide range of values, for the distributions considered the values are all very similar. The distributions that are most different from the others are Marshall and Olkin's, with its singularity in the p.d.f. at $y = x$, and Kimeldorf and Sampson's, with its oddly shaped support. With these exceptions, at $\tau = 0.5$, ρ_g lies in the range .667 to .707, even though it could theoretically take any value between .250 and .875. This suggests the following question to close with: is there some class of bivariate distributions which includes nearly all of those that occur, for which only a narrow range of ρ_g (for given τ) is possible? For instance, if every quantile of y for given x increases with x, and vice versa, can bounds for ρ_g in terms of τ be found?

A clue can be found from the Marshall and Olkin (1967a) and Kimeldorf and Sampson (1975a) distributions. Consider distributions whose marginals are uniform (0,1). Consider probability densities along

Table 13.1: Comparison of the values of ρ_g corresponding to four values of τ.

Distribution	$\tau = \frac{1}{5}$	$\tau = \frac{1}{3}$	$\tau = \frac{1}{2}$	$\tau = \frac{3}{4}$
(Lower bound)	−.200	.000	.250	.625
(Upper bound)	.680	.778	.875	.969
Normal	.296	.483	.690	.917
Farlie-Gumbel-Morgenstern	.300 *	-	-	-
Ali-Mikhail-Haq	.297	.478	-	-
Frank	.297 †	.484 †	.695 †	.922 †
Pareto	.295	.478	.682 ‡	?
Marshall and Olkin	.273	.429	.600	.818
Kimeldorf and Sampson	.300	.500	.750	.937
Weighted linear combination: exponential	.289	.467	.667	.900
Weighted linear combination: Laplace	.293 §	.473 §	.674	.904 §
Weighted linear combination: uniform	.295	.483	.700	.927
Trivariate reduction: exponential	.292	.472 ¶	.675	.906
Trivariate reduction: Laplace	.293 §	.473	.676 §	.906
Trivariate reduction: uniform	.298 §	.490	.707	.927
Section 13.10	.298	.486	.695	.919
Equation (13.24)	.291	.471	.673	.905
(New lower bound)	.265	.414	.581	.803
(New upper bound)	.300	.500	.750	.937

Notes:

* For the iterated F-G-M distribution with $\tau = 0.2$, ρ_g lies between .297 and .301, depending on what α and β are. The former corresponds to $\alpha = 0.446$, $\beta = 1.784$, the latter to $\alpha = 1$, $\beta = -0.385$.

† We are grateful to R B Nelsen of Lewis and Clark College for calculating these values.

‡ One way of finding this is to use equation (1) of Lavoie (1986).

§ We are grateful to M E Johnson of Los Alamos National Laboratory for calculating these values.

¶ This result is implicit in Table III of David and Fix (1961).

$y = 1 - x$ and lines parallel to it. High correlation corresponds to these densities being high close to $y = x$. But for a given degree of correlation as measured by τ,

1. A low ρ_g seems to correspond to the density having a high peak in the region of $x = y = 0.5$, low shoulders in the regions of $x = 0.25, y = 0.75$ and $x = 0.75, y = 0.25$, and thick tails as $x \to 0, y \to 1$ and $x \to 1, y \to 0$;

2. Conversely, a relatively high ρ_g corresponds to a low peak, high shoulders, and thin tails.

As we shall see shortly, the peak-shoulders-tails distinction is a red herring. But the contrast between the two distributions, with their respective peculiar features, does suggest the following conjecture.

A conjecture. We conjecture that the following is a correct interpretation of this observation—but would welcome seeing a rigorous proof.

A. Consider distributions on the unit square that consist of a uniform distribution, on which is superimposed (i) a set of rectangles of uniform and equal extra p.d.f., extending corner-to-corner diagonally upwards[5], and (ii) a monotonically increasing line of probability singularity joining the diagonally-opposite corners of the rectangles. For these distributions,

$$-1 + \sqrt{1 + 3\tau} \le \rho_g \le 2\tau - \tau^2 \qquad (13.25)$$

The upper bound is not necessarily attainable, and, in particular, $\rho_g \le 3\tau/2$ also.

B. The distributions described in part A of the conjecture are equivalent to those for which every quantile of y for given x is nondecreasing, and every quantile of x for given y is also nondecreasing. (The reason is that monotonic transformations of the axes leave rank correlations unaffected.) And this is what we called positive regression dependence in Chapter 12.

Remarks on the conjecture:

- The lower bound in (13.25) occurs when the distribution is that of Section 10.9.1, a mixture of perfect correlation ($Y \equiv X$) and independence (p.d.f. = 1 for the whole unit square). When these occur in the proportions $p : 1 - p$, $\rho_g = p$ and $\tau = p(2 + p)/3$, and hence $\rho_g = -1 + \sqrt{1 + 3\tau}$. (These results were previously obtained by Konijn, 1959, p. 277.) Values of this expression, and of $\min(2\tau - \tau^2, \frac{3}{2}\tau)$, are included in Table 13.1.

- The Kimeldorf-Sampson distribution of Section 10.7 provides the upper bound, though it is not necessary for the areas of constant p.d.f. to be specified so restrictively—indeed, all the probability mass could be concentrated at points (for 3 points, containing probabilities p, q, and $1 - p - q$ respectively, $\rho_g = \frac{3}{2}\tau - 3pq(1 - p - q)$).

- For a given value of τ:

  - A relatively high ρ_g (as with the Kimeldorf-Sampson distribution) indicates there is zero correlation locally but good correlation on the larger scale.

  - A relatively low ρ_g (as with the Marshall and Olkin distribution) indicates there is high correlation locally but poor correlation on the larger scale. It is true that over most of the support the local-scale correlation is poor—but it is very good on the singularity, and so it averages out to be relatively high.

  - As was known in the early 1950's, very high ρ_g only occurs with negative correlation locally contrasted with positive overall correlation, and very low ρ_g only with negative overall correlation contrasted with positive correlation locally.

[5]To clarify: let us have n values of X, arranged in increasing order, and n values of Y, also in increasing order. Then, for $1 \le i \le n - 1$, the ith rectangle of nonzero p.d.f. is defined by the points: (x_i, y_i), (x_i, y_{i+1}), (x_{i+1}, y_{i+1}), (x_{i+1}, y_i).

- As mentioned in Section 13.11, Professor R B Nelsen has found τ and ρ_g for some distributions for which the probability mass is concentrated on the two curves $y = a(x)$ and $y = a^{-1}(x)$. If, for instance, the marginals are uniform and $a(x) = \max\left(\frac{1-\alpha}{1+\alpha}x, x - \frac{\alpha}{1+\alpha}, \frac{1+\alpha}{1-\alpha}x - \frac{2\alpha}{1-\alpha}\right)$, then $\tau = 1 - \alpha$ and $\rho_g = 1 - 2\alpha^2 + \frac{2\alpha^4}{(1+\alpha)^2}$. For example, if $\alpha = \frac{1}{2}$, then $\tau = 0.5$ and $\rho_g = 0.556$. This distribution has a low centre around $x = y = 0.5$, high shoulders near $x = 0.25$, $y = 0.75$ and $x = 0.75$, $y = 0.25$, and thin tails as $x \to 0$, $y \to 1$ and $x \to 1$, $y \to 0$, yet ρ_g is low. So our first idea about peak-shoulders-tails was wrong, and the above interpretation (a low ρ_g meaning high local correlation but poor larger-scale correlation) is better.

- It is likely that the "local dependence function" of Holland and Wang (1987b)—see Section 11.14—could be helpful in taking this further. This function is defined as $\frac{\partial^2}{\partial x \partial y}\log[h(x,y)]$; Holland and Wang use the symbol $\gamma(x,y)$ for it. The bivariate normal distribution has constant γ (as, in fact, has been known for many years).[6] We conjecture that (for given τ) a relatively high ρ_g corresponds to a low γ (when this is averaged in some way, presumably by weighting according to the p.d.f.), and vice versa.

- Since ρ_g can exceed $3\tau/2$ for the iterated F-G-M distribution (13.9)—see footnote to Table 13.1—we suggest that there is something odd about it. Perhaps it is possible to show that the conditional quantiles are not all increasing.

- For both the bounds that have been suggested in (13.25),[7] and for all the distributions above for which it could easily be obtained, $\frac{d\rho_g}{d\tau} = \frac{3}{2}$ at $\tau = 0$. Hence it seems likely that if this is not the case, then the family of distributions is in some sense ill-behaved.

What we are claiming, then, is that PRD $\Rightarrow$ (13.25). (i) Is this true? (ii) Can bounds for ρ_g in terms of τ be derived for the other concepts of positive dependence in Chapter 12?

[6] For the bivariate normal, $\gamma = \rho/(1-\rho^2)$. Holland and Wang also give $\gamma = 4\alpha/[1 + \alpha(2x-1)(2y-1)]^2$ for the F-G-M distribution with uniform marginals, and $\gamma = 6xy/(c^2 + x^2 + y^2)^2$ for the Cauchy distribution. They point out that this last expression changes sign as (x,y) changes quadrant in the plane, X and Y being positively associated in the first and third quadrants, and negatively associated in the second and fourth—see also the contours illustrated in Barnett (1980).

[7] Remember the upper one is replaced by $3\tau/2$ when $\tau < \frac{1}{2}$.

Chapter 14

Diagonal Expansion

14.1 Introduction

Diagonal expansions of bivariate distributions are also known as the canonical forms of bivariate distributions. The method under consideration is essentially an extension to the theoretical parent population of the procedure used by Fisher (1940), Hirschfeld (1935), and Hotelling (1936) to derive certain results in connection with contingency tables. H O Lancaster is generally considered as a main contributor in formalising this extension to a general class of bivariate distributions. For a historical development of the subject, see Lancaster (1958).

This Chapter is arranged as follows:

14.2 A general expansion of a bivariate distribution.

14.3 ϕ^2-bounded bivariate distributions.

14.4 Biorthogonal property and diagonal expansion.

14.5 Conditions for polynomial biorthogonal property.

14.6 Canonical sequence and moment sequence.

14.7 Remarks.

14.8 Diagonal expansion and construction of Markov processes.

14.2 A General Expansion of a Bivariate Distribution

In this Section, we shall see how bivariate distributions may be expressed as diagonal expansions in terms of orthogonal polynomials and marginals F and G.

Let H be a bivariate distribution with given marginals F and G. Let $\{\xi_n\}$ and $\{\eta_n\}$ be complete sets of orthonormal functions[1] defined on F and G, respectively. Let

$$\rho_{ij} = \text{corr}[\xi_i(X), \eta_j(Y)] = \int \xi_i(x)\eta_j(y)\, dH(x,y). \tag{14.1}$$

If $\sum_{i=1}^{\infty}\sum_{j=1}^{\infty} \rho_{ij}^2$ is convergent, then from results in the theory of integral equations,

$$dH(x,y) = dF(x)dG(y)\sum_{i=0}^{\infty}\sum_{j=0}^{\infty} \rho_{ij}\xi_i(x)\eta_j(y), \tag{14.2}$$

[1]Orthonormal functions ξ satisfy

$$\int \xi_i(x)\xi_j(x)\, dF(x) = \delta_{ij},$$

where δ_{ij} is the Kronecker delta, as in Section 1.5.4.

A *complete* set of orthonormal functions is one such that there is no non-zero function outside it that is orthogonal to all functions in it.

where the double summation is convergent in the mean square sense[2] with respect to F and G.

Conversely, a bivariate distribution is completely determined by its marginal distributions and the matrix of correlations of any pair of complete sets of orthonormal functions on the marginal distributions. The background to the expansion of this type is set out in Lancaster's (1969) book as well as in his earlier publications (e.g., Lancaster, 1958, 1963a), and there is a later review by Lancaster (1980).

14.3 ϕ^2-Bounded Bivariate Distributions

For given F, G, and H, let us define

$$\phi^2 + 1 = \int \Omega(x,y)\,dH(x,y) = \int\int \Omega^2(x,y)\,dF(x)\,dG(y), \tag{14.3}$$

where[3] $\Omega(x,y) = h(x,y)/[f(x)g(y)]$. ϕ^2 is related to Karl Pearson's mean square contingency.[4] We say that H is ϕ^2-bounded if $\phi^2 < \infty$. On substituting (14.2) into (14.3) and using (14.1) and the orthogonality property, we have

$$\phi^2 = \sum_{i=1}^{\infty}\sum_{j=1}^{\infty} \rho_{ij}^2.$$

Hence, ϕ^2-bounded is equivalent to $\sum_{i=1}^{\infty}\sum_{j=1}^{\infty}\rho_{ij}^2 < \infty$. In the remainder of this Chapter, we shall assume that H is ϕ^2-bounded; this apparent restriction is unimportant as all distributions met with in practice satisfy it.

14.4 Biorthogonal Property and Diagonal Expansion

Lancaster (1969) has shown that the two sets of orthonormal functions $\{\xi_n\}$ and $\{\eta_n\}$ can be transformed,[5] respectively, to two new sets $\{\xi_n'\}$ and $\{\eta_n'\}$, which satisfy

$$E[\xi_i'(X)\eta_j'(Y)] = \delta_{ij}\rho_i \tag{14.4}$$

($1 \geq \rho_1^2 \geq \rho_2^2 \ldots$). We then say the two series of orthonormal functions $\{\xi_n'\}$ and $\{\eta_n'\}$ possess the biorthogonal property if (14.4) holds. It is now clear that the infinite matrix of correlation coefficients becomes diagonal and H can be represented as

$$dH(x,y) = dF(x)dG(y)\sum_{i=0}^{\infty}\rho_i\xi_i'(x)\eta_i'(y). \tag{14.5}$$

ξ_i' and η_i' are called the canonical variables and the ρ's are called the canonical correlations. We then say H has a diagonal expansion in terms of the marginals and the canonical variables.

14.5 Conditions for Polynomial Biorthogonal Property

From now on, we shall assume that both X and Y have finite moments of all orders, and hence we can construct unique[6] sets of orthonormal polynomials $\{P_n\}$ and $\{Q_n\}$ with respect to F and G, respectively,

---

[2]That is, there exists a function $\ell(x,y)$ such that $\int\int\left(\sum_{i=0}^{M}\sum_{j=0}^{N}\rho_{ij}\xi_i(x)\eta_j(y) - \ell(x,y)\right)^2 dF(x)\,dG(y)$ tends to 0 as $M, N \to \infty$. In this case, $\ell(x,y)$ is the $\Omega(x,y)$ which will be defined shortly.

[3]The definition of $\Omega(x,y)$ which now follows only applies when the joint density and the marginal densities exist. More generally, $\Omega(x,y)$ is the Radon-Nikodým derivative; that is, Ω is, roughly, the ratio of the joint measure induced by H to the product measure induced by F and G.

[4]Pearson (1904) defined mean square contingency as the ratio of the chi-squared calculated for a contingency table to the sample size n. He used it as a measure of association between X and Y. Divide the supports of X and Y into I and J regions respectively. Then ϕ^2 as we have defined it is the limit of $\frac{1}{n}\sum(O-E)^2/E$ (where the symbols take their usual meanings) as I, J, and n tend to infinity.

[5]The transformations are linear and orthogonal, $\xi_n' = \sum_{i=1}^{\infty} a_{ni}\xi_i$ and $\eta_n' = \sum_{i=1}^{\infty} b_{ni}\eta_i$, with $\sum_{i=1}^{\infty} a_{ni}^2 = \sum_{i=1}^{\infty} b_{ni}^2 = 1$.

[6]Apart from a constant factor in each polynomial.

i.e., $E[P_n(X)] = 0$, $E[P_m(X)P_n(X)] = \delta_{mn}$, and $P_0(X) = 1$, and similarly for the Q's. We further assume that these sets of orthonormal polynomials are complete on their respective marginals.[7]

Definition: a bivariate d.f. H with complete sets of orthonormal polynomials $\{P_n\}$ and $\{Q_n\}$ is said to satisfy the polynomial biorthogonal property if

$$E[P_m Q_n] = \delta_{mn}\rho_{mn}, \tag{14.6}$$

ρ_{mn} being the product-moment correlation between $P_m(X)$ and $Q_n(Y)$. Using the same argument as previously, we deduce that (14.6) is equivalent to the expression

$$dH(x,y) = dF(x)dG(y)\sum_{i=0}^{\infty} \rho_i P_i(x)Q_i(y). \tag{14.7}$$

Brown (1958) shows that (14.7) holds if and only if

$$\left.\begin{array}{rcl} E[X^n|Y=y] & = & \text{a polynomial in } y \text{ of degree} \le n \\ E[Y^n|X=x] & = & \text{a polynomial in } x \text{ of degree} \le n \end{array}\right\} \tag{14.8}$$

Now, (14.8) is equivalent to

$$\left.\begin{array}{rcl} E[P_n(X)|Y] & = & \rho_n Q_n(Y) \\ E[Q_n(Y)|X] & = & \rho_n P_n(X) \end{array}\right\} \tag{14.9}$$

Hence H has a diagonal expansion in terms of F and G and their respective orthonormal polynomials if either (14.8) or (14.9) holds.

The sequence $\{\rho_n\}$ in (14.7) or (14.5) is called the canonical correlation sequence (or, for brevity, just the canonical sequence).

For the orthogonal polynomials that correspond to certain marginals, see Table 14.1.

14.6 Canonical Sequence and Moment Sequence

14.6.1 Normal Marginals

One important special case of the formula (14.7) arises when the marginals are normal and hence the orthogonal functions are Hermite polynomials. In this case,

$$h(x,y) = \frac{1}{2\pi}\exp[-(x^2+y^2)/2]\sum_{i=0}^{\infty} c_i H_i(x)H_i(y), \tag{14.10}$$

where the series converges in the mean square sense if $\sum_{i=0}^{\infty} c_i^2 < \infty$. The Hermite polynomials H_i here have been normalised so that they are orthonormal with respect to the normal density as the weight function:

$$\frac{1}{\sqrt{2\pi}}\int_{-\infty}^{\infty} H_i(x)H_j(y)\exp(-x^2/2) = \begin{cases} 0 & \text{if } i \ne j \\ 1 & \text{if } i = j \end{cases} \tag{14.11}$$

In order to generate a family of bivariate density functions from (14.10), it is necessary to choose the coefficients c_i such that the series for $h(x,y)$ is non-negative. Then c_i would be the correlation coefficient between $H_i(X)$ and $H_i(Y)$; hence, only expansions of the type (14.10) for which $|c_i| \le 1$ $(i = 1, 2, \ldots)$ are of interest.

It has been shown by Sarmanov (1966) (also, Sarmanov and Bratoeva, 1967) that for the c's to be the canonical correlations, it is necessary and sufficient that the c's be the moments of some probability distribution concentrated within the interval $[-1, 1]$. Thus, for example, the trivial case $c_1 = 1$, with all other c's being 0, satisfies the above requirements. Of more interest, perhaps, is the case where there is

[7] The assumption of the completeness of the orthonormal system, though important, will not create any problem, as all the systems of polynomials used in this context are complete. However, it is not true that every system of orthonormal polynomials is complete (Szegö, 1975, p. 40).

Table 14.1: Marginal distributions and the corresponding orthogonal polynomials.

Orthogonal Polynomial*	Marginal Distribution	Support	References†
Hermite	Normal	$-\infty$ to ∞	[1]‡ [2]
Laguerre	Gamma	0 to ∞	[1]‡ [3]‡ [4] [5]
Chebyshev (first kind)	$f \propto (1-x^2)^{-1/2}$	-1 to 1	[1]
Ultraspherical (Gegenbauer)	Pearson type II	-1 to 1	[6] [7]
Jacobi (P_n)	Pearson type I	-1 to 1	[8]
Jacobi (G_n)	Beta	0 to 1	[9]
Legendre	Uniform	-1 to 1	[8] [10]
Shifted Legendre	Uniform	0 to 1	[11]
Pollaczek	Meixner hypergeometric	$-\infty$ to ∞	[12] [13]

Notes:

* All the orthogonal polynomials given here can be found in Abramowitz and Stegun (1964, pp. 774–775), except for the Pollaczek system, for which, see Szegö (1975, p. 393).

† [1] Barrett and Lampard (1955); [2] Sarmanov (1966); [3] Kibble (1941); [4] Griffiths (1969); [5] Jensen (1970); [6] McFadden (1966); [7] Derin et al (1977); [8] Wong and Thomas (1962); [9] Lee (1971); [10] Sarmanov (1968); [11] Sarmanov (1974); [12] Eagleson (1964); [13] Lancaster (1983).

‡ In these instances, the canonical correlation coefficients are of simple power only.

some simple pattern to the c's, such as c_i = either R^i or 0, for some R between -1 and 1. Then (14.10) is of the form

$$h(x,y) = \frac{1}{2\pi}\exp[-(x^2+y^2)/2]\sum_{j=0}^{\infty} R^{k_j} H_{k_j}(x) H_{k_j}(y), \tag{14.12}$$

where $0 < k_1 < k_2 < \ldots$ are integers. Only two choices of k_j for which the sum (14.12) is positive for all x and y are known:

- $k_j = j$. This yields the well-known expansion of the standard bivariate normal with zero means and unit variances.
- $k_j = 2j$. In this case, a linear combination of two bivariate normal distributions with correlations R and $-R$, respectively, is obtained:

$$h(x,y;R) = \tfrac{1}{2}\psi(x,y;R) + \tfrac{1}{2}\psi(x,y;-R), \tag{14.13}$$

where ψ is the standardised bivariate normal p.d.f.

14.6.2 Other Marginals

It is now clear that to generate a new bivariate distribution with a diagonal expansion in terms of given marginals and their associated orthonormal polynomials, it is necessary to find a sequence $\{\rho_n\}$ such that it is a canonical sequence.

Following Sarmanov's pioneering work on the characterisation of a canonical sequence by a moment sequence in the case of a bivariate normal distribution, many authors extended his idea to other bivariate distributions. For example, Griffiths (1969) (see also Askey, 1970) has shown that a sequence $\{\rho_n\}$ is a canonical sequence for a bivariate gamma distribution if and only if

$$\rho_n = \sqrt{\frac{(\alpha,n)}{(\beta,n)}}\int_0^1 t^n \, d\mu(t), \tag{14.14}$$

where (a, n) is the Appell symbol, i.e., $\Gamma(a+n)/\Gamma(n)$, and F and G are gamma distribution functions such that $f(x) = x^{\alpha-1}e^{-x}/\Gamma(\alpha)$ and $g(y) = y^{\beta-1}e^{-y}/\Gamma(\beta)$, with $\alpha < \beta$. Griffiths (1969) also gave several canonical sequences for which μ can be identified. We may consider (14.14) as a scaled moment sequence since it is a product of the nth moment of a distribution μ and a scale factor. Note that when $\alpha = \beta$, i.e., when $F \equiv G$, the canonical sequence is a moment sequence.

Consider now an arbitrary bivariate distribution H that has an expansion given by (14.7). Let S_X and S_Y denote the supports of F and G, respectively. We shall now divide the characterisation properties of a canonical sequence, according to whether or not F and G are the same.

14.6.3 The Case of Identical F and G

We assume here that the moment sequence $\{c_n\}$ satisfies $|c_n| \leq 1$, $n = 0, 1, \ldots$, and that its distribution is concentrated within the interval $[-1, 1]$. Then we have the following:

- The class C of canonical sequences of H is convex. That is, if $\{c_n\}$ and $\{d_n\}$ both belong to C, then $\{\alpha c_n + (1-\alpha)d_n\}$ is also in C, for $0 < \alpha \leq 1$.
- Furthermore, Tyan et al (1976) show that $\{c_n d_n\}$ is also a canonical sequence.
- If S_X is unbounded,[8] then every canonical sequence $\{\rho_n\}$ is a moment sequence. This can be proved by the idea given in Sarmanov (1966), which we discussed earlier in the context of normal marginals.
- Again if S_X is unbounded, a moment sequence $\{c_n\}$ is a canonical sequence if $\{\rho^n : |\rho| < 1\}$ is a canonical sequence. The reason is given below:
  - $\{\rho^n\}$ being a canonical sequence implies that $\sum_{n=0}^{\infty} \rho^n P_n(x)P_n(y) \geq 0$, $|\rho| \leq 1$.
  - $\{c_n\}$ being a moment sequence implies that $c_n = \int_{-1}^{1} t^n \, d\mu(t)$ for some distribution μ defined over $[-1, 1]$; hence

$$\begin{aligned} dH(x,y) &= dF(x)dG(y)\sum_{n=0}^{\infty} c_n P_n(x)P_n(y) \\ &= dF(x)dG(y)\int_{-1}^{1}\left[\sum_{n=0}^{\infty} \rho^n P_n(x)P_n(y)\right] d\mu(\rho) \qquad (14.15) \\ &\geq 0. \end{aligned}$$

  Tyan et al (1976) gave an example in which $\{\rho^n\}$ is not a canonical sequence. Let $f(x) = 1/\cosh \pi x$ $(-\infty < x < \infty)$, a hyperbolic secant distribution, which is a special case of the Meixner hypergeometric distribution (see Section 2.14.2). (In this case, the orthonormal polynomials defined on F, in the complete set $\{P_n\}$, are known as the Pollaczek polynomials.) Tyan et al showed that $\sum_{n=0}^{\infty} \rho^n P_n(x)P_n(y)$ may be negative for some choices of x and y. Hence the function $dH(x,y) = dF(x)dG(y)\sum_{n=0}^{\infty} \rho^n P_n(x)P_n(y)$ is not a bivariate distribution. This illustrates that $\{\rho^n\}$ is not always a canonical sequence.

- If S_X is bounded, then a canonical sequence need not be a moment sequence. In this case of bounded support, it is very simple to find a canonical sequence, and hence it is easy to generate a bivariate distribution. The reason is

$$\begin{aligned} dH(x,y) &= dF(x)dG(y)[1 + \rho_1 P_1(x)P_1(y)] \qquad (14.16) \\ &\geq 0 \end{aligned}$$

  for sufficiently small ρ_1—choose it such that $\rho_1 \leq 1/k$, where $|P_1(x)P_1(y)| \leq k$ for all x and y in S_X. Here the canonical sequence $(1, \rho_1, 0, \ldots)$ is clearly not a moment sequence.[9] Therefore, a canonical

[8] A set is bounded if there exists a number $K < \infty$ such that $s \leq K$ for every member s. A set is unbounded if it is not bounded.

[9] Suppose it were a moment sequence of a random variable Z, i.e., $\rho_n = E(Z^n)$. Then $E(Z^2) = \rho_2 = 0$ implies $Z = \rho_1 = 0$, which is a contradiction.

sequence for a bivariate distribution with bounded supports is not necessarily a moment sequence. Lai (1975) has noted the particular bivariate Pearson Type II distribution given in Wong and Thomas (1962, p. 515) has a canonical sequence $\{\rho_n = \rho^{n^2}\}$ which is clearly not a moment sequence.[10] Two other bivariate Pearson Type II distributions considered in Derin et al (1977) also have canonical sequences that are not moment sequences.

14.6.4 The Case of Different F and G

Before the results, some necessary notation:

- The two sequences of orthonormal polynomials are written as $P_n(x) = \sum_{k=0}^{n} a_{nk}\, x^k$ and $Q_m(y) = \sum_{k=0}^{m} b_{mk}\, y^k$. (The only use we make of the a's and b's is of the leading coefficients a_{nn} and b_{mm}.)
- RR: the family of subsets which are bounded below but unbounded above, e.g., $[0, \infty)$.
- LR: the family of subsets which are bounded above but unbounded below, e.g., $(-\infty, 0]$.
- BR: the family of subsets which have neither upper bound nor lower bound, e.g., $(-\infty, \infty)$.

The following results, concerning necessary conditions for a sequence $\{\rho_n\}$ to be a canonical sequence, are due to Tyan and Thomas (1975) and Tyan et al (1976).

- If S_X and S_Y are both contained in BR, then

$$\rho_n = \frac{a_{nn}}{b_{nn}} \int_{-\gamma}^{\gamma} t^n \, d\mu_1(t) = \frac{b_{nn}}{a_{nn}} \int_{-\zeta}^{\zeta} t^n \, d\mu_2(t), \tag{14.17}$$

  where μ_1 and μ_2 are distributions concentrated in $[-\gamma, \gamma]$ and $[-\zeta, \zeta]$, respectively, and[11]

$$0 < \liminf_{n\to\infty} \left|\frac{a_{2n\,2n}}{b_{2n\,2n}}\right|^{1/2n} = \zeta \le \gamma^{-1} = \limsup_{n\to\infty} \left|\frac{a_{2n\,2n}}{b_{2n\,2n}}\right|^{1/2n} < \infty. \tag{14.18}$$

- If S_X and S_Y are both contained in either RR or LR, then

$$\rho_n = \frac{a_{nn}}{b_{nn}} \int_{0}^{\gamma} t^n \, d\mu_1(t) = \frac{b_{nn}}{a_{nn}} \int_{0}^{\zeta} t^n \, d\mu_2(t), \tag{14.19}$$

  where μ_1 and μ_2 are distributions concentrated in $[0, \gamma]$ and $[0, \zeta]$, respectively, and

$$0 < \liminf_{n\to\infty} \left|\frac{a_{nn}}{b_{nn}}\right|^{1/n} = \zeta \le \gamma^{-1} = \limsup_{n\to\infty} \left|\frac{a_{nn}}{b_{nn}}\right|^{1/n} < \infty. \tag{14.20}$$

- If one of S_X, S_Y is in RR while the other is in LR, then

$$\rho_n = \frac{a_{nn}}{b_{nn}} \int_{-\gamma}^{0} t^n \, d\mu_1(t) = \frac{b_{nn}}{a_{nn}} \int_{-\zeta}^{0} t^n \, d\mu_2(t), \tag{14.21}$$

  where γ and ζ are defined as in (14.20), and μ_1 and μ_2 are distributions with supports contained in $[-\gamma, 0]$ and $[-\zeta, 0]$, respectively.

- If S_X is in RR or LR, while S_Y is in BR, then $\rho_n = 0$, for $n = 1, 2, \ldots$

[10] If it were a moment sequence of a random variable Z, then we would then have a negative variance: $\mathrm{var}(Z) = E(Z^2) - [E(Z)]^2 = \rho_2 - \rho_1^2 = \rho^4 - \rho^2 < 0$ for $|\rho| < 1$.

[11] $\liminf a_n$ and $\limsup a_n$ are the smallest and largest limit points of a sequence $\{a_n\}$. A point p is a limit point of the sequence $\{a_n\}$ if every interval, no matter how small, centred on p contains at least another point $q \neq p$ such that q belongs to $\{a_n\}$. Formally, p is a limit point of $\{a_n\}$ if, given $\epsilon > 0$, and given N, there exists $n \ge N$ such that $|a_n - p| < \epsilon$.

14.7 Remarks

- If a symmetric bivariate distribution with unbounded supports can be expanded as in (14.7), then $\text{cov}(X,Y)=0$ implies that X and Y are independent. This is because the canonical sequence $\{\rho_n\}$ is a moment sequence of a probability distribution defined over $[-1,1]$, and hence $\rho_0 = 1 \geq \rho_1^2 \geq \rho_2^2 \ldots$ Therefore, $\rho_1 = \text{corr}(X,Y) = 0$ would imply that $\rho_n = 0$ for all $n \geq 1$, and hence $dH(x,y) = dF(x)\,dG(y)$.

- It is considered by some—for example, Kotz (1975, p. 259)—that the method of the diagonal expansion provides the most powerful unified attack on the structure of bivariate distributions. The study of the canonical sequence enables us to construct bivariate distributions with specified marginals. For example, we can generate a bivariate distribution with normal marginals using Hermite polynomials and a moment sequence of a probability distribution concentrated within $[-1,1]$.

- However, it should not be thought that the set $\{\rho_n\}$ is insensitive to the marginals. (On the contrary, it is well-known that the ordinary product-moment correlation is changed by anything other than a linear transformation of the marginals.)

- Agterberg (1984) has applied the concept of diagonal expansion of a bivariate distribution to a mineral resource evaluation problem. Rocks and geological areas are often sampled by using cells. The effect of cell size on the resulting statistics is of interest. Let X represent the random variable for the average mineral concentration of small cells and Y that of large cells. Two assumptions were imposed as follows:

  1. X and Y can be transformed into Z_1 and Z_2, respectively, such that $X = T_1(Z_1)$ and $Y = T_2(Z_2)$. The joint distribution of Z_1 and Z_2 can be expanded diagonally in terms of the marginals and their respective orthonormal polynomials.
  2. The regression of X on Y satisfies the condition $E(X|Y) = Y$. (In other words, the cell size does not affect the expectation.)

  Consequently, the density of Y can be derived from that of X. Agterberg considered six types of transformations (T_1, T_2) and found that his models fitted well with the given geological data.

14.8 Diagonal Expansion and Construction of Markov Processes

The theory of diagonal expansion of a bivariate distribution is useful in modelling Markov processes, especially in the context of electrical engineering. Papers include Barrett and Lampard (1955), Wong and Thomas (1962), Wong (1964), McFadden (1966), Lampard (1968), Lee (1971), Cambanis and Liu (1971), Tyan and Thomas (1975), and Derin et al (1977). The transition probability density function (or the kernel function) of a Markov process may be defined via $k(x,y) = h(x,y)/f(x)$ (e.g., Wong, 1964) or $k(x,y) = h(x,y)/\sqrt{f(x)f(y)}$ (e.g., Thomas, 1978), in both cases assuming h to have a diagonal expansion and that the marginals are equal.

The diagonal expansion theory also appears in Markov processes in hydrology, e.g., Phatarfod (1976) and Sarmanov (1974).

Thomas (1978) gave a survey on applying bivariate distributions with diagonal expansions to engineering problems.

Chapter 15

Variables in Common

15.1 Introduction

"Trivariate reduction" or "variables in common"—these terms are used for schemes of construction of pairs of r.v.'s that start with three (or more) r.v.'s and perform some operations on them that reduce them to two in number.

Different authors have used these terms in slightly different ways. A narrow definition is:

$$\left.\begin{array}{rcl} X &=& X_1 + X_3 \\ Y &=& X_2 + X_3 \end{array}\right\} \tag{15.1}$$

with X_1, X_2, X_3 being i.i.d. A broad definition is:

$$\left.\begin{array}{rcl} X &=& T_1(X_1, X_2, X_3) \\ Y &=& T_2(X_1, X_2, X_3) \end{array}\right\} \tag{15.2}$$

with X_1, X_2, X_3 not necessarily identically distributed or independent. Another possible definition is:

$$\left.\begin{array}{rcl} X &=& T(X_1, X_3) \\ Y &=& T(X_2, X_3) \end{array}\right\} \tag{15.3}$$

with (i) the X_i being independently distributed and having d.f. $F_0(x_i; \lambda_i)$, and (ii) X and Y having distributions $F_0(x; \lambda_1 + \lambda_2)$ and $F_0(y; \lambda_1 + \lambda_3)$, respectively. This is the definition of trivariate reduction given by Mardia (1970a, p. 74), and of additive random elements in common given by Eagleson and Lancaster (1967).

We do not find it necessary to tie down exactly what these terms mean, or to distinguish them. The general idea is clear, and we will meet several examples of its application in this Chapter.

The majority of distributions discussed in this Chapter have already, to some extent, been dealt with in Chapters 5–10. The Sections are as follows:

15.2 General description.

15.3 Additive model.

15.4 Bivariate distributions having common denominator.

15.5 The transformation being the minimum.

15.6 Other forms of the variables in common technique.

15.2 General Description

Let X_i $(i = 1,2,3)$ be three independent random variables with distribution functions $F_i(x_i; \lambda_i)$. The F_i are often assumed to be the same, but the parameters λ_i may be different. Suppose there exists a function T such that

$$\left.\begin{array}{rcl} X & = & T(X_1, X_3) \\ Y & = & T(X_2, X_3) \end{array}\right\} \qquad (15.4)$$

Then X and Y are said to have a bivariate distribution generated by a trivariate reduction technique. Pearson (1896) generated the bivariate normal in this way and Cherian (1941) a bivariate gamma.

More generally, let us define

$$\left.\begin{array}{rcl} X & = & T_1(X_1, \ldots X_n) \\ Y & = & T_2(X_1, \ldots X_n) \end{array}\right\} \qquad (15.5)$$

where X and Y have one or more X_i in common, and the X_i $(i = 1, 2, \ldots n)$ may not be mutually independent. The structure of T is obviously important, but we consider only simple transforms of the X_i. Usually the X_i will be mutually independent, but occasionally they will be allowed to be dependent.

The following example is taken from Section 6 of Sumita and Kijima (1985). It is from the field of production engineering. Suppose a machine is alternately producing items or being maintained. Let a period of useful production (length X_3) followed by a maintenance period (length X_1) be referred to as a cycle (length $X_1 + X_3$). The cost incurred during a cycle consists of running costs during the production period (which are proportional to the length of the production period, X_3), and of such things as parts for maintenance (a random variable, X_2). Then

$$\begin{array}{rcl} \text{Length of cycle } X & = & X_1 + X_3 \\ \text{Total cost } Y & = & X_2 + cX_3. \end{array}$$

Sumita and Kijima assume X_1, X_2, and X_3 have exponential distributions.

An example motivated from plant breeding is the following. Suppose we are interested in the true values X of a particular characteristic, but we can only observe $Y = X + \varepsilon$, where ε is an error term. What is the distribution of X within the population selected by the requirement that $Y > y$? For the case of ε being normally distributed, see Curnow (1958).

15.3 Additive Model

15.3.1 Background

The first model we consider is:

$$T(X_1, X_3) = X_1 + X_3 \qquad (15.6)$$

The X_i are usually taken to come from the same family of distributions; it may happen that the family is closed under convolution, i.e., the sum $X_1 + X_3$ also belongs to the same family of distributions.

As we mentioned in Section 15.2, Pearson (1896) obtained the bivariate normal using the trivariate reduction technique. In his well-known dice problem, W F R Weldon first constructed a bivariate binomial distribution by using (15.1) and (15.3), X_i being independent binomial variables.

Cherian (1941) and David and Fix (1961) obtained a bivariate gamma distribution in the same manner. Let $X = X_1 + X_3$, $Y = X_2 + X_3$, where X_i are independent gamma r.v.'s with shape parameters λ_i. Then the joint p.d.f. of X and Y is given by (10.8).

Eagleson (1964) used a particular additive model in which the sums and the X_i belong to the same family of distributions to obtain a class of bivariate distributions whose marginals are members of Meixner classes, defined in Section 15.3.2 (see also Lancaster, 1975, 1983). Meixner's collection of distributions have often appeared in characterisation theorems because of their regression properties. Some of these characterisations and properties have been discussed by Lai (1982). The Meixner collection has also appeared in Morris (1982, 1983). We now give a brief account of the Meixner classes of bivariate distributions.

15.3.2 Meixner Classes

Suppose that X is a centred (i.e., zero mean) random variable possessing a moment generating function, with distribution function, G, on which can be defined an orthogonal polynomial system, $\{P_n\}$, where $P_n(x) = x^n +$ terms of lower order, such that $\int P_m P_n \, dG = \delta_{mn} b_m$. Here, δ_{mn} is the Kronecker delta, as in Section 1.5.4, and b_m is a normalising constant. Meixner (1934) considered these distributions, for which the generating function for their orthogonal polynomials is of the form

$$K(x,t) = \sum_{n=0}^{\infty} P_n(x)t^n/n! = \exp[xu(t)]/M[u(t)], \tag{15.7}$$

where

$$u(t) = t + \text{possibly terms in higher powers of } t$$

is a real power series in t, and $M(.)$ is necessarily the moment generating function.

It has been shown by Meixner (1934) (see also Lancaster, 1975) that there are precisely six statistical distributions for which (15.7) is satisfied:

- Positive binomial,
- Normal,
- Poisson,
- Gamma (transformed),
- Negative binomial, and
- Meixner hypergeometric.

The first five are in common use and the sixth distribution (i.e., Meixner hypergeometric) has been discussed in diverse literature. Its main properties are listed in Section 2.14.2.

Eagleson (1964) showed that if X_i belong to the same Meixner class and if they are mutually independent, then the X and Y obtained by (15.1) also belong to the same Meixner class, and their joint distribution satisfies the biorthogonal property (see also Section 14.4), i.e.,

$$dH(x,y) = dF(x)dG(y) \sum_{n=0}^{\infty} \rho_n P_n(x) P_n(y). \tag{15.8}$$

15.3.3 Correlation

It is easy to see that the correlation coefficient of X and Y in the additive model is given by

$$\frac{\text{var}(X_3)}{\sqrt{\text{var}(X_1 + X_3)\text{var}(X_2 + X_3)}}, \tag{15.9}$$

which is always positive. It follows at once that we cannot obtain bivariate distributions with negative correlations in such a scheme; independent marginals can only be obtained by letting $X_3 =$ a constant be included in the family.

X and Y obtained in this way will have linear regression on each other. This is a consequence of a theorem of Rao (1947), which was restated in Lancaster (1975). See also Eagleson and Lancaster (1967).

15.3.4 Symmetric Stable Distributions

A class of bivariate symmetric stable distributions can also be obtained via the additive model. Let the X_i be three mutually independent, symmetric stable[1] random variables with characteristic functions

[1] See Section 2.14.4. For $\alpha = 1$, the distribution is Cauchy, and for $\alpha = 2$, it is normal.

$\exp(-\lambda_i|t|^\alpha)$, $\lambda_i \geq 0$, $0 < \alpha \leq 2$. Consider the transformations $X = X_1 + X_3$, $Y = X_2 + X_3$; then the joint characteristic function of (X, Y) is

$$\varphi(s,t) = \exp(-\lambda_1|s|^\alpha - \lambda_3|t+s|^\alpha - \lambda_2|t|^\alpha). \tag{15.10}$$

de Silva and Griffiths (1980) constructed a test of independence for this class of bivariate distributions. For more on bivariate symmetric stable distributions, see Section 7.11.

15.3.5 Bivariate Triangular Distribution

Eagleson and Lancaster (1967) constructed a bivariate triangular distribution by letting the X_i have uniform distributions on $[0, 1]$.

- The marginal p.d.f.'s are $f(x) = x$ (for $0 \leq x \leq 1$), $f(x) = 2 - x$ (for $1 \leq x \leq 2$), and similarly for y.
- The regression is linear, $E(Y|X = x) = \frac{x+1}{2}$.
- $E(Y^2|X = x) = \frac{1}{12} + \frac{x}{2} + c(x)$, where $c(x) = x^2/2$ (for $0 \leq x \leq 1$), $c(x) = (x^2 - x + 1)/3$ (for $1 \leq x \leq 2$) (Eagleson and Lancaster, 1967). As this is not a polynomial of the second degree, the canonical variables (see Section 14.4) associated with the diagonal expansion of the bivariate triangular distribution are not polynomials (see Section 14.5). The example was constructed as a counterexample to the proposition that linear regression implies the canonical variables are polynomials. Griffiths (1978) showed that these canonical variables, though not polynomials themselves, have a relationship with the Legendre polynomials.

15.3.6 Summing Several i.i.d. Variables

What follows generalises the model we have been considering so far in that more than two variables are added together; but it is a specialisation also, as the variables are now i.i.d.

Let X_i $(i = 1, 2, \ldots)$ be a sequence of i.i.d. variables and define

$$\left.\begin{array}{rcl} X & = & \sum_{i \in A} X_i \\ Y & = & \sum_{i \in B} X_i \end{array}\right\} \tag{15.11}$$

where A and B are subsets of the positive integers. The joint distribution of X and Y has a correlation coefficient given by

$$\rho(X,Y) = \frac{n(A \cap B)}{[n(A)n(B)]^{1/2}}, \tag{15.12}$$

where $n(A)$ denotes the number of elements in the set A. Clearly, X and Y are independent if $A \cap B = \emptyset$ i.e., $\rho(X, Y) = 0$) and $\rho(X, Y) = 1$ if $A \equiv B$. For further details, see Lancaster (1982b).

15.4 Bivariate Distributions Having Common Denominator

15.4.1 Explanation

In this section, we let X_3, independent of X_1 and X_2, be the common denominator of X and Y, which are defined as

$$\left.\begin{array}{rcl} X & = & X_1/X_3 \\ Y & = & X_2/X_3 \end{array}\right\} \tag{15.13}$$

Many of the well-known bivariate distributions are generated in this way,[2] and we will give several examples. First, we note the close relationship of this method with compounding.

[2]Turning away from distribution construction for a moment, a similar pair of equations is often used in the data analysis context, with X_3 being some general measure of size—e.g., in economics, X_1 and X_2 might be measures of the total wealth of country, and X_3 its population, and, in biology, X_1 and X_2 might be lengths of parts of an animal's body, and X_3 its overall length. In any particular application, there might be controversy over whether an empirical positive correlation between the two ratios X_1/X_3 and X_2/X_3 is genuine, or spuriously results from dividing by the same factor, X_3. This subject is bound up with ideas of "neutrality" (for which, see Section 12.4.2). See also Pendleton (1986) and Prather (1988).

15.4.2 Compounding

As pointed out in Section 3.4.2, one may obtain bivariate distributions via the method of compounding. Let X and Y be two random variables having parameters θ_1 and θ_2, respectively. For given values of θ_1 and θ_2, X and Y are taken to be independent. The basic idea of compounding is to let θ_1 and θ_2 be random variables, and the observed distribution of X and Y results from integrating over the unobserved distribution of θ_1 and θ_2. Then the resulting distribution is a compound bivariate distribution. If we assume that θ_1 and θ_2 are identically equal, i.e., $\theta_1 \equiv \theta_2 = \theta$, and that θ is a scale parameter, then the method of compounding is equivalent to the trivariate reduction method (15.13). More precisely, X and Y can be expressed as $X = X_1/\theta$, $Y = X_2/\theta$, where X_1 and X_2 are independent for a given value of θ. Then

$$\begin{aligned} H(x,y) &= \Pr(X \le x, Y \le y) \\ &= \Pr(X_1 \le \theta x, X_2 \le \theta y) \\ &= \int \Pr(X_1 \le \theta x)\Pr(X_2 \le \theta y) f(\theta)\, d\theta \\ &= \int F_{X_1}(\theta x) F_{X_2}(\theta y) f(\theta)\, d\theta \end{aligned}$$

(where $f(\theta)$ is the p.d.f. of θ), which is the familiar equation for compounding of a scale parameter.

See also Lai (1987).

15.4.3 Example 1: Bivariate Cauchy Distribution

Let X_1 and X_2 be two independent normal variates, and X_3, independent of X_1 and X_2, be distributed as χ_1 (i.e., chi-distribution with 1 degree of freedom). Then the joint distribution of $X = X_1/X_3$ and $Y = X_2/X_3$ is a bivariate Cauchy distribution with probability density function given by (6.6).

15.4.4 Example 2: Bivariate *t*-Distribution

Let X_1 and X_2 have a joint standard bivariate normal density and X_3, independent of X_1 and X_2, be distributed as χ_ν. Then the joint distribution of $X = X_1(X_3/\sqrt{\nu})^{-1}$ and $Y = X_2(X_3/\sqrt{\nu})^{-1}$ is a bivariate t-distribution with ν degrees of freedom, whose density is given by (6.1).

15.4.5 Example 3: Bivariate *F*-Distribution

Let X_1, X_2, and X_3 be independent chi-squared random variables with ν_1, ν_2 and ν_3 degrees of freedom, respectively. Then $X = \frac{X_1/\nu_1}{X_3/\nu_3}$ and $Y = \frac{X_2/\nu_2}{X_3/\nu_3}$ have joint probability density given by (6.9). See Section 6.3 and Mardia (1970a, pp. 92–93). We may generalise the above distribution to the case where X_1 and X_2 have non-centrality parameters λ_1 and λ_2, respectively. The correlation structure for this generalised bivariate F-distribution is considered in detail by Feingold and Korsog (1986).

15.4.6 Example 4: Jensen's Bivariate *F*-Distribution

Suppose X_1 and X_2 have a correlated bivariate chi-squared distribution of Kibble's type with shape parameter $\alpha = n/2$, and X_3, independent of X_1 and X_2, is also a chi-squared variate with m degrees of freedom. Then $X = \frac{X_1/n}{X_3/m}$ and $Y = \frac{X_2/n}{X_3/m}$ have a bivariate F-distribution of Krishnaiah's (1964, 1965) type. See Section 7.8.6. More generally, let Q_1 and Q_2 follow Jensen's (1970) bivariate chi-squared distribution with degrees of freedom r and s, respectively, and V, independent of Q_1 and Q_2, is a chi-squared variate with ν degrees of freedom. Then $X = \frac{Q_1/r}{V/\nu}$ and $Y = \frac{Q_2/s}{V/\nu}$ follow Jensen's bivariate F-distribution. (See Section 7.8 for further details).

15.4.7 Example 5: Bivariate Pareto Distribution

Suppose X_1 and X_2 are independent unit exponential variates, and that X_3, independent of X_1 and X_2, has a gamma distribution. The joint distribution of X and Y is then bivariate Pareto (see especially

Section 20.1). More generally, if X_1 and X_2 have unit gamma distributions, instead, then a bivariate inverted beta distribution is the result (see Section 20.7.2).

If we suppose (X_1, X_2) has a Farlie-Gumbel-Morgenstern distribution with unit exponential marginals, and that X_3 has an independent gamma distribution with shape parameter c, then the pair $X = X_1/X_3$, $Y = X_2/X_3$ has a bivariate distribution with Pareto marginals and survival distribution function given by equation (20.11)—see Section 20.6 and Johnson (1987, pp. 170–171).

15.5 The Transformation Being the Minimum

Let X_i $(i = 1, 2, 3)$ belong to the same one-parameter family with distribution functions $F(x_i; \lambda_i)$. (We assume that other parameters, if present, are common to all X_i.) We now wish to find the family that is closed under the transformation $T(X_1, X_3) = \min(X_1, X_3)$, i.e., we want to find distribution functions $F(x; \lambda)$ such that

$$\bar{F}(x; \lambda_1)\bar{F}(x; \lambda_3) = \bar{F}(x; \lambda_1 + \lambda_3) \tag{15.14}$$

(where $\bar{F}$, as usual, is $1 - F$). This in turn implies that

$$\bar{F}(x; \lambda) = [\bar{F}(x)]^{\lambda}. \tag{15.15}$$

There are several continuous distributions satisfying the above (see Arnold, 1967)—the exponential, Pareto, and Weibull.

Marshall and Olkin (1967a) constructed their bivariate exponential distribution by taking F to be the exponential distribution and defining $X = \min(X_1, X_3)$ and $Y = \min(X_2, X_3)$, thus giving

$$\bar{H} = \exp[-\lambda_1 x - \lambda_3 y - \lambda_2 \max(x, y)]. \tag{15.16}$$

See Section 9.2 for details.

The case of T being the maximum is similar.

15.6 Other Forms of the Variables in Common Technique

15.6.1 Bivariate Chi-Squared Distribution

Let X_1, X_2, X_3 be independent univariate standard normal variates, and define

$$\left.\begin{aligned} X &= X_1^2 + X_3^2 \\ Y &= X_2^2 + X_3^2 \end{aligned}\right\} \tag{15.17}$$

Then X and Y have a joint bivariate chi-squared distribution (with two degrees of freedom) and the joint moment generating function is

$$E(e^{sX+tY}) = \{[1 - 2(s + t)](1 - 2s)(1 - 2t)\}^{-1/2}. \tag{15.18}$$

The joint density is not of simple form. (This is an example of Cherian's construction of a bivariate gamma distribution, see Section 10.3.)

Note that if X_1, X_2, X_3 are each supposed to have χ_2^2 distributions (i.e., exponential), the joint density function of X and Y takes a simple form:

$$h = (e^{-\max(x,y)/2} - e^{-(x+y)/2})/4. \tag{15.19}$$

See Johnson and Kotz, 1972, pp. 260–261.)

15.6.2 Bivariate Beta Distribution

This example illustrates that X and Y may have more than one variable in common.

Let X_i ($i = 1,2,3$) be independent, and have gamma distributions with shape parameters θ_i. Consider

$$\left.\begin{array}{rcl} X & = & X_1/(X_1 + X_2 + X_3) \\ Y & = & X_2/(X_1 + X_2 + X_3) \end{array}\right\} \tag{15.20}$$

Then X and Y have a bivariate beta distribution with probability density given by (6.12). We will obtain the same bivariate beta density if the X_i in (15.20) are three independent beta variates with parameters $(\theta_i, 1)$, respectively, conditional on $X_1 + X_2 + X_3 \le 1$.

15.6.3 Bivariate Z-Distribution

Consider three independent gamma variates X_1, X_2 and X_3 with shape parameters α, β and ν, respectively. Form two variables X and Y via the following definition:

$$\left.\begin{array}{rcl} X & = & \log X_3 - \log X_1 \\ Y & = & \log X_3 - \log X_2 \end{array}\right\} \tag{15.21}$$

The joint moment generating function of X and Y can be obtained in a straightforward manner, giving

$$M(s,t) = \frac{\Gamma(\nu + s + t)\Gamma(\alpha - s)\Gamma(\beta - t)}{\Gamma(\alpha)\Gamma(\beta)\Gamma(\nu)}. \tag{15.22}$$

By inverting the moment generating function (15.22), we obtain the joint density function of X and Y:

$$h = \frac{\Gamma(\nu + \alpha + \beta)}{\Gamma(\alpha)\Gamma(\beta)\Gamma(\nu)} \frac{e^{-\alpha x - \beta y}}{(1 + e^{-x} + e^{-y})^{\alpha+\beta+\gamma}}. \tag{15.23}$$

By writing $X = -\log(X_1/X_3)$ and $Y = -\log(X_2/X_3)$, we see that the distribution of (X,Y) is simply a logarithmic transformation of the bivariate inverted beta distribution considered in Section 6.5.15, Hutchinson (1979b, 1981b), and Lee (1981). As the marginals are Z-distributions (see the end of Section 2.5.3) we may call (15.23) a bivariate Z-distribution or a generalised bivariate logistic distribution. (See Malik and Abraham, 1973, and Lindley and Singpurwalla, 1986.)

15.6.4 Cross-References

- Multiplicative trivariate reduction, especially utilising Khintchine's theorem—Sections 10.2 and 10.5.
- Weighted linear combination—Section 10.4.
- Several methods (Tosch and Holmes, 1980; Lawrance and Lewis, 1983; Raftery, 1984, 1985) oriented towards the reliability context and exponential distributions—Sections 9.9, 9.10, 9.11.
- Drivers' injury severities—Section 21.2.

Chapter 16

Overview of Generating Random Variates

16.1 Introduction

Devroye (1986) has recently given the world xvi + 843 pages on generating random variates. We shall do no more than skim the surface of the subject. In view of the importance of simulation as a tool for studying both real-world systems using mathematical models of them, and (more specifically) the properties of summary statistics and hypothesis tests, it is important we do not avoid entirely the question of how to generate a sample from a specified probability function.

We have referred to published algorithms (some coded in a language such as FORTRAN, some not coded) at several points in Chapters 4–10. The structure of the present Chapter is as follows:

16.2 The univariate case—the major approaches.

16.3 The univariate case—some specific distributions.

16.4 The bivariate case—general methods.

16.5 The bivariate case—some specific distributions.

For further information, see especially the recent texts by Devroye (1986) and Johnson (1987), and also Schmeiser (1980) and Johnson et al (1984).

16.2 The Univariate Case—The Major Approaches

16.2.1 Contents of this Section

Our starting point is that we assume we can easily and efficiently generate independent uniform random numbers on a computer (see, e.g., Cohen, 1986); we wish to manipulate these in some way so as to obtain random variates satisfying specific distributional properties.

Valuable references include Bratley et al (1983, Chapter 5), Devroye (1981; 1986, especially Chapters VII–IX), Fishman (1978, Chapters 8 and 9), Hoaglin (1983b), Kennedy and Gentle (1980, Chapter 6), Knuth (1981, especially Section 3.4.1), Law and Kelton (1982), Leemis and Schmeiser (1985), Morgan (1984, Chapters 4 and 5), Ripley (1983; 1987, Section 3.4), and Rubinstein (1981, Chapter 3).

We will discuss the major approaches to generating random variates under the headings:

- Inverse probability integral transform,
- Composition,
- Acceptance/rejection,
- Ratio of uniform deviates, and

- Transformations.

We note that the best method of random variate generation for a given distribution may depend on the value of some parameter of that distribution, e.g., on its shape parameter. (Indeed, there are many other things that may affect what is the "best" method, such as the number of variates to be generated for each different set of parameters of the distribution, the availability or otherwise of fast generators for some related distribution, whether the algorithm is to be coded in a high or low level language, and the criteria that we use to make the assessment, such as speed, or portability, or simplicity.)

The notation $U(a,b)$ to mean the uniform variate over $[a,b]$ will be used. Sometimes $U(0,1)$ will be written simply as U, and sometimes a subscript will be affixed to denoted the first, second, etc., example; lower case u will be used for a particular value.

16.2.2 Inverse Probability Integral Transform

The use of the inverse of the (cumulative) distribution function is of wide applicability; in principle, it can be used whenever the distribution function is known.[1] It is based on the following well-known result. If X has a continuous distribution function F, then $U = F(X)$ is distributed with uniform density on $[0,1]$; conversely,[2] $X = F^{-1}(U)$ has distribution function $F(x)$. We can thus write the method as

1. Generate u from U.

2. Set $x = F^{-1}(u)$, or the modified expression (see footnote 2) if F^{-1} does not exist.

For discrete distributions, this method is easy, but straightforward application to continuous distributions is limited to variates having F^{-1} in a simple closed form. For example, we can obtain exponential variates from the transformation $X = -\log(1-U)$ or $X = -\log U$. For other members of the gamma family, i.e., for arbitrary shape parameter, this approach is inefficient since the evaluation of F^{-1} must be done iteratively.[3]

We note that, according to Hoaglin (1983b), this method is generally slower than other, seemingly more complicated, methods. However, Schmeiser (1980) gives three reasons for its use, even so—easy generation of order statistics, easy generation of truncated distributions, and easy implementation of "variance reduction" techniques in simulation models.

If only the density f is known explicitly, not the d.f. F, then the additional step of numerically integrating this is an added burden. The recent paper of Ulrich and Watson (1987) is directed at this situation.

16.2.3 Composition

The mixture method or composition technique is based upon representing the density f from which variates are to be generated as $f(x) = \sum_{i=1}^{n} p_i f_i(x)$, where $\sum_{i=1}^{n} p_i = 1$. The mixture algorithm is simply to generate variates from each f_i with probability p_i. See Peterson and Kronmal (1982, 1985) for detailed discussion.

Composition must not be confused with convolution. The linear combination of random variables, such as $X = \sum_{i=1}^{n} a_i X_i$, is a convolution—generate such a r.v. X directly from the X_i's and this formula, quite different from the composition construction.

[1] There are times when the distribution is defined through the characteristic function φ, and it is difficult or impossible to get F directly. In this situation, with some provisos about the integrability and continuity of φ and its first two derivatives, one may generate the random variates by means of the acceptance/rejection method of Section 16.2.4 (Devroye, 1986, pp. 695–716). Similarly, sometimes only a sequence of moments, or of Fourier coefficients, may be known—see Devroye (1989).

[2] If F is not continuous or not strictly increasing (or both), then the inverse does not exist. In this case, the definition $X = \inf\{x : F(x) \geq U\}$, comes to our rescue. This means that x assumes the infimum (or smallest value) for which $F(x)$ is at least u.

[3] There are three popular algorithms for numerical inversion, if this should be needed to obtain $F^{-1}(u)$: the bisection method, the secant method, and the Newton-Raphson method. See Devroye (1986, pp. 32–33).

16.2.4 Acceptance/Rejection

The acceptance/rejection method[4] has been a fruitful approach to developing new algorithms for generating univariate observations (see Tadikamalla, 1978a). The idea of this method is to generate a variate from a density function which somewhat resembles the desired density function f. First, select a function t such that $t(x) \geq f(x)$ for all values of x. (t is called a dominating function.) Let $g(x) = t(x)/c$, where $c = \int_{-\infty}^{\infty} t(x)\,dx$; hence g is a probability density.

The algorithm is:

1. Generate x from density function $g(x)$.

2. Generate u from U.

3. If $u \leq f(x)/t(x)$, then accept x, otherwise go to step 1.

The number c is the expected number of "trials" (or iterations) until an acceptance. $1/c$ is generally referred to as the "efficiency" of the procedure. The factors to be considered in selecting g are:

- Step 1 of the algorithm should be executed quickly.

- c should be close to 1.

- The acceptance/rejection test should be simple (i.e., $f(x)/t(x)$ should be easy to evaluate.)

Kronmal et al (1978) and Kronmal and Peterson (1979) proposed what they called the alias-rejection-mixture method. It is based on two ingredients: (i) Walker's (1974a,b) alias method, and (ii) the rejection-mixture method, which is a combination of the mixture and acceptance/rejection methods. Kronmal and Peterson (1984) have recently proposed an "acceptance-complement" method.

16.2.5 Ratio of Uniform Deviates

For a density f, if (U, V) is uniformly distributed over the region $0 \leq u \leq \sqrt{f(v/u)}$, then $X = V/U$ has the desired density f (Kinderman and Monahan, 1977; Hoaglin, 1983b; Devroye, 1986, Section IV.7).

16.2.6 Transformations

Many methods for obtaining r.v.'s are based on first generating some intermediate non-uniform random variates $Y_1, Y_2, \ldots Y_n$, and then setting $X = f(Y_1, Y_2, \ldots Y_n)$.

Among transformations of a single variate (i.e., $n = 1$), familiar examples are non-standard normal variates by $X = \mu + \sigma Z$, where Z is a standard normal variate; $U(a, b)$ by $X = a + (b - a)U$; and lognormal variates by $X = \exp(Y)$, where Y is the appropriate normal variate.

As to generating one random variate from two or more other non-uniform random variates, examples include gamma variates (having an integer shape parameter k) as the sum of k exponential variates, beta variates as a ratio of gammas, t from the standard normal and chi-squared, F from chi-squareds, chi-squareds from normals, and many others.

16.3 The Univariate Case—Some Specific Distributions

16.3.1 The Normal Distribution

A well-known exact method for generating normal variates is that of Box and Muller (1958). It gives two independent standard normal variates X_1, X_2, by means of the sequence of calculations $R = [-2\log(U_1)]^{1/2}$, $\alpha = 2\pi U_2$, $X_1 = R\sin\alpha$, and $X_2 = R\cos\alpha$. A change-of-variables argument demonstrates the validity of this algorithm. Alternatively, observe that (R, α) are the polar coordinates of (X_1, X_2). Let X_1 and X_2 be two independent standard normal variates. Then their density is symmetric about the origin. So α is

[4] Called in some literature simply the rejection method.

uniform on $(0, 2\pi)$ and $R^2 = X_1^2 + X_2^2$ has a chi-squared distribution with two degrees of freedom. Now, this chi-squared distribution is the exponential distribution with mean 2.

A modified polar method is due to Marsaglia and Bray (1964); it avoids the trigonometric functions. From variates V_1 and V_2 that are uniformly distributed over $[-1, 1]$, $W = V_1^2 + V_2^2$ is calculated. If $W > 1$, the pair (V_1, V_2) is rejected, and a new one is generated. Having an acceptable pair, we then calculate the normal variates:

$$\left.\begin{aligned} X_1 &= \left(\tfrac{-2\log W}{W}\right)^{1/2} V_1 \\ X_2 &= \left(\tfrac{-2\log W}{W}\right)^{1/2} V_2 \end{aligned}\right\} \qquad (16.1)$$

Later algorithms have been chiefly based on the composition and acceptance/rejection techniques. One example is the rectangle-wedge-tail method of Marsaglia et al (1964)—the normal distribution is seen as made up from rectangles, wedges between the rectangles and the true density, and tails.

Schmeiser (1980) provides a long list of references. See also Devroye (1986, especially Section IX.1), Rubinstein (1981, Section 3.6.4), and Ripley (1987, especially pp. 82–87). FORTRAN codes for the Box and Muller (1958) and Ahrens and Dieter (1972) methods are given by Bratley et al (1983, p. 297 and p. 318). There are two FORTRAN programs in Best (1978b), also. The comparison of algorithms made by Kinderman and Ramage (1976) is notable for paying attention to user-oriented features such as machine-independence, brevity, and implementation in a high-level language, rather than being confined to speed and accuracy.[5]

16.3.2 The Gamma Distribution

It is sufficient to generate the random variates from the standard gamma distribution having density

$$f = x^{\alpha-1}\exp(-x)/\Gamma(\alpha) \qquad (16.2)$$

$(x \geq 0)$; if the scale parameter is other than 1, and/or the lower end of the distribution is other than 0, a linear transformation can be applied.

The case $\alpha = 1$ (exponential variates): as mentioned earlier, the usual method is the inverse transformation $x = -\log(u)$. For other methods, see Schmeiser (1980, pp. 84–85).

The case $\alpha = k$, an integer (Erlang variates): these may be generated by $x = -\log(\prod_{i=1}^{k} u_i)$. However, the execution time grows linearly with k.

According to Schmeiser (1980), the earliest exact method for generating a gamma variate for any $\alpha > 0$ is due to Jöhnk (1964). However, this is based on the method for Erlang variates, so has the same disadvantage. Since the mid-1970's, many algorithms have been developed. Among them are those of are Ahrens and Dieter (1974), Atkinson and Pearce (1976), Cheng (1977), Best (1978), Tadikamalla (1978a,b), Tadikamalla and Johnson (1978), and Schmeiser and Lal (1980). Most of these algorithms are based on the acceptance/rejection method we discussed earlier. Several computer algorithms are listed in Fishman (1978, pp. 422–429). For extensive surveys on generating gamma variates, see Schmeiser (1980), Tadikamalla and Johnson (1981), Rubinstein (1981, Section 3.6.2), Devroye (1986, pp. 401–428), and Ripley (1987, pp. 88–90). FORTRAN listings of some leading algorithms are given in the second of these,[6] there are several in Văduva (1977), and there are two in Bratley et al (1983, pp. 312–313). Recent papers include those of Barbu (1987), Monahan (1987), and Minh (1988).

16.3.3 The Beta Distribution

We may obtain beta variates from $X = W/(W + Y)$, where W and Y are gamma variates with shape parameters a and b, respectively.

[5]In passing, we note that Kinderman and Ramage said "The algorithms discussed in this paper were coded in as comparable a manner as the authors could manage... we have experimented with several versions of the coding." This throws light on something that has concerned us from time to time: how can we know that a purported comparison of algorithms really is that, rather than a comparison of their codings? The tone of the above quotation suggests there is not—or was not in the mid-1970's—any great science in the step from algorithm to FORTRAN code.

[6]Minh (1988) notes a couple of misprints in algorithm G4PE.

A method due to Jöhnk (1964) uses the relationship between beta and uniform variates: set $Y = U_1^{1/a}$, $Z = U_2^{1/b}$. If $Y + Z \leq 1$, calculate $X = Y/(Y + Z)$. Then X has a beta distribution with parameters a and b. Several other methods are discussed in the references cited in Section 16.2.1.

Unfortunately, the execution time for Jöhnk's algorithm grows indefinitely with increasing a and/or b, because $U_1^{1/a} + U_2^{1/b}$ will then often be greater than 1. Cheng (1978) described an algorithm called BB whose execution time becomes constant as a or b increases. It involves using the acceptance/rejection method to generate an observation from the beta distribution of the second kind. According to Schmeiser (1980), algorithm B4PE developed in Schmeiser and Babu (1980) executes in about half the time of BB, but the set-up time is longer, and it requires more lines of code. Cheng's method is coded in FORTRAN at page 295 of Bratley et al (1983). Algorithms devised by Sakasegawa (1983) are shown by that author to be also very fast in execution, albeit (as with B4PE) with considerable set-up time.

"The bottom linc is that the choice of a method depends upon the user: if he is not willing to invest a lot of time, he should use the ratio of gamma variates. If he does not mind coding short programs, and a and/or b vary frequently, one of the rejection methods based upon analysis of the beta density or upon universal inequalities can be used. The method of Cheng is very robust. For special cases, such as symmetric beta densities, rejection from the normal density is very competitive. If the user does not foresee frequent changes in a and b, a strip table method or the algorithm of Schmeiser and Babu (1980) are recommended. Finally, when both parameters are smaller than one, it is possible to use rejection from polynomial densities or to apply Jöhnk's method."—Devroye (1986, p. 433).

16.3.4 The t-Distribution

Random variates from the t-distribution with ν degrees of freedom may be generated by the transformation method, from the ratio of a normal variate and the square root of an independent gamma variate having shape parameter $\alpha = \nu/2$. For other methods, see Devroye (1986, pp. 445–454).

Best (1978b) showed that t_ν-variates can be generated by the acceptance/rejection method. The density function of t_ν, for $\nu \geq 3$, is dominated by a multiple of the density of t_3. Hence the algorithm involves (i) generating a t_3 variate by the ratio-of-uniform method (for which, see Section 16.2.5 and Devroye, 1986, pp. 194–203), and (ii) generating t_ν (for $\nu > 3$) by the acceptance/rejection method based upon t_3. Best provided a FORTRAN program. Best's algorithm is summarised and slightly modified by Devroye (1986, pp. 449–451).

Several diverse methods are listed in the article by Hoaglin (1983b).

16.3.5 Other Distributions

The paper by Tadikamalla (1984) deals with the normal, gamma, beta, and t-distributions, plus the inverse Gaussian and exponential power distributions, and gives references for the stable and some other distributions. Section IX.2 of Devroye (1986) is on the exponential distribution, IX.3 on gamma distributions, IX.6 on stable densities, and IX.7.5 on the generalised inverse Gaussian. There is a FORTRAN routine for generating a stable variate at pp. 314–315 of Bratley et al (1983). For "phase-type" distributions, see Neuts and Pagano (1981).

16.3.6 IMSL and NAG

The well-known IMSL and NAG libraries include routines to generate random variates from many continuous univariate distributions,[7] and the multivariate normal. Morgan (1984, Appendix 1) describes the programs more fully. Lewis (1980) and Gentle (1986) discussed the use of the IMSL in simulation and statistical analysis. A package by Lewis et al (1986) (reviewed by Burn, 1987) includes routines for generating random variates from normal, Laplace, Cauchy, gamma, Pareto, and beta distributions.

[7] Beta, Cauchy, chi-squared, exponential, exponential mixture, F, gamma, inverted beta, logistic, lognormal, normal, stable, t, triangular, and Weibull variates can be obtained from programs in one or both these libraries.

16.3.7 Microcomputers and Random Variate Generation

The widespread availability of microcomputers has encouraged the implementation of software that was previously available only on larger computers. According to Thesen et al (1984), the relatively slow speed and small word size of microcomputers make the efficient generation of pseudorandom numbers more difficult. As a result, some algorithms on these machines are excessively time-consuming, while others gain speed by sacrificing "randomness". Thesen et al (1984) and Thesen (1985) presented several uniform random number generators claimed to be especially suitable for microcomputers. Ripley's (1983) paper, also, was oriented to the microcomputer context. The situation with microcomputers is rapidly changing, of course, and some aspects of these articles may already be outdated.

Taking up the issue of "randomness", we may mention that two of the more common tests of the statistical properties of a pseudorandom number generator are the frequency test and the serial correlation test. Gleason (1985) reported the results of these tests applied to random number generators available in BASIC on (i) IBM PC's, and (ii) compatible machines. ((i) and (ii) have different random number generators.) He concluded that the generators on the IBM compatibles perform better than the generators on the IBM PC's.

16.4 The Bivariate Case—General Methods

16.4.1 The Setting

We shall now turn our attention to generation of variates from continuous bivariate distributions. Much of our discussion applies to the multivariate case also.

To establish a framework for specific generation algorithms (given in Section 16.5 below), a discussion will be given of two general methods, the transformation and conditional distribution techniques. The usual context for application of the latter is where the conditional distribution function is explicitly given, whereas the transformation method does not need this, and is usually met in the trivariate reduction context.

Acceptance/rejection methods, which are extensively used in the univariate context, are not examined in this Section. Although, in principle, this method applies in multivariate situations, practical difficulties have stifled its use. According to Johnson et al (1984), these difficulties include:

- A lack of suitable dominating functions,
- Complications in optimising the choice of parameters in the dominating function, and
- Low efficiencies.

For a discussion of these difficulties, and an idea to overcome them, see Johnson (1987, pp. 46–48).

As in the univariate situation, the composition (or probability mixing) method is also used to generate bivariate random vectors. Schmeiser and Lal (1982) used this method to generate bivariate gamma random variables.

In what follows, we shall use (X_1, X_2), rather than (X, Y), to denote the pair of variates that we wish to generate.

16.4.2 The Conditional Distribution Method

The idea, usually attributed to Rosenblatt (1952), is:

1. Generate x_1 from the marginal distribution of X_1.
2. Generate x_2 from the conditional distribution of X_2 given $X_1 = x_1$.

The appropriateness of this method for a given bivariate distribution depends on there being an efficient method for generating the required univariate distributions.

One might ask which variate ought to be X_1 and which X_2. Rubinstein (1981, p. 61) states bluntly "Unfortunately, there is no way to find a priori the optimal order of representing the variates in the vector to minimise the CPU time."

16.4.3 The Transformation Method

This method is best presented in the multivariate context. The idea is that we let $\mathbf{X} = (X_1, X_2, \ldots X_p)'$ be the p-dimensional random vector that we want, whose distribution may be difficult to generate directly, and we let $\mathbf{Y} = (Y_1, Y_2, \ldots Y_q)'$ be a q-dimensional ($q \geq p$) random vector having a distribution that is easier to generate. Then if there exists a function $\mathbf{a}(\mathbf{Y}) = (a_1(\mathbf{Y}), \ldots a_p(\mathbf{Y}))$ such that $\mathbf{a}(\mathbf{Y})$ has the same distribution as $\mathbf{X}$, we can get a random realisation of $\mathbf{X}$ by first generating $\mathbf{Y}$ and then evaluating $\mathbf{a}(\mathbf{Y})$.

This method is most appealing when the specific transformation is already available. For some distributions, this is indeed the case.[8] However, for some arbitrary multivariate density $h(\mathbf{x})$, it is seldom obvious what transformation of what random vectors (that are themselves easy to generate) will give rise to it. The following advice (Johnson et al, 1984; Johnson, 1987, p. 46) may help:

- Carry out a thorough search of the literature for an appropriate construction scheme. Rarely does a distribution emerge from a vacuum—usually there is some derivation, such as by compounding, convolution, or transformation.
- Attempt (invertible) transformations of $\mathbf{X}$. Is there a recognisable result? Start with a component transformation such as a probability integral transformation (and compare the resulting expression with known bivariate uniform distributions), or a transformation to exponential marginals (and compare with known bivariate exponential distributions).
- Perhaps h can be recognised as a mixture, $ph_1 + (1-p)h_2$, with h_1 and h_2 being well-known.
- Check if the p.d.f. can be written as a function of the quadratic form $ax_1^2 + bx_1x_2 + cx_2^2$. Then, generation of random variates is easy, as the distribution would then belong to the elliptical class discussed in Sections 7.2 and 16.5.2.

16.4.4 Methods Reflecting the Distribution's Construction

Some bivariate distributions are more easily thought of in terms of how they are constructed, rather than in terms of a formula for a d.f. or p.d.f. It may happen that the method of construction can be directly adapted to random variate generation, as when the method is trivariate reduction or other form of transformation.

16.5 The Bivariate Case—Some Specific Distributions

For a variety of bivariate distributions, this Section describes specific generation algorithms that use the conditional and transformation methods.

16.5.1 Bivariate Normal Distribution

The conditional distribution and transformation methods for generating bivariate and multivariate normal random deviates have been available for some time (Scheuer and Stoller, 1962; Hurst and Knop, 1972). Let $(X_1, X_2)'$ denote the bivariate normal vector with covariance matrix $\mathbf{\Sigma}$. Define

$$\left.\begin{aligned} Y_1 &= (X_1 - \mu_1)/\sigma_1 \\ Y_2 &= \frac{(X_2-\mu_2) - \frac{\sigma_2}{\sigma_1}(X_1-\mu_1)\rho}{\sigma_2(1-\rho^2)^{1/2}} \end{aligned}\right\} \qquad (16.3)$$

Then Y_1 and Y_2 are two independent unit normals, and we can now write

$$\left.\begin{aligned} X_1 &= \sigma_1 Y_1 + \mu_1 \\ X_2 &= \sigma_2 \rho Y_1 + \sigma_2(1-\rho^2)^{1/2} Y_2 + \mu_2 \end{aligned}\right\} \qquad (16.4)$$

[8] For an example, we need look no further than bivariate t, which can be obtained as $X_i = Z_i/W^{1/2}$ (for $i = 1, 2$), where (Z_1, Z_2) has a standardised bivariate normal distribution, and W, independent of the Z's, has a chi-squared distribution.

Univariate normal generators are widely available (see Section 16.3.1).

More generally, let $\mathbf{X} \sim N_p(\boldsymbol{\mu}, \boldsymbol{\Sigma})$, i.e., $\mathbf{X}$ is a p-dimensional multivariate normal random vector with mean vector $\boldsymbol{\mu}$ and covariance matrix $\boldsymbol{\Sigma}$. Let $\mathbf{L}$ be the lower triangular matrix of the Choleski decomposition of $\boldsymbol{\Sigma}$, that is, a matrix such that $\boldsymbol{\Sigma} = \mathbf{LL}'$. (Routines for computing $\mathbf{L}$ are in many computer software packages.) Given p independent univariate standard normal variates, $\mathbf{Y}' = (Y_1, \ldots Y_p)$, transform them via

$$\mathbf{X} = \mathbf{LY} + \boldsymbol{\mu} \tag{16.5}$$

to achieve the desired $N_p(\boldsymbol{\mu}, \boldsymbol{\Sigma})$ distribution.

In 2 or 3 dimensions, $\mathbf{L}$ can easily be given. If $\boldsymbol{\Sigma} = \begin{pmatrix} \sigma_1^2 & \rho_{12}\sigma_1\sigma_2 & \rho_{13}\sigma_1\sigma_3 \\ \rho_{12}\sigma_1\sigma_2 & \sigma_2^2 & \rho_{23}\sigma_2\sigma_3 \\ \rho_{13}\sigma_1\sigma_3 & \rho_{23}\sigma_2\sigma_3 & \sigma_3^2 \end{pmatrix}$, then

$$\mathbf{L} = \begin{pmatrix} \sigma_1 & 0 & 0 \\ \sigma_2\rho_{12} & \sigma_2(1-\rho_{12}^2)^{1/2} & 0 \\ \sigma_3\rho_{13} & \frac{\sigma_3(\rho_{23}-\rho_{12}\rho_{13})}{(1-\rho_{12}^2)^{1/2}} & \sigma_3[(1-\rho_{12}^2)(1-\rho_{13}^2)-(\rho_{23}-\rho_{12}\rho_{13})^2]^{1/2} \end{pmatrix} \tag{16.6}$$

As to the bivariate case, delete the third row and column of $\mathbf{L}$.

The conditional distribution method is almost the same. In the first term on the right hand side of the second equation in (16.4), it would write $(X_1 - \mu_1)/\sigma_1$ instead of Y_1, thus interpreting the distribution of X_2 as being conditional upon a known value of X_1.

The trivariate reduction method also is simple, albeit at the expense of using three independent normal variates per (X_1, X_2) pair generated.

A FORTRAN program, that includes the Choleski decomposition procedure, has been published by Bedall and Zimmermann (1976). Routines for generating vectors from the multivariate normal distribution are in both the IMSL and the NAG collections (see Section 16.3.6 above). One for the bivariate normal is in STATLIB (Brelsford and Relles, 1981, p. 375). Ghosh and Kulatilake (1987) have published a listing of a FORTRAN program to generate random variates from the multivariate normal distribution. An APL listing for the bivariate case is in Bouver and Bargmann (1981).

See also Devroye (1986, Section XI.2), Johnson (1987, pp. 52–54), Kennedy and Gentle (1980, Section 6.5.9), Ripley (1987, pp. 98–99), Rubinstein (1981, Section 3.5.3), and Văduva (1985).

16.5.2 Elliptically-Contoured Bivariate Distributions

We are concerned here with the class of distributions described in Section 7.2, not with all those whose contours are ellipses. The class includes the bivariate normal, Cauchy, and t distributions.

Define $R^2 = (X_1^2 - 2\rho X_1 X_2 + X_2^2)/(1-\rho^2)$, where ρ is the off-diagonal entry in the scaling matrix $\boldsymbol{\Sigma}$, that is, $\boldsymbol{\Sigma} = \begin{pmatrix} 1 & \rho \\ \rho & 1 \end{pmatrix}$. Let $\mathbf{L}$ be the lower triangle (Choleski) decomposition of $\boldsymbol{\Sigma}$. Then, repeating equation (7.2), $\mathbf{X}$ may be represented as

$$(X_1, X_2)' = R\mathbf{L}\mathbf{U}^{(2)} + \boldsymbol{\mu}, \tag{16.7}$$

where $\mathbf{U}^{(2)}$ is uniformly distributed on the circumference of a unit circle and is independent of R.

Generation of $\mathbf{U}^{(2)}$ is easy. The expression for $\mathbf{L}$ in the bivariate case is simple. Choice of a particular member of this class of distributions determines what the distribution of R is. Generation of the vector $\mathbf{X}$ is therefore just as easy or just as difficult as generation of the single variate R.

For the p-dimensional case, an explicit expression for $\mathbf{L}$ would be more complicated, and we would need a random vector with a uniform distribution on the surface of the p-dimensional unit hypersphere, instead of on the circumference of the unit circle, but otherwise the method would be the same.

16.5.3 The Gumbel Type I Bivariate Exponential Distribution

Equation (5.34) gives the p.d.f. The marginals are exponential, and the conditional distribution of X_2 given $X_1 = x_1$ has density

$$g(x_2|x_1) = [(1+\theta x_1)(1+\theta x_2) - \theta]\exp[-x_2(1+\theta x_1)],$$

which can be rewritten as

$$g(x_2|x_1) = p\beta\exp(-\beta x_2) + (1-p)\beta^2 x_2 \exp(-\beta x_2), \tag{16.8}$$

where $\beta = 1 + \theta x_1$ and $p = (\beta - \theta)/\beta$. The form in (16.8) is a mixture density, arising by a mechanism that with probability p generates an exponential variate with mean β^{-1}, and with probability $1-p$ generates the sum of two independent exponential variates, each having mean β^{-1}. Generation is therefore straightforward. It is an example of the method of composition (Section 16.2.3).

16.5.4 Generation of Bivariate Distributions with Given Marginals

16.5.4.1 Background

Many simulation models require the specification of joint continuous bivariate distributions as input. If there is adequate theory or sufficient data on which to choose a specific bivariate distribution, the problem is well defined and can usually be solved by one of the methods discussed earlier.

In many situations, there is really no adequate theory or sufficient data to be able to specify a unique bivariate distribution. However, it may be realistic to specify the marginal distributions of the random variables and a measure of dependence between them. For example, Johnson and Tenenbein (1978, 1981) stated that in the context of an investment portfolio simulation, a joint distribution of stock and bond returns may have to be specified. Because of lack of data, it may be difficult to specify completely the joint distribution of stock and bond returns. However, it is possible to specify the marginal distributions and some measures of dependence between the variables concerned.

We shall write the marginal d.f.'s of X_1 and X_2 as F_1 and F_2, respectively.

16.5.4.2 Weighted Linear Combination and Trivariate Reduction

Johnson and Tenenbein (1978, 1979, 1981) considered two procedures to generate bivariate random variables when the marginals are specified and whose measures of dependence, Kendall's τ or grade correlation ρ_g, can be specified.

Let W_1 and W_2 be i.i.d. r.v.'s with common density function g (uniform, Laplace or exponential). Set

$$\left.\begin{aligned} Y_1 &= W_1 \\ Y_2 &= cW_1 + (1-c)W_2 \end{aligned}\right\} \tag{16.9}$$

($0 < c < 1$), and then find the marginals of Y_1 and Y_2. Suppose F_1 and F_2 are the marginals that we require, obtain X_1 and X_2 by the appropriate univariate transformations, $X_1 = F_1^{-1}[G_1(Y_1)]$ and $X_2 = F_2^{-1}[G_2(Y_2)]$, where G_1 and G_2 are the distribution functions of Y_1 and Y_2, respectively; these are determined by c, which in turn is determined by the chosen value of τ or ρ_g. (Notice that this method is not applicable if the product-moment correlation is specified, as it would be changed by the univariate transformations of Y_i to get X_i.)

Trivariate reduction is similar to weighted linear combination, except that Y_1 and Y_2 are now defined by

$$\left.\begin{aligned} Y_1 &= W_1 + \beta W_3 \\ Y_2 &= W_2 + \beta W_3 \end{aligned}\right\} \tag{16.10}$$

($0 < \beta < \infty$), where W_3, independent of W_1 and W_2, also has density g.

16.5.4.3 Schmeiser and Lal's Methods

Schmeiser and Lal (1982) pointed out that many random phenomenon may be modelled as dependent gamma variates with correlation coefficient specified. For given gamma marginals and correlation coefficient ρ, Schmeiser and Lal showed how (i) the trivariate reduction algorithm (i.e., Cherian's method) and (ii) the composition algorithm made up from the independence case and the upper Fréchet bound can be used to generate a pair of gamma variates. Moreover, they developed a family of algorithms which can produce bivariate gamma vectors having any parameters α_1, α_2 (shape parameters), β_1, β_2 (scaling parameters), and ρ (Pearson's product-moment correlation coefficient).

Let $Z \sim \text{Gamma}(\gamma, 1)$, $W_i \sim \text{Gamma}(\delta_i, 1)$, $U \sim U(0,1)$, and they be mutually independent. Then, because of the reproduciability of the gamma distribution,

$$X_1 = (\text{Ga}_1^{-1}(U) + Z + W_1)\beta_1 \tag{16.11}$$

and

$$X_2 = (\text{Ga}_2^{-1}(V) + Z + W_2)\beta_2 \tag{16.12}$$

are each gamma variates, with shape parameters $\alpha_1 = \lambda_1 + \gamma + \delta_1$ and $\alpha_2 = \lambda_2 + \gamma + \delta_2$, and scaling parameters β_1 and β_2, respectively, and either $V = U$ or $V = 1 - U$. (Ga_i denotes the distribution function of $\text{Gamma}(\lambda_i, 1)$, $i = 1, 2$.) Pearson's product-moment correlation is given by

$$\rho = \frac{E[\text{Ga}_1^{-1}(U)\text{Ga}_2^{-1}(V)] - \lambda_1\lambda_2 + \gamma}{\sqrt{\alpha_1\alpha_2}} \tag{16.13}$$

Given that we know what values of α_1, α_2, β_1, β_2, and ρ we want, we now wish to select values of λ_1, λ_2, γ, δ_1, δ_2, β_1, and β_2, such that:

$$\left.\begin{array}{lcl} \lambda_1 + \gamma + \delta_1 & = & \alpha_1 \\ \lambda_2 + \gamma + \delta_2 & = & \alpha_2 \\ E[G_1^{-1}(U)G_2^{-1}(V)] - \lambda_1\lambda_2 + \gamma & = & \rho\sqrt{\alpha_1\alpha_2} \\ \lambda_1, \lambda_2, \gamma, \delta_1, \delta_2 & \geq & 0 \end{array}\right\} \tag{16.14}$$

(β_1 and β_2 do not appear here, they can be set directly.) As we are using five variables to satisfy three equality conditions, finding a set of parameter values corresponds to finding a feasible solution, rather than an optimal solution, to a nonlinear programming problem.

Schmeiser and Lal (1982) gave some guidelines for an efficient solution procedure, and developed an algorithm called GBIV, which both determines the parameter values and generates the random vector (X_1, X_2). It takes $\gamma = \delta_2 = 0$. For scatterplots of data generated by this algorithm, see Schmeiser and Lal (1982) and Hsu and Nelson (1987).

16.5.4.4 Cubic Transformation of Normals

Fleishman (1978) proposed generating a random variable with given mean, standard deviation, skewness, and kurtosis by taking a standard normal random deviate Z and forming a new r.v. X from a cubic expression, $\sum_{i=0}^{3} a_i Z^i$. (The a's can be obtained by solving a system of equations involving the desired values of the first four moments.) Though the method was criticised by Tadikamalla (1980b), its simplicity makes it attractive, and it has been extended to the multivariate situation by Vale and Maurelli (1983). Their method is to set $X_1 = \sum_{i=0}^{3} a_i Z_1^i$ and $X_2 = \sum_{i=0}^{3} b_i Z_2^i$, where (i) the a's and the b's are determined by the desired univariate moments, and (ii) the correlation between Z_1 and Z_2 is determined by the desired correlation between X_1 and X_2—solution of the cubic equation $\rho_{X_1X_2} = \sum_{i=1}^{3} c_i \rho_{Z_1Z_2}^i$ is required, where the c's are given in terms of the a's and b's by Vale and Maurelli.

Chapter 17

Reliability Models Leading to Bivariate Distributions

17.1 Introduction

According to Barlow and Proschan (1975, Chapter 5), the starting point for many reliability analyses is the assumption that lifetimes of components are independent. However, this is often unrealistic, and it is desired to incorporate some form of positive dependence in the model, chiefly because of the common environmental conditions that the components are exposed to. Thus the study of bivariate distributions becomes an important aspect of reliability analysis.

Just as the univariate exponential distribution is important in describing the lifetime of a single component, so bivariate distributions with exponential marginals are also much used in describing the joint lifetimes of two components. According to Block (1985), some of the earliest examples of bivariate exponentials occurred implicitly as special cases of bivariate gamma distributions, by choosing parameters such that the marginals reduce to exponential. One such bivariate example is that studied in Downton (1970), a special case of Kibble's gamma. In Section 17.2, we show that the bivariate exponential distribution of Marshall and Olkin (1967a) arises from three different types of shock model. An account of the derivation of Downton's (1970) distribution using a successive damage model is also given there. Section 17.3 describes a bivariate geometric compounding mechanism from which many bivariate exponential distributions can be derived. Freund's bivariate distribution cannot be obtained from the scheme in Section 17.3, so a brief reminder of its derivation in relation to reliability is given in Section 17.4. In Section 17.5, we note how a bivariate Pareto distribution may arise in the context of two components sharing a common environment. Finally, we will also discuss three variants of a stress and strength model, in Section 17.6.

Concerning parameterised models in the multivariate lifetime context, Lawless (1982, p. 482) writes "At the present time there is not a lot of motivation for such models, either in the form of realistic failure processes that lead to specific models or in the form of bodies of data that certain models can be shown to fit well." Notwithstanding this note of caution, we hope the models below may have some appeal.

17.2 Two Bivariate Exponential Distributions

As an introduction to the geometric compounding scheme of Section 17.3, we describe here two distributions, with exponential marginals, that will later be seen to be special cases of a common model.

17.2.1 Marshall and Olkin's Shock Model

17.2.1.1 The Distribution

One of the most cited bivariate exponential distributions is that introduced by Marshall and Olkin (1967a)—see our Section 9.2—whose survival distribution is

$$\bar{H} = \exp[-\lambda_1 x - \lambda_2 y - \lambda_{12} \max(x, y)] \tag{17.1}$$

$(x, y \geq 0)$, where the λ's are positive[1] parameters. This distribution, denoted BVE, arises in ways now to be described.

17.2.1.2 Fatal Shocks

Consider a system of two components. Suppose there are three independent sources of shocks present in the environment:

- A shock from source 1 destroys component 1; it occurs at random time U, where $\Pr(U > t) = e^{-\lambda_1 t}$.
- A shock from source 2 destroys component 2; it occurs at random time V, where $\Pr(V > t) = e^{-\lambda_2 t}$.
- Finally, a shock from source 3 destroys both components; it occurs at random time W, where $\Pr(W > t) = e^{-\lambda_{12} t}$.

Thus, the life length of component 1 satisfies $X = \min(U, W)$, while the life length of component 2 satisfies $Y = \min(V, W)$. Hence, the joint survival probability is as given in (17.1).

We may equivalently think of this as three independent Poisson processes, $Z_1(t; \lambda_1)$, $Z_2(t; \lambda_2)$, and $Z_3(t; \lambda_{12})$, governing the occurrence of fatal shocks of the three types.

17.2.1.3 Non-Fatal Shocks

It could be that the shocks sometimes knock out a component, sometimes not.[2] Consider events in the Poisson process with rate θ that cause failure to the ith component (but not the other) with probability p_i $(i = 1, 2)$, and cause failure to both components with probability p_{12}, where $1 - p_1 - p_2 - p_{12} > 0$. If $\lambda_i = p_i\theta$, and $\lambda_{12} = p_{12}\theta$, then the respective times to failure X and Y of components 1 and 2 obey equation (17.1). See Marshall and Olkin (1985a) for such a representation as this.

17.2.1.4 A Random Sums Model

This distribution turns out to be a special case of a random-sums model (Marshall and Olkin, 1985a), to be discussed in Section 17.3.4.

17.2.2 The Lack of Memory Property of Marshall and Olkin's Distribution

The univariate exponential distribution is characterised by the functional equation

$$\bar{F}(s + \delta) = \bar{F}(s)\bar{F}(\delta) \tag{17.2}$$

$(s, \delta > 0)$. This is referred to as the lack of memory (LMP) property. The same equation in vectors $\mathbf{s}$, $\boldsymbol{\delta}$ leads to the case of independence.[3] Equation (17.2) can be rewritten as

$$\Pr(X > s + \delta | X > \delta) = \Pr(X > s), \tag{17.3}$$

so that the probability of surviving an additional time s for a component of age δ is the same as for a new component. A bivariate analogue of (17.3) is

$$\Pr(X > s_1 + \delta, Y > s_2 + \delta | X > \delta, Y > \delta) = \Pr(X > s_1, Y > s_2), \tag{17.4}$$

[1] Strictly, nonnegative parameters such that $\lambda_1 + \lambda_{12} > 0$, $\lambda_2 + \lambda_{12} > 0$.

[2] The usual term, "non-fatal shock" model, is perhaps unfortunate, as it may suggest the shocks are injurious, whereas in fact it is assumed they are either fatal or do not have any effect at all.

[3] It would become

$$\Pr(X > s_1 + \delta_1, Y > s_2 + \delta_2) = \Pr(X > s_1, Y > s_2) \Pr(X > \delta_1, Y > \delta_2).$$

Putting $\delta_1 = 0$ and $s_2 = 0$, we obtain $\Pr(X > s_1, Y > \delta_2) = \Pr(X > s_1) \Pr(Y > \delta_2)$, which implies that X and Y are independent. Further, it follows that

$$\Pr(X > s_1 + \delta_1) \Pr(Y > s_2 + \delta_2) = \Pr(X > s_1) \Pr(X > \delta_1) \Pr(Y > s_2) \Pr(Y > \delta_2),$$

and by virtue of the univariate LMP (i.e., (17.2)), we conclude that both X and Y are exponential variates.

which asserts that the joint survival probability of a pair of components, each of age δ, is the same as that of a pair of new components. We may write (17.4) equivalently as

$$\bar{H}(s_1 + \delta, s_2 + \delta) = \bar{H}(s_1, s_2)\bar{H}(\delta, \delta) \tag{17.5}$$

(where $s_1, s_2, \delta > 0$). This also is termed the LMP. The bivariate exponential distribution of Marshall and Olkin is the only distribution with exponential marginals that satisfies (17.5), see, for example, Barlow and Proschan (1981, pp. 130–131). This functional equation has many possible solutions if the requirment of exponential marginals is not imposed. The class of possible solutions of the equation is characterised in Ghurye and Marshall (1984). Apart from Marshall and Olkin's BVE, other known solutions of (17.5) are:

- Freund's distribution (described in Section 9.3),
- ACBVE of Block and Basu (see Section 9.5),
- PSE of Proschan and Sullo (1974), and
- BEE of Friday and Patil (1977), which includes Freund's, Marshall and Olkin's BVE, and the ACBVE as special cases. See Section 9.8.

The formula for the BEE is given in (9.31), and the survival function of the PSE (Proschan-Sullo extension) is as follows:

$$\bar{H} = \begin{cases} (\lambda_1 + \lambda_2 - \lambda_2')^{-1}\{\lambda_1 \exp[-(\lambda_1 + \lambda_2 - \lambda_2')x - (\lambda_0 + \lambda_2')y] + (\lambda_2 - \lambda_2')\exp(-\lambda y)\} & \text{for } x \le y \\ (\lambda_1 + \lambda_2 - \lambda_1')^{-1}\{\lambda_2 \exp[-(\lambda_0 + \lambda_1')x - (\lambda_1 + \lambda_2 - \lambda_1')y] + (\lambda_1 - \lambda_1')\exp(-\lambda x)\} & \text{for } x \ge y \end{cases}$$

where the parameters λ_0, λ_1, λ_2, λ_1', λ_2' are all positive, $\lambda = \lambda_0 + \lambda_1 + \lambda_2$, and $x, y \ge 0$. This distribution subsumes both the BVE of Marshall and Olkin (1967a) and Freund's (1961) distribution.

The LMP (17.5) is characterised by Block and Basu (1974) and Block (1977c). The characterisation theorem can be stated as follows. Let (X, Y) be a nonnegative bivariate random variable with absolutely continuous marginal distribution functions F and G, and let $U = \min(X, Y)$ and $V = X - Y$. Then the LMP holds if and only if there is a $\theta > 0$ such that

- U and V are independent, and
- U is an exponential variate with mean θ^{-1}.

Furthermore, if the LMP holds, the distribution of V is given by

$$\Pr(V \le v) = \left\{ \begin{array}{ll} F(v) + \theta^{-1}f(v) & \text{if } v \ge 0 \\ 1 - G(-v) - \theta^{-1}g(-v) & \text{if } v < 0 \end{array} \right\} \tag{17.6}$$

where f and g are the density functions corresponding to F and G. A final point is that if $\bar{H}(x, y)$ has the LMP, so also does the survival function $\frac{1}{2}[\bar{H}(x, y) + \bar{H}(y, x)]$.

Gumbel's bivariate exponential distribution, type 1, is characterised by another form of bivariate lack of memory (see Section 5.10.9). A survey of bivariate exponential distributions and their applications to reliability has been given recently by Basu (1988b).

17.2.3 Downton's Successive Damage Model

For some purposes, the property of Marshall and Olkin's shock model that it is made up of a singular distribution and a continuous one (see (9.3)) is a disadvantage. The singularity in the p.d.f. is caused by there being a non-zero probability that the lives of the two components are the same. It is perhaps more usual that failure of one component weakens a second, so shortening its life, without actually causing catastrophic failure.

With these points in view, Downton (1970) proposed a successive damage model such that the resulting bivariate distribution generalises a weaker property of the univariate exponential. Consider a two-component system such that each component is subject to non-fatal shocks which arise according to two independent Poisson processes. If each component fails after a random number of shocks, where these random numbers follow a certain correlated geometric distribution (see Section 17.3.6), the joint density of times to failure of the two components is given by

$$\frac{\mu_1\mu_2}{1-\rho}\exp\left(\frac{-\mu_1 x-\mu_2 y}{1-\rho}\right) I_0\left(\frac{2\sqrt{\rho\mu_1\mu_2 xy}}{1-\rho}\right) \tag{17.7}$$

$(x, y \geq 0)$, where I_0 is the modified Bessel function of the first kind of order 0. It may be noted that for the above equation to be a density function, the correlation ρ must be positive, a restriction which arises naturally from the model by which the density was obtained. In an equipment failure context, it is difficult to imagine a situation which could lead to negative correlation, so that this restriction, which is shared by Marshall and Olkin's distribution, is not a serious one.

17.3 Bivariate Exponentials and a Geometric Compounding Scheme

17.3.1 Background

Many bivariate exponential distributions may arise in one of the following two ways: first, as a consequence of a random shock model due to Arnold (1975b); second, from a characteristic function equation due to Paulson (1973) and Paulson and Uppuluri (1972a,b). Block (1977a) used a bivariate geometric compounding mechanism to unify the approaches of previous authors.[4] Before describing it, we say a few words about probability generating functions and the bivariate geometric distribution.

17.3.2 Probability Generating Function

Let N be a nonnegative integer-valued random variable. Then the probability generating function (p.g.f.) of N is defined as $P(s) = E(s^N)$. Similarly, the bivariate p.g.f. of (N_1, N_2) is defined as $P(s_1, s_2) = E(s_1^{N_1} s_2^{N_2})$.

It is easy to show that if $\psi(t)$ is the c.f. of the i.i.d. r.v.'s X_i, then $P[\psi(t)]$ is the c.f. of the compound r.v. $\sum_{i=1}^N X_i$, which is a sum of a random number of X_i's.

17.3.3 Bivariate Geometric Distribution

A random variable N has a geometric distribution if

$$\Pr(N = n) = p^{n-1}(1-p), \tag{17.8}$$

for all positive integers n and some probability p.

If X's are i.i.d. r.v.'s with an exponential distribution, then $\sum_{i=1}^N X_i$ also has an exponential distribution if N has a geometric distribution.[5]

We say that a random variable (N_1, N_2) has a bivariate geometric distribution if the marginals are geometric distributions. (We are not concerned with the specific bivariate structure.)

17.3.4 Bivariate Geometric Distribution Arising from a Shock Model

Suppose we have two components receiving shocks in discrete cycles. (No assumption is made at this stage concerning the time interval between successive shocks.) In each cycle, there is a shock to both

[4] Marshall and Olkin (1967a), Downton (1970), Hawkes (1972), Paulson (1973), Arnold (1975a,b). Another relevant paper is that of Ohi and Nishida (1978).

[5] The proof of this is simple. The p.g.f. of N is $\frac{(1-p)s}{1-ps}$ and the c.f. of X is $\psi(t) = (1-i\mu t)^{-1}$, where $\mu = E(X)$. It follows from Section 17.3.2 that the c.f. of $\sum_{i=1}^N X_i$ is simply $\varphi(t) = P[\psi(t)] = \left(1 - i\frac{\mu t}{1-p}\right)^{-1}$, which is the c.f. of the exponential distribution with mean $\mu/(1-p)$.

components: with probability p_{11} both components survive, with probability p_{10} the first survives and the second fails, with probability p_{01} the first fails and the second survives, and with probability p_{00} both fail. The outcomes of the shocks are independent from cycle to cycle. By conditioning on the outcome of the first cycle (Hawkes, 1972; Arnold, 1975b), we find that the number of shocks (N_1, N_2) to failure of components 1 and 2 satisfies the following functional equation in the p.g.f.:

$$g(s_1, s_2) = s_1 s_2[p_{00} + p_{01}g(1, s_2) + p_{10}g(s_1, 1) + p_{11}g(s_1, s_2)]. \tag{17.9}$$

The survival function $\bar{H}(n_1, n_2) = \Pr(N_1 > n_1, N_2 > n_2)$ associated with (17.9) is given by

$$\bar{H}(n_1, n_2) = \begin{Bmatrix} p_{11}^{n_1}(p_{01} + p_{11})^{n_2 - n_1} & \text{if } n_1 \leq n_2 \\ \\ p_{11}^{n_2}(p_{10} + p_{11})^{n_1 - n_2} & \text{if } n_2 \leq n_1 \end{Bmatrix} \tag{17.10}$$

where $p_{00} + p_{01} + p_{10} + p_{11} = 1$, $p_{10} + p_{11} < 1$, and $p_{01} + p_{11} < 1$. The p.g.f. of (N_1, N_2) is given by

$$g(s_1, s_2) = \frac{p_{00}s_1 s_2}{1 - p_{11}s_1 s_2} + \frac{p_{01}(p_{00} + p_{10})s_2}{1 - (p_{01} + p_{11})s_2} + \frac{p_{10}(p_{00} + p_{01})s_1}{1 - (p_{10} + p_{11})s_1}. \tag{17.11}$$

(Equation (17.11) was also derived in Esary and Marshall, 1973.)

If $s_1 s_2$ in (17.9) is replaced by the characteristic function of any bivariate distribution with exponential marginals, $\psi(t_1, t_2)$,[6] we obtain the c.f. $\varphi(t_1, t_2) = E[\exp(it_1 X + it_2 Y)]$, which satisfies the functional equation

$$\varphi(t_1, t_2) = \psi(t_1, t_2)[p_{00} + p_{01}\varphi(0, t_2) + p_{10}\varphi(t_1, 0) + p_{11}\varphi(t_1, t_2)]. \tag{17.12}$$

By using the idea developed in Section 17.3.2 (i.e., that the characteristic function of a random sum is the composition of ψ and P), we see that (17.12) corresponds to the compounding of the distribution with c.f. ψ with respect to the bivariate geometric distribution given in (17.10). In other words, (17.12) is the characteristic function equation of the bivariate random variable

$$(X, Y) = \left(\sum_{i=1}^{N_1} X_{i1}, \sum_{i=1}^{N_2} X_{i2}\right), \tag{17.13}$$

where (N_1, N_2) has the bivariate geometric distribution given in (17.10), and is independent of (X_{i1}, X_{i2}) $(i = 1, 2, \ldots)$, which are independent and identically distributed having exponential c.f. $\psi(t_1, t_2)$. (X, Y) in (17.13) has a bivariate exponential distribution, since univariate geometric sums of exponential variables are exponential (see Section 17.3.3). X can now be interpreted in the following way (with a similar interpretation for Y). Component 1 is subjected to shocks which arrive according to a Poisson process. The probability that a shock will cause failure is $p_{01} + p_{00}$. Then X represents the lifetime of the first component. For the bivariate exponential distribution of Marshall and Olkin, $X_{i1} = X_{i2}$, for all i.

The aim of Section 17.3.4 is to provide a common framework (which we call the compounding scheme) for constructing various well-known bivariate exponential distributions. Two ingredients are used: (i) an input bivariate exponential with c.f. ψ, and (ii) a set of non-negative parameters p_{ij} $(i, j = 0, 1)$ such that $p_{00} + p_{01} + p_{10} + p_{11} = 1$, $p_{01} + p_{11} < 1$, and $p_{10} + p_{11} < 1$. This leads to the formulation (17.12). Alternatively, in the formulation (17.13), we may regard the bivariate exponential distribution we want to construct as the compound distribution, the bivariate gamma distribution with discrete shape parameters N_1 and N_2 as the compounded distribution, and the bivariate geometric distribution as the compounding distribution.

Table 17.1 (adapted from Block, 1977a) summarises the ways in which the various bivariate distributions satisfy the equation and hence fit into compounding scheme. It gives the names of the distributions, the "input" bivariate exponential distributions, and the parameters that correspond to our notation. The distributions are arranged in order from simplest to most complex. (The first entry indicates that a pair of independent exponential variates may be obtained from a pair of mutually completely dependent and identical exponential variates.)

[6] Thus we can call this c.f. ψ a "generator" or an "input" function. In order to obtain a desired output φ, we need to choose an appropriate ψ, together with appropriate p_{00}, p_{01}, p_{10}, and p_{11}.

Table 17.1: Bivariate exponential distributions arising from a compounding scheme. (The p_{ij}'s in the first row, and ψ and φ, are as we have defined; otherwise, the notations are as in the original publications.)

Distribution	$\psi(t_1,t_2)$	p_{00}	p_{01}	p_{10}
Independent marginals	$[1-i\theta(t_1+t_2)]^{-1}$	0	θ/θ_1	θ/θ_2
Marshall and Olkin (1967a)	$[1-i\theta(t_1+t_2)]^{-1}$	$\theta\lambda_{12}$	$\theta\lambda_1$	$\theta\lambda_2$
Downton (1970)	$\left[1-\frac{it_1}{\mu_1(1+\gamma)}\right]^{-1}\left[1-\frac{it_2}{\mu_2(1+\gamma)}\right]^{-1}$	$(1+\gamma)^{-1}$	0	0
Hawkes (1972)	$(1-iP_1t_1/\mu_1)^{-1}(1-iP_2t_2/\mu_2)^{-1}$ $(P_1=p_{11}+p_{10}, P_2=p_{11}+p_{01})$	p_{11}	p_{10}	p_{01}
Paulson (1973)	$(1-i\theta_1t_1)^{-1}(1-i\theta_2t_2)^{-1}$	a	c	b
Arnold (1975b)	$\psi(t_1,t_2)\in E_{n-1}^{(2)}$	p_{11}	p_{10}	p_{01}

$E_n^{(2)}=\{\varphi(t_1,t_2):\psi(t_1,t_2)\in E_{n-1}^{(2)}\}$, with $E_0^{(2)}=\{\psi(t_1,t_2)=[1+i\theta(t_1+t_2)]^{-1}\}$

Some observations about the bivariate exponential distributions can be made from the table. First, it is clear that Downton's distribution is a special case of the Hawkes and Paulson distributions. The latter two distributions can be seen to be the same, but they were derived differently. Arnold's class given contains all of the exponential distributions given in the table. The first and second are in $E_1^{(2)}$; then, since the first is in $E_1^{(2)}$, it follows that the third, fourth, and fifth are in $E_2^{(2)}$.

17.3.5 Infinitely Divisible Bivariate Exponential Distributions

We now consider the infinite divisibility of the bivariate exponential distributions discussed above.[7]

Vere-Jones (1967) proved that Kibble's bivariate gamma distribution is infinitely divisible, for every shape parameter α; hence Downton's bivariate exponential distribution is also infinitely divisible.

Block and Paulson (1984) proved the following.

- The Marshall and Olkin distribution is not infinitely divisible when all three λ's are greater than 0; it is only infinitely divisible in the special cases when one of them is 0.
- Suppose ψ in (17.12) is given by $\psi(t_1,t_2)=[(1-i\alpha t_1)(1-i\beta t_2)]^{-1}$ (hence the first two cases in Table 17.1 are excluded). Then φ is infinitely divisible if (i) p_{01} or p_{10} is 0, with $p_{00},p_{11}>0$, or (ii) $p_{00}p_{11}=p_{01}p_{10}$. It is not infinitely divisible otherwise.

17.3.6 Bivariate Cumulative Damage Models

17.3.6.1 The Univariate Case

Esary et al (1973) have discussed certain univariate cumulative damage[8] models, as follows. A device is affected by shocks occurring randomly in time according to a Poisson process with rate λ. These shocks occur at times X_1, X_1+X_2, ..., where the X_i are the interarrival times for the Poisson process and consequently have independent exponential distributions whose mean is $1/\lambda$. The device fails when a random number N of shocks occurs. The time to failure of the device is then $T=\sum_{i=1}^N X_i$. If N is a nonnegative integer-valued random variable independent of the Poisson process, then T has the survival function

$$\bar{H}(t)=\sum_{k=0}^{\infty}\frac{e^{-\lambda t}(\lambda t)^k}{k!}\Pr(N>k). \tag{17.14}$$

[7]Recall from footnote 4 of Chapter 7 that a bivariate distribution with characteristic function φ is said to be infinitely divisible if $\varphi^{1/n}$ is a characteristic function for every integer n.

[8]This term, from Barlow and Proschan (1981, p. 91) seems more descriptive than the "shock models and wear processes" of Esary et al (1973). Griffith (1988b) regards shock models and cumulative damage models as special cases of the more general class of wear processes.

An alternative way of viewing this is as each shock causing a random amount of damage to the device, which fails when the accumulated damage exceeds a specified capacity or threshold. Hence N exceeding k corresponds to total damage from k shocks being below the threshold.

17.3.6.2 The Bivariate Case

A bivariate generalisation will now be described. A two-component device is subjected to shocks. Assuming the numbers of shocks until failure of these components are N_1 and N_2, respectively, and N_1 and N_2 are independent of the inter-shock times X_{ij} (which for $j = 1, 2$, are i.i.d. exponential variates), means that the times until failures are

$$(T_1, T_2) = \left(\sum_{i=1}^{N_1} X_{i1}, \sum_{i=1}^{N_2} X_{i2} \right). \tag{17.15}$$

Analogously to (17.14), these times have a bivariate survival function

$$\bar{H}(t_1, t_2) = \sum_{k_1=0}^{\infty} \sum_{k_2=0}^{\infty} \Pr[Z_1(t_1) = k_1, Z_2(t_2) = k_2] \Pr[N_1 > k_1, N_2 > k_2], \tag{17.16}$$

where $Z_i(t)$ denotes the number of shocks occurring in the interval $(0, t_i)$.

The variables $Z_1(t_1)$ and $Z_2(t_2)$ are univariate Poisson processes which are dependent if X_{i1} and X_{i2} are dependent. Assuming (N_1, N_2) has a bivariate geometric distribution given by (17.10), shock models of the Marshall and Olkin type and of the Downton-Hawkes-Paulson type can be derived by choosing an appropriate bivariate distribution and a set of parameters p_{ij} $(i, j = 0, 1)$. See Block (1977a) for details.

The model here is essentially the same as that of Section 17.3.4 except (i) this model is described in terms of accumulated damage, (ii) more emphasis is placed here on the bivariate counting process for shocks, namely, the bivariate Poisson process, and (iii) there is no mention of cycles here.

17.4 Freund's Bivariate Model

This was one of the first bivariate distributions to be obtained from reliability considerations—Freund (1961). It is based upon exponential distributions, but does not have exponential marginals. This model envisages a two-component system in which the failure of one component affects the lifetime of the other component. Initially, the unaffected lifetimes have exponential distributions; the affected lifetimes also have exponential distributions, though with an altered parameter. The joint survival distribution is given by equation (9.15). See Section 9.3 for details.

17.5 Bivariate Distributions Arising from Components Sharing a Common Environment

Consider a system of two components (in series, or in parallel), which operates in an environment whose characteristics may affect its reliability. Suppose that in environment i, the components' lifetimes are exponentially distributed, with mean lifetime $1/\lambda_i$. Assume that λ has a gamma distribution over the population of environments. Then the joint p.d.f. of the lifetimes x and y of the two components is of bivariate Pareto form,

$$h = \frac{(a+1)(a+2)b^{a+1}}{(b+x+y)^{a+3}}, \tag{17.17}$$

where a and b are the parameters of the gamma distribution of λ. (If the components are in series, we will be especially interested in $\bar{H}(t, t)$, whereas if they are in parallel, $1 - H(t, t)$ will be of chief concern.)

This subject is taken further at other points in this book. Section 5.8 gives the distribution function, correlation, conditional properties, etc. Section 15.4.7 puts the idea into the variables in common form. In Section 20.7, generalisations of (17.17) are given, including taking the compounded distributions to be gamma instead of exponential.

17.6 Results on $\Pr(X > Y)$: A Stress-Strength Model

17.6.1 The Basic Idea

Let X be the strength of a component, subjected to a stress Y. The component fails if at any moment the applied stress (or load) is greater than its strength. The stress is a function of the environment to which the component is subjected, whereas strength depends on material properties, manufacturing procedures, and so on. The reliability R of a component is the probability that its strength exceeds the stress,

$$R = \Pr(X > Y) \tag{17.18}$$

This model was considered by Birnbaum (1956), and has since found an increasing number of applications in many different areas, especially in the structural and aircraft industries. A recent review paper is that by Johnson (1988). A similar formulation occurs in hydrology—let X be the input of a pollutant into a river of flow Y, and assume that the flora and fauna of the river are sensitive to the concentration of the pollutant. Then $\Pr(X > cY)$ is the relevant quantity—Plate and Duckstein (1987, pp. 56–58).

In many situations, the distribution of Y (or of both X and Y) is completely known, except possibly for a few unknown parameters, and it is desired to obtain parametric solutions. Church and Harris (1970), Downton (1973), Owen et al (1964), Govindarajula (1968), and more recently, Reiser and Guttman (1986 1987), have considered the problem of stress and strength under the assumption that X and Y have independent normal distributions. Since in many physical situations, especially in reliability, exponential and related distributions provide more realistic models, it is desirable to obtain estimators of R for these cases. Some results for the exponential case are given by Tong (1974), Kelly et al (1976), and Basu (1981).

Most of the authors have assumed that X and Y are independent. However, it is more realistic to assume some form of dependence between X and Y, as they may be influenced by a common environmental factor. We shall now evaluate R for three different models in which X and Y are correlated. See also Sections 19.8.3, 19.8.4, and 24.2—4.

17.6.2 Marshall and Olkin's Model

Suppose X and Y have the joint bivariate exponential distribution of Marshall and Olkin (1967a), i.e. (17.1). In the notation of Section 17.2.1, $X > Y$ if and only if $V < \min(U, W)$. Hence

$$\begin{aligned} R &= \Pr[V < \min(U, W)] \\ &= \lambda_2/\lambda, \end{aligned}$$

where $\lambda = \lambda_1 + \lambda_2 + \lambda_{12}$ (see Basu, 1981). Similarly, $\Pr(X < Y) = \lambda_1/\lambda$. Also, $\Pr(X \geq Y) = R + \Pr(X = Y) = (\lambda_2 + \lambda_{12})/\lambda$. Awad et al (1981) derived three estimators for $\Pr(X < Y)$, $\Pr(X > Y)$, and $\Pr(X = Y)$.

17.6.3 Downton's Model

Suppose X and Y have a joint density function given by (17.7). Lai (1985) showed that

$$R = (1 - \rho^2) \sum_{i=0}^{\infty} \rho^{2i} B_\alpha(i+1, i+1)/B(i+1, i+1), \tag{17.19}$$

where $\alpha = \mu_2/(\mu_1 + \mu_2)$ and B_x is the incomplete beta function. Perhaps this infinite series could be replaced by an expression involving some sort of hypergeometric function, but we have not yet done so.

Note that for $\rho = 0$ (i.e., when X and Y are independent), $R = \mu_2/(\mu_1 + \mu_2)$, as expected (Tong, 1974).

17.6.4 Independent Truncated Bivariate Normal Model

Let X, the strength, have a truncated normal distribution, with the stress Y (before truncation of X) being normally distributed with mean μ_2 and standard deviation σ_2. (We think of proof-testing here. Assuming that X and Y are independent and that X is restricted to the range x_0 to x_1, Parikh and Shet (1966) showed that $\Pr(X > Y)$ can be expressed in terms of the bivariate normal integral.

Chapter 18

Reliability Classes: Concepts of Ageing

18.1 Introduction

This Chapter concentrates on "life" distributions, by which we mean those for which negative values do not occur, i.e., $F(x) = 0$ for $x < 0$. The variate X is thought of as the time to failure (or death) of a electrical or mechanical component (or organism), but other interpretations may be possible—an inter-event time is normally necessarily positive.

In this Chapter, we focus on classes of life distributions based on notions of ageing—IFR ("increasing failure rate") is perhaps the best-known, but we shall meet several others also, and study their interrelationships. The major parts of the Chapter are devoted to the naming and interpretation of univariate reliability classes (Section 18.2), and to how these notions may be extended to the bivariate case (Section 18.3). For further reading, we call attention especially to the book by Barlow and Proschan (1975)—see its Chapter 4 for the univariate classes and Section 5 of Chapter 5 for the multivariate case. A recent expository paper is by Bergman (1985).

Most readers will know that statistical theory applied to distributions of lifetime lengths plays an important part in both the reliability engineering literature and the biometrics literature. We may also note a third applications area: Heckman and Singer (1986) review econometric work on duration variables (e.g., lengths of periods of unemployment, or time intervals between purchases of a certain good), much of which, they say, has borrowed freely and often uncritically from reliability theory and biostatistics.

18.2 Reliability Classes—Univariate Ideas

18.2.1 Introducing the Jargon

The concepts of increasing and decreasing failure rates for univariate distributions have been found very useful in reliability theory.[1] The classes of distributions having these ageing properties are designated the IFR and DFR distributions, respectively,[2] and have been extensively studied. Other more general classes such as "increasing failure rate on average" (IFRA), "new better than used" (NBU), "new better than used in expectation" (NBUE), and "decreasing mean residual life" (DMRL) have also been of interest. For fuller accounts of these classes see, e.g., Bryson and Siddiqui (1969), Barlow and Proschan (1981), and Klefsjö (1983), and Hollander and Proschan (1984).

The notion of "harmonically new better than used in expectation" (HNBUE) was introduced by Rolski (1975) and studied by Klefsjö (1981, 1982). Further generalisations along this line were given by Basu and Ebrahimi (1984). A class of distributions denoted $\mathcal{L}$ has an ageing property that is based on the Laplace transform, and was put forward by Klefsjö (1983). Deshpande et al (1986) used stochastic dominance[3]

[1] The exponential distribution is characterised by having constant failure rate, and thus stands at the boundary between the IFR and DFR classes. For a review of goodness-of-fit tests of the exponential distribution, which gives some explicit attention to IFR and DFR alternatives, see Stephens (1986c).

[2] Some authors refer to them as IHR and DHR, H standing for hazard.

[3] Let X and Y be two random variables with distribution functions G and H, respectively. We say that X is stochastically larger than Y if $G(x) \leq H(x)$, for all x. In economic theory, this is known as first-order stochastic dominance. Higher orders

comparisons to describe positive ageing and suggested several new positive ageing criteria based on these ideas—see their paper for details. Two further classes, NBUFR ("new better than used in failure rate") and NBUFRA ("new better than used in failure rate average") require the absolute continuity of the distribution function, and have been discussed in Loh (1984a,b), Deshpande et al (1986), Kochar and Wiens (1987), and Abouammoh and Ahmed (1988).

Rather than $F(x)$, we often think of $\bar{F}(x) = \Pr(X > x) = 1 - F(x)$, which is known as the survival distribution or reliability function. The mean of X is denoted μ.

The function

$$\bar{F}(t|x) = \bar{F}(x+t)/\bar{F}(x) \tag{18.1}$$

represents the survival function of a unit of age x, i.e., the conditional probability that a unit of age x will survive for an additional t units of time. The mean remaining (residual) life, at age x, is $E(X - x|X > x)$, which may be shown to be $\int_0^\infty \bar{F}(t|x)\,dt$.

When $F'(x) = f(x)$ exists, we can define the failure rate (or hazard rate, or force of mortality) as

$$r(x) = f(x)/\bar{F}(x), \tag{18.2}$$

for x such that $\bar{F}(x) > 0$. It follows that, if $r(x)$ exists,

$$-\log \bar{F}(t) = \int_0^t r(x)\,dx, \tag{18.3}$$

which represents the cumulative failure rate.

We are now ready to define several reliability classes. Having done so, we shall then list the relationships between them.

18.2.2 Definitions

The nine reliability classes mentioned above are defined thus:

1. F is said to be IFR if $\bar{F}(t|x)$ is decreasing in $0 \leq x < \infty$ for each $t \geq 0$. When the density exists, this is equivalent to $r(x) = f(x)/\bar{F}(x)$ being increasing[4] in $x \geq 0$.

2. F is said to be IFRA if $-(1/t)\log \bar{F}(t)$ is increasing in t. (That is, if $-\log \bar{F}$ is a starshaped function; for this notion, see Dykstra, 1985.) As $-\log \bar{F}(t)$ given in (18.3) represents the cumulative failure rate, the name given to this class is appropriate. Block and Savits (1976) showed that IFRA is equivalent to $E^\alpha[h(X)] \leq E[h^\alpha(X/\alpha)]$ for all continuous nonnegative increasing functions h and all α such that $0 < \alpha \leq 1$. This class is perhaps the most important in reliability analysis. It is the smallest class containing the exponential distribution which is closed under the formation of coherent systems.[5] It has been shown that a device subject to shocks governed by a Poisson process, which

---

of stochastic dominance are defined in Deshpande et al (1986).

[4] We use the term "increasing" when perhaps "nondecreasing" would be more strictly correct. Similarly, we use "decreasing" to mean "nonincreasing".

[5] In order to define a coherent system, we need to define a structure function first. A structure function of a system is a binary function ϕ such that

$$\phi(x_1, x_2, \ldots x_n) = \begin{cases} 1 & \text{if the system is functioning} \\ 0 & \text{if the system has failed,} \end{cases}$$

where n is the number of components in the system and x_i indicates the state of the ith component,

$$x_i = \begin{cases} 1 & \text{if component } i \text{ is functioning} \\ 0 & \text{if component } i \text{ has failed,} \end{cases}$$

for $i = 1, 2, \ldots n$. The ith component is irrelevant to the structure if ϕ is constant in x_i (i.e., the state of the component does not affect the system). Otherwise, the ith component is relevant (essential).

A system is called *monotone* if its structure function ϕ is increasing in its components, and, in addition, $\phi(0, 0, \ldots 0) = 0$ and $\phi(1, 1, \ldots 1) = 1$.

A system is called *coherent* if ϕ is increasing in its arguments and if all the components are relevant.

It can be shown that a coherent system is always monotone.

fails when the accumulated damage exceeds a fixed threshold, has an IFRA distribution (Esary et al, 1973).

3. F is said to be DMRL if the mean remaining life function $\int_0^\infty \bar{F}(t|x)\,dt$ is decreasing in x. That is, the older the device is, the smaller is its mean residual life.

4. F is said to be NBU if $\bar{F}(t|x) \le \bar{F}(t)$ for $x \ge 0$, $t \ge 0$. This means that a device of any particular age has a stochastically smaller remaining lifetime than does a new device.

5. F is said to be NBUE if $\int_0^\infty \bar{F}(t|x)\,dt \le \mu$ for $x \ge 0$. This means that a device of any particular age has a smaller mean remaining lifetime than does a new device.

6. F is said to be HNBUE if $\int_x^\infty \bar{F}(t)\,dt \le \mu \exp(-x/\mu)$ for $x \ge 0$. There is an alternative definition in terms of the mean residual life.

7. F is said to be an $\mathcal{L}$-distribution if $\int_0^\infty \exp(-st)\bar{F}(t)\,dt \ge \mu/(1+s)$ for $s \ge 0$. The expression $\mu/(1+s)$ can be written as $\int_0^\infty \exp(-sx)\bar{G}(x)\,dx$, where $\bar{G}(x) = \exp(-x/\mu)$. This means that the inequality is one between the Laplace transforms of $\bar{F}$ and of an exponential survival function with the same mean as F. For more interpretations of this class, see Section 2 of Klefsjö (1983).

8. F is said to be NBUFR if $r(x) \ge r(0)$ for $x \ge 0$.

9. F is said to be NBUFRA if $r(0) \le \frac{1}{x}\int_0^x r(t)\,dt = \frac{-\log r(x)}{x}$.

Using Laplace transforms, Block and Savits (1980a) established necessary and sufficient conditions for the IFR, IFRA, DMRL, NBU, and NBUE properties to hold.

18.2.3 Interrelationships

The following chain of implications exists among the classes (Deshpande et al, 1986; Kochar and Wiens, 1987):

$$\begin{array}{ccccccccc}
\text{IFR} & \Rightarrow & \text{IFRA} & \Rightarrow & \text{NBU} & \Rightarrow & \text{NBUFR} & \Rightarrow & \text{NBUFRA} \\
\Downarrow & & & & \Downarrow & & & & \\
\text{DMRL} & & \Rightarrow & & \text{NBUE} & \Rightarrow & \text{HNBUE} & \Rightarrow & \mathcal{L}
\end{array}$$

In definitions 1–9 of Section 18.2.2, if we reverse the inequalities and interchange "increasing" and "decreasing", we obtain the classes DFR, DFRA, NWU, IMRL, NWUE, HNWUE, $\bar{\mathcal{L}}$, NWUFR, and NWUFRA. They satisfy the same chain of implications.

18.2.4 Properties of IFR and DFR

Patel (1983) gives the following properties of the IFR and DFR concepts:

1. If X_1 and X_2 are both IFR, so is $X_1 + X_2$; but the DFR property is not so preserved.

2. A mixture of DFR distributions is also DFR; but this is not necessarily true of IFR distributions.

3. Parallel systems of identical IFR units are IFR.

4. Series systems of (not necessarily identical) IFR units are IFR.

5. Order statistics from an IFR distribution have IFR distributions, but this is not true for spacings from an IFR distribution; order statistics from a DFR distribution do not necessarily have a DFR distribution, but spacings from a DFR distribution are DFR.

6. The p.d.f. of an IFR distribution need not be unimodal.

7. The p.d.f. of a DFR distribution is a decreasing function.

Further issues:

- Bounds for F, and moment inequalities—see Patel (1983).
- Classification of several well-known distributions[6] as IFR, DFR, neither, or possibly either—see Patel (1983).
- IFR is equivalent to $\log \bar{F}$ being concave, and DFR to $\log \bar{F}$ being convex—see Shaked (1983) and Steutel (1985).
- The concept of PF_2 (Pólya type 2 frequency) distributions and how it relates to the IFR and DFR properties—see Rao (1986, 1988) and Kim and Proschan (1988).
- The book by Kaufmann et al (1977) has chapters on lifetimes of components, increasing failure rate, structure functions and reliability networks, reliability of systems, redundance, two types of failures, and Pólya functions of order 2.

18.2.5 Applications

From the definitions of the life distribution classes, results may be derived concerning such things as properties of systems (based upon properties of components), bounds for survival functions, moment inequalities, and algorithms for use in maintenance policies (Hollander and Proschan, 1984).

Block and Sampson (1983) conclude their review of inequalities on multivariate distributions[7] by sketching applications of such ideas as PQD and IFR to reliability theory.

See Newby (1986) for some simple explanations and examples of analyses in terms of mean residual life, hazard, and hazard rate average.

18.2.6 Ordering

A distribution F is said to be more IFR than G if the ratio of hazard rates $\frac{r_F[F^{-1}(t)]}{r_G[G^{-1}(t)]}$ is nondecreasing in t. (This definition has assumed the failure rates exist.) Orderings of each of the other reliability classes may also be defined.

There is a chain of implications connecting the orderings of the classes, that is similar but not identical to that which connects the classes (Section 18.2.3) (Kochar and Wiens, 1987).

18.2.7 Further Classes

The classes of distributions defined above certainly do not exhaust the possibilities.

- SIFR (stochastically increasing failure rate) and SNBU (stochastically new better than used) classes were defined by Singh and Deshpande (1985) as follows. With Y_i being independent exponentially distributed r.v.'s with mean the same as that of X, SIFR means that the conditional probability of survival until random time $\sum_{i=1}^{k} Y_i$ given that the item has not failed before random time $\sum_{i=1}^{k-1} Y_i$ is decreasing in k; further, SNBU means that if the item has survived until $\sum_{i=1}^{k} Y_i$, the probability that it will survive the next Y_{k+1} units of time is no more than the probability that a new component will survive time Y_{k+1}. IFR implies SIFR as well as implying NBU, and both of these in turn imply SNBU.
- NBU-t_0 (new better than used of age t_0) class: the survival probability at time 0 is greater than or equal to the conditional survival probability at age t_0. See Hollander et al (1986).

[6] Exponential, gamma, Weibull, half-normal, modified extreme value, Pareto, lognormal, log-logistic, Birnbaum-Saunders, inverse Gaussian.

[7] A few words about the content of this article seem called for here. After an introduction, there is discussion of: positive and negative dependence; convex symmetric set inequalities; stochastic ordering; stochastic majorisation and inequalities obtained by majorisation; Chebyshev and Kolmogorov-type inequalities; multivariate moment inequalities; applications.

- BMRL-t_0 (better mean residual life at t_0) class: the mean residual life declines during the time 0 to t_0, and thereafter is no greater than what it was at t_0. See Kulasekera and Park (1987). In the chain of implications in Section 18.2.3, this class falls between DMRL and NBUE.

- Launer (1984) defined some classes in terms of the variance of residual life—for instance, DVRL, decreasing variance of residual life. This was as a preliminary to deriving some inequalities for the survival function $\bar{F}$ in terms of the moments.

- Joe and Proschan (1984) have studied classes defined in terms of the 100α percentile of residual life—examples are DPRL-α (decreasing 100α percentile residual life), and NBUP-α (new better than used with respect to the 100α percentile).[8]

18.3 Bivariate Reliability Classes

18.3.1 Alternative Requirements

In dealing with multicomponent systems, it is of great interest to obtain suitable bivariate and multivariate extensions of the univariate ageing properties mentioned above. In each case, however, several definitions are possible, because of different requirements imposed by various authors. This can be illustrated in the papers by Buchanan and Singpurwalla (1977) and Esary and Marshall (1979).

The multivariate definitions of the classes (IFR, IFRA, NBU, NBUE, and DMRL) given by Buchanan and Singpurwalla (1977) are motivated by the following requirements:

- The definitions should be based upon conditions imposed on the joint survival function, rather than on the corresponding random variables.

- The definitions should coincide with those that are accepted for a single variable.

- The definitions should lead to a chain of implications which is analogous to the chain of implications connecting the corresponding univariate classes (Section 18.2.3).

- The arguments which prompt the definitions should be natural extensions of those used for the various univariate definitions.

Specifically with regard to IFRA, Esary and Marshall (1979), in contrast to these requirements, gave six multivariate IFRA definitions which are not based upon requirements on the joint survival function, but on generalisations of those reliability properties which explain why the univariate IFRA concept is so important (e.g., that coherent systems have IFRA life distributions whenever their components have independent IFRA life distributions). Further, Esary and Marshall did not intend to meet the third requirement of Buchanan and Singpurwalla.

We note that if the marginal distributions of bivariate ∇ distributions are also univariate ∇, Buchanan and Singpurwalla (1977) call H jointly ∇, where ∇ denotes IFR, IFRA, NBU, NBUE, or DMRL.

18.3.2 Bivariate IFR

R Harris (1970) introduced the quantity

$$\bar{H}(x_1+t_1, x_2+t_2)/\bar{H}(x_1, x_2), \tag{18.4}$$

which is seen to be a direct bivariate analogue of the right hand side of (18.1). Then, one definition of H being a bivariate increasing failure rate distribution is that (18.4) decreases in $x_1 \geq 0$, $x_2 \geq 0$, for all $t_1 \geq 0$, $t_2 \geq 0$.

There are several variants of the above, and we shall list four of them.

1. $\bar{H}(x+t, x+t)/\bar{H}(x, x)$ decreases in $x \geq 0$ for all $t \geq 0$.

[8] F being NBUP-α for all α between 0 and 1 is equivalent to F being NBU. F being DPRL-α for all α between 0 and 1 is equivalent to F being IFR.

2. $\bar{H}(x+t_1, x+t_2)/\bar{H}(x,x)$ decreases in $x \geq 0$ for all $t_1, t_2 \geq 0$.

3. $\bar{H}(x_1+t, x_2+t)/\bar{H}(x_1,x_2)$ decreases in $x_1, x_2 \geq 0$ for all $t \geq 0$.

4. $\bar{H}(x_1+t_1, x_2+t_2)/\bar{H}(x_1,x_2)$ decreases in $x_1, x_2 \geq 0$ for all $t_1, t_2 \geq 0$.

Version (3) is due to R Harris (1970) and Brindley and Thompson (1972); (1), (2), and (4) are due to Marshall (1975b). Many other versions are possible, but it appears that of all of these, (3) is perhaps the most important. One reason for this is that this condition can be interpreted as meaning that the lifetime of a series system of two components of different ages decreases stochastically as the ages of the components increase. Other reasons for its importance, according to Barlow and Proschan (1981, pp. 152–154), are:

- The marginal lifetimes have the univariate IFR property.
- And so does $\min(X_1, X_2)$.
- There exists a sensible definition of multivariate IFR such that it is satisfied by the union of two mutually independent sets having property (3).
- The ratio $\bar{H}(x_1+t, x_2+t)/\bar{H}(x_1,x_2)$ depends on t but not on x_1 or x_2, if and only if the distribution is the bivariate exponential of Marshall and Olkin (1967a).

A notion upon which other versions of bivariate IFR have been based is a generalised form of the univariate hazard rate $r(x) = f(x)/\bar{F}(x)$. Basu (1971) defined the bivariate failure (hazard) rate to be $h/\bar{H}$. But Johnson and Kotz (1973b) opined that a multivariate generalisation ought appropriately to be a vector, not a scalar. So define

$$r_i(x_1,x_2) = \frac{-\partial}{\partial x_i}\log \bar{F}(x_1,x_2) = \frac{-\partial \bar{H}(x_1,x_2)}{\partial x_i} \Big/ \bar{H}(x_1,x_2) \tag{18.5}$$

for $i = 1,2$. The quantity $\mathbf{r}(x_1,x_2) = (r_1(x_1,x_2), r_2(x_1,x_2))$ is called the hazard gradient (see Block, 1977b, Marshall, 1975c, and Johnson and Kotz, 1975).[9] According to the idea of Johnson and Kotz (1973b), then, bivariate IFR would imply that for all (x_1,x_2), r_1 is an increasing function of x_1, and r_2 is an increasing function of x_2. Block (1977b) showed that H has property (3) if and only if $\mathbf{r}(x_1+t, x_2+t)$ increases in $t \geq 0$ for all $x_1, x_2 \geq 0$.

For other versions of bivariate IFR, see Savits (1985). For further developments with regard to both $h/\bar{H}$ and $\mathbf{r}$, see Shanbhag and Kotz (1987).

18.3.3 Bivariate IFRA

As with bivariate IFR, many possible definitions for bivariate IFRA have been proposed. Let $\mathbf{T} = (T_1, T_2)$ be a nonnegative vector with survival function $\bar{H}(t_1,t_2) = \Pr(T_1 > t_1, T_2 > t_2)$. The following definition is due to Block and Savits (1980b).

Definition: The vector $\mathbf{T}$ (or its distribution function H) is said to be bivariate IFRA if

$$E^{\alpha}[a(T_1,T_2)] \leq E[a^{\alpha}(T_1,T_2)/\alpha], \tag{18.6}$$

for all continuous nonnegative increasing functions a and all α such that $0 < \alpha \leq 1$. Although mathematically uncomplicated, the above definition seems to lack meaningful physical interpretation. As the class of all continuous nonnegative increasing functions in R is a subclass of all continuous nonnegative increasing functions in R^2, it is clear the two component lifetimes are individually IFRA.

Other conditions for bivariate IFRA are:

[9]The hazard gradient is sufficient to determine the distribution:

$$\bar{H}(\mathbf{x}) = \exp\left[-\int_{\mathbf{0}}^{\mathbf{x}} \mathbf{r}(\mathbf{z})\,d\mathbf{z}\right]$$

analogously to the univariate case (18.3), where the integral is a line integral over any sufficiently smooth continuous path beginning at $\mathbf{0}$ and terminating at $\mathbf{x}$ (Marshall, 1975c).

A. $\bar{H}^{\alpha}(t_1, t_2) \leq \bar{H}(\alpha t_1, \alpha t_2)$, where again $0 < \alpha \leq 1$, for all $t_1 \geq 0$, $t_2 \geq 0$.

B. $\mathbf{T}$ is such that each monotone system formed from $\mathbf{T}$ is univariate IFRA.

C. $\mathbf{T}$ is such that there exist independent random variables $X_1, X_2, \ldots X_k$ and monotone life functions[10] τ_1 and τ_2, such that $T_i = \tau_i(X_1, X_2, \ldots X_k)$.

Σ. $\mathbf{T}$ is such that there exist independent IFRA random variables $X_1, X_2, \ldots X_k$ and nonempty sets S_i of $\{1, 2, \ldots k\}$ such that $T_i = \sum_{j \in S_i} X_j$ (for $i = 1, 2$).

D. $\mathbf{T}$ is such that there exist independent IFRA random variables $X_1, X_2, \ldots X_k$ and nonempty sets S_i of $\{1, 2, \ldots k\}$ such that $T_i = \min_{j \in S_i} X_j$ (for $i = 1, 2$).

E. $\mathbf{T}$ is such that $\min(T_1, T_2)$ is IFRA.

F. $\mathbf{T}$ is such that $\min(a_1 T_1, a_2 T_2)$ is IFRA for all $a_1 \geq 0$, $a_2 \geq 0$.

Conditions A, B, C, D, E, and F have been given by Esary and Marshall (1979), and condition Σ was given by Block and Savits (1980b). See also Singpurwalla and Buchanan (1977). Note that T_i in conditions C, Σ, and D represents the lifetime of a subsystem, whereas X_i denotes the lifetime of a basic component.

The following relationships hold between BIFRA in the sense of Block and Savits and the seven conditions above:

$$\begin{array}{ccccc} \mathrm{D} & & \Sigma & & \mathrm{F} \\ \Downarrow & & \Downarrow & & \Updownarrow \\ \mathrm{C} & \Rightarrow & \mathrm{BIFRA} & \Rightarrow & \mathrm{A} \\ & & \Downarrow & & \Downarrow \\ & & \mathrm{B} & \Rightarrow & \mathrm{E} \end{array}$$

With one possible exception, the above chain is complete, i.e., no more implications are possible. The exception is that it is not known whether $\Sigma \Rightarrow \mathrm{C}$ holds, though Block and Savits (1982) conjectured that it does not.

An important property common to all these versions of bivariate IFRA is that the marginals are univariate IFRA (Block and Savits, 1982).

A recent review is by Shaked and Shanthikumar (1986). And a paper that includes another notion of BIFRA is that of Mukherjee and Chatterjee (1988).

18.3.4 Bivariate NBU

There are also several versions of bivariate NBU based on the inequality

$$\frac{\bar{H}(x_1 + t_1, x_2 + t_2)}{\bar{H}(x_1, x_2)} \leq \bar{H}(t_1, t_2). \tag{18.7}$$

This can be interpreted as the conditional survival probability of two components of different ages being less than the corresponding survival probability $\bar{H}(t_1, t_2)$ of two new components.

Consider the following five versions:

1. $\bar{H}(x + t, x + t) \leq \bar{H}(x, x)\bar{H}(t, t)$ for all $t, x \geq 0$.

2. $\bar{H}(x + t_1, x + t_2) \leq \bar{H}(x, x)\bar{H}(t_1, t_2)$ for all $t_1, t_2, x \geq 0$.

3. $\bar{H}(x_1 + t, x_2 + t) \leq \bar{H}(x_1, x_2)\bar{H}(t, t)$ for all $t, x_1, x_2 \geq 0$.

4. $\bar{H}(x_1 + t_1, x_2 + t_2) \leq \bar{H}(x_1, x_2)\bar{H}(t_1, t_2)$ for all $t_1, t_2, x_1, x_2 \geq 0$.

[10]The life function τ of a system expresses the lifetimes of a system in terms of lifetimes of its components. τ is called monotone if the structure function ϕ is monotone.

5. $\bar{H}(x_1+t_1, x_2+t_2) \leq \bar{H}(x_1, x_2)\bar{H}(t_1, t_2)$ for all $t_1, t_2, x_1, x_2 \geq 0$ which satisfy $(x_i - x_j)(t_i - t_j) \geq 0$ for $i, j = 1, 2$.

The first four were given by Buchanan and Singpurwalla (1977), and condition (5) by Marshall and Shaked (1979). Version (3) has perhaps the greatest intuitive appeal.

Another definition of bivariate NBU is given by Marshall and Shaked (1982): for a bivariate vector **T**, it requires that

$$\Pr[\mathbf{T} \in (\alpha+\beta)A] \leq \Pr(\mathbf{T} \in \alpha A)\Pr(\mathbf{T} \in \beta A), \tag{18.8}$$

for every $\alpha \geq 0$, $\beta \geq 0$ and every region A in the positive quadrant for which X exceeds some value x_A and Y exceeds some value y_A. Marshall and Shaked gave several conditions equivalent to it, and they showed that it is implied by Block and Savits' bivariate IFRA given in (18.6).

For several other definitions of multivariate NBU, see Marshall and Shaked (1986a), who also establish relationships between them. For a review, see Marshall and Shaked (1986b).

18.3.5 Bivariate NBUE, DMRL, and HNBUE

Less attention seems to have been given to bivariate NBUE, DMRL, and HNBUE concepts than to the three above. It was pointed out by Block and Savits (1981a) that various versions of bivariate NBUE can be obtained by integrating versions 1–4 of bivariate NBU. For instance, the third one will become the condition $\int_0^\infty \bar{H}(x_1+t, x_2+t)\,dt \leq \bar{H}(x_1, x_2)\int_0^\infty \bar{H}(t,t)\,dt$ for all $x_1, x_2 \geq 0$. This can be interpreted as a used series system of two components of ages x_1 and x_2, respectively, having a smaller remaining life than a new system.

Corresponding to, and implying, each of the four versions of bivariate NBUE is a version of bivariate DMRL. For instance, to that of the previous paragraph corresponds the condition that $\int_0^\infty \bar{H}(x_1+t, x_2+t)\,dt/\bar{H}(x_1, x_2)$ shall decrease in $x_1, x_2 \geq 0$.

For further details concerning these definitions, see Buchanan and Singpurwalla (1977).

Arnold and Zahedi (1988) introduced a new multivariate remaining life function which is different from that of Buchanan and Singpurwalla (1977). Based on this new definition, Zahedi (1985) put forward four classes of DMRL distributions; the relations among them were also established, but their relationships with the ideas of Buchanan and Singpurwalla (1977) were not given.

As to bivariate HNBUE, Basu et al (1983) gave eight definitions, the first group of four being due to Klefsjö (1980) and based upon a multivariate version of the first sentence in Section 18.2.2—6, and the second group of four being based upon a multivariate version of the mean residual life definition. Basu et al give some closure properties for the second group of classes. See also Basu and Ebrahimi (1986).

18.4 Concluding Remarks

It is clear from the material presented in this Chapter that research on ageing properties (univariate, bivariate, and multivariate) is currently being vigorously pursued. Surveys are given by Block and Savits (1981a,b, 1988).

At present, however, it seems that many of the bivariate definitions above lack clear physical interpretations, though it is true that some can be deduced from shock models—for details, see Marshall and Shaked (1979), Ghosh and Ebrahimi (1983), and Savits (1988).

Further, it is our impression that the univariate concepts of Section 18.2 exist in something of a mathematical limbo, and have not been applied by the mainstream of reliability practitioners. It may be that practical concern is very much with multi-component systems—in which case, appreciation and clarification of the bivariate concepts (and hence of the multivariate concepts) could lead to greater application of this whole body of work.

Part IV

Mostly Applications

Chapter 19

Applications of the Bivariate Normal Distribution

19.1 Organisation of Part IV

The subject of this Part is the use of bivariate distributions in modelling data. There are six Chapters. Chapter 19 gives brief accounts of applications of the bivariate normal distribution. Chapter 20 concentrates on the bivariate Pareto distribution, though actually the key point is the obtaining of bivariate distributions by the method of compounding. In Chapter 21, some data on severity of injury in road accidents is analysed; the distribution fitted is the bivariate normal, though this was largely a matter of convenience, and others obtained by the method of variables in common (if they were mathematically tractable) could be tried. Bivariate generalisations of probit analysis is the subject of Chapter 22; the chief areas of application are biology and econometrics. Psychophysics is that branch of psychology that deals with the relation between a sensory stimulus and its perception, and an important subject within it is signal detection theory; bivariate models in that context are discussed in Chapter 23. Finally, Chapter 24 is a collection of datasets which one might fit bivariate distributions to; there are 60 tables of data therein, and many more are referenced.

19.2 Introduction to Chapter 19

We all know what the bivariate normal distribution is used for, don't we? For haven't we all written down some pairs of numbers, given names to them, and asked our students to calculate an estimate of ρ? Nevertheless, the reader may find some examples here which are unfamiliar to him or her. The arrangement is by field of application. Chapters 20–23 deal with selected types of application in greater depth; with these, as well as with those in this Chapter, it is possible, or even customary, to employ the bivariate normal distribution.

Applications of variants of the bivariate normal, such as univariate transformations and mixtures, are included; indeed, the very first two examples refer to the bivariate lognormal and S_{BB} distributions, respectively.

Many multivariate statistical procedures—factor analysis, discriminant analysis, cluster analysis, and so on—often have as their basis an underlying multivariate normal model. These are not considered here, in part because they form a subject of their own which is dealt with well in other books, and in part because their greatest relevance is to the *multi*variate situation, as distinct from the *bi*variate.

The fields of application in this Chapter have been arranged as follows:

19.3 Biology.

19.4 Medicine.

19.5 Genetics: the multifactorial model of disease transmission.

19.6 Psychology.

19.7 Air, earth, water, and climate.

19.8 Physical and technological sciences.

19.9 Sociology and economics.

19.3 Biology

19.3.1 Size and Shape of Animals

The size and shape of different individuals of one species, and of different species, have long been studied by biologists. Simple-mindedly, one might suppose that if shape remained constant, mass would vary as the third power of body length, cross-sectional area of limbs would vary only as the second power, and larger animals would tend to be relatively more vulnerable to fractures. Perhaps the width of a limb bone should be expected to be proportional to the power 1.5 of its length? Thus attention focuses on power-law relationships, $Y \propto X^b$, and it is natural to linearise these by taking logarithms, $\log Y = a + b \log X$. An example of a paper in this tradition is that by Jolicoeur (1963), who had four measurements on the limbs of North American martens, and who took logarithms before proceding with his analysis.

Malmgren (1979) fitted both the bivariate normal and the bivariate lognormal distributions to several datasets consisting of measurements of a microfossil group, the planktonic foraminifera. He found that the log transformation generally improved conformity to normality, but not to the extent that the distributions could be said to be normal. A similar study involving fossil and living animals was that of Reyment (1971). Some views of this author are worth quoting: "The effect of logarithmically transforming the data appears to be a mixed blessing. Apart from the fact that the investigator is removed one step from the biological relationships in his data by carrying out a transformation of them"—by this, Reyment means that non-normality may be a symptom of some type of biological heterogeneity—"the betterment in the multivariate interconnections with respect to normality tends to be slight, despite the general improvement in the univariate values".

Two measurements, like length and weight, of an animal may have a bivariate lognormal distribution for animals of a particular age. Since animals grow, if a sample contains specimens of various ages, the distribution will be a mixture of bivariate lognormals. A recent paper on this was by Schnute (1984), and was mainly concerned with the length and weight of herring.

19.3.2 The Dimensions of Trees

One of the things that forest managers are interested in is the size of their trees—height, diameter, and volume. For an overview of statistical distributions applied here, see Warren (1979). There have been several papers in which the univariate S_B, bivariate S_{BB}, and trivariate S_{BBB} distributions have been applied to such variates. For example, Schreuder and Hafley (1977) reported having fitted the S_{BB} distribution to 21 datasets of diameter and height measurements. (See items 5 and 6 in Section 24.7.2 for examples.) They found that the (nonlinear) shape of the regression of height on diameter was satisfactory, as was the amount of variation in height for given values of the diameter. With respect to this latter point, they comment "This is in conformance with one hypothesis for even-aged stands of pine that the tallest (dominant) trees and the shortest (suppressed) trees are associated, respectively, with the largest and smallest diameters, and vary little in height within their respective range of diameters. Tree diameters more closely associated with the mean diameter (codominant and intermediate trees) of the stand exhibit much more variation in height." A later paper on this subject is that of Hafley and Buford (1985).

19.3.3 The Home Ranges of Animals

The subject of the "home ranges" or "territories" of wild animals is of interest to ecologists. One model is that of a bivariate normal spatial distribution of activity of an individual animal. There are many reasons, such as variations in landform or vegetation, why such a distribution might not have any smooth form, let alone be normal, but nevertheless the bivariate normal has achieved some popularity in this context. (The only other ones we have come across are a uniform distribution within a convex polygon, and a bivariate normal with a flat central area. Yang (1979) has said that the bivariate normal is indistinguishable from a large number of non-normal distributions, with the sample sizes typically used.) A recent paper, that has an extensive list of references, is that of W P Smith (1983).

Two points occur to us:

- *Home* range presumably implies something different from full range, namely, the exclusion of locations where the animal is found rarely. In this case, it should presumably (again) only ever contain, say, 90% of the animal's activity. So if the home range is supposed to have a smooth distribution extending to infinity in all directions like the bivariate normal does, and our statistical task is to estimate where the middle 90% (say) of this distribution lies, such an area would account for only 81% of the animal's activity. Alternatively, is the smooth bivariate distribution supposed to describe all of the animal's activity, with the middle 90%, say, defined to be the home range? Whatever the answer, there has been some interest in procedures that give low weight to observations that are far from the centre of activity, whether they are true observations or erroneous ones (e.g., Samuel and Garton, 1985).

- There has been little description in the publications we have seen of the precise process of sampling an animal's location. Presumably it is taken for granted that the observations are sufficiently separated in time for them to be in some sense statistically independent of each other.

We note that W P Smith (1983) included two small datasets on the location of deer, and Samuel and Garton (1985) one on the location of a bear.

19.3.4 Probit Analysis

See Chapter 22.

19.4 Medicine

19.4.1 Epidemiology

Epidemiological studies often find it necessary to make an assumption about the joint distribution of two (or more) variables considered to be possible risk factors for the disease under consideration. For instance, Halperin et al (1979) were concerned with the probability of death (from any cause) as a function of systolic blood pressure and number of cigarettes smoked per day. The interest of Halperin et al was primarily methodological: they demonstrated that if X and Y have a bivariate normal distribution, with the risk of death being a probit function $\Phi(\alpha+\beta_1 x+\beta_2 y)$, then not only do X and Y have different means in the group that die and the group that survive, but also the variances and covariances in the groups differ. Halperin et al also obtained analogous results when X and Y were assumed to have the Farlie-Gumbel-Morgenstern distribution with exponential marginals. This approach taken by Halperin et al contrasts with that of Truett et al (1967), who assumed that X and Y had a bivariate normal distribution in the survivor and nonsurvivor groups separately, and, moreover, the two variance-covariance matrices were the same.

19.4.2 Blood Pressure Measurements

el Lozy (1982) interpreted the bivariate distribution of (X,Y) as of $(X, X+\varepsilon)$, where X = true blood pressure, Y = observed blood pressure, and ε = error. el Lozy's concern was with the proportions of a population falling in each cell of the fourfold table (true blood pressure normal or high) $\times$ (observed blood pressure normal or high), and how these would be affected by different numbers of measurements of blood pressure (given that the variance of ε is known).

19.4.3 Mixtures—Graphical Methods

Tarter and Silvers (1975) graphically examined the hypothesis of a mixture of bivariate normal distributions for two datasets of medical significance, (i) triglyceride (log-transformed) and cholesterol in normal women from Modesto, California, and (ii) age and DDT level in migrant field workers.

19.4.4 Mixtures—Normals and Haemophilia Carriers

Basford and McLachlan (1985) used a mixture of bivariate lognormal distributions in analysing AHF activity and AHF-like antigen in normal women and in haemophilia A carriers. See also item 24.9.3—8.

19.4.5 Pharmacokinetics

In a simulation of individualised drug treatment, Katz and D'Argenio (1986) considered it appropriate to assume a bivariate lognormal distribution in the population of patients of the two parameters of the pharmacokinetic model used.

19.4.6 Length and Diameter of Airborne Fibres

This subject is included under Medicine because of the role of airborne fibres in lung disease. Schneider and Holst (1983) and Holst and Schneider (1985) have used the bivariate lognormal distribution to describe the diameter D and length L of airborne man-made mineral fibres. The first of these papers includes (i) some data showing the conditional mean and standard deviation of $\log L$, given $\log D$, have the properties required by the bivariate lognormal distribution, and (ii) the deduction that a bivariate lognormal distribution of D and L implies that the diameter of fibres selected with probability proportional to fibre length has a lognormal distribution. The second paper extends this work by considering the general variable $W = aD^r L^s$, and derives (i) its properties (its distribution is lognormal), and (ii) the properties of the joint distribution of D and L when fibres are selected with probability proportional to their value of W (their distribution is bivariate lognormal). Cheng (1986) has also discussed the above-defined W, and given a number of empirical results. Patil et al (1988) include a brief account of bivariate weighted distributions in their survey.

19.4.7 Leukemia

Survival time and white blood count: see Section 8.2.8.

19.4.8 Severity of Injury in Road Accidents

See Chapter 21.

19.4.9 Evaluation of X-Ray Images

See Section 23.3.

19.4.10 Remark

The applications in Sections 19.4.2, 19.5, and 23.3, and in Chapters 21 and 22, were briefly considered by Hutchinson (1984).

19.5 Genetics: The Multifactorial Model of Disease Transmission

19.5.1 General Description

"Disease transmission" refers to the inheritance of an abnormality. "Multifactorial" distinguishes the model from those which postulate genetic loci that are few in number, giving rise to inheritance probabilities that follow the Mendelian laws.

According to the multifactorial model of disease transmission there exists a single-dimensioned quantity called liability, which determines whether or not an individual succumbs to the disease in question, and which may be determined jointly by genes and by environment. Individuals whose liability is greater than some threshold value manifest the disease, whereas those whose liability is lower do not. An increased incidence of the disease in relatives of affected individuals is regarded as being due to positive correlation between the liabilities of people who are related to each other.

To make calculations with this model, it is necessary to assume the joint distribution of liability in two related persons to have some particular mathematical form. It is usual to assume it is a bivariate normal distribution.[1]

With this topic, the multivariate distribution of liability in three or more related people is usually of more concern than a bivariate distribution. This is because the incidences of the disease in the general population and in relatives of diseased individuals are constraints. That is, if we represent the probability of liability exceeding x by $\bar{F}(x)$ and the probability of the liability for person 1 exceeding x and that of person 2 exceeding y by $\bar{H}(x,y)$ (with margins $\bar{F}(x)$ and $\bar{F}(y)$), and we know the population incidence of the disease to be p_1 and the incidence in relatives of affected individuals to be p_2, then we have these equations:

$$\left.\begin{aligned} p_1 &= \bar{F}(a) \\ p_2 &= \bar{H}(a,a)/\bar{F}(a) \end{aligned}\right\} \tag{19.1}$$

where a is the threshold above which the disease manifests itself. Much of the work in this topic involves using (19.1) to determine a and an association parameter of the distribution $\bar{H}(x,y)$, and then using these quantities in multivariate extensions of $\bar{H}(x,y)$, representing the joint distribution of liability of a number of relatives.[2] The practical application is in estimating recurrence risks in families with known histories of the disease.

The specifically *bivariate* aspects of this topic arise when the disease can occur at two (or more) levels of severity. An example may be taken from psychiatry: Cloninger et al (1975) suggested that sociopathy and hysteria are related forms of the same syndrome. Specifically they hypothesised that women develop sociopathy if their liability is greater than a_1, and hysteria if it is between a_2 and a_1, and that men develop sociopathy if their liability exceeds a_3 (and do not manifest hysteria). This example will be taken further in Section 20.4, where a comparison of the applicability of the bivariate normal and bivariate Pareto distributions is made.

19.5.2 Correlated Traits

Another example in genetics of the bivariate normal assumption is provided by the paper of Dawson and Elston (1984). Quoting from the abstract, this paper refers to "the situation in which proband selection is made on the basis of one trait, but the desired object of analysis is another related trait for which the individuals in the pedigree also are measured". (The term "proband" usually means someone with the disease or defect concerned.) Dawson and Elston give several examples of how such a situation might arise in practice, one being selection of probands on the basis of high cholesterol or triglyceride values, but interest subsequently transferring to lipoprotein levels; another might occur with normally-distributed random error, if different measurement procedures are used to obtain the values used for the proband selection and analytic purposes. Dawson and Elston indicate how their results are modified when the distribution is not bivariate normal, but a mixture of bivariate normals; whereas the former might be an appropriate assumption under polygenic inheritance, the latter might apply when major gene segregation is of importance.

19.5.3 Mixtures—Twins of Unknown Zygosity

19.5.3.1 The Problem

This example is concerned with fitting a model to disease concordance data in twins, in circumstances where it is not known whether each twin pair is monozygotic (identical twins) or dizygotic (non-identical). The

[1] Frequently, the following approximation is made: liability in the general population is normally distributed, and liability in relatives of affected persons is also normally distributed, with mean and variance calculated on the assumption that the joint distribution is bivariate normal. If the latter is true, liability in relatives of affected persons is not exactly normally distributed.

[2] One of the simplest examples is the probability of a child of two affected parents also being affected. Represent the probability of $X > a$ and $Y > a$ and $Z > a$ by $\bar{H}(a,a,a)$. Then the conditional probability of $Z > a$ given that $X > a$ and $Y > a$ is $\bar{H}(a,a,a)/\bar{H}(a,a)$. We should really include the association parameter—representing how strongly correlated are liabilities of children and parents—in this expression, and write it as $\bar{H}(a,a,a;r,r,0)/[\bar{H}(a)]^2$. We have here assumed the child's liability to be equally correlated with that of each of its parents, and that there is no correlation of parents' liabilities.

bivariate normal distribution comes in again because of the multifactorial model of disease transmission, but in this case a mixture is required, as the two types of twin pair may possibly have very different correlations of their disease status. (One might say that a form of "latent structure" model is required. This is a term used when the observed data is made up of a mixture of two or more unobservable different populations.) Blischke (1978) mentions the same basic idea, of twins being a mixture of two types, though without reference to disease status.

19.5.3.2 The Model

The model may be built up as follows.

- We shall use the following notation. Let t_M be the threshold such that males manifest the disease if their liability exceeds it, and t_F be the corresponding threshold for females. Let ρ_1 be the correlation between liabilities of monozygotic twins, and ρ_2 be the correlation for dizygotic twins. As usual, $L(x, y; \rho)$ will be the upper right volume under the bivariate normal density function, and $\Phi(x)$ will be the cumulative univariate normal integral.

- First, consider same-sexed twin pairs.
  - The proportion for which both twins suffer the disease is $L(t, t; \rho)$, where t is t_M if we are thinking about males, and t_F for females, and ρ is ρ_1 for monozygotic twins, and ρ_2 for dizygotic twins.
  - The proportion that have exactly one of the pair suffering the disease is $2[1 - \Phi(t) - L(t, t; \rho)]$.
  - The proportion in which both twins are free from the disease is $2\Phi(t) - 1 + L(t, t; \rho)$.

- Next, consider unlike-sexed twin pairs. Obviously, these are all dizygotic.
  - The proportion in which both suffer the disease is $L(t_M, t_F; \rho_2)$.
  - The proportion in which the boy has the disease and the girl does not is $1 - \Phi(t_M) - L(t_M, t_F; \rho_2)$.
  - The proportion in which the reverse is the case is $1 - \Phi(t_F) - L(t_M, t_F; \rho_2)$.
  - The proportion in which neither has the disease is $\Phi(t_M) + \Phi(t_F) - 1 + L(t_M, t_F; \rho_2)$.

- Before applying the above expressions to data in which the zygosity of the twins is not known, we require a little more notation. Let p and q $(= 1 - p)$ be the proportions of twin pairs that are monozygotic and dizygotic, respectively. Let m and f $(= 1 - m)$ be the proportions of male and female fertilised ova, respectively.

- Considering sex and zygosity, there are five types of twin pair, and their proportions are:

male monozygotic	mp
female monozygotic	fp
male-male dizygotic	m^2q
female-female dizygotic	f^2q
male-female dizygotic	$2mfq$

- Considering the observable characteristics, sex and disease status, there are seven types of twin pair having at least one affected member. We now give formulae for their relative proportions. We write M and F for unaffected males and females, and $\bar{\text{M}}$ and $\bar{\text{F}}$ for males and females having the disease.

$\bar{M}\bar{M}$	$mpL(t_M, t_M; \rho_1) + m^2qL(t_M, t_M; \rho_2)$
$\bar{M}M$	$2mp[1 - \Phi(t_M) - L(t_M, t_M; \rho_1)] + 2m^2q[1 - \Phi(t_M) - L(t_M, t_M; \rho_2)]$
$\bar{F}\bar{F}$	$fpL(t_F, t_F; \rho_1) + f^2qL(t_F, t_F; \rho_2)$
$\bar{F}F$	$2fp[1 - \Phi(t_F) - L(t_F, t_F; \rho_1)] + 2f^2q[1 - \Phi(t_F) - L(t_F, t_F; \rho_2)]$
$\bar{M}\bar{F}$	$2mfqL(t_M, t_F; \rho_2)$
$\bar{M}F$	$2mfq[1 - \Phi(t_M) - L(t_M, t_F; \rho_2)]$
$M\bar{F}$	$2mfq[1 - \Phi(t_F) - L(t_M, t_F; \rho_2)]$

19.5.3.3 The Data

The data consisted of the numbers of the above-listed seven types of twin pair, for the following five conditions: mongolism, schizophrenia, coronary occlusion, congenital neural tube defects, and infantile hypertrophic pyloric stenosis. It was given in Table 5 of Marshall and Knox (1980).

References to the original sources of the datasets may be found in Marshall and Knox (1980). For two of the conditions (coronary occlusion and infantile pyloric stenosis), we have noticed minor discrepancies between the original data and that given by Marshall and Knox; the latter figures were used when fitting the above model.

The model fitted by Marshall and Knox was different from that above. Nevertheless, some of their interpretations of their results were in terms of polygenic inheritance; Hutchinson (1982c) felt this to be inappropriate, and therefore proposed the above model.

19.5.3.4 Some Notes on Fitting the Model to the Data

- It was assumed that $m = f = 0.5$, and $p = 0.38$ (or 0.263 for mongolism).
- The conditions $0 \leq \rho_2 \leq \rho_1 \leq 1$ were imposed.
- The condition $\rho_2 = \rho_1/2$ was not imposed. (This condition holds if the correlations are entirely genetic and the genetic variation entirely additive—Curnow, 1972.)
- Consequently, there are four parameters to be fitted to the data.
- The method of fitting was to minimise $\sum(O-E)^2/E$. The minimum value of this may be interpreted as having a chi-squared distribution with 2 degrees of freedom.

19.5.3.5 Results

In the following table, the estimated thresholds t_M and t_F have been expressed as the corresponding incidences I_M and I_F, these being the estimated proportions of the population having the condition concerned:

Condition	I_M	I_F	ρ_1	ρ_2	$\sum\frac{(O-E)^2}{E}$	$\sum\frac{(O-E_{MK})^2}{E_{MK}}$
Mongolism	0.35%	0.35%	1.00	0.00	4.8	4.7
Schizophrenia	7.9%	15.1%	0.98	0.00	4.6	1.8
Coronary occlusion	0.94%	0.40%	0.85	0.85	21.7	26.4
Neural tube defects	0.40%	0.62%	0.00	0.00	2.8	2.0
Infantile pyloric stenosis	23.6%	5.9%	0.07	0.07	2.3	2.3

(The 5% critical value for chi-squared with 2 degrees of freedom is 6.0. The final column gives the goodness-of-fit statistic for the Marshall and Knox model.)

There should be no surprise when we find (below) that the goodness-of-fit is not very sensitive to the incidences I_M and I_F—remember that no disease-free twin pairs were included in the datasets, only the seven types of pair having at least one affected member.

We now comment briefly on the results for the individual conditions.

- Mongolism:
  - The parameters imply a complete association between disease occurrence in monozygotic twins, but no association between dizygotic twins. This is entirely compatible with the current understanding of the condition. (Indeed, the multifactorial model is not a credible hypothesis, except at the boundary point $\rho_1 = 1$, $\rho_2 = 0$, where it may be interpreted in other terms.)

  - The estimated incidences are rather larger than the incidence of the condition in the general population, about 0.1%. However, on using $I_M = I_F = 0.001, \rho_1 = 1, \rho_2 = 0$, the goodness-of-fit statistic was found to be unchanged (to 1 decimal place).

- Schizophrenia:

  - The surprising finding of almost complete concordance between monozygotic twins and no correlation between dizygotic twins seems to be a feature of the data, rather than an artefact of the model, as an analogous result was obtained by Marshall and Knox.
  - On using the values (suggested by Curnow and Smith, 1975) $I_M = I_F = 0.01, \rho_1 = 2\rho_2 = 0.90$, the goodness-of-fit statistic was found to be 47.2, which with 6 degrees of freedom is highly significant. However, values that are quite close to these ($I_M = 0.0080, I_F = 0.0122, \rho_1 = 2\rho_2 = 0.98$) gave a goodness-of-fit statistic as low as 8.5.

- Concerning the results for coronary occlusion, the original source of the data gave the believed zygosity of the twin pairs; the reason for the poor fit of both models can be traced to there being 176 like-sexed dizygotic twins, and only 72 of unlike sex (rather than roughly equal numbers).

- Neural tube defects:

  - That the fitted correlations were zero is surprising.
  - On using the values (suggested by Curnow and Smith, 1975) $I_M = I_F = 0.003, \rho_1 = 2\rho_2 = 0.76$, the goodness-of-fit statistic was found to be 14.6, which with 6 degrees of freedom is significant at the 5% level. However, values that are quite close to these ($I_M = 0.0025$, $I_F = 0.0034$, $\rho_1 = 2\rho_2 = 0.70$) improved the goodness-of-fit statistic to 9.5.

- Infantile pyloric stenosis:

  - The estimated parameters are not credible—the population incidences are much too high, and the correlations too low.
  - On using the values (suggested by Curnow and Smith, 1975) $I_M = 0.005$, $I_F = 0.001$, $\rho_1 = \rho_2 = 0.74$, the goodness-of-fit statistic was found to be 13.1, which with 6 degrees of freedom is significant at the 5% level. However, a set of values that are quite close to these ($I_M = 0.0047$, $I_F = 0.0011$, $\rho_1 = 2\rho_2 = 0.78$) improved the goodness-of-fit statistic to 9.7.

19.5.4 Mixtures—A Paper by Tan

We do not understand it fully, but it appears that Section 4 of the paper by Tan (1978) provides another application in population genetics of mixtures of bivariate normal distributions. There is a pair of equations, similar to the variables-in-common form, connecting observable values of two quantitative characters to correlated genetic variables and to correlated environmental disturbances.

19.5.5 A Veterinary Example

Although the multifactorial model is most often encountered applied to humans, it is not restricted to that context. A paper by Foulley and Gianola (1984) was concerned with the difficulty of calving and the viability of the calves born. These authors had a number (s) of 2×2 contingency tables; they wished to explain the relative numbers of observations in the 4 cells by the explanatory variables associated with the s tables, each 2×2 table being regarded as a double-dichotomisation of a bivariate normal distribution. Their paper was somewhat preliminary in character, however, as the numerical example which they gave was of hypothetical data, and the value of the correlation coefficient between the two response variables was assumed to be both known and the same for all s tables.

19.6 Psychology

19.6.1 Mental Abilities

The intercorrelations between different types of abilities, especially of children, have long been of interest to psychologists and educationists. Usually, no form of the multivariate distribution is explicitly assumed, other than that it is sufficiently like the bivariate normal for the sample product-moment correlation to be meaningful. Burns (1982) has shown how tables of the bivariate normal integral—the variates being reading and IQ scores—can be used in the classification of children who have a reading disability (i.e., read much more poorly than would be expected from their intelligence). And in other papers, Burns (1983, 1984) has applied the idea in related fields.

Owen and Li (1980) were concerned with properties of academic selection procedures, in the sense of the joint distribution of a screening variable (e.g., score on the Scholastic Aptitude Test) and a performance variable (e.g., graduation grade points). Most of their paper was concerned with the truncated bivariate normal distribution, in the tradition of the work mentioned in Section 4.16.2 above, but at pp. 165–166 they suggested using Johnson's system of transformations if the marginals are not normal.

Huynh (1979) assumed that the true ability and the test score of a subject jointly have a bivariate normal distribution. The paper was concerned with selecting a cut-off score representing mastery of a test, given that the costs of the two types of error (false passing and false failing) are known.

For another model involving the bivariate normal distribution in mental test theory, see Section 23.5.

19.6.2 Replicated Paired Comparison Tests

At page 169 of Hutchinson (1979a), it is briefly mentioned that volumes under a bivariate normal surface could arise in a model for replicated paired comparison tests (for which subject, see Section 20.5). The idea is basically that of trivariate reduction—the difference between the utilities of the two products under comparison being $Z + \varepsilon_1$ in the first test, and $Z + \varepsilon_2$ in the second, with Z having a normal distribution over the population of people tested, and the ε's also having a normal distribution. Hutchinson's reason for suggesting this, rather than the models of our Section 20.5, was its ease of accommodating any effect of order of presentation—simply include a parameter Δ to represent the effect of presenting B before A rather than the reverse.

19.6.3 Signal Detection Theory

See Chapter 23 for applications of bivariate versions of signal detection theory to (i) the effect of signal probability on the slope of the ROC, (ii) the evaluation of X-ray images, (iii) information retrieval systems, and (iv) performance in two-alternative choice tests.

19.7 Air, Earth, Water, and Climate

19.7.1 Wind Speed

The two components into which wind velocity may be resolved are sometimes assumed to have a bivariate normal distribution. See, for example, Essenwanger (1976, pp. 105–112). If it is assumed that the two orthogonal components are i.i.d. normal variates with zero means, it can be deduced that the magnitude of wind speed, $\sqrt{X^2 + Y^2}$, has a Rayleigh distribution; hence the popularity of the Rayleigh distribution in this context. See Chou and Corotis (1983) for more general results. This paper argued that the Rayleigh distribution is still a good approximation even for a general bivariate normal distribution of wind velocity. Unfortunately, however, the value of the results is lessened by their presentation as sketches of the p.d.f.—a p.d.f. can look of roughly Rayleigh shape and yet the slope of $\log[-\log(1-F)]$ vs. $\log x$ still be noticeably different from 2.

The Weibull distribution arises if the above assumptions apply to some power of the orthogonal components of the wind velocity (Tuller and Brett, 1984).

19.7.2 Meteorological Conditions at Two Times

Gringorten (1971) has proposed the use of the bivariate normal distribution in meteorology, to describe the associations between conditions (temperature, cloud cover, etc.) at two moments time t apart. The correlation coefficient ρ is supposed to be related to the hour-to-hour correlation coefficient ρ_0 by $\rho = \rho_0^t$.

19.7.3 Rainfall

The gamma distribution is a popular choice for describing rainfall. Moran (1970) wanted a bivariate distribution with gamma marginals for use when rainfall observations are paired one with another (for instance, resulting from cloud-seeding experiments), and suggested starting with the bivariate normal and transforming its marginals to be gamma distributions.

19.7.4 Rainfall Duration and Amount

Hiemstra and Creese (1970) wanted to simulate chronological sequences of precipitation data. In doing this, they assumed bivariate lognormal distributions of several pairs of variables, including duration and amount of precipitation.

19.7.5 Cloud-Seeding Experiments

Cloud-seeding experiments commonly employ a target and a control area. An analysis of an experiment in Colorado was reported by Mielke et al (1977), who took X to be the precipitation in the target area and Y to be the precipitation in the control area, and assumed these variates to have a bivariate lognormal distribution. By comparing results when the target area was seeded with those when seeding did not take place, an estimate of the effectiveness of the treatment could be calculated. (The result was that seeding increased precipitation by about 50% when cloud-top temperature was estimated as at least −23°C, but had no effect at lower cloud-top temperatures.) The analysis was restricted to cases where $X \geq 0.015$ inches, and 0.01 was added on to the Y-measurements, because of the difficulty of using the lognormal distribution with observations of zero, a disadvantage also met with in other applications.

An interesting essay by Gabriel (1988) goes well beyond the specific subject of weather modification, and is thought-provoking about the interactions of statistics and statisticians with other members of a scientific team generally.

19.7.6 Errors—Tropical Cyclone Forecasts

The error in the 24-hour forecast position of the centre of a tropical cyclone was the subject treated by Crutcher et al (1982). The forecast position of each storm (979 in number) was plotted relative to its actual position. On the basis of previous work, these authors regarded the bivariate normal distribution as (i) the natural focus for their work, but (ii) likely to predict too few small and too few large errors and too many medium-sized ones. They therefore fitted a two-component model to the data. Note that the data points form a roughly elliptical cloud centred on $(0,0)$, there is no visual impression of clusters. Not surprisingly, then, the two components were both centred close to $(0,0)$, and differed chiefly in their standard deviations. If that were where the analysis stopped, one might doubt the validity of the two components, preferring to describe the data as a single component with greater kurtosis than the normal. But Crutcher et al went on to associate the storms that probably belonged to component 1, and those that probably belonged to component 2, with different geographical regions, and to give meteorological reasons for larger errors with the one as compared with the other.

19.7.7 River Flows and Rain

Clarke (1979) was concerned with augmenting a short series of annual streamflows by using a longer historical series of annual precipitations. One of the models used was a bivariate normal distribution of streamflow and precipitation. (Another was a bivariate gamma distribution, see Section 8.12.5.6.)

19.7.8 River Floods

In a study of river flow, Stănescu and Catană (1974) took $\log(W - c_1)$ and $\log(Q - c_2)$ to have a bivariate normal distribution, W being flood volume, Q the peak discharge at the same gauging station, and the c's being constants. The same method was used for $\log(W - c_1)$ and $\log(W' - c_1')$ (volumes at two different gauging stations), and for $\log(Q - c_2)$ and $\log(Q' - c_2')$ (peak discharges at two different gauging stations).

19.7.9 Tides at Two Times

Tides are sometimes described by a normal distribution, and the tide at two different moments of time by a bivariate normal distribution. An example of this occurs at equation (36) of Tayfun (1979)—in this case, the tide was being treated as a random variable because it was desired to superimpose upon it a storm surge or tsunami of random height occurring at a random time.

19.7.10 Height and Period of Waves

This subject is touched on at several other points in this book (Sections 8.12.5.1, 9.12.13, 10.8, 24.12.2—8, and 24.12.3—5). Ochi (1979) presented evidence that the marginal distributions of these variates is closely lognormal, and therefore the bivariate lognormal distribution is a natural choice to make. Burrows and Salih (1987) also included this distribution in their study (for which, see Section 8.12.5.1).

19.7.11 Rock Discontinuities

Kulatilake (1986) remarks that "the assessment of bedrock discontinuities and their orientations is often one of the most important geological inputs in engineering design of underground structures and structures founded on bedrock." Though recognising the dubious validity of applying the bivariate normal to a distribution on a sphere, Kulatilake fitted it to observed values of dip angle and dip direction of discontinuities (joints) in a metamorphic gneiss outcrop.

19.7.12 Earthquake Motion

Pinto and Nuti (1981) suggested that the three components of the intensity of ground motion in an earthquake (namely, acceleration, velocity, and displacement), jointly have a trivariate lognormal distribution.

19.7.13 Slope Stability

See Section 19.8.2.

19.8 Physical and Technological Sciences

19.8.1 Measuring Instrument Errors and a Calibration System

The following two applications are well-known because they appear in National Bureau of Standards (1959, Section 2.2).

- "A normal population of product items U with known mean μ_u and known standard deviation σ_u is to be measured by an instrument which is itself subject to an error Z which is normally distributed with mean zero (the instrument is properly calibrated) and with known standard deviation σ_z. The consumer's loss, abbreviated CL, is defined to be the joint probability that an item drawn at random from the population is nonconforming and that it is accepted by the inspection scheme. The producer's loss, abbreviated PL, is defined to be the joint probability that an item drawn at random from the population is conforming and that it is rejected by the inspection scheme ... The problem is to find CL and PL."

- "A calibration system starts with a device, the primary standard, that is always considered to be correct. The problem is to transfer a reading from the primary standard to another device, the secondary standard, located some distance from it and to make statements about the readings of the secondary standard. The transfer to the secondary standard will be made by a transfer device. After the transfer device is returned to the primary standard, its readings must not differ from the reading on the primary by more than a given amount if the readings of the secondary standard are to be certified. This process is called loop closure." It is wished to determine the probability that the reading of the secondary standard has an error less than a specified amount, given loop closure.

19.8.2 Structural Reliability

The wording here is adapted from Der Kiureghian and Liu (1986). The structural reliability problem is often formulated in terms of a joint distribution of basic random variables (which represent uncertain quantities such as loads, environmental factors, material properties, structural dimensions, and variables introduced to account for modelling and prediction errors), and a performance function such that if it is less than zero, the structure fails. For example, strength and load might be the random variables, and performance = strength − load. Unless we are prepared to make the very strong assumption that all the r.v.'s are independent, we immediately become concerned with bivariate or multivariate distributions.

It seems to us that most literature on this topic has been concerned with evaluating the multivariate (as distinct from bivariate) normal (as distinct from any other) distribution over the failure region. We can, however, mention the following:

- A-Grivas and Asaoka (1982) considered the joint distribution of two measures of soil strength, the cohesion and the coefficient of friction, in a slope stability analysis. They compared results from two alternatives, that these variables had a bivariate normal distribution, and that they had a bivariate beta distribution; the choice did not seem to affect the results greatly.

- Chowdhury and A-Grivas (1982) were concerned with how the failure of a soil slope progresses through that slope. They considered vertical slices through the slope, and assumed the safety margins of adjacent slices to have a bivariate normal distribution.

- Nguyen and Chowdhury (1985) were also concerned with slope stability. The hydraulic characteristics of soil or rock are an important factor in determining its stability, and Nguyen and Chowdhury assumed a bivariate normal distribution for the permeability of the aquifer and the storage coefficient.

- Example 1 of Der Kiureghian and Liu (1986) was concerned with two variables whose distribution was only bivariate normal after marginal transformations had been made. We have already summarised (in Section 4.14.5) their study of how the correlation coefficient is affected by marginal transformations.

- The approach of Grigoriu (1983), too, was to assume multivariate normality once the marginals had been transformed to be univariate normal.

We have already referred to the paper by Thomopoulos and Longinow (1984) in Section 4.14.1 above.

19.8.3 Strength of Lumber

Taylor and Bender (1988) have suggested that an appropriate procedure for modelling the joint distribution of modulus of elasticity and tensile strength of lumber is to transform their marginals to be normal and then assume bivariate normality. They give an example in which the original distributions were three-parameter Weibull. (It appears that Taylor and Bender calculated the correlation coefficient on the original observations, whereas the equivalent normally-transformed ones should have been used.)

19.8.4 Destructive Testing

Evans et al (1984) were concerned with estimating the correlation between two variables that, individually, can only be observed through destructive testing; proof loading a unit in one failure mode and then loading the survivors to destruction in the other mode is what is done. Having observations

$$\begin{array}{rl} x_j & \text{if } x_j \leq L \\ \text{or} \quad y_j & \text{if } x_j > L, \end{array}$$

a choice of joint distribution of X and Y was necessary in order to estimate the correlation. Evans et al assumed a bivariate normal distribution to obtain their maximum-likelihood equations; and they also used a bivariate normal distribution in a simulation of the performance of their estimation procedure. (The marginal means and standard deviations could be assumed to be known, having been found from two other series of tests, measuring X or Y alone.) Our suspicion is that there is some danger of any such procedure being quite sensitive to the distributional assumption, because the data itself is providing so little direct evidence about how X and Y co-vary.

A slightly more complex problem in the same vein was that of Bartlett and Lwin (1984); see Section 24.2—4. These authors assumed a trivariate lognormal distribution of their three variables.

19.8.5 Thermodynamics

Muthu (1982, pp. 304–307) writes about fluctuations of thermodynamic quantities characterising a system in equilibrium, saying they are very small, but important in the neighbourhood of critical points. He gives an equation for the probability of a fluctuation in terms of ΔE, ΔS, and ΔV, the deviations of the energy, entropy, and volume from their equilibrium values. By expanding ΔE as a Taylor series, he converts this into a form recognisable as a bivariate normal distribution of ΔS and ΔV.

19.8.6 Astronomy

Barton and David (1958) studied the joint distribution of the angular velocity and distance from the galactic centre of cepheid variable stars, and in Section 7 of their paper, entertained the possibility that it was made up of a mixture of two bivariate normal populations.

19.8.7 Radar

Bouver (1987) used the bivariate normal distribution to describe the difference between the positions of a target as estimated by two radar systems. The specific problem was to determine whether the two positions represented one target or two separate ones. Bouver expressed the probability of correctly accepting them as a single target in terms of the integral of a bivariate normal p.d.f. over an elliptical region.

19.8.8 Optics

The optical phase of light being focussed by a lens is commonly assumed to have a normal distribution, being the result of many small disturbances by the turbulent atmosphere, and the joint distribution of phases at two points across a pupil is assumed to be bivariate normal, the correlation coefficient being a function of the distance between the two points. See Frieden (1983, Sections 4.28.3 and 8.6.1).

19.8.9 Shooting at a Target

One of the areas that the bivariate normal distribution is classically supposed to be relevant to (e.g., Johnson and Kotz, 1972, p. 85) is the position a shot (or bomb, or missile, etc.) hits, relative to its target.

- Mason and Young (1985) use such data, which they attribute to Bates (1966), in evaluating the performance of a test of bivariate normality.

- Shephard (1989) notes that, late in 1944, Roland D Clarke fitted the bivariate normal distribution to the places of impact of flying bombs on London, Antwerp, and Brussels. (As can be imagined, it was a matter of considerable importance to determine whether they were landing at random or were clustering around specific targets.)

As well as the location of impact, the bivariate normal distribution is also used to describe (i) the probability of destroying a target at location (X, Y) with respect to the impact point, and (ii) the value of the target itself, at a point (X, Y) with respect to its centre.

Eckler (1988) gives a review of this subject.

19.8.10 Forces in Wood Grinding

In studying the force variation in the grinding of wood, DeVries et al (1978) were prepared to assume a bivariate normal distribution for the components of the force tangential to and at right angles to the wood-stone interface. They transformed the p.d.f. $\phi(x, y; \rho)$ into a p.d.f. $h(r, m)$, where $r = \sqrt{x^2 + y^2}$ and $m = y/x$; m is interpretable as a coefficient of friction.

19.9 Sociology and Economics

19.9.1 Household Size and Income

Kmietowicz (1984) applied the bivariate lognormal distribution to a cross-tabulation of household size and income in rural Iraq, and found it gave a satisfactory fit. Results on four other datasets (three from Iraq and one from Kenya) were more briefly reported.

19.9.2 Commodities Trading

Lien (1985) is concerned with the Houthakker effect, which "relates to the notion that the correlation between the spot and futures prices depends on the stocks of the commodity. Specifically, when inventories of a commodity are large, spot and futures prices of the commodity tend to be highly correlated, while at low inventory levels, spot and futures prices are less highly correlated. Since large inventories tend to be associated with low spot prices and vice versa, the correlation pattern between spot and futures prices will depend on the spot price level." From formulae for the moments of a truncated lognormal distribution, Lien shows that whereas $\rho_L = \rho_U$ for symmetric distributions like the bivariate normal, $\rho_L > \rho_U$ for the bivariate lognormal distribution with correlation between 0 and 0.54, where ρ_L is the correlation between X and Y given that X is less than its mean, and ρ_U is the correlation between X and Y given that X is greater than its mean.

19.9.3 Blackjack Gambling

Griffin (1976) reports that "The card-counter in a game of blackjack wishes to assess favorability for a particular action using his knowledge of a parameter which reflects, usually imperfectly, the composition of the unplayed subset of cards he is confronting. The Gauss-Markov theorem allows an intuitive approximation of multicard interactions by linear effects. A limit theorem and the results of extensive simulations justify the treatment of the distribution of favorability as approximately normal. The efficiency of a card-counting system depends on its ordinary correlation coefficient with the linear effects."

19.9.4 Policy Analysis

A one-page note by Wilson (1983) includes mention of a thesis (Martin, 1983) that used Johnson's system of transformations with 80 trivariate datasets arising in a large-scale policy analysis project.

19.9.5 Sample Selection Equations

Within econometrics, there is a body of work concerned with sample selection equations. By this, we mean that we want to estimate the parameters of $y = \boldsymbol{\beta}_1\mathbf{z}_1 + \varepsilon_1$, but we can only observe $(\mathbf{z}_1, y)$ if $\boldsymbol{\beta}_2\mathbf{z}_2 + \varepsilon_2 > 0$. Problems arise if the disturbances $(\varepsilon_1, \varepsilon_2)$ are not independent. For $(\varepsilon_1, \varepsilon_2)$ having a bivariate normal distribution, see, for example, Heckman (1976); for other cases, see Ray et al (1980). There is a good deal of concern that the consequences of failure of the assumption of (univariate) normality are considerably more serious in this context than with ordinary regression models. See, for example, Duncan (1983).

The sort of application to which this is relevant is where y is a wage offer to an individual, but is only observed if the individual accepts it, i.e., if it exceeds what economists call the individual's "reservation" wage. The following quotation from Heckman (1976, p. 477) illustrates this in a specific context: "If the presence of children affects the work decision but does not affect market wages, regression evidence from selected samples of working women that women with children earn lower wages is not necessarily evidence that there is market discrimination against such women."

An extended version of this was proposed as a unionism and wage rate model by Lee (1978):

$$\begin{aligned} \log y_1 &= \boldsymbol{\beta}_1\mathbf{z}_1 + \varepsilon_1 \\ \log y_2 &= \boldsymbol{\beta}_2\mathbf{z}_2 + \varepsilon_2 \\ y_3 &= \boldsymbol{\beta}_3\mathbf{z}_3 + \varepsilon_3 \end{aligned}$$

in which y_1 is the potential wage for a union member, y_2 is the potential wage for the same person if not a union member, and y_3 is a latent variable representing the utility gain from being a union member; either y_1 or y_2 is observed, depending on the value of y_3. A bivariate normal distribution is assumed for each of the pairs of disturbances, $(\varepsilon_1, \varepsilon_3)$ and $(\varepsilon_2, \varepsilon_3)$. Lee (1984) proposed a test for bivariate normality in this type of model within the wider class of bivariate Edgeworth series distributions, and applied it to this example.

Ham (1982) extends Heckman's approach by considering the operation of two correlated selection rules—specifically, referring to unemployment and underemployment (these conditions being constraints on the number of weeks of work per year and number of hours of work per week, respectively). For Ham's data, the disturbance terms in the two selection equations were estimated to have a small positive correlation, 0.2.

Related to sample selection is the idea that we may have a demand equation and a supply equation,

$$\left.\begin{aligned} d &= \boldsymbol{\beta}_1\mathbf{z}_1 + \varepsilon_1 \\ s &= \boldsymbol{\beta}_2\mathbf{z}_2 + \varepsilon_2 \end{aligned}\right\} \tag{19.2}$$

and observe only the amount transacted, $\min(d, s)$. See Lee (1982) for a test for the bivariate normality of $(\varepsilon_1, \varepsilon_2)$ with, as earlier, the bivariate Edgeworth series distribution being the wider class. See also Gilliland and Hannan (1983).

For further econometric applications of the bivariate normal distribution, see Chapter 22, especially Section 22.4.2. For reviews of this area of work, see Quandt (1982), Amemiya (1985, Chapter 10), Heckman and Macurdy (1986), and especially Maddala (1986).

Chapter 20

Construction of the Bivariate Pareto Distribution by Compounding, with Applications

20.1 Description of the Method

The derivation of a probability distribution by compounding involves (i) starting with one distribution, (ii) assuming that one parameter of that distribution is not constant but is itself a random variable having a certain distribution, and (iii) integrating over all possible values of this to produce the new distribution.

We shall concentrate in this Chapter on the familiar example of the Pareto distribution, derived by assuming that the reciprocal of the mean of an exponential distribution has a gamma distribution. In the univariate case,

$$\Pr(X > x) = \exp(-\theta x), \tag{20.1}$$

and we suppose that the total population consists of a large number of sub-populations, each with its own value of θ must necessarily be positive; if we assume it has a gamma distribution, we obtain

$$\begin{aligned} \Pr(X > x) &= \int_0^\infty \exp(-\theta x) \frac{\theta^{c-1} \exp(-\theta/b)}{b^c \Gamma(c)} \, d\theta \\ &= (1 + bx)^{-c}, \end{aligned} \tag{20.2}$$

the Pareto distribution (Section 2.9); b can be seen to be merely a scale parameter. For more on the relation between the distribution of θ and of X, see McNolty et al (1980) and Gupta (1982).

A bivariate version of this concept is to consider that in each subpopulation, two variables X and Y are independent and have the same distribution, and that a parameter of this distribution varies between subpopulations. We may readily obtain the bivariate Pareto distribution (Sections 5.8 and 5.9) in this way: if $\Pr(X > x) = \exp(-\theta x)$ and $\Pr(Y > y) = \exp(-\theta y)$, then $\Pr(X > x, Y > y) = \exp[-\theta(x + y)]$ assuming that θ has a gamma distribution,

$$\Pr(X > x, Y > y) = \int_0^\infty \exp[-\theta(x + y)] \frac{\theta^{c-1} \exp(-\theta/b)}{b^c \Gamma(c)} \, d\theta \tag{20.3}$$

$$= (1 + bx + by)^{-c}. \tag{20.4}$$

After a moment's thought, one realises that substitution of $A(x)$ for x, and $B(y)$ for y, in the integral expression (20.3) will simply lead to the same substitution in the result (20.4). What this means in terms of the distribution of X and Y given θ is that $\Pr(X > x) = \exp[-\theta A(x)]$ and $\Pr(Y > y) = \exp[-\theta B(y)]$. Alternatively, if one wants to write $\Pr(X > x) = [1 - C(x)]^\theta$ and $\Pr(Y > y) = [1 - D(y)]^\theta$, where C and D are distribution functions, then the result will be

$$\Pr(X > x, Y > y) = \{1 - b\log[1 - C(x)] - b\log[1 - D(y)]\}^{-c}. \tag{20.5}$$

Table 20.1: Proportions in which the four grades of certainty of diagnosis of MS were used.

	Certain	Probable	Possible	Unlikely
Winnipeg patients				
Winnipeg neurologist	.56	.25	.07	.11
New Orleans neurologist	.30	.32	.23	.15
New Orleans patients				
Winnipeg neurologist	.16	.42	.16	.26
New Orleans neurologist	.12	.26	.32	.30

In the previous paragraph, we have retained the bivariate Pareto copula, but let the marginals be arbitrary. If we, instead, are uninterested in the copula but insist that the marginals be Pareto, then we can start with any bivariate distribution having exponential marginals, and let the scale parameter have a gamma distribution. There will be an example of this in Section 20.6; see also Arnold (1983, Section 6.2.1).

The bivariate normal distribution may also be derived by compounding. If X and Y are independent and each is normally distributed with mean μ and unit standard deviation, their joint distribution is circular normal centred upon (μ, μ). Now, if μ itself has a normal distribution of zero mean and standard deviation σ, the joint distribution of X and Y becomes bivariate normal with mean $(0,0)$, variances each equal to $1+\sigma^2$, and correlation $\sqrt{\sigma^2/(1+\sigma^2)}$.

For the equivalence of compounding to the denominator-in-common version of trivariate reduction, see Section 15.4.2.

The Sections of this Chapter are as follows:

20.2 Ordinal-scale assessments by two observers.

20.3 Injury severity in road accidents.

20.4 The multifactorial model of disease transmission.

20.5 Replicated paired comparison tests.

20.6 Compounding of an F-G-M distribution, and application to trace metal concentrations.

20.7 Generalising the bivariate Pareto distribution.

20.2 Ordinal-Scale Assessments by Two Observers

20.2.1 The Topic, in a Medical Context

There are very many contexts in which some object is graded by two or more observers, and the process of assigning grades is not entirely an objective one. Medical examples are prominent among these contexts, and one such is that of Westlund and Kurland (1953): these authors collected data on the diagnosis of multiple sclerosis (MS). There were two series of patients (from different geographical areas), each classified independently by two neurologists, using a four-category scale from "certain" through "probable" and "possible" to "unlikely".

Perhaps the most important feature of this data is the differences in the marginal proportions. Table 20.1 immediately indicates that the Winnipeg neurologist is more inclined to diagnose MS than his New Orleans colleague, and that the incidence is higher in the Winnipeg series of patients than in the New Orleans series.

Table 20.2: Numbers of patients with each combination of grades assigned by the two neurologists.

		Winnipeg neurologist			
		1	2	3	4
Winnipeg patients					
	1	38	5	0	1
New Orleans	2	33	11	3	0
neurologist	3	10	14	5	6
	4	3	7	3	10
New Orleans patients					
	1	5	3	0	0
New Orleans	2	3	11	4	0
neurologist	3	2	13	3	4
	4	1	2	4	14

Note: Grade 1 indicates "certain" multiple sclerosis, grade 4 indicates "unlikely".

20.2.2 Application of the Model

Of more interest to us at present, however, is the joint distribution of classifications by the two neurologists—this is given in Table 20.2. Let us assume that the neurologists use identical scales in assessing patients, i.e., that they differ only in where they place the boundaries between the categories. (Some comments on this assumption are made later.) The difficult or ill-defined nature of their task means that each would not always give the same classification to any one particular patient. We suppose that there is a continuous scale representing the degree to which a patient's symptoms resemble multiple sclerosis, and that the probability distribution for any one patient follows equation (20.1). This variation can be regarded as being due to error in assessment[1]. Different patients will have different values of θ. Then, on assuming that θ has a gamma distribution, we arrive at equation (20.4) for the joint distribution of "similarity to multiple sclerosis" as assessed by the two neurologists.

20.2.3 Results

Fitting of this model to the data of Table 20.2 will necessitate determining at least eight parameters: three thresholds for each neurologist, a scale parameter b for one series of patients (that for the other may be taken as unity without loss of generality), and the shape parameter c. If it proved necessary to allow the two series to have different values of c, there would be nine parameters altogether.

The results obtained (the method of fitting was that of minimising Pearson's chi-squared, $\sum(O-E)^2/E$) were as follows:

- Having set the scale parameter b at 1.0 for the Winnipeg series, that for the New Orleans series was estimated to be 0.28, reflecting the lower incidence there.
- The shape parameter $c = 0.70$.
- The three thresholds of the Winnipeg neurologist were 1.9, 9.5, and 20.3, whereas those of the New Orleans neurologist were 0.7, 3.3, and 14.8, reflecting the latter's greater reluctance to diagnose MS.
- The goodness-of-fit statistic X^2 was 25.7, which compared with the χ^2 distribution with 22 degrees of freedom is not significant at the 5% level.

[1]The exponential may strike the reader as an unsuitable distribution for observer error. But in this application, the continuous scale is unobservable. Consequently, the shapes of the distributions are meaningless, as any monotonic transformation can be performed on the scale of "similarity to multiple sclerosis". See Figure 4 of Hutchinson (1976) for an illustration of a pair of distributions that appear roughly normal, and yet which is equivalent to a pair of negative exponential distributions.

Table 20.3: Severities of injury to vehicle drivers in head-on collisions between two vehicles of equal mass.

		Driver of first vehicle			
		Fatal	Serious	Slight	Uninjured
	Fatal	5	25	12	16
Driver of second vehicle	Serious		134	162	202
	Slight			177	532
	Uninjured				?

Notes:

- The collisions were those at sites where the speed limit in force was greater than 40 m.p.h., in Great Britain, during 1969-72.
- As the two vehicles were equivalent, entries below the diagonal have been combined with those above.
- Only accidents involving injury were reported, hence the number in which both drivers were uninjured is unknown.

20.2.1 Comments

If we had used the bivariate normal instead of the bivariate Pareto to model this data, the goodness-of-fit would probably have been about the same. The significance of the results lies not in the choice of distribution, but that at least one choice resulted in an acceptable fit. To see the importance of this, we can identify three reasons why the two neurologists might differ in their classification of a patient:

1. Random error, which would also manifest itself in the same neurologist, presented with the same case history a second time, sometimes giving a different classification from the first time.
2. Differences in their placement of the thresholds between categories, while using the same scale to measure the degree of similarity to multiple sclerosis.
3. The use of different criteria to construct their scales of assessment (that is, placing different levels of importance on the various symptoms present or absent, biochemical tests positive or negative, and so on).

The conclusion from the successful fit of the bivariate Pareto model is that Table 20.2 may be explained by disagreements of types (1) and (2) only, with no need to adduce type (3). This is quite important, as modifying the boundary positions that a clinician uses is likely to be much easier than changing his whole concept of the scale.

For perspectives on Table 20.2 from the observer-agreement standpoint (Cohen's kappa—see Section 10.9—and similar methods), see Landis and Koch (1977), Krauth (1984), Hutchinson (1985), Darroch and McCloud (1986), and Agresti (1988).

20.3 Injury Severity in Road Accidents

20.3.1 The Topic and the Model

We are here concerned with the joint distribution of the severities of injury to two people in the same road accident. We shall return to such data in Chapter 21; here, we may mention that one reason for its importance is that impact speed is one of the chief factors determining injury, yet is very difficult to determine, so an alternative approach is to use the injury to casualty no. 1 as a proxy for speed when interpreting the injury to casualty no. 2, and vice versa.

Table 20.3 gives the data. There are two special features of this: (i) it refers to injuries to the two drivers of vehicles in head-on collisions, the two vehicles being of equal mass and hence indistinguishable in the data, unlike most of the data in Chapter 21, where a heavier vehicle and a lighter vehicle could be distinguished; (ii) the number of cases in which both drivers escaped injury is not known, since such accidents are not reported.

We suppose that at any speed of impact, the severity of injury to each driver follows equation (20.1), and that the distribution of impact speeds is such that θ has a gamma distribution. The result is the bivariate Pareto distribution. (When fitting this to the data of Table 20.3, some slight modifications need to be made to the formulae to take account of the features mentioned in the previous paragraph.)

20.3.2 Results

The results obtained (by minimising Pearson's chi-squared) were as follows:

- The threshold of injury corresponding to death was fixed at 1.0; the thresholds for serious and slight injury were respectively 0.21 and 0.075.
- The scale parameter $b = 5.65$. (Alternatively, if the convention $b = 1.0$ is adopted, then the thresholds for fatal, serious, and slight injury become 5.65, 1.19, and 0.42 respectively.)
- The shape parameter $c = 2.08$.
- The goodness-of-fit statistic X^2 was 25.5, which compared with the χ^2 distribution with 4 degrees of freedom is highly statistically significant, so according to conventional criteria the model must be rejected.

20.3.3 A Further Model and Results

We can introduce another parameter in order to improve the fit in the following way. Consider the propositions:

1. The severity distribution in vehicle 1 is exponential with parameter θ_1; θ_1 has a gamma distribution.
2. The severity distribution in vehicle 2 is exponential with parameter θ_2; θ_2 has the same gamma distribution as in (1).
3. θ_1 and θ_2 are identically the same, $\theta_1 \equiv \theta_2$.
4. As an alternative to (3), θ_1 and θ_2 are independent.

We have so far been assuming (1), (2), and (3), thus arriving at the bivariate Pareto distribution. Undoubtedly (3) is more realistic than (4). But it may be that there is some departure from (3) in the direction of (4)—this could arise from different interior designs of the vehicles, or from different susceptibilities to injury of the drivers. So suppose that θ_1 and θ_2 jointly have Kibble's bivariate gamma distribution (for which, see Section 7.5), i.e., their marginal densities are of gamma form with shape parameter c and their joint p.d.f. is

$$h(\theta_1, \theta_2) = \frac{(\theta_1\theta_2)^{(c-1)/2}}{b^{c+1}(1-k)^{(c-1)/2}k\Gamma(c)} \exp[-(\theta_1+\theta_2)/bk] I_{c-1}\left(\frac{2\sqrt{(1-k)\theta_1\theta_2}}{bk}\right) \tag{20.6}$$

(where $k < 1$ and I_ν is the modified Bessel function of the first kind). Then on performing the integration

$$\Pr(X > x, Y > y) = \int_0^\infty \int_0^\infty \exp(-\theta_1 x - \theta_2 y) h(\theta_1, \theta_2)\, d\theta_1\, d\theta_2 \tag{20.7}$$

(by making use of equation (18) from page 197 of Erdelyi, 1954a), we find the result is

$$(1 + bx + by + kb^2xy)^{-c}. \tag{20.8}$$

This is what we have called the Durling-Pareto distribution at Section 5.8.13. We may note that $k = 0$ corresponds to proposition (3), and $k = 1$ to proposition (4). When this distribution was fitted to the data of Table 20.3, the results were:

Table 20.4: Data from Cloninger, Reich, and Guze (1975) on the percentage prevalences of sociopathy and hysteria.

Diagnosis and sex of probands, and prevalences in the general population		Prevalences of illnesses in first-degree relatives of the probands: Sociopathy (men)	Sociopathy (women)	Hysteria (women)
Sociopathic men	3.3	25.9	6.3	9.9
Sociopathic women	0.9	40.0	15.4	24.0
Hysteric women	2.4	15.0	0.0	25.8

Note: The sample sizes on which the above percentages were based were as follows:

329	54	79	79
318	5	13	13
153	20	56	56

(The reason why, for example, 2.4% of 153 is not an exact integer, is that an age correction was made to the prevalences of hysteria.)

- Fixing the threshold of injury corresponding to death at 1.0, the thresholds for serious and slight injury were respectively 0.090 and 0.027, and $b = 46$.
- $c = 1.1$
- $k = 0.12$
- Goodness-of-fit $X^2 = 16.7$.

However, scant importance should be attached to these particular parameter values, as it was found possible to change any one of them by a considerable amount, and yet find values of the others giving an almost unaltered fit. But it is noteworthy that X^2 shows a substantial reduction from the 25.5 obtained when k was assumed to be zero.

As c was so close to 1, fitting of the distribution (20.8) was repeated with c constrained to be 1. That is, the distribution of the exponential parameter is itself exponential: the bivariate distribution of θ_1 and θ_2 thus has the form studied by Downton (1970) and Nagao and Kadoya (1971) (see our Section 9.12). The goodness-of-fit was hardly changed from that in the previous paragraph.

Although we have achieved some improvement in the fit, proper resolution of this issue will not be made until Chapter 21. There, when Table 20.3 is considered in the context of data from collisions between vehicle of unequal masses, an important reason for the fit being poor is found to be variability between accidents in where the serious/slight and slight/none thresholds are located.

20.4 The Multifactorial Model of Disease Transmission

20.4.1 The Topic, and the Relevance of the Pareto Distribution

This topic has been introduced in Section 19.5.1. Passing reference was made there to data of Cloninger et al (1975) on the occurrence of sociopathy and hysteria in families. We take up this example in a moment.

The bivariate Pareto distribution may be justified in this context as follows. We may imagine that in each family the liabilities x and y of two relatives are independent and have the same distribution, which is negative exponential. Then the upper right probability volume is $\exp[-\theta(x+y)]$. There are many families, each with a different value of θ. So the question of what is a suitable distribution for θ arises. To answer this, we realise that we want the distribution over families of the probability of disease occurrence,

Table 20.5: Results of fitting the bivariate normal and Pareto models to the data of Table 20.4.

	Normal	Pareto	Reversed Pareto
Proportion of sociopathy in the male population	3.7%	3.8%	3.4%
Proportion of sociopathy in the female population	0.6%	0.5%	0.8%
Proportion of hysteria in the female population	2.9%	3.0%	2.5%
Rank correlation of liability between relatives	0.37	0.16	0.80
Goodness-of-fit	11.0	11.4	12.4

Note: The rank correlation given is Kendall's tau. The shape parameter c for the Pareto and reversed Pareto models was respectively 2.7 and 0.12. For the normal model, the product-moment correlation ρ was 0.55.

$\exp(-\theta x)$, to have the same form whether we are concerned with a severe form of the disease ($x = a_1$) or a milder form ($x = a_2$). This is true if θ has a gamma distribution, in which case equation (20.4) results.

20.4.2 Data and Results

Table 20.4 gives the data, taken from Table II of Cloninger et al (1975). The results of fitting the model to the data are in Table 20.5, taken from Hutchinson and Satterthwaite (1977). Both the Pareto and normal distributions were fitted, and in this instance the former was used both in the form described (in which high values of the variate correspond to high liability) and in reversed form (in which low values correspond to high liability).[2]

To facilitate comparison between the distributions, their parameters are presented in Table 20.5 not as the thresholds themselves, but as the proportions of the general population estimated as possessing the traits under consideration. It can be seen that these proportions are insensitive to choice of model.

20.4.3 Comments on Estimated Correlations

However, the correlation coefficient is sensitive to choice of model. If this statement only referred to the product-moment correlation, noone would be surprised—it is well-known that this can be altered greatly by monotonic distortion of the axes. (If we apply a non-linear transformation to either or both of two variables having a bivariate normal distribution, then the absolute value of the correlation in the new bivariate distribution is less than it was in the original one—see Section 4.14.5.) But it is true also of rank correlation, as shown in Table 20.5. A little further understanding may be gained from the following calculations. Let us use the extreme 2% of the population for our comparisons, as this is the approximate disease prevalence we are dealing with. Then:

- For the distribution $(1 + x + y)^{-2.7}$, given that X is in the top 2%, the probability that Y is also in the top 2% is 22%.
- For the same distribution—an unchanged correlation, of course—given that X is in the bottom 2%, the probability that Y is also in the bottom 2% is 2.7%.
- For the distribution $(1 + x + y)^{-0.12}$, given that X is in the top 2%, the probability that Y is also in the top 2% is 92%.
- For the same distribution, given that X is in the bottom 2%, the probability that Y is also in the bottom 2% is 16%.

[2] Hutchinson and Satterthwaite (1977) described these latter results in terms of the bivariate distribution $H = (1 - e^{-x} + e^{-y})^{-c}$ (whose properties they described in Satterthwaite and Hutchinson, 1978); they did not then realise that, when considering data that is merely ordinal, this distribution is effectively the Pareto in reversed form.

The first and fourth conditional probabilities calculated above are roughly similar to each other and to the probabilities shown in the three rightmost columns of Table 20.4, the other two conditional probabilities are quite different. Kendall's tau is calculated from the whole distribution, we only have information about the proportion in one tail, that is why tau is sensitive to the choice of distribution.

20.4.4 A Trivariate Approximation

Returning briefly to the trivariate case, we may mention a method for calculating an approximate value of $\bar{H}(a, a, a; r, r, 0)/[\bar{H}(a)]^2$ in the normal case. (See footnote 2 in Chapter 19 for why this ratio is important.) This is to (i) determine the parameters a and c of the Pareto distribution from disease incidences p_1 and p_2 using equations (19.1), and then (ii) calculate $\left(1 + \frac{a}{(1+a)^2+a^3}\right)^{-c}$. The accuracy of this method is demonstrated in Table 3 of Hutchinson (1980), but the mathematical reason for its success is not known.

20.5 Replicated Paired Comparison Tests

20.5.1 The Topic

In paired comparison tests, people are asked which of two products they prefer (which of two brands of sausage, or two detergents, perhaps). If someone is asked to choose between the same products twice, the answer may not be the same each time. This repetition of the test is what "replicated" refers to.

Frequently, the subjects are permitted to give "no preference" as a response. In this case, a 3×3 table with total sample size N would result if N people took part in the experiment. Some examples are given in Table 20.6.

Hutchinson's (1979a) paper was a comment on earlier ones by Wierenga (1974) and Horsnell (1977), an attempt to combine the best features of the two. In the Introduction, Hutchinson said "Although it contains useful criticisms of Wierenga's approach, Horsnell's paper seems to be in conflict with modern psychological ideas about the process of perceiving and deciding. Thus, Horsnell assumes that the population of subjects can be sharply divided into discriminators and non-discriminators and into A-preferrers and B-preferrers. The responses of the subjects in some of the groups which he envisages do not vary at all between different trials, and in another group choice between A and B is entirely random. I suggest that neither of these alternative characteristics of the groups is likely to represent realistically a subject's behaviour. Unless the task is very easy, no subject will be able to maintain absolute consistency of response. [On the other hand, even when he or she] claims not to know the answer to the question posed, the subject has some degree of information available and, if forced to choose, his or her guesses are correct in a higher-than-chance proportion of cases... The stochastic model of Wierenga (1974) is preferable to that of Horsnell because it recognises explicitly these aspects of the psychology of preference and choice. Nevertheless, the objections of Horsnell (1977) have force: the absence of any degrees of freedom in Wierenga's model for testing it, the failure to model the 'no preference' category, and the [failure to take account of the] fact that the probabilities of preferring A and of preferring B cannot vary independently."

20.5.2 The Model

One might attempt to model such data by assuming

- That the difference in the utilities of item A and item B has distribution function $F(u - \theta_i)$ for the ith person,
- That "prefer A", "no preference", and "prefer B" correspond to this difference being greater than some level t, being between $-t$ and t, and being less than $-t$ respectively, and
- That θ_i has some distribution over the population of people.

However, as people differ substantially in the frequency with which they use the "no preference" category, it would be necessary to write t as t_i and take it as varying over people, as θ does.

Table 20.6: Twelve datasets obtained in replicated paired comparison tests.

	Test 1			Test 2			Test 3			Test 4	
24	9	30	18	4	28	20	7	25	25	22	43
	4	13		2	7		9	11		14	28
		20			42			28			29
	Test 5			**Test 6**			**Test 7**			**Test 8**	
43	8	20	63	16	32	40	13	48	12	6	31
	5	9		15	13		14	10		5	10
		15			21			35			36
	Test 9			**Test 10**			**Test 11**			**Test 12**	
57	9	17	50	9	20	21	8	18	13	12	25
	2	6		4	4		23	3		14	5
		9			13			27			31

Notes:

- This data is taken from Wierenga (1974). It was also analysed by Horsnell (1977).
- For each test, the numbers in the table are the numbers of respondents giving a certain pattern of responses, as follows:
  - In the first row are the numbers preferring A in one test, and preferring A, having no preference, and preferring B in the other test.
  - In the second row are the numbers having no preference on both occasions, and having no preference on one occasion and preferring B on the other.
  - In the third row is the number preferring B on both occasions.

  Thus the format is analogous to that of Table 20.3.

What Hutchinson (1979a) did, instead, was to set $p =$ probability of preferring A, conditional upon a preference for A or B being stated, and $q =$ probability of stating no preference. The distributions of p and q were assumed to be independent, that of p being beta,[3] and that of $-\log(q)$ being gamma.[4]

The probability of preferring A both times is $(1-q)^2p^2$ averaged over all subjects. If p has probability density function $p^{u-1}(1-p)^{v-1}/B(u,v)$, and q has density function $[\log(1/q)]^{c-1}q^{d-1}d^c/\Gamma(c)$, then we find the probability of preferring A both times is

$$\frac{u(u+1)}{(u+v)(u+v+1)}\left[\left(\frac{d}{d+2}\right)^c - 2\left(\frac{d}{d+1}\right)^c + 1\right]. \tag{20.9}$$

This is obtained simply by evaluating the integrals $\int_0^1 p^2p^{u-1}(1-p)^{v-1}\,dp$ and $\int_0^1 (1-q)^2[\log(1/q)]^{c-1}q^{d-1}\,dq$, since p and q are assumed independent. Other probabilities may be obtained similarly and are given in Table 20.7.

20.5.3 Results and Comments

As this type of experiment has only six cells of data, there are only five free probabilities; the model has four parameters, leaving only one degree of freedom for testing the goodness of fit. Considering all 12

[3] Chosen for reasons of symmetry, $1-p$ also having a beta distribution.

[4] Chosen because then a power of q also has a distribution from this family, a property that turned out to be useful in an extended model.

Table 20.7: Probabilities for response patterns in experiments like Wierenga's.

Response pattern	Probability
Two A's	$\frac{u(u+1)}{(u+v)(u+v+1)}\left[\left(\frac{d}{d+2}\right)^c - 2\left(\frac{d}{d+1}\right)^c + 1\right]$
One A, one N	$\frac{2u}{u+v}\left[\left(\frac{d}{d+1}\right)^c - \left(\frac{d}{d+2}\right)^c\right]$
One A, one B	$\frac{2uv}{(u+v)(u+v+1)}\left[\left(\frac{d}{d+2}\right)^c - 2\left(\frac{d}{d+1}\right)^c + 1\right]$
Two N's	$\left(\frac{d}{d+2}\right)^c$
One N, one B	$\frac{2v}{u+v}\left[\left(\frac{d}{d+1}\right)^c - \left(\frac{d}{d+2}\right)^c\right]$
Two B's	$\frac{v(v+1)}{(u+v)(u+v+1)}\left[\left(\frac{d}{d+2}\right)^c - 2\left(\frac{d}{d+1}\right)^c + 1\right]$

Note: A = prefer A, B = prefer B, N = no preference.

datasets in Table 20.6, there will be 12 d.f.; the total goodness-of-fit statistic X^2 turned out to be 18.2, not statistically significant according to the usual criterion and thus tending to support this model.[5]

Some extensions to this model are discussed by Hutchinson (1979a). Horsnell (1969, 1977) and the discussants of the paper by Horsnell (1969) had suggested the need for models that accommodate (i) experiments in which a distinction between discrimination and preference is made, (ii) the finding (sometimes attributed to fatigue) of an increased proportion of "no preference" in the second trial, and (iii) the possible effect of order of presentation on preference. Hutchinson went on to re-examine Horsnell's (1969) data in the light of an extension of the above model incorporating parameters to reflect the first two points; a way of handling the third point was merely outlined.

A recent paper on this topic is by Buchanan (1988). In Buchanan's model, the subject always chooses his or her preferred product if he or she does discriminate; but discrimination does not always occur.

[5]In fact, the fit of the model to the data could be rejected ($P < 0.05$) for one dataset, no. 12. It is instructive to look at this a little more closely, as it points up a qualitative difference between Horsnell's model and that described here.

Consider those people who respond "no preference" (N) in one trial but give a preference (A or B) in the other. Horsnell's model says these people must be guessers, and therefore that there should be equal numbers preferring A in one trial as preferring B in one trial. The model proposed here, on the other hand, says that if the population of subjects on average prefers A, this will manifest itself both in the response pattern (A, A) being more common than (B, B), and in the response pattern (N, A) being more common than (N, B).

For dataset 12, (A, A) is less common than (B, B), but (N, A) is more common than (N, B). This leads to the fit of the present model being poorer than that of Horsnell's. If it were to be generally found in experiments of this sort that the numbers of (N, A) and (N, B) were equal, then the stochastic model would have to be rejected, but we are reluctant to do so on the basis of this one dataset, in view of the analogy with areas of psychology where partial knowledge has been shown to be a weakened form of full knowledge, rather than qualitatively different. The difference between the numbers of (N, A) and (N, B) is of the same sign as the difference between the numbers of (A, A) and (B, B) for 8 of the 12 datasets.

20.6 Compounding of an F-G-M Distribution, and Application to Trace Metal Concentrations

20.6.1 Derivation of the Distribution

Suppose, now, we start with the Farlie-Gumbel-Morgenstern distribution that has exponential marginals, i.e.,

$$\bar{H} = e^{-\theta x}e^{-\theta y}[1 + \alpha(1 - e^{-\theta x})(1 - e^{-\theta y})], \qquad (20.10)$$

instead of with independent exponentials. As before, let θ have a gamma distribution. Then the new distribution is

$$\bar{H} = [1 + \alpha](1 + bx + by)^{-c} - \alpha(1 + 2bx + by)^{-c} - \alpha(1 + bx + 2by)^{-c} + \alpha(1 + 2bx + 2by)^{-c}. \qquad (20.11)$$

This idea is due to Cook and Johnson (1986). The resulting distribution includes the F-G-M as a special case ($c \to \infty$) and also the Pareto ($\alpha = 0$). As with their 1981 paper, Cook and Johnson were more interested in the cases where the marginals had been transformed to be uniform or normal than in Pareto marginals. See also Johnson (1987, Section 9.2).

We could (as in Section 20.3.3) take the scale parameters of X and Y to be correlated rather than identical, and obtain a distribution like (20.11) but with extra terms kb^2xy, $2kb^2xy$, $2kb^2xy$, and $4kb^2xy$ within the four sets of round brackets, respectively—see Bagchi and Samanta (1984), who actually started with the Weibull-marginals version of (20.10); they also considered the case of the two θ's having an F-G-M distribution with exponential marginals.

20.6.2 Description of Some Properties

Cook and Johnson (1986) have given some properties of distribution (20.11) after transformation of the marginals to be normal:

- A table of correlation in terms of α and c (β and α in their notation). Correlation increases with α and decreases with c, as the following brief extract from their table shows:

	$\alpha = -1$	$\alpha = 0$	$\alpha = 1$
$c = 0$	1.00	1.00	1.00
$c = 1$	.35	.50	.65
$c = \infty$	−.32	.00	.32

- Illustrations of contours of probability density, showing how the outer contours are pointed at one end and blunt at the other, reflecting changing conditional variance. And if α is negative, it is possible for the inner contours to have a negative orientation and the outer contours a positive one.

20.6.3 Application

The distribution was fitted to the concentrations of metals in water samples collected in part of Colorado. The context was exploration for uranium: the presence or absence of certain trace metals can assist the assessment of whether the area is worth investigating further.

Scatterplots (655 points) of three pairs of metal concentrations were shown by Cook and Johnson (1986): logarithms of the concentrations were used, and normal distributions were then found to fit the marginals satisfactorily. Overlaid on the scatterplots were contours of four fitted distributions: the bivariate normal, (20.11) with normal marginals, the F-G-M with normal marginals, and Pareto with normal marginals. (The third and fourth are special cases of the second.)

Cook and Johnson also summarised the quality of fit by means of the log-likelihood. The F-G-M distribution could be rejected easily, as it can only accommodate weak dependence. As to the other three, the log-likelihoods were as follows:

	Dataset 1	Dataset 2	Dataset 3
Bivariate normal	−99 ($r = .60$)	645 ($r = .48$)	277 ($r = .36$)
Pareto	−64 ($c = 0.53$)	628 ($c = 1.48$)	293 ($c = 1.50$)
Equation (20.11)	−58 ($c = 0.79$, $\alpha = .93$)	651 ($c = 3.85$, $\alpha = .98$)	295 ($c = 1.92$, $\alpha = .37$)

(The estimates of the association parameters are shown in brackets.)

Comparing the two models that have a single association parameter, the normal is better in one case and the Pareto better in two. Distribution (20.11) is necessarily better than the Pareto, since it subsumes it, and is also better than the normal in each case.

20.7 Generalising the Bivariate Pareto Distribution

20.7.1 Introduction

The results herein were given originally in Section 6 of Hutchinson (1979b), and developed in more detail in Hutchinson (1981b). We consider the scheme

$$\Pr(X < x, Y < y) = \int_0^\infty F(\theta x)F(\theta y)g(\theta)\,d\theta, \tag{20.12}$$

or, more generally,

$$\Pr(X < x, Y < y) = \int_0^\infty \int_0^\infty F(\theta_X x)F(\theta_Y y)g(\theta_X, \theta_Y)\,d\theta_X\,d\theta_Y, \tag{20.13}$$

where F is a d.f. and $g(.)$ (or $g(.,.)$) is a p.d.f. We shall obtain expressions for distributions that we will call GG1 and GGD. These three-character codes, and others, will now be explained:

- The first character refers to the distribution of X and Y for given θ. The four alternatives, exponential, Weibull, gamma, and Stacy, are denoted E, W, G, and S, respectively.
- The second character refers to the distribution of θ—exponential or gamma (E, G).
- The third character refers to how θ for X is related to θ for Y. The figure 1 is used when they are identical (1 for unit correlation), and the letter D (for dependent) for the general case when they are positively correlated but not identical. The form that the dependence will take will be that of Kibble's bivariate gamma distribution, as in Section 20.3.3.

The following table shows how these codes correspond to distributions that we have met already.

Name	Code	Equation	F	g
Pareto	EG1	(20.4)	Exponential*	Gamma, $\theta_X \equiv \theta_Y$
Takahasi-Burr	WG1	Section 5.8.12	Weibull	Gamma, $\theta_X \equiv \theta_Y$
Durling-Pareto	EGD	(20.8)	Exponential†	Kibble's bivariate gamma
	EED	Section 5.8.13.4	Exponential†	Downton's bivariate exponential
Durling-Burr	WGD	Section 5.8.13.3	Weibull†	Kibble's bivariate gamma
Inverted beta (F)	GG1	(20.16)	Gamma	Gamma, $\theta_X \equiv \theta_Y$
	GGD	(20.19)	Gamma	Kibble's bivariate gamma
	SGD	Section 20.7.3	Stacy	Kibble's bivariate gamma

* Reminder: for, given θ, X and Y having the F-G-M form of dependence, see the first paragraph of Section 20.6.1.

† Reminder: for, given the respective θ's, X and Y having the F-G-M form of dependence, see the second paragraph of Section 20.6.1.

20.7.2 Distribution GG1

In equation (20.1), the distribution of X for a given value of θ has taken an elementary mathematical form. Now let us say it is a gamma distribution. In other words, we seek a bivariate analogue of the construction of the univariate beta distribution by Dubey (1970). Therefore, with $\Gamma_x(a)$ being the incomplete gamma function, as in footnote 2 of Chapter 2, H will be $\Gamma_{\theta x}(a)\Gamma_{\theta y}(a)[\Gamma(a)]^{-2}$ averaged over θ:

$$H = \int_0^\infty \frac{\Gamma_{\theta x}(a)\Gamma_{\theta y}(a)\theta^{c-1}\exp(-\theta/b)}{b^c\Gamma(c)[\Gamma(a)]^2}\,d\theta. \qquad (20.14)$$

We can write this in terms of the Whittaker function M as

$$H = \frac{(xy)^{(a-1)/2}}{a^2 b^c\Gamma(c)[\Gamma(a)]^2}\int_0^\infty \theta^{a+c-2}\exp\left[-\left(\tfrac{1}{b}+\tfrac{x}{2}+\tfrac{y}{2}\right)\theta\right] M_{\frac{a-1}{2},\frac{a}{2}}(x\theta)\,M_{\frac{a-1}{2},\frac{a}{2}}(y\theta)\,d\theta, \qquad (20.15)$$

and, using equation (14) on page 216 of Erdelyi (1954a), we find the result is

$$H = \frac{\Gamma(2a+c)(bx)^a(by)^a}{a^2\Gamma(c)[\Gamma(a)]^2(1+bx+by)^{2a+c}}F_2\left(2a+c;1,1;a+1,a+1;\frac{bx}{1+bx+by},\frac{by}{1+bx+by}\right). \qquad (20.16)$$

F_2 is one of Appell's hypergeometric functions of two variables[6]. The p.d.f. corresponding to (20.16) is

$$\frac{x^{a-1}y^{a-1}}{b^c\Gamma(c)[\Gamma(a)]^2}\int_0^\infty \theta^{2a+c-1}\exp[-(x+y+b^{-1})\theta]\,d\theta = \frac{b^2\Gamma(2a+c)}{\Gamma(c)[\Gamma(a)]^2}\frac{(bx)^{a-1}(by)^{a-1}}{(1+bx+by)^{2a+c}}. \qquad (20.17)$$

This density is not new—as Lee (1981) points out, it is effectively the bivariate F-distribution (our Section 6.3), also known as the bivariate inverted (or second kind of) beta (or Dirichlet) distribution. Amos and Bulgren (1972) seem to have been aware that the d.f. could be expressed in such a way as (20.16). See also Arnold (1983, p. 251).

20.7.3 Distributions GGD and SGD

For distribution GGD, we require the integral

$$\int_0^\infty\int_0^\infty \Gamma_{\theta_X x}(a)\Gamma_{\theta_Y y}(a)g(\theta_X,\theta_Y)\,d\theta_X\,d\theta_Y, \qquad (20.18)$$

where $g(\theta_X,\theta_Y)$ is Kibble's bivariate gamma density, equation (20.6).[7] The modified Bessel function may be written as

$$I_\nu(u) = \sum_{m=0}^\infty \frac{(u/2)^{\nu+2m}}{m!\Gamma(\nu+m+1)}.$$

Expression (20.18) may then be evaluated by using this series substituted for the modified Bessel function, then interchanging the order of summation and integration, and applying equation (11) on page 215 of Erdelyi (1954a) and equation (15) on page 308 of Erdelyi (1954b). The result is:

$$H = \frac{[\Gamma(q+c)]^2k^c(bkx)^a(bky)^a}{a^2\Gamma(a)[\Gamma(c)]^2(1+bkx)^{a+c}(1+bky)^{a+c}}K\left(1,a,c;\frac{bkx}{1+bkx},\frac{bky}{1+bky},\frac{1-k}{(1+bkx)(1+bky)}\right). \qquad (20.19)$$

Here, K is a special case of Saran's triple hypergeometric function F_K (for which, see Exton, 1976, Section 2.10), and is defined by:

$$K(\alpha,\beta,\gamma;u,v,w) = \sum_{i=0}^\infty\sum_{j=0}^\infty\sum_{k=0}^\infty \frac{(\alpha,i)(\alpha,j)(\beta+\gamma,i+k)(\beta+\gamma,j+k)}{(\beta+1,i)(\beta+1,j)(\gamma,k)}\frac{u^iv^jw^k}{i!j!k!}.$$

[6] Its definition is:

$$F_2(\alpha;\beta,\beta';\gamma,\gamma';u,v) = \sum_{i=0}^\infty\sum_{j=0}^\infty \frac{(\alpha,i+j)(\beta,i)(\beta',j)}{(\gamma,i)(\gamma',j)}\frac{u^iv^j}{i!j!},$$

where (ν,n) is the Appell symbol, $\Gamma(\nu+n)/\Gamma(\nu)$, as in footnote 3 of Chapter 13.

[7] Because g has powers and exponentials of θ_X (and θ_Y) in it, equation (20.18) may be regarded as a bivariate Laplace or a bivariate Mellin transform. Steece (1977) has given a readable introductory account of how integral transforms are useful in statistics, notable for giving almost as much attention to bivariate transforms (Fourier and Mellin) as to univariate ones.

Applying a certain identity involving F_K, expression (20.19) may alternatively be written as proportional to

$$(bkx)^a (bky)^a K(a, a, c; -bkx, -bky, 1-k),$$

but this is really only valid for sufficiently small x and y, because of the limited region of convergence of the hypergeometric series.

Distribution SGD may readily be obtained by substituting x^α for x, and y^α for y, throughout.

The p.d.f. corresponding to (20.19) is:

$$h = \frac{[\Gamma(a+c)]^2 b^2 k^{c+2} (bkx)^{a-1} (bky)^{a-1}}{[\Gamma(a)]^2 [\Gamma(c)]^2 (1+bkx)^{a+c} (1+bky)^{a+c}} \, {}_2F_1\left(a+c, a+c; c; \frac{1-k}{(1+bkx)(1+bky)}\right). \tag{20.20}$$

Lee (1981) confirmed these results, and called this the "correlated bivariate inverted beta distribution"— but this name might wrongly lead one to suppose there is no correlation in the usual bivariate inverted beta distribution.

20.7.4 Properties

Lee (1981) found that the conditional moments are

$$E(Y^n | X = x) = \frac{(-1)^n (a, n)}{b^n (1-c, n)} \left(\frac{1-\rho+bx}{1+bx}\right)^{a+c} {}_2F_1\left(a+c, c-n; c; \tfrac{\rho}{1+bx}\right) \tag{20.21}$$

(where $\rho = 1 - k$ is the product moment correlation between θ_X and θ_Y), that the joint moments are

$$E(X^m Y^n) = \frac{(-1)^{m+n} (1-\rho)^{m+n} (a, m)(a, n)}{b^{m+n} (1-c, m)(1-c, n)} {}_2F_1(m, n; c; \rho), \tag{20.22}$$

and that the correlation is

$$\frac{a(c-2)}{a+c-1} [{}_2F_1(1, 1; c; \rho) - 1]. \tag{20.23}$$

20.7.5 An Alternative

It is easy to suggest that, instead of θ having a gamma distribution, its inverse θ^{-1} might have a gamma distribution. It might be possible to obtain explicit results with no more difficulty than above.

20.7.6 A Bayesian Application

It is fairly rare to find examples of compounding over more than one variable. We therefore mention the following example, even though it is from an area we are not paying much attention to, that of Bayesian statistics. Specifically, it concerns getting a posterior probability function from a prior. Papadopoulos (1973) considered the case of the Weibull failure model $\bar{F} = \exp(-x^\xi / \theta)$, and let the joint distribution of θ and ξ be of Farlie-Gumbel-Morgenstern type. Most of his results were in a form needing numerical integration.

Chapter 21

Joint Distributions of Injury Severity: A Bivariate Normal Model

21.1 Introduction

In crashes between two vehicles, one of the most important factors affecting the severity of injury sustained by their occupants is the velocity change which the vehicles undergo at impact. This, in turn, is determined by the configuration of the impact, the speeds of the vehicles, and their relative masses. Routinely-collected data specifies the accident configuration (head-on, front-into-rear, etc.) and the masses of the vehicle (as make and model are known[1]), but not their speeds (as this is too lengthy and difficult a process to be done routinely). It is possible, therefore, to obtain a table of the frequency of occurrence of all combinations of injury severities to the two drivers, and to stratify this table according to mass ratio[2] of the vehicles and according to accident configuration.

Table 21.1 is an example of this type of data. Tables for 17 ranges of mass ratio are shown there; further stratification has been made according to whether the speed limit was at most 40 m.p.h. (denoted "urban") or higher ("rural"). Note that the number of accidents in which both drivers were uninjured is not known, since such accidents are not recorded (unless another person was injured). Also, as has been explained in Section 20.3, in the case of collisions between vehicles of equal mass, the table has in effect been folded about the diagonal. When mass ratio is close to unity, most of the crashes are between two cars; when the two masses are very unequal, the crashes are between a car and a commercial vehicle. It is debatable just what ranges of mass ratio are appropriately analysed together (as various factors may be different in car-truck crashes from car-car crashes). Table 21.1 excludes impacts where the mass ratio was more extreme than 5:1.

The processing of British road accident data in order to obtain tables like Table 21.1 was described by Grime and Hutchinson (1979, 1982); their analyses were confined to the marginal totals of such tables. In the remainder of this Chapter, a model for such data will be set up that is of variables-in-common type; results obtained when it was applied to the data of Table 21.1 will be presented and discussed; and then another dataset, giving the joint distribution of injury severities to driver and front seat passenger in single-vehicle crashes, will be analysed by the same method.

21.2 A Variables-in-Common Approach

The essence of the idea is that there are a number of variables that together determine the degree of injury sustained by a casualty, but one of these is common to both the drivers in a two-vehicle crash: the relative velocity of their vehicles. Repeating equations (15.1), it may happen that two observed random variables

[1] For no good reason that we know of, it is actually rather rare to find either the vehicle masses given directly in routine road accident data, or the makes and models. But make and model was given in British data in the late 1960's/early 1970's, which is what is analysed here.

[2] Vehicle mass as such is not used as a stratifying variable. The validity of this omission is discussed below, at the end of Section 21.2.

X and Y are created from three other independent random variables X_1, X_2, and X_3, thus:

$$\left.\begin{array}{rcl} X & = & X_1 + X_3 \\ Y & = & X_2 + X_3 \end{array}\right\} \tag{21.1}$$

In the present context, X and Y are the two severities of injury, X_3 is a factor deriving from crash speed, and X_1 and X_2 are random variables influenced by how well the vehicle is designed, the driver's susceptibility to injury, and so on.

Such a model would be expected to lead to a positive correlation between the two injury severities, and indeed that can be seen in Table 21.1.

We now specify this idea in more detail. Consider head-on collisions between vehicles, one of which is k times as massive as the other. (We take $k \leq 1$.) If their relative velocity at impact was v, the velocity changes of the two vehicles may be found to be (on applying conservation of momentum and assuming the collision to be inelastic) $v/(1+k)$ and $kv/(1+k)$. As stated earlier, v is unknown but k is known.

Because it will prove mathematically tractable, assume that injury severity (X) has a log-linear relation with velocity change (Δ_v):

$$X = a + b\log(\Delta_v) + \varepsilon, \tag{21.2}$$

where ε is a normally distributed random variable with mean 0 and variance σ_ε^2. Factors such as vehicle interior design and driver injury tolerance are subsumed within ε.

The severities of injury to the two drivers are therefore given by

$$\left.\begin{array}{rcl} X & = & a + b\log\left(\frac{v}{1+k}\right) + \varepsilon_1 \\ Y & = & a + b\log\left(\frac{kv}{1+k}\right) + \varepsilon_2 \end{array}\right\} \tag{21.3}$$

so that

$$\left.\begin{array}{rcl} X & = & a + b\log\left(\frac{1}{1+k}\right) + b\log(v) + \varepsilon_1 \\ Y & = & a + b\log\left(\frac{k}{1+k}\right) + b\log(v) + \varepsilon_2 \end{array}\right\} \tag{21.4}$$

Again for no better reason than it leads to a convenient result, assume impact speed v has a lognormal distribution. Taking $\log(v)$ to have mean μ and variance σ^2, X and Y will jointly have a bivariate normal distribution with the following properties:

$$E(X) = a + b\log\left(\frac{1}{1+k}\right) + b\mu \tag{21.5}$$

$$E(Y) = a + b\log\left(\frac{k}{1+k}\right) + b\mu \tag{21.6}$$

$$\mathrm{var}(X) = b^2\sigma^2 + \sigma_\varepsilon^2 \tag{21.7}$$

$$\mathrm{var}(Y) = b^2\sigma^2 + \sigma_\varepsilon^2 \tag{21.8}$$

$$\mathrm{corr}(X,Y) = \frac{b^2\sigma^2}{b^2\sigma^2 + \sigma_\varepsilon^2}. \tag{21.9}$$

Therefore, if an injury is called "severe" if X is greater than a threshold t and "minor" if it is less than t, the probability of both drivers being severely injured is

$$L\left(\frac{t - a - b\mu - b\log\left(\frac{1}{1+k}\right)}{\sqrt{b^2\sigma^2 + \sigma_\varepsilon^2}}, \frac{t - a - b\mu - b\log\left(\frac{k}{1+k}\right)}{\sqrt{b^2\sigma^2 + \sigma_\varepsilon^2}}; \frac{b^2\sigma^2}{b^2\sigma^2 + \sigma_\varepsilon^2}\right), \tag{21.10}$$

where the symbol $L(h,k;\rho)$ is, as usual, the upper right volume of the standardised bivariate normal distribution. In Table 21.1, there are four levels of injury severity, so there will be three thresholds—t_1, t_2, and t_3—to be estimated.

Some of the parameters in the model are confounded with others. In the presentation of the results below, the following conventions will be adopted: $a = 0$, $t_3 = 0$, $b^2\sigma^2 + \sigma_\varepsilon^2 = 1$. Further, $b^2\sigma^2/(b^2\sigma^2 + \sigma_\varepsilon^2)$ will be referred to as the correlation ρ. Thus there will be five parameters—two thresholds, μ, b, and σ.

In this model, it has been assumed that impact speed has the same distribution for all masses of vehicles, and that the relation between injury severity and velocity change is also independent of vehicle mass. British data examined by Grime and Hutchinson (1979, 1982) shows that mass of vehicle does not affect injury severity in single-vehicle accidents, nor in two-vehicle accidents in which the two vehicles are of almost equal weight. The simplest way to interpret this is to see it as confirmation of the aforementioned assumptions.[3] The contrary finding in North Carolina, that vehicle mass does affect injury severity, the greater injury being in smaller cars, should be mentioned (Campbell and Reinfurt, 1973; O'Neill et al, 1974); the reason for the divergence between British and U.S. data in this respect is an important unsolved question.

21.3 Results

21.3.1 First Model

Little time need be spent on the results obtained when the above model was fitted to the data. For one thing, the values of the goodness-of-fit statistic (maybe we ought to call it the badness-of-fit) were large. Perhaps more importantly, there was a particular pattern to the difference between observed and expected values: the numbers of cases in which both drivers were seriously injured or both were slightly injured were greater in the data than predicted, and the numbers of cases in which one driver was seriously injured and one slightly injured were less.

21.3.2 Second Model

We suggest that the reason for why and how the first model failed is variability in the positioning of the thresholds separating the four categories of injury. Although criteria are laid down for the categories, there is nevertheless some latitude in their interpretation.[4] Accordingly, let it be assumed that the thresholds are the same for both drivers in the same accident, because they are classified by the same policeman at the same time, but are different for different accidents.

Specifically, it is assumed that there is no variability in the position of the fatality threshold, t_1, but that thresholds t_2 and t_3 jointly have a bivariate normal distribution[5] with standard deviations σ_2 and σ_3, and correlation ρ_{23}. Mantel (1966, pp. 92–93) had this idea in a setting that will be more familiar to many readers, that of thresholds for response to drugs or poisons; see Section 22.3.1 for this.

The effect of this is that there appears to be a greater correlation in injury severities when t_2 or t_3 is used as a dividing line than when t_1 is.

The estimates of the parameters were as follows:

t_1	:	1.56	μ (rural data)	:	−0.53
t_2	:	0.68	μ (urban data)	:	−1.15
t_3	:	0.00 (defined to be so)	σ (rural)	:	1.13
σ_2	:	0.48	σ (urban)	:	1.16
σ_3	:	0.48	b	:	0.71
ρ_{23}	:	0.71	Hence σ_ε (rural)	:	0.45
			and σ_ε (urban)	:	0.42

Clearly, we could constrain σ to be the same in urban as in rural areas, and σ_2 to be equal to σ_3, without noticeable deterioration of the fit.

[3] The possibility should be admitted that bigger cars, for instance, might be involved in higher-speed crashes, with injury at a given speed being lower in them than in smaller vehicles by just the right amount that the two effects cancel out.

[4] See Sections 8.8–8.10 of Hutchinson (1987) for discussion of this.

[5] This is not an ideal assumption, as it permits t_3 to exceed t_2, and either to exceed t_1. We assume this is so rare as not to be a problem.

Table 21.1: The joint distribution of injury severities to the two drivers of vehicles involved in head-on collisions in Great Britain, 1969–72.

The four columns of each table refer to the severity of injury to the driver of the heavier vehicle: fatal, serious, slight, and none (from left to right). The four rows refer to the severity of injury to the driver of the lighter vehicle: fatal, serious, slight, and none (from top to bottom).

Observed-rural				Predicted-rural				Observed-urban				Predicted-urban			
Equal masses															
5	25	12	16	6	24	15	10	0	8	7	9	2	9	6	4
	134	162	202		121	153	193		75	87	133		71	99	135
		177	532			200	543			145	504			158	485
			?				2835				?				4488
Mass ratio = .95 to .99															
6	16	3	2	7	13	8	6	1	2	1	4	2	5	3	2
16	176	65	100	13	131	84	108	5	94	59	58	5	81	57	79
4	80	218	290	8	81	216	300	1	52	209	270	3	55	180	284
2	88	301	?	5	101	286	3064	1	70	276	?	2	74	269	5114
Mass ratio = .90 to .94															
6	20	7	5	6	12	8	6	1	10	1	1	2	5	3	2
18	134	78	106	11	117	78	103	5	84	61	71	5	77	57	81
3	63	205	305	6	70	193	282	2	43	193	306	3	50	170	284
1	78	195	?	4	84	244	2742	2	51	218	?	2	65	242	4861
Mass ratio = .85 to .89															
6	14	6	6	7	15	10	8	1	5	2	4	2	6	4	3
7	175	94	126	12	137	95	129	4	96	75	95	5	89	68	101
6	65	225	365	7	78	225	347	1	43	223	372	3	55	196	347
6	79	256	?	5	92	271	3215	0	64	231	?	2	70	264	5629
Mass ratio = .80 to .84															
4	18	6	3	5	13	8	7	0	10	6	4	2	6	4	3
9	109	93	118	9	107	78	110	0	73	73	94	4	78	63	97
0	39	182	356	5	59	177	288	0	28	217	330	2	47	174	327
2	50	148	?	3	67	202	2543	1	39	212	?	1	57	220	5023
Mass ratio = .75 to .79															
5	19	5	9	6	14	9	8	2	6	4	1	2	6	4	3
8	123	84	122	9	110	83	123	2	62	63	84	3	70	59	95
2	51	192	342	5	58	182	316	2	35	168	331	2	39	155	312
3	55	165	?	3	64	197	2646	2	45	174	?	1	46	184	4521
Mass ratio = .70 to .74															
6	6	8	4	4	11	8	7	1	7	5	3	2	5	4	3
9	109	92	103	6	86	68	104	2	53	68	102	3	63	56	94
1	48	150	211	3	43	142	263	0	36	145	295	2	34	140	303
3	41	147	?	2	46	144	2090	3	40	144	?	1	38	156	4152

Table 21.1: (continued).

Observed-rural				Predicted-rural				Observed-urban				Predicted-urban			
Mass ratio = .65 to .69															
2	19	13	19	7	19	14	12	1	10	21	6	2	8	6	5
6	123	104	176	9	135	112	181	3	80	102	170	4	88	83	147
3	64	199	449	5	64	222	443	4	42	170	462	2	44	195	459
10	80	238	?	3	66	213	3343	2	53	170	?	1	49	203	5925
Mass ratio = .60 to .64															
2	10	5	3	4	11	8	8	0	5	7	4	1	5	4	4
4	58	68	129	5	73	64	109	1	45	46	99	2	50	50	94
2	31	116	311	2	33	121	260	0	14	101	331	1	24	111	286
0	21	80	?	1	33	109	1859	1	15	97	?	1	25	108	3482
Mass ratio = .55 to .59															
0	12	8	6	2	6	5	5	1	2	2	4	1	3	2	2
0	32	37	74	2	37	34	62	2	23	28	48	1	24	26	51
1	5	35	176	1	15	61	143	0	7	48	165	0	11	53	151
0	21	32	?	1	15	51	967	0	16	37	?	0	11	48	1729
Mass ratio = .50 to .54															
2	11	9	11	2	8	7	7	1	5	3	2	1	3	3	3
3	40	45	76	3	48	48	91	1	30	30	72	1	29	33	72
1	17	79	200	1	19	79	206	0	16	62	183	0	12	65	206
0	28	75	?	1	17	62	1308	0	15	74	?	0	12	54	2210
Mass ratio = .45 to .49															
0	13	11	3	2	8	7	8	0	8	6	1	1	3	3	3
2	38	49	90	2	46	49	100	0	21	29	68	1	26	31	73
1	19	54	215	1	17	76	219	1	10	48	196	0	10	57	204
1	29	79	?	0	15	54	1313	0	11	64	?	0	9	43	2049
Mass ratio = .40 to .44															
0	7	6	6	1	4	4	5	0	5	4	2	0	2	2	2
0	21	17	38	1	21	24	54	0	15	21	38	0	11	15	39
1	13	26	119	0	7	35	114	0	3	17	96	0	4	25	105
0	11	33	?	0	6	23	642	0	2	23	?	0	3	17	987
Mass ratio = .35 to .39															
0	7	8	5	1	3	3	4	0	1	3	5	0	2	2	2
0	15	16	40	1	16	20	50	0	9	12	42	0	10	15	44
0	5	29	107	0	5	27	103	1	4	23	98	0	3	23	113
0	3	19	?	0	4	16	542	1	6	27	?	0	3	14	997
Mass ratio = .30 to .34															
0	9	7	8	1	4	4	6	0	0	2	1	0	1	1	2
0	9	20	64	1	16	22	62	0	4	8	42	0	7	12	39
0	6	18	114	0	4	27	121	0	1	18	94	0	2	16	97
1	7	21	?	0	3	14	598	0	2	18	?	0	2	9	801

Table 21.1: (continued).

Observed-rural				Predicted-rural				Observed-urban				Predicted-urban			
						Mass ratio = .25 to .29									
0	6	7	8	1	4	5	8	0	0	0	7	0	1	2	3
0	11	13	87	0	16	25	79	0	4	12	54	0	8	14	55
1	1	23	146	0	4	26	149	0	4	15	132	0	2	17	131
0	7	22	?	0	3	12	687	0	2	14	?	0	1	8	1005
						Mass ratio = .20 to .24									
0	3	6	12	0	4	5	9	0	2	4	6	0	1	2	3
0	2	13	89	0	12	21	84	1	3	5	50	0	5	11	53
0	2	16	158	0	2	20	150	0	2	11	108	0	1	12	119
0	2	15	?	0	2	8	645	0	4	16	?	0	1	5	850

The maximum-likelihood statistic $G^2 = 2\sum(O)\log(O/E)$ was the goodness-of-fit measure used in this case, and it turned out to be 928.9. There are 32 data tables with 14 free probabilities each, and 2 with 8 free probabilities, and we have estimated 10 parameters, so it would be natural to compare the goodness-of-fit with a chi-squared distribution having 454 degrees of freedom ($= 32 \times 14 + 2 \times 8 - 10$), in which case we would conclude that the model is a very poor fit to the data.

Finally, we note that although the model produces an estimate of the number of crashes in which both drivers are uninjured, we suspect that this estimate might be sensitive to the particular distributional assumptions made, as it is in effect obtained by extrapolation beyond the observable region.

21.4 Comments on the Fit

A conventional chi-squared test would reject the second model; however, a number of lines of argument suggest that this basis of assessing the model is inadequate.

1. Inspection of the predicted frequencies (Table 21.1) in comparison with the data shows reasonably close agreement between them.

2. When the data was split into two subsets, similar parameter estimates were obtained in both—they typically differed by .05. (The two subsets consisted of data tables for alternate ranges of mass ratios—i.e., 1.00, .90 to .94, .80 to .84, etc., in one subset, and 1.00, .95 to .99, .85 to .89, etc., in the other.)

3. Since there is evidence about the effect of velocity change on injury severity both by comparison of the two drivers at a given mass ratio and between the different mass ratio ranges, it is possible to fit a model in which μ varies with mass ratio. When a model was tried in which μ was made a quadratic function of k both in rural and in urban areas, the reduction of G^2 was only 6.9, for the addition of 4 extra parameters.

4. Suppose that σ, σ_2, σ_3, and ρ_{23} remain constants, and thresholds t_{11}, t_{12}, t_{13} for the driver of the lighter vehicle and t_{21}, t_{22}, t_{23} for the driver of the heavier vehicle are fitted to each dataset (now changing the convention to $\mu = 0$ rather than $t_3 = 0$). Then our model makes predictions about how the estimates should depend on mass ratio. These predictions are detailed on page 219 of Hutchinson (1983a), and graphical demonstrations of their degree of success are given on pp. 220–221.[6]

5. There are a very large number of observations in our data (25837 in 498 cells). With such a large number, it is possible that a model that is realistic to a useful degree might deviate significantly from

[6]The data analysed at that time was actually grouped into ranges of .10 of mass ratio, rather than the .05 of Table 21.1.

the data. (In addition, Hutchinson and Mayne (1977) and Nicholson (1985) have shown that there is rather more variability in road accident figures than would be expected on the Poisson assumption.)

6. Support for the previous point is given by predicting the proportion in each cell of a given range of mass ratios by the average of the corresponding proportions at the mass ratio ranges either side. This gives a value of G^2 about 1.7 times as large as that expected.

7. There are lots of reasons why we would expect any model to differ significantly from the data. These fall into three main groups: data errors, model structure inadequacies, model detail inadequacies. Consequently, we are not surprised or distressed when it does fail—the important question is whether it strikes the right balance between simplicity and exactness to be useful.

21.5 Practical Conclusions

The substantive conclusions are dealt with more extensively in Hutchinson (1983a). The principal ones are as follows.

1. The relationship between velocity change and injury severity. We must appreciate the limitations we are working under. First, our dependent variable (injury severity) is inherently unquantifiable. Second, there is no information in our data about the absolute values of our independent variable (velocity change). Given these limitations, our measure of the strength of dependence, b, was estimated to be 0.71.

2. The magnitudes of the variabilities of the injury thresholds. These were found to be about the same as the random variability in injury severity that occurs at any given velocity change.

3. The average impact speed in urban areas was estimated to be 56% of that in rural areas, this being obtained from the respective values of μ and σ.

4. The estimated numbers of fatal, serious, and slight injuries per 100 collisions within each range of mass ratios. For mass ratios greater than about 0.6, i.e., for predominantly car-car collisions, these are uninfluenced by mass ratio; but when the vehicles are more unequal in mass (i.e., predominantly collisions between cars and commercial vehicles), the decrease in fatalities in the heavier vehicle does not compensate for the increase in the lighter vehicle, and the number of fatalities per 100 collisions is higher.

21.6 Discussion of the Realism of the Model

Some factors that have been omitted from the model should be mentioned.

- Variables such as driver age (as a proxy for susceptibility to injury), the interior design of the vehicle, and the wearing of a seat belt are not included in the analysis. Although they are known to affect severity of injury, they appear in the model only through their contributions to the random variation of injury severity that is present at any velocity change.

- Not all injuries will result simply from the driver hitting the interior of the vehicle with relative velocity equal to the vehicle's velocity change. A proportion (probably small) of the injuries will arise from intrusion of the vehicle structure into the passenger compartment, or from impacts subsequent to the initial collision between the two vehicles. And in a (small) proportion of cases, a seat belt will have been worn.

- If one investigates accidents in detail, one finds that all sorts of peculiar things can happen, which are inadequately represented in the standard categories available on the reporting form. Although all accidents included in the present data are classified as head-on, there undoubtedly are included a considerable number in which the accident events and mechanisms of injury are far from what is usual.

- Another consideration that has been omitted is that of incompatibility between vehicle structures, which might lead to, for instance, a greater degree of intrusion into the passenger compartment when some combinations of vehicle types are in collision.

- From comparing hospital and police records, it is known that many accidents in which only slight injury occurs, and a substantial proportion of serious injury accidents, are not reported to the police and hence are omitted from the data. This factor is included in our analysis only by the variabilities of the thresholds.

Some other reasons for discrepancies between our model and the data include:

- Data errors, such as those resulting from forcing an unusual incident into the categories used on the standard reporting form, transcription errors, errors in the estimation of vehicle masses, and the exclusion of cases for which some data (e.g., accident type, or vehicle mass) was unknown.

- The particular assumptions made in Section 21.2 were chosen for the convenience of the model they give rise to—there is no reason to suppose that when injury severity is defined by its mean being linearly related to $\log(\Delta_v)$, then the variability is normal in shape and has a constant variance, or that $\log(v)$ has a normal distribution, or that t_2 and t_3 jointly have a bivariate normal distribution. And injuries that are qualitatively different are inherently difficult to compare on a single dimension of severity.

The way we handled variability in injury classification thresholds was arguably a rather optimistic one, in the sense that comparison was always being made between two quantities—a severity and a threshold—on the same scale; there was no real fog creeping in, as there would be if some classifications were being made on the basis of guesswork, for instance. A possible alternative way of handling the excess of observations in the (serious, serious) and (slight, slight) cells would be with a distribution having a singularity along $Y = X$. This would be a suitable choice if we thought that in some proportion of crashes, there was no perception of the injury of individual casualties, merely a perception of the severity of the accident as a unit, with both drivers being assigned the severity classification of the accident.

21.7 Two Ideas for the Future

21.7.1 Semiparametrics?

We have seen above that there are some quantities that can be estimated, and others which cannot be, with data of the given form. In our model, too, we see some features we are firm about (what we might call our core assumptions), and others which are made solely for mathematical tractability (peripheral assumptions). For instance, trivariate reduction is essential to the model, but the normality of the distributions is merely a convenience. Is there any statistical tool that would help direct our attention at the conclusions that are feasible from a given body of data, that would express these results independently of the peripheral assumptions, perhaps generating upper and lower limits within which the results of any model embodying the core assumptions must fall, whatever peripheral assumptions are made?

We suggest that the rapidly-developing area of "semiparametric" inference may provide such a tool. The idea behind this is that some aspects of a model should be parameterised, and others be nonparametric. Wellner (1985) refers to "this very large and rich class of models which spans the middle ground between parametric and nonparametric models". His paper includes an effort—one that is admitted to be a provisional one—to classify semiparametric methods and show how this concept helps unify a number of diverse strands of statistical work. See also Oakes (1988).

We are confident that the qualitative features of the model we have fitted to Table 21.1 are ones that ought to be present. We are not altogether confident of the detailed development, and certainly we know little of the robustness of the results with respect to the peripheral assumptions. It would be of great interest to see a semiparametric analysis of Table 21.1.

The core of the model is that the two injury severities are given by $a\left(\frac{v}{1+k}\right) + \varepsilon_1$ and $a\left(\frac{kv}{1+k}\right) + \varepsilon_2$, where v, ε_1, and ε_2 are independent random variables, and $a(.)$ is some increasing function. If modified as

Table 21.2: Severities of injury to driver and front seat passenger in single-vehicle crashes. The format is similar to that of Table 21.1, but with the columns referring to the driver's injury, and the rows to the passenger's.

Observed	- Rural	nonoverturning	-	Predicted			
61	156	52	43	50	167	82	52
114	2133	1317	1315	106	2048	1380	1250
32	923	3657	2918	39	912	3768	2866
441	5847	9203	?	456	5802	9234	28588

Observed	- Rural	overturning	-	Predicted			
33	119	58	41	35	105	52	35
114	1549	1071	1004	107	1696	1032	961
71	970	3763	2277	53	1044	3517	2435
276	4030	7547	?	264	4066	7522	27284

Observed	- Urban	nonoverturning	-	Predicted			
50	152	79	43	71	167	70	41
105	2052	1423	1209	103	1910	1382	1356
37	911	4562	4867	33	885	4790	4681
390	6483	16189	?	399	6457	16206	118011

Observed	- Urban	overturning	-	Predicted			
9	23	21	9	10	21	9	5
28	291	217	216	22	341	217	217
10	200	1057	781	9	220	972	824
43	813	2252	?	44	810	2254	19356

suggested in Section 21.3.2 to allow for between-accident variability of thresholds, it would say that these severities are compared with thresholds t_1, t_2, t_3, which are such that t_1 is a constant but t_2 and t_3 have some joint distribution over the support $t_3 < t_2 < t_1$. The peripheral assumptions are the choice of $a(.)$, and the various distributions chosen for v, the ε's, and (t_2, t_3).

21.7.2 Equations Adapted from Econometrics?

There is a body of work within econometrics that deals with sets of equations like the following:

$$
\begin{aligned}
x &= \boldsymbol{\beta}_1 \mathbf{z} + \varepsilon_1 \\
y &= \boldsymbol{\beta}_2 \mathbf{z} + \varepsilon_2
\end{aligned}
$$

where $\mathbf{z}$ is a vector of known regressor variables, the $\boldsymbol{\beta}$'s are vectors of parameters to be estimated, and the ε's are error terms that are usually taken to have a bivariate normal distribution.

Very often, the dependent variables are not directly observed:

- Perhaps x is observed only if $y > 0$. (This is referred to as sample selection. See Section 19.9.5.)
- Perhaps only $\min(x, y)$ is observed. (This occurs in disequilibrium models, of supply and demand for example. See Section 19.9.5.)
- Perhaps the dependent variables are condensed into binary variables. (See Section 22.4.2.)

There are two reasons why we feel injury severity data might be usefully handled with equations like the above:

- One might accept that injury severity is mostly a continuous variate, grouped only for convenience into a limited number of categories, but one might nevertheless want to distinguish death and/or no injury as qualitatively different states, with different equations determining their probabilities of occurrence.

- It is known there are many defects in mass accident data, routinely collected as it is.[7] One is that many accidents do not get reported to the police, and hence never get included in a databank. The likelihood of this happening varies with the severity of the accident, and other factors. Hence one might want a sample selection equation, as well as one for injury severity.

(These points apply to injury severity studies generally, and are not restricted to joint distributions as analysed in this Chapter.)

21.8 Appendix. Injury Severities of Two Occupants in Single-Vehicle Crashes

21.8.1 The Data

In this Appendix, we turn to the severities of injury to driver and front seat passenger in single-vehicle accidents. The data is given in Table 21.2; as is Table 21.1, it is from Great Britain, 1969–72.

An additional difficulty with this dataset is that it was not possible to reliably identify cases in which there was no front seat passenger in the vehicle. Consequently, the "no injury to passenger" category includes both cases where a passenger was present but uninjured and those where there was no passenger.

21.8.2 The Statistical Model

Mean injury severity can be expected to increase with speed of the crash. Let the relation be $a + b\zeta(v)$ for the vehicle driver and $c + d\zeta(v)$ for the front seat passenger. Assume that the statistical scatter about this is normal in shape, with constant variance, σ_ε^2. Further assume that the distribution of v is such that $\zeta(v)$ has a normal distribution with mean μ and variance σ^2. (Unlike Section 21.2, it is not necessary to specify that ζ is the logarithmic function.) As before, a bivariate normal distribution of injury severities is obtained. To take account of the unknown proportion of cases in which there actually was a passenger present, an extra parameter p is introduced.

21.8.3 Results

See Hutchinson (1986) for full details. In order to obtain a satisfactory fit of the model to the data, it was found necessary to allow $\zeta(v)$ to have a different mean when no passenger is present. The corresponding predicted numbers of observations are included in Table 21.2. Total goodness-of-fit $G^2 = 20.7$, which with 16 degrees of freedom is not unusually high. The model used permitted the thresholds and their variabilities to be different in the four types of accidents; when they were constrained to be the same, G^2 increased to 186.4. (That the fit is so much worse is disquieting, but Hutchinson (1986, paragraph (iii) of Section 7) was able to suggest possible explanations.)

Some parameters are equivalent in this model to those used for the two-vehicle crash data. These are t_1/t_2, σ_2/t_2, σ_3/t_2, and ρ_{23}. There turns out to be a gratifying degree of similarity in the estimated values:

	Single-vehicle	Two-vehicle
t_1/t_2	2.37	2.31
σ_2/t_2	0.65	0.71
σ_3/t_2	0.88	0.71
ρ_{23}	0.60	0.71

[7] See Chapter 4 of Hutchinson (1987).

Chapter 22

Bivariate Generalisations of Probit Analysis

22.1 Outline of Probit Analysis

When the proportion of organisms giving some particular response (such as death) following exposure to a certain concentration of stimulant is related to this concentration, probit analysis is a technique commonly employed. It is often found that the shape of the function connecting the logarithm of concentration and probability of response is that of the cumulative normal distribution.[1] An interpretation of this is that each organism has a threshold for response, and this threshold has a lognormal distribution in the population of organisms. The range of biological responses to which probit analysis can be applied is a wide one—examples are dead vs. alive, paralysed vs. not, asleep vs. awake, in oestrus vs. not, germinated vs. not, and recovered vs. ill.

Important books on the subject include ones by Finney (1971) and Hewlett and Plackett (1979). Relevant articles in the *Encyclopedia of Statistical Sciences* include those of Bhattacharyya (1985), Kiefer (1985), and Ashford (1986).

Various generalisations of probit analysis have been proposed. Some of them involve bivariate distributions, usually the normal. While many occur in biological settings, others are of an econometric nature. The arrangement of the Chapter is as follows:

22.2 Two stimulants.

22.3 One stimulus, three alternative outcomes.

22.4 Bivariate response.

22.2 Two Stimulants

22.2.1 Two Stimulants in the Conventional Setting

To determine what happens when two stimulants are applied, we may suppose that the log-thresholds fo[r] response to them jointly have a bivariate normal distribution, the correlation possibly being non-zero.

We also require a rule for determining the region of integration that gives the response probability a[s] the volume under the bivariate normal surface. Let the thresholds be u_1 and u_2 and the concentrations b[e] v_1 and v_2. Two alternative rules are as follows:

1. The probability of non-response is the probability that *each* threshold is larger than the respectiv[e] concentration of stimulant.

[1]The words "probit" and "normit" refer specifically to the normal distribution. Other distributions are also employed especially the logistic (when "logit" is the name used). "Quantal" or "all-or-none" or "binary" responses are terms whic[h] avoid the implication of any particular distribution.

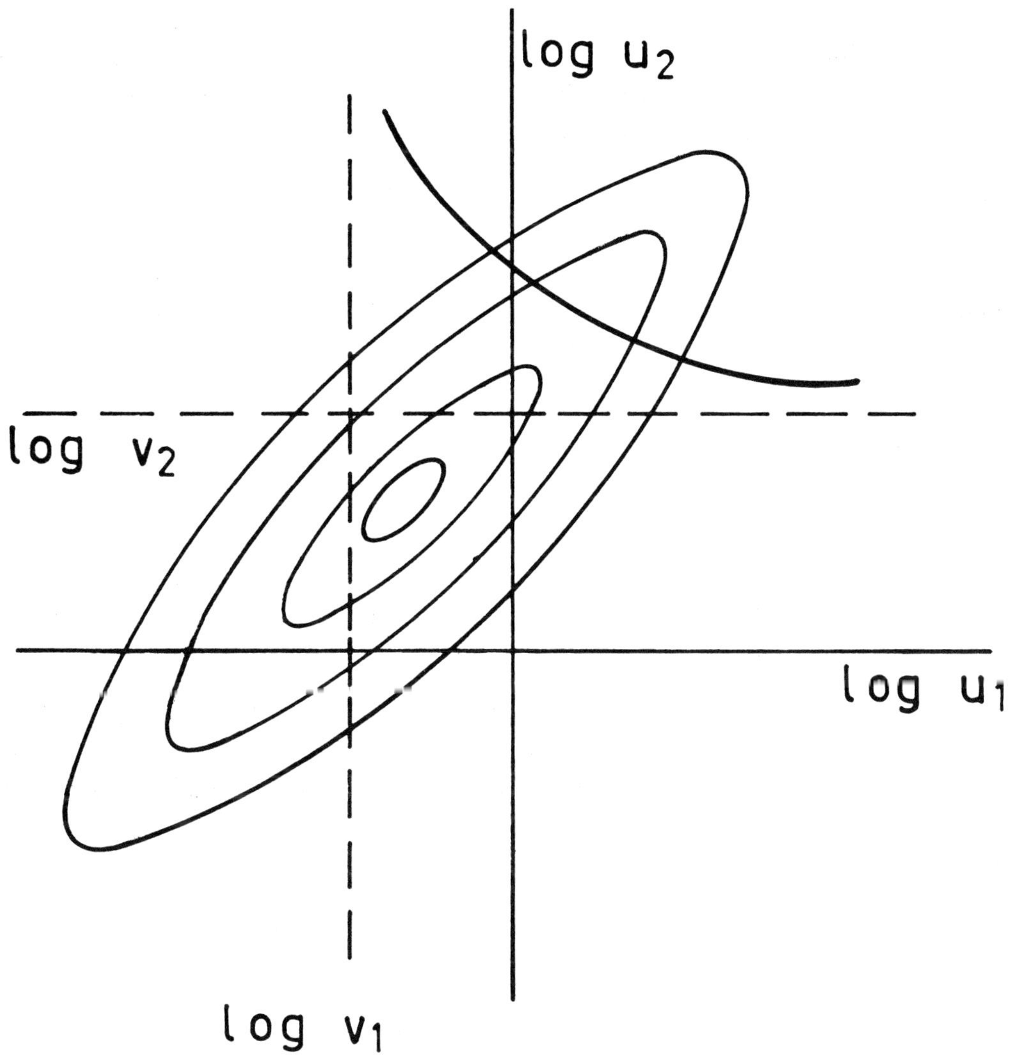

Figure 22.1: Models for the joint action of two stimulants: a bivariate distribution of the thresholds u_1 and u_2 is shown.

2. The concentrations add together according to their ratios to their separate thresholds, i.e., response occurs if $\frac{v_1}{u_1} + \frac{v_2}{u_2} > 1$.

To visualise this, perhaps the easiest way is as in Figure 22.1: imagine horizontal and vertical axes being the logarithms of u_1 and u_2; in this case, the variables will jointly have a bivariate normal distribution, and the contours of the p.d.f. will be ellipses. For concentrations v_1 and v_2 of the stimulants, draw lines parallel to the axes at $\log u_1 = \log v_1$ and $\log u_2 = \log v_2$. Then the probability of non-response is:

- In case (1), the volume under the bivariate normal surface in the upper right quadrant, $\log u_1 > \log v_1$ and $\log u_2 > \log v_2$;
- And in case (2), the appropriate region of integration is to the upper right of a curve which lies within that quadrant and has $\log v_1$ and $\log v_2$ as asymptotes, its equation being $v_1 \exp(-\log u_1) + v_2 \exp(-\log u_2) = 1$. (The equation has been written this way because the axes are $\log u_1$ and $\log u_2$.)

The following generalisation subsumes both the above cases: the concentrations add together according o the expression

$$\left(\frac{v_1}{u_1}\right)^{1/\lambda} + \left(\frac{v_2}{u_2}\right)^{1/\lambda}, \tag{22.1}$$

and response occurs if the result exceeds 1. Setting $\lambda = 1$ gives case (2), and letting $\lambda \to 0$ gives case (1).

All three cases are classified as "non-interactive" action of the two stimulants; case (1) is called "independent" action,[2] and case (2) "simple similar" action. Expression (22.1) may simply be thought of as a formula that includes both the others, or some justification for it provided by supposing that case (2) applies to the effective amounts of the stimulants acting, and that these are related to the dosages by a power function. The foregoing account is based upon those of Hewlett and Plackett (1959; 1979, Sections 7.7–7.11).

The bivariate normal model is not the only available tool here. One might suppose the bivariate normal only applies after a transformation of the marginals to be logistic or Weibull, as Christensen and Chen (1985) did. Or one might set up a log-linear model, as Piegorsch et al (1988) did.

Seven datasets in which combinations of stimulants were applied were reanalysed by Durling (1974). (Most concerned the effects of poisons on insects.) Durling used the independent action model, with several choices of bivariate distribution:

- Bivariate normal,
- Farlie-Gumbel-Morgenstern with logistic marginals, and
- Four varieties of the bivariate Burr distribution—with or without the third and fourth standardised central moments being set to the same values as for the normal distribution, and with or without the extra association term that Durling (1975) proposed.

For most of the datasets, it did not much matter which distribution was chosen, though the bivariate normal was best in most cases. (Recall that the Farlie-Gumbel-Morgenstern distribution is limited to describing weak dependence, so a logistic marginals model with stronger correlation might have done better.)

22.2.2 Econometrics: Two Utilities

A somewhat similar model is sometimes of use in econometrics, when the observed binary outcome of a random utility model does not reflect the binary choice of a single decision-maker, but rather the joint unobserved binary choices of two decision-makers. As an example, consider the probability that a trainee will be retained by a sponsoring company after training. In this situation, the employer must decide whether or not to make a job offer, and the applicant must decide whether or not to seek it. Each individual's decision is not observed, only whether or not the trainee continues working after the completion of training. The probability of this may be modelled by assuming the joint distribution of the utilities to the employer and to the trainee is bivariate normal, and both the employer and the trainee have a utility threshold, exceedance of which corresponds to wanting continuation of working (Poirier, 1980). Meng and Schmidt (1985) cite a study by Connolly (1983) of contracts between public employees' unions and municipalities in Michigan; these contracts may be negotiated or arbitrated; arbitration occurs if either side requests it; it may not be known to outsiders which party requested arbitration, or if both did.

22.2.3 Gap Acceptance by Drivers

In a study of drivers' gap-acceptance behaviour, Storr et al (1980) used a bivariate normal model. The driving manoeuvre which they studied involved first crossing one stream of traffic and then merging with another stream (in Britain, turning right from a minor road into a major road, or left in the U.S.). The probability of accepting a particular pair of gaps in the nearside and farside traffic streams was modelled by a bivariate normal distribution (on logarithmic time axes).

[2] A test for independent action has been proposed by Wahrendorf and Brown (1980). It does not rely upon any particular assumed bivariate distribution of tolerances. Instead, it is based on the idea that Fréchet's inequalities (see Section 3.1.2) for the bivariate distribution of tolerances must be obeyed for this model. (It could happen by chance that the proportion of responses when both stimulants are applied is inconsistent with these inequalities, but one can ask, in the normal manner of statistical testing, whether data like that observed would be likely to occur under the null hypothesis of the inequalities being obeyed.)

22.3 One Stimulus, Three Alternative Outcomes

22.3.1 Three Ordered Responses

It is straightforward to write down the following equations for probabilities of three ordered responses (0, 1, 2) when a single stimulus (at level z) is applied:

$$\left.\begin{array}{lcl} \Pr(Y=2) & = & \bar{F}(a_2 + b\log z) \\ \Pr(Y=2)+\Pr(Y=1) & = & \bar{F}(a_1 + b\log z) \end{array}\right\} \qquad (22.2)$$

where F is some distribution function, such as the normal or logistic, and where also sometimes some other function of z than the log is used. Thus, only a univariate distribution is involved. See also Cox (1970, Section 7.4). A model in which a bivariate distribution occurred was that of Mantel (1966, pp. 92–93), who suggested assuming the two thresholds (which might correspond to death and illness, for instance) might have a bivariate lognormal distribution. (Perfect correlation and equality of standard deviations correspond to (22.2) above.) This is essentially the same model as used in Section 21.3.2 for the joint distributions of two thresholds of injury severity, and was justified by Mantel on the same grounds as there—that although, in principle, it can lead to daft results (such as the death threshold being exceeded but not the illness threshold), it may in practice be sufficiently accurate to be useful.

22.3.2 Aitchison and Bennett's Idea

Simple probit analysis is for one stimulus and two alternative outcomes. The generalisation of Section 22.2 is to two stimuli and two possible outcomes. The generalisation to one stimulus and ordered responses is fairly trivial (equation (22.2)). The generalisation suggested by Aitchison and Bennett (1970) is to one stimulus and three (or, more generally still, several) unordered outcomes that are alternatives to one another, i.e., only one of them occurs.

Corresponding to each possible response is a normally distributed random variable, the mean of which is linearly related to the level of stimulant. The response which has the highest value of this random variable on a particular trial is the one that occurs. Thus when there are three possible responses, the probability of response 1 occurring is the probability that $X - Y$ exceeds 0 and $X - Z$ exceeds 0, where X, Y, Z are the three random variables, each normally distributed. $X - Y$ and $X - Z$ have, of course, a bivariate normal distribution. See Section 23.5 for a similar model in the context of ability testing.

According to Amemiya (1985, pp. 307–308), this model has rarely been put to practical use, because of the difficulty of the computations, which involve an integral in $(n-1)$ dimensions if there are n alternatives. More common—in econometrics, at any rate—is the multinomial logit model, which we shall now outline.

22.3.3 The Multinomial Logit Model

There has been considerable development of ideas like the above in the econometric literature, particularly in modelling choices made by travellers (e.g., Daganzo et al, 1977). The random variables are envisaged to be utilities associated with the various alternatives, and the responses are choices of one alternative. The review by Small (1988) of books by Ben-Akiva and Lerman (1985) and Train (1986) is helpful, especially for the succinct presentation of the basic models, and the identification of some differences in nomenclature and notation that may be found in the literature.

Suppose the utility associated with choice j is decomposable into deterministic and stochastic components, $\boldsymbol{\beta}\mathbf{z}_j + \varepsilon$. The vector $\mathbf{z}_j$ consists of known characteristics of j, the vector $\boldsymbol{\beta}$ consists of unknown parameters which need to be estimated, and the ε's have the type I extreme-value distribution, i.e., $\exp[-\exp(\varepsilon)]$. Then if the choice associated with the highest utility is chosen, the probability of choice j is $\exp(\boldsymbol{\beta}\mathbf{z}_j)/\sum_i \exp(\boldsymbol{\beta}\mathbf{z}_i)$, assuming the ε's to be mutually independent. See also Cox (1970, Section 7.5).

The assumption that the ε's are independent leads to the model having a property that is at once elegant and restrictive—that of "independence from irrelevant alternatives". By this is meant that choices j and k occur in the proportions $\exp(\boldsymbol{\beta}\mathbf{z}_j) : \exp(\boldsymbol{\beta}\mathbf{z}_k)$, i.e., their relative proportions depend only on the characteristics of j and k, not on the characteristics of any other choice that might or might not be available.

It is necessary to relax this assumption in order to account for similarity between the available choices. (The extreme example imagined in the transport literature is that choice between car, blue bus, and red bus as a mode of travel ought to be treated differently from the choice between car, bus, and train.) The following example shows what can be done in the case of three available choices.

- Retain the distribution of ε_0 as $\exp[-\exp(-\varepsilon_0)]$.
- Let ε_1 and ε_2 have the joint distribution $\exp\{-[\exp(-\varepsilon_1/\rho) + \exp(-\varepsilon_2/\rho)]^\rho\}$, i.e., the bivariate extreme-value distribution from Sections 5.6.9.4 and 9.17.5.2.
- Then the probability of selecting choice 0 is $\exp(\boldsymbol{\beta}\mathbf{z}_0)/\{\exp(\boldsymbol{\beta}\mathbf{z}_0) + [\exp(\boldsymbol{\beta}\mathbf{z}_1/\rho) + \exp(\boldsymbol{\beta}\mathbf{z}_2/\rho)]^\rho\}$, and the other two choices occur in the relative proportions $\exp(\boldsymbol{\beta}\mathbf{z}_1/\rho) : \exp(\boldsymbol{\beta}\mathbf{z}_2/\rho)$.

A multivariate generalised extreme-value distribution, $\exp\{-A[\exp(-\varepsilon_0), \exp(-\varepsilon_1), \exp(-\varepsilon_2)]\}$, leads to a more general formulation still. McFadden (1978) and Ameniya (1985, Section 9.3.7) give the necessary properties of the function A, and the explicit expression for the probability of choice j which can be derived.

For more on this, see Amemiya (1985, Chapter 9, especially pp. 296–302).

22.4 Bivariate Response

22.4.1 Medical Applications

We now take our examination of generalisations of probit analysis one step further, by specifying that the response observed may be multivariate. For instance, Ashford and Sowden (1970) consider data in which two systems of an organism are observed, and it is possible that either or both or neither respond. If the joint distribution of tolerances of the two systems to the stimulus applied is bivariate normal, the probabilities of each combination of outcomes are given by appropriate volumes under a bivariate normal density function. The example considered by Ashford and Sowden is to the two symptoms of breathlessness (present or absent) and wheeze (present or absent) in coal miners of different ages. Ashford (1981) sets out a general framework within which variant forms of probit analysis can be discussed. There are three main concepts: the system and its subsystems, the sites of action (of the drugs or other stimuli), and the components of the stimulus.

Mantel and Brown (1973) applied the logistic model, equation (22.5) below, to the data of Ashford and Sowden. In their Model 5, h, k, and ρ were linear functions of age, and the fit was not as good as that obtained by Ashford and Sowden with their bivariate normal model; Mantel and Brown employed other forms of dependency of these parameters on age in their other models.

Freeman et al (1978) fitted what might be called a conditional Weibull model to Ashford and Sowden's data:

$$\left.\begin{array}{rcl}\Pr(X=1,Y=1) & = & h(1-\rho) \\ \Pr(X=1,Y=0) & = & k-h(1-\rho) \\ \Pr(X=0,Y=1) & = & h\rho\end{array}\right\} \qquad (22.3)$$

This is an asymmetric model, so we need to mention that X refers to wheeze, and Y to breathlessness; in this case, h, k, and ρ were Weibull functions of age, $h = 1 - \exp(-a \text{ age}^b)$, for instance.

We have already noted in Section 5.2.11 an application of the 6-variate Farlie-Gumbel-Morgenstern distribution to tumour occurrence at six sites (Chinchilli and Breen, 1985).

We have already noted in Section 9.2.8 an application of the Marshall and Olkin distribution to tumour occurrence at two sites (Rai and Van Ryzin, 1984). A model with exponential marginals can be justified in terms of a one-hit model with Poisson shocks,[3] and the model with Weibull marginals which Rai and Van Ryzin actually used might be called a multi-hit model.

[3] That is, we reinterpret the probability of a Poisson variate, whose mean is proportional to dose, exceeding zero, in terms of an exponential tolerance density.

22.4.2 Econometric and Sociological Applications

22.4.2.1 Morimune

In the study by Morimune (1979), the response data was a set of 2×2 tables of whether a family owned a house and whether their house had at least five rooms; there were nine such tables, referring to different levels of income. The first model fitted to this data utilised the bivariate normal distribution, so that

$$\left.\begin{array}{rcl} \Pr(X=1,Y=1) & = & \Psi(h,k;\rho) \\ \Pr(X=1,Y=1)+\Pr(X=1,Y=0) & = & \Phi(h) \\ \Pr(X=1,Y=1)+\Pr(X=0,Y=1) & = & \Phi(k) \end{array}\right\} \quad (22.4)$$

Ψ and Φ being the cumulative bivariate and univariate normal distributions, respectively, and h, k, and ρ each being a linear function of income. (Notice that ρ was not constant in this application.) The second model was of a type described under the heading of the Pareto distribution, in Section 5.8.13.4:

$$\left.\begin{array}{rcl} \Pr(X=1,Y=1) & = & D^{-1}\exp(h+k+\rho) \\ \Pr(X=1,Y=0) & = & D^{-1}\exp(h) \\ \Pr(X=0,Y=1) & = & D^{-1}\exp(k) \end{array}\right\} \quad (22.5)$$

in which $D = 1+\exp(h)+\exp(k)+\exp(h+k+\rho)$, and the symbols h, k, and ρ have been used to point up the analogy with the bivariate normal model (and are, again, linear functions of income). Section 7.6 of Cox (1970) is relevant to this sort of model. The third model wrote the four probabilities as linear functions of income, and gave a much worse fit than the others.

With some types of explanatory variable, the normal and logistic models are equivalent. See Lee (1979) for some remarks clarifying this.

22.4.2.2 Amemiya

The following is Example 6 from Amemiya (1975). Suppose the two classifications to be purchase of car ($x = 1$ if yes, $x = 0$ if no) and purchase of house ($y = 1$ if yes, $y = 0$ if no). Suppose the probabilities of these purchases are logistic functions of known external variables $\mathbf{z}$:

$$\left.\begin{array}{rcl} \Pr(X=1|y) & = & [1+\exp(-\boldsymbol{\beta}_1\mathbf{z}-\beta_{12}y)]^{-1} \\ \\ \Pr(Y=1|x) & = & [1+\exp(-\boldsymbol{\beta}_2\mathbf{z}-\beta_{21}x)]^{-1} \end{array}\right\} \quad (22.6)$$

The dependence of $\Pr(X = 1)$ on y and of $\Pr(Y = 1)$ on x might arise because purchase of house uses money that might otherwise be used for car purchase, and vice versa. Amemiya shows that β_{12} and β_{21} must be equal, and that

$$\Pr(X=x,Y=y) = \frac{\exp(\boldsymbol{\beta}_1\mathbf{z}x+\boldsymbol{\beta}_2\mathbf{z}y+\beta_{12}xy)}{1+\exp(\boldsymbol{\beta}_1\mathbf{z})+\exp(\boldsymbol{\beta}_2\mathbf{z})+\exp(\boldsymbol{\beta}_1\mathbf{z}+\boldsymbol{\beta}_2\mathbf{z}+\beta_{12})}. \quad (22.7)$$

Thus this is essentially the same model as (22.5).

22.4.2.3 Sample Selection Equations

These ideas are related to what are referred to as sample selection equations in Section 19.9.5. Write a two-equation model as

$$\left.\begin{array}{rcl} x & = & \boldsymbol{\beta}_1\mathbf{z}_1+\varepsilon_1 \\ y & = & \boldsymbol{\beta}_2\mathbf{z}_2+\varepsilon_2 \end{array}\right\} \quad (22.8)$$

It is usual to assume the disturbances $(\varepsilon_1, \varepsilon_2)$ have a bivariate normal distribution. Now note three cases:

- The bivariate probit model refers to our observing the two indicators, $I_{\{x>0\}}$ and $I_{\{y>0\}}$.
- Sample selection occurs when we observe $xI_{\{y>0\}}$.
- A disequilibrium model (of demand and supply, for instance) refers to our observing $\min(x, y)$.

22.4.2.4 Three Other Studies

- Greene (1984) utilised a similar idea in a public policy setting. The response data was a 2×2 table of whether a respondent voted yes or no in a school tax referendum and whether he or she had at least one child enrolled in a public school (in the American sense of this term). Several relevant facts were known about the respondents, such as the number of years of living in the community, income, and yearly property taxes; these acted essentially as regressors for the marginal proportions, the correlation coefficient between the two response variables being a further parameter to be estimated.

- Muthén (1979) applied a model similar to that of Ashford and Sowden to some sociological data relating to sex-typing of chores.

- The importance of health in an individual's decision whether or not to retire is, so Butler et al (1985) tell us, the subject of some dispute in the economics literature. Their paper—and some others which they refer to—uses a bivariate normal model; the concern is with whether retired people stay retired, and the two methods of exiting this state are re-employment and death. (Death may serve as an indicator of health, as an alternative to self-reported health, either of these being capable of acting as a proxy for a general health variable that may affect the retirement decision.) The paper was presented at a conference, and is somewhat abbreviated in nature,[4] and we class it under probit models more by guesswork than by real understanding. The conclusions as to the correlation between work and health were that "The probabilities of exiting the sample via reentry or death are not independent events. Tastes for work and poor health appear to be positively correlated in this example suggesting that, controlling for income and demographic factors, those who continue to work the longest will die at an earlier age than those who consume leisure during retirement." See Butler et al (1989) for related work. These authors have been influenced by the "competing risks" literature (see Section 9.18.3). However, we have to say that several important explanations about their model and about the analogy between the competing risks and the econometric viewpoints appear to be missing from their paper. Perhaps they have written a more accessible account that we are not aware of. Butler et al (1989) found that the unmeasured propensities to die and to return to work may be positively related (non-significant correlation = 0.36), and that taking account of this changes somewhat the estimated effects of policy-relevant variables.

22.4.2.5 The Cost of Partial Observability

Meng and Schmidt (1985) integrate bivariate response with the partially-observable choices of Section 22.2.2. They also discuss intermediate cases: (i) whether or not the second type of response occurred is known if and only if the first type of response did occur; (ii) when it is known that at least one of the types of response failed to occur, either it is also known that the first response did not occur, or it is known that the second response did not occur. Thus the distinct conditions that can be observed are as follows.

Model	Number of distinct conditions	List of distinct conditions
Poirier	2	(1,1); (1,0) or (0,1) or (0,0)
Meng and Schmidt's Case 3	3	(1,1); (1,0); (0,1) or (0,0)
Meng and Schmidt's Case 4	3	(1,1); (0,1) or (0,0); (1,0) or (0,0)
Full observability	4	(1,1); (1,0); (0,1); (0,0)

As an example of their Case 3, Meng and Schmidt cited a study by Farber (1983) in which workers were classified as unionised, non-union but desiring union representation, or non-union and not desiring union representation. (A positive estimated correlation coefficient of 0.78 in this study was interpreted as indicating that "unobserved factors which make workers more likely to work on union jobs also make these workers more likely to prefer union representation.") As a potential example of their Case 4, they mention Connolly's (1983) study cited in Section 22.2.2 above; Connolly did not use this model as the extra classification was not considered reliable.

[4] The paper by Butler et al (1986) is even further condensed.

Meng and Schmidt were concerned with the cost of the dependent variables being only partially observable, "cost" referring to the loss of asymptotic efficiency of the parameter estimates. They were unable to draw any general quantitative conclusions, because these depended so strongly on the parameters of the model, but said "We find the cost of partial observability to be fairly high, so that extra observability information may be worth collecting, if possible."

Chapter 23

Bivariate Models in Signal Detection Theory

23.1 Signal Detection Theory in Psychophysics

Our first task in this Chapter is to outline what "signal detection theory" (SDT) means in the context of psychophysics. We will see that the concept of univariate distributions is needed, but normally bivariate distributions do not play any part. Then we will describe a number of variant models of perception and response in which bivariate distributions do become relevant. It will be the bivariate normal that we will concentrate on.

23.1.1 Background

Signal detection theory is a methodology which has been applied increasingly widely since its introduction during the 1950's in the analysis of psychophysical data. SDT concepts have led to major advances in the analysis of information-retrieval systems, medical decision-making, and quality control in industry, among other areas.

In its early days, SDT in the psychophysical sense was connected with the electronics usage of the same term. We shall not be concerned with the latter here, though we can mention there is occasional mention of bivariate distributions in that literature—e.g., a paper by Ehrenberg (1976), in which "a series approximation is obtained for the bivariate density of two filtered Poisson processes with orthogonal response functions. The approximation is derived by first using a bivariate Edgeworth series to obtain a bivariate density conditioned on a fixed number of terms in the Poisson sum. Then, the conditioning is removed using the density for the Poisson counting variable."

The basic idea of SDT is that in a binary identification task, two types of error can be made—the subject might say "A" when the correct answer is "B", or "B" when the correct answer is "A". (The identification task, may, for example, be such that A = no signal, and B = signal present.) The frequencies of these two types of error will be determined by two factors—how good the subject is at the task, and his or her relative predispositions towards the two alternative answers. The first of these factors will be influenced by the ability of the subject and how physically distinct the two stimuli A and B are, whereas the second will be influenced by the rewards or penalties for answering A or B correctly or incorrectly, and the relative frequencies of presentation.

SDT was first used in psychophysical experiments into the detection by human subjects of faint visual or aural stimuli. It has the advantage of combining the frequencies of the two types of error—reporting the presence of the stimulus when one is not presented ("false alarm"), and reporting no stimulus when there is one ("miss")—into one quantity. How this is done, and how the methodology is applied to variants of this experiment, will now be explained. A detailed review will not be given, however, as there have been many papers and books on the topic (e.g., Swets, 1964; Green and Swets, 1966; McNichol, 1972; Laming, 1973, Chapters 5–7; Egan, 1975). SDT has been tried out in many fields of application, far removed from its origins; for reviews, see Swets and Green (1978) and Hutchinson (1981a).

To give an idea of the place of SDT in psychophysics, we may quote from the review by McGill and Teich (1989) of Laming's (1986) book. "This is a very important book, perhaps the most significant advance in psychophysical detection theory since the now-classical collection of papers, [Swets, 1964]... The principal virtue here is the prospect of reviving an immensely important part of sensory psychology. The original work pioneered by W P Tanner, J A Swets, and D M Green had this bold quality. It challenged the very idea of a threshold, pointed to the critical importance of outcomes or payoffs, and gave us a vision of sensory processes as approximations to a mathematically defined 'ideal observer'... The early vitality of signal detection theory, a powerful driving force in auditory research of the 1960's, dissipated slowly into a methodological literature not unlike Thurstone scaling—an important tool but essentially a curiosity... Intensity discrimination couched in the framework of a single broad detection theory simply lost its fastball... Then along came Laming in 1986 with his breath of fresh air regenerating the excitement of the search for missing pieces of the puzzle surrounding sensory communication."

23.1.2 Mathematical Formulation

The basis of the analysis is to consider that there is some variable X which the subject uses for deciding whether or not to report detecting a signal. The variable might be for example the level of activity in some part of the nervous system. In any situation, X varies from moment to moment, but on average it is higher when a signal is present than when one is not. It is further supposed that the subject reports a signal if X is greater than some response threshold T and reports no signal when X is less than T (the alternative, that X is on average lower when a signal is present than when one is not, with the subject reporting a signal when X is less than some response threshold, can easily be changed into the present formulation by making the transformation $X \to -X$).

The probabilities of reporting a signal are $\bar{F}_A(t) = \int_t^\infty f_A(x)\,dx$ and $\bar{F}_B(t) = \int_t^\infty f_B(x)\,dx$, where f_A and f_B are the p.d.f.'s of X in the "noise" and "signal plus noise" conditions, $\bar{F}_A$ is the probability of a "false alarm", and $\bar{F}_B$ the probability of a correct detection.

23.1.3 The ROC Curve

The relationship between $\bar{F}_A$ and $\bar{F}_B$ is fundamental to signal detection theory. A function connecting them is traced out as the response threshold T moves along the X-axis. This may be done experimentally, because the location of T is to some extent under the control of the subject, and his or her placing of it may be influenced by the experimenter providing information on the a priori likelihood of a stimulus, or manipulating the pay-off matrix (amount of reward or punishment for each combination of stimulus presented/not presented and detection reported/no detection reported). For instance, if the subject believes that in nine trials out of ten a stimulus will be presented, a higher rate of detection will be reported than if that the probability is believed to be only one in ten. The contribution of signal detection theory to this situation is to emphasise that the probability of false alarm will increase along with the probability of correct detection, and that the relation between these two is of interest.

This relation will be referred to as an ROC curve. This originally meant Receiver Operating Characteristic because of early application to the analysis of radar receiver performance, but Swets (1973) prefers Relative Operating Characteristic because of an intimate connection with the Operating Characteristic (complement of power function) of statistical hypothesis testing.

For examples of empirical ROC's, we may refer to Swets et al (1961), who experimentally determined 13 points $(\bar{F}_A, \bar{F}_B)$ for each of four observers, using a visual stimulus. Each point was based on a two-hour experimental session consisting of 200 presentations of noise and 200 of signal. Changes in the response threshold T were produced by changes in the payoff matrix. Experimental determination of a positive relation between $\bar{F}_A$ and $\bar{F}_B$ is the first reason for the popularity of signal detection theory; success in predicting performance in other types of experiment is the second reason, and this will be discussed in Section 23.1.5 below. The third, and perhaps most important, reason is the elegance of the theory as a framework for integrating a field of study.

Graphs similar to the ROC find uses in a variety of other contexts: see Hutchinson (1985b).

23.1.4 Alternative Shapes of ROC

One specific assumption often made is that f_A and f_B are two normal p.d.f.'s having equal variance. The shape of the ROC curve then depends solely on the difference between the means of these distributions expressed in units of the common standard deviation, this difference being known as d'. A generalisation of this hypothesis involves relaxing the assumption that the standard deviations are equal, this being referred to as the normal unequal-variance model. The difference d' measures the strength of the stimulus relative to the sensitivity of the subject, increasing when the former is made physically stronger or the subject's sensitivity is increased in some way, whereas the criterion (usually expressed as $\beta = f_B(t)/f_A(t)$) depends on the response strategy of the subject. Together they determine the point $(\bar{F}_A, \bar{F}_B)$ at which the subject operates.

It is worth pointing out that we do not need to adhere strictly to the proposition that the threshold T is at a fixed point: if the p.d.f.'s f_A and f_B are separated by d' and each has a variance of unity, and the threshold is variable, its position being normally distributed with variance σ^2, this is equivalent to the threshold being fixed but the distributions having variance $1 + \sigma^2$. The normal model thus retains its validity, but, since the difference between the means of f_A and f_B is expressed in units of the standard deviation, d' is reduced to $d'/\sqrt{1+\sigma^2}$.

Most attention in the literature has been given to the normal models, but the chief contribution of signal detection theory lies in the assertion that a positive relation should exist between the probabilities of false alarm and correct detection; the particular functional form is of secondary importance, and a matter for empirical investigation. What shape must the ROC be? It must go through $(0,0)$ and $(1,1)$, be monotonically increasing, and its first derivative must be monotonically decreasing (see, for instance, Laming, 1973, pp. 98–103). The simplest functions for which this is true are the power curves, $\bar{F}_B = \bar{F}_A^k$. For such curves to arise, the decision axis must be monotonically transformable so that f_A and f_B are exponential curves: if $f_A(x) = \lambda \exp(-\lambda x)$ and $f_B(x) = \mu \exp(-\mu x)$, $\bar{F}_A(t) = \exp(-\lambda t)$ and $\bar{F}_B(t) = \exp(-\mu t)$, so that $\bar{F}_B = \bar{F}_A^{\mu/\lambda}$. The general shapes of the power-curve ROC's are necessarily similar to those from the normal equal-variance model, but the latter are symmetric about the diagonal from $(0,1)$ to $(1,0)$ and are tangential to the axes both at $(0,0)$ and at $(1,1)$, whereas the former are not symmetrical and are tangential only at $(0,0)$.

Another functional form for the relation between $\bar{F}_A$ and $\bar{F}_B$ may be found by assuming f_A and f_B to be logistic p.d.f.'s. This leads to ROC's which are symmetric about the diagonal but not tangential to the axis, neither at $(0,0)$ nor at $(1,1)$.

It should, however, be noted at this point that whereas any two distributions f_A and f_B define a corresponding ROC, to each ROC an infinite number of pairs of distributions correspond, since, given one pair that gives rise to a particular ROC, another pair may be generated by distorting—stretching or compressing—the decision axis X in any monotonic way. In other words, given an ROC and any distribution f_A, a distribution f_B can be found to correspond.

23.1.5 Other Experimental Situations

From the model described above, predictions may be made of the subject's performance in other types of experiment, and the success of such predictions is one of the reasons for the popularity of signal detection theory. The main alternative experiments are the four below.

23.1.5.1 Rating

The rating experiment has the advantage of reducing the number of trials necessary to obtain an empirical ROC: instead of being asked to respond "Yes" or "No" to the question "Was a signal presented?", the subject has a number of response categories available, typically five or six, from "certainly present" through "probably present" and "probably not present" to "certainly not present". Each boundary between adjacent categories may be treated as if it were the dividing line between "Yes" and "No", so that six response categories yield five pairs of operating probabilities from the one sequence of stimulus presentations.

23.1.5.2 Forced-Choice

In the forced-choice experiment, the signal is presented in exactly one of n observation intervals, and the subject is required to choose which. If the subject's value of d' is known from a Yes/No experiment using the same signal intensity, then the probability of him or her being correct in the forced-choice experiment may be predicted. Swets (1959) performed a comparison of this sort and found that several forced-choice procedures (with different n) gave very similar results to the Yes/No experiment.

23.1.5.3 Second-Choice

The second-choice experiment consists of asking a participant in a forced-choice experiment to make a second choice when his or her first choice is wrong. If detection is an all-or-nothing matter, getting the first choice wrong means that the stimulus failed to excite the detection mechanism and therefore that second choices will be correct only at the chance level (independent of the signal strength). Signal detection theory, on the other hand, implies that the subject is capable of ordering the several alternatives according to their likelihood of containing the signal, and therefore that the proportion of correct second choices should be above the chance level and be positively related to signal strength. Swets et al (1961) report an experiment in which this was found to be the case.

23.1.5.4 Same-Different

The same-different experiment is more often used when the subject is required to *discriminate* between two signals than in *detection* tasks. As its name suggests, the subject has to say whether the two presentations were the same (without having to identify them as of type A or type B) or different (without having to say which was which).

23.1.6 Remark

The degree of empirical success of signal detection theory is still a matter of some controversy. But there is no denying its usefulness as a unifying conceptualisation.

That concludes our introduction to conventional signal detection theory. The next four Sections of this Chapter give examples of SDT concepts in bivariate settings.

23.2 The Effect of Signal Probability on the Slope of the ROC

Results from an auditory discrimination task in which the subjects used a rating procedure with four response categories were reported by Schulman and Greenberg (1970). The value of the probability of signal presentation was varied, and it was found that as this probability increased, so did the slope of the ROC when plotted on normalised coordinates. Hutchinson (1982a) used a two-stage decision model in a possible explanation of this finding. (This paper concentrated on the relatively narrow issue of the particular result of Schulman and Greenberg, and the explanation proposed by Treisman (1977), and did not attempt to go into the extensive literature on intertrial dependencies in perceptual experiments.)

The basis of the model is that the threshold criteria according to which the subject classifies each stimulus are not fixed. They have some random trial-to-trial variability. This variability is hypothesised to be smallest when the threshold is at the position on the decision axis where the odds of the stimulus being signal or noise are equal. The variability is larger when the threshold is located towards either extreme of the decision axis.

The subject's task is to put the stimulus into one of four categories. There are therefore three thresholds in use. It is assumed that the stimulus effect is compared with the middle threshold first, and then with one of the others. For the stimulus to be placed in the highest category, therefore, it is necessary that (i) X must have exceeded t_2, and (ii) X must have exceeded t_1, where:

$$\begin{aligned} X &= \text{the stimulus effect on this trial,} \\ t_2 &= \text{the position of the middle threshold on this trial,} \\ t_1 &= \text{the position of the highest threshold on this trial.} \end{aligned}$$

Horizontal axis: difference between stimulus effect and second threshold.
Vertical axis: difference between stimulus effect and first threshold.

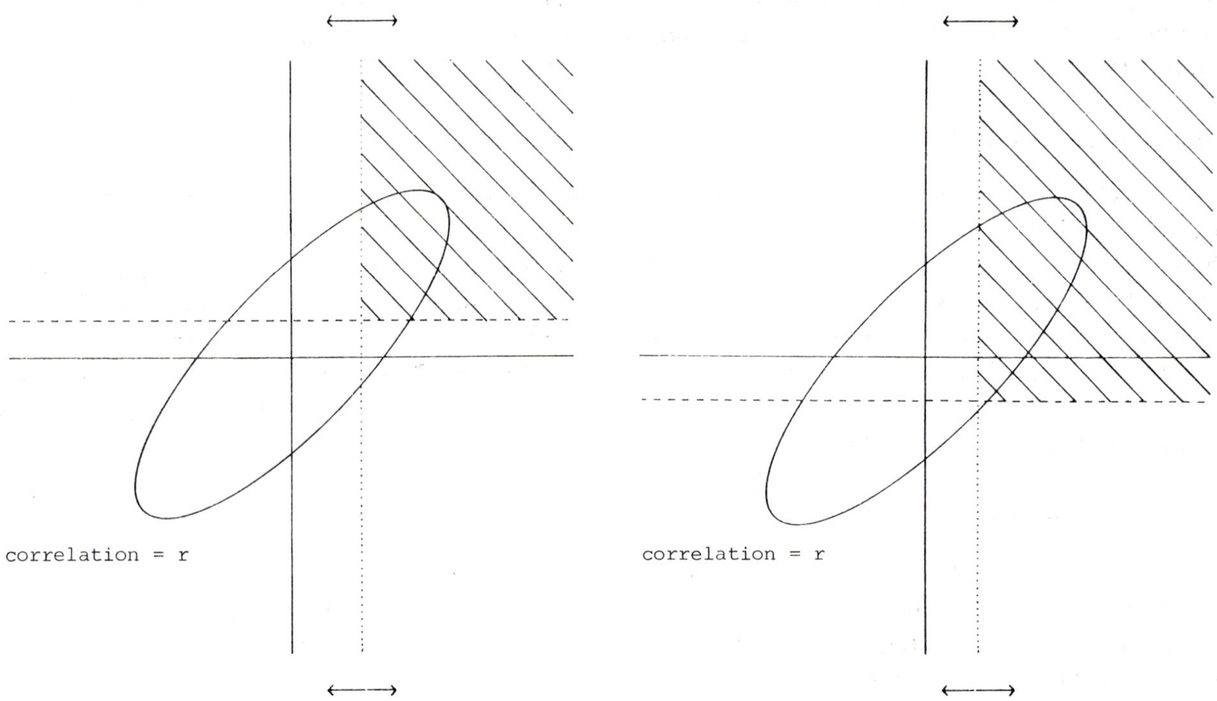

The relevant probabilities are given by the volumes under the density surfaces in the shaded regions.

Figure 23.1: Two-stage models of perception.
Note that the following conventions are observed in Figures 23.1, 23.2, and 23.4: all distributions are represented by an ellipse, standardised to zero means and unit standard deviations; a dashed line is used to represent some fixed criterion; a dotted line is used to represent a moveable threshold.

If X, T_2, and T_1 each have a normal distribution, the two variables $X - T_2$ and $X - T_1$ jointly have a bivariate normal distribution with a positive correlation. The probability of the highest category being used is therefore $\Pr(X - T_2 > 0$ and $X - T_1 > 0)$. The variables $X - T_2$ and $X - T_1$ will have different means when a signal is presented from when it is not. This is represented in Figure 23.1 (in which, in common with other Figures, the distributions are shown as having zero means and the different probabilities shown using adjusted thresholds). Part (a) of the Figure represents the case when noise is presented, and part (b) is for signal presentation. The details of how the two volumes are calculated depend on the function connecting the threshold variance with position on the decision axis—see Hutchinson (1982a).

The dependence of ROC slope on signal presentation probability arises as follows. The position of the point of equal odds of signal and of noise is dependent on signal presentation probability. Hence the function relating threshold variability to position on the decision axis is similarly dependent. And a threshold variability that is different for different threshold positions affects ROC slope.

The following example was given by Hutchinson (1982a). Let us take X to have zero mean and unit variance in the noise condition, and to have mean = 1 and unit variance in the signal condition, and the thresholds to be located at -1, $\frac{1}{2}$, and 2. Let us further assume that the threshold variance, σ_H^2, is an increasing function of distance from the point of equal odds, that is, of $|x - \xi|$, ξ denoting the point of equal odds. ξ is given by $(1-p)\exp(-\xi^2/2) = p\exp[-(\xi - 1)^2/2]$, so that if p is 0.1 (for example), $\xi = 2.70$, and if p is 0.9, $\xi = -1.70$.

Now, assume specifically that σ_H^2 is actually equal to $|x - \xi|$ and not merely some function of it.

Table 23.1: Signal probability and the slope of the ROC: results from the theory, certain parameter values having been assumed.

Threshold	$p = .1$ Noise	$p = .1$ Signal	$p = .9$ Noise	$p = .9$ Signal
(a) Probabilities				
3 (lowest)	.768	.904	.818	.954
2 (middle)	.390	.610	.390	.610
1 (highest)	.046	.182	.096	.232
(b) z values				
3 (lowest)	−.73	−1.30	−.91	−1.69
2 (middle)	.28	−.28	.28	−.28
1 (highest)	1.69	.91	1.30	.73
(c) $\Delta z_S/\Delta z_N$				
	.91		1.10	

Therefore, total effective variance (including that of the threshold) is $1 + |x - 2.7|$ if $p = 0.1$, and $1 + |x + 1.7|$ if $p = 0.9$.

We shall assume that the stimulus effect is compared first with T_2 and then with one of the other criteria. For an example of how to obtain theoretical expressions for the probabilities of exceeding the criteria, suppose that $p = 0.1$ and consider the probability of T_1 being exceeded in a trial in which noise is presented. For this to happen, it is necessary that (i) $X - T_2$ must have exceeded 0, and that (ii) $X - T_1$ must have exceeded 0. The two variables $X - T_2$ and $X - T_1$ jointly have a bivariate normal distribution: the former has a mean of $-.5$ and a variance of 3.2, and the latter a mean of -2 and a variance of 1.7. The correlation between the two variables is, as in equation (15.9),

$$\frac{\text{variance of the variable in common, i.e., } I_N}{\sqrt{[\text{variance of } (X - T_2)] \times [\text{variance of } (X - T_1)]}}$$

which is $1/\sqrt{5.44}$. Consequently, the probability is $L\left(\frac{0.5}{\sqrt{3.2}}, \frac{2.0}{\sqrt{1.7}}; \frac{1.0}{\sqrt{5.44}}\right)$, where L is the upper right volume under the bivariate normal density surface; this works out to be 0.046. Similar expressions may be obtained for the other probabilities. The results are given in part (a) of Table 23.1. Part (b) of the Table shows the corresponding standard normal deviates (z values). Finally, part (c) gives a summary statistic, the ratio of the change in z for the signal condition to that for the noise condition; this is bigger for the bigger value of p, in accord with Schulman and Greenberg's result.

See Treisman and Faulkner (1984) for further developments.

23.3 Evaluation of X-ray Images

Swensson (1980) considers the process of evaluating X-ray images for the possible presence of abnormality. He suggests this is a two-stage process: "To report a target, a radiologist must consider a specific pattern feature and decide that it cannot be interpreted as part of the normal anatomy. This final decision stage must logically follow some earlier stage of visual search, during which that particular feature of the pattern was selected for attention and specific consideration."

Swensson models this by assuming the two variables—the feature's salience as represented by the observer's preattentive filter, and that measured by his or her cognitive evaluation stage—jointly have a bivariate normal distribution. This bivariate distribution is different for true abnormalities from that for normal features on the X-ray. In order for a feature to be reported, it is necessary that it pass both a

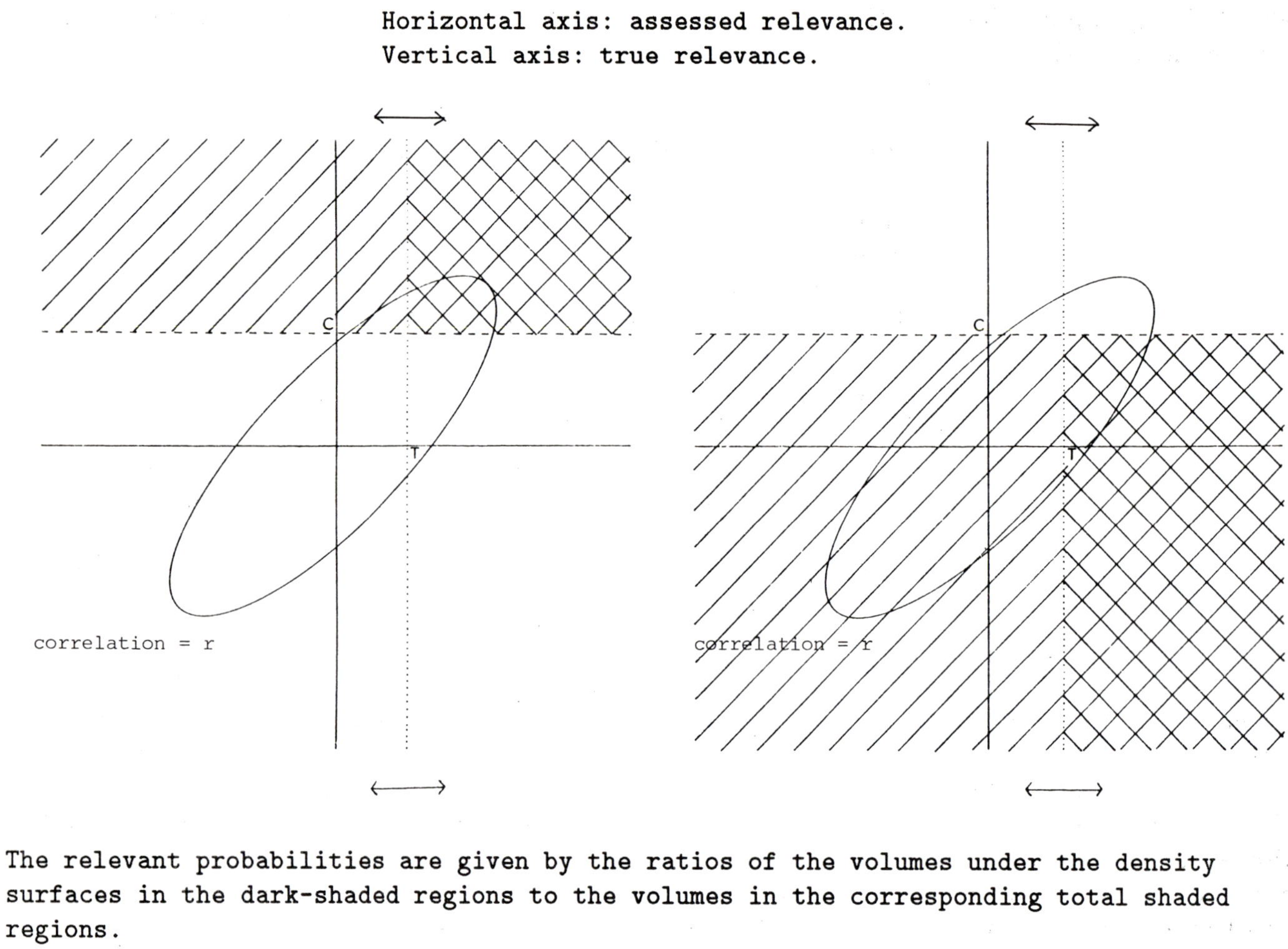

Figure 23.2: A model for information retrieval. (See also the note to Figure 23.1.)

salience criterion and a threshold in the evaluation. As the threshold position is changed, so an ROC curve is traced out.

In the absence of an abnormality, the two variables may be taken to have zero means and unit standard deviations, and correlation ρ_1. When one is present, the means and standard deviations will take values μ_X, μ_Y, σ_X, σ_Y, with the correlation being ρ_2. The salience criterion C is a seventh parameter. In order to be useful, the number of free parameters needs to be reduced. For instance, one might assume $\rho_1 = \rho_2$, $\sigma_X = \sigma_Y = 1$. Another approach to constraining the model's parameters is to measure performance in circumstances designed to bypass the initial filter, or to vary its selection threshold.

Two series of experiments are reported by Swensson. Both involve a comparison between a search task (the observer reported all the abnormal findings he detected on each radiograph) and a nonsearch task (the observer reported only on precise locations on the radiographs). The nonsearch conditions were assumed to bypass the radiologists' initial filter without affecting the explicit cognitive evaluation of each test feature.

23.4 Information Retrieval Systems

We ask a computerised information retrieval (IR) system for references to a certain topic. As well as giving us some references that are relevant to our topic, it will usually retrieve some references that are not, and will fail to retrieve some that are. Swets (1963) proposed that signal detection theory be used to measure the performance of IR systems. We may consider that the IR system has a decision axis: notionally, we can envisage it calculating a "relevance score" for each item in the databank, and that those passing a certain threshold are output by the system, and those falling below the threshold are not. The user of the system decides which items are relevant and which are not. These two populations will each have a distribution

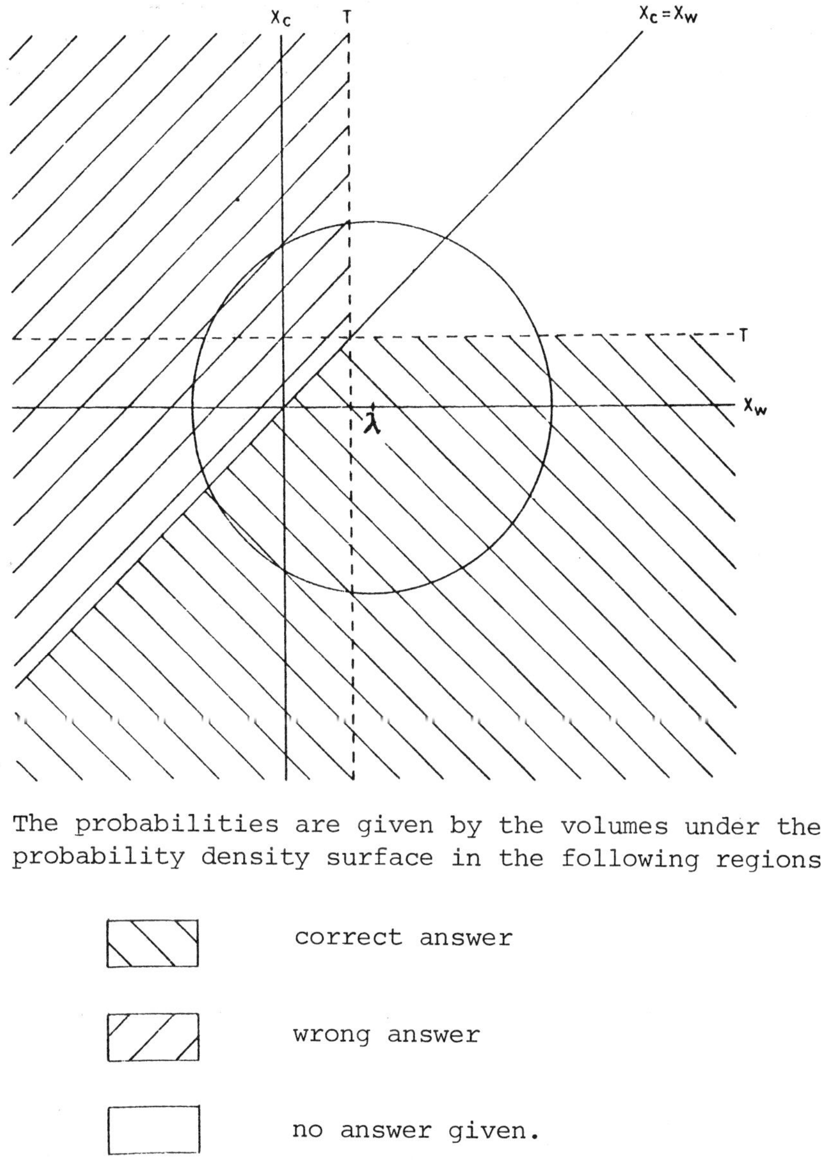

Figure 23.3: Showing the probabilities of the three possible responses in a two-alternative choice test.

of relevance scores, that for the relevant items having a higher mean.

Hutchinson (1978) made the point that the true relevance of an item in the databank is likely to be an ordered variable, not a dichotomy. He suggested that this true relevance and the relevance as assessed by the IR system might jointly have a bivariate normal distribution, and that their intercorrelation would be a suitable measure of the effectiveness of the IR system. The ROC curve for the system would be generated as shown in Figure 23.2. The dashed line at C divides the population of items into two groups, truly relevant and truly non-relevant. The dotted line at T shows the corresponding division by the IR system. As one adjusts the position of T so as to get a greater or lesser number of items output from the system, so the ROC curve will be traced out by the proportions shown in the Figure.

23.5 Performance in Two-Alternative Choice Tests

Assessment of intelligence, aptitude, ability, or educational attainment frequently uses multiple-choice tests. Subjects' responses to the questions fall into three classes—correct, wrong, no answer given. This last category causes a problem in scoring the test. It is inferior to giving the correct answer, and (presumably) superior to giving a wrong answer, but where exactly does it stand compared with these? The importance

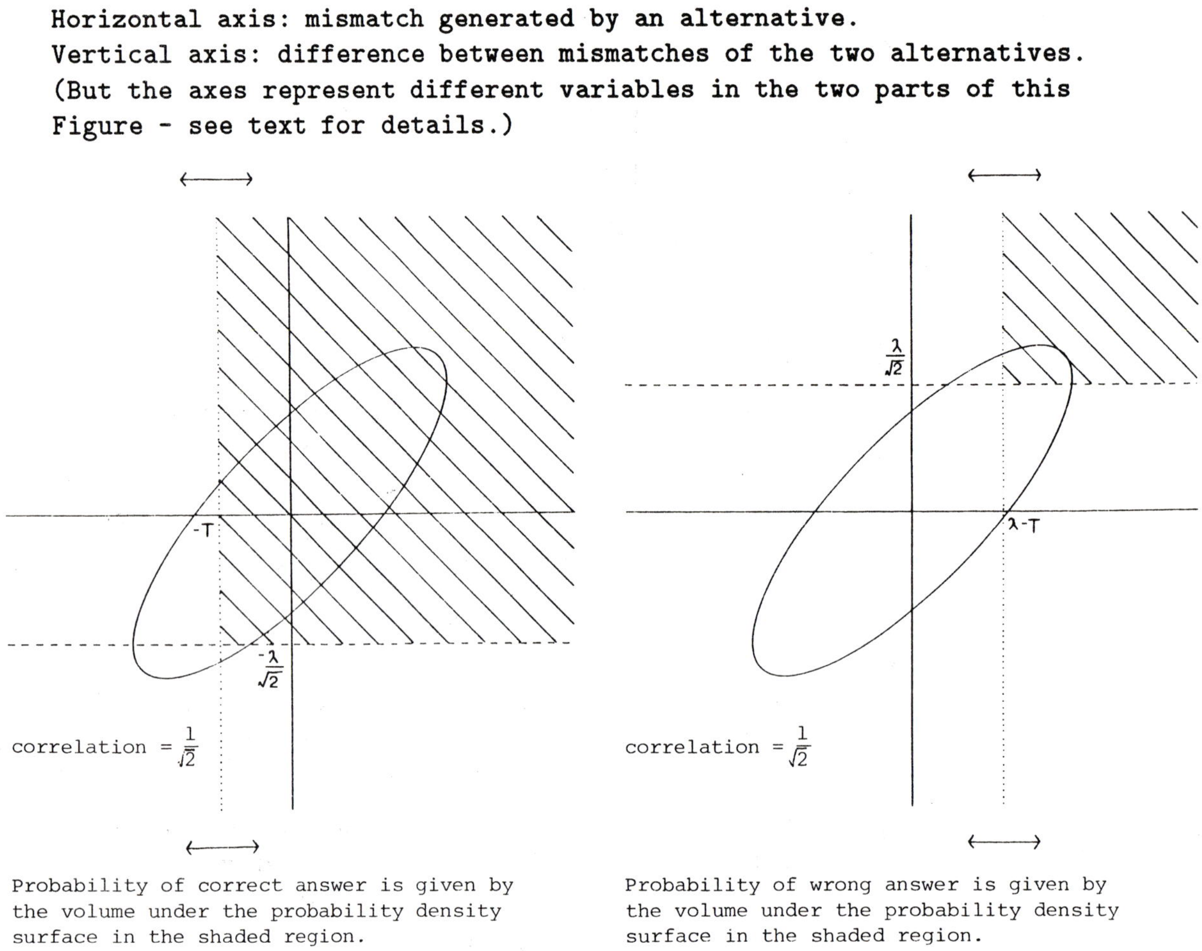

Figure 23.4: Performance in two-alternative choice tests. (See also the note to Figure 23.1.)

of this question arises from the finding that if subjects are forced to make a choice on questions they initially left unanswered, they answer correctly in a higher-than-chance proportion of cases; and, furthermore, that this tendency is greater in the more timid or anxious type of personality (Votaw, 1936; Sherriffs and Boomer, 1954). This is undesirable, since it is intended that the test be personality-free.

Hutchinson (1977) formulated this as a response-threshold problem, and showed the connexion between signal detection theory and formulae for calculating ability from the proportions of items answered correctly, answered wrongly, and unanswered. A greater variety of models of this type are described in Hutchinson (1982b). One of them involves the bivariate normal distribution.

This is for the special case in which the subject has only two alternatives to choose from. (These alternatives may perhaps be "True" vs. "False", or "Yes" vs. "No".) It is envisaged that each of the two alternatives gives rise in the subject to a variable corresponding to the degree to which it fails to fit the answer ("mismatch"). The mismatch has a distribution with a smaller mean in the case of the correct answer than in the case of the wrong answer. The subject selects the alternative corresponding to the lower mismatch. But if the lower mismatch is greater than some threshold, then the question is left unanswered. Different degrees of encouragement to the subject to respond would correspond to different positions of this threshold.

If the mismatch for the correct alternative has a normal distribution with mean 0 and standard deviation 1, and for the incorrect alternative has mean λ and standard deviation 1, then the probability of a correct answer is

$$\Pr(X_C < T \text{ and } X_C - X_W < 0),$$

and the probability of a wrong answer is

$$\Pr(X_W < T \text{ and } X_W - X_C < 0),$$

where X_C is the mismatch for the correct alternative, X_W is the mismatch for the wrong alternative, and T is the response threshold.

These probabilities are shown in Figure 23.3. X_C and X_W jointly have a circular bivariate normal distribution—that is, they have equal standard deviations and are independent. This is represented by a circle centred on $(\lambda, 0)$. The probability of no answer being given is the probability that both X_C and X_W exceed T. This is the unshaded area in Figure 23.3. The probabilities of selecting the correct answer and the wrong answer are also shown.

In order to get expressions for the probabilities that may be determined using tables of, or an algorithm for, the cumulative bivariate normal distribution, we proceed as follows. The joint distribution of the variables X_C and $X_C - X_W$ is bivariate normal, with the following parameters: $E(X_C) = 0$, s.d.$(X_C) = 1$, $E(X_C - X_W) = -\lambda$, s.d.$(X_C - X_W) = \sqrt{2}$, correlation $= 2^{-1/2}$.

Similarily, the joint distribution of the variables X_W and $X_W - X_C$ is bivariate normal, with the following parameters: $E(X_W) = \lambda$, s.d.$(X_W) = 1$, $E(X_W - X_C) = \lambda$, s.d.$(X_W - X_C) = \sqrt{2}$, correlation $= 2^{-1/2}$.

Therefore,

$$c = L\left(-T, \frac{-\lambda}{\sqrt{2}}; 2^{-1/2}\right) \quad (23.1)$$

$$w = L\left(\lambda - T, \frac{\lambda}{\sqrt{2}}; 2^{-1/2}\right) \quad (23.2)$$

and these quantities are shown in Figure 23.4. In order to solve these equations for λ, it appears necessary to prepare a special table or algorithm.

We may also recall from Section 22.3.2 the generalisation of probit analysis suggested by Aitchison and Bennett (1970). This model bears some resemblance to that for performance in two-alternative choice tests, the more so because Aitchison and Bennett regard their model as "particularly appropriate to situations where the category of outcome determined by the experimental unit may be regarded as the result of some reasoning, or possibly subconscious, process of assessing the effects of the choices of the various response categories". Indeed, it is possible that by permitting variability in the response threshold T in Figure 23.4, a model formally identical to that of Aitchison and Bennett could be obtained (and could be re-interpreted in terms of randomly-distributed utilities being estimated for each of the alternative courses of action, i.e., answer 1, answer 2, no answer).

23.6 The ABX Experiment

Brief mention should be made of a discussion of various signal detection paradigms by Macmillan et al (1977). What is called the ABX experiment involves the four possible stimuli ABA, ABB, BAA, and BAB. The subject's task is to respond "1" to the first and fourth of these (in which the third simulus is the same as the first) and "2" to the second and third. The *optimal* decision rule may be formulated in terms of $X_1 - X_2$ (the difference between the effects of the first two stimuli) and X_3. The probabilities of responding "1" or "2" are obtained by integrating circular bivariate normal distributions over regions with curved boundaries. It is therefore not possible to convert the problem to one of integrating a bivariate normal distribution with non-zero correlation over a rectangular area. Macmillan et al obtained their results using a Monte Carlo sampling technique. The more *plausible* decision rule based upon comparing $|X_1 - X_3|$ with $|X_2 - X_3|$ leads to probabilities expressible in terms of the univariate normal distribution (Pierce and Gilbert, 1958).

23.7 Graham et al

We have only recently noticed the paper by Graham et al (1987), so have not had time to study it thoroughly, but we believe it does not consider such a wide variety of models as we have done in this

Chapter. (It does have much greater psychological content, though.) The detection rules which it considers are (i) the decision variable will be below a criterion if and only if every one of the monitored random variables is (maximum-output rule), and (ii) the decision variable is the sum of the monitored random variables (sum-of-outputs rule). It may be that these are equivalent to two models for response to mixtures of stimuli in the quantal response context, where the "stimuli" are usually drugs or poisons, discussed in Section 22.2.1.

23.8 Overview

There are a number of features which the applications described above have in common. The *bivariate normal distribution* and *signal detection theory* are two. In connexion with the latter, let us make explicit mention of the emphasis on the *relation between two probabilities*, as when constructing the ROC. Both the analysis of radiologists' performance and the explanation of ROC slope increasing with the probability of signal presentation relied upon a *two-stage model* of what was happening: in the first case there were two stages for perception, in the second the categorisation of the stimulus was in two stages. Both the ROC slope example and the consideration of performance in two-alternative choice tests relied upon the bivariate normal distribution being obtained by from combinations of univariate distributions by the variables-in-common (trivariate reduction) method—for which, see especially Chapter 15.

In no case was there any fundamental reason for choosing the bivariate normal distribution as the bivariate distribution to base the analysis upon. It was chosen because it is the one that is most widely understood and has convenient properties. So any success which it has in explaining a pattern of data should not deter us from trying other models also, preferably ones which include the bivariate normal as a special case.

Chapter 24

Examples of Data

24.1 Introduction

24.1.1 Purpose of the Chapter

We believe that many readers will welcome having a number of datasets—some well known, others not—reproduced in this Chapter, along with a brief description of the variables and how the information was obtained.

Pressure of space being what it is, many journals discourage printing of raw data. We think that is misguided, as

- The data may yield answers to some substantive questions with relatively crude, pencil-and-paper, analysis,
- The data may be a suitable testing-ground for sophisticated analysis that the original author did not think of, and
- Academic staff like to give their students data that is real rather than contrived.

Thanks are due to the originators of the various datasets herein; we hope no-one is annoyed at not being asked for permission—that would have been a misuse of our time and energy, we felt. How did we choose the data to be included? Well, the sample is largely one of convenience, determined by what came to hand while writing this book. Thus it is dominated by datasets that have been discussed in the statistical literature recently. Our impression is that in some fields, there is a wealth of available examples from which to choose, whereas in others, a lack of published data has led to the same one or two datasets being treated time and time again. Among the best sources of further examples are the early volumes of *Biometrika.*

The reader may wish we had carefully read each source reference to check there was nothing odd in the data collection method, and had performed some simple analyses to check there is nothing weird in the data itself. Yes, but again the expenditure of effort would have been disproportionate to the benefit.

24.1.2 Reader, Beware!

24.1.2.1 Some Quotations

Having said how welcome it is when someone publishes observations in a relatively unprocessed form, we must immediately acknowledge that there are dangers with reusing such data. In a book review, Dewey (1988) says "Although the book contains several useful example datasets, too few have a proper provenance. No health statistician would analyse data without knowing where it came from, and what it was collected for." Similarly, the Preface to the collection of datasets published by Andrews and Herzberg (1985) mentions "reservations about the encouragement of unthinking manipulation of numbers" and goes on to express the view "that statistical methods should not be applied to numbers but rather to the situations giving rise to the data". Other writers—e.g., Preece (1986), Margolin and Collings (1987)—have given the world whole papers on the subject. The latter authors, for example, say that if there is to be meaningful reanalysis "from which new insights are possible, as opposed to mere numerical illustrations of the evaluation of formulae, reusers need to appreciate fully the true context of the data they are intending to adapt... One finds that some datasets, including a few that are well known and have been analyzed by numerous authors, may not be as simple and straightforward as many thought."

24.1.2.2 A Response

While we have much sympathy with these sentiments, they provide an incomplete picture; many data analysts—expert statistical craftsmen and tyros alike—have their thinking stimulated and their understanding of both the data at hand and the calculations they are performing on it deepened by number-manipulation that is initially "unthinking".

24.1.2.3 Alternatives to Analysis as a Distribution

In some cases, consideration of the data as a bivariate distribution may be less appropriate than some other style of analysis—as a contingency table (if there are few different values of X and Y), as a regression problem (if there is little scatter about a trend line connecting X and Y), as a paired t-test (if interest centres on the difference between X and Y), or as an analysis of variance (if supplementary information is given on other variables). In a few cases, it is doubtful if the data is (at least) ordinal, rather than merely nominal. And sometimes it may be important to take account of the order in which the observations occurred.

24.1.2.4 Conclusion

To sum up, while we wouldn't encourage *analysis* of data of unknown origin, nevertheless we do encourage *play* with it.

24.1.3 Organisation of the Material

There are five Sections referring to particular statistical frameworks:

- Reliability,
- Extremes,
- Two measurements of the same thing,
- Compositions (totalling 100%),
- A few discrete examples.

Then come eight Sections referring to particular types of subject matter:

- Shapes of living things,
- Biological sciences,
- Medicine and human biology,
- Mental abilities,
- Sociology and economics,
- Air, earth, water, and climate,
- Physical and technological sciences,
- Miscellaneous.

Table 24.1: Hours to failure of the transmission (x) and transmission pump (y) of caterpillar tractors.

Tractors 1-5		Tractors 6-10		Tractors 11-20	
x	y	x	y	x	y
1641	850	6367	3832	3953	6310
5556	1607	9460	3871	6922	6310
5421	2225	6679	4142	4210	6378
3168	3223	6142	4300	5161	6449
1534	3379	5995	4789	4732	6949

Within the Sections, four sub-section headings are used repeatedly. These chiefly refer to the format in which the data is presented:

- Data presented as exact values. The examples under this are of pairs of (x, y) values.
- Data presented as frequency tables. Here, the frequencies with which combinations of ranges (or ordinal categories) of x and y occurred are given.
- References to scatterplots. Here, some references to published scatterplots are given. Pairs of (x, y) values could possibly be obtained by writing to the authors of the respective papers, or by measurement from the graphs. The product-moment correlation coefficient, r, is sometimes given.
- Cross-references. There are some examples which could have been included under more than one heading. These are noted here.

The reader will appreciate that we have space for only the briefest of descriptions. As to the data itself, in a few cases we have condensed the categories somewhat, as compared to the source publication.

24.1.4 Data in Other Chapters

We draw attention here to data appearing in earlier Chapters of the book:

1. Table 20.2, on degree of certainty of diagnosis of multiple sclerosis by two neurologists.
2. Table 20.4, on the prevalences of sociopathy and hysteria.
3. Table 20.6, on replicated paired comparison tests.
4. Table 21.1 (and also Table 20.3), on the joint distribution of injury severities to vehicle drivers in head-on collisions.
5. Table 21.2, on the joint distribution of injury severities to driver and front seat passenger in single-vehicle crashes.

24.2 Reliability

We give here four Tables of paired values (x, y) that are in some sense times of survival until failure.

1. The first (Table 24.1) is of the times to first failure of the transmission and the transmission pump of D9G-66A caterpillar tractors. It comes from Barlow and Proschan (1977), and has also been considered by O'Neill (1985). A recent paper by Nayak (1988) utilises this data in a hypothesis-testing context.

Table 24.2: Hours to failure of three parts of the insulation of ten motors.

Motor	Turn	Phase	Ground
1	1764	2436	2436
2	2436	2436	2490
3	2436	2436	2436
4	2436 +	2772 +	2772
5	2436	2436 +	2436
6	2436	4116 +	4116 +
7	3108	4116 +	4116 +
8	3108	4116 +	4116 +
9	3108	3108	3108 +
10	3108	4116 +	4116 +

+ Denotes time without failure.

Table 24.3: Time (minutes) to inception of a defect (x) and further time until failure (y), of cable insulation.

Specimens 1-5		Specimens 6-10		Specimens 11-15		Specimens 16-20	
x	y	x	y	x	y	x	y
228	30	1740 +	?	1227	39	2600 +	?
106	8	155	7	254	46	195	27
246	66	414	30	2440 +	?	117	27
700	72	1374	90	435	85	724	21
473	25	128	4	1155	85	300	96

+ Denotes time without defect occurring.

2. Table 24.2 is a trivariate dataset from Nelson (1982, pp. 187–188) that was also examined by Hougaard (1988?). It gives the time to failure of the insulation of three parts of 10 motors. The times are midpoints of intervals—hence the repetition of values; Nelson states that the rounding "affects the data little". A number of observations were censored.

3. Table 24.3 gives information about the failure of electrical cable insulation. The data recorded are the time until a defect can first be detected microscopically, and the additional time until failure of the insulation. In three cases, no defect occurred before observation was terminated. The source is Lawless (1982, pp. 476–477), who credits Stone (1978).

4. A dataset of unusual structure was the subject of Bartlett and Lwin's (1984) paper. Tests on timber of bending strength B and tension strength T are destructive, so it is impossible to measure both; the measurement of modulus of elasticity E is not destructive, so can be carried out in addition to that of B or T. Bartlett and Lwin had the following data:

   - T and E on 74 pieces of timber,
   - B and E on another 74 pieces of timber,
   - B and E on 60 pieces of timber such that $B \leq 31.08$ MPa,
   - T and E on 86 pieces of timber which survived a bending load of 31.08, i.e., such that $B > 31.08$.

Table 24.4: Floods of the Ocmulgee River, at Macon (x) and Hawkinsville (y).

1910-1919		1920-1929		1930-1939		1940-1949	
x	y	x	y	x	y	x	y
28.8	18.8	66.2	45.2	64.4	50.0	14.2	13.3
8.5	5.9	37.0	30.0	10.7	12.2	7.3	6.9
44.8	44.4	48.6	44.0	19.6	16.2	73.4	57.0
51.0	52.0	28.3	30.3	19.0	19.9	44.8	41.6
4.8	5.9	21.0	15.2	16.9	17.4	50.2	46.8
19.1	20.1	72.5	79.0	22.7	13.5	40.4	26.2
47.8	40.4	28.3	19.3	65.3	61.0	57.6	35.4
25.4	27.0	7.9	7.6	33.3	25.8	32.6	34.8
14.3	14.3	47.1	42.4	31.0	33.0	24.0	28.2
31.0	40.0	73.4	70.5	33.9	37.9	84.0	68.0

They assumed that, after logarithmic transformation, the three variables had a trivariate normal distribution, and estimated the three means, the three variances, and the three correlations. Of special interest was the correlation between $\log B$ and $\log T$, the two variables which cannot be observed simultaneously.

In view of the indirectness of their evidence about this, it would be of interest to know how sensitive were the results obtained to the assumption of trivariate normality.

The data is not given in full in the paper, only (i) summary parameters for the 74 + 74 pieces of timber on which E and either T or B were measured, and (ii) scatterplots of $\log B$ vs. $\log E$ (given that $B \leq 31.08$) and of $\log T$ vs. $\log E$ (given that $B > 31.08$).

A rather simpler example of similar type was given by Evans et al (1984). In this case, the strength of lumber in tension and in bending was again the issue, but modulus of elasticity was not measured. Proof loading in bending followed by test to failure in tension was again envisaged. Evans et al reported a simulation study of estimating by maximum likelihood the correlation between the two strengths—assuming known univariate means and variances, a realistic assumption as large samples broken in tension testing alone or bending alone are typically available.

Incidentally—perhaps more than incidentally from a practical viewpoint—both papers indicated that the assumption that proof loading does not damage the timber which survives it is only approximately true.

24.3 Extremes

We give here some Tables of paired values (x, y) that have appeared in the literature on bivariate extreme values.

1. Table 24.4 refers to the sizes of the annual floods on the Ocmulgee River (Georgia), downstream at Macon, and upstream at Hawkinsville, in the 40 consecutive years, 1910–1949. It is taken from Gumbel and Goldstein (1964), who cite Carter (1951) as their source.

2. Table 24.5 refers to floods on the Fox River (Wisconsin), upstream at Berlin, and downstream at Wrightstown, in the 33 consecutive years, 1918–1950. It is taken from Gumbel and Mustafi (1967). (Differences between the two readings were analysed by Bain and Engelhardt, 1973, and constitute dataset BAEN in the book edited by D'Agostino and Stephens, 1986.)

3. Tawn (1988a) analysed the annual maximum sea levels for Lowestoft and Sheerness, two ports on the east coast of Britain. The sample size was 74. The data was only given in the form of a scatterplot,

Table 24.5: Floods of the Fox River, at Berlin (x) and Wrightstown (y).

1918-1919		1920-1929		1930-1939		1940-1950	
x	y	x	y	x	y	x	y
		5.15	16.6	3.00	6.6	4.72	17.5
		2.45	14.2	1.14	3.1	3.54	16.6
		5.92	20.1	1.91	9.9	2.74	19.8
		6.05	13.7	2.60	8.9	5.08	21.3
		4.02	15.5	1.91	6.7	2.29	10.8
		2.52	8.3	4.34	11.1	3.46	15.8
		3.44	9.1	4.34	6.3	6.90	21.3
		3.17	13.3	3.26	13.5	3.16	11.0
6.05	16.3	5.92	15.1	6.19	18.0	4.54	10.3
2.67	13.1	6.62	20.6	4.91	18.2	2.60	6.4
						4.63	10.9

Table 24.6: Extreme values of the command and weakness-detection channels (x and y), from an experiment with spacecraft command receivers.

Data pairs 1-10		Data pairs 11-20		Data pairs 21-30	
x	y	x	y	x	y
-95	-157	-159	-333	-172	-293
-138	-199	-204	-300	-112	-282
-181	-321	-197	-327	-112	-336
-158	-355	-157	-321	-121	-238
-146	-209	-185	-304	-171	-215
-179	-331	-173	-274	-190	-272
-192	-273	-19	-216	-151	-326
-211	-299	-103	-366	-110	-185
-169	-274	-108	-253	-174	-255
-198	-322	-153	-265	-170	-240

and is not reproduced here.

4. The data of Table 24.6 is taken from Posner et al (1969), and refers to a spacecraft command receiver. The reason for a bivariate analysis is that there are two channels, one being for the command, the other for detecting whether the command signal is weak (and therefore should be rejected) or not. A bit of data may be

   - Received correctly (if X is less than some value x_0) and accepted (if $Y < y_0$),
   - Received correctly but rejected ($Y > y_0$),
   - Received incorrectly ($X > x_0$) but accepted, or
   - Received incorrectly and rejected.

   Posner et al reported extremes from 100 values of the command channel, and 100 values of the weakness-detection channel, the experiment having been repeated 30 times.

24.4 Two Measurements of the Same Thing

While some of the examples in this Section actually are what the title says, in other cases, "assessment" would be a better word than "measurement", and "similar concept" a better phrase than "same thing".

24.4.1 Data Presented as Exact Values

1. In Appendix C of Rodriguez and Taniguchi (1980), there is some data on octane number requirements, as determined by customers and by expert raters. See Table 24.7. The source of the data is CRC (1980). There is a scatterplot for an earlier year's data (CRC, 1977), having 169 data points, and for which $r = 0.56$ (which happens to be the same as for the data in Table 24.7), in Rodriguez (1982b).

2. Trivariate data: the burning times of powder train fuses, as measured by three observers on 30 rounds of ammunition which were fired from a gun, are given in Table 24.8. This is taken from Grubbs (1948), who cites Simon (1946). The data for observers A and C has been used by Thompson (1963) and Draper and Guttman (1975). (Running times of 20 mechanical time fuses, measured by operators stopping two independent clocks, are given by Grubbs (1983), but it is not clear whether it is real or artificial data; not reproduced here.)

3. Trivariate data: velocities of rounds fired from a 155mm gun, as recorded by two standard velocity chronographs (the Fotobalk and the Counter) and one test instrument (the Terma) are given in Table 24.9, taken from Grubbs (1973).

4. Trivariate data, of observations obtained with a standard measuring device, and duplicated observations made with a new device, was given by Hahn and Nelson (1970). However, it is not clear whether it is real or artificial data. Not reproduced here.

5. Barnett (1969) published some quadrivariate data, of measurements of the vital capacity of the lungs of 72 patients: the four variates were the combinations of either a standard or an experimental instrument, and either a skilled or an unskilled operative. Not reproduced here.

6. Kaplow et al (1986) give a table relating to the stability over time of stained slides used to measure leucocyte alkaline phosphatase activity. Four measurements are given, two having been made at the same time by different methods, and two at other times. Not reproduced here.

7. Several datasets have been published of the classification into ordinal categories of a number of items by two or more observers. There are two reasons why we do not give space to these. The first is that there are very often several observers, so the data is multivariate rather than bivariate. The second is that we question whether the samples are usually randomly chosen from a meaningful distribution. In particular, we suspect that the items included are those that are difficult to classify—there is under-representation of those that every observer would put in the lowest category, and of those that every observer would put in the highest category. However, for the well-known data of Holmquist et al (1967), we note that Agresti (1988) has cast the data for the first two pathologists into a 5×5 ordinal contingency table, and proceeds with an analysis on that basis, so presumably he feels differently.

24.4.2 Data Presented as Frequency Tables

1. The level of diphtheria antitoxin was determined in 124 serum samples, using two methods, the haemagglutination test and the rabbit intradermal test. The data is given in Table 24.10, taken from van Belle and Cornell (1971), who cite Schubert and Cornell (1958) as their source. See also Rosenzweig (1979, 198?).

2. The following is taken from the summary of the paper of Whitehead et al (1986): "All histopathological cases completed in the British Army are sent to the Army Histopathology Registry (AHR) for archiving; prior to placing cases in the archive both microscopic material and submitted reports are reviewed by staff of the AHR. A 'nonagreed' report is produced for those cases in which the

Table 24.7: Octane requirements of 248 cars, as determined by customers and by raters.

Cars 1-41		Cars 42-81		Cars 82-123		Cars 124-158		Cars 159-195		Cars 196-229	
R	C	R	C	R	C	R	C	R	C	R	C
94.5	92.0	95.7	86.4	96.0	94.0	94.0	89.0	94.0	84.9	90.0	87.0
94.0	88.0	96.6	88.2	90.0	86.0	86.0	80.0	96.0	90.0	94.0	89.0
94.0	90.0	93.9	91.6	91.0	84.0	88.0	82.0	88.3	84.9	96.0	89.0
93.0	93.0	94.8	85.4	82.0	82.0	91.0	86.0	96.5	92.8	90.0	83.0
88.0	88.0	96.6	83.6	95.0	93.0	88.0	87.0	98.7	91.5	97.0	91.0
91.0	89.0	93.9	89.1	89.0	83.0	91.5	87.0	94.0	78.7	98.0	86.0
93.0	89.0	94.8	83.6	96.0	93.0	85.0	87.0	86.7	86.7	94.0	89.0
90.5	84.0	89.1	76.6	94.0	89.0	95.0	86.0	90.0	84.9	99.0	91.0
96.0	88.0	94.8	91.6	96.0	79.0	86.7	84.9	91.0	89.0	89.0	87.0
92.0	86.0	93.9	87.3	88.0	84.0	91.5	84.9	95.0	92.0	94.5	90.0
94.0	88.0	94.0	87.0	91.0	79.0	94.0	91.5	95.0	91.0	91.0	83.0
92.0	82.0	96.0	88.0	96.0	86.0	86.7	76.6	94.0	86.0	92.0	85.0
92.0	82.0	90.0	77.0	92.0	80.0	96.0	91.5	94.0	82.0	89.0	83.0
92.0	84.0	93.0	89.0	90.0	85.0	88.3	84.9	94.0	91.0	93.0	93.0
95.0	89.0	94.0	90.0	88.5	88.0	95.0	84.9	92.0	85.0	95.0	87.0
96.0	90.0	96.0	84.0	90.0	82.0	91.5	90.0	88.0	80.0	88.0	89.0
92.0	88.0	96.0	88.0	93.0	83.0	84.9	86.7	89.0	87.0	89.0	87.0
86.0	80.0	96.0	84.0	93.0	84.0	88.3	86.7	92.5	83.0	89.0	86.0
96.0	90.0	92.0	86.0	98.0	94.0	97.5	98.2	93.5	82.0	89.0	87.0
93.0	85.0	97.0	91.0	95.5	92.0	90.0	84.9	96.0	90.0	92.0	84.0
91.0	87.0	88.0	87.0	94.0	90.0	91.5	88.3	88.0	86.0	89.0	90.0
92.0	87.0	91.0	90.0	95.0	87.0	91.5	80.8	93.0	84.0	96.0	92.0
91.0	86.0	87.0	77.0	90.0	81.0	91.5	80.8	92.0	86.0	95.0	86.0
95.0	90.0	93.0	86.0	98.5	94.0	92.8	88.3	90.0	85.0	95.0	87.0
84.0	80.0	92.0	88.0	90.0	88.0	92.8	83.0	91.0	86.0	96.0	90.0
93.0	87.0	92.0	86.0	91.0	85.0	95.0	84.9	93.0	87.0	93.0	79.0
88.0	82.0	99.0	88.0	96.0	88.0	88.3	84.9	90.0	82.0	94.0	90.0
87.0	87.0	89.0	80.0	95.0	94.0	95.0	91.5	96.0	88.0	86.0	85.0
94.0	85.0	88.0	77.0	94.0	87.0	94.0	84.9	92.0	85.0	96.0	95.0
92.0	86.0	91.0	83.0	92.0	88.0	91.5	84.9	91.0	85.0	88.0	83.0
94.0	89.0	87.0	83.0	89.0	84.0	91.5	84.9	93.0	90.0	83.0	82.0
94.8	89.1	99.5	93.0	97.0	90.0	84.9	83.0	93.0	90.0	89.0	78.0
95.7	90.8	92.0	86.0	94.0	80.0	88.3	80.8	94.0	86.0	93.0	87.0
92.8	88.2	90.0	86.0	88.0	89.0	86.7	86.7	89.5	85.0	88.0	84.0
96.6	88.2	96.0	88.0	90.0	86.0	91.5	88.3	90.0	88.0	96.0	L
96.6	91.6	85.0	77.0	90.0	86.0	87.0	L	91.0	88.0	94.0	L
89.1	85.4	94.0	81.0	88.0	86.0	92.0	L	95.0	82.0	H	90.0
89.1	80.8	95.0	88.0	96.0	91.0	89.0	L	86.7	L		
91.6	85.4	91.0	79.0	87.0	79.0	86.0	L	H	84.9		
94.8	90.0	96.0	89.0	91.0	83.0	90.0	L				
95.7	90.0			89.0	86.0	91.5	L				
				94.0	88.0	86.5	L				
				90.0	L	83.0	L				
				86.0	L	84.9	L				
				86.0	L	80.8	L				
				90.0	L						

H means >99.2, L means <76.6; the slightly irregular column lengths make sense in terms of the code numbers used for the cars by Rodriguez and Taniguchi.

Table 24.8: Burning times of fuses, each set for 10 seconds, as recorded by three observers, A, B, and C.

Fuses 1-10			Fuses 11-20			Fuses 21-30		
A	B	C	A	B	C	A	B	C
10.10	10.07	10.07	9.62	9.61	9.64	9.50	9.49	9.50
9.98	9.90	9.90	10.24	10.23	10.24	9.56	9.56	9.55
9.89	9.85	9.86	9.84	9.83	9.86	9.54	9.53	9.54
9.79	9.71	9.70	9.62	9.58	9.63	9.89	9.89	9.88
9.67	9.65	9.65	9.60	9.60	9.65	9.53	9.52	9.51
9.89	9.83	9.83	9.74	9.73	9.74	9.52	9.52	9.53
9.82	9.75	9.79	10.32	10.32	10.34	9.44	9.43	9.45
9.59	9.56	9.59	9.86	9.86	9.86	9.67	9.67	9.67
9.76	9.68	9.72	10.01	lost	10.03	9.77	9.76	9.78
9.93	9.89	9.92	9.65	9.64	9.65	9.86	9.84	9.86

Table 24.9: Velocities (m/s) of twelve successive rounds, as measured by three chronographs.

Rounds 20-25			Rounds 26-31		
Foto	Counter	Terma	Foto	Counter	Terma
793.8	794.6	793.2	791.7	792.4	791.6
793.1	793.9	793.3	792.3	792.8	792.4
792.4	793.2	792.6	789.6	790.2	788.5
794.0	794.0	793.8	794.4	795.0	794.7
791.4	792.2	791.6	790.9	791.6	791.3
792.4	793.1	791.6	793.5	793.8	793.5

Table 24.10: Cross-tabulation of the numbers of serum samples (from a total of 124) having various levels of diphtheria antitoxin, as determined by the haemagglutination and rabbit intradermal methods.

		Units (rabbit intradermal test method)						
		.003	.01	.03	.1	.3	1.	3.
	.003	9	17	7	9			
Units	.01	1	7	4	8			
(haem-	.03		3	4	13			
agglutination	.1		1	1	14			
test	.3				11	1		
method)	1.				4	4	3	
	3.					1	1	1

Table 24.11: The pathological and clinical significance of differences of opinion: numbers of cases.

		Clinical significance of difference of opinion			
		0	1	2N	2U
Pathological	A	174	66	13	18
significance	B	63	77	24	9
of difference	C	17	8	16	27
of opinion	D	0	0	0	9

The four categories of clinical significance are: none; intermediate; definite/substantial (not urgent); definite/substantial (urgent).

The four categories of pathological significance as regarding diagnosis or guidance to the clinician are: none; minor; significant; very significant.

receiving pathologist has a dissenting opinion or for which he thinks other comments may be helpful." Resulting from an examination of all nonagreed reports produced over a 19-month period, Table 24.11 compares the significances of the difference of opinion, when judged with emphasis on the pathological or, alternatively, the clinical aspects.

3. Radiology is a field that for a long time has paid attention to intra- and inter-observer agreement. The following example, taken from Cookson et al (1985), concerns pulmonary asbestosis. Table 24.12 shows the categories assigned to radiographs by two film readers. (Three slight typographical errors in the table given by Cookson et al have been corrected in the present version, thanks to Dr A W Musk.) The population is 354 claimants for compensation among former workers at the crocidolite mine and mill, Wittenoom Gorge, Western Australia, less 74 cases for which no radiographs were found; the film graded was the latest available.

4. Sherman et al (1984) made a study of the performance of humans and a computer system in the classification of photographs of urothelial cells as negative, atypical, suspicious, or malignant. There are two tables in their paper that are relevant to us—of classification by observer 1 at session 1 vs. consensus classification, and of classification by observer 1 at session 1 vs. classification by the same observer at session 2. Not reproduced here, as the photographs were not a random sample, being selected to have 50 examples in each of the four categories, according to the consensus.

5. Table 24.13 compares the cytologic and histopathologic evaluations of 42 renal-cell carcinomas, for a series of patients seen at Umeå's urologic clinic (Ljungberg et al, 1987).

6. Table 24.14 shows results from histologic and colposcopic investigations of a series of patients seen at Montréal's colposcopy clinic (Lambert et al, 1985). The histologic result is considered to be the correct one.

7. Kaplow et al (1986) present a table comparing staining intensity (five-point scale), as rated by an instrument and subjectively by a highly-trained technologist. Not reproduced here.

8. Koch (1983) gives a table of 1000 subjects according to the standard version and a modified version of a diagnostic procedure, each of these having four ordered categories, but the data is fictitious. Not given here.

9. Table 24.15 gives two observers' assessments of the severity of secondary caries attack of 194 occlusal fillings in 50 adult patients seen at the Royal School of Dentistry, Stockholm. (The number of fillings which both observers considered as caries-free is not recorded.) The source is Table 74 of Markén (1962).

Table 24.12: Observer agreement on final film for each subject for whom films were located.

Reader 1 \ Reader 2	0/0	0/1	1/0	1/1	1/2	2/1	2/2	2/3	3/2	3/3	3/4
0/0	36	4	4	1							
0/1	8	0	12	5							
1/0	9	0	19	13	9	2	0	0	0	1	
1/1	0	0	5	17	6	3	3	0	0	0	
1/2	1	0	0	5	6	5	3	1	0	0	
2/1			1	2	3	8	3	1	0	0	
2/2				1	2	5	14	6	4	1	
2/3						3	8	5	2	3	
3/2							1	3	4	1	
3/3								1	1	14	1
3/4											2

The categories used are the standard ones recommended by the International Labour Organization for classifying radiographs for pneumoconiosis.

Table 24.13: Renal-cell carcinomas, cross-classified by their cytologic and histopathologic evaluations.

		Histopathologic grade 1	2	3	4
Cytologic grade	1	0	2	0	1
	2	1	6	6	1
	3	0	0	16	5
	4	0	0	2	2

Among the carcinoma grades, 1 is the most normal and 4 the most abnormal.

Table 24.14: Numbers of patients, cross-classified by colposcopy and histology results.

		Histology N	C	I	II	IIIA	IIIB
Colposcopy grade of whiteness	I (=benign)	19	5	5	1	10	8
	II (=malignant)	14	13	6	14	14	32
	III (=malignant)	2	0	1	5	4	12

oding of histology: N and C are benign results (negative and condyloma, respectively), hereas the others are malignant (grades of cervical intraepithelial neoplasia).

f this table should be analysed by some scheme of scoring the errors, the most mportant distinction is between a benign and a malignant diagnosis.

Table 24.15: Severity of secondary caries, as assessed by two observers.

Obs. B	Observer A 0	1	2	3	*
0	?	22	1	0	1
1	24	67	6	0	2
2	12	20	15	3	0
3	0	1	5	6	0
*	0	0	0	1	8

The symbol * represents a recording of secondary caries, but of a different type to that represented by the grades 1-3.

Table 24.16: Cross-tabulation of the number of dyes, according to lightfastness as measured by two methods.

	High level of dye Glasshouse method					Medium level of dye Glasshouse method					Low level of dye Glasshouse method							
MBTF method	7+	7	6-7	6	5-6	7+	7	6-7	6	5-6	7+	7	6-7	6	5-6	5	4	3
7+	0	1	1	0	0													
7	2	6	5	5	1	0	7	7	2	2	1	2	8	3	0	1	1	0
6-7	1	3	14	6	0	1	1	7	10	1	0	2	5	7	1	1	0	0
6	0	1	2	2	1	0	0	6	5	2	0	0	4	5	3	0	0	1
5-6											0	0	0	3	0	0	0	0
5											0	0	0	2	1	0	0	0

10. Lightfastness trials of a number of wool dyes were carried out using two methods, exposure to sun in a glasshouse, and exposure to artificial light (a 500W MBTF lamp). The dyes were used in three strengths (usually, 1%, 0.5%, and 0.1%). Table 24.16 was prepared from the data in Ingham and Smith (1986).

    This is an example of data to which several of the reservations of Section 24.1.2 apply. In particular:

    - It is not clear that in some sense the dyes included in the trials are representative of a population of dyes, about which it is meaningful to speak of a distribution of lightfastness.
    - Other information about the dyes is given by Ingham and Smith, including their colour and their manufacturer. So an analysis-of-variance style of procedure might be appropriate.

11. Hogben and Lumb's (1967) book of tables of sea wave heights and periods will be referred to at item 24.12.2—8 below; Section 7 of the text which introduces it compares visual observations of wave height with spectral analysis measurements. The data consisted of about 400 wave records obtained over a seven-year period, from British Ocean Weather Ships in the North Atlantic; we do not reproduce it as we think it must have been selected, in some ill-defined fashion, to be representative of rough seas. Hogben and Lumb also compare visual observations and spectral analysis measurements of wave period; the correlation is substantially weaker than for height. D L Harris (1970, 1972b) gives similar tables.

12. Section 6 of Hogben and Lumb (1967) compares two sources of visual observations of wave height and period: Ocean Weather Ships, and Voluntary Observing Ships, the latter being in the same general

Table 24.17: Cross-tabulation of truck GVW, as rated by the manufacturer and as reported in use.

	Rating by manufacturer							
Reported use	1	2	3	4	5	6	7	8
1	15576	5356	21	262	49	49	46	39
2	2304	5856	74	415	1040	147	13	22
3	224	1242	71	372	1204	1725	29	27
4	64	445	64	336	899	1368	31	35
5	71	428	58	275	1989	3086	272	94
6	36	63	13	152	2386	6931	566	587
7	14	20	2	15	529	2821	934	687
8	62	28	4	11	79	1374	1676	11434

The eight categories of GVW have the following upper class boundaries, in thousands of pounds: 6, 10, 14, 16, 19.5, 26, 33, no upper limit.

Table 24.18: Age at which finished full time education, as reported by subjects and by proxies.

		Response by subject (husband/child)					
		-14	15	16	17	18	19+
	-14	296	20	4			
Response	15	38	241	42	3	2	2
by proxy	16	6	30	247	16	6	2
(wife/	17		2	9	30	7	1
parent)	18		1	3	14	36	7
	19+		1	2	2	8	60

area of sea as the former, but not usually in the immediate vicinity. Data not reproduced here.

13. An estimate of the cloud cover of a region of the earth may be made from visible or infrared radiance data obtained by a geostationary satellite. For a stratocumulus cloud field off the coast of Chile, Figure 8 of Minnis and Harrison (1984) compares the two methods on a pixel-by-pixel basis. Not reproduced here. Another report on this subject is by Platt (1981), but the data is presented in the form of contour diagrams. The paper by Xie and Mitsuta (1988) brings ground-observed cloud cover into consideration also.

14. Table 24.17 is of the gross vehicle weight (GVW) of trucks, as rated by the manufacturer and as actually used (as reported in the Truck Inventory and Use Survey). The data is from Robison and Mesenbourg (1980). Further tables in that reference show manufacturer-rated GVW by State-registered GVW, the latter both unrecoded and after a process of recoding.

15. Waite (1911) published data on the judgments by schoolmasters of their pupils' abilities, using a five-point scale. This included two tables in which judgments by different masters were cross-tabulated, and one of judgments by the same master at different times. Not reproduced here.

16. Proxy responses are often used in interview surveys. That is, to keep costs down, a question about someone unavailable for interview is answered by someone else (a proxy), instead of searching out the subject personally. How accurate are proxy responses? A paper on this is by Martin and Butcher (1982). Data of theirs on age at which people finished full time education, as reported by themselves and by a proxy, is given here as Table 24.18.

Table 24.19: Cross-tabulation of the number of people within various bands of salaries or wages, according to the Census Bureau's Current Population Survey and the Internal Revenue.

	Internal Revenue Service				
	None	Under $3000	$3000-$4999	$5000-$7999	$8000 or more
Census Bureau					
None	1659	128	14	15	12
Under $3000	100	1060	67	6	2
$3000-$4999	11	77	630	99	4
$5000-$7999	10	8	77	810	69
$8000 or more	6	0	2	25	344

17. Scheuren and Oh (1975) report a cross-check of U.S. wages and salaries in 1963 between two data sources, the Census Bureau and the Internal Revenue Service. Their data is given in Table 24.19.

18. Another economic example is given by Whiteman (1981): for the nonfarm self-employed, there is a table (total frequency = 336) comparing salary or money taken out of the business for living expenses, as ascertained in a personal interview, and net income from own business, as reported in a mail-back self-administered questionnaire. Not reproduced here.

19. Hauser and Massagli (1983) cross-tabulate people's occupation (major group classification), as recorded in the U.S. Census and the Current Population Survey. One might be prepared to take the groups, or some of them, as being ordered with respect to some such construct as social class or status. Not reproduced here.

20. Agresti (1984, Table 11.2), citing Radelet and Pierce (1983) as the source, gives four 3×3 tables of homicides in Florida: the two dimensions are whether the homicide was committed concurrently with another felony (yes, possible, no), as classified (i) by the police, and (ii) by the prosecutor; the four tables refer to the races of defendant and victim. Not reproduced here.

24.4.3 References to Scatterplots

1. Glucose concentration in potatoes, as measured by enzymatic and Reflocheck methods. About 117 data points. $r = 0.75$. Louwes and Neele (1987).

2. Percent total body fat and kidney fat index of white-tailed deer. 40 data points. $r = 0.87$. Finger et al (1981).

3. Femur marrow fat and kidney fat index of white-tailed deer. 96 + 19 data points (adults and fawns, respectively). Kie et al (1983).

4. (a) Concentration of thyroxine, measured twice. 68 data points. $r = 0.52$.
(b) Concentration of AFP, measured twice. 102 data points. $r = 0.84$.
Yamada et al (1984).

5. Protein content of wheat flour, measured by two methods. 50 data points. Martens and Naes (1983).

6. Comparison of cytologic and histologic assessments of patients with cervical intraepithelial neoplasia. 40 data points. Wied et al (1984).

7. Comparison of certain measurements relevant to intraocular melanoma by two technicians. 100 data points. $r = 0.91$. Gamel et al (1985).

8. Comparison of instrumental and manual assays of leucocyte alkaline phosphatase activity. 81 data points. $r = 0.97$. Kaplow et al (1986).

9. Comparison of instrumental and manual counts of the three types of leucocytes (each being expressed as a percentage of the total leucocytes). Numerals on the scatterplots denote the number of data points at each coordinate; however, "9" is used to mean "9 or more", and this occurs frequently, so exact numbers are not discernible. 500 data points.
   (a) Lyphocytes. $r = 0.94$.
   (b) Monocytes. $r = 0.81$.
   (c) Granulocytes. $r = 0.94$.
   Peebles et al (1986).

10. Predicted seismic performances of buildings, all three pairs of three methods. About 74, 36, and 29 data points in the three graphs. $r = 0.81$, $r = 0.78$, and $r = 0.57$. Endo et al (1984).

11. (a) Grades of uranium ore, as measured by chemical and radiometric methods. About 221 data points. $r = 0.77$.
   (b) The same data, log-transformed. $r = 0.82$.
   Camisani-Calzolari (1984).

24.4.4 Cross-References

Some of the examples in Section 24.6 below (discrete data) could be considered as replicate counts of the same thing. In particular, see 24.6—9 (galaxies) and 24.6—20 (working bullocks). Indeed, the notion of repeated measurements is sufficiently broad that some would say it encompasses a number of other examples, also: bark thicknesses in four directions (Table 24.34), chromosome viability at two temperatures (Table 24.35), underemployment at present and in previous year (item 24.11.2—14), and so on.

24.5 Compositions (Totalling 100%)

24.5.1 Aitchison's Book and Datasets

The distinctive feature of compositional data is that it consists of proportions, which must sum to unity. (See our Sections 4.15 and 12.4.2 above.) An important recent book on the subject is by Aitchison (1986). As this devotes the 52 pages of its Appendix D to listings of 40 datasets of compositions, and as such data is frequently multivariate rather than bivariate, we need give no more than a single example. (However, in a number of cases, Aitchison's data appears to be contrived, and in a number more it does not appear suited to distributional fitting.) To illustrate the range of problems that may come under this heading, we briefly describe the subject-matter of some of Aitchison's datasets.

1. Sand, silt, clay compositions of 39 sediment samples at different water depths in an Arctic lake.

2. Compositions of lavas from Skye. (See Table 24.20 and item 24.5.2—3 below.)

3. Household expenditures on four groups of commodities.

4. Urinary excretions of steroid metabolites.

5. Daily activity patterns of a statistician.

6. Flesh, skin, stone proportions of fruit.

7. Colour-size compositions of clam colonies.

8. Serum protein compositions of blood samples.

9. Subjective diagnostic probabilities.

Table 24.20: Percentages of three oxides in 23 aphyric Skye lavas.

A	F	M	A	F	M	A	F	M	A	F	M	A	F	M
52	42	6	37	54	9	21	60	19	20	58	22	13	52	35
52	44	4	27	58	15	25	53	22	16	62	22	14	47	39
47	48	5	27	54	19	24	54	22	17	57	26	24	56	20
45	49	6	23	59	18	22	55	23	14	54	32			
40	50	10	22	59	19	22	56	22	13	55	32			

A = sodium and potassium oxides, F = iron oxide, M = magnesium oxide.

10. Compositions and total pebble counts of glacial tills.
11. Sand, silt, clay compositions in sediments.
12. Activities during shifts of machine operators.
13. Percentages of different leucocytes in blood samples of ten patients, as determined by four different methods.
14. Measurements of three angles (totalling 180°) for 102 skulls.
15. Areal compositions by abundance of vegetation and of animals.
16. Composition of fossil pollens.
17. Heart and total body weights of domestic cats.
18. Heart and total thorax widths.
19. Predator-prey compositions.
20. Percentages of boys in each hair-colour category, in each of the counties of Scotland. Similarly for eye colour.

24.5.2 Further References

In the following cases, the data was presented as exact values:

1. Five components (quartz, microcline, plagioclase, biotite, others) in twelve thin sections of granite from Bradford, Rhode Island. Aitchison and Shen (1984), and Aitchison (1986), who cite Chayes (1971).
2. 16 oxides in 45 specimens of effusive rocks of the Taupo volcanic association. Steiner (1958); see also Darroch and Ratcliff (1970, 1978), Snow (1975), and Aitchison (1981, 1984, 1986).
3. 10 major oxides in 32 specimens of Eocene lavas (basalts) of the Isle of Skye. Thompson et al (1972); see also Aitchison (1981, 1982, 1984, 1985, 1986). The simplified version which Aitchison (1986, p. 360) gives of this data is shown as Table 24.20.
4. 11 major oxides in 60 specimens of lavas of the Belingwe Greenstone Belt, Zimbabwe. Nisbet et al (1977); see also Aitchison (1984, 1986).
5. Aitchison (1984, 1986) also re-examined the data of Carr (1981) on 10 major oxides in 102 specimens of igneous rocks of the southern Sydney Basin, New South Wales. But the data is not actually listed by Carr (1981), being said to be available on request.

Table 24.21: Cross-tabulation of the numbers of accidents to a sample of bus drivers in parts of Northern Ireland, in two consecutive two-year periods.

		Number of accidents (1952-1953)								
		0	1	2	3	4	5	6	7	Total
	0	117	96	55	19	2	2	0	0	291
Number of	1	61	69	47	27	8	5	1	0	218
	2	34	42	31	13	7	2	3	0	132
accidents	3	7	15	16	7	3	1	0	0	49
	4	3	3	1	1	2	1	1	1	13
(1954-1955)	5	2	1	0	0	0	0	0	0	3
	6	0	0	0	0	1	0	0	0	1
	7	0	0	0	1	0	0	0	0	1
Total		224	226	150	68	23	11	5	1	708

Table 24.22: Cross-tabulation of the numbers of accidents to London bus drivers in two years.

		Number of accidents (year 2)								
		0	1	2	3	4	5	6	7	Total
	0	15	15	4	2	0	0	1	0	37
Number of	1	17	18	9	3	3	0	1	1	52
	2	4	16	12	6	2	3	0	0	43
	3	2	6	5	2	4	0	0	0	19
accidents	4	1	4	4	0	1	0	0	0	10
	5	1	0	0	0	1	0	0	0	2
	6	0	0	1	0	0	0	0	0	1
(year 3)	7	0	0	0	0	0	0	0	0	0
	8	0	0	0	0	1	0	0	0	1
	9	0	0	0	0	0	0	1	0	1
Total		40	59	35	13	12	3	3	1	166

6. 4 pollen types in 73 sets of fossil pollens in the Bellas Artes core from the Valley of Mexico. Clisby and Sears (1955); see also Mosimann (1962), Snow (1975), Darroch and Ratcliff (1978), and Aitchison (1981).

7. A cross-tabulation of the fat content and protein content of 115 different varieties of soy beans is given as Example XV—6b of Bliss and Calhoun (1954), who cite Vessereau (1948).

In addition, we should draw attention to Waite's fingerprint data (see 24.6—19 below).

There are two scatterplots arising from chemical analyses of 61 samples of Devonian shales from Wise County, Virginia, in Hohn and Nuhfer (1980).

There is more data from the earth sciences in Section 24.12 below.

24.6 A Few Discrete Examples

In this Section, the axes of the data tables are inherently discrete, being counts of events, as distinct from divisions of what can be imagined to be a continuous scale. It is natural, therefore, to use discrete bivariate

Table 24.23: Numbers of births when aged 20–30 years and when aged 30–35 years, 977 women.

		Number of births when aged 30-35			
		0	1	2	3+
Number	0	41	3	2	0
of	1	22	15	5	2
births	2	33	60	46	3
when	3	31	109	127	24
aged	4	25	119	158	24
20-30	5+	15	45	61	7

distributions to analyse them—but it might be that something useful would come out of fitting continuous bivariate distributions to them.

1. Table 24.21 is of the number of bus driver accidents in two consecutive two-year periods. It is from Table 6.18 of Cresswell and Froggatt (1963); there are five other tables of this type in that source. See also Kemp and Papageorgiou (1982), Papageorgiou et al (1983), and Xekalaki (1984).

2. Table 24.22 is also of numbers of accidents. It is taken from Edwards and Gurland (1961), who attribute it to Farmer and Chambers (1939), and refers to the numbers of accidents sustained by 166 London bus drivers in two consecutive years. It has also been considered by Arbous and Kerrich (1951), Hamdan and Al-Bayyati (1971), Khatri (1971), and Subrahmaniam and Subrahmaniam (1973).

3. Fitzpatrick (1958) attributed the idea of cross-tabulating accident data obtained in two different time periods to E M Newbold. Other tables of this type of data include:

   - That concerning 122 railway shunters of the Western Transvaal System, from Adelstein (1952). (Judging by the univariate tables which appear in that paper, there is a slight misprint in the bivariate table, the version in Arbous and Kerrich (1951), Gupta and Jain (1978), Barnett (1979), and Ferreri (1983) being correct. Fitzpatrick (1958) and Holgate (1964) have also discussed this data. The table in Maritz (1950), and also in Hamdan (1972), is based upon the same investigation but is slightly different.) Adelstein gives numerous other bivariate tables, also.
   - That concerning 29531 drivers in Connecticut, in the periods 1931–1933 and 1934–1936 (Xekalaki, 1984).
   - That concerning 3112 sites on a road network. This was given by Senn and Collie (1988); McGuigan (1985) had used the same data, though in a different form.
   - That concerning 621 white boys in California, when aged 4–7 and when aged 8–11. Mellinger et al (1965) published this, and it has also been treated by Ord (1972, Section 7.10), Crockett (1979), Lee and Ong (1986), and Loukas and Kemp (1986).
   - That concerning flight aborts—not exactly accidents—for 109 aircraft, in two consecutive six-month periods (Mitchell and Paulson, 1981).

   These are not reproduced here.

4. Table 24.23 is taken from Bhattacharya and Nath (1985); for a sample of married women in India, it gives their number of births when they were aged 20–30 and when aged 30–35.

5. There is a table of the numbers of families containing b boys and c girls (total number of families = 2133) in Rao et al (1973), who cite Reed and Reed (1965), and in Papageorgiou and Kemp (1988). Not reproduced here.

Table 24.24: Cross-tabulation of the numbers of diatom specimens collected by two methods, 137 species.

		Number of specimens collected on styrofoam							
		0	1	2	3-9	10-99	100-999	1000+	Total
Number of	0	?	13	7	2	0	0	0	22
	1	19	1	2	2	1	0	0	25
specimens	2	6	2	0	1	1	0	0	10
	3-9	4	1	6	16	2	0	0	29
collected on	10-99	0	0	1	5	29	3	0	38
	100-999	0	0	0	0	3	8	1	12
glass slides	1000+	0	0	0	0	0	0	1	1
	Total	29	17	16	26	36	11	2	137

6. A table of the first and second lambing records of a flock of 227 Merino ewes (0, 1, or 2 lambs in 1952 and 1953) was considered by Tallis (1962), who cited Turner et al (1958). See also Caussinus (1965) and Drasgow (1986). Not reproduced here.

7. Chatfield (1975) gives a table concerning purchases of two brands of breakfast cereals in a 48-week period. It has 13 rows and 8 columns (but many of the cell frequencies are combined, so there are only 51 cells), and a total frequency of 909. Not reproduced here.

8. The numbers of specimens of diatoms collected on glass slides and on styrofoam, in the Potomac River, for various species: Table 24.24 has been constructed from the 137 paired (x, y) values in Taillie et al (1979), who cite Hohn and Hellerman (1963). It is not known how many species were missed by both methods, so the number in the $(0, 0)$ cell is unknown.

9. In Section 3 of their paper, Mayall et al (1958) were concerned with the chance variation in the number of galaxies counted in regularly placed small squares on photographic plates. Table 24.25 shows that despite taking much care, the numbers of images counted differ considerably, not only for counts on two different plates exposed on a region of space (bottom of Table), but also for two counts on the same plate by the same observer separated by a lapse of time (top).

10. Weldon (1901) cross-tabulated the numbers of pistils and of stamens in flowers of the lesser celandine, attributing the data to MacLeod (1899). See also Rahman (1972, pp. 204–205). Not reproduced here.

11. Holgate (1966) cross-tabulated some data collected by Mr P Grieg-Smith on the numbers of plants of two species (*Lacistema aggregatum* and *Protium guianense*) found in 100 systematically laid out contiguous quadrats in a Trinidad rain forest. Further study of this is by Gillings (1974), Crockett (1979), Loukas and Kemp (1986), Ong and Lee (1986), and Paul and Ho (1989). Not reproduced here.

12. Example 5.1 of Mardia (1970a) is of the numbers of Müllerian glands on the right and left forelegs of 2000 male pigs. The likely source of the data (and also similar data for female pigs) is Davenport and Bullard (1896); see Pearson and Filon (1898). Mardia cites Jørgensen (1916) and Steffensen (1922). Not reproduced here.

13. Table 24.26 shows students classified according to number of newspapers read and number looked at. It is taken from Kendall and Stuart (1977, p. 29), who attribute it to H S Booker. See also Wahrendorf (1980), Goodman (1981a), Dale (1983, Section 3.5.3; 1984), and Upton (1985).

Table 24.25: Numbers of $10' \times 10'$ squares on photographic plates of a region of space in which the given numbers of galaxies were counted on two occasions.

		First count on plate 3340											
		0	1	2	3	4	5	6	7	8	9	13	14
	0	66	34	1	2	1							
	1	37	100	32	7	3*							
	2	3	40	50	20	7*	3						
Second	3	1	10	16	31	13	4	1					
	4		3	6	15	13	6	2					
count	5			1	3	6	5	1	1				
	6			3	1	1	3	4	1				
on	7					2	0	4	3	0			
	8					1	0	1	1	0	1		
plate	9								1	0	1		
	10									1	0		
3340	11									0	0		
	12									1	0		
	13										0	1*	0*
	14											1*	0*
	0	64	35	3	2								
	1	33	90	33	9	3	3						
	2	6	42	36	20	12	3						
Count	3	2	14	24	27	13	5	3					
	4	1	3	6	13	6	4	1	1				
on	5	1	1	5	4	6	2	2	1				
	6		0	1	2	5	2	3	3	0*	1*		
plate	7		1	0	1	1	0	1	2	1	0		
	8		1	1	1	1	1	2	0	0	0		
3341	9						1	1	0	1	1		
	10										0	0	
	11											1	
	12											1	

* There seem to be minor misprints in Tables 2 and 3 of Mayall et al (1958). The eight frequencies marked with a * have been adjusted slightly from the values they print.

Table 24.26: Numbers of newspapers read and looked at, on a particular day.

		Number looked at only					
		0	1	2	3	4	5+
	0	77	75	19	10	1	2
Number	1	179	136	65	20	15	2
	2	86	70	45	18	1	3
read	3	17	21	13	3	4	0
	4	4	2	2	2	0	0
	5+	2	2	0	2	0	0

Table 24.27: Numbers of letters and of syllables in German words.

	Number of syllables								
No. of letters	1	2	3	4	5	6	7	8	9
1	9								
2	114	4							
3	645	101							
4	962	1256	7						
5	444	1492	253						
6	69	1603	1105	22					
7	2	1394	1840	204	1				
8		430	1893	643	12				
9		91	1125	883	43				
10		25	566	896	167	3			
11			157	572	201	10			
12			28	268	200	31			
13			4	99	154	52	1		
14			1	43	81	50	4		
15				6	44	29	8		
16				4	14	20	8	4	
17					3	10	11	1	
18						5	2	2	1
19						2	3	0	1
20						2	3	3	0
21							2	1	3
22									3

14. Table 24.27 gives the joint distribution of the number of syllables and the number of phonemes (letters) in 20453 headwords in Vietor-Meyer's *Deutsches Aussprachewörterbuch.* It is taken from Table E.4 of Derman et al (1973), who cite Herdan (1966).

15. For the numbers of occupations to which various combinations of numbers of men and women belong, see Patil and Bildikar (1967), who took their data (a sample obtained in Ann Arbor, Michigan) from Clark et al (1964). It has also been analysed by Gupta and Jain (1976). Not reproduced here.

16. Arbous and Sichel (1954) gave data on the number of absences from work of 248 shift workers in a large industrial plant in two successive one-year periods; it has also been considered by Mitchell and Paulson (1981). Not reproduced here.

17. Table 24.28, taken from Talwalker (1975), shows the numbers of mild and of moderate/severe side effects observed per patient in a certain clinical trial of an antipsychotic drug.

18. Breslow (1982) and Agresti (1984, p. 217) give a table of the number (0, 1, 2, or 3) of beverages drunk at burning hot temperatures by 80 (Singaporean Chinese male) oesophageal cancer patients and matched controls. (The original study was by de Jong et al, 1974.) Not reproduced here.

19. There are numerous two-way tables relating to fingerprints given by Waite (1915). For instance, the first one is of the number of fingers having arches vs. the number having small loops, for the right hand, the sample size being 2000 people. See also Pearson (1930), Goodman (1968), Darroch and Ratcliff (1973), Plackett (1981, Examples 3.6 and 6.7), and Fienberg (1982). Methods used for compositional data (see Section 24.5 above) are relevant. Not given here. Another example of discrete compositional data occurs when a predetermined number of pollen grains, say 100, are

Table 24.28: Numbers of mild side effects (x) and of moderate/severe side effects (y).

	x						
	0	1	2	3	4	5	6
y							
0	13	1	2	0	1	0	0
1	3	1	1	1	2	1	0
2	2	4	0	2	1	2	0
3	4	3	1	0	2	0	0
4	2	2	2	0	0	0	1
5	0	0	2	1	1	1	1
6	3	0	0	0	2	0	0

Table 24.29: Numbers of households tabulated according to the numbers of working bullocks recorded by a census and by a sample verification survey—rural India.

		Number, as in the census											
		0	1	2	3	4	5	6	7	8	9	10+	Unknown
	0	13110	199	260	8	13	3	5					1135
Number,	1	240	2308	261	19	12	1	0					327
	2	337	196	5650	177	106	7	5	0	1	0	3	793
as in the	3	17	7	117	472	62	8	3	2	2	0	1	66
	4	23	4	102	52	849	33	27	0	2	0	1	99
National	5	0	0	10	2	31	91	17	1	1	0	0	6
	6	5	0	8	5	19	7	172	5	12	0	3	13
Sample	7	0	1	0	0	0	0	7	26	3	0	1	0
	8	1	0	1	0	5	1	14	2	44	2	6	5
Survey	9	0	0	1	0	1	1	1	1	1	7	1	1
	10+	1	1	1	0	1	0	2	0	2	1	48	5

Table 24.30: Frequency distribution of digits supplied by guesswork—first two digits.

		Second missing digit									
		0	1	2	3	4	5	6	7	8	9
	0	9	7	9	4	6	5	1	2	3	3
First	1	14	9	29	13	10	10	7	7	6	3
	2	12	8	6	22	16	13	18	12	9	7
	3	6	4	14	6	33	20	15	18	11	14
missing	4	11	4	19	24	5	38	32	13	7	8
	5	9	5	12	14	13	9	24	17	14	6
	6	5	5	16	14	7	10	7	40	11	7
digit	7	5	5	7	5	7	11	13	7	21	13
	8	5	3	11	6	9	10	4	16	3	21
	9	10	9	10	9	9	8	6	8	9	13

Table 24.31: Lengths, widths, and heights (mm) of the carapaces of 24 male and 24 female painted turtles.

Males			Females			Males			Females		
L	W	H	L	W	H	L	W	H	L	W	H
93	74	37	98	81	38	116	90	43	137	98	51
94	78	35	103	84	38	117	90	41	138	99	51
96	80	35	103	86	42	117	91	41	141	105	53
101	84	39	105	86	40	119	93	41	147	108	57
102	85	38	109	88	44	120	89	40	149	107	55
103	81	37	123	92	50	120	93	44	153	107	56
104	83	39	123	95	46	121	95	42	155	115	63
106	83	39	133	99	51	125	93	45	155	117	60
107	82	38	133	102	51	127	96	45	158	115	62
112	89	40	133	102	51	128	95	45	159	118	63
113	88	40	134	100	48	131	95	46	162	124	61
114	86	40	136	102	49	135	106	47	177	132	67

counted and classified into species. Returning to fingerprints, an example of dermal ridge counts of parents and children has been classified to 24.9.2—12 below, there being so many categories that it seems unnatural to use discrete methods.

20. Some Indian data on the number of working bullocks per household, as reported in the livestock census and in the National Sample Survey, was given by Mahalanobis and Lahiri (1960). See Table 24.29.

21. Digit preference error is a ubiquitous feature of many types of survey. To get an idea of the magnitude of biases that could be attributed to an isolated effect of digit preference, 220 employees of the Indian Statistical Institute were asked to fill in three missing middle digits in a group of seven-digit numbers. See Table 24.30 for part of the results. The source of this is Zarkovich (1966, Table 51), who cites Das Gupta and Mitra (1958).

22. One of the more unusual applications of bivariate distributions is by Midlarsky (1983), who sought to use them in understanding international alliances in nineteenth-century Europe. Table 4(a) in that paper gives the numbers of countries with i alliances in existence in 1816–1857 and j alliances in existence in 1858–1899. But, of course, the number of countries (15) is too small for anything subtle to appear. Not reproduced here.

24.7 Shapes of Living Things

24.7.1 Data Presented as Exact Values

1. A trivariate example: Jolicoeur and Mosimann (1960) gave data on the length, width, and height of painted turtles, and we reprint it as Table 24.31. That for the 24 females has also been considered by Pimentel (1979, Sections 4.3–4.4).

2. Four morphological measurements (length of 8th tergite, length of superuncus, length of right valva, and length of right forewing) on swallowtail butterflies: data for four species, having about 30 specimens each, was given by Brower (1959). The first two measurements are in Sokal and Rohlf (1969, Exercise 15.3). Not given here.

3. Three quadrivariate examples: four measurements (sepal length, sepal width, petal length, petal width) of 50 plants of each of three types of iris. Data collected by Dr Edgar Anderson (see Anderson, 1935), but published by Fisher (1936). Reproduced by Kendall (1966, Table I; 1975, Table 3.4),

Table 24.32: Length and breadth of beans.

Length (mm)	Breadth (mm) 6.375	6.625	6.875	7.125	7.375	7.625	7.875	8.125	8.375	8.625	8.875	9.125	Total
9.5	1												1
10.0	1	3	1	1	1								7
10.5	1	4	7	6	0								18
11.0	0	1	13	11	11								36
11.5	0	0	12	32	22	4							70
12.0	0	2	21	25	37	27	3						115
12.5	1	0	8	35	78	55	19	3					199
13.0			9	28	124	175	89	12					437
13.5			1	18	137	361	330	73	9				929
14.0				13	91	469	794	362	56	2			1787
14.5				1	23	236	871	913	227	23			2294
15.0					6	65	385	956	574	93	0	3	2082
15.5						4	81	375	494	156	19	0	1129
16.0						1	7	44	105	101	17	0	275
16.5								4	18	23	8	2	55
17.0										2	4	0	6
Total	4	10	72	170	530	1397	2579	2742	1483	400	48	5	9440

Kendall and Stuart (1976, Table 44.1), Gnanadesikan (1977, Exhibit 34a), Mardia et al (1979, Table 1.2.2), Chambers et al (1983, dataset 14), Andrews and Herzberg (1985, Dataset 1), and in part by Johnson and Wichern (1982, Table 10.4); other studies of it include Friedman and Rubin (1967), Wolfe (1970), Scott and Symons (1971), Duda and Hart (1973, p. 218), Gnanadesikan (1973), Friedman and Tukey (1974), Dagnelie (1975, Chapter 14), Devlin et al (1975), Binder (1978), O'Neill (1978), Rincon-Gallardo et al (1979), Tukey and Tukey (1981a,b), Everitt and Hand (1981, pp. 43–44 and 115–116), Andrews (1982), Isogai (1982), Koziol (1982, 1983), Nicholson and Littlefield (1983), Royston (1983), Tukey (1983), Barnett and Lewis (1984, Section 9.4.5), Gabriel (1985), Titterington et al (1985, pp. 173–174), Csörgő (1986), Becker, Cleveland, and Weil (1987), Becker, Cleveland, and Wilks (1987), Baringhaus and Henze (1988), and Lesaffre and Albert (1989). Not given here.

4. A 5-variate example: five measurements on sea cucumbers were given in Simon (1977), who took them from the data published by Edwards (1908); not given here.

5. A 7-variate example: seven measurements of molluscs (genus *Astarte*) from the Pliocene deposits north of Antwerp were given by Symons and Ringele (1976). The sample size was 150. Not given here.

6. McNeil (1979) listed the length and diameter of jellyfish from two locations near Sydney (22 + 24 specimens), attributing the data to K Pulley. Not given here.

7. Six 18-variate examples: 18 measurements of skulls of both sexes of three species of kangaroo, with 25 individuals in most of the six groups. This is Dataset No. 53 in Andrews and Herzberg (1985); not given here.

8. Greenacre (1984, Section 9.10) summarises an analysis of 17 measurements on 132 fossil skulls of cave bears, and states that the data matrix and a fuller report are in Greenacre (1974). Not given here.

Table 24.33: Cross-tabulation of the diameter and height of shortleaf pine trees.

		Height (feet)										
		15	20	25	30	35	40	45	50	55	60	65
	Stand 1											
	2	8	11	0	1							
	3	1	1	2	2	1						
	4		1	2	4	6						
Diameter	5				1	2	3	1				
	6				2	1	3	7	3			
	7						0	3	7	1		
(inches)	8						1	4	4	6	2	
	9								4	4	0	
	10								1	1	0	1
	11									2	0	0
	Stand 2											
	2	12	33	26								
	3	1	11	32	30	15	2	0	2			
	4		1	10	36	38	30	0	2			
Diameter	5				7	20	35	45	6			
	6				2	3	15	41	41	6	3	
	7						2	9	31	15	4	
(inches)	8						1	6	14	20	10	2
	9							2	6	14	6	0
	10								1	5	6	1
	11									2	0	0

Note also the various other datasets of biological measurements mentioned by Greenacre (1984, pp. 318–319).

9. The tail length and wing length of 15 examples of each of four subspecies of birds are listed by Brown and Hettmansperger (1987); not given here.

10. Manly (1986a, Table 1.1) gives 5 measurements on 49 female sparrows, collected moribund after a storm. There were 21 survivors and 28 nonsurvivors. Data taken from Bumpus (1898). See also Manly (1986b, 1988). Data not reproduced here.

11. Mardia et al (1979, p. 329) give two measurements on 18 males of each of two species of flea-beetles. The data is a subset of that in Lubischew (1962) (113 males, in five species, up to six measurements each). Not reproduced here.

24.7.2 Data Presented as Frequency Tables

1. The lengths and breadths of beans were cross-tabulated by Pretorius (1930), who took the data from Wicksell (1917); the ultimate source of it may be Johannsen (1903). Other studies of it include Johnson (1949), Gayen (1951), Bhatt and Dave (1965), Kendall and Stuart (1977, pp. 20–21, 172), Mardia (1970a, Examples 3.1, 4.1, and 8.1; 1970c), Blæsild (1981), and Blæsild and Jensen (1981). See our Table 24.32. (In Blæsild's paper, the entries in cells (11.0, 6.375) and (12.5, 6.375) are respectively 1 and 0, with corresponding changes in the row totals.)

2. Alexander (1961, Example 45.1) gave the lengths and weights of 157 lima beans. Not reproduced here.

3. Tienstra (1956, Table 1) gave the lengths and thicknesses of 350 pears, attributing the data to R Roelofs. Not reproduced here.

4. Lengths and girths of 955 eggs of the common tern: not reproduced here, see Watson et al (1923), in which there are numerous other tables, Tippett (1952b), Rahman (1972, pp. 206–207), and Chambers et al (1983, dataset 21).

5. Table 24.33 is of the diameter and height of shortleaf pine trees, in natural stands, and is taken from Schreuder and Hafley (1977).[1] (A comment on the data is made by Seheult, 1978.)

6. Trivariate datasets of the diameter, height, and volume of trees. Data on 321 loblolly pines, in 20 size categories: Schreuder et al (1982a). Data on 3105 loblolly pines, in 22 size categories: Schreuder et al (1982b). These datasets are not reproduced here. of 408 180-year-old pine trees. Data not reproduced here.

7. For the diameter and weight of 327 ears of maize, see Snedecor and Cochran (1967, Table 7.10.1), who cite Lindstrom (1935).

8. A 6-variate example: 6 bone measurements on 276 leghorn chickens were reported by Wright (1954), and are also considered by Pimentel (1979, Chapter 1). Not reproduced here.

9. For the femur length and humerus length of rabbits, see Waugh (1952, Table 15.5), who cites Castle (1916, p. 68). Not reproduced here.

10. For the diameters (antero-posterior and dorso-ventral) of the lower valve of *Pecten opercularis*, sample size = 537, see Yule and Kendall (1937, Table 11.1), who attribute the data to Davenport (1903).

11. The size of cell and size of body length in *Daphnia magna* were correlated by Warren (1903); see also Pearson (1905). Not reproduced here.

12. (a) Weight and stature of British women.
(b) Bust girth and stature of British women.
These datasets are not reproduced here. They may be found in Kendall and Stuart (1979, Tables 26.1 and 26.2), who took them from Kemsley (1957).

13. Weight and height of 285 Boston University women students: not reproduced here, see Mode (1966, Table 18.2).

14. Weight and height of military recruits: not reproduced here, see Martin (1960).

15. Sitting height and stature of 6999 men: not reproduced here, see Ruger (1932–33), in which there are numerous other bivariate tables, and Rahman (1968, Example 16.2).

16. Height and forearm length of 348 men: not reproduced here, see Galton (1888) and Snedecor and Cochran (1967, Table 7.4.1).

17. Hip and chest girths of 4157 six-year-old boys: not reproduced here, see Griffin (1962, Chart 13.3).

18. Weight and height of Glasgow schoolchildren of particular ages. Besides these two dimensions, there were three others in the data: sex, age (9 categories), and socioeconomic class of the district of the city (4 categories). Not reproduced here, see Elderton (1914) and also Isserlis (1915).

[1] For the first stand, the total of the frequencies is 104, but Schreuder and Hafley say there were 105 trees. For the second stand, the total of the frequencies is 662, but Schreuder and Hafley say there were 566 trees. We do not know the reason for these discrepancies.

Table 24.34: Weights of cork borings, representing the bark thickness of cork trees, measured in four directions.

Trees 1-10				Trees 11-20				Trees 21-28			
N	E	S	W	N	E	S	W	N	E	S	W
72	66	76	77	32	30	34	28	39	35	34	37
60	53	66	63	63	45	74	63	32	30	30	32
56	57	64	58	54	46	60	52	60	50	67	54
41	29	36	38	47	51	52	43	35	37	48	39
32	32	35	36	91	79	100	75	39	36	39	31
30	35	34	26	56	68	47	50	50	34	37	40
39	39	31	27	79	65	70	61	43	37	39	50
42	43	31	25	81	80	68	58	48	54	57	43
37	40	31	25	78	55	67	60				
33	29	27	36	46	38	37	38				

24.7.3 References to Scatterplots

1. Length and weight of herring (both log-transformed). 279 data points. $r = 0.99$. Schnute (1984).

2. Mantle length and cube root of body weight of Pacific squid. 28 data points. Barnett (1985b).

3. (a) Length and width of young snapping turtles. About 53 data points.
   (b) Lengths of radius and humerus of adult human skeletons. About 111 data points.
   (c) Lengths of femur and humerus of adult human skeletons. About 111 data points.
   Mosimann (1975a), the human skeleton measurements being from Schultz (1953).

4. (a) Length and width of humerus (log-transformed) of specimens of the North American marten. 92 + 47 data points (males and females, respectively).
   (b) Length and width of femur (log-transformed) of specimens of the North American marten. 92 + 47 data points (males and females, respectively).
   Jolicoeur (1963); see also Hills (1982).

5. No datasets are given in full, but nevertheless attention should be drawn to papers by Reyment (1971, 1985) and Bookstein (1986).

24.8 Biological Sciences

24.8.1 Data Presented as Exact Values

1. A quadrivariate example: Rao (1948a) gave data on the thickness of bark of 28 cork trees, measured from the north, east, south, and west directions—reproduced in our Table 24.34. This came to our notice via the recent paper by Best and Rayner (1988), who consider the N and E directions only. See also Pearson (1956) and Pettitt (1979).

2. Best and Rayner (1988) also gave 26 measurements, by different laboratories, of the fat and protein contents of a sample of milk. Not reproduced here.

3. Widths of cottonwood leaves when fresh (X) and when dry (Y) were given by Sokal and Rohlf (1969, Exercise 15.7), who attribute the data to T J Crovello. Not reproduced here. There was little shrinkage, and hence the correlation was high. We therefore surmise that any subtleties in the bivariate distribution should be sought in the joint distribution of X and either $X - Y$ or X/Y, rather than X and Y.

4. Total length of aphid stem mothers and the mean thorax length of their parthenogenetic offspring (based on measurement of four alates, or winged forms), sample size = 15: see Sokal and Rohlf (1969, Box 15.6). Not given here.

5. Gill weight and body weight in the crab *Pachygrapsus crassipes*, sample size = 12: see Sokal and Rohlf (1969, Box 15.1), who attribute the data to L Miller. Not given here.

6. A few datasets have been published relating to the bivariate normal model for the home range or territory of individual animals (see Section 19.3.3). Smith (1983) gives two datasets, both of the locations of a deer over a two or three month period (20 and 26 observations). Samuel and Garton (1985) give one of the locations of a black bear (25 observations). Not reproduced here.

7. An experiment was performed on a number of rats to determine their sensitivity to a harmless tactile stimulus. For each animal, X and Y were measures of sensitivity without and with having been dosed with an analgesic drug. Pairs (X, Y) may be regarded as having a bivariate distribution over the population of rats. The data is in Table 1 of Crowder (1989). In fact, there are 15 such distributions there—5 dose levels $\times$ 3 levels of lead. (The number of rats was about 8 in each case.) Crowder chose to (i) treat the data for the 3 lead levels separately, (ii) to parameterise the effect of dose as either a linear or quadratic function, and (iii) to use the distribution described in Section 5.6.5.

24.8.2 Data Presented as Frequency Tables

1. Over the winter of 1980–81, blackbirds were caught in mist nets and ringed and released as they entered two winter roosts in Fife, Scotland. The colours of the orbital ring around the eye and of the upper mandible were recorded; these change over the winter from black to yellow, and were recorded on 3- and 5-point scales, respectively. The results are given by Dale (1983, Section 3.4.2; 1985). Not reproduced here.

2. The age and yield of milk per week in 4912 Ayrshire cows is one of the examples that Kendall and Stuart (1977, p. 29) give. It also appears as Table 11.4 of Yule and Kendall (1937). The data comes from Tocher (1928), wherein there are many other tables. Not reproduced here.

3. Table 24.35, taken from Band and Ives (1963), shows the viability distribution of chromosomes of *Drosophila melanogaster* at 25°C relative to performance at 17°C, separately for heterozygotes and homozygotes. Viability refers to the percentage of the wild-type progeny which emerge among the four possibilities; 25% is the baseline, though the overall proportion for the heterozygotes in these experiments was a little greater, 27%, as is commonly found. The total number of chromosomes or combinations tested—the total table frequency—was 136 for the heterozygotes and 149 for the homozygotes. Part (a) of Table 24.35 was included in Exercise 15.2 of Sokal and Rohlf (1969).

4. For the lengths of the median cross vein in the right and left forewings of 500 worker bees, see Sokal and Rohlf (1969, Box 15.2), who cite Phillips (1929). Not reproduced here.

5. For a cross-tabulation of the length of a mother-frond of duckweed (when the daughter frond separated from it) and the length of the daughter frond (when its first daughter frond separated), with sample size 266, see Table 11.9 of Yule and Kendall (1937). Not reproduced here.

24.8.3 References to Scatterplots

1. Density of scarlet mites and rate of leaf deterioration. 81 data points. $r = 0.81$. Oomen (1984).

2. Mildew of wheat: proportion of leaves visibly diseased and mean number of pustules per leaf (i.e., incidence and severity of disease), both transformed. 42 data points. $r = 0.93$. Daamen (1986).

Table 24.35: Distribution of chromosome viability at two temperatures. (See text for further explanation.)

(a) Heterozygotes

At 25 deg. \ At 17 deg.	0	0	2	4	6	8	10	12	14	16	18	20	22	24	26	28	30	32	34	36	38	40
40																			1	0	0	1
38																		1	1			
36															1	0	0	0	1			
34													1	1	0	1	3	1	2			
32										1	0	0	0	0	3	5	3	3	0	0	1	
30												2	2	1	6	4	4	3	0	1	1	1
28												1	0	4	7	7	9	3	2	1		
26													3	2	7	4	3	2	2	1		
24													1	0	5	1	1					
22										1	0	0	0	2	2	2	0	0	2			
20									1	0	0	0	0	0	1	0	1					
18																	1					
16														1								

(b) Homozygotes

At 25 deg. \ At 17 deg.	0	0	2	4	6	8	10	12	14	16	18	20	22	24	26	28	30	32	34	36	38	40
42																	1					
40																						
38																						
36																						
34														1	1	0	1					
32														3	0	1	0	1				
30													1	1	1	2	0	0	0	0	0	1
28											1	1	4	4	3	1	0	0	1			
26										1	1	1	5	4	2							
24							1	0	0	1	0	0	2	2	3	0	2	2				
22					1	0	0	1	0	1	2	2	5	2	5	1	2					
20												1	4	0	1	0	0	1				
18							1	0	0	1	0	1	1	1	0	1						
16									1	0	0	0	0	2								
14								1	0	0	0	0	1									
12					1	0	0	1	1	0	1	1	0	1	0	1						
10		1	0	0	1	0	0	1	0	0	0	0	2									
8		1	0	1	0	0	0	1	0	0	0	0	1									
6		1	0	0	1	0	0	0	0	1												
4			1																			
2		3	2	0	0	1	0	0	1	0	1	0	0	1								
0		19	3	0	1																	

The row and column labels are upper class boundaries. Thus the entry at upper right of part (a) indicates that there was one chromosome for which the viability fell in the range 38-40% at both temperatures.

24.8.4 Cross-References

Sections 24.7 (Shapes of Living Things) and 24.9 (Medicine and Human Biology) provide additional examples from the biological sciences. In addition, there are the following:

- 24.4.3—1,2,3,4,5 (scatterplots only),
- 24.5.1—6,7,15,16,17,18,19 (compositional data),
- 24.5.2—6,7 (compositional data),
- 24.6—6 (lambing),
- 24.6—8 (species of diatoms),
- 24.6—10 (pistils and stamens),
- 24.6—11 (two species in a rain forest),
- 24.6—12 (Müllerian glands, pigs),
- 24.12.1—1 (biological quality of water),
- 24.13.2—1 (physical properties of spruce),
- 24.13.1—1 (strength of timber).

24.9 Medicine and Human Biology

24.9.1 Data Presented as Exact Values

1. Vital capacity of 200 patients, before and after oral surgery, is listed by Chilton (1967, Table 8.9), who cites Frank and Frasher (1940). Not reproduced here.

2. A 6-variate dataset, six blood chemical measurements on 30 healthy blood donors at Princeton Hospital, New Jersey, is listed by Simon (1974). Not reproduced here.

3. A dataset with 16 measured and 16 categorical variables on 117 male coronary patients and 117 age-matched healthy middle-aged men is given by Kronmal and Tarter (1974), the data being from the study by Kasser and Bruce (1969), though not printed therein. Kronmal and Tarter were concerned with a technique for producing estimates of univariate and bivariate densities without assuming a functional form for the distribution under consideration. Among their graphs are (lineprinter) contours of maximum vs. resting heart rate for the patient and the normal groups, and diastolic vs. systolic blood pressure, similarly. Not reproduced here.

4. Observations of blood non-protein nitrogen and serum chloride, in 37 patients with diabetic acidosis, are given by Bliss and Calhoun (1954, Example XV—6a), who cite Danowski et al (1946). Not reproduced here.

5. Winer and Cossette (1979) list two datasets of the length and width (and also their ratio) of chrysotile (asbestos) fibres, there being approximately 100 fibres in each dataset. Noted here because of the relevance to lung diseases. Not reproduced here.

6. If we understand the French correctly, there are two trivariate datasets in Romeder (1973, p. 127), respectively referring to three characteristics of the voices of 26 young female schizophrenics and of 27 young females considered normal. Not reproduced here.

Table 24.36: Periodontal condition and calcium level intake of 135 women.

		Calcium intake level (gm/day)						
		0-.25	.25-.40	.40-.55	.55-.70	.70-.85	.85-1.00	1.00+
Periodontal condition	A	0	5	3	10	4	4	3
	B	0	4	5	8	5	1	0
	C	7	19	11	3	5	0	1
	D	3	12	10	1	2	0	0
	E	3	5	1	0	0	0	0

Periodontal condition is ordered from A = healthy to E = very advanced disease.

7. Two variables (scores on psychological tests) are listed for 25 psychotics and 25 normals by Smith (1947). Two bivariate normal distributions, each with zero correlation, appears a reasonable hypothesis. See also Bartlett (1965, p. 208) and Mardia et al (1979, Example 11.3.3). Smith also gives two scores derived from intelligence testing, for 50 psychotics and 349 normals. Datasets not reproduced here.

8. Data on the urinary excretion rates for cortisol and cortisone of 27 patients with a form of Cushing's syndrome are listed by Aitchison and Dunsmore (1975, Table 4.1); see also Guttman (1988), who calculates tolerance regions based on the assumption of bivariate normality of the log-transformed rates. Data not reproduced here.

9. Royston (1983) lists six haematological variables measured on 103 West Indian or African paint sprayers in a car assembly plant. Not reproduced here.

24.9.2 Data Presented as Frequency Tables

1. Table 24.36 relates to the periodontal condition and calcium intake level of 135 women. It is taken from Radusch (1939), wherein are three other tables of similar type; condensed versions have been examined by Williams (1952) (who attributes it to Treloar, 1939, p. 228), Kendall and Stuart (1979, pp. 606–607), and Goodman (1981b, 1982).

2. The calcium and milk intake levels of 135 women were given as a scatterplot by Radusch (1939), and as a table of frequencies by Chilton (1967, Table 8.14). Not reproduced here.

3. The amounts of phosphate and of cholesterol in the peripheral blood of 136 patients with marked periodontal disease were cross-tabulated by Chilton (1967, Table 8.13), who cites Karshan et al (1946), though the data is not tabulated there. Not reproduced here.

4. Cross-tabulations of the length and diameter of amosite fibres found in lung tissue specimens and sampled at a mine in South Africa are given by Cheng (1986), who cites Pooley and Clark (1979). Not reproduced here.

5. The data of Table 24.37 is from Bishop and Fienberg (1969), who describe it as follows: "The following unpublished data were collected from patients at the Massachusetts General Hospital by Jones and Poskanzer. On admission, 121 hospitalized patients were graded on a five-point scale A–E of increasing severity according to their physical disability following a stroke. On discharge, they were re-rated on the same scale. As none of the patients had a second stroke their score on the second examination could only be the same or better than on the original examination." The data has also been considered by Altham (1975), Bishop and Fienberg (1969), Haberman (1974b, pp. 268–269), and Goodman (1979a).

Table 24.37: Initial and final ratings of stroke patients.

		Final rating				
		A	B	C	D	E
	A	5				
Initial	B	4	5			
	C	6	4	4		
rating	D	9	10	4	1	
	E	11	23	12	15	8

Table 24.38: School children in Caithness and Aberdeen classified by eye colour and hair colour.

	Hair colour (Caithness)					Hair colour (Aberdeen)				
	Fair	Red	Medium	Dark	Black	Fair	Red	Medium	Dark	Black
Eye colour										
Blue	326	38	241	110	3	1368	170	1041	398	1
Light	688	116	584	188	4	2577	474	2703	932	11
Medium	343	84	909	412	26	1390	420	3826	1842	33
Dark	98	48	403	681	85	454	255	1848	2506	112

6. Data on the eye and hair colour of Scottish children, Table 24.38, has recently been considered by Goodman (1981b, 1982, 1986—Reply to Discussion), Hill (1982), Greenacre (1984, Section 9.1), Kent (1986), and Gilula and Haberman (1986). It comes from Maung (1941), in which there are numerous similar tables, referring to different schools. The original survey was that reported by Tocher (1908). Other papers dealing with this data include Fisher (1940) and Lancaster (1958).

7. A table of eye and hair colour (3 and 4 categories, respectively) of 6800 people of Germanic stock is in Kendall and Stuart (1979, p. 616), who attribute it to O Ammon. See also Goodman and Kruskal (1954), Anderson and Sclove (1974, Table 5.21; 1978, Table 5.21), and Haberman (1982). Another table of such data (4×4, sample size = 592) is given by Snee (1974); see also Snee and Pfeifer (1983), and Diaconis and Efron (1985). A 5×5 table of hair colour in brothers is in Pearson (1904). A table, due to Galton, of eye colours in fathers and sons is in Pearson (1900); in condensed form, it appears as Table 5.4 of Yule and Kendall (1937). These tables are not reproduced here.

8. Table 24.39, data on left-eye and right-eye vision of men and women employed in ordnance factories in 1943–46, published by Stuart (1953), has been much discussed. No doubt there have been other studies besides those listed below.

   - Regarding the data for women, see Stuart (1953, 1955), Caussinus (1965), Bhapkar (1966, 1979), Ireland et al (1969), Bishop et al (1975, p. 284), Gokhale and Kullback (1978, Example 3 in Chapter 8), McCullagh (1978, 1980), Kendall and Stuart (1979, Table 33.5), Agresti (1980, 1983; 1984, p. 215), Plackett (1981, Table 3.5 and Example 6.6), Goodman (1982), Hettmansperger (1982), Upton (1985), and Margolin and Collings (1987).
   - As to the data for men, see Stuart (1953), Caussinus (1965), McCullagh (1978, 1980), Kendall and Stuart (1979, Table 33.3), Agresti (1980), Anscombe (1981), Plackett (1981, Table 7.3), Dale (1983, Section 3.5.3), Nguyen and Sampson (1987), and Margolin and Collings (1987).

See Tomizawa (1985, 1986) for a 4×4 table of right eye and left eye grades of 4746 students in the Faculty of Science and Technology, Science University of Tokyo, in April 1982.

Table 24.39: Unaided distance vision (1 = highest grade, 4 = lowest) in men and women aged 30–39.

		Left eye (Men)				Left eye (Women)			
		1	2	3	4	1	2	3	4
	1	821	112	85	35	1520	266	124	66
Right	2	116	494	145	27	234	1512	432	78
eye	3	72	151	583	87	117	362	1772	205
	4	43	34	106	331	36	82	179	492

9. Various measures of alcohol usage by 10th–12th grade adolescents are cross-tabulated by Rachal et al (1981). One example is an overall classification of the respondents according to quantity and frequency of drinking, vs. the highest amount ever used. Not reproduced here.

10. Two bivariate distributions of blood pressure and serum cholesterol, for people having or without coronary heart disease, originated from Cornfield (1962) and are given by Ku and Kullback (1974), Gokhale and Kullback (1978, Example 5 in Chapter 4), and Agresti (1984, Table 6.1). Not reproduced here.

11. Pearson and Lee (1903) give numerous two-way tables of measurements of parents and their children, among them the statures of mothers and sons (see also Derman et al, 1973, Example 5.1), the statures of fathers and daughters (see also Lancaster and Hamdan, 1964), and the statures of fathers and sons (but there is a slight printing error, the table being corrected in Pearson, 1904; see also Table 11.3 of Yule and Kendall, 1937, and Table 5.5 of Chakravarti et al, 1967). Not reproduced here.

12. Dermal ridge counts of parents and their children: not reproduced here, see Holt (1952), in which there are numerous other bivariate tables, and Acton (1959, Table 1). Strictly, the two variates are discrete.

24.9.3 References to Scatterplots

1. Concentrations of ethanol and methanol in blood of suspected drinking drivers. 519 data points. $r = 0.47$. Jones and Löwinger (1988).

2. Inside and outside diameters of nuclear rings. 424 data points. Arechaga et al (1987).

3. Rate of flow and acidity of the saliva. 85 data points. Anderson (1949); see also Chilton (1967, Figure 8.9).

4. Dental height in occlusion and in physiologic rest. 28 data points. Chilton (1967, Figure 11.1), who cites Wylie (1944).

5. Diameters of cerebral arteries.
   (a) Left posterior cerebral artery and left posterior communicating artery. 100 data points. $r = -0.48$.
   (b) As above, but right side. $r = -0.57$.
   Kamath (1981).

6. Concentrations of triglyceride and cholesterol in 320 diseased patients in a heart disease study. Silverman (1986, Figure 4.6), who cites Scott et al (1978). Silverman uses this data to illustrate estimating bivariate contours by kernel methods.

7. Serum calcium and serum parathyroid hormone (transformed). 544 data points. Brunden (1978).

8. AHF activity and AHF-like antigen (both log-transformed) in normal women and in haemophilia A carriers. 30 + 45 data points. McLachlan and Basford (1988, Figure 1 in Chapter 3), who cite

Habbema et al (1974). See also Ganesalingam and McLachlan (1979), Makov (1980), Smith and Spiegelhalter (1981), and Basford and McLachlan (1985).

9. Maximal ouabain binding capacity and serum T4 in thyrotoxic patients with a history of paralysis. 60 data points. $r = -0.57$. Lam et al (1989).

24.9.4 Cross-References

The following examples might have been included under the Medicine and Human Biology heading:

- 24.4.1—5 (lung vital capacity),
- 24.4.1—6 (stability of stained slides),
- 24.4.1—7 (agreement between pathologists),
- 24.4.2—1 (diphtheria antitoxin),
- 24.4.2—2 (histopathology disagreements),
- 24.4.2—3 (asbestosis),
- 24.4.2—4 (urothelial cell classification),
- 24.4.2—5 (renal-cell carcinoma),
- 24.4.2—6 (comparison of histology and colposcopy),
- 24.4.2—7 (staining intensity),
- 24.4.2—8 (diagnostic procedure),
- 24.4.2—9 (assessment of caries),
- 24.4.3—6,7,8,9 (references to scatterplots only),
- 24.6—1,2,3 (numbers of accidents),
- 24.6—4 (numbers of births),
- 24.6—5 (gender of children),
- 24.6—17 (drug side effects),
- 24.6—18 (oesophageal cancer),
- 24.6—19 (fingerprints),
- 24.7.2—13,14,15,16,17,18 (anthropometric data),
- 24.11.1—3 (heights of husbands and wives),
- 24.11.2—4 (mental health in Manhattan).

In addition, there are some medical examples in Section 24.5 (compositions, totalling 100%).

24.10 Mental Abilities

24.10.1 Data Presented as Exact Values

A 7-variate dataset: age, I.Q., and performance on tests of five subjects, of 109 fourth-grade pupils, are given in Appendix A of Walker (1943). Not reproduced here.

Table 24.40: Grade on the Armed Forces Qualification Test and years of school completed—whites, U.S.A.

		None	Elementary school			High school		College		
			1-4	5-6	7-8	1-2	3-4	1-2	3-4	5+
	I	0	2	1	31	228	3608	4420	1469	24
Mental	II	6	6	18	423	1892	15449	9252	1976	24
	III	25	44	155	2355	6317	22445	6270	941	9
group	IV	56	292	883	4331	5029	6363	721	72	0
	V	483	961	937	1428	757	280	14	3	0

The entries in this table are the expected numbers per 100,000 of population.

Mental groups I-V refer to percentile scores 100-93, 92-65, 64-31, 30-10, and 9 or below, respectively.

Table 24.41: Classification of 1725 schoolboys according to clothing and intelligence.

	Intelligence					
	A,B	C	D	E	F	G
Clothing						
I	33	48	113	209	194	39
II	41	100	202	255	138	15
III	39	58	70	61	33	4
IV,V	17	13	22	10	10	1

A denotes the mentally defective, G the very able.
I denotes the best clothing, V the worst.

24.10.2 Data Presented as Frequency Tables

1. Male whites, aged 19–21, were cross-classified by educational attainment and performance on the Armed Forces Qualification Test. The data, Table 24.40, comes from Karpinos (1966). There are actually 30 similar tables in that reference—3 racial groups (white, negro, total) $\times$ 10 geographical categories (continental U.S.A. + 9 regions).

2. Hayashi (1951) cross-tabulates amount of schooling and score on a literacy test, for 1000 Japanese of all ages. Not reproduced here.

3. Table 24.41, from Gilby (1911), shows the result when schoolboys from eight primary schools were classified according to quality of their clothing and the teacher's estimate of their intelligence. Among later authors who have considered this data are Pearson (1913b), Quensel (1966), Kendall and Stuart (1979, Table 33.2), Anscombe (1981, Chapter 12), and Dale (1983, Section 3.5.3; 1984).

4. A subject dear to the hearts of our academic readers will be the relationship between students' performances at school and at university. A cross-tabulation of A-level points score by degree classification was given by Green (1984), and is reproduced here as Table 24.42.

5. A bivariate frequency distribution of the scores on the California Test of Mental Maturity and the California Achievement Test of 182 sixth-graders in a public school is given by Games and Klare (1967, Figure XIII.12). Not reproduced here.

6. A bivariate frequency distribution of the scores of 500 Scottish schoolboys on a verbal intelligence test and the Binet intelligence test is given by Ferguson (1971, Table 8.2), who cites Macmeeken

Table 24.42: Points score at A-level and degree classification.

		Degree classification				
		I	IIa	IIb	III	Pass
	15	22	13	10	3	0
Points	14	20	21	31	9	2
	13	13	43	31	16	10
score	12	7	21	35	18	5
	11	3	21	26	32	8
at	10	3	17	25	20	12
	9	1	10	9	15	11
A-level	8	1	2	4	12	6
	7	0	1	2	6	1
	6	0	0	2	1	0

(1940). Not reproduced here.

7. Verbal and mathematical scores on the Scholastic Aptitude Test of the Princeton economics class of 1938, a sample size of 67, were given in Wilks (1948, Exercise 13.3.3). Not reproduced here.

24.10.3 Further References

1. A scatterplot of course grade vs. percent of classes missed by students, with about 117 data points, was given by Winkler (1980).

2. See also 24.4.2—15 (schoolmasters' judgments of pupils' abilities), 24.9.1—7 (intelligence testing of psychotics and normals), and 24.11.1—2 (National Child Development Study).

24.11 Sociology and Economics

24.11.1 Data Presented as Exact Values

The examples below are all from the recent book by Marsh (1988). None are reproduced here.

1. A dataset obtained from the 1979 General Household Survey (U.K.) lists 24 variables on 200 men. A number of the variables are ordinal, e.g., current social class and highest educational qualification. Marsh (1988, pp. 298–304).

2. A dataset obtained from the National Child Development Study (U.K.) records 10 variables on 238 individuals born in 1958. Several of the variables are ability scores, others reflect social class. Marsh (1988, pp. 309–314).

3. A dataset that includes the social classes and heights of husbands and wives (200 couples) is listed by Marsh at pp. 315–319.

4. At pp. 320–321, Marsh gives a dataset of family income and social class (50 family units).

5. A dataset of 10 features (population, unemployment rate, geographical area, etc.) of British towns (280 in number) is listed by Marsh (1988, pp. 324–330).

Table 24.43: Male subjects classified by their own and by their father's occupational status category.

(a) British data

		Subject's status							
		I	II	III	IV	Va	Vb	VI	VII
	I	50	19	26	8	7	11	6	2
	II	16	40	34	18	11	20	8	3
Father's	III	12	35	65	66	35	88	23	21
	IV	11	20	58	110	40	183	64	32
	Va	2	8	12	23	25	46	28	12
status	Vb	12	28	102	162	90	554	230	177
	VI	0*	6	19	40	21	158	143	71
	VII	0*	3	14	32	15	126	91	106

(b) Danish data

		I-IV	V	VI	VII	VIII -IX
	I-IV	18	17	16	4	2
Father's	V	24	105	109	59	21
	VI	23	84	289	217	95
status	VII	8	49	175	348	198
	VIII-IX	6	8	69	201	246

* These entries totalled 3 in the (condensed) table used by Wahrendorf (1980).
The classes I-IX are ordered from highest to lowest. The criteria for determining occupational status were different in the two studies.

24.11.2 Data Presented as Frequency Tables

1. Table 24.43 shows male subjects classified by their own and by their father's occupational status category. Points about this:

   - The source of the British data is Glass and Hall (1954), though the data there is in the form of a 7×7 table, and the first publication of the 8×8 table was by Miller (1960).
   - The source of the Danish data is Svalastoga (1959).
   - There are 20 similar tables, from various countries, in Miller (1960).
   - In the same book as Glass and Hall (1954), are contributions by Mukherjee (1954) that includes four tables of this type, and by Berent (1954) that includes cross-tabulations of the social origins of wives and husbands and of the education levels of wives and husbands. (Further tables of the last type, of U.S. origin, are the major examples in Chapters 8 and 9 of Haberman, 1979.)
   - Even our cursory examination of the literature has turned up the following studies of the British and/or the Danish data: White (1963), Goodman (1965, 1968, 1969, 1972, 1979b, 1981a), Mosteller (1968), Haberman (1974b, Chapter 6), Bishop et al (1975, Example 3.6—3), Everitt (1977, Sections 6.3–6.4), Morgan and Titterington (1977), Payne et al (1977), McCullagh (1978), Simon (1978), Duncan (1979), Hauser (1980), Wahrendorf (1980), Anscombe (1981, Chapter 12), Kimeldorf et al (1982), Jones and Pittelkow (1983), Agresti (1984, Table 11.1, and also p. 215), and Clogg (1986). No doubt there are others.

2. Table 24.44 is another cross-tabulation of son's and father's occupational group. Freeman (1981) reconstructed this data from information in Scudder and Anderson (1954). The two parts of the Table refer to sons who remained in a community and those who migrated.

Table 24.44: Comparison of occupational group of fathers and sons.

		Resident sons					Migrant sons				
		P	B	C	S	O	P	B	C	S	O
	P	16	15	0	1	0	12	9	1	1	0
Father's	B	7	85	37	27	12	21	66	23	18	6
	C	0	10	23	18	17	3	22	13	15	3
occup'n	S	0	9	12	72	36	3	27	12	51	19
	O	0	9	9	25	51	1	7	9	9	12

The five occupational groups are professional, business, clerical, skilled, and other.

Table 24.45: Voters in 1976 Wisconsin primaries cross-classified by party affiliation and political ideology.

	Conservative	Moderate	Liberal
Democrat	100	156	143
Independent	141	210	119
Republican	127	72	15

3. Table 24.45 shows voters in 1976 Wisconsin primary elections classified according to party affiliation and political ideology. It is taken from Agresti (1983; 1984, Table 5.3), who reconstructed it from Table 6 of Hedlund (1977–78).

4. Table 24.46, on mental health in Manhattan, has been much analysed. It is taken from Srole et al (1962, Table 12.1). See also Haberman (1974a; 1979, Section 6.1), Goodman (1979b, 1985a,b, 1986), Escoufier (1982), Agresti (1984, Table 7.6), Upton (1985), Gilula (1985, 1986), Agresti and Chuang (1986), and Gilula and Haberman (1986).

5. Table 24.47 concerns reported happiness as related to years of schooling; its origin was the (U.S.) General Social Survey, and it was published and analysed by Clogg (1980, 1982). See also Chuang-Stein and Francom (1988).

Table 24.46: Subjects in midtown Manhattan cross-classified by mental health status and parental socio-economic status.

	Parental socioeconomic stratum					
	A	B	C	D	E	F
Well	64	57	57	72	36	21
Mild symptom formation	94	94	105	141	97	71
Moderate symptom formation	58	54	65	77	54	54
Marked symptom formation	31	21	34	51	43	39
Severe symptom formation	10	11	23	32	27	22
Incapacitated	5	8	3	11	8	10

The strata A-F are ordered from highest to lowest.

Table 24.47: Cross-tabulation of reported happiness and years of schooling.

(a) Stratified by sex.

	Males				Females			
	<12	12	13-16	17+	<12	12	13-16	17+
N	40	21	14	3	62	26	12	3
H	131	116	112	27	155	156	95	15
V	82	61	55	27	87	127	76	15

(b) Stratified by number of siblings.

	0-1 siblings				2-3 siblings				4-5 siblings			
N	15	31	35	18	34	60	45	14	36	46	30	3
H	17	60	63	15	53	96	74	15	70	45	39	9
V	7	5	5	2	20	12	10	1	23	11	4	2

	6-7 siblings				8+ siblings			
N	22	25	13	3	61	26	8	4
H	67	40	24	2	79	31	7	1
V	16	12	4	0	36	7	3	1

Coding of reported happiness: N denotes not too happy, H pretty happy, and V very happy.

6. Further data from the General Social Survey—a trivariate dataset, satisfaction with three aspects of life (residence, hobbies, and family)—is given in Table 24.48. Again, Clogg (1980, 1982) is the source.

7. Among the datasets in Pretorius (1930), and considered by Bhatt and Dave (1965), are cross-tabulations of the ages of bride and bridegroom at marriage, and of mother and father at babies' births. (Another example of the former type of table is Table 11.2 of Yule and Kendall, 1937, which is reproduced as Table 5.1 of Zelditch, 1959.) Similar data is routinely published in the demographic statistics of some countries. See Cox (1985) for a discussion. No examples are reproduced here.

8. A trivariate dataset: attitude to abortion (3 categories), years of schooling (3 categories), ideal number of children (3 categories). Data from the 1972 General Social Survey used by Smith (1976) and Agresti (1977). Not reproduced here.

9. A trivariate dataset: aptitude level (5 categories), educational level (4 categories), occupational group (4 categories). Data in Gilula (1986), who cites Beaton (1975) and Fienberg (1980, pp. 44–48). Not reproduced here.

10. Table 24.49 relates total miles driven and family income. It is from Guerrero and Johnson (1980, 1988), who cite Holmes (1974, pp. 197–198).

11. Regarding poverty in India, Rao (1977) gives a table in which 11468 households are cross-classified according to calorie intake (per consumer unit) and monthly expenditure (per consumer unit), 6 categories and 10 categories respectively. Not reproduced here.

12. Kmietowicz (1984) has published a cross-tabulation of household size and income in rural Iraq. Not reproduced here.

Table 24.48: Three indicators of satisfaction with life.

L	R	C=1	C=2	C=3	C=4
1	1	76	14	15	4
1	2	32	17	7	3
1	3	64	23	28	15
1	4	41	11	27	16
2	1	15	2	7	4
2	2	27	20	9	5
2	3	57	31	24	15
2	4	27	9	22	16
3	1	13	6	13	5
3	2	12	13	10	6
3	3	46	32	75	20
3	4	54	26	58	55
4	1	7	6	7	6
4	2	7	2	3	6
4	3	12	11	31	15
4	4	52	36	80	101

L refers to hobbies, R to family, and C to residence, and responses 1-4 are in order of increasing satisfaction.

Table 24.49: Cross-tabulation of thousands of miles driven and family income of car owners (in 1973).

	Upper class limit of range of family income (dollars)									
Thousands of miles	3050	4900	6550	8700	10850	12900	15350	18500	23500	Higher
0-5	123	141	99	89	82	41	46	18	18	6
5-10	42	91	109	134	102	69	68	44	51	39
10-15	31	50	102	109	130	146	116	100	76	61
15-20	15	16	32	46	60	75	80	97	68	58
20-25	7	18	20	44	51	53	51	60	62	60
25-30	2	6	7	12	19	30	42	33	44	37
30-35	1	6	10	16	18	25	23	28	19	43
35+	6	9	13	16	25	36	33	39	44	64

13. Call discount rates and percentage of reserves on deposits in New York associated banks (weekly returns, sample size = 780) are cross-tabulated in Table 11.5 of Yule and Kendall (1937), who cite Norton (1902). See also Mardia (1970b), Reyment (1971), Rincon-Gallardo et al (1979), and Csörgő (1986). Not reproduced here.

14. Ham and Hsiao (1984) give a table of weeks worked in the previous year and hours per week usually worked (at the time of the interview), for Canadian prime-aged[2] males, as recorded in the Census. Not reproduced here.

15. The age and salary of 55 members of the Southern Graduate and Student Section of the Institute of Electrical Engineers are listed by Barnett and Lewis (1984, Table 9.1), the data being used to illustrate outlier-detection techniques. Not reproduced here.

24.11.3 Cross-References

The following examples might have been included under the Sociology and Economics heading:

- 24.4.2—16 (proxy responses),
- 24.4.2—17 (wages cross-check),
- 24.4.2—18 (income cross-check),
- 24.4.2—19 (occupational classification),
- 24.4.2—20 (homicide),
- 24.6—7 (purchases of breakfast cereals),
- 24.6—13 (newspaper reading),
- 24.6—15 (occupations of men and women),
- 24.6—16 (absences from work),
- 24.6—20 (working bullocks per household),
- 24.6—22 (nineteenth-century alliances),
- 24.9.2—9 (adolescent alcohol usage).

24.12 Air, Earth, Water, and Climate

24.12.1 Data Presented as Exact Values

1. Table 24.50 is a trivariate dataset: it gives three values of an index of biological quality of the water at 35 sites on 13 streams of the Taranaki ringplain in New Zealand. The index is called the MCI—macroinvertebrate community index—and reflects whether the macroinvertebrates found are pollution-tolerant types or include pollution-intolerant ones. High values indicate the presence of the latter, and thus good biological quality of the water. The data is taken from Stark (1985).

2. A 10-variate dataset concerning certain characteristics of seawater is given by Yamamoto (1976). Among the characteristics are salinity, median size of mineral grains, and combustible organic matter concentration. There were 164 seawater samples altogether; as listed, they could be assigned to three geographical areas. Not reproduced here.

[2]What a nice way with words the Canadians have! This refers to those aged 25–50.

Table 24.50: MCI values in winter, spring, and summer, at 35 sites on streams of the Taranaki ringplain.

Win	Spr	Sum	Win	Spr	Sum	Win	Spr	Sum	Win	Spr	Sum
154	156	143	126	162	120	129	113	123	92	95	78
146	143	132	130	142	133	129	124	121	92	92	92
137	123	148	133	120	127	132	138	110	93	84	75
150	133	154	132	137	130	103	118	118	43	70	76
132	140	132	140	143	122	105	110	121	63	51	51
136	137	142	126	141	122	128	106	86			
139	144	134	109	122	125	113	95	112			
140	146	132	140	125	128	106	93	109			
160	129	139	115	104	116	110	100	103			
139	140	143	128	122	120	86	57	95			

3. A 6-variate dataset is given in Table 24.51. It lists the sulphur dioxide concentrations in air at two points in Helsinki, for the 31 days of March 1983. For each day, the maximum and minimum (1-hour averages) are given along with the mean. The data is reported in Annex II of UNSC and ECE (1984); in Annex III of the same publication is a more extensive dataset (seven sites) from Budapest for the same month.

4. A dataset of 34 5-variate observations is given in Table 24.52. Four of the variates are summary measures of the climate for the months of April to December—sum of the average daily temperatures, hours of sunshine, number of very hot days, and level of rainfall. The fifth is the quality of Bordeaux wines (on a seven-point scale). The 34 years are 1924–1957. The data is from Table 1 of Dacunha-Castelle and Duflo (1986).

5. Rain volumes from clouds which were seeded and matched clouds which were not are given in Table 24.53. (The seeded cloud was randomly chosen from each pair.) The source of the data is Salsburg (1983, pp. 104–105).

6. Another cloud-seeding example is given by David and Fix (1961)—the rain (in inches) for two areas in southern California. (i) Data for 1957, "target" and "comparison", 13 pairs of observations in the seeded condition and 12 pairs in the not seeded condition. (ii) Data for 1958, when seeding operations were also taking place in other areas nearby. 12 pairs of observations in the seeded condition and 14 pairs in the not seeded condition. Not reproduced here.

7. A trivariate dataset: responses to "How pleased are you with the local climate?" and "How pleased are you with the construction of your home?" (both of these being on a 7-point scale), and the usage of gas during a winter, for a sample of homeowners in the planned community of Twin Rivers, New Jersey, are given by Mayer and Robinson (1974, 1975). Not reproduced here.

8. Table LXXVI of Brooks and Carruthers (1953) gives the mean atmospheric pressure and the total rainfall in January of the years 1881–1920, at Greenwich. Not reproduced here.

24.12.2 Data Presented as Frequency Tables

1. Table 24.54 refers to two nondimensional measures of wind gusting, denoted by T1 and T2, observed at Cape Canaveral, Florida. It is from Table 2 of Smith et al (1982).

2. There are three tables of the zonal (westerly) and meridional (southerly) components of wind speeds in the book by Essenwanger (1976, Figures 2.4–2.6). Not reproduced here.

Table 24.51: Sulphur dioxide concentrations (micrograms per cubic metre) at two points in Helsinki.

Days 1-16						Days 17-31					
At Hermanni			At Lansi-Pakila			At Hermanni			At Lansi-Pakila		
Mean	Max	Min	Mean	Max	Min	Mean	Max	Min	Mean	Max	Min
34	91	0	37	107	0	55	117	5	8	18	0
164	278	80	71	101	16	50	139	27	11	37	0
84	139	32	44	80	11	23	59	5	6	32	0
63	128	27	22	32	11	12	32	0	2	11	0
19	37	5	3	11	0	38	75	0	16	48	0
13	43	0	3	16	0	11	16	0			
64	123	0	19	69	0	11	32	0	3	16	0
25	64	0	24	59	0	32	72	8	3	24	0
39	117	5	17	48	0	24	51	3	8	40	0
130	203	84	43	91	0	16	32	8			
97	203	21	30	69	0	32	67	0	3	24	0
24	64	5	6	48	0	48	61	16		19	0
30	69	11	10	27	0	40	101	8		16	0
56	91	11	10	27	0	16	37	0	4	19	0
39	123	5	7	10	0	32	100	3	3	32	0
52	123	18	27	59	5						

Hermanni is an industrial and residential urban area, Lansi-Pakila is a suburban area.

3. Among the datasets in Pretorius (1930), and considered by Gayen (1951) and Bhatt and Dave (1965), are cross-tabulations of barometric pressure at Greenwich on alternate days for the period 1848–1926: three tables (whole year, summer months, winter months), the largest of which has 25 rows and 25 columns, and a total of 28855 observations. In Pearson (1925) (and also in Pearson and Lee, 1897), there is a table of contemporaneous barometric heights at Southampton and at Laudale. See also Pearson (1913b), Rhodes (1923), and Gayen (1951). This table has 30 rows and 26 columns, with 2922 observations. These tables are not reproduced here.

4. Gringorten (1971) cross-tabulates the temperatures at midnights 24 hours apart, for the month of July, the place being Minneapolis and the time period 1943–1952. See Table 24.55.

5. Table LIX of Brooks and Carruthers (1953) cross-tabulates the temperature and the relative humidity at 1pm for the 310 days of July 1931–1940, at Calshot, England. Not reproduced here.

6. Fisher and Hoblyn (1928) cross-tabulate daily maximum and minimum temperatures occurring at Rothamsted. There are separate tables for the twelve months, and the period covered is 1876–1926. Rahman (1972, p. 203) gives one of the tables. Not reproduced here.

7. Cornish and Evans (1964) cross-tabulate daily maximum and minimum temperatures occurring at Adelaide. There are twelve tables in that reference; the first refers to the first 30 days of the years 1861–1944, and has 46 rows, 57 columns, and a total of 2520 observations; the others are similar, referring to other times of the year. Not reproduced here.

8. There have been many tables published of the joint distribution of the heights and periods of water waves—we have mentioned some theoretical distributions used in this context in Section 10.8—and we have come across the following:

   - Figure 8(a) of Kimura (1981) is given here as Table 24.56; the data was collected in a wave tank experiment. (The variates are expressed in units of their respective root-mean-squares, that is

Table 24.52: Four climatic variables and the quality of Bordeaux wines, 1924–1957.

Yr	SADT	HS	NVHD	RF	Q	Yr	SADT	HS	NVHD	RF	Q
47	3478	1317	42	259	1	46	3061	1175	12	261	4
49	3458	1508	43	286	1	40	3094	1329	11	339	4
45	3381	1444	25	253	1	38	3019	1239	16	275	4
29	3267	1386	35	225	1	39	3022	1285	9	303	4
43	3308	1368	24	282	2	35	3182	1711	28	326	5
34	3317	1362	25	326	2	57	3043	1208	14	371	5
52	3270	1399	24	306	2	25	3000	1053	11	338	5
28	3245	1258	36	294	2						
55	3247	1277	19	375	2	30	3080	966	13	417	6
37	3221	1424	21	382	2	36	2998	1102	9	349	6
53	3198	1299	20	367	2	56	3083	1195	5	441	6
						27	3085	970	4	467	6
44	3212	1289	17	302	3	51	3052	1186	14	443	6
50	3252	1361	26	346	3	54	2904	1164	6	311	6
33	3318	1310	29	427	3	31	2974	1185	12	488	6
42	3227	1331	21	414	3	41	3009	1210	15	536	6
48	3126	1248	11	315	3						
26	3155	1133	19	393	3	32	3038	1103	14	677	7
24	3064	1201	10	361	3						

The variables are labelled as follows. Yr: Year. SADT: Sum of average daily temperatures. HS: Hours of sunshine. NVHD: Number of very hot days. RF: Level of rainfall (mm). Q: Quality of wines (1=best, 7=worst).

Table 24.53: Rain volume from seeded (x) and unseeded (y) clouds.

Trials 1-5		Trials 6-10		Trials 11-16	
x	y	x	y	x	y
129.6	26.1	302.8	0.0	200.7	68.5
31.4	26.3	119.0	17.3	274.7	81.2
2745.6	87.0	4.1	24.4	283.7	47.3
489.1	95.0	92.4	11.5	7.7	28.6
430.0	372.4	17.5	321.2	1656.0	830.1
				978.0	345.5

Table 24.54: Joint distribution of two nondimensional measures of wind gusting.

	T2				
	0-2	2-4	4-6	6-8	8+
T1					
0-2	18	28	23	7	2
2-4	0	22	17	6	6
4-6	0	6	6	4	0
6+	0	1	4	0	0

Table 24.55: Temperatures (Fahrenheit scale) at successive midnights, July in Minneapolis.

		Temperature 24 hr later						
		51-55	56-60	61-65	66-70	71-75	76-80	81-85
	51-55	0	3	1	1			
	56-60	2	3	11	7	2	1	
Midnight	61-65	0	8	29	25	10	3	
	66-70	1	4	27	30	24	7	
temperature	71-75	1	1	8	29	31	11	2
	76-80		1	3	7	9	9	2
	81-85				1	3	0	2

the normalisation referred to.)

- The book by Hogben and Lumb (1967), already referred to at 24.4.2—11,12 above, is perhaps the classic reference. In it are more than 3000 tables obtained by visual estimation by Voluntary Observing Ships sailing in different seas of the world, at different seasons of the year.
- Harris (1972a) gives three tables from sites in the U.S.A.
- Longuet-Higgins (1975) gives a scatter diagram of wave height and wavelength, for 399 consecutive waves from the Gulf of Mexico, attributing the data to C L Bretschneider.
- Chang et al (1976) give a scatter diagram of instrumentally-measured wave heights and periods during typhoons.
- Chakrabarti and Cooley (1977) measured 1624 waves from a North Atlantic storm, over an interval of $3\frac{1}{2}$ days. (See also Longuet-Higgins, 1983.)
- Thornton and Schaeffer (1979) show contours of the joint distributions of each pair from: wave height, wave period, and particle velocity. Their data referred to breaking waves, on a California beach. (Some researchers have a tough life.)
- Ochi (1979) reproduces three tables previously published by others.
- Goda (1979) analyses 89 series of observations, from five places in Japan; the method of presentation is interesting, as it involves grouping the series according to their correlation coefficients, resulting in five tables which refer to correlations −0.25 to 0.19, 0.20 to 0.39, 0.40 to 0.59, 0.60 to 0.69, and 0.70 to 0.81. Longuet-Higgins (1983) reconsidered some of this data.
- Bretschneider and Rocheleau (1979) give three tables of data from Hawaii.
- A table from an Ocean Weather Station in the eastern North Atlantic is given by Draper and Bownass (1983).
- There are 54 tables of data obtained at a point on the Japanese coast in Hotta et al (1983). A photographic method was used in this case.
- Shum and Melville (1984) give contour diagrams for two datasets, one being for calm conditions in the Pacific Ocean, the other for a hurricane in the Gulf of Mexico. The first of these is analysed using three different band passes.
- The paper by Myrhaug and Kjeldsen (1984) is concerned with the joint distribution of height and the following variables: crest front steepness, vertical asymmetry factor, and total wave steepness, as well as period. The data was from the North Sea.
- Krogstad (1985) gives a table referring to a site off the coast of Norway.
- Hameed and Baba (1985) give a table referring to the west coast of India during the monsoon season.
- Burrows and Salih (1987) give three tables from sites in Britain.

Table 24.56: The joint distribution of wave heights and periods, from a wave tank experiment.

Normalised period \ Normalised height	0.1	0.2	0.3	0.4	0.5	0.6	0.7	0.8	0.9	1.0	1.1	1.2	1.3	1.4	1.5	1.6	1.7	1.8	1.9	2.0	2.1	2.2	2.3	2.4+
0.1	1																							
0.2	8																							
0.3	12	6	1																					
0.4	1	12	7	4	1																			
0.5	1	7	13	12	12	1																		
0.6	1	6	12	11	17	10	10	2	1	1														
0.7		0	7	13	14	14	13	9	4	2	1	1	2	1										
0.8		0	7	4	9	13	19	9	5	13	6	3	2	1	1	2								
0.9		1	2	6	12	10	17	14	13	9	12	9	4	5	1	2	1							
1.0				2	8	7	11	6	5	15	9	11	7	7	5	3	2	4	2	2	0	0	1	2
1.1				4	2	8	8	7	7	12	15	8	8	9	8	3	1	4	2	2	0	1	1	0
1.2				0	2	1	5	6	10	9	9	12	6	13	5	6	6	2	4	3	1	0	1	2
1.3				0	1	2	5	3	8	4	10	11	10	5	4	8	5	3	3	1	2	3	0	1
1.4				1	0	2	4	4	7	10	10	12	10	7	3	5	4	1	1	0	1			
1.5					2	1	1	3	4	6	3	5	4	2	4	5	0	1	2					
1.6					1	0	3	2	2	3	3	3	2	2	0	0								
1.7							0	1	1	4	1	0	1	0	0	1								
1.8							2	1	0	0	1	0	1	0	0	1								
1.9									1															

9. Liu (1987) cross-tabulated wind speed and wave height on Lake Superior—see Table 24.57.

10. Kulatilake (1986) cross-tabulated dip angle and dip direction of a cluster of joint discontinuities in a metamorphic gneiss outcrop. The sample size was 37. Not reproduced here.

11. Yang et al (1984) compared the degree of damage to multi-storey brick buildings in the Tangshan and Haichen earthquakes with that predicted; two methods of prediction were used, the comparisons with actual damage being shown in the two parts of Table 24.58. (If further tables of predictions vs. observations are required, the weather forecasting literature is a good place to look.)

24.12.3 References to Scatterplots

1. Magnitude and length of wind gusts. 150 data points for each of 11 conditions—2 components × 6 altitudes, with one combination missing; the two components are referred to as zonal and meridional, which presumably mean westerly and southerly. Tubbs (1983b). There are also two examples in Smith and Adelfang (1981).

2. Wind speed and wave height during North Sea storms, both transformed. About 34 data points. Rosbjerg and Knudsen (1984).

3. High water levels at two points on the German North Sea coast during storm surges. About 50 data points. Christiansen and Seifert (1979).

4. In Figure 2 of Castano et al (1978), heights of floods at two points on a tributary river are given—at its confluence with the major river (X) and some 60 km upstream (Y). The form of the data is unusual: it is "L-shaped" truncated. That is, only those floods are shown for which $X > x_L$ or $Y > y_L$. Castano et al were concerned with how to choose a bivariate distribution $h(x, y)$ for use in

Table 24.57: Joint distribution of wind speed and significant wave height.

		Wind speed (m/s)								
		0-2	2-4	4-6	6-8	8-10	10-12	12-14	14-16	16-18
	20-21								3	
Significant	19-20								0	
	18-19							1	1	
	17-18						1	0	4	
wave	16-17						1	2	2	
	15-16						0	4	3	
	14-15						4	15	10	
height	13-14						3	6	6	
	12-13						6	6	2	
	11-12					5	2	4	1	1
(in units	10-11					3	10	9	0	
	9-10				1	11	28	5	1	
	8-9			2	8	43	56	22	0	
of thirds	7-8			2	22	70	41	9	0	
	6-7		1	7	59	88	45	3	2	
	5-6	2	5	50	209	129	26	4		
of a	4-5	6	30	138	278	111	13	1		
	3-4	24	88	315	385	76	5			
	2-3	82	491	1007	428	30	2			
metre)	1-2	331	1191	1146	125	1				
	0-1	1236	2229	750	21					

Table 24.58: Predicted and actual degree of earthquake damage.

	Prediction method: strength discrimination only — Actual damage						Prediction method: synthetical discrimination — Actual damage					
	B	Sl	M	Se	P	T	B	Sl	M	Se	P	T
Predicted												
B	19	5	3				19	1	1			
Sl	3	25	7	5	1		3	29	1			
M	2	7	51	25	3	2	2	8	59	4		
Se		2	9	73	3	2		1	9	102		
P				1	17	13				0	27	2
T				6	7	57				4	4	72

The grades of damage are: basically intact, slightly damaged, moderately damaged, seriously damaged, partial collapse, total collapse.

levee design studies. They proceeded by considering two conditional distributions, of (X, Y) given (i) $Y > y_L$, or (ii) $X > x_L$ and $Y < y_L$. It so happened that within both (i) and (ii), the correlation between X and Y was practically zero, so the bivariate aspects of the work from there on get less interesting.

5. Heights of long- and short-period waves. $r = 0.88$. Botes et al (1983).

6. Monthly total rainfall and monthly total rainfall energy index, storms greater than 0.51 cm, at Watkinsville, Georgia, 1940–1979. 480 data points. $r = 0.66$. Snyder and Thomas (1987).

7. Monthly total rainfall and monthly average temperature, at Watkinsville, Georgia, March–May 1937–1983. 141 data points. $r = 0.33$. Snyder and Thomas (1987).

8. Total depth of rainfall and (a) duration, (b) peak intensity, of rainstorms during July in Osaka. 163 data points. Etoh and Murota (1986).

9. Peak rainfall intensity and maximum storm surge of the sea level, typhoons, Osaka. About 117 data points. Hashino (1985).

10. Mass and the velocity of fall of snowflakes. Four scatterplots, totalling about 273 data points. Sasyo and Matsuo (1980).

11. Maximum acceleration of ground during an earthquake (as simulated by a computer program) and thickness of alluvium, for parts of Nagoya city. About 154 data points. Ichihara and Yamada (1979).

12. Percentages of magnesium and aluminium in a nickel ore. About 116 data points. Chiles (1984).

13. Two parameters of rock mechanical quality. 122 data points. Girardi and Matar de Sarquís (1984).

24.12.4 Cross-References

See Section 24.3 for data on floods of rivers, and sea levels, items 24.4.2—11 and 24.4.2—12 for wave height and period assessments by two methods, item 24.4.2—13 for cloud cover assessment by two methods, item 24.4.3—10 for seismic performances of buildings, predicted by three methods, and item 24.4.3—11 for grades of uranium ore, measured by two methods. Many of the examples in Section 24.5 (compositions, totalling 100%) are geological ones.

24.13 Physical and Technological Sciences

24.13.1 Data Presented as Exact Values

1. Six measurements of the strength characteristics of beams made from 55 red pine trees are given by Bliss and Calhoun (1954, Example XV—7), who cite a 1943 Yale University dissertation by J H Kraemer. The six measurements were the load applied at the proportional limit, the load at the breaking point, and the deflection at the proportional limit, for beams from two different parts of each tree. Not reproduced here.

2. For the density and porosity of 25 pellets of uranium dioxide, see Table 24.59. This is taken from Ainscough and Wheeler (1968), and is also in Lentner (1972, pp. 251–252).

3. Wilks (1948, Table 13.5) gives the thickness before and after rolling, at 88 positions of a coil of sheet metal, attributing the data to Milbourn. Not reproduced here.

4. Wilks (1948, Exercise 13.4.7) gives some trivariate data, the hardness, density, and tensile strength of 10 aluminium die castings. At Exercise 13.2.3, he attributes the data to Shewhart. Not reproduced here.

Table 24.59: The density and porosity of 25 pellets of sintered uranium dioxide.

Specimens 1-10		Specimens 11-20		Specimens 21-25	
Density (gm/cc)	Porosity (%)	Density (gm/cc)	Porosity (%)	Density (gm/cc)	Porosity (%)
10.673	2.55	10.600	3.21	10.639	2.86
10.675	2.53	10.614	3.09	10.651	2.75
10.661	2.66	10.614	3.09	10.638	2.87
10.651	2.75	10.612	3.10	10.642	2.83
10.661	2.66	10.624	2.99	10.653	2.73
10.660	2.67	10.630	2.94		
10.648	2.78	10.612	3.10		
10.662*	3.01	10.592	3.29		
10.600	3.21	10.560	3.58		
10.616	3.07	10.649	2.77		

* Printed as 10.612 in Lentner (1972).

5. The percentage of iron in the form of pig iron (as distinct from scrap iron) and the lime consumption are listed and plotted for 100 casts of steel by Tippett (1952a, Chapter 12), who cites the British Iron and Steel Federation (pre-1952). See also Chakravarti et al (1967, Table 5.13). Not reproduced here.

6. Decays of neutral V-particles observed in a cloud chamber: Page and Newth (1954) listed 26 observations, which were also given by Neyman (1955) and considered by Steck (c. 1958), of the decay time (t) and the potential period of observation (T). Not reproduced here, as T seems to be treated as an uninteresting feature of the experimental arrangement, whose importance is merely to limit the range of observable t ($0 < t < T$).

7. An astronomical example: two estimated quantities, distance from the galactic centre and angular velocity about the galactic centre, are given for 65 cepheid variable stars by Barton and David (1958), who took the data from Weaver (1955). Not reproduced here.

24.13.2 Data Presented as Frequency Tables

1. The modulus of elasticity and the density of spruce are cross-tabulated by Mitropol'skii (1966, Table 1), the sample size being 400; not given here.

2. The strength and the stretching of cotton thread, for a sample size of 1000, are cross-tabulated by Mitropol'skii (1966, Table 3), who attributes the data to E A Sankov; not given here.

3. Table 6 of Mitropol'skii (1966) refers to the yield strength and the toughness of steel, the sample size being 791; it is given here as Table 24.60.

24.13.3 Cross-References

Various of the examples in Sections 24.5 and 24.12 might have been included under the Physical and Technological Sciences heading. In addition, there are the following:

- 24.2—1,2,3,4 (mostly failure times),
- 24.3—4 (extreme values, spacecraft communication),
- 24.4.1—1,2,3,4 (replicated observations),

Table 24.60: The yield strength and the toughness of steel.

		Yield strength (kg/sqmm)								
		48	50	52	54	56	58	60	62	64
	4.5								1	1
	5.5					2	3	3	2	0
Toughness	6.5			5	7	21	27	21	10	1
	7.5	1	4	11	28	48	43	29	5	0
	8.5	0	4	22	60	63	38	23	1	3
	9.5	0	5	15	32	47	26	6	1	
(kgm/sqcm)	10.5	1	2	14	37	28	12	6	3	
	11.5	1	2	4	22	16	6	2		
	12.5	0	0	4	8	3	0	1		

- 24.4.2—10 (lightfastness of dyes),
- 24.4.3—8,9 (scatterplots only),
- 24.6—9 (counts of galaxies),
- 24.9.1—5 (asbestos fibres),
- 24.9.2—4 (amosite fibres),
- 24.11.1—2 (air pollution in Helsinki),
- 24.11.3—6,7 (scatterplots only).

24.14 Miscellaneous

24.14.1 Data Presented as Exact Values

1. Nine variables (price, length, weight, etc.) for 56 automobiles were listed in Table 1 of Fowlkes et al (1987), being a subset of dataset 7 of Chambers et al (1983). Not reproduced here.
2. The scores of 15 expert riflemen, firing when kneeling and when standing, are listed by Wilks (1948, Exercise 13.3.1), who attributes the data to Scarborough and Wagner. Not reproduced here.
3. For 15 law schools, the average score of new admissions on two tests is given by Efron (1979, 1982b) and Hinkley (1983). Not reproduced here.
4. Rounds were fired from a certain military weapon oriented at a fixed elevation angle and located at a fixed distance from the target. The deviations in deflection and range from the target were measured. Data for 30 rounds (out of a total of 150) are listed by Mason and Young (1985), who cite Bates (1966). Not reproduced here.

24.14.2 References to Scatterplots

There are numerous scatterplots in Thayer et al (1978) of some measure of computer software reliability vs. some explanatory variable. For example, their Figure 4.11 plots number of actual problems against number of statements in the routine, for 249 routines in a large command and control software package. However, it is perhaps more natural to think of this type of data in a regression context, rather than as a bivariate distribution.

References

In ordering these references, letters like ø and ö have been treated as if they were simply o.

Note also that there are so many works listed that in a number of instances there are two or more primary authors with the same surname; initials have taken precedence over year of publication in deciding the ordering, though where it appears that a single author sometimes uses and sometimes does not use a secondary initial, a single sequence is used.

The location(s) where each reference is cited is shown in square brackets. In the case of a citation within a Table, Figure, or footnote, the abbreviation T, F, or fn is used.

A

Abdel-Hameed, M, and Sampson, A R (1978). Positive dependence of the bivariate and trivariate absolute normal, t, χ^2, and F distributions. Annals of Statistics, 6, 1360–1368. [12.3.3, 12.3.5]

Abouammoh, A M, and Ahmed, A N (1988). The new better than used failure rate class of life distribution. Advances in Applied Probability, 20, 237–240. [18.2.1]

Abrahams, J, and Thomas, J B (1984). A note on the characterization of bivariate densities by conditional densities. Communications in Statistics—Theory and Methods, 13, 395–400. [8.1.3, 8.2.7, 8.4.5, 8.5]

Abramowitz, M, and Stegun, I A (1964). Handbook of Mathematical Functions. Applied Mathematics Series, No. 55, National Bureau of Standards. Washington, D.C.: U.S. Government Printing Office. (Republished subsequently—New York: Dover.) [T14.1]

Acton, F S (1959). Analysis of Straight-Line Data. New York: Wiley. [24.9.2—12]

Adachi, K, and Kodama, M (1980). Availability analysis of two-unit warm standby system with inspection time. Microelectronics and Reliability, 20, 449–455. [9.3.10]

Adegboye, O S, and Gupta, A K (1981). On a multivariate F-distribution and its application. American Statistical Association, 1981 Proceedings of the Social Statistics Section, pp. 314–317. [6.3.9]

Adelstein, A M (1952). Accident proneness: A criticism of the concept based upon an analysis of shunters' accidents. Journal of the Royal Statistical Society, Series A, 115, 354–400 (Discussion, 401–410). [24.6—3]

Agresti, A (1977). Considerations in measuring partial association for ordinal categorical data. Journal of the American Statistical Association, 72, 37–45. [24.11.2—8]

Agresti, A (1980). Generalized odds ratios for ordinal data. Biometrics, 36, 59–67. [24.9.2—8]

Agresti, A (1983). A survey of strategies for modeling cross-classifications having ordinal variables. Journal of the American Statistical Association, 78, 184–198. [24.9.2—8, 24.11.2—2]

Agresti, A (1984). Analysis of Ordinal Categorical Data. New York: Wiley. [24.4.2—20, 24.6—18, 24.9.2—8, 24.9.2—10, 24.11.2—1, 24.11.2—3, 24.11.2—4]

Agresti, A (1988). A model for agreement between ratings on an ordinal scale. Biometrics, 44, 539–548. [20.2.4, 24.4.1—7]

Agresti, A (1989). An agreement model with kappa as parameter. Statistics and Probability Letters, 7, 271–273. [10.9.4]

Agresti, A, and Chuang, C (1986). Bayesian and maximum likelihood approaches to order-restricted inference for models for ordinal categorical data. In R Dykstra, T Robertson, and F T Wright (Editors), Advances in Order Restricted Statistical Inference, pp. 6–27. (Lecture Notes in Statistics, No. 37.) Berlin: Springer. [24.11.2—4]

A Grivas, D, and Asaoka, A (1982). Slope safety prediction under static and seismic loads. Journal of the Geotechnical Engineering Division, Proceedings of the American Society of Civil Engineers, 108, 713–729. [6.5.12, 19.8.2]

Agterberg, F P (1984). Use of spatial analysis in mineral resource evaluation. Mathematical Geology (Journal of the International Association for Mathematical Geology), 16, 565–588 (Comments by J W Tukey, 591–594, and Reply by F P Agterberg, 595–600). [14.7]

Ahrens, J H, and Dieter, U (1972). Computer methods for sampling from the exponential and normal distributions. Communications of the Association for Computing Machinery, 15, 873–882. [16.3.1]

Ahrens, J H, and Dieter, U (1974). Computer methods for sampling from gamma, beta, Poisson and binomial distributions. Computing, 12, 223–246. [16.3.2]

Ainscough, J B, and Wheeler, M J (1968). The high-temperature thermal conductivity of sintered uranium dioxide. British Journal of Applied Physics (Journal of Physics, D), Series 2, 1, 859–868. [24.13.1—2]

Aitchison, J (1981). A new approach to null correlations of proportions. Mathematical Geology (Journal of the International Association for Mathematical Geology), 13, 175–189. [24.5.2—2, 24.5.2—3, 24.5.2—6]

Aitchison, J (1982). The statistical analysis of compositional data. Journal of the Royal Statistical Society, Series B, 44, 139–160 (Discussion, 161–177). [4.15, 24.5.2—3]

Aitchison, J (1984). Reducing the dimensionality of compositional data sets. Mathematical Geology (Journal of the International Association for Mathematical Geology), 16, 617–635. [24.5.2—2, 24.5.2—3, 24.5.2—4, 24.5.2—5]

Aitchison, J (1985). A general class of distributions on the simplex. Journal of the Royal Statistical Society, Series B, 47, 136–146. [4.15, 24.5.2—3]

Aitchison, J (1986). The Statistical Analysis of Compositional Data. London: Chapman and Hall. [4.15, 24.5.1, 24.5.2—2, 24.5.2—3, 24.5.2—4, 24.5.2—5]

Aitchison, J, and Bennett, J A (1970). Polychotomous quantal response by maximum indicant. Biometrika, 57, 253–262. [22.3.2,

23.5]

Aitchison, J, and Dunsmore, I R (1975). Statistical Prediction Analysis. Cambridge: Cambridge University Press. [24.9.1—8]

Aitchison, J, and Shen, S M (1980). Logistic-normal distributions: Some properties and uses. Biometrika, 67, 261–272. [4.15]

Aitchison, J, and Shen, S M (1984). Measurement error in compositional data. Mathematical Geology (Journal of the International Association for Mathematical Geology), 16, 637–650. [24.5.2—1]

Aitkin, M A, and Hume, M W (1965). Correlation in a singly truncated bivariate normal distribution. II. Rank correlation. Biometrika, 52, 639–643. [13.12]

Alexander, H W (1961). Elements of Mathematical Statistics. New York: Wiley. [24.7.2—2]

Al-Hussaini, E K, and Abd-el-Hakim, N S (1981). Bivariate inverse Gaussian distribution. Annals of the Institute of Statistical Mathematics, Tokyo, 33, 57–66. [7.9.8]

Ali, M M, Mikhail, N N, and Haq, M S (1978). A class of bivariate distributions including the bivariate logistic. Journal of Multivariate Analysis, 8, 405–412. [5.3.6, 5.3.9, 13.3]

Al-Saadi, S D, and Young, D H (1980). Estimators for the correlation coefficient in a bivariate exponential distribution. Journal of Statistical Computation and Simulation, 11, 13–20. [9.12.5]

Altham, P M E (1975). Quasi-independent triangular contingency tables. Biometrics, 31, 233–238. [24.9.2—5]

Amemiya, T (1975). Qualitative response models. Annals of Economic and Social Measurement, 4, 363–372. [5.8.13.4, 22.4.2.2]

Amemiya, T (1985). Advanced Econometrics. Oxford: Blackwell. (Parts of Chapter 9 appeared in "Qualitative response models: A survey", in Journal of Economic Literature, 19, 1483–1536 (1981). Chapter 10 is very similar to "Tobit models: A survey", in Journal of Econometrics, 24, 3–61 (1984).) [5.7.10, 19.9.5, 22.3.2, 22.3.3]

Amos, D E (1969). On computation of bivariate normal distribution. Mathematics of Computation, 23, 655–659. [4.12.4]

Amos, D E (1980). Algorithm 556. Exponential integrals. ACM (Association for Computing Machinery) Transactions on Mathematical Software, 6, 420–428 (Remark, 9, 525). [8.2.2]

Amos, D E, and Bulgren, W G (1972). Computation of a multivariate F distribution. Mathematics of Computation, 26, 255–264. [6.3.3, 6.3.11, 7.8.8, 20.7.2]

Andersen, P K , and Væth, M (1988). Survival analysis. In Encyclopedia of Statistical Sciences. Volume 9, pp. 119–129. New York: Wiley. [2.15.3—8]

Anderson, D J (1949). The hydrogen-ion concentration of the saliva, II. The relationship between hydrogen-ion concentration and rate of flow of saliva. Journal of Dental Research, 28, 583–588. [24.9.3—3]

Anderson, E (1935). The irises of the Gaspe Peninsula. Bulletin of the American Iris Society, 59, 2–5. [24.7.1—3]

Anderson, T W (1958). An Introduction to Multivariate Analysis. New York: Wiley. [4.18.1]

Anderson, T W, and Ghurye, S G (1978). Unique factorization of products of bivariate normal cumulative distribution functions. Annals of the Institute of Statistical Mathematics, Tokyo, 30, 63–70. [4.9]

Anderson, T W, and Sclove, S L (1974). Introductory Statistical Analysis. Boston: Houghton Mifflin. [24.9.2—7]

Anderson, T W, and Sclove, S L (1978). An Introduction to the Statistical Analysis of Data. Boston: Houghton Mifflin. [24.9.2—7]

Andrews, D F (1982). Andrews function plots. In Encyclopedia of Statistical Sciences. Volume 1, pp. 85–87. New York: Wiley. [24.7.1—3]

Andrews, D F, Gnanadesikan, R, and Warner, J L (1971). Transformations of multivariate data. Biometrics, 27, 825–840. [4.13.6]

Andrews, D F, Gnanadesikan, R, and Warner, J L (1973). Methods for assessing multivariate normality. In P R Krishnaiah (Editor), Multivariate Analysis—III, pp. 95–116. New York: Academic Press. [4.13.1]

Andrews, D F, and Herzberg, A M (1985). Data. A Collection of Problems from Many Fields for the Student and Research Worker. New York: Springer. [24.1.2.1, 24.7.1—3, 24.7.1—7]

Anscombe, F J (1981). Computing in Statistical Science through APL. New York: Springer. [4.2, 5.7.7, 5.7.9, 5.7.10, 24.9.2—8, 24.10.2—3, 24.11.2—1]

Antle, C E (1985). Lognormal distribution. In Encyclopedia of Statistical Sciences. Volume 5, pp. 134–136. New York: Wiley. [2.15.1—21]

Antle, C E, and Bain, L J (1988). Weibull distribution. In Encyclopedia of Statistical Sciences. Volume 9, pp. 549–556. New York: Wiley. [2.15.1—51]

Apostolakis, G, and Moieni, P (1987). The foundations of models of dependence in probabilistic safety assessment. Reliability Engineering, 18, 177–195. [6.5.12]

Arbous, A G, and Kerrich, J E (1951). Accident statistics and the concept of accident-proneness. Biometrics, 7, 340–432. [24.6—2, 24.6—3]

Arbous, A G, and Sichel, H S (1954). New techniques for the analysis of absenteeism data. Biometrika, 41, 77–90, plus a Table. [24.6—16]

Arechaga, J, Diaz, J, and Bahr, G F (1987). Morphometry of nuclear rings at high electron microscopic resolution. Analytical and Quantitative Cytology and Histology, 9, 335–340. [24.9.3—2]

Arnold, B C (1967). A note on multivariate distributions with specified marginals. Journal of the American Statistical Association, 62, 1460–1461. [15.3.1, 15.5]

Arnold, B C (1975a). A characterization of the exponential distribution by geometric compounding. Sankhyā, Series A, 37, 164–173. [9.12.12, fn17.4]

Arnold, B C (1975b). Multivariate exponential distributions based on hierarchical successive damage. Journal of Applied Probability, 12, 142–147. [T17.1, 17.3.1, 17.3.4, fn17.4]

Arnold, B C (1983). Pareto distributions. (Statistical Distributions in Scientific Work, Volume 5.) Fairland, Maryland: International Co-operative Publishing House. [2.9, 5.2.12, 5.7.9, 5.8.12, 5.8.13.2, 9.2.12, 20.1, 20.7.2]

Arnold, B C (1985). Pareto distribution. In Encyclopedia of Statistical Sciences. Volume 6, pp. 568–574. New York: Wiley. [2.15.1—31]

Arnold, B C (1987a). Bivariate distributions with Pareto conditionals. Statistics and Probability Letters, 5, 263–266. [8.3.4]

Arnold, B C (1987b). Dependence in conditionally specified distributions. Presented at the Hidden Valley Conference on Dependence. (Proceedings to be published by the Institute of Mathematical Statistics as Topics in Statistical Dependence.) [8.2.6]

Arnold, B C, and Strauss, D J (1987). Bivariate distributions with conditionals in prescribed exponential families. Technical Report No. 151, Department of Statistics, University of California, Riverside. [8.2.6]

Arnold, B C, and Strauss, D (1988). Bivariate distributions with exponential conditionals. Journal of the American Statistical Association, 83, 522–527. [8.2.7]

Arnold, B C, and Zahedi, H (1988). On multivariate mean remaining life functions. Journal of Multivariate Analysis, 25, 1–9. [18.3.5]

Arnold, S F (1988). Union-intersection principle. In Encyclopedia of Statistical Sciences. Volume 9, pp. 417–420. New York: Wiley. [4.13.5.4]

Aroian, L A (1947). The probability function of the product of two normally distributed variables. Annals of Mathematical Statistics, 18, 265–271. [4.9]

Aroian, L A (1978). Mathematical forms of the distribution of the product of two normal variables. Communications in Statistics, Part A—Theory and Methods, 7, 165–172. [4.9]

Aroian, L A (1986). The distribution of the quotient of two correlated normal random variables. American Statistical Association, 1986 Proceedings of the Business and Economic Statistics Section, pp. 612–613. [4.9]

Ashford, J R (1981). General models for the joint action of mixtures of drugs. Biometrics, 37, 457–474. [22.4.1]

Ashford, J R (1986). Quantal response analysis. In Encyclopedia of Statistical Sciences. Volume 7, pp. 402–408. New York: Wiley. [22.1]

Ashford, J R, and Sowden, R R (1970). Multi-variate probit analysis. Biometrics, 26, 535–546. [5.7.10, 22.4.1]

Askey, R (1970). Orthogonal polynomials and positivity. In Studies in Applied Mathematics. 6: Wave Propagation and Special Functions, pp. 64–85. Philadelphia: Society for Industrial and Applied Mathematics. [14.6.2]

Assaf, D, Langberg, N A, Savits, T H, and Shaked, M (1984). Multivariate phase-type distributions. Operations Research, 32, 688–702. [9.7.11, 9.13]

Atkinson, A C, and Pearce, M C (1976). The computer generation of beta, gamma and normal random variables. Journal of the Royal Statistical Society, Series A, 139, 431–448 (Discussion, 448-461). [16.3.2]

Awad, A M, Azzam, M M, and Hamdan, M A (1981). Some inference results on Pr(X < Y) in the bivariate exponential model. Communications in Statistics, Part A—Theory and Methods, 10, 2515–2525. [17.6.2]

Azlarov, T A, and Volodin, N A (1986). Characterization Problems Associated with the Exponential Distribution. New York: Springer. (Original Soviet edition was dated 1982.) [9.2.7]

B

Bacon-Shone, J (1985). Algorithm AS 208. Fitting a multivariate logistic normal distribution by the method of moments. Applied Statistics, 34, 81–86. [4.15]

Bacon-Shone, J, and Fung, W K (1987). A new graphical method for detecting single and multiple outliers in univariate and multivariate data. Applied Statistics, 36, 153–162. [4.13.2]

Bagchi, S B, and Samanta, K C (1984). On some compound bivariate Weibull distributions in life testing. IAPQR (Indian Association for Productivity, Quality and Reliability) Transactions, 9, 41–49. [20.6.1]

Bagchi, S B, and Samanta, K C (1985). A study of some properties of bivariate Burr distributions. Bulletin of the Calcutta Mathematical Society, 77, 370–383. [5.2.12, 5.8.13.3, 6.8]

Bain, L (1983). Gamma distribution. In Encyclopedia of Statistical Sciences. Volume 3, pp. 292–298. New York: Wiley. [2.15.1—8]

Bain, L J, and Engelhardt, M (1973). Interval estimation for the two-parametric double exponential distribution. Technometrics, 15, 875–887. [24.3—2]

Band, H T, and Ives, P T (1963). Genetic structure of populations. I. On the nature of the genetic load in the South Amherst population of *Drosophila melanogaster*. Evolution, 17, 198–215. [24.8.2—3]

Banerjee, A K (1977). A bivariate inverse Gaussian distribution (Preliminary report). (Abstract only.) Institute of Mathematical Statistics Bulletin, 6, 138–139. [7.9.8]

Barbu, Gh (1987). A new fast method for computer generation of gamma and beta random variables by transformations of uniform variables. Statistics, 18, 453–464. [16.3.2]

Baringhaus, L, and Henze, N (1988). A consistent test for multivariate normality based on the empirical characteristic function. Metrika, 35, 339–348. [4.13.5.3, 24.7.1—3]

Barlow, R E, and Proschan, F (1975). Statistical Theory of Reliability and Life Testing. Probability Models. New York: Holt, Rinehart and Winston. [12.2.1, 17.1, 18.1]

Barlow, R E, and Proschan, F (1977). Techniques for analyzing multivariate failure data. In C P Tsokos and I N Shimi (Editors), The Theory and Applications of Reliability. Volume 1, pp. 373–396. New York: Academic Press. [9.3.11, 24.2—1]

Barlow, R E, and Proschan, F (1981). Statistical Theory of Reliability and Life Testing. Silver Springs, Maryland: To Begin With. [fn12.1, 12.2.1, 12.3.1, 12.3.3, 17.2.2, fn17.8, 18.2.1, 18.3.2]

Barndorff-Nielsen, O (1977). Exponentially decreasing distributions for the logarithm of particle size. Proceedings of the Royal Society, Series A, 353, 401–419. [7.10.6]

Barndorff-Nielsen, O (1978). Hyperbolic distributions and distributions on hyperbolae. Scandinavian Journal of Statistics, 5, 151–157. [7.10.6]

Barndorff-Nielsen, O (1982). Exponential families. In Encyclopedia of Statistical Sciences. Volume 2, pp. 587–596. New York: Wiley. [2.15.1—5]

Barndorff-Nielsen, O (1983). On a formula for the distribution of the maximum likelihood estimator. Biometrika, 70, 343–365. [7.11.6, 8.10]

Barndorff-Nielsen, O, and Blæsild, P (1983a). Hyperbolic distributions. In Encyclopedia of Statistical Sciences. Volume 3, pp. 700–707. New York: Wiley. [2.15.1—12, 7.10.6]

Barndorff-Nielsen, O, and Blæsild, P (1983b). Reproductive exponential families. Annals of Statistics, 11, 770–782. [8.10]

Barndorff-Nielsen, O E, and Blæsild, P (1988). Reproductive models. In Encyclopedia of Statistical Sciences. Volume 8, pp. 86–89. New York: Wiley. [2.15.2—49]

Barndorff-Nielsen, O, Kent, J, and Sørensen, M (1982). Normal variance-mean mixtures and *z* distributions. International Statistical Review, 50, 145–159. [2.4.8]

Barnett, V D (1969). Simultaneous pairwise linear structural relationships. Biometrics, 25, 129–142. [24.4.1—5]

Barnett, V (1979). Some outlier tests for multivariate samples. South African Statistical Journal, 13, 29–52. [4.2, 5.8.10. 5.9.3, 5.10.9, 5.11.5, 11.5.2, 24.6—3]

Barnett, V (1980). Some bivariate uniform distributions. Communications in Statistics, Part A—Theory and Methods, 9, 453–461 (Correction, 10, 1457). [1.6, 4.7, 4.14.3, 5.1.3, 5.2.7, 5.7.7, 5.11.1, 5.11.5, 6.2.13, fn13.6]

Barnett, V (1983a). Reduced distance measures and transformation in processing multivariate outliers. Australian Journal of Statistics, 25, 64–75. [4.13.2, 5.8.10, 5.11.5]

Barnett, V (1983b). Principles and methods for handling outliers in data sets. In T Wright (Editor), Statistical Methods and the Improvement of Data Quality, pp. 131–166. New York: Academic Press. [4.13.2, 5.8.10, 5.10.9]

Barnett, V (1985a). The bivariate exponential distribution; A review and some new results. Statistica Neerlandica, 39, 343–357. [5.10.5, 9.1, 11.5.2]

Barnett, V (1985b). Detection and testing of different types of outlier in linear structural relationships. Australian Journal of Statistics, 27, 151–162. [4.13.2, 24.7.3—2]

Barnett, V, and Lewis, T (1984). Outliers in Statistical Data. 2nd Edition. Chichester: Wiley. [4.13.2, 5.8.10, 5.10.8, 24.7.1—3, 24.11.2—15]

Barrett, J F, and Lampard, D G (1955). An expansion of some second-order probability distributions and its applications to noise problems. IEEE (Institute of Electrical and Electronics Engineers) Transactions on Information Theory, IT-1, 10–15. [T14.1, 14.8]

Bartlett, M S (1965). Multivariate statistics. In T H Waterman and H J Morowitz (Editors), Theoretical and Mathematical Biology, pp. ???–??? New York: Blaisdell. [24.9.1—7]

Bartlett, N R, and Lwin, T (1984). Estimating a relationship between different destructive tests on timber. Applied Statistics, 33, 65–72. [19.8.4, 24.2—4]

Barton, D E, and David, F N (1958). Models of functional relationship illustrated on astronomical data. Bulletin de l'Institut International de Statistique, 37, Book 3, 9–33 (Discussion, Book 1, 68–71). [19.8.6, 24.13.1—7]

Basford, K E, and McLachlan, G J (1985). Likelihood estimation with normal mixture models. Applied Statistics, 34, 282–289. [19.4.4, 24.9.3—8]

Basu, A P (1971). Bivariate failure rate. Journal of the American Statistical Association, 66, 103–104. [18.3.2]

Basu, A P (1981). The estimation of $P(X < Y)$ for distributions useful in life testing. Naval Research Logistics Quarterly, 28, 383–392. [17.6.1, 17.6.2]

Basu, A P (1988a). Reliability, probabilistic. In Encyclopedia of Statistical Sciences. Volume 8, pp. 24–29. New York: Wiley. [2.15.2—48]

Basu, A P (1988b). Multivariate exponential distributions and their applications in reliability. In P R Krishnaiah and C R Rao (Editors), Handbook of Statistics. Volume 7, Quality Control and Reliability, pp. 467–476. Amsterdam: North-Holland. [9.1, 17.2.2]

Basu, A P, and Block, H W (1975). On characterizing univariate and multivariate exponential distributions with applications. In G P Patil, S Kotz, and J K Ord (Editors), A Modern Course on Distributions in Scientific Work. Volume 3—Characterizations and Applications, pp. 399–421. Dordrecht: Reidel. [9.2.7]

Basu, A P, and Ebrahimi, N (1984). On k-order harmonic new better than used in expectation distributions. Annals of the Institute of Statistical Mathematics, Tokyo, 36, 87–100. [18.2.1]

Basu, A P, and Ebrahimi, N (1986). HNBUE and HNWUE distributions—A survey. In A P Basu (Editor), Reliability and Quality Control, pp. 33–46. Amsterdam: North–Holland. [18.3.5]

Basu, A P, and Ebrahimi, N (1987). On a bivariate accelerated life test. Journal of Statistical Planning and Inference, 16, 297–304.

Basu, A P, Ebrahimi, N, and Klefsjö, B (1983). Multivariate harmonic new better than used in expectation distributions. Scandinavian Journal of Statistics, 10, 19–25. [18.3.5]

Bates, C (1966). The chi-square test of goodness of fit for a bivariate normal distribution. Technical Memorandum No. K-77/66, U.S. Naval Weapons Laboratory, Dahlgren, Virginia. [19.8.9, 24.14.1—4]

Battjes, J A (1971). Run-up distributions of waves breaking on slopes. Journal of the Waterways, Harbors and Coastal Engineering Division, Transactions of the American Society of Civil Engineers, 97, 91–114. [9.12.13]

Bauer, H (1972). Probability Theory and Elements of Measure Theory. New York: Holt, Rinehart and Winston. [11.5.1]

Baughman, A L (1988). A FORTRAN function for the bivariate normal integral. Computer Methods and Programs in Biomedicine, 27, 169–174. [4.12.1, 4.12.8]

Baxter, L A (1988). Relevation. In Encyclopedia of Statistical Sciences. Volume 8, pp. 14–16. New York: Wiley. [2.15.2—46, fn5.1]

Beaton, A E (1975). The influence of education and ability on salary and attitudes. In F T Jaster (Editor), Education, Income, and Human Behavior, pp. 365–396. New York: McGraw-Hill. [24.11.2—9]

Bechhofer, R E, and Dunnett, C W (1987). Four Sets of Tables. Percentage Points of Multivariate Student t Distributions. (Selected Tables in Mathematical Statistics, Volume 11.) Providence, Rhode Island: American Mathematical Society. [6.2.12]

Becker, P J, and Roux, J J J (1981). A bivariate extension of the gamma distribution. South African Statistical Journal, 15, 1–12. [9.3.14, 9.13]

Becker, R A, Cleveland, W S, and Weil, G (1987). The use of brushing and rotation for data analysis. Bulletin of the International Statistical Institute, 52, Book 3, 369–386 (Discussion, 401-407). [24.7.1—3]

Becker, R A, Cleveland, W S, and Wilks, A R (1987). Dynamic graphics for data analysis. Statistical Science, 2, 355–383 (Discussion, 383–395). [24.7.1—3]

Bedall, F K, and Zimmermann, H (1976). On the generation of $N(\mu, \Sigma)$-distributed random vectors by $N(0, 1)$-distributed random numbers. Biometrische Zeitschrift, 18, 467–471. [16.5.1]

Ben-Akiva, M, and Lerman, S R (1985). Discrete Choice Analysis: Theory and Application to Travel Demand. Cambridge, Massachusetts: MIT Press. [22.3.3]

Bentler, P M, and Berkane, M (1985). Developments in the elliptical theory generalization of normal multivariate analysis. American Statistical Association, 1985 Proceedings of the Social Statistics Section, pp. 291–295. [7.2.1]

Bera, A, and John, S (1983). Tests for multivariate normality with Pearson alternatives. Communications in Statistics—Theory and Methods, 12, 103–117. [4.13.5.4, 4.13.5.8, 4.13.5.8]

Berent, J (1954). Social mobility and marriage: A study of trends in England and Wales. In D V Glass (Editor), Social Mobility in Britain, pp. 321–338. London: Routledge and Kegan Paul. [24.11.2—1]

Bergman, B (1985). On reliability theory and its applications. Scandinavian Journal of Statistics, 12, 1–30 (Discussion, 30–41). [18.1]

Bergman, B, and Klefsjö, B (1988). Total time on test transform. In Encyclopedia of Statistical Sciences. Volume 9, pp. 297–300. New York: Wiley. [2.15.2—61]

Berkane, M, and Bentler, P M (1986a). Moments of elliptically distributed random variates. Statistics and Probability Letters, 4, 333–335. [7.2.10]

Berkane, M, and Bentler, P M (1986b). Characterizing parameters of elliptical distributions. American Statistical Association, 1986 Proceedings of the Social Statistics Section, pp. 278–279. [7.2.10]

Berkane, M, and Bentler, P M (1987). Characterizing parameters of multivariate elliptical distributions. Communications in Statistics—Simulation and Computation, 16, 193–198. [7.2.10]

Berland, R, and Dussauchoy, A (1973). Aspects statistiques des régimes de microdécharges electriques entre électrodes métalliques placées dans un vide industriel. Vacuum, 23, 415–421. [8.6.10]

Best, D J (1978a). Letter to the editor. Applied Statistics, 27, 181. [16.3.2]

Best, D J (1978b). A simple algorithm for the computer generation of random samples from a Student's t or symmetric beta distribution. In L C A Corsten and J Hermans (Editors), COMPSTAT 1978. Proceedings in Computational Statistics, pp. 341–347. Wien:

Physica-Verlag. [16.3.1, 16.3.4]

Best, D J, and Rayner, J C W (1988). A test for bivariate normality. Statistics and Probability Letters, 6, 407–412. [4.13.5.9, 24.8.1—1, 24.8.1—2]

Bhapkar, V P (1966). A note on the equivalence of two criteria for hypotheses in categorical data. Journal of the American Statistical Association, 61, 228–235. [24.9.2—8]

Bhapkar, V P (1979). On tests of marginal symmetry and quasi-symmetry in two and three-dimensional contingency tables. Biometrics, 35, 417–426. [24.9.2—8]

Bhaskara Rao, M, Krishnaiah, P R, and Subramanyam, K (1987). A structure theorem on bivariate positive quadrant dependent distributions and tests for independence in two-way contingency tables. Journal of Multivariate Analysis, 23, 93–118. [12.3.1]

Bhatt, N M, and Dave, P H (1964). A note on the correlation between polynomial transformations of normal variates. Journal of the Indian Statistical Association, 2, 177–181. [4.14.5]

Bhatt, N M, and Dave, P H (1965). Change in normal correlation due to exponential transformations of standard normal variates. Journal of the Indian Statistical Association, 3, 46–54. [24.7.2—1, 24.11.2—7, 24.12.2—3]

Bhattacharya, B N, and Nath, D C (1985). On some bivariate distributions of number of births. Sankhyā, Series B, 47, 372–384. [24.6—4]

Bhattacharya, S K (1966). A modified Bessel function model in life testing. Metrika, 11, 131–144. [2.6.7]

Bhattacharya, S K, and Holla, M S (1963). Bivariate life-testing models for two component systems. Annals of the Institute of Statistical Mathematics, Tokyo, 15, 37–43. [9.3.15]

Bhattacharya, S K, and Holla, M S (1965). On a life distribution with stochastic deviations in the mean. Annals of the Institute of Statistical Mathematics, Tokyo, 17, 97–104. [2.6.7]

Bhattacharya, S K, and Kumar, S (1986). E-IG model in life testing. Calcutta Statistical Association Bulletin, 35, 85–90. [2.6.7]

Bhattacharyya, B B (1985). Multinomial probit: Multinomial logit. In Encyclopedia of Statistical Sciences. Volume 5, pp. 665–668. New York: Wiley. [22.1]

Bildikar, S, and Patil, G P (1968). Multivariate exponential-type distributions. Annals of Mathematical Statistics, 39, 1316–1326. [fn4.1]

Bilodeau, M (1989). On the monotone regression dependence for Archimedian bivariate uniform. Communications in Statistics—Theory and Methods, 18, 981–988. [5.4.9, 5.9.5, 10.4.4, fn12.8]

Binder, D A (1978). Bayesian cluster analysis. Biometrika, 65, 31–38. [24.7.1—3]

Birnbaum, Z W (1956). On a use of the Mann-Whitney statistic. In J Neyman (Editor), Proceedings of the 3rd Berkeley Symposium on Mathematical Statistics and Probability. Volume 1, pp. 13–17. Berkeley: University of California Press. [17.6.1]

Birnbaum, Z W, and Saunders, S C (1969). A new family of life distributions. Journal of Applied Probability, 6, 319–327.

Bishop, Y M M, and Fienberg, S E (1969). Incomplete two-dimensional contingency tables. Biometrics, 25, 119–128. [24.9.2—5]

Bishop, Y M M, Fienberg, S E, and Holland, P W (1975). Discrete Multivariate Analysis: Theory and Practice. Cambridge, Massachusetts: MIT Press. [24.9.2—8, 24.11.2—1]

Biswas, S, and Nair, G (1984). A generalization of Freund's model for a repairable paired component based on a bivariate Geiger Muller (G. M.) counter. Microelectronics and Reliability, 24, 671–675. [9.3.10]

Bjerager, P, and Skov, K (1982). A simple formula approximating the normal distribution function. In Euromech 155: Reliability Theory of Structural Engineering Systems, pp. 217–231. Lyngby: Danish Engineering Academy. [4.12.2]

Blæsild, P (1981). The two-dimensional hyperbolic distribution and related distributions, with an application to Johannsen's bean data. Biometrika, 68, 251–263. [7.10.5, 7.10.6, 7.10.7, 24.7.2—1]

Blæsild, P, and Jensen, J L (1981). Multivariate distributions of hyperbolic type. In C Taillie, G P Patil, and B A Baldessari (Editors), Statistical Distributions in Scientific Work. Volume 4—Models, Structures, and Characterizations, pp. 45–66. Dordrecht: Reidel. [7.10.5, 7.10.7, 24.7.2—1]

Blake, I F (1979). An Introduction to Applied Probability. New York: Wiley. [3.2.1, 11.5.2]

Blischke, W R (1978). Mixtures of distributions. In W H Kruskal and J M Tanur (Editors), International Encyclopedia of Statistics. Volume 1, pp. 174–180. New York: Free Press. [19.5.3.1]

Bliss, C I, and Calhoun, D W (1954). An Outline of Biometry. New Haven, Connecticut: Yale Co-Operative Corporation. [24.5.2—7, 24.9.1—4, 24.13.1—1]

Block, H W (1975). Continuous multivariate exponential extensions. In R E Barlow, J R Fussell, and N D Singpurwalla (Editors), Reliability and Fault Tree Analysis. Theoretical and Applied Aspects of System Reliability and Safety Assessment, pp. 285–306. Philadelphia: Society for Industrial and Applied Mathematics. [9.5.7]

Block, H W (1977a). A family of bivariate life distributions. In C P Tsokos and I N Shimi (Editors), The Theory and Applications of Reliability. Volume 1, pp. 349–371. New York: Academic Press. [17.3.1, 17.3.4, 17.3.6.2]

Block, H W (1977b). Multivariate reliability classes. In P R Krishnaiah (Editor), Applications of Statistics, pp. 79–88. Amsterdam: North-Holland. [18.3.2]

Block, H W (1977c). A characterization of a bivariate exponential distribution. Annals of Statistics, 5, 808–812. [17.2.2]

Block, H W (1985). Multivariate exponential distribution. In Encyclopedia of Statistical Sciences. Volume 6, pp. 55–59. New York: Wiley. [9.1, 17.1]

Block, H W, and Basu, A P (1974). A continuous bivariate exponential extension. Journal of the American Statistical Association, 69, 1031–1037. [9.3.8, 9.5.5, 9.5.7, 17.2.2]

Block, H W, and Paulson, A S (1984). A note on infinite divisibility of some bivariate exponential and geometric distributions arising from a compound process. Sankhyā, Series A, 46, 102–109. [17.3.5]

Block, H W, and Rao, B R (1973). A beta warning-time distribution and a distended beta distribution. Sankhyā, Series B, 35, 79–84. [8.6.11]

Block, H W, and Sampson, A R (1983). Inequalities on distributions: Bivariate and multivariate. In Encyclopedia of Statistical Sciences. Volume 4, pp. 76–82. New York: Wiley. [18.2.5]

Block, H W, and Savits, T H (1976). The IFRA closure problem. Annals of Probability, 4, 1030–1032. [18.2.2]

Block, H W, and Savits, T H (1980a). Laplace transforms for classes of life distributions. Annals of Probability, 8, 465–474. [18.2.2]

Block, H W, and Savits, T H (1980b). Multivariate increasing failure rate distributions. Annals of Probability, 8, 793–801. [9.16, 18.3.3]

Block, H W, and Savits, T H (1981a). Multivariate classes in reliability theory. Mathematics of Operations Research, 6, 453–461. [18.3.5, 18.4]

Block, H W, and Savits, T H (1981b). Multivariate distributions in reliability theory and life testing. In C Taillie, G P Patil, and B A Baldessari (Editors), Statistical Distributions in Scientific Work. Volume 5—Inferential Problems and Properties, pp. 271–288. Dordrecht: Reidel. [18.4]

Block, H W, and Savits, T H (1982). The class of MIFRA lifetime and its relation to other classes. Naval Research Logistics Quarterly, 29, 55–61. [18.3.3]

Block, H W, and Savits, T H (1988). Multivariate nonparametric classes in reliability. In P R Krishnaiah and C R Rao (Editors), Handbook of Statistics. Volume 7, Quality Control and Reliability, pp. 121–129. Amsterdam: North-Holland. [18.4]

Block, H W, Savits, T H, and Shaked, M (1982a). Some concepts of negative dependence. Annals of Probability, 10, 765–772. [12.4.1]

Block, H W, Savits, T H, and Shaked, M (1982b). Negative dependence. In J Crowley and R A Johnson (Editors), Survival Analysis, pp. 206–215. Hayward, California: Institute of Mathematical Statistics. [12.4.1]

Block, H W, Savits, T H, and Shaked, M (1985). A concept of negative dependence using stochastic ordering. Statistics and Probability Letters, 3, 81–86. [12.4.1]

Block, H W, and Ting, M L (1981). Some concepts of multivariate dependence. Communications in Statistics, Part A—Theory and Methods, 10, 749–762. [12.2.1]

Blomqvist, N (1950). On a measure of dependence between two random variables. Annals of Mathematical Statistics, 21, 593–600. [12.3.1]

Blumen, I, and Ypelaar, M A (1980). A family of bivariate distributions for rank-based statistics with an application to Kendall's tau. American Statistical Association, 1980 Proceedings of the Business and Economic Statistics Section, pp. 386–390. [8.8.5]

Bondesson, L (1988a). Shot-noise processes and distributions. In Encyclopedia of Statistical Sciences. Volume 8, pp. 448–452. New York: Wiley. [2.15.2—54]

Bondesson, L (1988b). T_1- and T_2-classes of distributions. In Encyclopedia of Statistical Sciences. Volume 9, pp. 157–159. New York: Wiley. [2.15.1—46]

Booker, J M, Johnson, M E, and Beckman, R J (1984). Investigation of an empirical probability measure based test for multivariate normality. American Statistical Association, 1984 Proceedings of the Statistical Computing Section, pp. 208–213. [4.13.5.8]

Bookstein, F L (1986). Size and shape spaces for landmark data in two dimensions. Statistical Science, 1, 181–222 (Discussion, 222–242). [24.7.3—5]

Borth, D M (1973). A modification of Owen's method for computing the bivariate normal integral. Applied Statistics, 22, 82–85. [4.12.4]

Botes, W A M, Russell, K S, and Huizinga, P (1983). Resonance in South African harbours. In B L Edge (Editor), Proceedings of the Eighteenth Coastal Engineering Conference. Volume I, pp. 439–453. New York: American Society of Civil Engineers. [24.12.3—5]

Bouver, H (1987). Application of the noncentral chi-square distribution to the probability of correlation between radar track data. American Statistical Association, 1987 Proceedings of the Statistical Computing Section, pp. 367–373. [19.8.7]

Bouver, H, and Bargmann, R E (1979). Comparison of computational algorithms for the evaluation of the univariate and bivariate normal distribution. In J F Gentleman (Editor), Proceedings of Computer Science and Statistics: 12th Annual Symposium on the Interface, pp. 344–348. [4.12.2, 4.12.8]

Bouver, H, and Bargmann, R E (1981). Evaluation and graphical application of probability contours for the bivariate normal distribution. American Statistical Association, 1981 Proceedings of the Statistical Computing Section, pp. 272–277. [4.13.3.2, 16.5.1]

Bowman, K O, and Shenton, L R (1982). Approximations to distributions. In Encyclopedia of Statistical Sciences. Volume 1, pp. 107–118. New York: Wiley. [2.15.2—1]

Bowman, K O, and Shenton, L R (1983). Johnson's system of distributions. In Encyclopedia of Statistical Sciences. Volume 4, pp. 303–314. New York: Wiley. [2.4.3, 2.15.1—16]

Bowman, K O, and Shenton, L R (1986). Moment ($\sqrt{b_1}$, b_2) techniques. In R B D'Agostino and M A Stephens (Editors), Goodness-of-Fit Techniques, pp. 279–329. New York: Dekker. [4.13.4.3, 8.12.2]

Box, G E P, and Cox, D R (1964). An analysis of transformations. Journal of the Royal Statistical Society, Series B, 26, 211–243 (Discussion, 244–252). [4.13.6]

Box, G E P, and Muller, M E (1958). A note on the generation of random normal deviates. Annals of Mathematical Statistics, 29, 610–611. [16.3.1]

Box, G E P, and Tiao, G C (1973). Bayesian Inference in Statistical Analysis. Reading, Massachusetts: Addison-Wesley. [4.13.6]

Branston, D (1979). A method of estimating the free speed distribution for a road. Transportation Science, 13, 130–145. [9.18.3]

Bratley, P, Fox, B L, and Schrage, L E (1983). A Guide to Simulation. New York: Springer. [16.2.1, 16.3.1, 16.3.2, 16.3.3, 16.3.5]

Brelsford, W M, and Relles, D A (1981). STATLIB. A Statistical Computing Library. Englewood Cliffs, New Jersey: Prentice-Hall. [4.2.8, 16.5.1]

Breslow, N (1982). Covariance adjustment of relative-risk estimates in matched studies. Biometrics, 38, 661–672. [24.6—18]

Bretschneider, C L, and Rocheleau, R E (1979). An evaluation of extreme wave climate at Keahole Point, Hawaii. In Proceedings of the Sixteenth Coastal Engineering Conference. Volume I, pp. 152–173. New York: American Society of Civil Engineers. [24.12.2—8]

Brewer, D W, Tubbs, J D, and Smith, O E (1987). A differential equations approach to the modal location for a family of bivariate gamma distributions. Journal of Multivariate Analysis, 21, 53–66. (Also appears as Chapter III of J D Tubbs, D W Brewer, and O E Smith (1983), Some Properties of a Five-Parameter Bivariate Probability Distribution. NASA Technical Memorandum 82550, Marshall Space Flight Center, Alabama.) [7.5.12, 7.7.11]

Brindley, E C, and Thompson, W A (1972). Dependence and aging aspects of multivariate survival. Journal of the American Statistical Association, 67, 822–830. [12.2.1, 18.3.2]

British Iron and Steel Federation (pre-1952). Statistical Methods in Industry. [24.13.1—5]

Broffitt, J D (1986). Zero correlation, independence, and normality. The American Statistician, 40, 276–277. [4.18.1]

Brooks, C E P, and Carruthers, N (1953). Handbook of Statistical Methods in Meteorology. London: HMSO. [24.12.1—8, 24.12.2—5]

Brower, L P (1959). Speciation in butterflies of the *Papilio glaucus* group. I. Morphological relationships and hybridization. Evolution, 13, 40–63. [24.7.1—2]

Brown, B M, and Hettmansperger, T P (1987). Invariant tests in bivariate models and the L_1 criterion. In Y Dodge (Editor), Statistical Data Analysis Based on the L_1 Norm and Related Methods, pp. 333–344. Amsterdam: North-Holland. [24.7.1—9]

Brown, C C, and Fears, T R (1981). Exact significance levels for multiple binomial testing with application to carcinogenicity screens. Biometrics, 37, 763–774. [5.2.11]

Brown, J L (1958). A criterion for the diagonal expansion of a second-order probability distribution in orthogonal polynomials. IRE (Institute of Radio Engineers) Transactions on Information Theory, IT-4, 172. [14.5]

Brucker, J (1979). A note on the bivariate normal distribution. Communications in Statistics, Part A—Theory and Methods, 8, 175–177. [4.5, 8.5]

Brunden, M N (1978). The probability contours and a goodness-of-fit test for the singly truncated bivariate normal distribution. Communications in Statistics, Part A—Theory and Methods, 7, 557–572. [4.16.1, 24.9.3—7]

Bryan-Jones, J (1987). A tutorial in computer algebra for statisticians. The Professional Statistician, 6(10), 5–8. [3.1.4]

Bryson, M C (1983). Heavy-tailed distributions. In Encyclopedia of Statistical Sciences. Volume 3, pp. 598–601. New York: Wiley. [2.15.2—13]

Bryson, M C, and Johnson, M E (1982). Constructing and simulating multivariate distributions using Khintchine's theorem. Journal of Statistical Computation and Simulation, 16, 129–137. [10.2.2, 10.5.1, 10.5.3]

Bryson, M C, and Siddiqui, M M (1969). Some criteria for aging. Journal of the American Statistical Association, 64, 1472–1483. [18.2.1]

Buchanan, B (1988). A model for repeat paired comparison preference tests. Psychometrika, 53, 209–221. [20.5.3]

Buchanan, W B, and Singpurwalla, N D (1977). Some stochastic characterizations of multivariate survival. In C P Tsokos and I N Shimi (Editors), The Theory and Applications of Reliability, pp. 329–348. New York: Academic Press. [18.3.1, 18.3.4, 18.3.5]

Bucklew, J A, and Gallagher, N C (1978). Quantization of bivariate circularly symmetric densities. In Proceedings of the Sixteenth Annual Allerton Conference on Communication, Control, and Computing, pp. 982–990. Urbana-Champaign: Department of Electrical Engineering and Coordinated Science Laboratory, University of Illinois. [4.10]

Buishand, T A (1984). Bivariate extreme-value data and the station-year method. Journal of Hydrology, 69, 77–95. [9.17.11]

Bumpus, H C (1898). The elimination of the unfit as illustrated by the introduced sparrow, *Passer domesticus*. Biological Lectures from the Marine Biology Laboratory, Woods Hole. Eleventh lecture, pp. 209–226. [24.7.1—10]

Burn, D A (1987). Software review of "Advanced Simulation and Statistics Package: IBM Professional FORTRAN Version" by P A W Lewis, E J Orav, and L Uribe, 1986. The American Statistician, 41, 324–327. [16.3.6]

Burnaby, T P (1966). Growth-invariant discriminant functions and generalised distances. Biometrics, 22, 96–110. [4.2]

Burns, E (1982). Bivariate normal distribution estimates of the prevalence of reading disabilities. Journal of Special Education, 16, 431–437. [19.6.1]

Burns, E (1983). Bivariate estimates of giftedness. Journal of School Psychology, 21, 261–262. [19.6.1]

Burns, E (1984). The bivariate normal-distribution and the IQ of learning-disability samples. Journal of Learning Disabilities, 17, 294–295. [19.6.1]

Burrows, R, and Salih, B A (1987). Statistical modelling of long-term wave climates. In B L Edge (Editor), Twentieth Coastal Engineering Conference, Proceedings. Volume I, pp. 42–56. New York: American Society of Civil Engineers. [8.12.5.1, 9.12.13, 19.7.10, 24.12.2—8]

Butler, J S, Anderson, K H, and Burkhauser, R V (1985). Work and health: A bivariate hazard model with correlated heterogeneity components. American Statistical Association, 1985 Proceedings of the Business and Economic Statistics Section, pp. 65–73. [22.4.2.4]

Butler, J S, Anderson, K H, and Burkhauser, R V (1986). Testing the relationship between work and health. A bivariate hazard model. Economics Letters, 20, 383–386. [fn22.4]

Butler, J S, Anderson, K H, and Burkhauser, R V (1989). Work and health after retirement: A competing risks model with semiparametric unobserved heterogeneity. Review of Economics and Statistics, 71, 46–53. [9.18.3, 22.4.2.4]

C

Cadwell, J H (1951). The bivariate normal integral. Biometrika, 38, 475–479. [4.12.6]

Cambanis, S (1977). Some properties and generalizations of multivariate Eyraud-Gumbel-Morgenstern distributions. Journal of Multivariate Analysis, 7, 551–559. [5.2.16]

Cambanis, S, Huang, S, and Simons, G (1981). On the theory of elliptically contoured distributions. Journal of Multivariate Analysis, 11, 368–385. [7.2.3]

Cambanis, S, and Liu, B (1971). On the expansion of a bivariate distribution and its relationship to output of a nonlinearity. IEEE (Institute of Electrical and Electronics Engineers) Transactions on Information Theory, IT-17, 17–25. [14.8]

Cambanis, S, Simons, G, and Stout, W (1976). Inequalities for $Ek(X,Y)$ when marginals are fixed. Zeitschrift für Wahrscheinlichkeitstheorie und verwandte Gebiete, 36, 285–294. [12.6]

Camisani-Calzolari, F A G M (1984). Geostatistical appraisal of a tabular uranium deposit in South Africa. In G Verly, M David, A G Journel, and A Marechal (Editors), Geostatistics for Natural Resources Characterization. Part 2, pp. 935–949. Dordrecht: Reidel. [24.4.3—11]

Campbell, B J, and Reinfurt, D W (1973). Relationship between driver crash injury and passenger car weight. Report from the Highway Safety Research Center, University of North Carolina at Chapel Hill. [21.2]

Carr, P F (1981). Distinction between Permian and post-Permian igneous rocks in the southern Sydney Basin, New South Wales, on the basis of major-element geochemistry. Mathematical Geology (Journal of the International Association for Mathematical Geology), 13, 193–200. [24.5.2—5]

Carter, R W (1951). Floods in Georgia. Geological Survey Circular, No. 100. [24.3—1]

Castano, E, Duckstein, L, and Bogardi, I (1978). Choice of distribution functions for hydrologic design. Water Resources Research, 14, 643–652. [24.12.3—4]

Castillo, E, and Galambos, J (1987a). Bivariate distributions with normal conditionals. In M H Hamza (Editor), Proceedings of the IASTED International Symposium, Cairo, pp. 59–62. Anaheim, California: Acta Press. [8.5]

Castillo, E, and Galambos, J (1987b). Bivariate distributions with Weibull conditionals. Technical Report from the Department of Mathematics, Temple University, Philadelphia. [8.2.7]

Castle, W E (1916). Genetics and Eugenics. Cambridge, Massachusetts: Harvard University Press. [24.7.2—10]

Caussinus, H (1965). Contribution à l'analyse statistique des tableaux de corrélation. Annales de la Faculté des Sciences de l'Université de Toulouse, 29, 77–182. [24.6—6, 24.9.2—8]

Chakrabarti, S K, and Cooley, R P (1977). Statistical distribution of periods and heights of ocean waves. Journal of Geophysical Research, 82, 1363–1368. [24.12.2—8]

Chakravarti, I M, Laha, R G, and Roy, J (1967). Handbook of Methods of Applied Statistics. Volume I. Techniques of Computation, Descriptive Methods, and Statistical Inference. New York: Wiley. [24.9.2—11, 24.13.1—5]

Chambers, J M, Cleveland, W S, Kleiner, B, and Tukey, P A (1983). Graphical Methods for Data Analysis. Belmont, California: Wadsworth, and Boston: Duxbury. [4.13.3.2, 4.13.3.5, 4.17.6, 11.5.2, 24.7.1—3, 24.7.2—4, 24.14.1—1]

Chang, C C C, Wang, M H, and Juang, J T (1976). Waves off

Taichung coast of Taiwan. In Proceedings of the Fifteenth Coastal Engineering Conference, pp. 129–150. New York: American Society of Civil Engineers. [24.12.2—8]

Chatfield, C (1975). A marketing application of a characterization theorem. In G P Patil, S Kotz, and J K Ord (Editors), A Modern Course on Distributions in Scientific Work. Volume 2—Model Building and Model Selection, pp. 175–185. Dordrecht: Reidel. [6.5.12, 24.6—7]

Chayes, F (1971). Ratio Correlation. Chicago: University of Chicago Press. [24.5.2—1]

Chen, H J (1979). Percentage points of multivariate t distribution with zero correlations and their application. Biometrical Journal, 21, 347–359. [6.2.11, 6.2.12]

Cheng, R C H (1977). The generation of gamma variables with non-integer shape parameter. Applied Statistics, 26, 71–75. [16.3.2]

Cheng, R C H (1978). Generating beta variates with non-integral shape parameters. Communications of the Association for Computing Machinery, 21, 317–322. [16.3.3]

Cheng, Y-S (1986). Bivariate lognormal distribution for characterizing asbestos fiber aerosols. Aerosol Science and Technology, 5, 359–368. [19.4.6, 24.9.2—4]

Cherian, K C (1941). A bi-variate correlated gamma-type distribution function. Journal of the Indian Mathematical Society, 5, 133–144. [10.3.8, 15.2, 15.3.1]

Chernick, M R (1983). Influence functions, outlier detection, and data editing. In T Wright (Editor), Statistical Methods and the Improvement of Data Quality, pp. 167–176. New York: Academic Press. [4.13.2]

Chew, V (1986). Radial error. In Encyclopedia of Statistical Sciences. Volume 7, pp. 499–500. New York: Wiley. [2.15.2—43]

Chiles, J P (1984). Simulation of a nickel deposit: Problems encountered and practical solutions. In G Verly, M David, A G Journel, and A Marechal (Editors), Geostatistics for Natural Resources Characterization. Part 2, pp. 1015–1030. Dordrecht: Reidel. [24.12.3—12]

Chilton, N W (1967). Design and Analysis in Dental and Oral Research. Philadelphia: Lippincott. [24.9.1—1, 24.9.2—2, 24.9.2—3, 24.9.3—3, 24.9.3—4]

Chinchilli, V M, and Breen, T J (1985). Testing the independence of q binary random variables. Report from the Department of Biostatistics, Medical College of Virginia, Richmond. (Abstract in Biometrics, 41, 578.) [5.2.11, 22.4.1]

Chmielewski, M A (1981). Elliptically symmetric distributions: A review and bibliography. International Statistical Review, 49, 67–74. [7.2.1]

Chou, K C, and Corotis, R B (1983). Generalized wind speed probability distribution. Journal of Engineering Mechanics, 109, 14–29. [4.9, 19.7.1]

Chou, Y-M, and Owen, D B (1984). An approximation to percentiles of a variable of the bivariate normal distribution when the other variable is truncated, with applications. Communications in Statistics—Theory and Methods, 13, 2535–2547. [4.16.1, 4.16.2]

Chowdhury, R N, and A-Grivas, D (1982). Probabilistic model of progressive failure of slopes. Journal of the Geotechnical Engineering Division, Proceedings of the American Society of Civil Engineers, 108, 803–819. [19.8.2]

Christensen, E R, and Chen, C-Y (1985). A general noninteractive multiple toxicity model including probit, logit, and Weibull transformations. Biometrics, 41, 711–725. [22.2.1]

Christiansen, H, and Siefert, W (1979). Storm surge prediction by combined wind and tide data. In Proceedings of the Sixteenth Coastal Engineering Conference. Volume I, pp. 965–974. New York: American Society of Civil Engineers. [24.12.3—3]

Chuang-Stein, C, and Francom, S F (1988). Some simple log-linear models for the analysis of ordinal response data. Journal of Applied Statistics, 15, 285–294. [24.11.2—5]

Church, J D, and Harris, B (1970). The estimation of reliability from stress-strength relationship. Technometrics, 12, 49–54. [17.6.1]

Clark, L A, Denby, L, Pregibon, D, Harshfield, G A, Pickering, T G, Blank, S, and Laragh, J H (1987). A data-based method for bivariate outlier detection: Application to automatic blood pressure recording devices. Psychophysiology, 24, 119–125. [4.13.2]

Clark, P J, Eckstrom, P T, and Linden, L C (1964). On the number of individuals per occupation in a human society. Ecology, 45, 367–372. [24.6—15]

Clarke, L E (1975). On marginal density functions of continuous densities. American Mathematical Monthly, 82, 845–846. [8.11]

Clarke, R T (1979). Extension of annual streamflow record by correlation with precipitation subject to heterogeneous errors. Water Resources Research, 15, 1081–1088. [8.12.5.6, 19.7.7]

Clarke, R T (1980). Bivariate gamma distributions for extending annual streamflow records from precipitation: Some large-sample results. Water Resources Research, 16, 863–870. [fn8.4, 8.12.5.6]

Clayton, D G (1978). A model for association in bivariate life tables and its application in epidemiological studies of familial tendency in chronic disease incidence. Biometrika, 65, 141–151. [5.9.6]

Clayton, D, and Cuzick, J (1985a). Multivariate generalizations of the proportional hazards model. Journal of the Royal Statistical Society, Series A, 148, 82–108 (Discussion, 108–117). [5.9.6]

Clayton, D, and Cuzick, J (1985b). The semi-parametric Pareto model for regression analysis of survival times. Bulletin of the International Statistical Institute, 51, Book 4, Paper 23.3 (Discussion, Book 5, 175–180). [5.9.6]

Clisby, K H, and Sears, P B (1955). Palynology in southern North America. Part III. Microfossil profiles under Mexico City correlated with sedimentary profiles. Geological Society of America Bulletin, 66, 511–520. [24.5.2—6]

Clogg, C C (1980). Analysis of association (ANOAS) in multi-way cross-classifications with ordered categories. American Statistical Association, 1980 Proceedings of the Social Statistics Section, pp. 438–443. [24.11.2—5, 24.11.2—6]

Clogg, C C (1982). Some models for the analysis of association in multiway cross-classifications having ordered categories. Journal of the American Statistical Association, 77, 803–815. [24.11.2—5, 24.11.2—6]

Clogg, C C (1986). Quasi-independence. In Encyclopedia of Statistical Sciences. Volume 7, pp. 460–464. New York: Wiley. [24.11.2—1]

Cloninger, C R, Reich, T, and Guze, S B (1975). The multifactorial model of disease transmission—III: Familial relationship between sociopathy and hysteria (Briquet's syndrome). British Journal of Psychiatry, 127, 23–32. [19.5.1, T20.4, 20.4.1, 20.4.2]

Cobb, L (1988). Statistical catastrophe theory. In Encyclopedia of Statistical Sciences. Volume 8, pp. 634–640. New York: Wiley. [2.15.2—56]

Cohen, J (1960). A coefficient of agreement for nominal scales. Educational and Psychological Measurement, 20, 37–46. [10.9.1]

Cohen, J P (1985). Normal extremes. In Encyclopedia of Statistical Sciences. Volume 6, pp. 359–361. New York: Wiley. [2.15.2—30]

Cohen, L and Zaparovanny, Y I (1980). Positive quantum joint distributions. Journal of Mathematical Physics, 21, 794–796. [3.1.3]

Cohen, M-D (1986). Pseudo-random number generators. In Encyclopedia of Statistical Sciences. Volume 7, pp. 327–333. New York: Wiley. [16.2.1]

Conde, S, and Kalla, S L (1979?). A Table of Gauss' Hypergeometric Function ${}_2F_1(a, b; c; x)$. Prepared at the Facultad de Ingeniería, División de Postgrado, Universidad del Zulia, Maracaibo, Venezuela. (Text of 8 pages plus computer output of tables, deposited in the file of unpublished mathematical tables held at the editorial office of Mathematics of Computation.) [fn13.3]

Connolly, M (1983). The impact of final offer arbitration on the bargaining process and wage outcomes. Ph.D. Thesis, Michigan State University. [22.2.2, 22.4.2.5]

Connor, R J, and Mosimann, J E (1969). Concepts of independence for proportions with a generalization of the Dirichlet distribution. Journal of the American Statistical Association, 64, 194–206. [6.5.14, 12.4.2]

Conolly, B W, and Choo, Q H (1979). The waiting time process for a generalized correlated queue with exponential demand and service. SIAM (Society for Industrial and Applied Mathematics) Journal on Applied Mathematics, 37, 263–275. [9.12.13]

Conway, D (1979). Multivariate distributions with specified marginals. Technical Report No. 145, Department of Statistics, Stanford University. [5.2.12, 5.3.5, 5.3.9, 5.7.9]

Conway, D (1981). Bivariate distribution contours. American Statistical Association, 1981 Proceedings of the Business and Economic Statistics Section, pp. 475–480. [5.1.3, 5.2.7, 5.3.7, 5.4.7, 5.6.6, 5.7.7, 9.2.9]

Conway, D (1983). Farlie-Gumbel-Morgenstern distributions. In Encyclopedia of Statistical Sciences. Volume 3, pp. 28–31. New York: Wiley. [5.2.10, 9.2.10]

Conway, D (1986). Plackett family of distributions. In Encyclopedia of Statistical Sciences. Volume 7, pp. 1–5. New York: Wiley. [5.7.9]

Cook, I D, and Barnes, J W (1981). Evaluation of the H-function integral and of probability distributions of functions of independent random variables. American Journal of Mathematical and Management Sciences, 1, 293–339. [10.6]

Cook, R D, and Johnson, M E (1981). A family of distributions for modelling non-elliptically symmetric multivariate data. Journal of the Royal Statistical Society, Series B, 43, 210–218. [5.8.12, 5.9.3, 5.9.6, 13.5]

Cook, R D, and Johnson, M E (1986). Generalized Burr-Pareto-logistic distributi on with applications to a uranium exploration data set. Technometrics, 28, 12 3–131. [5.2.7, 5.2.11, 5.2.16, 5.8.7, 5.8.12, 5.9.3, 5.9.6, 20.6.1, 20.6.2, 20.6.3]

Cookson, W O C M, de Klerk, N H, Musk, A W, Glancy, J J, Armstrong, B K, and Hobbs, M S T (1985). The natural history of asbestosis in former crocidolite workers of Wittenoom Gorge. In Statcomp-Medstat 85, Symposium Proceedings of the Medical Statistics and Statistical Computing Sections of the Statistical Society of Australia, pp. 144–165. Sydney: Statistical Society of Australia. [24.4.2—3]

Cordóva, J R, and Rodríguez-Iturbe, I (1985). On the probabilistic structure of storm surface runoff. Water Resources Research, 21, 755–763. [8.12.5.7, 9.12.13]

Cornfield, J (1962). Joint dependence of risk of coronary heart disease on serum cholesterol and systolic blood pressure: A discriminant function analysis. Federation Proceedings, 21, Supplement No. 11, 58–61. [24.9.2—10]

Cornish, E A, and Evans, M J (1964). An analysis of daily temperatures at Adelaide, South Australia. Division of Mathematical Statistics Technical Paper No. 17. Melbourne: Commonwealth Scientific and Industrial Research Organization. (Reprinted in Published Papers of E A Cornish. A Memorial Volume, pp. 581–627. Adelaide: E A Cornish Memorial Appeal, 1974.) [24.12.2—7]

Correia, F N (1987). Engineering risk in flood studies using multivariate partial duration series. In L Duckstein and E J Plate (Editors), Engineering Reliability and Risk in Water Resources, pp. 307–332. Dordrecht: Nijhoff. [8.12.5.5]

Cowan, R (1987). A bivariate exponential distribution arising in random geometry. Annals of the Institute of Statistical Mathematics, Tokyo, 39, 103–111. [9.14.6, 9.14.7, 9.14.8, 9.14.9]

Cox, C, and Munk, W (1954). Measurement of the roughness of the sea surface from photographs of the sun's glitter. Journal of the Optical Society of America, 44, 838–850. [3.11.3]

Cox, D R (1970). Analysis of Binary Data. London: Methuen. [22.3.1, 22.3.3, 22.4.2.1]

Cox, D R, and Lewis, P A W (1966). The Statistical Analysis of Series of Events. London: Chapman and Hall. [9.11.8]

Cox, D R, and Miller, H D (1965). The Theory of Stochastic Processes. London: Chapman and Hall. [3.9.2]

Cox, D R, and Oakes, D (1984). Analysis of Survival Data. London: Chapman and Hall. [5.9.2, 9.18.2]

Cox, D R, and Small, N J H (1978). Testing multivariate normality. Biometrika, 65, 263–272. [4.13.5.5]

Cox, P R (1985). Marriage. In Encyclopedia of Statistical Sciences. Volume 5, pp. 273–274. New York: Wiley. [24.11.2—7]

Craig, C C (1936). On the frequency function of xy. Annals of Mathematical Statistics, 7, 1–15. [4.9]

Cramér, H (1946). Mathematical Methods of Statistics. Princeton, New Jersey: Princeton University Press. [4.9, 11.5.1]

CRC (1977). 1975 CRC customer/rater knock perception study. Report No. 492, Coordinating Research Council, Atlanta. [24.4.1—1]

CRC (1980). 1978 CRC customer versus rater octane number requirement program. Report No. 514, Coordinating Research Council, Atlanta. [24.4.1—1]

Cresswell, W L, and Froggatt, P (1963). The Causation of Bus Driver Accidents. An Epidemiological Study. London: Oxford University Press. [24.6—1]

Critchlow, D (1988). Ulam's metric. In Encyclopedia of Statistical Sciences. Volume 9, pp. 379–380. New York: Wiley. [fn11.9]

Crockett, N G (1979). A quick test of fit of a bivariate distribution. In D McNeil (Editor), Interactive Statistics, pp. 185–191. Amsterdam: North-Holland. [24.6—3, 24.6—11]

Crofts, A E (1969). An investigation of normal-lognormal distributions. Technical Report No. 32, Department of Statistics, Southern Methodist University, Dallas. [4.14.2.1]

Crofts, A E and Owen, D B (1972). Large sample maximum likelihood estimation in a normal-lognormal distribution. South African Statistical Journal, 6, 1–10. [4.14.2.1]

Crowder, M (1985). A distributional model for repeated failure time measurement s. Journal of the Royal Statistical Society, Series B, 47, 447–452. [5.8.12, 5.9.6, 6.8]

Crowder, M (1989). A multivariate distribution with Weibull connections. Journal of the Royal Statistical Society, Series B, 51, 93–107. [5.6.5, 5.6.6, 5.6.8, 24.8.1—7]

Crutcher, H L, Neumann, C J, and Pelissier, J M (1982). Tropical cyclone forecast errors and the multimodal bivariate normal distribution. Journal of Applied Meteorology, 21, 978–987. [19.7.6]

Csörgő, S (1984). Testing by the empirical characteristic function: A survey. In P Mandl and M Hušková (Editors), Asymptotic Statistics 2. Proceedings of the Third Prague Symposium on Asymptotic Statistics, pp. 45–56. Amsterdam: Elsevier. [4.13.5.3]

Csörgő, S (1986). Testing for normality in arbitrary dimension. Annals of Statistics, 14, 708–723. [4.13.1, 4.13.5.3, 24.7.1—3, 24.11.2—13]

Csörgő, S (1989). Consistency of some tests for multivariate normality. Metrika, 36, 107–116. [4.13.5.3]

Cuadras C M, and Augé, J (1981). A continuous general multivariate distribution and its properties. Communications in Statistics, Part A—Theory and Methods, 10, 339–353. [5.5, 9.2.9, 13.6]

Cunnane, C (1978). Unbiased plotting positions—A review. Journal of Hydrology, 37, 205–222. [4.13.3.1]

Curnow, R N (1958). The consequences of errors of measurement for selection from certain non-normal distributions. Bulletin de l'Institut International de Statistique, 37, Book 3, 291–308 (Discussion, Book 1, 59–60). [15.2]

Curnow, R N (1972). The multifactorial model for the inheritance of liability to disease and its implications for relatives at risk. Biometrics, 28, 931–946. [19.5.3.4]

Curnow, R N, and Smith, C (1975). Multifactorial models for familial diseases in man. Journal of the Royal Statistical Society, Series A, 138, 131–156 (Discussion, 156–169). [19.5.3.5]

D

Daamen, R A (1986). Measures of disease intensity in powdery mildew (*Erysiphe graminis*) of winter wheat. 2. Relationships and errors of estimation of pustule number, incidence and severity. Netherlands Journal of Plant Pathology, 92, 207–222. [24.8.3—2]

Dabrowska, D (1982). Parametric and nonparametric models with special schemes of stochastic dependence. In B V Gnedenko, M L Puri, and I Vincze (Editors), Nonparametric Statistical Inference. Volume I, pp. 171–182. Amsterdam: North-Holland. (Colloquia Mathematica Societatis János Bolyai, 32.) [10.3.3]

Dacunha-Castelle, D, and Duflo, M (1986). Probability and Statistics. Volume 1. New York: Springer. (Original French edition was dated 1983.) [24.12.1—4]

Daganzo, C F, Bouthelier, F, and Sheffi, Y (1977). Multinomial probit and qualitative choice: A computationally efficient algorithm. Transportation Science, 11, 338–358. [22.3.3]

Dagnelie, P (1975). Analyse Statistique à Plusieurs Variables. Brussels: Vander. [24.7.1—3]

D'Agostino, R B (1982). Departures from normality, tests for. In Encyclopedia of Statistical Sciences. Volume 2, pp. 315–324. New York: Wiley. [2.15.2—5, 4.13.4.1]

D'Agostino, R B (1986a). Graphical analysis. In R B D'Agostino and M A Stephens (Editors), Goodness-of-Fit Techniques, pp. 7–62. New York: Dekker. [4.13.3.1]

D'Agostino, R B (1986b). Tests for the normal distribution. In R B D'Agostino and M A Stephens (Editors), Goodness-of-Fit Techniques, pp. 367–419. New York: Dekker. [4.13.4.1, 4.13.4.3]

D'Agostino, R, and Pearson, E S (1973). Tests for departure from normality. Empirical results for the distributions of b_2 and $\sqrt{b_1}$. Biometrika, 60, 613–622. [4.13.4.3]

D'Agostino, R B, and Stephens, M A (1986) (Editors). Goodness-of-Fit Techniques. New York: Dekker. [24.3—2. Individual papers are cited at various places.]

Dagum, C (1983). Income distribution models. In Encyclopedia of Statistical Sciences. Volume 4, pp. 27–34. New York: Wiley. [2.15.3—3]

Dagum, C (1985). Lorenz curve. In Encyclopedia of Statistical Sciences. Volume 5, pp. 156–161. New York: Wiley. [2.15.2–24]

Dale, J R (1983). Statistical methods for ordered categorical responses. Report No. 114, Applied Mathematics Division, Department of Scientific and Industrial Research, Wellington, New Zealand. [5.7.10, 24.6—13, 24.8.2—1, 24.9.2—8, 24.10.2—3]

Dale, J R (1984). Local versus global association for bivariate ordered responses. Biometrika, 71, 507–514. [5.7.10, 24.6—13, 24.10.2—3]

Dale, J R (1985). A bivariate discrete model of changing colour in blackbirds. In B J T Morgan and P M North (Editors), Statistics in Ornithology, pp. 25–36. Berlin: Springer. [5.7.10, 24.8.2—1]

Dale, J R (1986). Global cross-ratio models for bivariate, discrete, ordered responses. Biometrics, 42, 909–917. [5.7.10]

Daley, D J (1974). Computation of bi- and tri-variate normal integrals. Applied Statistics, 23, 435–438. [4.12.4]

Damsleth, E, and El-Shaarawi, A H (1989). ARMA models with double-exponentially distributed noise. Journal of the Royal Statistical Society, Series B, 51, 61–69. [10.10]

Daniels, H E (1950). Rank correlation and population models. Journal of the Royal Statistical Society, Series B, 12, 171–181 (Discussion, 182–191). [13.1]

Danowski, T S, Winkler, A W, and Peters, J P (1946). Salt depletion, peripheral vascular collapse, and the treatment of diabetic acidosis. Yale Journal of Biology and Medicine, 18, 405–417. [24.9.1—4]

Darroch, J N, and McCloud, P I (1986). Category distinguishability and observer agreement. Australian Journal of Statistics, 28, 371–388. [20.2.4]

Darroch, J N, and Ratcliff, D (1970). Null correlation for proportions—II. Mathematical Geology (Journal of the International Association for Mathematical Geology), 2, 307–312. [24.5.2—2]

Darroch, J N, and Ratcliff, D (1973). Tests of F-independence with reference to quasi-independence and Waite's fingerprint data. Biometrika, 60, 395–401. [24.6—19]

Darroch, J N, and Ratcliff, D (1978). No-association of proportions. Mathematical Geology (Journal of the International Association for Mathematical Geology), 10, 361–368. [24.5.2—2, 24.5.2—6]

Das Gupta, A, and Mitra, S N (1958). A technical note on age grouping. Report No. 12, The National Sample Survey, Cabinet Secretariat, New Delhi. [24.6—21]

Davenport, C B (1903). (Title not known.) Proceedings of the American Academy of Arts and Sciences, 39, 149–???. [24.7.2—11]

Davenport, C B, and Bullard, C (1896). (Title not known.) Proceedings of the American Academy of Arts and Sciences, 32, 87–??. [24.6—12]

Davenport, J H (1985). Symbolic and numeric manipulation of integrals. In W L Liranker and R A Toupin (Editors), Accurate Scientific Computations, pp. 168–180. Berlin: Springer. [3.1.4]

David, F N, and Fix, E (1961). Rank correlation and regression in a nonnormal surface. In J Neyman (Editor), Proceedings of the 4th Berkeley Symposium on Mathematical Statistics and Probability. Volume 1, pp. 177–197. Berkeley: University of California Press. [T13.1, 13.9, 15.3.1, 24.12.1—6]

David, H A (1985). Order statistics. In Encyclopedia of Statistical Sciences. Volume 6, pp. 504–510. New York: Wiley. [2.15.2—34]

David, H A, and Moeschberger, M L (1978). The Theory of Competing Risks. London: Griffin. [4.10, 9.2.7, 9.12.12, 9.18.2]

Davis, B M, and Jalkanen, G J (1988). Nonparametric estimation of multivariate joint and conditional spatial distributions. Mathematical Geology (Journal of the International Association for Mathematical Geology), 20, 367–381. [4.16.2]

Davis, C E, and Steinberg, S M (1986). Quantile estimation. In Encyclopedia of Statistical Sciences. Volume 7, pp. 408–412. New York: Wiley. [2.15.2—41]

Dawson, D V, and Elston, R C (1984). A bivariate problem in human genetics: Ascertainment of families through a correlated trait. American Journal of Medical Genetics, 18, 435–448. [19.5.2]

DeBrota, D J, Dittus, R S, Roberts, S D, Wilson, J R, Swain, J J, and Venkatraman, S (1988). Input modeling with the Johnson system of distributions. In M A Abrams, P L Haigh, and J C Comfort (Editors), 1988 Winter Simulation Conference Proceedings, pp. 165–179. Piscataway, New Jersey: Institute of Electrical and Electronics Engineers. [2.4.3, 4.14.2.5]

Deddens, J A, and Koch, G G (1988). Survival analysis, grouped data. In Encyclopedia of Statistical Sciences. Volume 9, pp. 129–134. New York: Wiley. [2.15.3—9]

de Haan, L, and Resnick, S I (1977). Limit theory for multivariate sample extremes. Zeitschrift für Wahrscheinlichkeitstheorie und verwandte Gebiete, 40, 317–337. [9.17.1]

Deheuvels, P (1980). The decomposition of infinite order and extreme multivariate distributions. In I M Chakravarti (Editor), Asymptotic Theory of Statistical Tests and Estimation. In Honor of Wassily Hoeffding, pp. 259–286. New York: Academic Press. [5.1.2]

Deheuvels, P (1982). Some applications of the dependence functions to statistical inference: Nonparametric estimates of extreme values distributions, and a Kiefer type universal bound for the uniform test of independence. In B V Gnedenko, M L Puri, and I Vincze (Editors), Nonparametric Statistical Inference. Volume I, pp. 183–201. Amsterdam: North-Holland. (Colloquia Mathematica Societatis János Bolyai, 32.) [5.1.2]

Deheuvels, P (1983). Point processes and multivariate sample extreme values. Journal of Multivariate Analysis, 13, 257–271. [9.17.1]

Deheuvels, P (1984). Probabilistic aspects of multivariate extremes. In J Tiago de Oliveira (Editor), Statistical Extremes and Applications, pp. 117–130. Dordrecht: Reidel. [9.17.3]

Deheuvels, P (1985). Point processes and multivariate extreme values (II). In P R Krishnaiah (Editor), Multivariate Analysis—VI, pp. 145–164. Amsterdam: North-Holland. [9.17.1]

de Jong, U W, Breslow, N, Goh Ewe Hong, J, Sridharan, M, and Shanmugaratnam, K (1974). Aetiological factors in oesophageal cancer in Singapore Chinese. International Journal of Cancer, 13, 291–303. [24.6—18]

Derin, H, Wise, G L, and Thomas, J B (1977). Bivariate densities with diagonal expansions in Gegenbauer polynomials. Journal of the Franklin Institute, 304, 243–249. [T14.1, 14.6.3, 14.8]

Der Kiureghian, A, and Liu, P-L (1986). Structural reliability under incomplete probability information. Journal of Engineering Mechanics, 112, 85–104. [4.14.5, 5.10.9, 19.8.2]

Derman, C, Gleser, L J, and Olkin, I (1973). A Guide to Probability Theory and Application. New York: Holt, Rinehart and Winston. [24.6—14, 24.9.2—11]

Deshpande, J V, Kochar, S C, and Singh, H (1986). Aspects of positive ageing. Journal of Applied Probability, 23, 748–758. [18.2.1, 18.2.3, fn18.3]

de Silva, B M (1978). A class of multivariate symmetric stable distributions. Journal of Multivariate Analysis, 8, 335–345. [7.11.4, 7.11.5.2, 7.11.6]

de Silva, B M (1979). The quotient of certain stable random variables. Sankhyā, Series B, 40, 279–281. [7.11.4]

de Silva, B M, and Griffiths, R C (1980). A test of independence for bivariate symmetric stable distributions. Australian Journal of Statistics, 22, 172–177. [7.11.4, 7.11.5.2, 15.3.4]

D'Este, G M (1981). A Morgenstern-type bivariate gamma distribution. Biometrika, 68, 339–340. [5.2.12]

Devlin, S J, Gnanadesikan, R, and Kettenring, J R (1975). Robust estimation and outlier detection with correlation coefficients. Biometrika, 62, 531–545. [11.5.1, 24.7.1—3]

Devlin, S J, Gnanadesikan, R, and Kettenring, J R (1976). Some multivariate applications of elliptical distributions. In S Ikeda, T Hayakawa, H Hudimoto, M Okamoto, M Siotani, and S Yamamoto (Editors), Essays in Probability and Statistics, pp. 365–393. Tokyo: Shinko Tsucho. [6.2.8, 6.2.13, 7.2.2, 7.2.10, 7.2.11, 7.2.12]

DeVries, W R, Dornfeld, D A, and Wu, S M (1978). Bivariate time series analysis of the effective forcee variation and friction coefficient distribution in wood grinding. Journal of Engineering for Industry (Transactions of the American Society of Mechanical Engineers), 100, 181–185. [19.8.10]

Devroye, L (1981). Recent results in non-uniform random variate generation. In T I Ören, C M Delfosse, and C M Shub (Editors), 1981 Winter Simulation Conference Proceedings. Volume 2, pp. 517–521. Piscataway, New Jersey: Institute of Electrical and Electronics Engineers. [16.2.1]

Devroye, L (1986). Non-Uniform Random Variate Generation. New York: Springer. [2.5.3, 3.1.2, 5.1.4, 5.2.8, 5.7.8, 5.8.12, 5.9.4, 5.9.5, 6.2.13, 6.5.7, 6.5.10, 6.6, 7.2.10, 10.3.7, 10.5.4, 16.1, fn16.1, 16.2.1, 16.2.5, fn16.3, 16.3.1, 16.3.2, 16.3.3, 16.3.4, 16.3.5, 16.5.1]

Devroye, L (1989). On random variate generation when only moments or Fourier coefficients are known. Mathematics and Computers in Simulation, 31, 71–89. [fn16.1]

Dewey, M (1988). Review of "Teaching Health Statistics: Twenty Lesson and Seminar Outlines" by S K Lwanga and C-Y Tye. Journal of the Royal Statistical Society, Series A, 151, 232. [24.1.2.1]

Dharmadhikari, S W, and Joag-Dev, K (1985). Multivariate unimodality. In Encyclopedia of Statistical Sciences. Volume 6, pp. 130–132. New York: Wiley. [5.1.3]

Diaconis, P, and Efron, B (1985). Testing for independence in a two-way table: New interpretations of the chi-square statistic. Annals of Statistics, 13, 845–874 (Discussion, 875–913). [24.9.2—7]

Dickey, J M (1983). Multiple hypergeometric functions: Probabilistic interpretations and statistical uses. Journal of the American Statistical Association, 78, 628–637. [6.5.14]

DiDonato, A R, and Hageman, R K (1982). A method for computing the integral of the bivariate normal distribution over an arbitrary polygon. SIAM (Society for Industrial and Applied Mathematics) Journal on Scientific and Statistical Computing, 3, 434–446. [4.12.6, 4.12.8]

DiDonato, A R, Jarnagin, M P, and Hageman, R K (1980). Computation of the integral of the bivariate normal distribution over convex polygons. SIAM (Society for Industrial and Applied Mathematics) Journal on Scientific and Statistical Computing, 1, 179–186. [4.12.6]

DiDonato, A R, and Morris, A H (1986). Computation of the incomplete gamma function ratios and their inverse. ACM (Association for Computing Machinery) Transactions on Mathematical Software, 12, 377–393 and 13, 318–319. [fn2.2]

Divgi, D R (1979a). Calculation of univariate and bivariate normal probability functions. Annals of Statistics, 7, 903–910. [4.12.2, 4.12.8]

Divgi, D R (1979b). Calculation of the tetrachoric correlation coefficient. Psychometrika, 44, 169–172. [11.11]

Donnelly, T G (1973). Algorithm 462. Bivariate normal distribution. Communications of the Association for Computing Machinery, 16, 638. [4.12.1, 4.12.8]

Downton, F (1970). Bivariate exponential distributions in reliability theory. Journal of the Royal Statistical Society, Series B, 32, 408–417. [5.2.9, 9.12.9, 17.1, T17.1, 17.2.3, fn17.4, 20.3.3]

Downton, F (1973). The estimation of $\Pr(Y < X)$ in the normal case. Technometrics, 15, 551–558. [17.6.1]

Draper, L, and Bownass, T M (1983). Unusual waves on European coasts, February 1979. In B L Edge (Editor), Proceedings of the Eighteenth Coastal Engineering Conference. Volume I, pp. 270–281. New York: American Society of Civil Engineers. [24.12.2—8]

Draper, N, and Guttman, I (1975). Two simultaneous measurement procedures: A Bayesian approach. Journal of the American Statistical Association, 70, 43–46. [24.4.1—2]

Drasgow, F (1986). Polychoric and polyserial correlations. In Encyclopedia of Statistical Sciences. Volume 7, pp. 68–74. New York: Wiley. [11.11, 24.6—6]

Drezner, Z (1978). Computation of the bivariate normal integral. Mathematics of Computation, 32, 277–279. [4.12.2]

Dubey, S D (1970). Compound gamma, beta, and F distributions. Metrika, 16, 27–31. [8.7, 20.7.2]

Duda, R O, and Hart, P E (1973). Pattern Classification and Scene Analysis. New York: Wiley. [24.7.1—3]

Duncan, G M (1983). Sample selectivity as a proxy variable problem: On the use and misuse of Gaussian selectivity corrections. In J D Reid (Editor), Research in Labor Economics. Supplement 2, New Approaches to Labor Unions, pp. 333–345. Greenwich, Connecticut: Jai Press. [19.9.5]

Duncan, O D (1979). How destination depends on origin in the occupational mobility table. American Journal of Sociology, 84, 793–803. [24.11.2—1]

Dunn, J E, and Houston, H A (1975). Automated decoding of an arbitrary probability distribution from its generating function. In J W Frane (Editor), Proceedings of the Computer Science and Statistics Eighth Annual Symposium on the Interface, pp. 77–84. Los Angeles: Health Sciences Computing Facility, University of California. [3.1.4]

Dunnett, C W, and Sobel, M (1954). A bivariate generalization of Student's t-distribution, with tables for certain special cases. Biometrika, 41, 153–169. [6.2.11]

Durbin, J, and Stuart, A (1951). Inversions and rank correlation coefficients. Journal of the Royal Statistical Society, Series B, 13, 303–309. [13.1]

Durling, F C (1974). Bivariate normit, logit, and Burrit analysis. Research Report No. 24, Department of Mathematics, University of Waikato, Hamilton, New Zealand. [5.2.11, 5.9.6, 22.2.1]

Durling, F C (1975). The bivariate Burr distribution. In G P Patil, S Kotz, and J K Ord (Editors), A Modern Course on Statistical Distributions in Scientific Work. Volume 1—Models and Structures, pp. 329–335. Dordrecht: Reidel. [5.8.13.3, 5.9.6, 6.8, 22.2.1]

Dussauchoy, A, and Berland, R (1972). Lois gamma à deux dimensions. Comptes Rendus de l'Academie des Sciences, Paris, Série A, 274, 1946–1949. [8.6.10]

Dussauchoy, A, and Berland, R (1975). A multivariate gamma type distribution whose marginal laws are gamma, and which has a property similar to a characteristic property of the normal case. In G P Patil, S Kotz, and J K Ord (Editors), A Modern Course on Distributions in Scientific Work. Volume 1—Models and Structures, pp. 319–328. Dordrecht: Reidel. [8.6.10]

Dykstra, R L (1985). Ordering, starshaped. In Encyclopedia of Statistical Sciences. Volume 6, pp. 499–501. New York: Wiley. [2.15.2—33, 18.2.2]

E

Eagleson, G K (1964). Polynomial expansions of bivariate distributions. Annals of Mathematical Statistics, 35, 1208–1215. [10.3.8, T14.1, 15.3.1, 15.3.2]

Eagleson, G K, and Lancaster, H O (1967). The regression system of sums with random elements in common. Australian Journal of Statistics, 9, 119–125. [15.1, 15.3.3, 15.3.5]

Ebrahimi, N (1987). Analysis of bivariate accelerated life test data for the bivariate exponential distribution. American Journal of Mathematical and Management Sciences, 7, 175–190. [9.2.7]

Ebrahimi, N, and Ghosh, M (1981). Multivariate negative dependence. Communications in Statistics—Theory and Methods, 11, 307–337. [12.4.1]

Ebrahimi, N, and Ghosh, M (1982). The ordering of negative quadrant dependence. Communications in Statistics—Theory and Methods, 11, 2389–2399. [12.4.1]

Eckler, A R (1988). Target coverage. In Encyclopedia of Statistical Sciences. Volume 9, pp. 180–183. New York: Wiley. [19.8.9]

Editors of the ESS (1982a). Bessel function distributions. In Encyclopedia of Statistical Sciences. Volume 1, pp. 224–225. New York: Wiley. [2.15.1—1]

Editors of the ESS (1982b). Dirichlet distribution. In Encyclopedia of Statistical Sciences. Volume 2, pp. 386–387. New York: Wiley. [6.5.11]

Editors of the ESS (1982c). Doubly noncentral distributions. In Encyclopedia of Statistical Sciences. Volume 2, pp. 423–424. New York: Wiley. [2.15.1—3b]

Editors of the ESS (1982d). Ellipsoid of concentration. In Encyclopedia of Statistical Sciences. Volume 2, p. 485. New York: Wiley. [7.2.10]

Editors of the ESS (1983). Lambda distributions. In Encyclopedia of Statistical Sciences. Volume 4, p. 466. New York: Wiley. [2.15.1—17b]

Editors of the ESS (1985a). Log-Laplace distribution. In Encyclopedia of Statistical Sciences. Volume 5, pp. 133–134. New York: Wiley. [2.15.1—20b]

Editors of the ESS (1985b). MacDonald distribution. In Encyclopedia of Statistical Sciences. Volume 5, pp. 171–172. New York: Wiley. [2.15.1—22]

Editors of the ESS (1985c). Modified normal distributions. In Encyclopedia of Statistical Sciences. Volume 5, pp. 590–591. New York: Wiley. [2.15.1—24]

Editors of the ESS (1985d). Moment-ratio diagrams. In Encyclopedia of Statistical Sciences. Volume 5, pp. 602–603. New York: Wiley. [2.15.2—29]

Editors of the ESS (1985e). Partially truncated distributions. In Encyclopedia of Statistical Sciences. Volume 6, pp. 610–611. New York: Wiley. [2.15.1—32]

Editors of the ESS (1985f). Perks' distributions. In Encyclopedia of Statistical Sciences. Volume 6, pp. 682–683. New York: Wiley. [2.15.1—34]

Editors of the ESS (1986a). Planck distribution. In Encyclopedia of Statistical Sciences. Volume 7, p. 6. New York: Wiley. [2.15.1—36]

Editors of the ESS (1986b). Poly-t distributions. In Encyclopedia of Statistical Sciences. Volume 7, p. 80. New York: Wiley. [2.15.1—37, 6.2.14]

Editors of the ESS (1986c). Quartic exponential distribution. In Encyclopedia of Statistical Sciences. Volume 7, p. 456. New York: Wiley. [2.15.1—38]

Editors of the ESS (1986d). Radon transformation. In Encyclopedia of Statistical Sciences. Volume 7, pp. 502–503. New York: Wiley. [2.15.2—44]

Editors of the ESS (1988a). Sargan distributions. In Encyclopedia of Statistical Sciences. Volume 8, pp. 260–261. New York: Wiley. [2.15.1—41]

Editors of the ESS (1988b). Self-reciprocal distributions. In Encyclopedia of Statistical Sciences. Volume 8, p. 352. New York: Wiley. [2.15.2—51]

Editors of the ESS (1988c). Skew-normal distributions. In Encyclopedia of Statistical Sciences. Volume 8, p. 507. New York: Wiley. [2.15.1—42]

Editors of the ESS (1988d). Tail ordering. In Encyclopedia of Statistical Sciences. Volume 9, pp. 169–170. New York: Wiley. [2.15.2—58]

Editors of the ESS (1988e). Trapezoidal distributions. In Encyclopedia of Statistical Sciences. Volume 9, p. 319. New York: Wiley. [2.15.1—48]

Editors of the ESS (1988f). Unimodality. In Encyclopedia of Statistical Sciences. Volume 9, pp. 415–416. New York: Wiley. [2.15.2—63]

Editors of the ESS (1988g). Van Zwet tail ordering. In Encyclopedia of Statistical Sciences. Volume 9, p. 465. New York: Wiley. [2.15.2—65]

Editors of the ESS (1988h). Wakeby distributions. In Encyclopedia of Statistical Sciences. Volume 9, pp. 513–514. New York: Wiley. [2.15.1—50]

Editors of the ESS (1988i). Weibull-exponential distributions. In Encyclopedia of Statistical Sciences. Volume 9, p. 556. New York: Wiley. [2.15.1—52]

Editors of the ESS (1988j). Whittaker-type distributions. In Encyclopedia of Statistical Sciences. Volume 9, p. 600. New York: Wiley. [2.15.1—54]

Edwards, C B, and Gurland, J (1961). A class of distributions applicable to accidents. Journal of the American Statistical Association, 56, 503–517. [24.6—2]

Edwards, C L (1908). Variation, development and growth in Holothuria Floridana Pourtalès and in Holothuria Atra Jäger. Biometrika, 6, 236–301. [24.7.1—4]

Efron, B (1979). Computers and the theory of statistics: Thinking the unthinkable. SIAM (Society for Industrial and Applied Mathematics) Review, 21, 460–480. [24.14.1—3]

Efron, B (1982a). Transformation theory: How normal is a family of distributions? Annals of Statistics, 10, 323–339. [2.4.6]

Efron, B (1982b). The Jackknife, the Bootstrap, and Other Resampling Plans. Philadelphia: Society for Industrial and Applied Mathematics. [24.14.1—3]

Egan, J P (1975). Signal Detection Theory and ROC Analysis. New York: Academic Press. [23.1.1]

Ehrenberg, J E (1976). An approximate bivariate density function for orthogonal filtered Poisson processes with applications to volume reverberation problems. Journal of the Acoustical Society of America, 59, 374–380. [23.1.1]

Eisenhart, C (1983a). Laws of error I: Development of the concept. In Encyclopedia of Statistical Sciences. Volume 4, pp. 530–547. New York: Wiley. [2.15.2—17]

Eisenhart, C (1983b). Laws of error II: The Gaussian distribution. In Encyclopedia of Statistical Sciences. Volume 4, pp. 547–562. New York: Wiley. [2.15.2—18]

Eisenhart, C (1983c). Laws of error III: Later (non-Gaussian) distributions. In Encyclopedia of Statistical Sciences. Volume 4, pp. 562–566. New York: Wiley. [2.15.2—19]

Elandt-Johnson, R C (1978). Some properties of bivariate Gumbel Type A distributions with proportional hazard rates. Journal of Multivariate Analysis, 8, 244–254. [9.17.5.1]

Elderton, E M (1914). Height and weight of school children in Glasgow. Biometrika, 10, 288–339. [24.7.2—19]

Elderton, W P, and Johnson, N L (1969). Systems of Frequency Curves. Cambridge: Cambridge University Press. [6.1]

Elffers, H (1980). On interpreting the product moment correlation coefficient. Statistica Neerlandica, 34, 3–11. [11.5.2]

el Lozy, M (1982). Simple computation of a bivariate normal integral arising from a problem of misclassification with applications to the diagnosis of hypertension. Communications in Statistics—Theory and Methods, 11, 2195–2205. [19.4.2]

Endo, T, Okifuji, A, Sugano, S, Hayashi, T, Shimizu, T, Takahara, K, Saito, H, and Yoneyama, Y (1984). Practices of seismic retrofit of existing concrete structures in Japan. In Proceedings of the Eighth World Conference on Earthquake Engineering. Volume 1, pp. 469–476. Englewood Cliffs, New Jersey: Prentice-Hall. [24.4.3—10]

Erdelyi, A (1954a) (Editor). Tables of Integral Transforms. Volume 1. New York: McGraw-Hill. [20.3.3, 20.7.2, 20.7.3]

Erdelyi, A (1954b) (Editor). Tables of Integral Transforms. Volume 2. New York: McGraw-Hill. [20.7.3]

Esary, J D, and Marshall, A W (1973). Multivariate geometric distributions generated by a cumulative damage process. Report NP555EY73041A, Naval Postgraduate School, Monterey, California. [17.3.4]

Esary, J D, and Marshall, A W (1979). Multivariate distributions with increasing hazard rate average. Annals of Probability, 7, 359–370. [18.3.1, 18.3.3]

Esary, J D, Marshall, A W, and Proschan, F (1973). Shock models and wear processes. Annals of Probability, 1, 627–647. [17.3.6.1, fn17.8, 18.2.2]

Esary, J D, and Proschan, F (1972). Relationships among some concepts of bivariate dependence. Annals of Mathematical Statistics, 43, 651–655. [12.2.1, 12.3.4]

Esary, J D, Proschan, F, and Walkup, D W (1967). Association of random variables, with applications. Annals of Mathematical Statistics, 38, 1466–1474. [12.2.1]

Escoufier, Y (1982). L'Analyse des tableaux de contingence simples et multiples. Metron, 40(1–2), 53–77. [24.11.2—4]

Essenwanger, O (1976). Applied Statistics in Atmospheric Science. Part A. Frequencies and Curve Fitting. Amsterdam: Elsevier. [19.7.1, 24.12.2—2]

Etoh, T, and Murota, A (1986). Probabilistic model of rainfall of a single storm. Journal of Hydroscience and Hydraulic Engineering, 4, 65–77. [8.12.5.7, 24.12.3—8]

Eubank, R L (1986). Quantiles. In Encyclopedia of Statistical Sciences. Volume 7, pp. 424–432. New York: Wiley. [2.15.2—42]

Evans, J W, Johnson, R A, and Green, D W (1984). Estimating the correlation between variables under destructive testing, or how to break the same board twice. Technometrics, 26, 285–290. [19.8.4, 24.2—4]

Everitt, B S (1977). The Analysis of Contingency Tables. London: Chapman and Hall. [24.11.2—1]

Everitt, B S (1985). Mixture distributions. In Encyclopedia of Statistical Sciences. Volume 5, pp. 559–569. New York: Wiley. [2.15.1—23, 3.4.1, 4.17.2]

Everitt, B S, and Hand, D J (1981). Finite Mixture Distributions. London: Chapman and Hall. [4.13.3.5, 4.17.1, 24.7.1—3]

Exton, H (1976). Multiple hypergeometric functions and applications. Chichester: Ellis Horwood. [20.7.3]

Eyraud, H (1936). Les principes de la mesure des corrélations. Annales de l'Université de Lyon, Section A, 1, 30–47. [5.2.10]

F

Fabius, J (1973). Two characterizations of the Dirichlet distribution. Annals of Statistics, 1, 583–587. [12.4.2]

Farber, H S (1983). Worker preferences for union representation. In J D Reid (Editor), Research in Labor Economics. Supplement 2, New Approaches to Labor Unions, pp. 171–205. Greenwich, Connecticut: Jai Press. [22.4.2.5]

Farlie, D J G (1960). The performance of some correlation coefficients for a general bivariate distribution. Biometrika, 47, 307–323. [5.2.10, 5.2.16, 13.2.1]

Farmer, E, and Chambers, E G (1939). A study of accident proneness among motor drivers. Industrial Health Research Board Report No. 84. London: HMSO. [24.6—2]

Feingold, M, and Korsog, P E (1986). The correlation and dependence between two F statistics with the same denominator. The American Statistician, 40, 218–220. [6.3.9, 15.4.5]

Feller, W (1971). An Introduction to Probability Theory and its Applications, Volume 2. 2nd Edition. New York: Wiley. [10.2.2, 15.7.3]

Ferguson, G A (1971). Statistical Analysis in Psychology and Education. 3rd Edition. New York: McGraw-Hill. [24.10.2—6]

Ferguson, T S (1962). A representation of the symmetric bivariate Cauchy distribution. Annals of Mathematical Statistics, 33, 1256–1266. [6.2.13]

Ferreri, C (1983). On the multivariate extended Pólya distribution and its interpretations. Metron, 41(3–4), 3–17. [24.6—3]

Fieller, E C, Hartley, H O, and Pearson, E S (1957). Tests for rank correlation coefficients. I. Biometrika, 44, 470–481. [13.1]

Fienberg, S E (1980). The Analysis of Cross-Classified Categorical Data. 2nd Edition. Cambridge, Massachusetts: MIT Press. [24.11.2—9]

Fienberg, S E (1982). Contingency tables. In Encyclopedia of Statistical Sciences. Volume 2, pp. 161–171. New York: Wiley. [24.6—19]

Finch, P D, and Groblicki, R (1984). Bivariate probability densities with given margins. Foundations of Physics, 14, 549–552. [3.1.3]

Finger, S E, Brisbin, I L, Smith, M H, and Urbston, D F (1981). Kidney fat as a predictor of body condition in white-tailed deer. Journal of Wildlife Management, 45, 964–968. [24.4.3—2]

Finney, D J (1971). Probit Analysis. 3rd Edition. Cambridge: Cambridge University Press. [22.1]

Fishburn, P C (1988). Utility theory. In Encyclopedia of Statistical Sciences. Volume 9, pp. 445–452. New York: Wiley. [2.15.2—64]

Fisher, R A (1936). The use of multiple measurements in taxonomic problems. Annals of Eugenics, London, 7, 179–188. (Reprinted in R A Fisher, Contributions to Mathematical Statistics, Paper 32. New York: Wiley, 1950. Also reprinted in J H Bennett (Editor), Collected Papers of R A Fisher. Volume III, pp. 465–475. Adelaide: University of Adelaide, 1973. And again reprinted in W R Atchley and E H Bryant (Editors), Multivariate Statistical Methods: Among-Groups Covariation, pp. 114–123. Stroudsburg, Pennsylvania: Dowden, Hutchinson and Ross, 1975.) [24.7.1—3]

Fisher, R A (1940). The precision of discriminant functions. Annals of Eugenics, London, 10, 422–429. (Reprinted in R A Fisher, Contributions to Mathematical Statistics, Paper 34. New York: Wiley, 1950. Also reprinted in J H Bennett (Editor), Collected Papers of R A Fisher. Volume IV, pp. 283–290. Adelaide: University of Adelaide, 1974.) [14.1, 24.9.2—6]

Fisher, R A, and Hoblyn, T N (1928). Maximum- and minimum-correlation tables in comparative climatology. Geografiska Annaler, 10, 267–281. (Reprinted in J H Bennett (Editor), Collected Papers of R A Fisher. Volume II, pp. 225–239. Adelaide: University of Adelaide, 1972.) [24.12.2—6]

Fishman, G S (1978). Principles of Discrete Event Simulation. New York: Wiley. [16.2.1, 16.3.2]

Fitzpatrick, R (1958). The detection of individual differences in accident susceptibility. Biometrics, 14, 50–68. [24.6—3]

Fleishman, A I (1978). A method for simulating non-normal distributions. Psychometrika, 43, 521–532. [16.5.4.4]

Flury, B K (1986). On sums of random variables and independence. The American Statistician, 40, 214–215. [4.18.1]

Folks, J L (1983a). Inverse distributions. In Encyclopedia of Statistical Sciences. Volume 4, pp. 244–246. New York: Wiley. [2.15.1—13, 2.15.2—16]

Folks, J L (1983b). Inverse Gaussian distribution. In Encyclopedia of Statistical Sciences. Volume 4, pp. 246–249. New York: Wiley. [2.15.1—14]

Foulley, J L, and Gianola, D (1984). Estimation of genetic merit from bivariate "all or none" responses. Génétique, Sélection, Evolution, 16, 285–306. [4.14.2, 19.5.5]

Fowlkes, E B, Gnanadesikan, R, and Kettenring, J R (1987). Variable selection in clustering and other contexts. In C L Mallows (Editor), Design, Data, and Analysis by Some Friends of Cuthbert Daniel, pp. 13–34. New York: Wiley. [24.14.1—1]

Frank, M J (1979). On the simultaneous associativity of $F(x, y)$ and $x + y - F(x, y)$. Aequationes Mathematicae, 19, 194–226. [5.4.6]

Frank, V H, and Frasher, C B (1940). Evaluation of pre-operative tests for anesthetics: Study of 200 cases. American Journal of Orthodontics and Oral Surgery, 26, 473–486. [24.9.1—1]

Fraser, D A S, and Streit, (1980). A further note on the bivariate normal distribution. Communications in Statistics, Part A—Theory and Methods, 10, 1097–1099. [4.5, 8.5]

Fréchet, M (1951). Sur les tableaux de corrélation dont les marges sont données. Annales de l'Université de Lyon, Série 3, 14, 53–77. [3.1.2, 3.4.1]

Freeman, D H (1981). The analysis of twice classified data. American Statistical Association, 1981 Proceedings of the Social Statistics Section, pp. 178–182. [24.11.2—2]

Freeman, D H, Freeman, J L, and Koch, G G (1978). A modified chi-squared approach for fitting Weibull models to synthetic life tables. Biometrical Journal, 20, 29–40. [22.4.1]

Freund, J E (1961). A bivariate extension of the exponential distribution. Journal of the American Statistical Association, 56, 971–977. [9.3.8, 9.13, 17.2.2, 17.4]

Friday, D S, and Patil, G P (1977). A bivariate exponential model with applications to reliability and computer generation of random variables. In C P Tsokos and I N Shimi (Editors), The Theory and Applications of Reliability. Volume 1, pp. 527–549. New York: Academic Press. [9.8, 17.2.2]

Frieden, B R (1983). Probability, Statistical Optics, and Data Testing. A Problem Solving Approach. Berlin: Springer. [3.11.3, 19.8.8]

Friedman, H P, and Rubin, J (1967). On some invariant criteria for grouping data. Journal of the American Statistical Association, 62, 1159–1178. [24.7.1—3]

Friedman, J H, and Stuetzle, W (1982). Projection pursuit methods for data analysis. In R L Launer and A F Siegel (Editors), Modern Data Analysis, pp. 123–147. New York: Academic Press. [4.13.3.4]

Friedman, J H, and Tukey, J W (1974). A projection pursuit algorithm for exploratory data analysis. IEEE (Institute of Electrical and Electronics Engineers) Transactions on Computing, C-23, 881–890. [4.13.3.4, 24.7.1—3]

G

Gabriel, K R (1985). Multivariate graphics. In Encyclopedia of Statistical Sciences. Volume 6, pp. 66–79. New York: Wiley. [24.7.1—3]

Gabriel, K R (1988). Weather modification—I. In Encyclopedia of Statistical Sciences. Volume 9, pp. 542–547. New York: Wiley. [19.7.5]

Gajjar, A V, and Subrahmaniam, K (1978). On the sample correlation coefficient in the truncated bivariate normal population. Communications in Statistics, Part B—Simulation and Computation, 7, 455–477. [4.16.1]

Galambos, J (1978). The Asymptotic Theory of Extreme Order Statistics. New York: Wiley. [5.1.2, 9.17.1]

Galambos, J (1982a). Characterizations of distributions. In Encyclopedia of Statistical Sciences. Volume 1, pp. 422–428. New York: Wiley. [2.15.2—3]

Galambos, J (1982b). Exponential distribution. In Encyclopedia of Statistical Sciences. Volume 2, pp. 583–587. New York: Wiley. [2.15.1—4]

Galambos, J (1985). Multivariate stable distributions. In Encyclopedia of Statistical Sciences. Volume 6, pp. 125–129. New York: Wiley. [7.11.3, 7.11.6]

Galambos, J, and Kotz, S (1978). Characterizations of Probability Distributions. A Unified Approach with an Emphasis on Exponential and Related Models. (Lecture Notes in Mathematics, No. 675.) Berlin: Springer. [9.17.1]

Galton, F (1888). Co-relations and their measurement, chiefly from anthropometric data. Proceedings of the Royal Society of London, 45, 135–145. [24.7.2—17]

Gamel, J W, Gleason, J, Williams, H, and Greenberg, R (1985). Reproducibility of nucleolar measurements in human intraocular melanoma cells on standard histologic microslides. Analytical and Quantitative Cytology and Histology, 7, 174–177. [24.4.3—7]

Games, P A, and Klare, G R (1967). Elementary Statistics. Data Analysis for the Behavioral Sciences. New York: McGraw-Hill. [24.10.2—5]

Ganesalingam, S, and McLachlan, G J (1979). A case study of two clustering methods based on maximum likelihood. Statistica Neerlandica, 33, 81–90. [24.9.3—8]

Gaver, D P (1970). Multivariate gamma distributions generated by mixture. Sankhyā, Series A, 32, 123–126. [3.4.2]

Gaver, D P (1972). Point process problems in reliability. In P A W Lewis (Editor), Stochastic Point Processes: Statistical Analysis, Theory, and Applications, pp. 774–800. New York: Wiley. [fn9.3, 9.12.9]

Gayen, A K (1951). The frequency distribution of the product-moment correlation coefficient in random samples of any size drawn from non-normal universes. Biometrika, 38, 219–247. [3.11.2, 24.7.2—1, 24.12.2—3]

Gebelein, H (1941). Das statistische Problem der Korrelation als Variations und Eigenwertproblem und sein Zusammenhang mit der Ausgleichsrechnung. Zeitschrift für Angewandte Mathematik und Mechanik, 21, 364–379. [11.6]

Geffroy, J (1958/59). Contributions à la théorie des valeurs extrême. Publications de l'Institut de Statistique de l'Université de Paris, 7, 37–121, and 8, 123–184. [fn9.5]

Genest, C (1987). Frank's family of bivariate distributions. Biometrika, 74, 549–555. [5.4.7, 5.4.8, 5.4.9, 13.4]

Genest, C, and MacKay, R J (1986a). Copules archimédiennes et familles de lois bidimensionnelles dont les marges sont données. Canadian Journal of Statistics, 14, 145–159. [5.2.10, 5.4.9, 5.6.4, 5.6.7, 5.9.5, 5.11.7, 13.3, 13.5]

Genest, C, and MacKay, J (1986b). The joy of copulas: Bivariate distributions with uniform marginals. The American Statistician, 40, 280–283. [5.3.9, 13.5]

Gentle, J E (1986). Simulation and analysis with IMSL routines. In J R Wilson, J O Henriksen, and S D Roberts (Editors), 1986 Winter Simulation Conference Proceedings, pp. 223–226. Piscataway, New Jersey: Institute of Electrical and Electronics Engineers. [16.3.6]

Ghirtis, G C (1967). Some problems of statistical inference relating to the double-gamma distribution. Trabajos de Estadística, 18, 67–87. [10.3.8]

Ghosh, A, and Kulatilake, P H S W (1987). A FORTRAN program for generation of multivariate normally distributed random variables. Computers and Geosciences, 13, 221–233. [16.5.1]

Ghosh, M, and Ebrahimi, N (1983). Shock models leading to multivariate NBU and NBUE distributions. In P K Sen (Editor), Contributions to Statistics: Essays in Honour of Norman Lloyd Johnson, pp. 175–184. Amsterdam: North-Holland. [18.4]

Ghurye, S G, and Marshall, A W (1984). Shock processes with aftershocks and multivariate lack of memory. Journal of Applied Probability, 21, 786–801. [17.2.2]

Gibbons, J D (1971). Nonparametric Statistical Inference. New York: McGraw-Hill. [11.4]

Gideon, R A, and Gurland, J (1978). A polynomial type approximation for bivariate normal variates. SIAM (Society for Industrial and Applied Mathematics) Journal on Applied Mathematics, 34, 681–684. [4.12.5]

Gideon, R A, and Hollister, R A (1987). A rank correlation coefficient resistant to outliers. Journal of the American Statistical Association, 82, 656–666. [11.8.4]

Gilby, W H (1911). On the significance of the teacher's appreciation of general intelligence. Biometrika, 8, 94–108. [24.10.2—3]

Gilliland, D C, and Hannan, J (1983). Fair-Jaffee model. In Encyclopedia of Statistical Sciences. Volume 3, pp. 22–24. New York: Wiley. [19.9.5]

Gillings, D B (1974). Some further results for bivariate generalizations of the Neyman type A distribution. Biometrics, 30, 619–628. [24.6—11]

Gilula, Z (1984). On some similarities between canonical correlation models and latent class models for two-way contingency tables. Biometrika, 71, 523–529. [5.2.15]

Gilula, Z (1985). On the analysis of heterogeneity among populations. Journal of the Royal Statistical Society, Series B, 76–83. [24.11.2—4]

Gilula, Z (1986). Grouping and association in contingency tables: An exploratory canonical correlation approach. Journal of the American Statistical Association, 81, 773–779. [24.11.2—4, 24.11.2—9]

Gilula, Z, and Haberman, S J (1986). Canonical analysis of contingency tables by maximum likelihood. Journal of the American Statistical Association, 81, 780–788. [24.9.2—6, 24.11.2—4]

Gilula, Z, Krieger, A M, and Ritov, Y (1988). Ordinal association in contingency tables: Some interpretive aspects. Journal of the American Statistical Association, 83, 540–545. [5.2.15]

Ginsberg, E S, and Zaborowski, D (1975). Algorithm 490. The dilogarithm function of a real argument. Communications of the Association for Computing Machinery, 18, 200–202 (Remark, ACM Transactions on Mathematical Software, 2, 112). [fn9.4]

Girardi, J P, and Matar de Sarquís, M (1984). Selection of a minimum fracturation density area by geostatistics in crystalline rock. In G Verly, M David, A G Journel, and A Marechal (Editors), Geostatistics for Natural Resources Characterization. Part 2, pp. 759–773. Dordrecht: Reidel. [24.12.3—13]

Glass, D V, and Hall, J R (1954). Social mobility in Great Britain: A study of inter-generation changes in status. In D V Glass (Editor), Social Mobility in Britain, pp. 177–217. London: Routledge and Kegan Paul. [24.11.2—1]

Gleason, J M (1985). Screening tests of pseudorandom number generators on IBM PCs and compatibles. In D T Gantz, G C Blais, and S L Solomon (Editors), 1985 Winter Simulation Conference Proceedings, pp. 505–507. Piscataway, New Jersey: Institute of Electrical and Electronics Engineers. [16.3.7]

Gnanadesikan, R (1973). Graphical methods for informal inference in multivariate data analysis. Bulletin of the International Statistical Institute, 45, 195- 206. [24.7.1—3]

Gnanadesikan, R (1977). Methods for Statistical Data Analysis of Multivariate Observations. New York: Wiley. [4.13.1, 4.13.3.3, 4.13.3.4, 24.7.1—3]

Goda, Y (1979). The observed joint distribution of periods and heights of sea waves. In Proceedings of the Sixteenth Coastal Engineering Conference. Volume I, pp. 227–246. New York: American Society of Civil Engineers. [10.8, 24.12.2—8]

Goel, L R, Gupta, P, and Singh, S K (1984). A two-unit parallel redundant system with three modes and bivariate exponential lifetimes. Microelectronics and Reliability, 24, 25–28. [9.3.10]

Goel, L R, Gupta, R, and Singh, S K (1985). Availability analysis of a two-unit (dissimilar) parallel system with inspection and bivariate exponential life times. Microelectronics and Reliability, 25, 77–80. [9.2.7]

Gokhale, D V (1973). On bivariate distributions with beta marginals. Metron, 31(1–4), 268–277. [10.2.3]

Gokhale, D V, and Kullback, S (1978). The Information in Contingency Tables. New York: Dekker. [24.9.2—8, 24.9.2—10]

Goldman, A S (1985). Particle-size statistics. In Encyclopedia of Statistical Sciences. Volume 6, pp. 619–622. New York: Wiley. [2.15.3—5]

Goodhardt, G J, Ehrenberg, A S C, and Chatfield, C (1984). The Dirichlet: A comprehensive model of buying behaviour. Journal of the Royal Statistical Society, Series A, 147, 621–643 (Discussion, 643–655). [6.5.12]

Goodman, L A (1965). On the statistical analysis of mobility data. American Journal of Sociology, 70, 564–585. [24.11.2—1]

Goodman, L A (1968). The analysis of cross-classified data: Independence, quasi-independence, and interactions in contingency tables with or without missing entries. Journal of the American Statistical Association, 63, 1091–1131. [24.6—19, 24.11.2—1]

Goodman, L A (1969). How to ransack social mobility tables and other kinds of cross-classification tables. American Journal of Sociology, 75, 1–40. [11.14.2, 24.11.2—1]

Goodman, L A (1972). Some multiplicative models for the analysis of crossclassified data. In L M Le Cam, J Neyman, and E L Scott (Editors), Proceedings of the Sixth Berkeley Symposium on Mathematical Statistics and Probability. Volume I, Theory of Statistics, pp. 649–696. Berkeley: University of California Press. [24.11.2—1]

Goodman, L A (1979a). On quasi-independence in triangular contingency tables. Biometrics, 35, 651–655. [24.9.2—5]

Goodman, L A (1979b). Simple models for the analysis of association in cross-classifications having ordered categories. Journal of the American Statistical Association, 74, 537–552. [24.11.2—1, 24.11.2—4]

Goodman, L A (1981a). Association models and the bivariate normal for contingency tables with ordered categories. Biometrika, 68, 347–355. [5.7.9, 5.7.10, 24.6—13, 24.11.2—1]

Goodman, L A (1981b). Association models and canonical correlation in the analysis of cross-classifications having ordered categories. Journal of the American Statistical Association, 76, 320–334. [24.9.2—1, 24.9.2—6]

Goodman, L A (1982). The analysis of contingency tables having ordered categories using log-linear and log-nonlinear models. Metron, 40(1–2), 37–52. [24.9.2—1, 24.9.2—6, 24.9.2—8]

Goodman, L A (1985a). The analysis of cross-classified data having ordered and/or unordered categories: Association models, correlation models, and asymmetry models for contingency tables with or without missing entries. Annals of Statistics, 13, 10–69. [24.11.2—4]

Goodman, L A (1985b). Correspondence analysis models, log-linear models, and log-bilinear models for the analysis of contingency tables. Bulletin of the International Statistical Institute, 51, Book 4, Paper 28 (Discussion, Book 5, 199–203). [24.11.2—4]

Goodman, L A (1986). Some useful extensions of the usual correspondence analysis approach and the usual log-linear models approach in the analysis of contingency tables. International Statistical Review, 54, 243–270 (Discussion, 271–309). [5.2.15, 24.9.2—6, 24.11.2—4]

Goodman, L A, and Kruskal, W H (1954). Measures of association for cross classifications. Journal of the American Statistical Association, 49, 732–764. [24.9.2—7]

Gordon, Y (1987). Elliptically contoured distributions. Probability Theory and Related Fields, 76, 429–438. [7.2.10]

Govindarajulu, Z (1968). Distribution-free confidence bounds for $P(X < Y)$. Annals of the Institute of Statistical Mathematics, Tokyo, 20, 229–238. [17.6.1]

Gradshteyn, I S, and Ryzhik, I M (1965). Tables of Integrals, Series, and Products. 4th Edition. New York: Academic Press. [13.5]

Graham, N, Kramer, P, and Yager, D (1987). Signal-detection models for multidimensional stimuli: Probability distributions and combination rules. Journal of Mathematical Psychology, 31, 366–409. [23.7]

Grauslund, H, and Lind, N C (1986). A normal probability integral and some applications. Structural Safety, 4, 31–40. [4.12.5]

Gray, J S (1981). Detecting pollution induced changes in communities using the log-normal distribution of individuals among species. Marine Pollution Bulletin, 12, 173–176.

Green, D M, and Swets, J A (1966). Signal Detection Theory and Psychophysics. New York: Wiley. Reprinted with a classified bibliography, 1974—Huntingdon, New York: Krieger. [23.1.1]

Green, P J (1981). Peeling bivariate data. In V Barnett (Editor), Interpreting Multivariate Data, pp. 3–19. Chichester: Wiley. [4.13.3.2]

Green, P J (1984). Contribution to Discussion of Professor Anderson's paper. Journal of the Royal Statistical Society, Series B, 46, 26–27. [24.10.2—4]

Green, P J (1985). Peeling data. In Encyclopedia of Statistical Sciences. Volume 6, pp. 660–664. New York: Wiley. [4.13.3.2]

Green, P J (1988). Sharpening data. In Encyclopedia of Statistical Sciences. Volume 8, pp. 431–433. New York: Wiley. [4.17.6]

Green, R F (1976). Outlier-prone and outlier-resistant distributions. Journal of the American Statistical Association, 71, 502–505. [4.13.2]

Green, R F (1985). Outlier-prone distribution. In Encyclopedia of Statistical Sciences. Volume 6, pp. 538–539. New York: Wiley. [2.15.2—36]

Greenacre, M J (1974). Analyse des correspondances: Les crânes d'ours des cavernes. Unpublished dissertation, Diplôme d'Etudes Approfondies, Université de Paris VI. [24.7.1—8]

Greenacre, M J (1984). Theory and Applications of Correspondence Analysis. London: Academic Press. [24.7.1—8, 24.9.2—6]

Greene, W H (1984). Estimation of the correlation coefficient in a bivariate probit model using the method of moments. Economics Letters, 16, 285–291. [22.4.2.4]

Greenwood, J A, and Hartley, H O (1962). Guide to Tables in Mathematical Statistics. Princeton, New Jersey: Princeton University Press. [4.12.7]

Griffin, J I (1962). Statistics. Methods and Applications. New York: Holt, Rinehart and Winston. [24.7.2—18]

Griffin, P A (1976). Use of bivariate normal approximations to evaluate card-counting systems in blackjack. Abstract in Notices of the American Mathematical Society, 23, A402. [19.9.3]

Griffith, W S (1988a). Shock models. In Encyclopedia of Statistical Sciences. Volume 8, pp. 442–446. New York: Wiley. [2.15.2—53]

Griffith, W S (1988b). Wear processes. In Encyclopedia of Statistical Sciences. Volume 9, pp. 540–542. New York: Wiley. [fn17.8]

Griffiths, R C (1969). The canonical correlation coefficients of bivariate gamma distributions. Annals of Mathematical Statistics, 40, 1401–1408. [10.3.8, T14.1, 14.6.2]

Griffiths, R C (1972). Linear dependence in bivariate distributions. Australian Journal of Statistics, 14, 182–187. [7.11.5.2]

Griffiths, R C (1978). On a bivariate triangular distribution. Australian Journal of Statistics, 20, 183–185. [15.3.5]

Griffiths, R C (1985). Orthogonal expansions. In Encyclopedia of Statistical Sciences. Volume 6, pp. 530–536. New York: Wiley. [2.15.2—35]

Grigoriu, M (1983). Approximate analysis of complex reliability problems. Structural Safety, 1, 277–288. [5.10.9, 19.8.2]

Grime, G, and Hutchinson, T P (1979). Vehicle mass and driver injury. Ergonomics, 22, 93–104. [21.1, 21.2]

Grime, G, and Hutchinson, T P (1982). The influence of vehicle weight on the risk of injury to drivers. In Ninth International Conference on Experimental Safety Vehicles, held in Kyoto, pp. 726–741. Washington, D.C.: National Highway Traffic Safety Administration. (Some misprints have been noted by the authors.) [21.1, 21.2]

Gringorten, I I (1971). Modelling conditional probability. Journal of Applied Meteorology, 10, 646–657. [19.7.2, 24.12.2—4]

Groenewoud, C, Hoaglin, D C, and Vitalis, J A (1967). Bivariate Normal Offset Circle Probability Tables with Offset Ellipse Transformations. Buffalo, New York: Cornell Aeronautical Laboratory. [4.12.1]

Groner, L H, and Cook, J (1986). Modeling the arithmetic of statistical distributions. In D Ziemann (Editor), APL86 Conference Proceedings: APL in Action, pp. 78–85. London: British Informatics Society. [fn3.3]

Gross, A J, and Clark, V A (1975). Survival Distributions: Reliability Applications in the Biomedical Sciences. New York: Wiley. [8.2.8]

Gross, A J, and Lam, C F (1981). Paired observations from a survival distribution. Biometrics, 37, 505–511. [9.5.9]

Gross, A J, and Miller, M C (1981). Some applications of statistical distribution theory to biology and medicine. In C Taillie, G P Patil, and B A Baldessari (Editors), Statistical Distributions in Scientific Work. Volume 6—Applications in Physical, Social, and Life Sciences, pp. 317–336. Dordrecht: Reidel. [9.1]

Grubbs, F E (1948). On estimating precision of measuring instruments and product variability. Journal of the American Statistical Association, 43, 243–264. [24.4.1—2]

Grubbs, F E (1973). Errors of measurement, precision, accuracy and the statistical comparison of measuring instruments. Technometrics, 15, 53–66. [24.4.1—3]

Grubbs, F E (1983). Grubbs' estimators (precision and accuracy of measurement). In Encyclopedia of Statistical Sciences. Volume 3, pp. 542–549. New York: Wiley. [24.4.1—2]

Guerrero, V M, and Johnson, R A (1980). Transforming grouped bivariate data to near normality. American Statistical Association, 1980 Proceedings of the Business and Economic Statistics Section, pp. 395–400. [24.11.2—10]

Guerrero, V M, and Johnson, R A (1988). Transforming grouped bivariate data to near normality. Statistics and Probability Letters, 6, 213–224. [24.11.2—10]

Gumbel, E J (1958). Distributions à plusieurs variables dont les marges sont données. Comptes Rendus de l'Académie des Sciences, Paris, 246, 2717–2719. [5.2.12]

Gumbel, E J (1960a). Bivariate exponential distributions. Journal of the American Statistical Association, 55, 698–707. [5.2.10, 5.2.12, 5.6.7, 5.6.9.1, 5.10.5, 5.10.6, 5.11.1, 9.17.5.2]

Gumbel, E J (1960b). Multivariate distributions with given margins and analytical examples. Bulletin de l'Institut International de Statistique, 37, Book 3, 363–373 (Discussion, Book 1, 114–115). [5.2.5, 5.2.10, 5.2.12, 5.6.9.1, 5.11.1]

Gumbel, E J (1961). Bivariate logistic distributions. Journal of the American Statistical Association, 56, 335–349. [5.2.10, 5.2.12, 5.6.7, 5.6.9.3, 5.8.12, 6.7]

Gumbel, E J (1962a). Statistical theory of extreme values (main results). In A E Sarhan and B G Greenberg (Editors), Contributions to Order Statistics, pp. 56–93. New York: Wiley. [fn9.5, 9.17.1]

Gumbel, E J (1962b). Multivariate extremal distributions. Bulletin de l'Institut International de Statistique, 39, Book 2, 471–475. [fn9.5, 9.17.1]

Gumbel, E J (1965). Two systems of bivariate extremal distributions. Bulletin of the International Statistical Institute, 41, Book 2, 749–763 (Discussion, 763). [9.17.1]

Gumbel, E J, and Goldstein, N (1964). Analysis of empirical bivariate extremal distributions. Journal of the American Statistical Association, 59, 794–816. [9.17.1, 9.17.11, 24.3—1]

Gumbel, E J, and Mustafi, C K (1967). Some analytical properties of bivariate extremal distributions. Journal of the American Statistical Association, 62, 569–588. [5.6.8, 5.6.9.4, 9.17.1, 9.17.5.1, 9.17.5.2, 9.17.8, 9.17.10, 9.17.11, 24.3—2]

Gunst, R F, and Webster, J T (1973). Density functions of the bivariate chi-square distribution. Journal of Statistical Computation and Simulation, 2, 275–288. [7.5.14, 7.7.10. 7.7.11]

Gupta, R C (1982). On the determination of mixing density in reliability studies. Journal of Statistical Planning and Inference, 7, 101–105. [20.1]

Gupta, R D, and Richards, D S P (1987). Multivariate Liouville distributions. Journal of Multivariate Analysis, 23, 233–256. [6.6]

Gupta, R P, and Jain, G C (1976). A generalized bivariate logarithmic series distribution. Biometrische Zeitschrift, 18, 169–173. [24.6—15]

Gupta, R P, and Jain, G C (1978). Some applications of a bivariate Borel-Tanner distribution. Biometrical Journal, 20, 99–105. [24.6—3]

Gupta, S S (1963a). Probability integrals of multivariate normal and multivariate t. Annals of Mathematical Statistics, 34, 792–828. [4.9, 4.12.9, 6.2.11]

Gupta, S S (1963b). Bibliography on the multivariate normal integrals and related topics. Annals of Mathematical Statistics, 34, 829–838. [4.12.9]

Gupta, S S, Panchapakesan, S, and Sohn, J K (1985). On the distribution of the Studentized maximum of equally correlated normal random variables. Communications in Statistics—Simulation and Computation, 14, 103–135. [6.2.12]

Guttman, I (1988). Tolerance regions, statistical. In Encyclopedia of Statistical Sciences. Volume 9, pp. 272–287. New York: Wiley. [24.9.1—8]

Guttman, L (1986). Polytonicity and monotonicity, coefficients of. In Encyclopedia of Statistical Sciences. Volume 7, pp. 80–87. New York: Wiley. [11.12]

H

Habbema, J D F, Hermans, J, and van den Broek, K (1974). A stepwise discriminant analysis program using density estimation. In G Bruckmann, F Ferschl, and L Schmetterer (Editors), COMPSTAT 1974. Proceedings in Computational Statistics, pp. 101–110. Wien: Physica-Verlag. [24.9.3—8]

Haberman, S J (1974a). Log-linear models for frequency tables with ordered classifications. Biometrics, 30, 589–600. [24.11.2—4]

Haberman, S J (1974b). The Analysis of Frequency Data. Chicago: University of Chicago Press. [24.9.2—5, 24.11.2—1]

Haberman, S J (1979). Analysis of Qualitative Data. Volume 2: New Developments. New York: Academic Press. [24.11.2—1, 24.11.2—4]

Haberman, S J (1982). Association, measures of. In Encyclopedia of Statistical Sciences. Volume 1, pp. 130–137. New York: Wiley. [24.9.2—7]

Hafley, W L, and Buford, M A (1985). A bivariate model for growth and yield prediction. Forest Science, 31, 237–247. [19.3.2]

Hagen, E W (1980). Common-mode/common-cause failure: A review. Annals of Nuclear Energy, 7, 509–517. [9.2.8]

Hahn, G J, and Nelson, W (1970). A problem in the statistical comparison of measuring devices. Technometrics, 12, 95–102. [24.4.1—4]

Haight, F A (1961). Index to the distributions of mathematical statistics. Journal of Research of the National Bureau of Standards—B. Mathematics and Mathematical Physics, 65B, 23–60. [2.1, 4.1]

Haldane, J B S (1942). Moments of the distribution of powers and products of normal variates. Biometrika, 32, 226–242. [4.9]

Hall, W J (1970). On characterizing dependence in joint distributions. In R C Bose, I M Chakravarti, P C Mahalanobis, C R Rao, and K J C Smith (Editors), Essays in Probability and Statistics, pp. 339–376. Chapel Hill: University of North Carolina Press. [11.6]

Halperin, M, Wu, M, and Gordon, T (1979). Genesis and interpretation of differences in distribution of baseline characteristics between cases and non-cases in cohort studies. Journal of Chronic Diseases, 32, 483–491. [5.2.11, 19.4.1]

Ham, J C (1982). Estimation of a labour supply model with censoring due to unemployment and underemployment. Review of Economic Studies, 49, 335–354. [19.9.5]

Ham, J, and Hsiao, C (1984). Two-stage estimation of structural labor supply parameters using interval data from the 1971 Canadian census. Journal of Econometrics, 24, 133–158. [24.11.2—14]

Hamdan, M A (1972). Estimation in the truncated bivariate Poisson distribution. Technometrics, 14, 37–45. [24.6—3]

Hamdan, M A, and Al-Bayyati, H A (1971). Canonical expansion of the compound correlated bivariate Poisson distribution. Journal of the American Statistical Association, 66, 390–393. [24.6—2]

Hamdy, H I, Son, M S, and Al-Mahmeed, M (1988). On the distribution of bivariate F-variable. Computational Statistics and Data Analysis, 6, 157–164. [6.3.11]

Hameed, T S S, and Baba, M (1985). Wave height distribution in shallow water. Ocean Engineering, 12, 309–319. [24.12.2—8]

Harkness, W L, and Harkness, M L (1968). Generalized hyperbolic secant distributions. Journal of the American Statistical Association, 63, 329–337. [2.15]

Harris, B (1988). Tetrachoric correlation coefficient. In Encyclopedia of Statistical Sciences. Volume 9, pp. 223–225. New York: Wiley. [11.11]

Harris, D L (1970). The analysis of wave records. In Proceedings of the Twelfth Coastal Engineering Conference. Volume 1, pp. 85–100. [24.4.2—11]

Harris, D L (1972a). Characteristics of wave records in the coastal zone. In Waves on Beaches and Resulting Sediment Transport, pp. 1–51. New York: Academic Press. [24.12.2—8]

Harris, D L (1972b). Wave estimates for coastal regions. In D J P Swift, D B Duane, and O H Pilkey (Editors), Shelf Sediment Transport: Process and Pattern, pp. 99–125. Stroudsburg, Pennsylvania: Dowden, Hutchinson and Ross. [24.4.2—11]

Harris, R (1968). Reliability applications of a bivariate exponential distribution. Operations Research, 16, 18–27. [9.2.7]

Harris, R (1970). A multivariate definition for increasing hazard rate distribution. Annals of Mathematical Statistics, 41, 713–717. [12.2.1, 12.3.5, 18.3.2]

Harter, H L (1984). Another look at plotting positions. Communications in Statistics—Theory and Methods, 13, 1613–1633. [4.13.3.1]

Hashino, M (1985). Formulation of the joint return period of two hydrologic variates associated with a Poisson process. Journal of Hydroscience and Hydraulic Engineering, 3, 73–84. [9.3.12, 9.4.3, 24.12.3—9]

Hashino, M, and Sugi, Y (1984). Study on combination of bivariate exponential distributions and its application. (In Japanese.) Scientific Papers of the Faculty of Engineering, University of Tokushima, No. 29, 49–57. [9.3.12, 9.4.3]

Hastings, N A J, and Peacock, J B (1975). Statistical Distributions. London: Butterworths. [2.1]

Hauptman, H (1985). Phase problem of X-ray crystallography, probabilistic methods in the. In Encyclopedia of Statistical Sciences. Volume 6, pp. 702–709. New York: Wiley. [2.15.3—6]

Hauser, R M (1980). Some exploratory methods for modeling mobility tables and other cross-classified data. In K F Schuessler (Editor), Sociological Methodology 1980, pp. 413–458. San Francisco: Jossey-Bass. [24.11.2—1]

Hauser, R M, and Massagli, M P (1983). Some models of agreement and disagreement in repeated measurements of occupation. Demography, 20, 449–460. [24.4.2—19]

Haver, S (1985). Wave climate off northern Norway. Applied Ocean Research, 7, 85–92. [8.12.5.1]

Hawkes, A G (1972). A bivariate exponential distribution with applications to reliability. Journal of the Royal Statistical Society, Series B, 34, 129–131. [9.12.12, T17.1, 17.3.4, fn17.4]

Hawkins, D M (1980). Identification of Outliers. London: Chapman and Hall. [4.13.2]

Hawkins, D M (1981). A new test for multivariate normality and homoscedasticity. Technometrics, 23, 105–110. [4.13.2, 4.13.5.6]

Hayashi, C (1951). On the prediction of phenomena from qualitative data and the quantification of qualitative data from the mathematico-statistical point of view. Annals of the Institute of Statistical Mathematics, Tokyo, 3, 69–98. [24.10.2—2]

Heckman, J J (1976). The common structure of statistical models of truncation, sample selection and limited dependent variables and a simple estimator for such models. Annals of Economic and Social Measurement, 5, 475–492. [19.9.5]

Heckman, J J, and Macurdy, T E (1986). Labor econometrics. In Z Griliches and M D Intriligator (Editors), Handbook of Econometrics. Volume III, pp. 1917–1977. Amsterdam: North-Holland. [19.9.5]

Heckman, J J, and Singer, B (1986). Econometric analysis of longitudinal data. In Z Griliches and M D Intriligator (Editors), Handbook of Econometrics. Volume III, pp. 1689–1763. Amsterdam: North-Holland. (A very similar paper appeared under the title "Econometric duration analysis" in Journal of Econometrics, 24,

64–132 (1984).) [18.1]

Hedlund, R D (1977–78). Cross-over voting in a 1976 open presidential primary. Public Opinion Quarterly, 41, 498–514. [24.11.2—3]

Herdan, G (1966). The Advanced Theory of Language as Choice and Chance. New York: Springer. [24.6—14]

Hettmansperger, T P (1982). Bowker's test for symmetry. In Encyclopedia of Statistical Sciences. Volume 1, pp. 304–306. New York: Wiley. [24.9.2—8]

Hewett, J, and Bulgren, W G (1971). Inequalities for some multivariate f-distributions with applications. Technometrics, 13, 397–402. [6.3.11]

Hewlett, P S, and Plackett, R L (1959). A unified theory for quantal responses to mixtures of drugs: Non-interactive action. Biometrics, 15, 591–610. [22.2.1]

Hewlett, P S, and Plackett, R L (1979). An Introduction to the Interpretation of Quantal Responses in Biology. London: Arnold. [22.1, 22.2.1]

Heyde, C C (1985). Multidimensional central limit theorems. In Encyclopedia of Statistical Sciences. Volume 5, pp. 643–646. New York: Wiley. [2.15.2—21, 4.9]

Hiemstra, L A V, and Creese, R C (1970). Synthetic generation of seasonal precipitation. Journal of Hydrology, 11, 30–46. [19.7.4]

Hill, B M (1988). Tail probabilities. In Encyclopedia of Statistical Sciences. Volume 9, pp. 170–173. New York: Wiley. [2.15.2—59]

Hill, M O (1982). Correspondence analysis. In Encyclopedia of Statistical Sciences. Volume 2, pp. 204–210. New York: Wiley. [24.9.2—6]

Hills, M (1982). Allometry. In Encyclopedia of Statistical Sciences. Volume 1, pp. 48–54. New York: Wiley. [24.7.3—4]

Hinkley, D (1983). Jackknife methods. In Encyclopedia of Statistical Sciences. Volume 4, pp. 280–287. New York: Wiley. [24.14.1—3]

Hinkley, D V (1969). On the ratio of two correlated normal random variables. Biometrika, 56, 635–639 (Correction, 57, 683). [4.9]

Hirano, K (1986a). Randomized gamma distribution. In Encyclopedia of Statistical Sciences. Volume 7, pp. 538–540. New York: Wiley. [2.15.1—39]

Hirano, K (1986b). Rayleigh distribution. In Encyclopedia of Statistical Sciences. Volume 7, pp. 647–649. New York: Wiley. [2.15.1—40]

Hirano, K, Kuboki, H, Aki, S, and Kuribayashi, A (1983). Figures of probability density functions in statistics. I—Continuous univariate case. Computer Science Monograph No. 19, Institute of Statistical Mathematics, Tokyo. [2.1]

Hirschfeld, H O (1935). A connection between correlation and contingency. Proceedings of the Cambridge Philosophical Society, 31, 520–524. [14.1]

Hoaglin, D C (1983a). g-and-h-distributions. In Encyclopedia of Statistical Sciences. Volume 3, pp. 298–301. New York: Wiley. [2.15.1—9]

Hoaglin, D C (1983b). Generation of random variables. In Encyclopedia of Statistical Sciences. Volume 3, pp. 376–382. New York: Wiley. [2.15.2—10, 16.2.1, 16.2.2, 16.2.5, 16.3.4]

Hoeffding, W (1940). Masstabinvariante Korrelationstheorie. Schriften des Mathematischen Instituts und des Instituts für Angewandte Mathematik der Universität Berlin, 5, 179–233. [3.1.2, 11.9, 12.2.2]

Hogben, N, and Lumb, F E (1967). Ocean Wave Statistics. A Statistical Survey of Wave Characteristics Estimated Visually from Voluntary Observing Ships Sailing Along the Shipping Routes of the World. London: HMSO. [24.4.2—11, 24.4.2—12, 24.12.2—8]

Hohn, M E, and Nuhfer, E B (1980). Asymmetric measures of association, closed data, and multivariate analysis. Mathematical Geology (Journal of the International Association for Mathematical Geology), 12, 235–246. [24.5.2]

Hohn, M H, and Hellerman, J (1963). The taxonomy and structure of diatom populations from three eastern North American rivers using three sampling methods. Transactions of the American Microscopical Society, 82, 250–329. [24.6—8]

Holgate, P (1964). Estimation for the bivariate Poisson distribution. Biometrika, 51, 241–245. [24.6—3]

Holgate, P (1966). Bivariate generalizations of Neyman's Type A distribution. Biometrika, 53, 241–245. [24.6—11]

Holla, M S, and Bhattacharya, S K (1965). A bivariate gamma distribution in life testing. Defence Science Journal (India), 15, 65–74. [9.3.10]

Holland, P W, and Wang, Y J (1987a). Regional dependence for continuous bivariate densities. Communications in Statistics—Theory and Methods, 16, 193–206. [11.14.1, 11.15.1, 11.15.2]

Holland, P W, and Wang, Y J (1987b). Dependence function for continuous bivariate densities. Communications in Statistics—Theory and Methods, 16, 863–876. [11.14.1, 11.14.2, 11.15.3, 13.3]

Hollander, M, Park, D H, and Proschan, F (1986). A class of life distributions for aging. Journal of the American Statistical Association, 81, 91–95. [18.2.7]

Hollander, M, and Proschan, F (1984). Nonparametric concepts and methods in reliability. In P R Krishnaiah and P K Sen (Editors), Handbook of Statistics. Volume 4. Nonparametric Methods, pp. 613–655. Amsterdam: North-Holland. [18.2.1, 18.2.5]

Holmes, J (1974). The relative burden of higher gasoline prices. In Five Thousand American Families—Patterns of Economic Progress. Volume IV. Ann Arbor: Institute for Social Research. [24.11.2—10]

Holmquist, N D, McMahan, C A, and Williams, O D (1967). Variability in classification of carcinoma in situ of the uterine cervix. Archives of Pathology, 84, 334–345. [24.4.1—7]

Holst, E, and Schneider, T (1985). Fibre size characterization and size analysis using general and bivariate log-normal distributions. Journal of Aerosol Science, 16, 407–413. [19.4.6]

Holt, S B (1952). Genetics of dermal ridges: Inheritance of total finger ridge-count. Annals of Eugenics, London, 17, 140–161. [24.9.2—12]

Horsnell, G (1969). A theory of consumer behaviour derived from repeat paired preference testing. Journal of the Royal Statistical Society, Series A, 132, 164–184 (Discussion, 185–192). [20.5.3]

Horsnell, G (1977). Paired comparison product testing when individual preferences are stochastic: An alternative model. Applied Statistics, 26, 162–172. [20.5.1, 20.5.3, T20.6]

Hotelling, H (1936). Relations between two sets of variables. Biometrika, 28, 321–377. [14.1]

Hotta, S, Mizuguchi, M, and Isobe, M (1983). A field study of waves in the nearshore zone. In B L Edge (Editor), Proceedings of the Eighteenth Coastal Engineering Conference. Volume I, pp. 38–57. New York: American Society of Civil Engineers. [24.12.2—8]

Hougaard, P (1986). A class of multivariate failure time distributions. Biometrika, 73, 671–678 (Correction, 75, 395). [5.6.5, 5.6.7, 5.6.8, 9.17.11]

Hougaard, P (1987). Modelling multivariate survival. Scandinavian Journal of Statistics, 14, 291–304. [1.6]

Hougaard, P (1988?). Fitting a multivariate failure time distribution. Report from the Biostatistical Department, Novo Research Institute, Denmark. [5.6.5, 5.6.7, 5.6.8, 24.2—2]

Howarth, R J, and Earle, S A M (1979). Application of a generalized power transformation to geochemical data. Mathematical

Geology (Journal of the International Association for Mathematical Geology), 11, 45–62. [4.13.6]

Hoyer, R W, and Mayer, L S (1976). The equivalence of various objective functions in a stochastic model of electoral competition. Technical Report No. 114, Series 2, Department of Statistics, Princeton University. [6.5.9, 6.5.12]

Hsu, J C, and Nelson, B L (1987). Control variates for quantile estimation. In A Thesen, H Grant, and W D Kelton (Editors), 1987 Winter Simulation Conference Proceedings, pp. 434–444. Piscataway, New Jersey: Institute of Electrical and Electronics Engineers. [16.5.4.3]

Huang, J S, and Kotz, S (1984). Correlation structure in iterated Farlie-Gumbel-Morgenstern distributions. Biometrika, 71, 633–636. [13.2.2]

Hurst, R L, and Knop, R E (1972). Algorithm 425. Generation of random correlated normal variables. Communications of the Association for Computing Machinery, 15, 355–357. [16.5.1]

Hutchinson, T P (1976). Statistical aspects of injury severity. Part I: Comparison of two populations when there are several grades of injury. Transportation Science, 10, 269–284. [fn20.1]

Hutchinson, T P (1977). On the relevance of signal detection theory to the correction for guessing. Contemporary Educational Psychology, 2, 50–54. [23.5]

Hutchinson, T P (1978). An extension of the signal detection model of information retrieval. Journal of Documentation, 34, 51–54. [23.4]

Hutchinson, T P (1979a). A comment on replicated paired comparisons. Applied Statistics, 28, 163–169. [19.6.2, 20.5.1, 20.5.2, 20.5.3]

Hutchinson, T P (1979b). Four applications of a bivariate Pareto distribution. Biometrical Journal, 21, 553–563. [5.8.13.2, 5.9.6, 6.3.3, 6.8, 15.6.3, 20.7.1]

Hutchinson, T P (1980). An easy method of calculating approximate recurrence risks using a multifactorial model of disease transmission. Annals of Human Genetics, 43, 285–293. [20.4.4]

Hutchinson, T P (1981a). A review of some unusual applications of signal detection theory. Quality and Quantity, 15, 71–98. [23.1.1]

Hutchinson, T P (1981b). Compound gamma bivariate distributions. Metrika, 28, 263–271. [5.8.13.2, 6.3.3, 6.8, 15.6.3, 20.7.1]

Hutchinson, T P (1982a). Signal probability and the slope of the ROC: A comment on Treisman. Psychological Bulletin, 91, 193–197. [23.2]

Hutchinson, T P (1982b). Some theories of performance in multiple choice tests, and their implications for variants of the task. British Journal of Mathematical and Statistical Psychology, 35, 71–89. [23.5]

Hutchinson, T P (1982c). Disease concordance and sex similarity in twins: Application of a multifactorial model with latent structure. Journal of Epidemiology and Community Health, 36, 155–156. [19.5.3.3]

Hutchinson, T P (1983a). A bivariate normal model for intra-accident correlations of driver injury with application to the effect of mass ratio. Accident Analysis and Prevention, 15, 215–224. [21.4, 21.5]

Hutchinson, T P (1983b). A note on applications of the competing risks model. Accident Analysis and Prevention, 15, 225–226. [9.18.3]

Hutchinson, T P (1984). Using the bivariate normal distribution to construct probability models in the health sciences. Communications in Statistics—Theory and Methods, 13, 1723–1733. [19.4.10]

Hutchinson, T P (1985a). Analysing severity data when assessors differ in their usage of the categories. The Statistician, 34, 183–195. [20.2.4]

Hutchinson, T P (1985b). Presenting one probability distribution as a function of another—Some applications. American Journal of Mathematical and Management Sciences, 5, 103–123. [23.1.3]

Hutchinson, T P (1986). Statistical modelling of injury severity, with special reference to driver and front seat passenger in single-vehicle crashes. Accident Analysis and Prevention, 18, 157–167. [21.8.3]

Hutchinson, T P (1987). Road Accident Statistics. Adelaide: Rumsby Scientific Publishing. [fn21.5, fn21.7]

Hutchinson, T P, and Mayne, A J (1977). The year-to-year variability in the numbers of road accidents. Traffic Engineering and Control, 18, 432–433. [21.4]

Hutchinson, T P, and Satterthwaite, S P (1977). Mathematical models for describing the clustering of sociopathy and hysteria in families: A comment on the recent paper by Cloninger et al. British Journal of Psychiatry, 130, 294–297. [fn20.2, 20.4.2]

Huynh, H (1979). A class of passing scores based on the bivariate normal model. American Statistical Association, 1979 Proceedings of the Social Statistics Section, pp. 310–313. [19.6.1]

I

Ichihara, M, and Yamada, K (1979). Earthquake damage prediction for water-supply pipes. Natural Disaster Science, 1, 57–80. [24.12.3—11]

Inaba, T, and Shirahata, S (1986). Measures of dependence in normal models and exponential models by information gain. Biometrika, 73, 345–352. [8.2.7, 8.2.8]

Ingham, P E, and Smith, R J (1986). Lightfastness of wool dyes. Report No. R129, Wool Research Organisation of New Zealand, Christchurch. [24.4.2—10]

Ireland, C T, Ku, H H, and Kullback, S (1969). Symmetry and marginal homogeneity of an $r \times r$ contingency table. Journal of the American Statistical Association, 64, 1323–1341. [24.9.2—8]

Isogai, T (1982). On a measure of multivariate skewness and a test for multivariate normality. Annals of the Institute of Statistical Mathematics, Tokyo, 34, 531–541. [24.7.1—3]

Isogai, T (1983a). Monte Carlo study on some measures for evaluating multinormality. Reports of Statistical Application Research, Union of Japanese Scientists and Engineers, 30, 1–10. [4.13.5.8]

Isogai, T (1983b). On measures of multivariate skewness and kurtosis. Mathematica Japonica, 28, 251–261. [4.13.5.8]

Isserlis, L (1915). On the partial correlation-ratio. Part II. Numerical. Biometrika, 11, 50–66. [24.7.2—19]

Itoi, T, Kodama, M, and Nishida, T (1978). Behaviour of a two correlated units redundant system with many types of failure. Microelectronics and Reliability, 17, 517–522. [9.8]

Itoi, T, Murakami, T, Kodama, M, and Nishida, T (1976). Reliability analysis of a parallel redundant system with two dissimilar correlated units. Technology Reports of Osaka University, 26, 403–409. [9.2.13]

Iyengar, S (1985). Hitting lines with two-dimensional Brownian motion. SIAM (Society for Industrial and Applied Mathematics) Journal on Applied Mathematics, 45, 983–989. [7.9.8]

Iyengar, S, and Patwardhan, G (1988). Recent developments in the inverse Gaussian distribution. In P R Krishnaiah and C R Rao (Editors), Handbook of Statistics. Volume 7, Quality Control and Reliability, pp. 479–490. Amsterdam: North-Holland. [7.9.8]

Izawa, T (1965). Two or multi-dimensional gamma-type distribution and its application to rainfall data. Papers in Meteorology and Geophysics, 15, 167–200. [7.5.11, 7.5.12, 7.5.13, 7.5.15]

J

Jain, G C, and Khan, M S H (1979). On an exponential family. Statistics, 10, 153–168. [7.5.9]

James, I R (1975). Multivariate distributions which have beta conditional distributions. Journal of the American Statistical Association, 70, 681–684. [6.5.14, 8.4.3, 8.4.4]

James, I (1981). Distributions associated with neutrality properties for random proportions. In C Taillie, G P Patil, and B A Baldessari (Editors), Statistical Distributions in Scientific Work. Volume 4—Models, Structures, and Characterizations, pp. 125–136. Dordrecht: Reidel. [8.4.4]

Japanese Standards Association (1972). Statistical Tables and Formulas with Computer Applications. Tokyo: Japanese Standards Association. [4.12.1, 4.12.7]

Jensen, D R (1969). An inequality for a class of bivariate chi-square distributions. Journal of the American Statistical Association, 64, 333–336. [7.5.12]

Jensen, D R (1970). The joint distribution of quadratic forms and related distributions. Australian Journal of Statistics, 12, 13–22. [7.7.2, 7.7.8, 7.7.12, 7.8.5, 7.8.7, T14.1, 15.4.6]

Jensen, D R (1985). Multivariate Weibull distributions. In Encyclopedia of Statistical Sciences. Volume 6, pp. 132–133. New York: Wiley. [9.16]

Jensen, D R (1988). Semi-independence. In Encyclopedia of Statistical Sciences. Volume 8, pp. 358–359. New York: Wiley. [11.5.1]

Joag-Dev, K (1984). Measures of dependence. In P R Krishnaiah and P K Sen (Editors), Handbook of Statistics. Volume 4. Nonparametric Methods, pp. 79–88. Amsterdam: North-Holland. [11.8.4, 12.2.1]

Joe, H, and Proschan, F (1984). Percentile residual life functions. Operations Research, 32, 668–678. [18.2.7]

Jogdeo, K (1964). Nonparametric methods for regression. Report S330, Mathematics Centre, Amsterdam. [3.7]

Jogdeo, K (1975). Dependence concepts and probability inequalities. In G P Patil, S Kotz, and J K Ord (Editors), A Modern Course on Distributions in Scientific Work. Volume 1—Models and Structures, pp. 271–279. Dordrecht: Reidel. [12.2.1]

Jogdeo, K (1982). Dependence, concepts of. In Encyclopedia of Statistical Sciences. Volume 2, pp. 324–334. [11.3.4, 12.2.1, 12.2.2, 12.3.1, 12.3.3]

Johannsen, F (1903). Ueber Erblichkeit in Populationen und in reinem Linien. Jena: Fischer. [24.7.2—1]

Jöhnk, M D (1964). Erzeugung von betaverteilten und gammaverteilten Zufallszahlen. Metrika, 8, 5–15. [16.3.2, 16.3.3]

Johnson, M E (1986). Distribution selection in statistical simulation studies. In J R Wilson, J O Henriksen, and S D Roberts (Editors), 1986 Winter Simulation Conference Proceedings, pp. 253–259. Piscataway, New Jersey: Institute of Electrical and Electronics Engineers. [5.9.3, 6.4.7]

Johnson, M E (1987). Multivariate Statistical Simulation. New York: Wiley. [1.6, 3.4.1, 4.7, 4.14.2.1, 4.14.2.5, 4.14.4, 4.14.5, 4.17.1, 4.17.2, 5.1.3, 5.1.4, 5.2.7, 5.2.8, 5.2.16, 5.3.5, 5.3.7, 5.3.8, 5.3.9, 5.7.7, 5.7.8, 5.8.12, 5.9.3, 5.9.5, 5.11.5, 5.11.6, 6.2.8, 6.2.9, 6.4.7, 6.4.8, 6.4.10, 7.2.1, 7.2.6, 7.3, 10.2.2, 10.5.1, 10.5.2, 10.5.3, 15.4.7, 16.1, 16.4.1, 16.4.3, 16.5.1, 20.6.1]

Johnson, M E, Bryson, M C, and Mills, C F (1981). Some new multivariate distributions with enhanced comparisons via contour and three-dimensional plots. Report LA-8903-MS, Los Alamos Scientific Laboratory, New Mexico. [4.7, 4.14.3, 5.1.3, 5.2.7, 5.7.7, 5.11.5, 10.5.2, 10.5.3]

Johnson, M E, and Tenenbein, A (1978). Bivariate distributions with given marginals and fixed measures of dependence. American Statistical Association, 1978 Proceedings of the Statistical Computing Section, pp. 261–263. [16.5.4.1, 16.5.4.2]

Johnson, M E, and Tenenbein, A (1979). Bivariate distributions with given marginals and fixed measures of dependence. Informal Report LA-7700-MS, Los Alamos Scientific Laboratory, New Mexico. [10.4.2, 10.4.3, 13.7, 13.8, 13.9, 16.5.4.2]

Johnson, M E, and Tenenbein, A (1981). A bivariate distribution family with specified marginals. Journal of the American Statistical Association, 76, 198-201. [3.7, 5.11.5, 10.4.2, 10.4.3, 13.8, 16.5.4.1, 16.5.4.2]

Johnson, M E, Wang, C, and Ramberg, J (1984). Generation of continuous multivariate distributions for statistical applications. American Journal of Mathematical and Management Sciences, 4, 225–248. [3.5, 4.13.7, 5.1.3, 5.1.4, 5.2.7, 5.2.8, 5.3.7, 5.3.8, 5.7.7, 5.7.8, 5.9.3, 5.9.4, 5.11.5, 5.11.6, 6.2.9, 6.2.13, 6.4.8, 6.4.9, 7.2.3, 7.2.9, 8.12.1, 10.5.4, 16.1, 16.4.1, 16.4.3]

Johnson, N L (1949). Bivariate distributions based on simple translation systems. Biometrika, 36, 297–304. [2.4.3, 3.2.2, 4.13.6, 4.14.2.1, 24.7.2—1]

Johnson, N L (1954). Systems of frequency curves derived from the first law of Laplace. Trabajos de Estadística, 5, 283–291. [2.7.2]

Johnson, N L, and Kotz, S (1970a). Distributions in Statistics: Continuous Univariate Distributions—1. New York: Wiley. [2.1, 2.4.3, 2.9, 3.9.1.1, 3.11.1, 9.17.10, 9.17.11]

Johnson, N L, and Kotz, S (1970b). Distributions in Statistics: Continuous Univariate Distributions—2. New York: Wiley. [9.17.10, 9.17.11]

Johnson, N L, and Kotz, S (1972). Distributions in Statistics: Continuous Multivariate Distributions. New York: Wiley. [1.1, 3.2.2, 3.9.1.2, 4.1, 4.2, 4.3.8, 4.5, 4.7, 4.9, 4.12.6, 4.12.9, 4.14.1, 4.14.2.4, 4.15, 4.16.1, 6.1, 6.2.5, 6.2.11, 6.2.12, 6.2.14, 6.3.10, 6.9.4, 7.5.3, 8.6.7, 9.2.13, 9.3.9, 9.17.10, 10.3.5, 15.6.1, 19.8.9]

Johnson, N L, and Kotz, S (1973a). Extended and multivariate Tukey lambda distributions. Biometrika, 60, 655–661. [6.5.14]

Johnson, N L, and Kotz, S (1973b). A vector valued multivariate hazard rate. Bulletin of the International Statistical Institute, 45, Book 1, 570–574. [18.3.2]

Johnson, N L, and Kotz, S (1975). A vector multivariate hazard rate. Journal of Multivariate Analysis, 5, 53–66. [18.3.2]

Johnson, N L, and Kotz, S (1977). On some generalized Farlie-Gumbel-Morgenstern distributions—II. Regression, correlation and further generalizations. Communications in Statistics, Part A—Theory and Methods, 6, 485–496. [5.2.13]

Johnson, N L, and Kotz, S (1981). Dependent relevations: Time-to-failure under dependence. American Journal of Mathematical and Management Sciences, 1, 155–165. [fn5.1]

Johnson, N L, and Kotz, S (1983). Multivariate dependent relevations with applications, and related topics. Naval Research Logistics Quarterly, 30, 163–169. [5.2.13]

Johnson, R A (1988). Stress-strength models for reliability. In P R Krishnaiah and C R Rao (Editors), Handbook of Statistics. Volume 7, Quality Control and Reliability, pp. 27–54. Amsterdam: North-Holland. [17.6.1]

Johnson, R A, and Wichern, D W (1982). Applied Multivariate Statistical Analysis. Englewood Cliffs, New Jersey: Prentice-Hall. [24.7.1—3]

Johnson, R E (1986). A convolution algorithm for calculating coefficients of Jensen's bivariate F distribution. Communications in Statistics—Simulation and Computation, 15, 1209–1213. [7.8.8]

Jolicoeur, P (1963). The degree of generality of robustness in *Martes americana*. Growth, 27, 1–27. [19.3.1, 24.7.3—4]

Jolicoeur, P, and Mosimann, J E (1960). Size and shape variation in the painted turtle. A principal component analysis. Growth, 24,

339–354. (Reprinted in E H Bryant and W R Atchley, Multivariate Statistical Methods: Within-Groups Covariation, pp. 86–101. Stroudsburg, Pennsylvania: Dowden, Hutchinson and Ross, 1975.) [24.7.1—1]

Jones, A W, and Löwinger, H (1988). Relationship between the concentration of ethanol and methanol in blood samples from Swedish drinking drivers. Forensic Science International, 37, 277–285. [24.9.3—1]

Jones, F L, and Pittelkow, Y E (1983). Analysis of occupational mobility tables using GLIM. The GLIM Newsletter, No. 7, 34–36. [24.11.2—1]

Jørgensen, B (1987). Exponential dispersion models. Journal of the Royal Statistical Society, Series B, 49, 127–145 (Discussion, 145–162). [8.9]

Jørgensen, N R (1916). Undersøgelser over Frequensflader og Korrelation. Copenhagen: Busck. [24.6—12]

Joshi, P C, and Lalitha, S (1985). A recurrence formula for the evaluation of a bivariate probability. Journal of the Indian Statistical Association, 23, 159–169. [6.4.10]

K

Kafadar, K (1988). Slash distribution. In Encyclopedia of Statistical Sciences. Volume 8, pp. 510–511. New York: Wiley. [2.15.1—43]

Kamath, S (1981). Observations on the length and diameter of vessels forming the circle of Willis. Journal of Anatomy, 133, 419–423. [24.9.3—5]

Kaminsky, F C, and Kirchhoff, R H (1988). Bivariate probability models for the description of average wind speed at two heights. Solar Energy, 40, 49–56. [8.12.5.2]

Kaplow, L S, Crouch, J Y, Meyers, J A, Kunz, H, and Garcia, G L (1986). Assessment of leukocyte alkaline phosphatase by image analysis. In M Andreeff (Editor), Clinical Cytometry, pp. 85–92. (Annals of the New York Academy of Sciences, 468.) [24.4.1—6, 24.4.2—7, 24.4.3—8]

Kappenman, R F (1985). Estimation for the three-parameter Weibull, lognormal, and gamma distributions. Computational Statistics and Data Analysis, 3, 11–23. [2.6.6]

Karlin, S, and Rinott, Y (1980). Classes of orderings of measures and related correlation inequalities, II. Multivariate reverse rule distributions. Journal of Multivariate Analysis, 10, 499–516. [12.4.1]

Karpinos, B D (1966). The mental test qualification of American youths for military service and its relation to educational attainment. American Statistical Association, Proceedings of the Social Statistics Section, 1966, pp. 92–111. [24.10.2—1]

Karshan, M, Tenenbaum, B, Karlan, F R, and Leonard, H J (1946). Blood studies in periodontoclasia. II. Journal of Dental Research, 25, 247–252. [24.9.2—3]

Kasser, I S, and Bruce, R A (1969). Comparative effects of aging and coronary heart disease on submaximal and maximal exercise. Circulation, 39, 759–774. [24.9.1—3]

Katz, D, and D'Argenio, D Z (1986). Implementation and evaluation of control strategies for individualizing dosage regimens, with application to the aminoglycoside antibiotics. Journal of Pharmokinetics and Biopharmaceutics, 14, 523–537. [19.4.5]

Kaufmann, A, Grouchko, D, and Cruon, R (1977). Mathematical Models for the Study of the Reliability of Systems. New York: Academic Press. [18.2.4]

Keating, J P, and Tripathi, R C (1985). Percentiles, estimation of. In Encyclopedia of Statistical Sciences. Volume 6, pp. 668–674. New York: Wiley. [2.15.2—37]

Kelker, D (1970). Distribution theory of spherical distributions and a location-scale parameter generalization. Sankhyā, Series A, 32, 419–430. [7.2.1, 7.2.6]

Kelley, G D, Kelley, J A, and Schucany, W R (1976). Efficient estimation of $P(Y < X)$ in the exponential case. Technometrics, 18, 359–360. [17.6.1]

Kellogg, S D, and Barnes, J W (1987). The distribution of products, quotients, and powers of two dependent H-function variates. Mathematics and Computers in Simulation, 29, 209–221. [10.6]

Kellogg, S D, and Barnes, J W (1989). The bivariate H-function distribution. Mathematics and Computers in Simulation, 31, 91–111. [6.5.9, 8.6.3, 8.6.8, 10.6]

Kemp, C D, and Papageorgiou, H (1982). Bivariate Hermite distributions. Sankhyā, Series A, 44, 269–280. [24.6—1]

Kemsley, W F F (1957). Women's Measurements and Sizes; A Study. London: HMSO. [24.7.2—13]

Kendall, M G (1966). Discrimination and classification. In P R Krishnaiah (Editor), Multivariate Analysis, pp. 165–185. New York: Academic Press. [24.7.1—3]

Kendall, M G (1975). Multivariate Analysis. London: Griffin. [24.7.1—3]

Kendall, M G, and Stuart, A (1976). The Advanced Theory of Statistics. Volume 3. Design and Analysis, and Time-Series. 3rd Edition. London: Griffin. [24.7—3]

Kendall, M G, and Stuart, A (1977). The Advanced Theory of Statistics. Volume 1. Distribution Theory. 4th Edition. London: Griffin. [3.11.1, 4.2, 4.5, 24.6—13, 4.7.2—1, 24.8.2—2]

Kendall, M G, and Stuart, A (1979). The Advanced Theory of Statistics. Volume 2. Inference and Relationship. 4th Edition. London: Griffin. [4.9, 11.5.2, 11.6, 24.7.2—13, 24.9.2—1, 24.9.2—7, 24.9.2—8, 24.10.2—3]

Kennedy, W J, and Gentle, J E (1980). Statistical Computing. New York: Dekker. [16.2.1, 16.5.1]

Kent, J T (1986). Contribution to Discussion of Professor Goodman's paper. International Statistical Review, 54, 273–274. [24.9.2—6]

Khan, M S H, and Jain, G C (1978). A class of distributions in the first emptiness of a semi-infinite reservoir. Biometrical Journal, 20, 243–252. [7.5.9]

Khatri, C G (1971). On multivariate contagious distributions. Sankhyā, Series B, 33, 197–216. [24.6—2]

Khatri, C G, and Rao, C R (1976). Characterizations of multivariate normality. I. Through independence of some statistics. Journal of Multivariate Analysis, 6, 81–94. [4.5]

Kibble, W F (1941). A two-variate gamma type distribution. Sankhyā, 5, 137–150. [7.5.6, 7.7.12, 9.12.12, T14.1]

Kie, J G, White, M, and Drawe, D L (1983). Condition parameters of white-tailed deer in Texas. Journal of Wildlife Management, 47, 583–594. [24.4.3—3]

Kiefer, N M (1985). Multivariate probit. In Encyclopedia of Statistical Sciences. Volume 6, pp. 108–110. New York: Wiley. [22.1]

Kim, J S, and Proschan, F (1988). Total positivity. In Encyclopedia of Statistical Sciences. Volume 9, pp. 289–297. New York: Wiley. [2.15.2—60, 12.3.3, fn12.7, 18.2.4]

Kimeldorf, G, May, J H, and Sampson, A R (1981). MONCOR—A program to compute concordant and other monotone correlations. In W F Eddy (Editor), Computer Science and Statistics: Proceedings of the 13th Symposium on the Interface, pp. 348–351. New York: Springer. [11.7.2]

Kimeldorf, G, May, J H, and Sampson, A R (1982). Concordant and discordant monotone correlations and their evaluation by nonlinear optimization. In S H Zanakis and J S Rustagi (Editors), Optimization in Statistics, With a View Towards Applications in Management Science and Operations Research (TIMS Studies in the Management Sciences, 19), pp. 117–130. Amsterdam: North-Holland. [11.7.2, 24.11.2—1]

Kimeldorf, G, and Sampson, A (1975a). One-parameter families of bivariate distributions with fixed marginals. Communications in Statistics, 4, 293–301. [3.2.4, 3.4.1, 5.1.1, 10.7.2, 13.13, 13.7]

Kimeldorf, G, and Sampson, A (1975b). Uniform representations of bivariate distributions. Communications in Statistics, 4, 617–627. [5.2.16, 13.2.3]

Kimeldorf, G, and Sampson, A (1978). Monotone dependence. Annals of Statistics, 5, 895–903. [11.3.3, 11.6, 11.7.1]

Kimeldorf, G, and Sampson, A R (1987). Positive dependence orderings. Annals of the Institute of Statistical Mathematics, Tokyo, 39, 113–128. [12.6]

Kimura, A (1981). Joint distribution of the wave heights and periods of random sea waves. Coastal Engineering in Japan, 24, 77–92. [9.12.13, 9.12.15, 24.12.2—8]

Kimura, A, and Seyama, A (1985). Statistical properties of short-term overtopping. In B L Edge (Editor), Nineteenth Coastal Engineering Conference, Proceedings. Volume I, pp. 532–546. New York: American Society of Civil Engineers. [9.12.13]

Kinderman, A J, and Monahan, J F (1977). Computer generation of random variables using the ratio of uniform deviates. ACM (Association for Computing Machinery) Transactions on Mathematical Software, 3, 257–260. [16.2.5]

Kinderman, A J, and Ramage, J G (1976). Computer generation of normal random variables. Journal of the American Statistical Association, 71, 893–896. [16.3.1]

Klefsjö, B (1980). On some classes of bivariate life distributions. Statistical Research Report 1980-9, Department of Mathematical Statistics, University of Umeå. [18.3.5]

Klefsjö, B (1981). HNBUE survival under some shock models. Scandinavian Journal of Statistics, 8, 39–47. [18.2.1]

Klefsjö, B (1982). The HNBUE and HNWUE classes of life distributions. Naval Research Logistics Quarterly, 29, 331–344. [18.2.1]

Klefsjö, B (1983). A useful ageing property based on the Laplace transform. Journal of Applied Probability, 20, 615–626. [18.2.1, 18.2.2]

Klein, J P (1986). A test for independence based on data from a bivariate series system. In A P Basu (Editor), Reliability and Quality Control, pp. 235–244. Amsterdam: North-Holland. [9.18.2]

Klein, J P, and Basu, A P (1985). Estimating reliability for bivariate exponential distributions. Sankhyā, Series B, 47, 346–353. [9.2.7, 9.3.10]

Klein, J P, and Moeschberger, M L (1988). Bounds on net survival probabilities for dependent competing risks. Biometrics, 44, 529–538. [5.9.6, 9.2.7, 9.3.10, 9.18.2]

Klemeš, V (1987). Empirical and causal models in hydrologic reliability analysis. In L Duckstein and E J Plate (Editors), Engineering Reliability and Risk in Water Resources, pp. 391–403. Dordrecht: Nijhoff. [9.17.11]

Kmietowicz, Z W (1984). The bivariate lognormal model for the distribution of household size and income. The Manchester School of Economic and Social Studies, 52, 196–210. [19.9.1, 24.11.2—12]

Knibbs, G H (1911). Studies in statistical representation. On the nature of the curve $y = Ax^m e^{nx^p}$. Journal of the Royal Society of New South Wales, 44, 341–367. [2.6.5]

Knuth, D E (1981). The Art of Computer Programming. Volume 2. Seminumerical Algorithms. 2nd Edition. Reading, Massachusetts: Addison-Wesley. [16.2.1]

Koch, G G (1983). Hierarchical kappa statistics. In Encyclopedia of Statistical Sciences. Volume 3, pp. 630–631. New York: Wiley. [24.4.2—8]

Kochar, S C, and Wiens, D P (1987). Partial orderings of life distributions with respect to their aging properties. Naval Research Logistics, 34, 823–829. [18.2.1, 18.2.3, 18.2.6]

Kocherlakota, S (1986). The bivariate inverse Gaussian distribution: An introduction. Communications in Statistics—Theory and Methods, 15, 1081–1112. [7.9.7, 7.9.8]

Kocherlakota, S, Kocherlakota, K, and Balakrishnan, N (1985). Effects of nonnormality on the SPRT for the correlation coefficient: Bivariate Edgeworth series distribution. Journal of Statistical Computation and Simulation, 23, 41–51. [3.11.2]

Konijn, H S (1959). Positive and negative dependence of two random variables. Sankhyā, 21, 269–280. [10.9, 13.13]

Kota, V K B (1984). Bivariate distributions in statistical spectroscopy studies: Fixed-J level densities, fixed-J averages and spin cut-off factors. Zeitschrift für Physik, Section A—Atoms and Nuclei, 315, 91–98. [3.11.4]

Kota, V K B, and Potbhare, V (1985). Bivariate distributions in statistical spectroscopy studies—II: Orthogonal polynomials in two variables and fixed E, J expectation values. Zeitschrift für Physik, Section A—Atoms and Nuclei, 322, 129–136. [3.11.4]

Kottas, J F, and Lau, H-S (1978). On handling dependent random variables in risk analysis. Journal of the Operational Research Society, 29, 1209–1217. [8.12.3]

Kotz, S (1975). Multivariate distributions at a cross road. In G P Patil, S Kotz, and J K Ord (Editors), A Modern Course on Distributions in Scientific Work. Volume 1—Models and Structures, pp. 247–270. Dordrecht: Reidel. [14.7]

Kotz, S, and Johnson, N (1977). Propriétés de dépendance des distributions itérées, généralisées à deux variables Farlie-Gumbel-Morgenstern. Comptes Rendus de l'Académie des Sciences, Paris, Série A, 285, 277–280. [5.2.13, 11.9, 13.2.2]

Kotz, S, and Johnson, N L (1984). Some replacement-times distributions in two-component systems. Reliability Engineering, 7, 151–157. [5.2.9]

Kovner, J L, and Patil, S A (1973). On the moments of the doubly truncated bivariate normal population with application to ratio estimate. Journal of the Indian Society of Agricultural Statistics, 25, 131–140. [4.16.1]

Kowalczyk, T (1977). General definition and sample counterparts of monotonic dependence functions of bivariate distributions. Statistics, 8, 351–365. [12.5.4]

Kowalczyk, T (1982). Shape of the monotone dependence function. Statistics, 13, 183–192. [12.5.4]

Kowalczyk, T, and Ledwina, T (1982). Some properties of chosen grade parameters and their rank counterparts. Statistics, 13, 547–553. [12.5.4]

Kowalczyk, T, and Pleszczyńska, E (1977). Monotonic dependence functions of bivariate distributions. Annals of Statistics, 5, 1221–1227. [12.5.1, 12.6]

Kowalski, C J (1970). The performance of some rough tests for bivariate normality before and after coordinate transformations to normality. Technometrics, 12, 517–544. [4.13.1, fn4.6]

Kowalski, C J (1973). Non-normal bivariate distributions with normal marginals. The American Statistician, 27, 103–106. [4.17.5, 4.18.1, 4.18.2]

Koziol, J A (1982). A class of invariant procedures for assessing multivariate normality. Biometrika, 69, 423–427. [24.7.1—3]

Koziol, J A (1983). On assessing multivariate normality. Journal of the Royal Statistical Society, Series B, 45, 358–361. [24.7.1—3]

Koziol, J A (1986). Assessing multivariate normality: A compendium. Communications in Statistics—Theory and Methods, 15, 2763–2783. [4.13.4.1, 4.13.5.9]

Koziol, J A (1987). An alternative formulation of Neyman's smooth goodness of fit tests under composite alternative. Metrika, 34, 17–24. [4.13.5.9]

Krauth, J (1984). A modification of kappa for interobserver bias. Biometrical Journal, 26, 435–445. [20.2.4]

Krishnaiah, P R (1963). Simultaneous tests and the efficiency of balanced incomplete block designs. Report No. ARL 63-174, Aerospace Research Laboratories, Wright-Patterson Air Force Base, Ohio. [7.5.2]

Krishnaiah, P R (1964). Multiple comparison tests in multivariate cases. Report ARL 64-124, Aerospace Research Laboratories, Wright-Patterson Air Force Base, Ohio. [6.3.9, 7.8.6, 15.4.6]

Krishnaiah, P R (1965). On the simultaneous ANOVA and MANOVA tests. Annals of the Institute of Statistical Mathematics, Tokyo, 17, 35–53. [6.3.9, 7.8.6, 7.8.7, 15.4.6]

Krishnaiah, P R (1980). Computations of some multivariate distributions. In P R Krishnaiah (Editor), Handbook of Statistics. Volume 1, pp. 745–971. Amsterdam: North-Holland. [7.5.14]

Krishnaiah, P R (1983). Multivariate gamma distributions and their applications in reliability. Technical Report No. 83-09, Center for Multivariate Analysis, University of Pittsburgh. [7.5.2, 7.7.11, 7.7.12]

Krishnaiah, P R (1985). Multivariate gamma distributions. In Encyclopedia of Statistical Sciences. Volume 6, pp. 63–66. New York: Wiley. [7.5.12]

Krishnaiah, P R, and Armitage, J V (1965). Probability integrals of the multivariate F-distribution, with tables and applications. Report ARL 65-236, Aerospace Research Laboratories, Wright-Patterson Air Force Base, Ohio. [7.8.8]

Krishnaiah, P R, Hagis, P, and Steinberg, L (1963). A note on the bivariate chi-distribution. SIAM (Society for Industrial and Applied Mathematics) Review, 5, 140–144. [7.5.15]

Krishnaiah, P R, and Rao, M M (1961). Remarks on a multivariate gamma distribution. American Mathematical Monthly, 68, 342–346. [7.5.8, 7.7.11]

Krishnamoorthy, A S, and Parthasarathy, M (1951). A multivariate gamma-type distribution. Annals of Mathematical Statistics, 22, 549–557 (Correction, 31, 229). [7.5.8, 7.7.2]

Krishnan, M (1976). The noncentral bivariate chi-squared distribution and extensions. Communications in Statistics, Part A—Theory and Methods, 5, 647–660. [7.6]

Krogstad, H E (1985). Height and period distributions of extreme waves. Applied Ocean Research, 7, 158–165. [8.12.5.1, 24.12.2—8]

Kronmal, R A, and Peterson, A V (1979). The alias and alias-rejection-mixture methods for generating random variables from probability distributions. In H J Highland, M G Spiegel, and R Shannon (Editors), 1979 Winter Simulation Conference. Volume 1, pp. 269–280. New York: Institute of Electrical and Electronics Engineers. [16.2.4]

Kronmal, R A, and Peterson, A V (1984). An acceptance-complement analogue of the mixture-plus-acceptance-rejection method for generating random variables. ACM (Association for Computing Machinery) Transactions on Mathematical Software, 10, 271–281. [16.2.4]

Kronmal, R A, Peterson, A V, and Lundberg, E D (1978). The alias-rejection-mixture method for generating random variables from continuous distributions. American Statistical Association, 1978 Proceedings of the Statistical Computing Section, pp. 106–110. [16.2.4]

Kronmal, R A, and Tarter, M (1974). The use of density estimates based on orthogonal expansions. In W J Dixon and W L Nicholson (Editors), Exploring Data Analysis. The Computer Revolution in Statistics, pp. 365–395 (Discussion, 396–400). Berkeley: University of California Press. [24.9.1—3]

Ku, H H, and Kullback, S (1974). Loglinear models in contingency table analysis. The American Statistician, 28, 115–122. [24.9.2—10]

Kulasekera, K B, and Park, D H (1987). The class of better mean residual life at age t_0. Microelectronics and Reliability, 27, 725–735. [18.2.7]

Kulatilake, P H S W (1986). Bivariate normal distribution fitting on discontinuity orientation clusters. Mathematical Geology (Journal of the International Association for Mathematical Geology), 18, 181–195. [19.7.11, 24.12.2—10]

Kwerel, S M (1983). Fréchet bounds. In Encyclopedia of Statistical Sciences. Volume 3, pp. 202–209. New York: Wiley. [3.1.2]

L

Laha, R G (1982). Characteristic functions. In Encyclopedia of Statistical Sciences. Volume 1, pp. 415–422. New York: Wiley. [2.2.2.2, 2.15.2—2]

Lai, C D (1975). Diagonal Expansions of Bivariate Probability Distributions and Their Application to Point Processes. Ph.D. Thesis, Victoria University of Wellington. [14.6.3]

Lai, C D (1978). An example of Wold's point processes with Markov-dependent intervals. Journal of Applied Probability, 15, 748–758. [9.12.13]

Lai, C D (1982). Meixner classes and Meixner hypergeometric distributions. Australian Journal of Statistics, 24, 221–233. [15.3.1]

Lai, C D (1985). On the reliability of a standby system composed of two dependent exponential components. Communications in Statistics—Theory and Methods, 14, 851–860. [9.12.12, 17.6.3]

Lai, C D (1986). Bounds on reliability of a coherent system with positively correlated components. IEEE (Institute of Electrical and Electronics Engineers) Transactions on Reliability, R-35, 508–511. [12.3.1]

Lai, C D (1987). Letter to the editor. Journal of Applied Probability, 24, 288–289. [15.4.2]

Lai, C D, and Moore, T (1984). Probability integrals of a bivariate gamma distribution. Journal of Statistical Computation and Simulation, 19, 205–213. [7.5.3, 7.5.14, 9.12.14]

Lam, K S L, Yeung, R T T, Benson, E A, and Wang, C (1989). Erythrocyte sodium-potassium pump in thyrotoxic periodic paralysis. Australian and New Zealand Journal of Medicine, 19, 6–10. [24.9.3—9]

Lambert, B, Rousseau, D, Lepage, Y, and St-Onge, M (1985). Colpohistologic correlations. A computerized study. Analytical and Quantitative Cytology and Histology, 7, 47–51. [24.4.2—6]

Laming, D (1973). Mathematical Psychology. London: Academic Press. [23.1.1, 23.1.4]

Laming, D (1986). Sensory Analysis. London: Academic Press. [23.1.1]

Lampard, D G (1968). A stochastic process whose successive intervals between events form a first order Markov chain—1. Journal of Applied Probability, 5, 648–668. [7.5.13, 14.8]

Lancaster, H O (1958). The structure of bivariate distributions. Annals of Mathematical Statistics, 29, 719–736. [14.1, 14.2, 24.9.2—6]

Lancaster, H O (1959). Zero correlation and independence. Australian Journal of Statistics, 21, 53–56. [4.17.5, 4.18.1]

Lancaster, H O (1963a). Correlations and canonical forms of bivariate distributions. Annals of Mathematical Statistics, 34, 532–538. [11.10, 14.2]

Lancaster, H O (1963b). Correlation and complete dependence of random variables. Annals of Mathematical Statistics, 34, 1315–1321. [11.3.2, 11.6, 11.10]

Lancaster, H O (1969). The Chi-Squared Distribution. New York: Wiley. [14.2, 14.4]

Lancaster, H O (1972). Development of notion of statistical dependence. Mathematical Chronicle, 2, 1–16. [4.2, 11.4.3]

Lancaster, H O (1975). Joint distributions in the Meixner classes. Journal of the Royal Statistical Society, Series B, 37, 434–443. [15.3.1, 15.3.2, 15.3.3]

Lancaster, H O (1980). Orthogonal models for contingency tables. In P R Krishnaiah (Editor), Developments in Statistics. Volume 3, pp. 99–157. New York: Academic Press. [14.2]

Lancaster, H O (1982a). Chi-square distribution. In Encyclopedia of Statistical Sciences. Volume 1, pp. 439–442. New York: Wiley. [2.15.1—3a]

Lancaster, H O (1982b). Dependence, measures and indices of. In Encyclopedia of Statistical Sciences. Volume 2, pp. 334–339. [fn11.3, 11.4, 11.9, 11.10, 11.13, 15.3.6]

Lancaster, H O (1983). Special joint distributions of Meixner variables. Australian Journal of Statistics, 25, 298–309. [T14.1, 15.3.1]

Lancaster, H O, and Hamdan, M A (1964). Estimation of the correlation coefficient in contingency tables with possibly nonmetrical characters. Psychometrika, 29, 383–391. [24.9.2—11]

Landis, J R, and Koch, G G (1977). The measurement of observer agreement for categorical data. Biometrics, 33, 159–174. [20.2.4]

Langaris, C (1986). A correlated queue with infinitely many servers. Journal of Applied Probability, 23, 155–165. [9.12.13]

Launer, R L (1984). Inequalities for NBUE and NWUE life distributions. Operations Research, 32, 660–667. [18.2.7]

Lavoie, J L (1986). Some evaluations for the generalized hypergeometric series. Mathematics of Computation, 46, 215–218. [T13.1]

Law, A M, and Kelton, W D (1982). Simulation Modeling and Analysis. New York: McGraw-Hill. [16.2.1]

Lawless, J F (1982). Statistical Models and Methods for Lifetime Data. New York: Wiley. [17.1, 24.2—3]

Lawless, J F (1988). Reliability, nonparametric methods in. In Encyclopedia of Statistical Sciences. Volume 8, pp. 20–24. New York: Wiley. [2.15.2—47]

Lawrance, A J, and Lewis, P A W (1983). Simple dependent pairs of exponential and uniform random variables. Operations Research, 31, 1179–1197. [9.10.1, 13.11, 15.6.4]

Lee, L (1979). Multivariate distributions having Weibull properties. Journal of Multivariate Analysis, 9, 267–277. [5.6.9.2, 9.2.10, 9.16]

Lee, L-F (1978). Unionism and wage rates: A simultaneous equations model with qualitative and limited dependent variables. International Economic Review, 19, 415–433. [19.9.5]

Lee, L-F (1979). On comparisons of normal and logistic models in the bivariate dichotomous analysis. Economics Letters, 4, 151–155. [22.4.2.1]

Lee, L-F (1982). A bivariate logit model. Report from the Center for Econometrics and Decision Sciences, University of Florida. [5.7.10, 19.9.5]

Lee, L-F (1984). Tests for the bivariate normal distribution in econometric models with selectivity. Econometrica, 52, 843–863. [19.9.5]

Lee, M L T (1985a). Dependence by total positivity. Annals of Probability, 13, 572–582. [12.3.5]

Lee, M L T (1985b). Dependence by reverse regular rule. Annals of Probability, 13, 583–591. [12.4.1]

Lee, P A (1971). A diagonal expansion for the 2-variate Dirichlet probability density function. SIAM (Society for Industrial and Applied Mathematics) Journal on Applied Mathematics, 21, 155–163. [6.5.11, T14.1, 14.8]

Lee, P A (1981). The correlated bivariate inverted beta distribution. Biometrical Journal, 23, 693–703. [14.8, 15.6.3, 20.7.2, 20.7.3, 20.7.4]

Lee, P A, and Ong, S H (1986). The bivariate non-central negative binomial distributions. Metrika, 33, 1–28. [24.6—3]

Lee, R-Y, Holland, B S, and Flueck, J A (1979). Distribution of a ratio of correlated gamma random variables. SIAM (Society for Industrial and Applied Mathematics) Journal of Applied Mathematics, 36, 304–320. [10.3.8]

Lee, S Y (1985). Maximum likelihood estimation of polychoric correlations in $r \times s \times t$ contingency tables. Journal of Statistical Computation and Simulation, 23, 53–67. [11.11]

Lee, S-Y, and Poon, W-Y (1987a). Two-step estimation of multivariate polychoric correlation. Communications in Statistics—Theory and Methods, 16, 307–320. [11.11]

Lee, S-Y, and Poon, W-Y (1987b). Some algorithms in computing GLS estimates of multivariate polychoric correlations. American Statistical Association, 1987 Proceedings of the Statistical Computing Section, pp. 444–447. [11.11]

Lee, S-Y, and Poon, W-Y (1987c). (See Li Xiqin and Pan Weixian, 1987.) [11.11]

Leemis, L, and Schmeiser, B (1985). Random variate generation for Monte Carlo experiments. IEEE (Institute of Electrical and Electronics Engineers) Transactions on Reliability, R-34, 81–85. [16.2.1]

Lehmann, E L (1966). Some concepts of dependence. Annals of Mathematical Statistics, 37, 1137–1153. [12.2.1, 12.3.1, 12.3.4, 12.4.1, 12.4.2]

Lentner, M (1972). Elementary Applied Statistics. Tarrytown-on-Hudson, New York: Bogden and Quigley. [24.13.1—2, T24.59]

Lesaffre, E, and Albert, A (1989). Partial separation in logistic discrimination. Journal of the Royal Statistical Society, Series B, 51, 109–119. [24.7.1—3]

Lew, E A (1982). Actuarial statistics—Life. In Encyclopedia of Statistical Sciences. Volume 1, pp. 16–20. New York: Wiley. [2.15.3—1]

Lewin, L (1981). Polylogarithms and Associated Functions. New York: North Holland. [fn9.4]

Lewis, P A W (1980). Chapter G of the IMSL library—Generation and testing of random deviates: Simulation. In T I Ören, C M Shub, and P F Roth (Editors), Proceedings of the 1980 Winter Simulation Conference, pp. 357–360. New York: Institute of Electrical and Electronics Engineers. [16.3.6]

Lewis, P A W, Orav, E J, and Uribe, L (1986). Advanced Simulation and Statistics Package: IBM Professional FORTRAN Version. (Software.) Monterey, California: Wadsworth and Brooks/Cole. [16.3.6]

Lewis, T (1975). Contribution to Discussion of paper by Tiago de Oliveira. Bulletin of the International Statistical Institute, 46, Book 2, 253. [9.17.11]

Li, L, and Owen, D B (1979). Two-sided screening procedures in the bivariate case. Technometrics, 21, 79–85. [4.16.2]

Liem, T C (1980). A computer program for Box-Cox transformations in regression modeels with heteroscedastic and autoregressive residuals. The American Statistician, 34, 121. [4.13.6]

Lien, D-H D (1985). Moments of truncated bivariate log-normal distributions. Economics Letters, 19, 243–247. [4.16.1, 19.9.2]

Lin, G D (1987). Relationships between two extensions of Farlie-Gumbel-Morgenstern distribution. Annals of the Institute of Statistical Mathematics, Tokyo, 39, 129–140. [5.2.13, 13.2.2]

Lindley, D V, and Singpurwalla, N D (1986). Multivariate distributions for the life lengths of components of a system sharing a common environment. Journal of Applied Probability, 23, 418–431. [5.9.3, 15.6.3]

Lindqvist, L (1976). SELLO, a FORTRAN IV program for the transformation of skewed distributions to normality. Computers and Geosciences, 1, 129–145. [4.14.5]

Lindstrom, E W (1935?). (Title not known.) American Naturalist, 49?, 311–???. [24.7.2—8]

Lingappaiah, G S (1983). Generalized Liouville distribution and some related multivariate distributions. Metron, 41(1–2), 135–146. [6.6]

Lingappaiah, G S (1984). Bivariate gamma distribution as a life test model. Aplikace Matematiky, 29, 182–188. [5.2.9]

Liu, P C (1987). Estimating long-term wave statistics from long-term wind statistics. In B L Edge (Editor), Twentieth Coastal Engineering Conference, Proceedings. Volume I, pp. 512–521. New York: American Society of Civil Engineers. [8.12.5.3, 24.12.2—9]

Li Xiqin and Pan Weixian (1987). Generalized least squares and maximum likelihood estimations of multivariate polychoric correlations. Acta Mathematicae Applicatae Sinica, 3, 351–357. [Cited in 11.11 as Lee and Poon (1987c).]

Ljungberg, B, Stenling, R, and Roos, G (1987). Flow cytometric DNA analysis of renal-cell carcinoma. A study of fine needle aspiration biopsies in comparison with multiple surgical samples. Analytical and Quantitative Cytology and Histology, 9, 505–508. [24.4.2—5]

Lochner, R H (1975). A generalized Dirichlet distribution in Bayesian life testing. Journal of the Royal Statistical Society, Series B, 37, 103–113. [6.5.14]

Loh, W Y (1984a). A new generalization of NBU distributions. IEEE (Institute of Electrical and Electronics Engineers) Transactions on Reliability, 33, 419–422. [18.2.1]

Loh, W Y (1984b). Bounds on ARE's for restricted classes of distributions defined via tail-orderings. Annals of Statistics, 12, 685–701. [18.2.1]

Longuet-Higgins, M S (1975). On the joint distribution of the periods and amplitudes of sea waves. Journal of Geophysical Research, 80, 2688–2694. [10.8, 24.12.2—8]

Longuet-Higgins, M S (1983). On the joint distribution of wave periods and amplitudes in a random wave field. Proceedings of the Royal Society of London, Series A, 389, 241–258. [10.8, 24.12.2—8]

Looney, S W (1986). A review of techniques for assessing multivariate normality. American Statistical Association, 1986 Proceedings of the Statistical Computing Section, pp. 280–285. [4.13.1, 4.13.5.8]

Loukas, S (1984). Simple methods for computer generation of bivariate beta random variables. Journal of Statistical Computation and Simulation, 20, 145–152. [6.5.7]

Loukas, S, and Kemp, C D (1986). The index of dispersion test for the bivariate Poisson distribution. Biometrics, 42, 941–948. [24.6—3, 24.6—11]

Louwes, K M, and Neele, A E F (1987). Selection for chip quality and specific gravity of potato clones: Possibilities for early generation selection. Potato Research, 30, 241–251. [24.4.3—1]

Lubischew, A A (1962). On the use of discriminant functions in taxonomy. Biometrics, 18, 455–477. [24.7.1—11]

Lukacs, E (1970). Characteristic Functions. 2nd Edition. London: Griffin. [2.2.2.2]

Lukacs, E (1983). Developments in Characteristic Function Theory. High Wycombe: Griffin. [2.2.2.2]

M

MacGillivray, H L (1985). Mean, median, mode, and skewness. In Encyclopedia of Statistical Sciences. Volume 5, pp. 364–367. New York: Wiley. [2.15.2—26]

MacKay, D B (1988). Thurstone's theory of comparative judgment. In Encyclopedia of Statistical Sciences. Volume 9, pp. 237–241. New York: Wiley. [2.15.3—10]

MacLeod, J (1899). Over de correlatie tusschen het aantal meeldraden en het aantal stampers bij het speenkruid (*Ficaria ranunculoides*). Botanisch Jaarboek, 11. [24.6—10]

Macmeeken, A M (1940). The Intelligence of a Representative Group of Scottish Children. London: University of London Press. [24.10.2—6]

MacMillan, N A, Kaplan, H L, and Creelman, C D (1977). The psychophysics of categorical perception. Psychological Review, 84, 452–471. [23.6]

Macomber, J H, and Myers, B L (1978). The bivariate beta distribution: Comparison of Monte Carlo generators and evaluation of parameter estimates. In H J Highland, N R Nielsen, and L G Hull (Editors), 1978 Winter Simulation Conference. Volume 1, pp. 142–152. New York: Institute of Electrical and Electronics Engineers. [6.5.10]

Maddala, G S (1986). Disequilibrium, self-selection, and switching models. In Z Griliches and M D Intriligator (Editors), Handbook of Econometrics. Volume III, pp. 1633–1688. Amsterdam: North-Holland. [19.9.5]

Madsen, R W (1982). A selection procedure using a screening variate. Technometrics, 24, 301–306. [4.16.2]

Mahalanobis, P C, and Lahiri, D B (1960). Analysis of errors in censuses and surveys with special reference to experience in India. Bulletin de l'Institut International de Statistique, 38, Book 2, 401–433. [24.6—20]

Makov, U E (1980). The statistical problems of unconfirmed cases in medical statistics. In E Girelli-Bruni (Editor), Teoria delle Decisioni in Medicina, pp. 149–162. Verona: Bertani. [24.9.3—8]

Malik, H J (1985). Logistic distribution. In Encyclopedia of Statistical Sciences. Volume 5, pp. 123–128. New York: Wiley. [2.15.1—20a]

Malik, H J, and Abraham, B (1973). Multivariate logistic distributions. Annals of Statistics, 1, 588–590. [15.6.3]

Malik, H J, and Trudel, R (1983). A generalization of a bivariate chi-square distribution. International Journal of Mathematical Education in Science and Technology, 14, 679–683. [7.5.10, 7.7.12]

Malik, H J, and Trudel, R (1985). Distributions of the product and the quotient from bivariates t, F and Pareto distribution. Communications in Statistics—Theory and Methods, 14, 2951–2962. [5.8.10, 6.2.10, 6.3.9]

Malkovich, J F, and Afifi, A A (1973). On tests for multivariate normality. Journal of the American Statistical Association, 68, 176–179. [4.13.5.4]

Malmgren, B A (1979). Multivariate normality tests of planktonic foraminiferal data. Mathematical Geology (Journal of the International Association for Mathematical Geology), 11, 285–297. [19.3.1]

Manly, B F J (1986a). Multivariate Statistical Methods: A Primer. London: Chapman and Hall. [24.7.1—10]

Manly, B F J (1986b). The survival of the fittest. In R J Brook, G C Arnold, T H Hassard, and R M Pringle (Editors), The Fascination of Statistics, pp. 331–350. New York: Dekker. [24.7.1—10]

Manly, B F J (1988). Van Valen's test. In Encyclopedia of Statistical Sciences. Volume 9, pp. 462–465. New York: Wiley. [24.7.1—10]

Mann, N R, and Singpurwalla, N D (1982). Extreme-value distributions. In Encyclopedia of Statistical Sciences. Volume 2, pp. 606–613. New York: Wiley. [2.15.1—6]

Mann, N R, and Singpurwalla, N D (1983). Life testing. In Encyclopedia of Statistical Sciences. Volume 4, pp. 632–639. New York: Wiley. [2.15.3—4]

Manoukian, E B (1986). Modern Concepts and Theorems of Mathematical Statistics. New York: Springer. [2.1]

Mantel, N (1966). Models for complex contingency tables and polychotomous dosage response curves. Biometrics, 22, 83–95. [21.3.2, 22.3.1]

Mantel, N, and Brown, C (1973). A logistic reanalysis of Ashford and Sowden's data on respiratory symptoms in British coal miners. Biometrics, 29, 649–665. [22.4.1]

Mardia, K V (1962). Multivariate Pareto distributions. Annals of Mathematical Statistics, 33, 1008–1015 (Correction, 34, 1603). [5.8.10, 7.5.15]

Mardia, K V (1967). Some contributions to contingency-type bivariate distributions. Biometrika, 54, 235–249 (Correction, 55, 597) [5.7.5, 5.7.9]

Mardia, K V (1970a). Families of Bivariate Distributions. London: Griffin. [1.1, 3.4.1, 3.5, 3.9.1.1, 3.9.1.5, 3.11.2, 4.1, 4.13.7, 4.14.2.4, 5.7.10, 6.2.6, 6.3.7, 6.5.14, 6.9.4, 7.5.7, 10.3.8, 15.1, 15.4.5, 24.6—12, 24.7.2—1]

Mardia, K V (1970b). Measures of multivariate skewness and kurtosis with applications. Biometrika, 57, 519–530. [4.13.5.7, 4.13.5.8, 24.11.2—13]

Mardia, K V (1970c). Some problems of fitting for contingency-type bivariate distributions. Journal of the Royal Statistical Society, Series B, 32, 254–264. [5.7.10, 24.7.2—1]

Mardia, K V (1974). Applications of some measures of multivariate skewness and kurtosis for testing normality and robustness studies. Sankhyā, Series B, 36, 115–128. [4.13.5.8]

Mardia, K V (1980). Tests of univariate and multivariate normality. In P R Krishnaiah (Editor), Handbook of Statistics. Volume 1. Analysis of Variance, pp. 279–320. Amsterdam: North-Holland. [4.13.1]

Mardia, K V (1985). Mardia's test of multinormality. In Encyclopedia of Statistical Sciences. Volume 5, pp. 217–221. New York: Wiley. [4.2, 4.13.5.8]

Mardia, K V, and Foster, K (1983). Omnibus tests for multinormality based on skewness and kurtosis. Communications in Statistics—Theory and Methods, 12, 207–221. [4.13.5.8]

Mardia, K V, Kent, J T, and Bibby, J M (1979). Multivariate Analysis. London: Academic Press. [24.7.1—3, 24.7.1—11, 24.9.1—7]

Mardia, K V, and Zemroch, P J (1975). Algorithm AS 84. Measures of multivariate skewness and kurtosis. Applied Statistics, 24, 262–265. [4.13.5.8]

Margolin, B H, and Collings, B J (1987). Reusing published examples: A case of caveat emptor. In C L Mallows (Editor), Design, Data, and Analysis by Some Friends of Cuthbert Daniel, pp. 261–273. New York: Wiley. [24.1.2.1, 24.9.2—8]

Maritz, J S (1950). On the validity of inferences drawn from the fitting of Poisson and negative binomial distributions to observed accident data. Psychological Bulletin, 47, 434–443. [24.6—3]

Markén, K-E (1962). Studies of deviations between observers in clinico-odontological recording. Selected Publications of the University Institute of Statistics, Uppsala, 19, Paper No. 2. [24.4.2—9]

Marsaglia, G, and Bray, T A (1964). A convenient method for generating normal variables. SIAM (Society for Industrial and Applied Mathematics) Review, 6, 260–264. [16.3.1]

Marsaglia, G, MacLaren, M D, and Bray, T A (1964). A fast procedure for generating normal random variables. Communications of the Association for Computing Machinery, 7, 4–10. [16.3.1]

Marsh, C (1988). Exploring Data. An Introduction to Data Analysis for Social Scientists. Cambridge, U.K.: Polity Press. [24.11.1, 24.11.1—1, 24.11.1—2, 24.11.1—3, 24.11.1—4, 24.11.1—5]

Marshall, A W (1975a). Multivariate distributions with monotone hazard rate. In F Proschan and R J Serfling (Editors), Reliability and Biometry, pp. 259–284. Philadelphia: Society for Industrial and Applied Mathematics.

Marshall, A W (1975b). Multivariate distributions with monotone hazard rate. In R E Barlow, J R Fussell, and N D Singpurwalla (Editors), Reliability and Fault Tree Analysis. Theoretical and Applied Aspects of System Reliability and Safety Assessment, pp. 259–284. Philadelphia: Society for Industrial and Applied Mathematics. [18.3.2]

Marshall, A W (1975c). Some comments on the hazard gradient. Stochastic Processes and Their Applications, 3, 293–300. [18.3.2, fn18.9]

Marshall, A W, and Olkin, I (1967a). A multivariate exponential distribution. Journal of the American Statistical Association, 62, 30–44. [3.8.2, 4.13.5.3, 5.2.9, 9.2.2, 9.2.7, 9.2.10, 9.13, 9.17.6.2, 13.6, 13.13, 15.5, 17.1, T17.1, 17.2.1.1, 17.2.2, fn17.4, 17.6.2, 18.3.2]

Marshall, A W, and Olkin, I (1967b). A generalized bivariate exponential distribution. Journal of Applied Probability, 4, 291–302. [9.2.13, 13.6]

Marshall, A W, and Olkin, I (1979). Inequalities: Theory of Majorization and Its Applications. New York: Academic Press. [6.6]

Marshall, A W, and Olkin, I (1985a). Multivariate exponential distributions, Marshall-Olkin. In Encyclopedia of Statistical Sciences. Volume 6, pp. 59–62. New York: Wiley. [9.2.7, 17.2.1.3, 17.2.1.4]

Marshall, A W, and Olkin, I (1985b). A family of bivariate distributions generated by the bivariate Bernoulli distributions. Journal of the American Statistical Association, 80, 332–338. [3.8.1, 3.8.3]

Marshall, A W, and Shaked, M (1979). Multivariate shock models for distribution with increasing hazard rate average. Annals of Probability, 7, 343–358. [18.3.4, 18.4]

Marshall, A W, and Shaked, M (1982). A class of multivariate new better than used distributions. Annals of Probability, 10, 259–264. [18.3.4]

Marshall, A W, and Shaked, M (1986a). Multivariate new better than used distributions. Mathematics of Operations Research, 11, 110–116. [18.3.4]

Marshall, A W, and Shaked, M (1986b). Multivariate new better than used distributions: A survey. Scandinavian Journal of Statistics, 13, 277–290. [18.3.4]

Marshall, T, and Knox, E G (1980). Disease concordance and sex similarity in twins. Journal of Epidemiology and Community Health, 34, 1–8. [19.5.3.3]

Martens, H, and Naes, T (1983). Multivariate calibration. In B R Kowalski (Editor), Chemometrics, pp. 147–156. Dordrecht: Reidel. [24.4.3—5]

Martin, J, and Butcher, B (1982). The quality of proxy information—Some results from a large-scale study. The Statistician, 31, 293–319. [24.4.2—16]

Martin, L (1960). Homométrie, allométrie et cograduation en biométrie généale. Biometrische Zeitschrift, 2, 73–97. [24.7.2—15]

Martin, O J (1983). ADSSIM: Aged and disabled services simulation. Master's Report, Department of Mechanical Engineering, University of Texas, Austin. [19.9.4]

Martinson, E O, and Hamdan, M A (1975). Algorithm AS 87: Calculation of the polychoric estimate of correlation in contingency tables. Applied Statistics, 24, 272–278. [11.11]

Martynov, G V (1981). Evaluation of the normal distribution function. Journal of Soviet Mathematics, 17, 1857–1875. (Original Russian article was dated 1979. [4.12.9]

Mason, R L, and Young, J C (1985). Re-examining two tests for bivariate normality. Communications in Statistics—Theory and Methods, 14, 1531–1546. [4.13.5.10, 19.8.9, 24.14.1—4]

Mathai, A M, and Pederzoli, G (1977). Characterizations of the Normal Probability Law. New York: Wiley. [4.5]

Mathar, R (1985). Outlier-prone and outlier-resistant multidimensional distributions. Statistics, 16, 451–456. [4.13.2]

Maung, K (1941). Measurement of association in a contingency table with special reference to the pigmentation of hair and eye colours of Scottish school children. Annals of Eugenics, London, 11, 189–223. [24.9.2—6]

Mayall, N U, Scott, E L, and Shane, C D (1958). Statistical problems in the study of galaxies. Bulletin de l'Institut International de Statistique, 37, Book 3, 35–53. [24.6—9, T24.25]

Mayer, L S, and Robinson, J A (1974). The monotone coefficient as an estimator of the correlation ratio between an ordinal variable and interval variable. Technical Report No. 72, Series 2, Department of Statistics, Princeton University. [24.12.1—6]

Mayer, L S, and Robinson, J A (1975). Multiple and partial monotone coefficients for regression models with ordinal predictor variables. Technical Report No. 95, Series 2, Department of Statistics, Princeton University. [24.12.1—6]

McAllister, P R (1983). A system for computing joint probabilities from Jensen's bivariate *F* distribution. Communications in Statistics—Simulation and Computation, 12, 223–238. [7.8.7, 7.8.8]

McAllister, P R, and Holland, B S (1980). Procedures for application of Jensen's bivariate F distribution to two way ANOVA with out interaction. American Statistical Association, 1980 Proceedings of the Biometrics Section, ???–???. [7.8.7]

McAllister, P R, Lee, R Y, and Holland, B S (1981). Computation of probabilities from Jensen's bivariate *F* distribution. Communications in Statistics, Part B—Simulation and Computation, 10, 249–263. [7.8.3, 7.8.8]

McCullagh, P (1978). A class of parametric models for the analysis of square contingency tables with ordered categories. Biometrika, 65, 413–418. [24.9.2—8, 24.11.2—1]

McCullagh, P (1980). Regression models for ordinal data. Journal of the Royal Statistical Society, Series B, 42, 109–127 (Discussion, 127–142). [24.9.2—8]

McCune, E D, and Gray, H L (1982). Cornish-Fisher and Edgeworth expansions. In Encyclopedia of Statistical Sciences. Volume 2, pp. 188–193. New York: Wiley. [2.15.2—4, 3.11.2]

McDonald, J E (1960). Remarks on correlation methods in geophysics. Tellus, 12, 176–183. [4.14.5, fn11.5]

McFadden, D (1978). Modelling the choice of residential location. In A Karlqvist, L Lundqvist, F Snickars, and J W Weibull (Editors), Spatial Interaction Theory and Planning Models, pp. 75–96. Amsterdam: North-Holland. [22.3.3]

McFadden, J A (1966). A diagonal expansion in Gegenbauer polynomials for a class of second-order probability densities. SIAM (Society for Industrial and Applied Mathematics) Journal on Applied Mathematics, 14, 1433–1436. [6.4.9, T14.1, 14.8]

McGill, W J, and Teich, M C (1989). A unique approach to stimulus detection theory in psychophysics based upon the properties of zero-mean Gaussian noise. (Review of "Sensory Analysis" by D Laming.) Journal of Mathematical Psychology, 33, 99–108. [23.1.1]

McGraw, D K, and Wagner, J F (1968). Elliptically symmetric distributions. IEEE (Institute of Electrical and Electronics Engineers) Transactions on Information Theory, IT-14, 110–120. [7.2.1]

McGuigan, D R D (1985). Accident migration—Or a flight of fancy? Traffic Engineering and Control, 26, 229–233. [24.6—3]

McKay, A T (1934). Sampling from batches. Journal of the Royal Statistical Society, Supplement, 1, 207–216. [3.9.1.4, 8.6.7, 8.6.11]

McLachlan, G J, and Basford, K E (1988). Mixture Models. Inference and Applications to Clustering. New York: Dekker. [3.4.1, 4.13.5.6, 4.17.1, 24.9.3—8]

McNeil, D (1979). Analysis of jellyfish data. In D McNeil (Editor), Interactive Statistics, pp. 153–158. Amsterdam: North-Holland. [24.7.1—6]

McNichol, D (1972). A Primer of Signal Detection Theory. London: Allen and Unwin. [23.1.1]

McNolty, F, Doyle, J, and Hansen, E (1980). Properties of the mixed exponential failure process. Technometrics, 22, 555–565. [20.1]

Mead, R, Riley, J, Dear, K, and Singh, S P (1986). Stability comparison of intercropping and monocropping systems. Biometrics, 42, 253–266. [8.12.4]

Mee, R W, and Owen, D B (1983). A simple approximation for bivariate normal probabilities. Journal of Quality Technology, 15, 72–75. [4.12.2]

Meixner, J (1934). Orthogonale Polynomsysteme mit einer besonderen Gestalt der erzeugenden Funktion. Journal of the London Mathematical Society, 9, 6–13. [15.3.2]

Mellinger, G D, Sylwester, D L, Gaffey, W R, and Manheimer, D I (1965). A mathematical model with applications to a study of accident repeatedness among children. Journal of the American Statistical Association, 60, 1046–1059. [24.6—3]

Melnick, E L, and Tenenbein, A (1982). Mispecifications of the normal distribution. The American Statistician, 36, 372–373. [4.18.3, 4.18.4]

Mendoza, G A, and Iglewicz, B (1983). A comparative study of systems of univariate frequency distributions. American Statistical Association, 1983 Proceedings of the Statistical Computing Section, pp. 249–254. [2.3.4]

Meng, C-L, and Schmidt, P (1985). On the cost of partial observability in the bivariate probit model. International Economic Review, 26, 71–85. [22.2.2, 22.4.2.5]

Michael, J R, and Schucany, W R (1986). Analysis of data from censored samples. In R B D'Agostino and M A Stephens (Editors), Goodness-of-Fit Techniques, pp. 461–496. New York: Dekker. [4.13.3.1, 4.13.4.1]

Midlarsky, M I (1983). Absence of memory in the nineteenth-century alliance system: Perspectives from queuing theory and bivariate probability distributions. American Journal of Political Science, 27, 762–784. [24.6—22]

Mielke, P W (1985). Omega distribution. In Encyclopedia of Statistical Sciences. Volume 6, pp. 416–417. New York: Wiley. [2.15.1—30]

Mielke, P W, and Flueck, J A (1976). Distributions of ratios for some selected bivariate probability functions. American Statistical Association, Proceedings of the Social Statistics Section, 1976, 608–613. [10.3.8]

Mielke, P W, Williams, J S, and Wu, S-c (1977). Covariance analysis technique based on bivariate log-normal distribution with weather modification applications. Journal of Applied Meteorology, 16, 183–187. [19.7.5]

Mihram, G A, and Hultquist, A R (1967). A bivariate warning-time/failure-time distribution. Journal of the American Statistical Association, 62, 589–599. [3.5, 8.6.7, 8.6.11, 8.12.1]

Mikhail, N N, Chasnov, R, and Wooldridge, T S (1987). Regression curves for Farlie-Gumbel-Morgenstern class of bivariate distribution. In R M Heiberger (Editor), Computer Science and Statistics: Proceedings of the 19th Symposium on the Interface, pp. 531–532. Alexandria, Virginia: American Statistical Association. [5.2.9]

Mikhail, N N, Falwell, J A, Bogue, A, and Weaver, T L (1987). Regression curves to a class of bivariate distributions including the bivariate logistic with application. In R M Heiberger (Editor), Computer Science and Statistics: Proceedings of the 19th Symposium on the Interface, pp. 525–530. Alexandria, Virginia: American Statistical Association. [5.3.9]

Miller, S M (1960). Comparative social mobility. Current Sociology, 9, 1–89. [24.11.2—1]

Minh, D L (1988). Generating gamma variates. ACM (Association for Computing Machinery) Transactions on Mathematical Software, 14, 261–266. [16.3.2, fn16.6]

Minnis, P, and Harrison, E F (1984). Diurnal variability of regional cloud and clear-sky radiative parameters derived from GOES data. Part I: Analysis method. Journal of Climate and Applied Meteorology, 23, 993–1011. [24.4.2—13]

Mitchell, C R, and Paulson, A S (1981). A new bivariate negative binomial distribution. Naval Research Logistics Quarterly, 28, 359–374. [24.6—3, 24.6—16]

Mitropol'skii, A K (1966). Correlation Equations for Statistical Computations. New York: Consultants Bureau. [3.11.2, 24.7.2—7, 24.13.2—1, 24.13.2—2, 24.13.2—3]

Mode, E B (1966). Elements of Probability and Statistics. Englewood Cliffs, New Jersey: Prentice-Hall. [24.7.2—14]

Moeschberger, M L (1974). Life tests under dependent competing causes of failure. Technometrics, 16, 39–47. [9.2.10]

Monahan, J F (1987). An algorithm for generating chi random variables. ACM (Association for Computing Machinery) Transactions on Mathematical Software, 13, 168–172 (Correction, 320). [16.3.2]

Monrad, D, and Stout, W (1988). Stable distributions. In Encyclopedia of Statistical Sciences. Volume 8, pp. 617–621. New York: Wiley. [2.15.1—44]

Moore, D S (1986). Tests of chi-squared type. In R B D'Agostino and M A Stephens (Editors), Goodness-of-Fit Techniques, pp. 63–95. New York: Dekker. [4.13.4.2]

Moore, R J, and Clarke, R T (1981). A distribution function approach to rainfall runoff modeling. Water Resources Research, 17, 1367–1382. [5.10.10]

Moothathu, T S K (1985). On a characterization property of multivariate symmetric stable distributions. Journal of the Indian Statistical Association, 23, 83–88. [7.11.6]

Moran, P A P (1967). Testing for correlation between non-negative variates. Biometrika, 54, 385–394. [9.1, 10.3.8, 11.5.1]

Moran, P A P (1970). The methodology of rain-making experiments. Review of the International Statistical Institute, 38, 105–115 (Discussion, 115–119). [19.7.3]

Morgan, B J T (1984). Elements of Simulation. London: Chapman and Hall. [5.7.7, 16.2.1, 16.3.6]

Morgan, B J T, and Titterington, D M (1977). A comparison of iterative methods for obtaining maximum likelihood estimates in contingency tables with a missing diagonal. Biometrika, 64, 265–269. [24.11.2—1]

Morgenstern, D (1956). Einfache Beispiele zweidimensionaler Verteilungen. Mitteilungsblatt für Mathematische Statistik, 8, 234–235. [5.2.10]

Morimune, K (1979). Comparisons of normal and logistic models in the bivariate dichotomous analysis. Econometrica, 47, 957–975. [5.7.10, 5.8.13.4, 22.4.2.1]

Morris, C N (1982). Natural exponential families with quadratic variance functions. Annals of Statistics, 10, 65–80. [15.3.1]

Morris, C N (1983). Natural exponential families with quadratic variance functions: Statistical theory. Annals of Statistics, 11, 515–529. [15.3.1]

Morris, C N (1985). Natural exponential families. In Encyclopedia of Statistical Sciences. Volume 6, pp. 157–159. New York: Wiley. [2.15.1—25]

Mosimann, J E (1962). On the compound multinomial distribution, the multivariate β-distribution and correlations among proportions. Biometrika, 49, 65–82. [24.5.2—6]

Mosimann, J E (1975a). Statistical problems of size and shape. I. Biological applications and basic theorems. In G P Patil, S Kotz, and J K Ord (Editors), A Modern Course on Distributions in Scientific Work. Volume 2—Model Building and Model Selection, pp. 187–217. Dordrecht: Reidel. [24.7.3—3]

Mosimann, J E (1975b). Statistical problems of size and shape. II. Characterizations of the lognormal, gamma and Dirichlet distributions. In G P Patil, S Kotz, and J K Ord (Editors), A Modern Course on Distributions in Scientific Work. Volume 2—Model Building and Model Selection, pp. 219–239. Dordrecht: Reidel. [12.4.2]

Mosimann, J E (1988). Size and shape analysis. In Encyclopedia of Statistical Sciences. Volume 8, pp. 497–507. New York: Wiley. [12.4.2]

Mosimann, J E, and Malley, J D (1981). The independence of size and shape before and after scale change. In C Taillie, G P Patil, and B A Baldessari (Editors), Statistical Distributions in Scientific Work. Volume 4—Models, Structures, and Characterizations, pp. 137–145. Dordrecht: Reidel. [12.4.2]

Mosteller, F (1968). Association and estimation in contingency tables. Journal of the American Statistical Association, 63, 1–28. [24.11.2—1]

Mudholkar, G S (1983). Fisher's z-transformation. In Encyclopedia of Statistical Sciences. Volume 3, pp. 130–135. New York: Wiley. [11.5.1]

Mukherjee, R (1954). A study of social mobility between three generations. In D V Glass (Editor), Social Mobility in Britain, pp. 266–290. London: Routledge and Kegan Paul. [24.11.2—1]

Mukherjee, S P, and Chatterjee, A (1988). A new MIFRA class of life distributions. Calcutta Statistical Association Bulletin, 37, 67–80. [18.3.3]

Mukherjee, S P, and Saran L K (1984). Bivariate inverse Rayleigh distributions in reliability studies. Journal of the Indian Statistical Association, 22, 23–31. [5.2.12]

Mukherjee, S P, and Sasmal, B C (1977). Life distributions of coherent dependent systems. Calcutta Statistical Association Bulletin, 26, 39–52. [5.2.9]

Muliere, P, and Scarsini, M (1987). Characterization of a Marshall-Olkin type class of distributions. Annals of the Institute of Statistical Mathematics, Tokyo, 39, 429–441. [9.2.12]

Mullooly, J P (1988). The variance of left-truncated continuous nonnegative distributions. The American Statistician, 42, 208–210. [fn6.9]

Mustonen, S (1982). Digression analysis. In Encyclopedia of Statistical Sciences. Volume 2, pp. 373–374. New York: Wiley. [4.17.7]

Muthén, B (1979). A structured probit model with latent variables. Journal of the American Statistical Association, 74, 807–811. [22.4.2.4]

Muthu, S K (1982). Probability and Errors for the Physical Sciences. New Delhi: Orient Longman. [19.8.5]

Myrhaug, D, and Kjeldsen, S P (1984). Parametric modelling of joint probability density distributions for steepness and asymmetry in deep water waves. Applied Ocean Research, 6, 207–220. [8.12.5.1, 24.12.2—8]

N

Nabar, S P (1988). A note on the bounds for the probability integral of bivariate Dirichlet distribution. Journal of the Indian Statistical Association, 26, 39–43. [6.5.13]

Nabeya, S (1951). Absolute moments in 2-dimensional normal distribution. Annals of the Institute of Statistical Mathematics, Tokyo, 3, 2–6. [4.3.8]

Nachtsheim, C J, and Johnson, M E (1988). A new family of multivariate distributions with applications to Monte Carlo studies. Journal of the American Statistical Association, 83, 984–989. [7.3]

Nagao, M, and Kadoya, M (1971). Two-variate exponential distribution and its numerical table for engineering application. Bulletin of the Disaster Prevention Research Institute, Kyoto University, 20, 183–215. [9.12.5, 9.12.6, 9.12.10, 9.12.12, 9.12.13, 9.12.14, 20.3.3]

Nagarsenker, B N (1970). A generalisation of beta densities and their multivariate analogues. Metron, 28(1–4), 156–168. [6.5.14, 6.5.15.4]

Nair, K R M, and Nair, N U (1988). On characterizing the bivariate exponential and geometric distributions. Annals of the Institute of Statistical Mathematics, Tokyo, 40, 267–271. [5.10.9]

Nair, N U, and Nair, V K R (1988). A characterization of the bivariate exponential distribution. Biometrical Journal, 30, 107–112. [5.10.9]

Narumi, S (1923a). On the general forms of bivariate frequency distributions which are mathematically possible when regression and variations are subjected to limiting conditions—I. Biometrika, 15, 77–88. [3.5]

Narumi, S (1923b). On the general forms of bivariate frequency distributions which are mathematically possible when regression and variations are subjected to limiting conditions—II. Biometrika, 15, 209–211. [3.5]

Nataf, A (1962). Détermination des distributions de probabilités dont les marges sont données. Comptes Rendus de l'Académie des Sciences, Paris, 255, 42–43. [3.2.1]

Nath, G B (1972). Moments of a linearly truncated bivariate normal distribution. Australian Journal of Statistics, 14, 97–102. [4.16.1]

National Bureau of Standards (1959). Tables of the Bivariate Normal Distribution Function and Related Functions. Applied Mathematics Series, No. 50. Washington, D.C.: U.S. Government Printing Office. [4.12.1, 4.12.6, 4.12.7, 4.12.9, 4.16.2, 19.8.1]

Nayak, T K (1987). Multivariate Lomax distribution: Properties and usefulness in reliability theory. Journal of Applied Probability, 24, 170–177. [5.8.10, 6.3.6, 6.3.7]

Nayak, T K (1988). Testing equality of conditionally independent exponential distributions. Communications in Statistics—Theory and Methods, 17, 807–820. [24.2—1]

Nelsen, R B (1986). Properties of a one-parameter family of bivariate distributions with specified marginals. Communications in Statistics—Theory and Methods, 15, 3277–3285. [5.4.5, 5.4.6, 5.4.8, 13.4]

Nelsen, R B (1989). Some two-parameter families of bivariate distributions with arbitrary margins. (Abstract only.) Abstracts of Papers Presented to the American Mathematical Society, 10, 115. [5.5, 13.6, 13.11]

Nelson, W (1982). Applied Life Data Analysis. New York: Wiley. [24.2—2]

Neuts, M F, and Pagano, M E (1981). Generating random variates from a distribution of phase type. In T I Ören, C M Delfosse, and C M Shub (Editors), 1981 Winter Simulation Conference Proceedings. Volume 2, pp. 381–387. Piscataway, New Jersey: Institute of Electrical and Electronics Engineers. [16.3.5]

Newby, M (1986). Applications of concepts of ageing in reliability data analysis. Reliability Engineering, 14, 291–308. [18.2.5]

Neyman, J (1955). The problem of inductive inference. Communications on Pure and Applied Mathematics, 8, 13–46. [24.13.1—6]

Nguyen, T T, and Sampson, A R (1987). Testing for positive quadrant dependence in ordinal contingency tables. Naval Research Logistics, 34, 859–877. [24.9.2—8]

Nguyen, V U, and Chowdhury, R N (1985). Simulation for risk analysis with correlated variables. Géotechnique, 35, 47–58. [19.8.2]

Nicholson, A J (1985). The variability of accident counts. Accident Analysis and Prevention, 17, 47–56. [21.4]

Nicholson, C (1943). The probability integral for two variables. Biometrika, 33, 59–72. [4.12.3, 4.12.7]

Nicholson, W L, and Littlefield, R J (1983). Interactive color graphics for multivariate data. In K W Heiner, R S Sacher, and J W Wilkinson (Editors), Computer Science and Statistics: Proceedings of the 14th Symposium on the Interface, pp. 211–219. New York: Springer. [24.7.1—3]

Nisbet, E G, Bickle, M J, and Martin, A (1977). The mafic and ultramafic lavas of the Belingwe Greenstone Belt, Rhodesia. Journal of Petrology, 18, 521–566. [24.5.2—4]

Norton, J P (1902). Statistical Studies in the New York Money Market. New York: Macmillan. [24.11.2—13]

O

Oakes, D (1982). A model for association in bivariate survival data. Journal of the Royal Statistical Society, Series B, 44, 414–422. [5.9.2, 5.9.6, 13.5]

Oakes, D (1986). Semiparametric inference in a model for association in bivariate survival data. Biometrika, 73, 353–361. [5.9.2, 5.9.6]

Oakes, D (1988). Semi-parametric models. In Encyclopedia of Statistical Sciences. Volume 8, pp. 367–369. New York: Wiley. [2.15.2—52, 21.7.1]

Obretenov, A (1985). Characterization of the multivariate Marshall-Olkin exponential distribution. Probability and Mathematical Statistics, 6, 51–56. [9.2.7]

Ochi, M K (1979). On long-term statistics for ocean and coastal waves. Proceedings of the Sixteenth Coastal Engineering Conference. Volume I, pp. 59–75. New York: American Society of Civil Engineers. [19.7.10, 24.12.2—8]

O'Cinneide, C A, and Raftery, A E (1989). A continuous multivariate exponential distribution that is multivariate phase type. Statistics and Probability Letters, 7, 323–325. [9.11.7]

Odeh, R E, and Owen, D B (1980). Tables for Normal Tolerance Limits, Sampling Plans, and Screening. New York: Dekker. [4.16.2]

Ohi, F, and Nishida, T (1978). Bivariate shock models and its application to the system reliability analysis. Mathematica Japonica, 23, 109–122. [fn17.4]

Ohi, F, and Nishida, T (1979). Bivariate Erlang distribution functions. Journal of the Japan Statistical Society, 9, 103–108. [9.2.13, 12.3.6]

Oja, H (1985). Ordering of distributions, partial. In Encyclopedia of Statistical Sciences. Volume 6, pp. 490–494. New York: Wiley. [2.15.2—32]

Olsson, U (1979). Maximum likelihood estimation of the polychoric correlation coefficient. Psychometrika, 44, 443–460. [11.11]

Olsson, U (1980). Measuring correlation in ordered two-way contingency tables. Journal of Marketing Research, 17, 391–394. [11.11]

O'Neill, B, Joksch, H, and Haddon, W (1974). Relationship between car size, car weight, and crash injuries in car-to-car crashes. Proceedings of the 3rd International Congress on Automotive Safety, held in San Francisco. Washington, D.C.: National Highway Traffic Safety Administration. [21.2]

O'Neill, T J (1978). Normal discrimination with unclassified observations. Journal of the American Statistical Association, 73, 821–826. [24.7.1—3]

O'Neill, T J (1985). Testing for symmetry and independence in a bivariate exponential distribution. Statistics and Probability Letters, 3, 269–274. [9.3.10, 9.3.11, 24.2—1]

Ong, S H, and Lee, P A (1986). Bivariate non-central negative binomial distribution: Another generalisation. Metrika, 33, 29–46. [24.6—11]

Oomen, P A (1984). Relation of scarlet mite (*Brevipalpus phoenicis*) density in tea with injury and yield. Netherlands Journal of Plant Pathology, 90, 199–212. [24.8.3—1]

Ord, J K (1972). Families of Frequency Distributions. London: Griffin. [2.1, 24.6—3]

Ord, J K (1983). Laplace distribution. In Encyclopedia of Statistical Sciences. Volume 4, pp. 473–475. New York: Wiley. [2.15.1—17a]

Ord, J K (1985). Pearson system of distributions. In Encyclopedia of Statistical Sciences. Volume 6, pp. 655–659. New York: Wiley. [2.15.1—33]

Osaki, S (1980). A two-unit parallel redundant system with bivariate exponential lifetimes. Microelectronics and Reliability, 20, 521–523. [9.2.7]

Osaki, S, Yamada, S, and Hishitani, J (1989). Availability theory for two-unit nonindependent series systems subject to shut-off rules. Reliability Engineering and System Safety, 25, 33-42. [9.2.7]

Owen, D B (1956). Tables for computing bivariate normal probabilities. Annals of Mathematical Statistics, 27, 1075–1090. [4.12.4, 4.12.7]

Owen, D B (1962). Handbook of Statistical Tables. Reading, Massachusetts: Addison-Wesley. [4.12.7]

Owen, D B (1980). A table of normal integrals. Communications in Statistics, Part B—Simulation and Computation, 9, 389–419 (Additions and Corrections, 10, 537–541). [4.12.4]

Owen, D B (1985). Noncentral t-distribution. In Encyclopedia of Statistical Sciences. Volume 6, pp. 286–290. New York: Wiley. [2.15.1—28]

Owen, D B (1988). Screening by correlated variates. In Encyclopedia of Statistical Sciences. Volume 8, pp. 309–312. New York: Wiley. [4.16.2]

Owen, D B, Crasewell, R J, and Hanson, D L (1964). Nonparametric upper confidence bounds for $\Pr\{Y < X\}$ when X and Y are normal. Journal of the American Statistical Association, 59, 906–924. [17.6.1]

Owen, D B, and Haas, R W (1978). Tables of the normal conditioned on t-distribution. In H A David (Editor), Contributions to Survey Sampling and Applied Statistics. Papers in Honor of H O Hartley, pp. 295–318. New York: Academic Press. [4.16.2]

Owen, D B, and Li, L (1980). The use of cutting scores in selection procedures. Journal of Educational Statistics, 5, 157-168. [19.6.1]

Owen, D B, Li, L, and Chou, Y M (1981). Prediction intervals for screening using a measured correlated variable. Technometrics, 23, 165–170. [4.16.2]

P

Page, D I, and Newth, J A (1954). The mean lifetime of the V_1^0-particles. Philosophical Magazine, 45, 38–46. [24.13.1—6]

Papadopoulos, A S (1983). Bayesian reliability of the Weibull failure model with bivariate characterization of the parameters. Metron, 41(1–2), 95–112. [20.7.2]

Papageorgiou, H, and Kemp, C D (1988). A method of estimation for some bivariate discrete distributions. Biometrical Journal, 30, 993–1001. [24.6—5]

Papageorgiou, H, Kemp, C D, and Loukas, S (1983). Some methods of estimation for the bivariate Hermite distribution. Biometrika, 70, 479–484. [24.6—1]

Parikh, N T, and Sheth, R J (1966). Applications of bivariate normal distribution: To a stress vs. strength problem in reliability analysis. Journal of the Indian Statistical Association, 4, 105–107. [17.6.4]

Parrish, R S (1981). Evaluation of bivariate cumulative probabilities using moments to fourth order. Journal of Statistical Computation and Simulation, 13, 181–194. [6.4.11, 8.1.2]

Parrish, R S, and Bargmann, R E (1981). A method for the evaluation of cumulative probabilities of bivariate distributions using the Pearson family. In C Taillie, G P Patil, and B A Baldessari (Editors), Statistical Distributions in Scientific Work. Volume 5—Inferential Problems and Properties, pp. 241–257. Dordrecht: Reidel. [6.4.10, 6.4.11, 6.5.13, 8.1.2, 8.6.9]

Parzen, E (1979). Nonparametric statistical data modeling. Journal of the American Statistical Association, 74, 105–121. [2.3.3]

Patel, J K (1983). Hazard rate and other classifications of distributions. In Encyclopedia of Statistical Sciences. Volume 3, pp. 590–594. New York: Wiley. [2.15.2—12, 18.2.4]

Patel, J K, Kapadia, C H, and Owen, D B (1976). Handbook of Statistical Distributions. New York: Dekker. [2.1]

Patel, J K, and Read, C B (1982). Handbook of the Normal Distribution. New York: Dekker. [4.2, 4.3.8, 4.12.1, 4.12.9]

Patel, K (1973). A catalog of failure distributions. Communications in Statistics, 1, 281–284. [2.3.5]

Patil, G P (1965). On a characterization of multivariate distribution by a set of its conditional distributions. Bulletin of the International Statistical Institute, 41, Book 2, 768–769. [8.1.3]

Patil, G P (1985). Linear exponential family. In Encyclopedia of Statistical Sciences. Volume 5, pp. 22–23. New York: Wiley. [2.15.1—18]

Patil, G P, and Bildikar, S (1967). Multivariate logarithmic series distribution as a probability model in population and community ecology and some of its statistical properties. Journal of the American Statistical Association, 62, 655–674. [24.6—15]

Patil, G P, Boswell, M T, and Ratnaparkhi, M V (1984). Dictionary and Classified Bibliography of Statistical Distributions in Scientific Work. Volume 2. Continuous Univariate Models. (Statistical Distributions in Scientific Work, Volume 7.) Fairland, Maryland: International Co-operative Publishing House. [2.1, 2.16]

Patil, G P, Boswell, M T, Ratnaparkhi, M V, and Roux, J J J (1984). Dictionary and Classified Bibliography of Statistical Distributions in Scientific Work. Volume 3. Multivariate Models. (Statistical Distributions in Scientific Work, Volume 8.) Fairland, Maryland: International Co-operative Publishing House. [4.1]

Patil, G P, Rao, C R, and Zelen, M (1988). Weighted distributions.

In Encyclopedia of Statistical Sciences. Volume 9, pp. 565–571. New York: Wiley. [2.15.1—53, 19.4.6]

Paul, S R, and Ho, N I (1989). Estimation in the bivariate Poisson distribution and hypothesis testing concerning independence. Communications in Statistics—Theory and Methods, 18, 1123–1133. [24.6—11]

Paulson, A S (1973). A characterization of the exponential distribution and a bivariate exponential distribution. Sankhyā, Series A, 35, 69–78. [9.12.12, T17.1, 17.3.1, fn17.4]

Paulson, A S, and Uppuluri, V R R (1972a). A characterization of the geometric distribution and a bivariate geometric distribution. Sankhyā, Series A, 34, 297–300. [17.3.1]

Paulson, A S, and Uppuluri, V R R (1972b). Limits laws of a sequence determined by a random difference equation governing a one-component system. Mathematical Biosciences, 13, 325–333. [17.3.1]

Payne, G, Ford, G, and Robertson, C (1977). A reappraisal of social mobility in Britain. Sociology, 11, 289–310. [24.11.2—1]

Pearson, E S (1956). Some aspects of the geometry of statistics. The use of visual presentation in understanding the theory and application of mathematical statistics. Journal of the Royal Statistical Society, Series A, 119, 125–146 (Proceedings of the Meeting, 146–149). (Reprinted in The Selected Papers of E S Pearson, pp. 252–275. Cambridge: Cambridge University Press.) [24.8.1—1]

Pearson, E S, Johnson, N L, and Burr, I W (1979). Comparisons of the percentage points of distributions with the same first four moments, chosen from eight different systems of frequency curves. Communications in Statistics, Part B—Simulation and Computation, 8, 191–229. [2.3.4]

Pearson, K (1896). Mathematical contributions to the theory of evolution—III. Regression, heredity and panmixia. Philosophical Transactions of the Royal Society of London, Series A, 187, 253–318. (Reprinted in Karl Pearson's Early Statistical Papers, pp. 113–178. Cambridge: Cambridge University Press, 1948.) [15.2, 15.3.1]

Pearson, K (1900). Mathematical contributions to the theory of evolution on the inheritance of characters not capable of exact quantitative measurement, VIII. Philosophical Transactions of the Royal Society of London, Series A, 195, 79–150. [24.9.2—7]

Pearson, K (1901). Mathematical contributions to the theory of evolution—VII. On the correlation of characters not quantitatively measurable. Philosophical Transactions of the Royal Society of London, Series A, 195, 1–47. [4.12.2]

Pearson, K (1904). Mathematical contributions to the theory of evolution—XIII. On the theory of contingency and its relation to association and normal correlation. Drapers' Company Research Memoirs, Biometric Series, I. (Reprinted in Karl Pearson's Early Statistical Papers, pp. 443–475. Cambridge: Cambridge University Press, 1948.) [fn14.4, 24.9.2—7, 24.9.2—11]

Pearson, K (1905). Mathematical contributions to the theory of evolution—XIV. On the general theory of skew correlation and nonlinear regression. Drapers' Company Research Memoirs, Biometric Series, II. (Reprinted in Karl Pearson's Early Statistical Papers, pp. 477–528. Cambridge: Cambridge University Press, 1948.) [11.5.3, 24.7.2—12]

Pearson, K (1913a). Note on the surface of constant association. Biometrika, 9, 534–537. [5.7.6, 5.7.7, 5.7.9]

Pearson, K (1913b). On the measurement of the influence of "broad categories" on correlation. Biometrika, 9, 116–139. [24.10.2—3, 24.12.2—3]

Pearson, K (1920). Notes on the history of correlation. Biometrika, 13, 25–45. [11.4.3]

Pearson, K (1923). Notes on skew frequency surfaces. Biometrika, 15, 220–230. [3.9.1.1]

Pearson, K (1924). On a certain double hypergeometric series and its representation by continuous frequency surfaces. Biometrika, 16, 172–188. [6.2.11]

Pearson, K (1925). The fifteen constant bivariate frequency surface. Biometrika, 17, 268–313. [1.6, 3.11.2, 24.12.2—3]

Pearson, K (1930). On the theory of contingency. I. Note on Professor J Arthur Harris' papers on the limitation in the applicability of the contingency coefficient. Journal of the American Statistical Association, 25, 320–323. (With a reply by J A Harris, A E Treloar, and M Wilder, pp. 323–327, and a postscript by K Pearson, p. 327.) [24.6—19]

Pearson, K (1931). Tables for Statisticians and Biometricians. Part II. Cambridge: Cambridge University Press. [4.12.7]

Pearson, K, and Filon, L N G (1898). Mathematical contributions to the theory of evolution—IV. On the probable errors of frequency constants and on the influence of random selection on variation and correlation. Philosophical Transactions of the Royal Society of London, Series A, 191, 229–311. (Reprinted in Karl Pearson's Early Statistical Papers, pp. 179–261. Cambridge: Cambridge University Press, 1948.) [24.6—12]

Pearson, K, and Lee, A (1897). On the distribution of frequency (variation and correlation) of the barometric height at divers stations. Philosophical Transactions of the Royal Society of London, Series A, 190, 423–469. [6.9.4, 24.12.2—3]

Pearson, K, and Lee, A (1903). On the laws of inheritance in man. I. Inheritance of physical characters. Biometrika, 2, 357–462. [24.9.2—11]

Pearson, K, and Young, A W (1918). On the product-moments of various orders of the normal correlation surface of two variates. Biometrika, 12, 86–92. [4.3.8]

Peebles, D, Kirchanski, S, Brown, M, and Hoffman, R (1986). Extended leukocyte parameters for hematology analyzers. In M Andreeff (Editor), Clinical Cytometry, pp. 104–112. (Annals of the New York Academy of Sciences, 468.) [24.4.3—9]

Pendleton, B F (1986). Ratio correlation. In Encyclopedia of Statistical Sciences. Volume 7, pp. 636–639. New York: Wiley. [12.4.2, fn15.2]

Peterson, A V, and Kronmal, R A (1982). On mixture methods for the computer generation of random variables. The American Statistician, 36, 184–191. [16.2.3]

Peterson, A V, and Kronmal, R A (1985). Mixture method. In Encyclopedia of Statistical Sciences. Volume 5, pp. 579–583. New York: Wiley. [16.2.3]

Pettitt, A N (1979). Testing for bivariate normality using the empirical distribution function. Communications in Statistics, Part A—Theory and Methods, 8, 699–712. [4.13.5.2, 24.8.1—1]

Pfanzagl, J (1982). Contributions to a General Asymptotic Statistical Theory. New York: Springer (Lecture Notes in Statistics, No. 13).

Phatarfod, R M (1976). Some aspects of stochastic reservoir theory. Journal of Hydrology, 30, 199–217. [7.5.13, 14.8]

Phillips, E F (1929). Variation and correlation in the appendages of the honeybee. Memoir No. 121, Cornell University Agricultural Experimental Station. [24.8.2—4]

Phillips, M J (1981). A preventive maintenance plan for a system subject to revealed and unrevealed faults. Reliability Engineering, 2, 221–231. [5.2.9]

Pickands, J (1981). Multivariate extreme value distributions. Bulletin of the International Statistical Institute, 49, Book 2, 859–878 (Discussion, 894–902). [9.17.1, 9.17.3, 9.17.9]

Piegorsch, W W, Weinberg, C R, and Margolin, B H (1988). Exploring simple independent action in multifactor tables of proportions. Biometrics, 44, 595–603. [22.2.1]

Pierce, J R, and Gilbert, E N (1958). On AX and ABX limens. Journal of the Acoustical Society of America, 30, 593–595. [23.6]

Pimentel, R A (1979). Morphometrics. The Multivariate Analysis of Biological Data. Dubuque, Iowa: Kendall/Hunt. [24.7.1—1, 24.7.2—9]

Pinto, P E, and Nuti, C (1981). Seismic structural risk model with multiple correlated ground motion parameters. In T Moan and M Shinozuka (Editors), Structural Safety and Reliability, pp. 725–737. Amsterdam: Elsevier. [19.7.12]

Plackett, R L (1965). A class of bivariate distributions. Journal of the American Statistical Association, 60, 516–522. [5.7.9]

Plackett, R L (1981). The Analysis of Categorical Data. 2nd Edition. London: Griffin. [24.6—19, 24.9.2—8]

Plate, E J, and Duckstein, L (1987). Reliability in hydraulic design. In L Duckstein and E J Plate (Editors), Engineering Reliability and Risk in Water Resources, pp. 27–60. Dordrecht: Nijhoff. [17.6.1]

Platt, C M R (1981). Two dimensional histograms of GMS-1 satellite visible albedo and infrared temperature for selected cloud systems. Technical Paper No. 40, Division of Atmospheric Physics, Commonwealth Scientific and Industrial Research Organization, Australia. [24.4.2—13]

Platz, O (1984). A Markov model for common-cause failures. Reliability Engineering, 9, 25–31. [9.8]

Poirier, D J (1980). Partial observability in bivariate probit models. Journal of Econometrics, 12, 209–217. [22.2.2]

Pooley, F D, and Clark, N (1979). Fiber dimensions and aspect ratio of crocidolite, chrysotile and amosite particles detected in lung tissue specimens. In I J Selikoff and E C Hammond (Editors), Health Hazards of Asbestos Exposure, pp. 711–716. (Annals of the New York Academy of Sciences, 330.) [24.9.2—4]

Posner, E C, Rodemich, E R, Ashlock, J C, and Lurie, S (1969). Application of an estimator of high efficiency in bivariate extreme value theory. Journal of the American Statistical Association, 64, 1403–1414. [9.17.11, 24.3—4]

Prather, J E (1988). Spurious correlation. In Encyclopedia of Statistical Sciences. Volume 8, pp. 613–614. New York: Wiley. [12.4.2, fn15.2]

Preece, D A (1986). Illustrative examples: Illustrative of what? The Statistician, 35, 33–44. [24.1.2.1]

Pregibon, D (1985). Link tests. In Encyclopedia of Statistical Sciences. Volume 5, pp. 82–85. New York: Wiley. [2.15.2—22]

Prékopa, A, and Szántai, T (1978). A new multivariate gamma distribution and its fitting to empirical streamflow data. Water Resources Research, 14, 19–24. [10.3.9]

Press, S J (1972a). Multivariate stable distributions. Journal of Multivariate Analysis, 2, 444–462. [7.11.2, 7.11.5.1, 7.11.6, 7.11.7]

Prêss, S J (1972b). Applied Multivariate Analysis. New York: Holt, Reinhart and Winston. [6.2.14, 7.11.2, 7.11.7]

Pretorius, S J (1930). Skew bivariate frequency surfaces, examined in the light of numerical illustrations. Biometrika, 22, 109–223. [24.7.2—1, 24.11.2—7, 24.12.2—3]

Proschan, F, and Sullo, P (1974). Estimating the parameters of a bivariate exponential distribution in several sampling situations. In F Proschan and R J Serfling (Editors), Reliability and Biometry. Statistical Analysis of Life Lengths, pp. 423–440. Philadelphia: Society for Industrial and Applied Mathematics. [17.2.2]

Prucha, I R, and Kelejian, H H (1984). The structure of simultaneous equation estimators: A generalization towards nonnormal disturbances. Econometrica, 52, 721–736. [6.2.11]

Q

Quandt, R E (1982). Econometric disequilibrium models. Communications in Statistics—Econometric Reviews, 1, 1–63 (Discussion, 65–96). [19.9.5]

Quandt, R E, and Ramsey, J B (1978). Estimating mixtures of normal distributions and switching regressions. Journal of the American Statistical Association, 73, 730–738 (Discussion, 738–752). [4.17.1]

Quensel, C-E (1966). On tests of independence and evaluation of the magnitude of the dependence between two categorized data. Skandinavisk Aktuarietidskrift, 1966, 199–217. [24.10.2—3]

Quesenberry, C P (1985). Model construction: Selection of distributions. In Encyclopedia of Statistical Sciences. Volume 5, pp. 583–589. New York: Wiley. [2.15.2—27]

Quesenberry, C P (1986a). Probability integral transformations. In Encyclopedia of Statistical Sciences. Volume 7, pp. 225–231. New York: Wiley. [2.15.2—39, 4.13.4.6, 4.13.5.2]

Quesenberry, C P (1986b). Some transformation methods in goodness-of-fit. In R B D'Agostino and M A Stephens (Editors), Goodness-of-Fit Techniques, pp. 235–277. New York: Dekker. [4.13.4.6]

R

Rachal, J V, Hubbard, R L, and Cavanaugh, E R (1981). Reliability and validity of the data and measures from two national surveys of adolescent drinking. American Statistical Association, 1981 Proceedings of the Section on Survey Research Methods, pp. 323–328. [24.9.2—9]

Radelet, M, and Pierce, G L (1983). Race and prosecutorial discretion in homicide cases. Presented at the Annual Meeting of the American Sociological Association, held in Detroit. [24.4.2—20]

Radusch, D F (1939). The relationship between periodontal conditions and certain dietary factors. Journal of Dental Research, 18, 305–327. [24.9.2—1, 24.9.2—2]

Raftery, A E (1984). A continuous multivariate exponential distribution. Communications in Statistics—Theory and Methods, 13, 947–965. [9.11.1, 9.11.4, 9.11.6, 9.11.8, 9.13, 13.11, 15.6.4]

Raftery, A E (1985). Some properties of a new continuous bivariate exponential distribution. Statistics and Decisions, Supplement Issue No. 2, 53–58. [9.11.1, 9.13, 13.11, 15.6.4]

Rahman, N A (1968). A Course in Theoretical Statistics. London: Griffin. [24.7.2—16]

Rahman, N A (1972). Practical Exercises in Probability and Statistics. London: Griffin. [24.6—10, 24.7.2—4, 24.12.2—6]

Rai, K, and Van Ryzin, J (1984). Multihit models for bivariate quantal responses. Statistics and Decisions, 2, 111–129. [9.2.8, 22.4.1]

Ramanarayanan, R, and Subramanian, A (1981). A 2-unit cold standby system with Marshall-Olkin bivariate exponential life and repair times. IEEE (Institute of Electrical and Electronics Engineers) Transactions on Reliability, R-30, 489–490. [9.2.7]

Ramig, P F, and Nelson, P R (1980). The probability integral for a bivariate generalization of the non-central t. Communications in Statistics, Part B—Simulation and Computation, 9, 621–631. [6.2.14]

Rand, R H (1984). Computer Algebra in Applied Mathematics: An Introduction to MACSYMA. London: Pitman. [fn3.1]

Rao, B R (1983). A general portfolio model for multivariate symmetric stable distributions. Metron, 41(1–2), 29–42. [7.11.7]

Rao, B R (1986). Pólya type 2 frequency (PF_2) distributions. In Encyclopedia of Statistical Sciences. Volume 7, pp. 65–68. New York: Wiley. [2.15.2—38, 18.2.4]

Rao, B R (1988). SSR_2 (strictly sign reverse regular functions of order 2). In Encyclopedia of Statistical Sciences. Volume 8, pp. 615–617. New York: Wiley. [18.2.4]

Rao, B R, Garg, M L, and Li, C C (1968). Correlation between the sample variances in a singly truncated bivariate normal distribution. Biometrika, 55, 433–436. [4.16.1]

Rao, B R, Mazumdar, S, Waller, J H, and Li, C C (1973). Correlation between the numbers of two types of children in a family. Biometrics, 29, 271–279. [24.6—5]

Rao, B V, and Sinha, B K (1988). A characterization of Dirichlet distributions. Journal of Multivariate Analysis, 25, 25–30. [6.5.11]

Rao, C R (1947). Note on a problem of Ragnar Frisch. Econometrica, 15, 245–249. [15.3.3]

Rao, C R (1948a). Tests of significance in multivariate analysis. Biometrika, 35, 58–79. [24.8.1—1]

Rao, C R (1948b). Large sample tests of statistical hypotheses concerning several parameters with applications to problems of estimation. Proceedings of the Cambridge Philosophical Society, 44, 50–57. [4.13.5.7]

Rao, C R (1975). Some problems in the characterization of the multivariate normal distribution. In G P Patil, S Kotz, and J K Ord (Editors), A Modern Course on Distributions in Scientific Work. Volume 3—Characterizations and Applications, pp. 1–13. Dordrecht: Reidel. [4.5]

Rao, V K R V (1977). Nutritional norms by calorie intake and measurement of poverty. Bulletin of the International Statistical Institute, 47, Book 1, 645–654 (Discussion, 671–679). [24.11.2—11]

Ratnaparkhi, M V (1981). Some bivariate distributions of (X,Y) where the conditional distribution of Y, given X, is either beta or unit-gamma. In C Taillie, G P Patil, and B A Baldessari (Editors), Statistical Distributions in Scientific Work. Volume 4—Models, Structures, and Characterizations, pp. 389–400. Dordrecht: Reidel. [8.6.11]

Ratnaparkhi, M V (1983a). Inverted beta distribution. In Encyclopedia of Statistical Sciences. Volume 4, pp. 253–256. New York: Wiley. [2.15.1—15]

Ratnaparkhi, M V (1983b). Inverted Dirichlet distribution. In Encyclopedia of Statistical Sciences. Volume 4, pp. 256–259. New York: Wiley. [6.3.10, 6.5.15.5]

Ratnaparkhi, M V (1985). Liouville-Dirichlet distribution. In Encyclopedia of Statistical Sciences. Volume 5, pp. 86–87. New York: Wiley. [6.6]

Ray, S C, Berk, R A, and Bielby, W T (1980). Correcting sample selection bias for bivariate logistic distribution of disturbances. American Statistical Association, 1980 Proceedings of the Business and Economic Statistics Section, pp. 456–459. [5.2.9, 5.9.5, 19.9.5]

Rayna, G (1987). REDUCE. Software for Algebraic Computation. New York: Springer. [fn3.2]

Read, C B (1985a). Linear hazard rate distribution. In Encyclopedia of Statistical Sciences. Volume 5, p. 24. New York: Wiley. [2.15.1—19]

Read, C B (1985b). Normal distribution. In Encyclopedia of Statistical Sciences. Volume 6, pp. 347–359. New York: Wiley. [2.15.1—29]

Read, C B (1988a). Spacings. In Encyclopedia of Statistical Sciences. Volume 8, pp. 566–569. New York: Wiley. [2.15.2—55]

Read, C B (1988b). *t* distribution. In Encyclopedia of Statistical Sciences. Volume 9, pp. 162–164. New York: Wiley. [2.15.1—47]

Read, C B (1988c). Uniform (or rectangular) distributions. In Encyclopedia of Statistical Sciences. Volume 9, pp. 411–414. New York: Wiley. [2.15.1—49]

Reed, S W, and Reed, S C (1965). Mental Retardation: A Family Study. Philadelphia: Saunders. [24.6—5]

Regier, M H, and Hamdan, M A (1971). Correlation in a bivariate normal distribution with truncation in both variables. Australian Journal of Statistics, 13, 77–82. [4.16.1]

Reiser, B, and Guttman, I (1986). Statistical inference for $\Pr(Y < X)$: The normal case. Technometrics, 28, 253–257. [17.6.1]

Reiser, B, and Guttman, I (1987). A comparison of three point estimators for $P(Y < X)$ in the normal case. Computational Statistics and Data Analysis, 5, 59–66. [17.6.1]

Rényi, A (1959). On measures of dependence. Acta Mathematica Academia Scientia Hungarica, 10, 441–451. [11.3.4, 11.4, 11.6, 11.10]

Rényi, A (1970). Probability Theory. Amsterdam: North Holland. [11.3.4, 11.6]

Resnick, S I (1982). Extremal processes. In Encyclopedia of Statistical Sciences. Volume 2, pp. 600–605. New York: Wiley. [2.15.2—8]

Reyment, R A (1971). Multivariate normality in morphometric analysis. Mathematical Geology (Journal of the International Association for Mathematical Geology), 3, 357–368. [4.13.5.8, 7.3, 19.3.1, 24.7.3—5, 24.11.2—13]

Reyment, R A (1985). Multivariate morphometrics and analysis of shape. Mathematical Geology (Journal of the International Association for Mathematical Geology), 17, 591–609. [24.7.3—5]

Rhodes, E C (1923). On a certain skew correlation surface. Biometrika, 14, 355–377. [6.9.4, 24.12.2—3]

Rincon-Gallardo, S, Quesenberry, C P, and O'Reilly, F J (1979). Conditional probability integral transformations and goodness-of-fit tests for multivariate normal distributions. Annals of Statistics, 7, 1052–1057. [24.7.1—3, 24.11.2—13]

Ripley, B D (1983). Computer generation of random variables: A tutorial. International Statistical Review, 51, 301–319. [16.2.1, 16.3.7]

Ripley, B D (1987). Stochastic Simulation. New York: Wiley. [16.2.1, 16.3.1, 16.3.2, 16.5.1]

Robison, E L, and Mesenbourg, T L (1980). Comparison of three measures of gross vehicle weight of trucks. American Statistical Association, 1980 Proceedings of the Social Statistics Section, pp. 312–317. [24.4.2—14]

Rodriguez, R N (1980). Multivariate Burr III distributions, Part I. Theoretical properties. Research Publication GMR-3232, General Motors Research Laboratories, Warren, Michigan. [5.2.7, 5.2.12, 5.2.16, 5.8.12, 5.8.13.5, 6.8]

Rodriguez, R N (1982a). Burr distributions. In Encyclopedia of Statistical Sciences. Volume 1, pp. 335–340. New York: Wiley. [2.15.1—2]

Rodriguez, R N (1982b). Correlation. In Encyclopedia of Statistical Sciences. Volume 2, pp. 193–204. New York: Wiley. [4.7, 11.5.1, 11.5.2, 24.4.1—1]

Rodriguez, R N (1983a). Frequency curves, systems of. In Encyclopedia of Statistical Sciences. Volume 3, pp. 212–225. New York: Wiley. [2.15.1—7]

Rodriguez, R N (1983b). Frequency surfaces, systems of. In Encyclopedia of Statistical Sciences. Volume 3, pp. 232–247. New York: Wiley. [3.11.2, 4.14.2.1, 4.12.2.4, 6.1, 6.8]

Rodriguez, R N, and Taniguchi, B Y (1980). A new statistical model for predicting customer octane satisfaction using trained-rater observations. (With Discussion.) Paper No. 801356, Society of Automotive Engineers. [5.8.13.5, 6.8, 24.4.1—1]

Rogers, W H, and Tukey, J W (1972). Understanding some long-tailed symmetrical distributions. Statistica Neerlandica, 26, 211–226. [2.4.7]

Rohatgi, V K (1985). Moment problem. In Encyclopedia of Statistical Sciences. Volume 5, pp. 600–602. New York: Wiley. [2.15.2—28]

Rolski, T (1975). Mean residual life. Bulletin of the International Statistical Institute, 46, 266–270. [18.2.1]

Romano, J P, and Siegel, A F (1986). Counterexamples in Probability and Statistics. Monterey, California: Wadsworth and Brooks/Cole. [4.14.5, 4.18.1, 8.11, fn11.4]

Romeder, J-M (1973). Méthodes et Programmes d'Analyse Discriminante. Paris: Dunod. [24.9.1—6]

Rosbjerg, D, and Knudsen, J (1984). POT-estimation of extreme sea states and the benefit of using wind data. In J Tiago de Oliveira (Editor), Statistical Extremes and Applications, pp. 611–620. Dordrecht: Reidel. [24.12.3—2]

Rosenblatt, M (1952). Remarks on a multivariate transformation. Annals of Mathematical Statistics, 23, 470–472. [4.13.5.2, 16.4.2]

Rosenblueth, E (1985). On computing normal reliabilities. Structural Safety, 2, 165–167 (Corrections, 3, 67). [4.12.1]

Rosenzweig, M (1979). Tests of symmetry in two-way contingency tables. Report from Bryant College, Rhode Island. [24.4.2—1]

Rosenzweig, M (198?). Testing for patterns in two-way contingency tables. Report from Bryant College, Rhode Island. [24.4.2—1]

Roux, J J J, and Becker, P J (1981). Compound distributions relevant to life testing. In C Taillie, G P Patil, and B A Baldessari (Editors), Statistical Distributions in Scientific Work. Volume 4—Models, Structures, and Characterizations, pp. 111-124. Dordrecht: Reidel. [9.3.13, 9.3.14]

Roy, D, and Mukherjee, S P (1989). Characterizations of some bivariate life-distributions. Journal of Multivariate Analysis, 28, 1–8. [9.2.7]

Roy, S N (1953). On a heuristic method of test construction and its use in multivariate analysis. Annals of Mathematical Statistics, 24, 220–238. [4.13.5.4]

Royston, J P (1983). Some techniques for assessing multivariate normality based on the Shapiro-Wilk W. Applied Statistics, 32, 121–133. [24.7.1—3, 24.9.1—9]

Ruben, H (1964). An asymptotic expansion for the multivariate normal distribution and Mills' ratio. Journal of Research of the National Bureau of Standards—B. Mathematics and Mathematical Physics, 68B, 3–11. [4.12.2]

Rubinstein, R Y (1981). Simulation and the Monte Carlo Method. New York: Wiley. [16.2.1, 16.3.1, 16.3.2, 16.4.2, 16.5.1]

Ruger, H A (1932–33). On the interrelationship of certain characters in man (males). Annals of Eugenics, London, 5, 59–104. [24.7.2—16]

Ruppert, D (1988). Trimming and Winsorization. In Encyclopedia of Statistical Sciences. Volume 9, pp. 348–353. New York: Wiley. [4.13.2, 11.5.1]

Ruymgaart, F H (1973). Non-normal bivariate densities with normal marginals and linear regression functions. Statistica Neerlandica, 27, 11–17. [4.17.5, 4.18.2]

S

Sakasegawa, H (1983). Stratified rejection and squeeze method for generating beta random numbers. Annals of the Institute of Statistical Mathematics, Tokyo, 35, 291–302. [16.3.3]

Salsburg, D (1983). Understanding Randomness. Exercises for Statisticians. New York: Dekker. [24.12.1—5]

Samanta, M (1975). A characterization of the bivariate exponential distribution. Journal of the Indian Society of Agricultural Statistics, 27, 67–70. [9.2.7]

Samuel, M D, and Garton, E O (1985). Home range: A weighted normal estimate and tests of underlying assumptions. Journal of Wildlife Management, 49, 513–519. [19.3.3, 24.8.1—6]

Sarkar, S K (1987). A continuous bivariate exponential distribution. Journal of the American Statistical Association, 82, 667–675. [9.6.5]

Sarmanov, I O (1968). Construction of correlation between uniformly distributed variables. (In Russian.) Trudy Gosundarstevennogo Gidrologicheskogo Instituta, 160, 82–89. [T14.1]

Sarmanov, I O (1974). New forms of correlation relationships between positive quantities applied in hydrology. In Mathematical Models in Hydrology Symposium, pp. 104–109. IAHS Publication No. 100, International Association of Hydrological Sciences. [5.2.16, 7.7.11, 13.2.3, T14.1, 14.8]

Sarmanov, O V (1962). Maximum correlation coefficient (nonsymmetric case). Selected Translations in Mathematical Statistics and Probability, 2, 207–210. (Original Russian article was dated 1958.) [11.6]

Sarmanov, O V (1963). Maximum correlation coefficient (symmetric case). Selected Translations in Mathematical Statistics and Probability, 4, 271–275. (Original Russian article was dated 1959.) [11.6]

Sarmanov, O V (1966). Generalized normal correlation and two-dimensional Fréchet classes. Soviet Mathematics, 7, 596–599. (Original article was in Russian.) [4.17.5, T14.1, 14.6.1, 14.6.3]

Sarmanov, O V, and Bratoeva, Z N (1967). Probabilistic properties of bilinear expansions of Hermite polynomials. Theory of Probability and its Applications, 12, 470–481. (Original article was in Russian.) [14.6.1]

Sasyo, Y, and Matsuo, T (1980). On the statistical investigation of fall velocity of snowflakes. Papers in Meteorology and Geophysics, 31, 61–79. [24.12.3—10]

Satterthwaite, S P, and Hutchinson, T P (1978). A generalisation of Gumbel's bivariate logistic distribution. Metrika, 25, 163–170. [5.8.12, 5.9.3, 6.7, fn20.2]

Saunders, S C (1983). Fatigue models. In Encyclopedia of Statistical Sciences. Volume 3, pp. 31–34. New York: Wiley. [2.15.3—2]

Savage, I R (1962). Mills' ratio for multivariate normal distributions. Journal of Research of the National Bureau of Standards—B. Mathematics and Mathematical Physics, 66B, 93–96. [4.12.2]

Savits, T H (1985). A multivariate IFR class. Journal of Applied Probability, 22, 197–204. [18.3.2]

Savits, T H (1988). Some multivariate distributions derived from a non-fatal shock model. Journal of Applied Probability, 25, 383–390. [18.4]

Saw, J G (1969). A bivariate exponential density and a test that two identical components in parallel behave independently. Technical Report No. 22, Department of Industrial and Systems Engineering, University of Florida, Gainesville. [9.2.13]

Scheuer, E M, and Stoller, D S (1962). On the generation of normal random vectors. Technometrics, 4, 278–281. [16.5.1]

Scheuren, F J, and Oh, H L (1975). A data analysis approach to fitting square tables. Communications in Statistics, 4, 595–615. [24.4.2—17]

Schmeiser, B W (1980). Random variate generation: A survey. In T I Oren, C M Shub, and F Roth (Editors), Simulation with Discrete Models: A State-of-the-Art View, pp. 79–104. New York: Institute of Electrical and Electronics Engineers. (Updated in T I Ören, C M Delfosse, and C M Shub (Editors), 1981 Winter Simulation

Conference Proceedings. Volume 1, pp. 227–242. Piscataway, New Jersey: Institute of Electrical and Electronics Engineers.) [16.1, 16.2.2, 16.3.1, 16.3.2, 16.3.3]

Schmeiser, B W, and Babu, A J G (1980). Beta variate generation via exponential majorizing functions. Operations Research, 28, 917–926. [16.3.3]

Schmeiser, B W, and Lal, R (1980). Squeeze methods for generating gamma variates. Journal of the American Statistical Association, 75, 679–682. [16.3.2, 16.4.1]

Schmeiser, B W, and Lal, R (1982). Bivariate gamma random vectors. Operations Research, 30, 355–374. [10.3.7, 16.5.4.3]

Schneider, T, and Holst, E (1983). Man-made mineral fibre size distributions utilizing unbiased and fibre length biased counting methods and the bivariate log-normal distribution. Journal of Aerosol Science, 14, 139–146. (The Authors have noted misprints in their equations (20) and (Appendix 1).) [19.4.6]

Schnute, J (1984). Linear mixtures: A new approach to bivariate trend lines. Journal of the American Statistical Association, 79, 1–8. [19.3.1, 24.7.3—1]

Schoenberg, I J (1938). Metric spaces and completely monotone functions. Annals of Mathematics, 39, 811–841. [16.5.2]

Schreiber, B M (1983). *L* class laws. In Encyclopedia of Statistical Sciences. Volume 4, pp. 578–579. New York: Wiley. [2.15.2—20]

Schreuder, H T, Bhattacharyya, H T, and McClure, J P (1982a). Towards a unified distribution theory for stand variables using the S_{BBB} distribution. Biometrics, 38, 137–142. [24.7.2—6]

Schreuder, H T, Bhattacharyya, H T, and McClure, J P (1982b). The S_{BBB} distribution: A potentially useful trivariate distribution. Canadian Journal of Forest Research, 12, 641–645. [24.7.2—6]

Schreuder, H T, and Hafley, W L (1977). A useful bivariate distribution for describing stand structure of tree heights and diameters. Biometrics, 33, 471–478. [19.3.2, 24.7.2—5]

Schriever, B F (1987a). Monotonicity of rank statistics in some non-parametric testing problems. Statistica Neerlandica, 41, 99–109. [12.6]

Schriever, B F (1987b). An ordering for positive dependence. Annals of Statistics, 15, 1208–1214. [12.6]

Schubert, J H, and Cornell, R G (1958). Determination of diphtheria and tetanus antitoxin by the hemagglutination test in comparison with tests in vivo. Journal of Laboratory and Clinical Medicine, 52, 737–743. [24.4.2—1]

Schucany, W R, Parr, W C, and Boyer, J E (1978). Correlation structure in Farlie-Gumbel-Morgenstern distributions. Biometrika, 65, 650–653. [5.2.5, 13.2.1]

Schulman, A I, and Greenberg, G Z (1970). Operating characteristics and a priori probability of the signal. Perception and Psychophysics, 8, 317–320. [23.2]

Schultz, A H (1953). (Title not known.) American Journal of Physical Anthropology, 11, 277–311. [24.7.3—3]

Schuster, E F (1984). Classification of probability laws by tail behavior. Journal of the American Statistical Association, 79, 936–939. [2.3.3]

Schuurmann, F J, Krishnaiah, P R, and Chattopadhyay, A K (1975). Tables of multivariate *F* distribution. Sankyā, Series B, 37, 308–331. [7.8.8]

Schwager, S J (1985). Multivariate skewness and kurtosis. In Encyclopedia of Statistical Sciences. Volume 6, pp. 122–125. New York: Wiley. [4.13.5.8]

Schweizer, B, and Wolff, E F (1976). Sur une mesure de dépendence pour les variables aléatories. Comptes Rendus de l'Académie des Sciences, Paris, Série A, 283, 659–661. [11.8]

Schweizer, B, and Wolff, E F (1981). On nonparametric measure of dependence for random variables. Annals of Statistics, 9, 879–885. [11.4, 11.9]

Scott, A J, and Symons, M J (1971). Clustering methods based on likelihood ratio criteria. Biometrics, 27, 387–397. [24.7.1—3]

Scott, D W, Gotto, A M, Cole, J S, and Gorry, G A (1978). Plasma lipids as collateral risk factors in coronary artery disease—A study of 371 males with chest pain. Journal of Chronic Diseases, 31, 337–345. [24.9.3—6]

Scudder, R, and Anderson, C A (1954). Migration and vertical occupational mobility. American Social Review, 19, 329–334. [24.11.2—2]

Seal, H L (1967). Studies in the history of probability and statistics. XV. The historical development of the Gauss linear model. Biometrika, 54, 1–24. [4.2]

Seal, H L (1988). Risk theory. In Encyclopedia of Statistical Sciences. Volume 8, pp. 152–156. New York: Wiley. [2.15.3—7]

Seaman, J W, and Odell, P L (1988). Variance, upper bounds. In Encyclopedia of Statistical Sciences. Volume 9, pp. 480–484. New York: Wiley. [2.15.2—67]

Searle, J A, Bethell, J, and Baggaley, G (1978). The variation of human tolerance to impact and its effect on the design and testing of automotive impact performance. Society of Automotive Engineers Paper No. 780885, presented at the 22nd Stapp Car Crash Conference. [9.18.3]

Searle, S R (1971). Linear Models. New York: Wiley. [7.8.7]

Seheult, A (1978). A note on stand structure of trees. Biometrics, 34, 522–523. [24.7.2—5]

Senn, S J, and Collie, G S (1988). Accident blackspots and the bivariate negative binomial. Traffic Engineering and Control, 29, 168–169. [24.6—3]

Seshadri, V (1987). Contribution to Discussion of paper by Jørgensen. Journal of the Royal Statistical Society, Series B, 49, 156. [8.9]

Seshadri, V, and Patil, G P (1963). A characterization of a bivariate distribution by the marginal and the conditional distributions of the same component. Annals of the Institute of Statistical Mathematics, Tokyo, 15, 215–221. [fn1.4, 8.1.4]

Shah, S M, and Parikh, N T (1964). Moments of singly and doubly truncated standard bivariate normal distribution. Vidya (Gujarat University), 7, 82–91. [4.16.1]

Shaked, M (1977). A family of concepts of dependence for bivariate distributions. Journal of the American Statistical Association, 72, 642–650. [12.2.1, 12.3.5, 12.3.6]

Shaked, M (1979). Some concepts of positive dependence for bivariate interchangeable distributions. Annals of the Institute of Statistical Mathematics, Tokyo, 31, 67–84. [12.2.1, 12.3.7]

Shaked, M (1982). A general theory of some positive dependence notions. Journal of Multivariate Analysis, 12, 199–218. [12.2.1]

Shaked, M (1983). Geometry in statistics: Convexity. In Encyclopedia of Statistical Sciences. Volume 3, pp. 419–424. New York: Wiley. [2.15.2—11, 18.2.4]

Shaked, M (1985). Ordering distributions by dispersion. In Encyclopedia of Statistical Sciences. Volume 6, pp. 485–490. New York: Wiley. [2.15.2—31]

Shaked, M (1988). Variance dilation. In Encyclopedia of Statistical Sciences. Volume 9, pp. 472–473. New York: Wiley. [2.15.2—66]

Shaked, M, and Shanthikumar, J G (1985). Phase type distributions. In Encyclopedia of Statistical Sciences. Volume 6, pp. 709–715. New York: Wiley. [2.15.1—35, 9.13]

Shaked, M, and Shanthikumar, J G (1986). IFRA processes. In

A P Basu (Editor), Reliability and Quality Control, pp. 345–352. Amsterdam: North–Holland. [18.3.3]

Shaked, M, and Tong, Y L (1988). Inequalities for probability contents of convex sets via geometric average. Journal of Multivariate Analysis, 24, 330–340. [7.2.10]

Shamseldin, A A, and Press, S J (1984). Bayesian parameter and reliability estimation for a bivariate exponential distribution. Parallel sampling. Journal of Econometrics, 24, 363–378. [9.2.7]

Shanbhag, D N, and Kotz, S (1987). Some new approaches to multivariate probability distributions. Journal of Multivariate Analysis, 22, 189–211. [18.3.2]

Shapiro, S S, and Wilk, M B (1965). An analysis of variance test for normality (complete samples). Biometrika, 52, 591–611. [4.13.4.5, 4.13.5.4]

Shea, B L (1988). Algorithm AS 239. Chi-squared and incomplete gamma integral. Applied Statistics, 37, 466–473. [fn2.2]

Shea, G A (1983). Hoeffding's lemma. In Encyclopedia of Statistical Sciences. Volume 3, pp. 648–649. New York: Wiley. [12.2.2]

Sheikh, A K, Ahmed, M, and Zulfiqar Ali (1987). Some remarks on the hazard functions of the inverted distributions. Reliability Engineering, 19, 255–261. [2.3.5]

Shenton, L R, and Bowman, K O (1977). A bivariate model for the distribution of $\sqrt{b_1}$ and b_2. Journal of the American Statistical Association, 72, 206–211. [8.12.2]

Shephard, R W (1989). Letter to the Editor. Journal of the Operational Research Society, 40, 113–114. [19.8.9]

Sherman, A B, Koss, L G, and Adams, S E (1984). Interobserver and intraobserver differences in the diagnosis of urothelial cells. Comparison with classification by computer. Analytical and Quantitative Cytology, 6, 112–120. [24.4.2—4]

Sherriffs, A C, and Boomer, D S (1954). Who is penalized by the penalty for guessing? Journal of Educational Psychology, 45, 81-90. [23.5]

Sherwood, H, and Taylor, M D (1988). Doubly stochastic measures with hairpin support. Probability Theory and Related Fields, 78, 617–626. [13.11]

Shum, K T, and Melville, W K (1984). Estimates of the joint statistics of amplitudes and periods of ocean waves using an integral transform technique. Journal of Geophysical Research, 89, 6467–6476. [24.12.2—8]

Sibuya, M (1960). Bivariate extreme statistics, I. Annals of the Institute of Statistical Mathematics, Tokyo, 11, 195–210. [9.17.1]

Siegel, A F (1988). Zero degrees of freedom. In Encyclopedia of Statistical Sciences. Volume 9, pp. 672–674. New York: Wiley. [2.15.1—55]

Siegel, J B (1985). Statistical Software for Microcomputers. A Guide to 40 Programs. New York: North-Holland. [4.12.8]

Sievers, G L (1986). Probability plotting. In Encyclopedia of Statistical Sciences. Volume 7, pp. 232–237. New York: Wiley. [2.15.2—40, 4.13.3.1]

Silverman, B W (1981). Density estimation for univariate and bivariate data. In V Barnett (Editor), Interpreting Multivariate Data, pp. 37–53. Chichester: Wiley. [4.13.3.5]

Silverman, B W (1986). Density Estimation for Statistics and Data Analysis. London: Chapman and Hall. [4.13.3.5, 24.9.3—6]

Simon, G (1974). An explicit relationship between multidimensional scaling and principal components—Use of nonmetric multidimensional scaling to find principal components. Technical Report No. 62, Series 2, Department of Statistics, Princeton University. [24.9.1—2]

Simon, G (1977). A non-parametric test of total independence based on Kendall's tau. Biometrika, 64, 277–282. [24.7.1—4]

Simon, G A (1978). Efficacies of measures of association for ordinal contingency tables. Journal of the American Statistical Association, 73, 545–551. [24.11.2—1]

Simon, L E (1946). On the relation of instrumentation to quality control. Instruments, 19, 654–656. [24.4.1—2]

Singh, H, and Deshpande, J V (1985). On some new ageing properties. Scandinavian Journal of Statistics, 12, 213–220. [18.2.7]

Sivazlian, B D (1981a). On a multivariate extension of the gamma and beta distributions. SIAM (Society for Industrial and Applied Mathematics) Journal of Applied Mathematics, 41, 205–209. [6.6]

Sivazlian, B D (1981b). A class of multivariate distributions. Australian Journal of Statistics, 23, 251–255. [6.6]

Small, K A (1988). Discrete choice econometrics. (Review of "Discrete Choice Analysis: Theory and Application to Travel Demand" by M Ben-Akiva and S R Lerman, and "Qualitative Choice Analysis: Theory, Econometrics, and an Application" by K Train.) Journal of Mathematical Psychology, 32, 80–87. [22.3.3]

Small, N J H (1985). Multivariate normality, testing for. In Encyclopedia of Statistical Sciences. Volume 6, pp. 95–100. New York: Wiley. [4.13.1, 4.13.4.1]

Smirnov, N V, and Bol'shev, L N (1962). Tables for Evaluating a Function of a Bivariate Normal Distribution. (In Russian.) Moscow: Is'datel'stov Akademii Nauk SSSR. [4.12.7]

Smith, A F M, and Spiegelhalter, D J (1981). Bayesian approaches to multivariate structure. In V Barnett (Editor), Interpreting Multivariate Data, pp. 335–348. Chichester: Wiley. [24.9.3—8]

Smith, C A B (1947). Some examples of discrimination. Annals of Eugenics, 13, 272–282. [24.9.1—7]

Smith, K W (1976). Marginal standardization and table shrinking: Aids in the traditional analysis of contingency tables. Social Forces, 54, 669–693. [24.11.2—8]

Smith, O E, and Adelfang, S I (1981). Gust model based on the bivariate gamma probability distribution. Journal of Spacecraft and Rockets, 18, 545–549. [7.5.13, 24.12.3—1]

Smith, O E, Adelfang, S I, and Tubbs, J D (1982). A bivariate gamma probability distribution with application to gust modeling. NASA Technical Memorandum 82483, Marshall Space Flight Center, Alabama. [7.5.11, 7.7.7, 7.7.9, 7.7.10, 7.7.11, 7.7.12, 24.12.2—1]

Smith, R L (1985). Statistics of extreme values. Bulletin of the International Statistical Institute, 51, Book 4, Paper 26 (Discussion, Book 5, 185–192). [9.17.1]

Smith, R L (1986). Extreme value theory based on the r largest annual events. Journal of Hydrology, 86, 27–43. [9.17.11]

Smith, W (1983). Inequalities for bivariate distributions with $X \leq Y$ and marginals given. Communications in Statistics—Theory and Methods, 12, 1371–1379. [3.1.2, 19.3.3, 24.8.1—6]

Smith, W L (1983). Generating functions. In Encyclopedia of Statistical Sciences. Volume 3, pp. 372–376. New York: Wiley. [2.15.2—9]

Smith, W P (1983). A bivariate normal test for elliptical home-range models: Biological implications and recommendations. Journal of Wildlife Management, 47, 611–619. [19.3.3]

Snedecor, G W, and Cochran, W G (1967). Statistical Methods. 6th Edition. Ames, Iowa: Iowa State University Press. [24.7.2—8, 24.7.2—17]

Snee, R D (1974). Graphical display of two-way contingency tables. The American Statistician, 28, 9–12. [24.9.2—7]

Snee, R D, and Pfeifer, C G (1983). Graphical representation of data. In Encyclopedia of Statistical Sciences. Volume 3, pp. 488–511. New York: Wiley. [24.9.2—7]

Snow, J W (1975). Association of proportions. Mathematical Geology (Journal of the International Association for Mathematical Geology), 7, 63–73. [24.5.2—2, 24.5.2—6]

Snyder, W M, and Thomas, A W (1987). Stochastic impacts on farming: IV. A form-free bivariate distribution to generate inputs to agricultural models. Transactions of the American Society of Agricultural Engineers, 30, 946–952. [24.12.3—6, 24.12.3—7]

Sogawa, N, Araki, M, and Imai, T (1986). Studies on multivariate conditional maximum entropy distribution and its characteristics. Journal of Hydroscience and Hydraulic Engineering, 4(1), 79–97. [8.12.5.7]

Sogawa, N, Araki, M, and Terashima, A (1984). Study on bivariate MEP distribution and its characteristics. (In Japanese.) Proceedings of the 28th Japanese Conference on Hydraulics, pp. 397–402. [8.12.5.7]

Sokal, R R, and Rohlf, F J (1969). Biometry. The Principles and Practice of Statistics in Biological Research. San Francisco: Freeman. [24.7.1—2, 24.8.1—3, 24.8.1—4, 24.8.1—5, 24.8.2—3, 24.8.2—4]

Sokolov, A A, Rantz, S E, and Roche, M (1976). Floodflow Computation. Methods Compiled from World Experience. Paris: UNESCO Press. [2.6.5]

Sowden, R R, and Ashford, J R (1969). Computation of bi-variate normal integral. Applied Statistics, 18, 169–180. [4.12.4]

Spanier, J, and Oldham, K B (1987). An Atlas of Functions. Washington, D.C.: Hemisphere, and Berlin: Springer. [1.5.4, fn9.4]

Spearman, C (1904). The proof and measurement of association between two things. American Journal of Psychology, 15, 72–101. [11.4.2]

Springer, M D (1979). The Algebra of Random Variables. New York: Wiley. [2.12.2, 4.9]

Springer, M D (1983). H-function distribution. In Encyclopedia of Statistical Sciences. Volume 3, pp. 619–621. New York: Wiley. [2.15.1—11, 10.6]

Springer, M D (1983). Integral transforms. In Encyclopedia of Statistical Sciences. Volume 4, pp. 169–176. New York: Wiley. [2.15.1—11, 2.15.2—15]

Spurrier, J D, and Weier, D R (1981). Bivariate survival model derived from a Weibull distribution. IEEE (Institute of Electrical and Electronics Engineers) Transactions on Reliability, R-30, 194–197. [9.3.12]

Srokosz, M A, and Challenor, P G (1987). Joint distributions of wave height and period: A critical comparison. Ocean Engineering, 14, 295–311. [10.8]

Srole, L, Langner, T S, Michael, S T, Opler, M K, and Rennie, T A C (1962). Mental Health in the Metropolis. The Midtown Manhattan Study. New York: McGraw-Hill. Revised edition, 1974—New York: NYU Press. [24.11.2—4]

Stacy, E W (1962). A generalization of gamma distribution. Annals of Mathematical Statistics, 33, 1187–1192. [2.6.5, 3.5]

Stănescu, V, and Catană, C (1974). Statistical analysis applied to the study of the components of a flood wave. In Mathematical Models in Hydrology Symposium, pp. 253–265. IAHS Publication No. 100, International Association of Hydrological Sciences. [19.7.8]

Stark, J D (1985). A macroinvertebrate community index of water quality for stony streams. Water and Soil Miscellaneous Publication No. 87, National Water and Soil Conservation Authority, New Zealand. [24.12.1—1]

Steck, G P (c. 1958). Limit theorems for conditional distributions. University of California Publications in Statistics, 2, 237–284. [24.13.1—6]

Steece, B M (1977). Integral transforms in statistics. Estadística, 31, 349–358. [fn20.7]

Steel, S J, and le Roux, N J (1987). A reparameterisation of a bivariate gamma extension. Communications in Statistics—Theory and Methods, 16, 293–305. [9.3.14]

Steele, J M (1985). MACSYMA as a tool for statisticians. American Statistical Association, 1985 Proceedings of the Statistical Computing Section, pp. 1–4. [3.1.4]

Steffensen, J F (1922). A correlation-formula. Skandinavisk Aktuarietidskrift, 5, 73–91. [24.6—12]

Steiner, A (1958). Petrogenetic implications of the 1954 Ngauruhoe lava and its xenoliths. New Zealand Journal of Geology and Geophysics, 1, 325–363. [24.5.2—2]

Stephens, M A (1982). EDF statistics. In Encyclopedia of Statistical Sciences. Volume 2, pp. 451–455. New York: Wiley. [2.15.2—7]

Stephens, M A (1986a). Tests based on EDF statistics. In R B D'Agostino and M A Stephens (Editors), Goodness-of-Fit Techniques, pp. 97–193. New York: Dekker. [4.13.4.4]

Stephens, M A (1986b). Tests based on regression and correlation. In R B D'Agostino and M A Stephens (Editors), Goodness-of-Fit Techniques, pp. 195–233. New York: Dekker. [4.13.4.5]

Stephens, M A (1986c). Tests for the exponential distribution. In R B D'Agostino and M A Stephens (Editors), Goodness-of-Fit Techniques, pp. 421–459. New York: Dekker. [fn48.1]

Stephens, M A (1988). Uniformity, tests of. In Encyclopedia of Statistical Sciences. Volume 9, pp. 402–410. New York: Wiley. [2.15.2—62]

Steutel, F W (1983). Infinite divisibility. In Encyclopedia of Statistical Sciences. Volume 4, pp. 114–116. New York: Wiley. [2.15.2—14]

Steutel, F W (1985). Log-concave and log-convex distributions. In Encyclopedia of Statistical Sciences. Volume 5, pp. 116–117. New York: Wiley. [2.15.2—23, 18.2.4]

Stone, G C (1978). Statistical analysis of accelerated aging tests on solid electrical insulation. M.A.Sc. Thesis, University of Waterloo. [24.2—3]

Storr, P A, Cooper, D F, McDowell, M R C (1980). Analysis of gap acceptance in a complex traffic manoeuvre. European Journal of Operational Research, 5, 94–101. [22.2.3]

Stout, W (1982). Domain of attraction. In Encyclopedia of Statistical Sciences. Volume 2, pp. 416–418. New York: Wiley. [2.15.2—6]

Strauss, D J (1985). Luce's choice axiom and generalizations. In Encyclopedia of Statistical Sciences. Volume 5, pp. 167–170. New York: Wiley. [2.15.2—25]

Strauss, D J (1986). Random utility models. In Encyclopedia of Statistical Sciences. Volume 7, pp. 572–573. New York: Wiley. [2.15.2—45]

Stuart, A (1953). The estimation and comparison of strengths of association in contingency tables. Biometrika, 40, 105–110. [24.9.2—8]

Stuart, A (1955). A test for homogeneity of the marginal distributions in a two-way classification. Biometrika, 42, 412–416. [24.9.2—8]

Subrahmaniam, K, and Gajjar, A V (1980). Robustness to non-normality of some transformations of the sample correlation coefficient. Journal of Multivariate Analysis, 10, 60–77. [3.11.2]

Subrahmaniam, K, and Subrahmaniam, K (1973). On the estimation of the parameters in the bivariate negative binomial distribution. Journal of the Royal Statistical Society, Series B, 35, 131–146. [24.6—2]

Sugasaw, Y, and Kaji, I (1981). Light maintenance for a two-unit parallel redundant system with bivariate exponential lifetimes. Microelectronics and Reliability, 21, 661–670. [9.2.7]

Sumita, U, and Kijima, M (1985). The bivariate Laguerre transform and its applications: Numerical exploration of bivariate processes. Advances in Applied Probability, 17, 683–708. [15.2]

Sutradhar, B C (1986). On the characteristic function of multivariate Student *t*-distribution. Canadian Journal of Statistics, 14, 329–337. [6.2.5]

Suzuki, M (1983). Estimation in a bivariate semi-lognormal distribution. Behaviormetrika, 13, 59–68. [4.14.2.1]

Suzuki, M, Iwase, K, and Shimizu, K (1984). Uniformly minimum variance unbiased estimation in a semi-lognormal distribution. Journal of the Japan Statistical Society, 14, 63–68. [4.14.2.1]

Svalastoga, K (1959). Prestige, Class, and Mobility. London: Heinemann. [24.11.2—1]

Swensson, R G (1980). A two-stage detection model applied to skilled visual search by radiologists. Perception and Psychophysics, 27, 11–16. [23.3]

Swets, J A (1959). Indices of signal discriminability obtained with various psychophysical procedures. Journal of the Acoustical Society of America, 31, 511–513. [23.1.5.2]

Swets, J A (1963). Information retrieval systems. Science, 141, 245–250. [23.4]

Swets, J A (1964) (Editor). Signal Detection and Recognition by Human Observers. New York: Wiley. [23.1.1]

Swets, J A (1973). The Relative Operating Characteristic in psychology. Science, 182, 990–1000. [23.1.3]

Swets, J A, and Green, D M (1978). Applications of signal detection theory. In H L Pick, H W Leibowitz, J E Singer, A Steinschneider, and H W Stevenson (Editors), Psychology: From Research to Practice, pp. 311–331. New York: Plenum Press. [23.1.1]

Swets, J A, Tanner, W P, and Birdsall, T G (1961). Decision processes in perception. Psychological Review, 68, 301–340. [23.1.3, 23.1.5.3]

Symons, F, and Ringele, A (1976). Study of time-related variability within the genus *Astarte* (bivalvia): A multivariate approach. Mathematical Geology (Journal of the International Association for Mathematical Geology), 8, 113–136. [24.7.1—5]

Szegö, G (1975). Orthogonal Polynomials. 4th Edition. Providence, Rhode Island: American Mathematical Society. [T14.1, fn14.7]

Székely, G J (1986). Paradoxes in Probability Theory and Mathematical Statistics. Dordrecht: Reidel. [8.11]

T

Tadikamalla, P R (1978a). Computer generation of gamma variables. Communications of the Association for Computing Machinery, 21, 419–422. [16.2.4, 16.3.2]

Tadikamalla, P R (1978b). Computer generation of gamma random variables—II. Communications of the Association for Computing Machinery, 21, 925–928. [16.3.2]

Tadikamalla, P R (1980). On simulating nonnormal distributions. Psychometrika, 45, 273–279. [16.5.4.4]

Tadikamalla, P R (1984). Modeling and generating stochastic inputs for simulation studies. American Journal of Mathematical and Management Sciences, 4, 203–223. [16.3.5]

Tadikamalla, P R, and Johnson, M E (1978). A survey of methods for sampling from the gamma distribution. In H J Highland, N R Nielsen, and L G Hull (Editors), 1978 Winter Simulation Conference, pp. 130–134. New York: Institute of Electrical and Electronics Engineers. [16.3.2]

Tadikamalla, P R, and Johnson, M E (1981). A complete guide to gamma variate generation. American Journal of Mathematical and Management Sciences, 1, 213–236. [16.3.2]

Tadikamalla, P R, and Johnson, N L (1982). Systems of frequency curves generated by transformations of logistic variables. Biometrika, 69, 461–465. [2.7.1]

Taguchi, T (1978). On a generalization of Gaussian distribution. Annals of the Institute of Statistical Mathematics, Tokyo, 30, 211–242 (Corrections and Supplements, 473–475). [7.4.2, 7.4.4]

Taillie, C, Ord, J K, Mosimann, J E, and Patil, G P (1979). Chance mechanisms underlying multivariate distributions. In J K Ord, G P Patil, and C Taillie (Editors), Statistical Distributions in Ecological Work, pp. 157–191. Fairland, Maryland: International Cooperative Publishing House. [24.6—8]

Takahasi, K (1965). Note on the multivariate Burr's distribution. Annals of the Institute of Statistical Mathematics, Tokyo, 17, 257–260. [5.8.12, 6.8]

Tallis, G M (1962). The maximum likelihood estimation of correlation from contingency tables. Biometrics, 18, 342–353. [24.6—6]

Talwalker, S (1975). Models in medicine and toxicology. In G P Patil, S Kotz, and J K Ord (Editors), A Modern Course on Distributions in Scientific Work. Volume 2—Model Building and Model Selection, pp. 263–274. Dordrecht: Reidel. [24.6—17]

Tan, W Y (1978). Some aspects of distribution theories in quantitative genetics, I. Some general results. Biometrical Journal, 20, 779–793. [19.5.4]

Tanabe, K, Sagae, M, and Ueda, S (1988). BNDE, Fortran subroutines for computing Bayesian nonparametric univariate and bivariate density estimator. Computer Science Monograph No. 24, Institute of Statistical Mathematics, Tokyo. [4.13.3.5]

Tarter, M, and Silvers, A (1975). Implementation and applications of a bivariate Gaussian mixture decomposition. Journal of the American Statistical Association, 70, 47–55. [4.17.1, 19.4.3]

Tawn, J A (1988a). Bivariate extreme value theory: Models and estimation. Biometrika, 75, 397–415. [9.17.1, 9.17.3, 9.17.4, 9.17.7.1, 9.17.8, 9.17.11, 24.3—3]

Tawn, J A (1988b). An extreme-value theory for dependent observations. Journal of Hydrology, 101, 227–250. [9.17.11]

Tayfun, M A (1979). Joint occurrences in coastal flooding. Journal of the Waterway, Port, Coastal and Ocean Division, Proceedings of the American Society of Civil Engineers, 105, 107–123. [19.7.9]

Taylor, S E, and Bender, D E (1988). Simulating correlated lumber properties using a modified multivariate normal approach. Transactions of the American Society of Agricultural Engineers, 31, 182–186. [19.8.3]

Tchen, A H (1980). Inequalities for distributions with given marginals. Annals of Probability, 8, 814–827. [11.8.4, 12.6]

Teichmann, T (1986). Joint probabilities of partially coupled events. Reliability Engineering, 14, 133–148. [5.2.11]

Thayer, T A, Lipow, M, and Nelson, E C (1978). Software Reliability. A Study of Large Project Reality. Amsterdam: North-Holland. [24.14.2]

Thesen, A (1985). An efficient generator of uniformly distributed random variates between zero and one. Simulation, 44, 17–22. [16.3.7]

Thesen, A, Sun, Z, and Wang, T-J (1984). Some efficient random number generators for micro-computers. In S Sheppard, U W Pooch, and C D Pegden (Editors), 1984 Winter Simulation Conference Proceedings, pp. 186–196. Piscataway, New Jersey: Institute of Electrical and Electronics Engineers. [16.3.7]

Thomas, J B (1978). Representation of bivariate distribution with applications. In J K Skwirzynski (Editor), Communication Systems and Random Process Theory, pp. 541–560. Alphen aan den Rijn:

Sijthoff and Noordhoff. [14.8]

Thomopoulos, N T, and Longinow, A (1984). Bivariate lognormal probability distribution. Journal of Structural Engineering, 110, 3045-3049. [4.14.1, 19.8.2]

Thompson, R N, Esson, J, and Dunham, A C (1972). Major element chemical variation in the Eocene lavas of the Isle of Skye, Scotland. Journal of Petrology, 13, 219–253. [24.5.2—3]

Thompson, W A (1963). Precision of simultaneous measurement procedures. Journal of the American Statistical Association, 58, 474–479. [24.4.1—2]

Thornton, E B, and Schaeffer, G (1979). Probability density functions of breaking waves. Proceedings of the Sixteenth Coastal Engineering Conference. Volume I, pp. 507–519. New York: American Society of Civil Engineers. [24.12.2—8]

Tiago de Oliveira, J (1961). La représentation des distributions extrêmales bivariées. Bulletin de l'Institut International de Statistique, 39, Book 2, 477–480. [9.17.1]

Tiago de Oliveira, J (1962/63). Structure theory of bivariate extremes, extensions. Estudos de Matematica, Estatistica, e Economicos, 7, 165–195. [9.17.1]

Tiago de Oliveira, J (1975a). Bivariate and multivariate extreme distributions. In G P Patil, S Kotz, and J K Ord (Editors), A Modern Course on Distributions in Scientific Work. Volume 1—Models and Structures, pp. 355–361. Dordrecht: Reidel. [9.17.1]

Tiago de Oliveira, J (1975b). Bivariate extremes; Extensions. Bulletin of the International Statistical Institute, 46, Book 2, 241–252 (Discussion, 253–254). [9.17.1, 9.17.8, 9.17.11]

Tiago de Oliveira, J (1980). Bivariate extremes: Foundations and statistics. In P R Krishnaiah (Editor), Multivariate Analysis—V, pp. 349–366. Amsterdam: North-Holland. [9.17.1, 9.17.8, 9.17.11]

Tiago de Oliveira, J (1983). Gumbel distribution. In Encyclopedia of Statistical Sciences. Volume 3, pp. 552–558. New York: Wiley. [2.15.1—10]

Tiago de Oliveira, J (1984). Bivariate models for extremes; Statistical decision. In J Tiago de Oliveira (Editor), Statistical Extremes and Applications, pp. 131–153. Dordrecht: Reidel. [9.17.1, 9.17.8]

Tiao, G G, and Guttman, I (1965). The inverted Dirichlet distribution with applications. Journal of the American Statistical Association, 60, 793–805 (Correction, 60, 1251–1252). [6.3.11]

Tienstra, J M (1956). Theory of the Adjustment of Normally Distributed Observations. Amsterdam: N V Uitgeverij "Argus". [24.7.2—3]

Tiku, M (1985a). Noncentral chi-square distribution. In Encyclopedia of Statistical Sciences. Volume 6, pp. 276–280. New York: Wiley. [2.15.1—26]

Tiku, M (1985b). Noncentral F-distribution. In Encyclopedia of Statistical Sciences. Volume 6, pp. 280–284. New York: Wiley. [2.15.1—27]

Tippett, L H C (1952a). Technological Applications of Statistics. London: Williams and Norgate. [24.13.1—5]

Tippett, L H C (1952b). The Methods of Statistics. 4th Edition. New York: Wiley. [24.7.2—4]

Titterington, D M, Smith, A F M, and Makov, U E (1985). Statistical Analysis of Finite Mixture Distributions. New York: Wiley. [4.17.1, 24.7.1—3]

Tocher, J F (1908). Pigmentation survey of school children in Scotland. Biometrika, 6, 129–235, plus an Appendix of 67 pages and 26 plates issued as a Supplement to Biometrika. [24.9.2—6]

Tocher, J F (1928). An investigation of the milk yield of dairy cows. Being a statistical analysis of the data of the Scottish Milk Records Association for the years 1908, 1909, 1911, 1912, 1920 and 1923. Biometrika, 20B, 106–244. [24.8.2—2]

Tolley, H D, and Norman, J E (1979). Time on trial estimates with bivariate risks. Biometrika, 66, 285–291. [5.2.9, 9.15]

Tomizawa, S (1985). The decompositions for point symmetry models in two-way contingency tables. Biometrical Journal, 27, 895–905. [24.9.2—8]

Tomizawa, S (1986). Four kinds of symmetry models and their decompositions in a square contingency table with ordered categories. Biometrical Journal, 28, 387–393. [24.9.2—8]

Tong, H (1974). A note on the estimation of $\Pr(Y < X)$ in the exponential case. Technometrics, 16, 625 (Correction, 17, 395). [17.6.1, 17.6.3]

Tong, Y L (1980). Probability Inequalities in Multivariate Distributions. New York: Academic Press. [4.9, 6.2.10, 7.2.10, 11.5.4, 12.2.1]

Tosch, T J, and Holmes, P T (1980). A bivariate failure model. Journal of the American Statistical Association, 75, 415–417. [9.9, 15.6.4]

Train, K (1986). Qualitative Choice Analysis: Theory, Econometrics, and an Application. Cambridge, Massachusetts: MIT Press. [22.3.3]

Treisman, M (1977). On the stability of d_s. Psychological Bulletin, 84, 235–243. [23.2]

Treisman, M, and Faulkner, A (1984). The effect of signal probability on the slope of the receiver operating characteristic given by the rating procedure. British Journal of Mathematical and Statistical Psychology, 37, 199–215. [23.2]

Treloar, A E (1939). Elements of Statistical Reasoning. New York: Wiley. [24.9.2—1]

Truett, J, Cornfield, J, and Kannel, W (1967). A multivariate analysis of the risk of coronary heart disease in Framingham. Journal of Chronic Diseases, 20, 511–524. [19.4.1]

Tubbs, J D (1983a). A method for determining if unequal shape parameters are necessary in a bivariate gamma distribution. Chapter II of J D Tubbs, D W Brewer, and O E Smith, Some Properties of a Five-Parameter Bivariate Probability Distribution. NASA Technical Memorandum 82550, Marshall Space Flight Center, Alabama. [7.7.11]

Tubbs, J D (1983b). Analysis of wind gust data. Chapter IV of J D Tubbs, D W Brewer, and O E Smith, Some Properties of a Five-Parameter Bivariate Probability Distribution. NASA Technical Memorandum 82550, Marshall Space Flight Center, Alabama. [7.7.7, 7.7.9, 24.12.3—1]

Tubbs, J D, and Smith, O E (1985). A note on the ratio of positively correlated gamma variates. Communications in Statistics—Theory and Methods, 14, 13–23. (Also appears as Chapter I of J D Tubbs, D W Brewer, and O E Smith (1983), Some Properties of a Five-Parameter Bivariate Probability Distribution. NASA Technical Memorandum 82550, Marshall Space Flight Center, Alabama.) [7.7.11]

Tukey, J W (1977). Exploratory Data Analysis. Reading, Massachusetts: Addison-Wesley. [9.11.8]

Tukey, P A (1983). Graphical methods. In Statistical Data Analysis (Proceedings of Symposia in Applied Mathematics, 28), pp. 8–48. Providence, Rhode Island: American Mathematical Society. [24.7.1—3]

Tukey, P A, and Tukey, J W (1981a). Preparation; Prechosen sequences of views. In V Barnett (Editor), Interpreting Multivariate Data, pp. 189–213. Chichester: Wiley. [24.7.1—3]

Tukey, P A, and Tukey, J W (1981b). Data-driven view selection; Agglomeration and sharpening. In V Barnett (Editor), Interpreting Multivariate Data, pp. 215–243. Chichester: Wiley. [4.13.3.4, 4.17.6, 24.7.1—3]

Tuller, S E, and Brett, A C (1984). Characteristics of wind velocity that favor the fitting of a Weibull distribution in windspeed analysis. Journal of Climate and Applied Meteorology, 23, 124–134. [19.7.1]

Turner, H N, Hayman, R H, and Prunster, R W (1958). Repeatability of twin births. Proceedings of the Australian Society of Animal Production, 2, 106–107. [24.6—6]

Tyan, S, Derin, H, and Thomas, J B (1976). Two necessary conditions on the representation of bivariate distributions by polynomials. Annals of Statistics, 4, 216–222. [14.6.3, 14.6.4]

Tyan, S, and Thomas, J B (1975). Characterization of a class of bivariate distribution functions. Journal of Multivariate Analysis, 5, 227–235. [14.6.4, 14.8]

U

Ulrich, G (1984). A class of multivariate distributions with applications in Monte Carlo and simulation. American Statistical Association, 1984 Proceedings of the Statistical Computing Section, pp. 185–188. [4.13.5.8, 6.5.14, 10.2.4]

Ulrich, G, and Chen, C-C (1987). A bivariate double exponential distribution and its generalizations. American Statistical Association, 1987 Proceedings of the Statistical Computing Section, pp. 127–129. [9.12.16]

Ulrich, G, and Watson, L T (1987). A method for computer generation of variates from arbitrary continuous distributions. SIAM (Society for Industrial and Applied Mathematics) Journal on Scientific and Statistical Computing, 8, 185–197. [16.2.2]

UNSC and ECE (1984). Statistics of Air Quality: Some Methods. Conference of European Statisticians Statistical Standards and Studies, No. 49. Geneva: United Nations Statistical Commission and Economic Commission for Europe. [24.12.1—3]

Upton, G J G (1985). A survey of loglinear models for ordinal variables in an $I \times J$ contingency table. Guru Nanak Journal of Sociology, 6, 1–18. [24.6—13, 24.9.2—8, 24.11.2—4]

V

Văduva, I (1977). On computer generation of gamma random variables by rejection and composition procedures. Mathematische Operationsforschung und Statistik, Series Statistics, 8, 545–576. [16.3.2]

Văduva, I (1985). Computer generation of random vectors based on transformation of uniformly distributed vectors. Proceedings of the Seventh Conference on Probability Theory, pp. 589–598. Bucarest: Editura Academiei, and Utrecht: VNU Science Press. [6.2.9, 6.5.10, 16.5.1]

Vale, C D, and Maurelli, V A (1983). Simulating multivariate nonnormal distributions. Psychometrika, 48, 465–471. [4.14.5, 16.5.4.4]

van Belle, G, and Cornell, R G (1971). Strengthening tests of symmetry in contingency tables. Biometrics, 27, 1074–1078. [24.4.2—1]

van der Vaart, H R (1973). On the existence of bivariate moments of lower order given the existence of moments of higher order. Statistica Neerlandica, 27, 97–102. [fn2.1]

van Praag, B M S, and Wesselman, B M (1989). Elliptical multivariate analysis. Journal of Econometrics, 41, 189–203. [7.2.1]

van Uven, M J (1947a). Extension of Pearson's probability distributions to two variables—I. Proceedings of the Koninklijke Nederlandse Akademie van Wetenschappen, 50, 1063–1070 (Indagationes Mathematicae, 9, 477–484). [3.9.1.1]

van Uven, M J (1947b). Extension of Pearson's probability distributions to two variables—II. Proceedings of the Koninklijke Nederlandse Akademie van Wetenschappen, 50, 1252–1264 (Indagationes Mathematicae, 9, 578–590). [3.9.1.1]

van Uven, M J (1948a). Extension of Pearson's probability distributions to two variables—III. Proceedings of the Koninklijke Nederlandse Akademie van Wetenschappen, 51, 41–52 (Indagationes Mathematicae, 10, 12–23). [3.9.1.1]

van Uven, M J (1948b). Extension of Pearson's probability distributions to two variables—IV. Proceedings of the Koninklijke Nederlandse Akademie van Wetenschappen, 51, 191–196 (Indagationes Mathematicae, 10, 62–67). [3.9.1.1]

van Yzeren, J (1972). A bivariate distribution with instructive properties as to normality, correlation and dependence. Statistica Neerlandica, 26, 55–56. [4.18.1]

van Zyl, J M (1987). Planar random movement and the bivariate normal density. South African Statistical Journal, 21, 1–12. [4.10]

Vere-Jones, D (1967). The infinite divisibility of a bivariate gamma distribution. Sankhyā, Series A, 29, 421–422. [7.5.6, 17.3.5]

Vessereau, A (1948). Methodes Statistiques en Biologie et en Agronomie. Paris: Bailliere. [24.5.2—7]

Votaw, D F (1936). The effect of do-not-guess directions upon the validity of true-false or multiple-choice tests. Journal of Educational Psychology, 27, 698–703. [23.5]

Vrijling, J K (1987). Probabilistic design of water-retaining structures. In L Duckstein and E J Plate (Editors), Engineering Reliability and Risk in Water Resources, pp. 115–134. Dordrecht: Nijhoff. [8.12.5.4]

W

Wachter, K W (1983). Haar distributions. In Encyclopedia of Statistical Sciences. Volume 3, pp. 559–562. New York: Wiley. [4.13.3.6]

Wahrendorf, J (1980). Inference in contingency tables with ordered categories using Plackett's coefficient of association for bivariate distributions. Biometrika, 67, 15–21. [5.7.10, 24.6—13, 24.11.2—1, T24.43]

Wahrendorf, J, and Brown, C C (1980). Bootstrapping a basic inequality in the analysis of joint action of two drugs. Biometrics, 36, 653–657. [fn22.2]

Waite, H (1911). The teacher's estimation of the general intelligence of school children. Biometrika, 8, 79–93. [24.4.2—15]

Waite, H (1915). Association of finger-prints. Biometrika, 10, 421–478. [24.6—19]

Walker, A J (1974a). New fast method for generating discrete random numbers with arbitrary frequency distributions. Electronics Letters, 10, 127–128. [16.2.4]

Walker, A J (1974b). Fast generation of uniformly distributed pseudo-random numbers with floating point representation. Electronics Letters, 10, 553–554. [16.2.4]

Walker, H M (1943). Elementary Statistical Methods. New York: Holt. [24.10.1]

Wang, Y (1987). The probability integrals of bivariate normal distributions: A contingency table approach. Biometrika, 74, 185–190. [4.12.2]

Warren, E (1903). A preliminary attempt to ascertain the relationship between the size of cell and the size of body in *Daphnia magna* straus. Biometrika, 2, 255–259. [24.7.2—12]

Warren, W G (1979). Some recent developments relating to statistical distributions in forestry and forest products research. In J K Ord, G P Patil, and C Taillie (Editors), Statistical Distributions in Ecological Work, pp. 247–250. Fairland, Maryland: International Co-operative Publishing House. [9.16, 19.3.2]

Wasan, M T (1968). On an inverse Gaussian process. Skandinavisk Aktuarietidskrift, 1968, 69–96. [7.9.8]

Wasan, M T, and Buckholtz, P (1973). Differential representation of a bivariate inverse Gaussian process. Journal of Multivariate Analysis, 3, 243–247. [7.9.8]

Watson, D M S, Watson, K M, Pearson, H S, Karn, M N, Irwin, J O, and Pearson, K (1923). On the nest and eggs of the common tern (*S. fluviatilis*). A third cooperative study. Biometrika, 15, 294–345. [24.7.2—4]

Waugh, A E (1952). Elements of Statistical Method. 3rd Edition. New York: McGraw-Hill. [24.7.2—10]

Weaver, H (1955). The value of the Oort A parameter derived from radial velocities of cepheid variable stars. Astronomical Journal, 60, 202–208. [24.13.1—7]

Weiss, G H (1988). Saddle point approximations. In Encyclopedia of Statistical Sciences. Volume 8, pp. 231–234. New York: Wiley. [2.15.2—50]

Weissman, I (1985). Multidimensional extreme value theory (Discussion). Bulletin of the International Statistical Institute, 51, Book 5, 187–190. [9.17.3]

Weldon, W F R (1901). Change in organic correlation of *Ficaria ranunculoides* during the flowering season. Biometrika, 1, 125–128. [24.6—10]

Wellner, J A (1985). Semiparametric models: Progress and problems. Bulletin of the International Statistical Institute, 51, Book 4, Paper 23.1 (Discussion, Book 5, 175–180). (Revised version in CWI (Centrum voor Wiskunde en Informatica, Amsterdam) Newsletter, No. 9, 2–24 (1985).) [21.7.1]

Westlund, K B, and Kurland, L T (1953). Studies on multiple sclerosis in Winnipeg, Manitoba, and New Orleans, Louisiana. I: Prevalence. Comparison between the patient groups in Winnipeg and New Orleans. American Journal of Hygiene, 57, 380–396. [20.2.1]

White, H C (1963). Cause and effect in social mobility tables. Behavioral Science, 7, 14–27. [24.11.2—1]

Whitehead, M E, Grieve, J H K, Payne, M J, and Ross, M S (1986). Quality assurance of histopathologic diagnosis in the British Army: Role of the Army Histopathology Registry in completed case review. Journal of the Royal Army Medical Corps, 132, 71–75. [24.4.2—2]

Whiteman, T C (1981). The measurement of nonfarm self-employment income: The 1979 ISDP research panel. American Statistical Association, 1981 Proceedings of the Section on Survey Research Methods, pp. 372–376. [24.4.2—18]

Whitt, W (1976). Bivariate distributions with given marginals. Annals of Statistics, 4, 1280–1289. [3.1.2]

Whitt, W (1988). Stochastic ordering. In Encyclopedia of Statistical Sciences. Volume 8, pp. 832–836. New York: Wiley. [2.15.2—57]

Wicksell, S D (1917). The correlation function of type A, and the regression of its characteristics. Kungliga Svenska Vetenskapsakademiens Handlinger, 58(3), 1–48. [24.7.2—1]

Wicksell, S D (1933). On correlation functions of type III. Biometrika, 25, 121–133. [7.5.6]

Wied, G L, Bibbo, M, Dytch, H E, and Bartels, P H (1984). Computer grading of cervical intraepithelial neoplastic lesions. I. Cytologic indices. Analytical and Quantitative Cytology and Histology, 7, 52–60. [24.4.3—6]

Wierenga, B (1974). Paired comparison product testing when individual preferences are stochastic. Applied Statistics, 23, 384–396. [20.5.1, T20.6]

Wilcox, R R (1985). Percentage points of the product of two correlated t variates. Communications in Statistics—Simulation and Computation, 14, 143–157. [6.2.10]

Wilcox, R R (1986). Percentage points of the bivariate t distribution for non-positive correlations. Metron, 44(1–4), 115–119. [6.2.12]

Wilks, S S (1948). Elementary Statistical Analysis. 10th Printing, 1958. Princeton, New Jersey: Princeton University Press. [24.10.2—7, 24.13.1—3, 24.13.1—4, 24.14.1—2]

Wilks, S S (1963). Mathematical Statistics. 2nd edition. New York: Wiley. [6.5.5, 6.5.7]

Willekens, E (1988). Subexponential distributions. In Encyclopedia of Statistical Sciences. Volume 9, pp. 53–54. New York: Wiley. [2.15.1—45]

Willett, P K, and Thomas, J B (1985). A simple bivariate density representation. In Proceedings of the 23rd Annual Allerton Conference on Communication, Control, and Computing, pp. 888–897. Urbana-Champaign: Coordinated Science Laboratory and Department of Electrical and Computer Engineering, University of Illinois. [5.2.14]

Willett, P K, and Thomas, J B (1987). Mixture models for underwater burst noise and their relationship to a simple bivariate density representation. IEEE (Institute of Electrical and Electronics Engineers) Journal of Oceanic Engineering, OE-12, 29–37. [5.2.14]

Williams, E J (1952). Use of scores for the analysis of association in contingency tables. Biometrika, 39, 274–289. [24.9.2—1]

Wilson, J R (1983). Fitting Johnson curves to univariate and multivariate data. In S Roberts, J Banks, and B Schmeiser (Editors), 1983 Winter Simulation Conference Proceedings. Volume 1, pp. 114–115. Piscataway, New Jersey: Institute of Electrical and Electronics Engineers. [19.9.4]

Wimp, J (1981). The computation of ${}_3F_2(1)$. International Journal of Computer Mathematics, 10, 55–62. [fn13.4]

Winer, A A, and Cossette, M (1979). The effect of aspect ratio on fiber counts: A preliminary study. In I J Selikoff and E C Hammond (Editors), Health Hazards of Asbestos Exposure, pp. 661–672. (Annals of the New York Academy of Sciences, 330.) [24.9.1—5]

Winkler, O W (1980). On triangular correlation. American Statistical Association, 1980 Proceedings of the Business and Economic Statistics Section, pp. 606–611. [24.10.3—1]

Wolfe, J E (1970). Pattern clustering by multivariate mixture analysis. Multivariate Behavioral Research, 5, 329–350. [4.17.1, 24.7.1—3]

Wong, E (1964). The construction of a class of stationary Markoff processes. In Stochastic Processes in Mathematical Physics and Engineering (Proceedings of Symposia in Applied Mathematics, 16), pp. 264–272. Providence, Rhode Island: American Mathematical Society. [3.9.2, 14.8]

Wong, E, and Thomas, J B (1962). On polynomial expansions of second-order distributions. SIAM (Society for Industrial and Applied Mathematics) Journal on Applied Mathematics, 10, 507–516. [3.9.2, T14.1, 14.6.3, 14.8]

Wright, S (1954). The interpretation of multivariate systems. In O Kempthorne, T A Bancroft, J W Gowen, and J L Lush (Editors), Statistics and Mathematics in Biology. Ames: Iowa State University Press. [24.7.2—9]

Wylie, W L (1944). Overbite and vertical facial dimensions in terms of muscle balance. Angle Orthodontist, 14, 13–17. [24.9.3—4]

X

Xekalaki, E (1984). The bivariate generalized Waring distribution and its application to accident theory. Journal of the Royal Statistical Society, Series A, 147, 488–498. [24.6—1, 24.6—3]

Xie, P, and Mitsuta, Y (1988). Rainfall estimation from GMS infrared and visible imagery data. (In Japanese.) Annuals of the Disaster Prevention Research Institute, Kyoto University, No. 31B—1, 201–217. [24.4.2—13]

Y

Yamada, H, Kuroda, A, Inaba, T, Yatabe, T, Nagashima, J, Muraki, T, Tanno, M, and Chiba, K (1984). Quality control of radioimmunoassay by bivariate analysis. (In Japanese.) Radioisotopes, 33, 370–375. [24.4.3—4]

Yamamoto, S (1976). Multivariate correlation analysis of suspended-sediment characteristics. Mathematical Geology (Journal of the International Association for Mathematical Geology), 8, 57–74. [24.12.1—2]

Yanagimoto, T (1972). Families of positive dependent random variables. Annals of the Institute of Statistical Mathematics, Tokyo, 24, 559–573. [12.2.1]

Yanagimoto, T, and Okamoto, M (1969). Partial orderings of permutations and monotonicity of a rank correlation statistic. Annals of the Institute of Statistical Mathematics, Tokyo, 21, 489–506. [12.6]

Yang, M C K (1979). Some results on size and utilization distribution of home range. In J K Ord, G P Patil, and C Taillie (Editors), Statistical Distributions in Ecological Work, pp. 429–449. Fairland, Maryland: International Co-operative Publishing House. [19.3.3]

Yang, Y, Yang, L, Gao, Y, and Yang, Y (1984). Prediction of earthquake damage to existing brick buildings in China. In Proceedings of the Eighth World Conference on Earthquake Engineering. Volume 1, pp. 401–408. Englewood Cliffs, New Jersey: Prentice-Hall. [24.12.2—11]

Young, J C, and Minder, C E (1974). Algorithm AS 76: An integral useful in calculating non-central t and bivariate normal probabilities. Applied Statistics, 23, 455–457 (Corrections, 27, 379; 28, 113; 28, 336; 34, 100–101; and 35, 310–312). (Reprinted with revisions in P Griffiths and I D Hill (Editors), Applied Statistics Algorithms, pp. 145–148. Chichester: Ellis Horwood, 1985.) [4.12.4, 4.12.8]

Yuan, P-T (1933). On the logarithmic frequency distribution and the semi-logarithmic correlation surface. Annals of Mathematical Statistics, 4, 30–74. [4.14.2.1]

Yule, G U (1926). Why do we sometimes get nonsense-correlations between time-series?—A study in sampling and the nature of time-series. Journal of the Royal Statistical Society, 89, 1–64 (Proceedings of the Meeting, 64–69). (Reprinted in A Stuart and M G Kendall (Selectors), Statistical Papers of George Udny Yule, pp. 325–388. London: Griffin.) [11.5.2]

Yule, G U, and Kendall, M G (1937). An Introduction to the Theory of Statistics. 11th Edition. London: Griffin. [24.7.2—11, 24,8.2—2, 24.8.2—5, 24.9.2—7, 24.9.2—11, 24.11.2—7, 24.11.2—13]

Z

Zahedi, H (1985). Some new classes of multivariate survival distribution functions. Journal of Statistical Planning and Inference, 11, 171–188. [18.3.5]

Zarkovich, S S (1966). Quality of Statistical Data. Rome: Food and Agriculture Organization of the United Nations. [24.6—21]

Zelditch, M (1959). A Basic Course in Sociological Statistics. New York: Holt. [24.11.2—7]

Zelen, M, and Severo, N C (1960). Graphs for bivariate normal probabilities. Annals of Mathematical Statistics, 31, 619–624. [4.7]

Index

Topics are referred to by the Section in which they appear; Chapters, Tables, Figures, and footnotes are identified by the abbreviations C, T, F, and fn. Citations of a particular author should be traced with the Section (or other) number at the end of the entry in the list of References.

Most of the applications have been grouped in this Index under one or other of the following entries: **Air, earth, water, and climate**; **Biology**; **Medicine**; **Physical and technological sciences**; **Psychology**; **Reliability**; **Sociology, economics, and business**.

Named bivariate distributions have been listed under **Bivariate distribution**, and univariate ones under **Univariate distribution**.

Abbreviations 1.5.1
Absolute moments, bivariate normal distribution 4.3.8
Accidents
Datasets 24.6—1, 24.6—2, 24.6—3
Injury severity 20.3, C21
Ageing, concepts of 2.15.2—12, C18
Applications 18.2.5
Bivariate 18.3
DMRL 18.3.5
HNBUE 18.3.5
IFR 18.3.2
IFRA 18.3.3
NBU 18.3.4
NBUE 18.3.5
Orderings 18.2.6
Reliability classes, univariate 18.2
Definitions of IFR, IFRA, DMRL, HNBUE, HNBUE, NBU, NBUE, HNBUE, $\mathcal{L}$, NBUFR and NBUFRA 18.2.2
Interrelationships 18.2.3
Air, earth, water, and climate
Barometric pressure 24.12.1—8, 24.12.2—3
Cloud seeding 19.7.3, 19.7.5
Datasets 24.12.1—5, 24.12.1—6
Cloud cover, replicated observations, datasets 24.4.2—13
Composition of rocks 4.15, 24.5
Factor analysis 4.14.5
Discontinuities in rocks 19.7.11
Datasets 24.12.2—10
Earthquakes 19.7.12
Datasets 24.4.2—10, 24.4.3—10, 24.12.2—11, 24.12.3—11
Mineral resources sampling
Gold and silver correlation 4.16.2
Magnesium and aluminium in a nickel ore 24.12.3—12
Small cells and large cells 14.7
Trace metal concentration, uranium exploration 5.2.11, 5.9.6, 20.6.3
Pollution
Datasets 24.12.1—1, 24.12.1—3
$\Pr(X > cY)$ being the relevant quantity 17.6.1
Sulphate concentration, Ontario, ARMA model 10.10
Rain
Agriculture-related climate variables 8.12.5.7, 24.12.3—6, 24.12.3—7
Annual and daily rainfall 8.12.5.7
Cloud-seeding 19.7.3, 19.7.5, 24.12.1—5, 24.12.1—6
Correlation, effect of transformations 4.14.5
Datasets 24.12.1—5, 24.12.1—6, 24.12.1—8, 24.12.3—6, 24.12.3—7, 24.12.3—8, 24.12.3—9
Duration and amount 8.12.5.7, 9.12.13, 19.7.4, 24.12.3—8
Emptiness of reservoirs 7.5.9
Intensity and storm surge 9.4.4, 24.12.3—9
Nonnormal distribution fn11.5
Paired observations 19.7.2, 19.7.3
Space-separated observations 7.5.13, 9.12.13
Station-year method 9.17.11
Storage-translation model 5.10.10
Streamflow 8.12.5.6, 19.7.7
River flows and floods
Datasets 24.3—1, 24.3—2, 24.12.3—4
Duration and peak discharge 8.12.5.5
Extreme-values context 5.6.8, 9.17.11
Flood volume and peak discharge 19.7.8
Nonnormal distribution fn11.5
Space-separated observations 5.6.8, 9.12.13, 9.17.11, 19.7.8
Streamflow and rain 8.12.5.6, 19.7.7
Time-separated observations 7.5.13, 10.3.9
Rock discontinuities 19.7.11
Datasets 24.12.2—10
Sea level 9.17.11, 19.7.9
Datasets 24.3—3, 24.12.3—3, 24.12.3—9
Skew distributions, effect on correlation fn11.5
Slope stability 19.8.2
Soil strength 6.5.12
Storms
Position errors 19.7.6
Surge and rainfall 9.4.4, 24.12.3—9
Surge and wave height 8.12.5.4
Surge (sea level) at two points 24.12.3—3
Surge superimposed on tide 19.7.9
Temperature
Datasets 24.12.1—4, 24.12.2—4, 24.12.2—5, 24.12.2—6, 24.12.2—7
Tides 19.7.9. See also **Air, earth, water, and climate**—Sea level
Time-separated observations of meteorological conditions 19.7.2
Waves
Height, repeated observations 24.4.2—11, 24.4.2—12
Height, successive waves 9.12.13
Height and period 8.12.5.1, 9.12.13, 10.8, 19.7.10, 24.12.2—8, 24.12.3—5
Height and storm surge 8.12.5.4
Height and wind speed 8.12.5.3, 24.12.2—9, 24.12.3—2
Period, repeated observations 24.4.2—11, 24.4.2—12
Slope 3.11.3
Wind
Datasets 24.12.2—1, 24.12.2—2, 24.12.2—9, 24.12.3—1, 24.12.3—2
Energy 8.12.5.2
Gust modelling 7.5.13, 7.7.9, 9.17.11
Speed 3.11.3, 8.12.5.2, 8.12.5.3, 19.7.1

Appell symbol, explanation fn13.3
Appell's hypergeometric function, definition fn20.6
Archimedean copula, explanation 5.1.2
ARMA model 10.10
Association. See Dependence, concepts of
Associativity and Frank's distribution 5.4.6
Autoregressive moving average model 10.10

Bayesian statistics 6.5.12, 20.7.6
Beach, Californian, and arduousness of researcher's life 24.12.2—8
Bessel function fn7.7
Beta function, definition fn2.2
Better mean residual life at t_0 (BMRL-t_0) 18.2.7
Biology. See also Dentistry; Medicine
Agriculture-related climate variables 8.12.5.7, 24.12.3—6, 24.12.3—7
Allometry. See **Biology**—Size and shape of living things
Bivariate normal distribution, applications of 19.3
Datasets 24.8
Forestry 19.3.2
Home ranges of animals 19.3.3
Datasets 24.8.1—6
Intercropping 8.12.4
Plant breeding, selection 15.2
Probit analysis 22.2.1
Size and shape of living things 19.3.1
Datasets 24.7
Fisher's iris data 24.7.1—3
Johannsen's bean data 3.11.2, 5.7.10, 7.10.7, 24.7.2—1
Trees, height and diameter 3.11.2, 19.3.2
Datasets 24.7.2—5, 24.7.2—6, 24.7.2—7
Biorthogonal property 14.4, 14.5
Bivariate distribution
AA 3.11.2
Ali-Mikhail-Haq 5.3
Rank correlations 13.3
ACBVE (Block and Basu) 9.5, 9.7, 17.2.2
Barndorff-Nielsen and Blæsild 8.10
BEE (Friday and Patil) 9.8, 17.2.2
Bernoulli 3.8.1
Bessel (Roux and Becker) 9.3.13
Beta T3.1, 3.9.1.2, 4.15, 6.5, 6.6, 10.6, 12.4.2, 15.6.2
Diagonal expansion T14.1
Beta conditionals 8.4
Beta-Stacy 3.5, 8.6.7, 8.6.11, 8.12.1
Block and Rao 8.6.11
Block's geometric compounding scheme 17.3.4
Blumen and Ypelaar 8.8
Bryson and Johnson 10.2.2
Burr 5.8.12, 5.8.13.3, 5.8.13.5, 5.9.5, 5.9.6, 6.8, 20.7.1
Bivariate probit analysis 22.2.1
Cauchy T3.1, 3.9.1.3, 6.2.13, 7.2.4, 15.4.3
Local dependence function fn13.5
Cavanié 10.8
Central limit theorem 4.9
Cherian's gamma 10.3, 13.9, 15.3.1, 15.6.1
Rank correlations 13.9
Chi 7.5.15
Chi-squared. See **Bivariate distribution**—Gamma marginals
Connor and Mosimann 6.5.14
Correlated bivariate inverted beta 20.7.3
Crowder 5.6.5
Cuadras and Augé 5.5, 9.2.9, 13.6. See also **Bivariate distribution**—Marshall and Olkin
Cubic transformation of normals 16.5.4.4
Dirichlet. See **Bivariate distribution**—Beta
Discontinuous marginals 8.11
Downton's exponential 5.2.9, 9.12, T17.1, 17.2.3, 17.3.4, 17.3.5, 17.3.6.2, 17.6.3, 20.3.3, 20.7.1
Dubey 8.7
Durling-Pareto 5.8.13, 20.3.3, 20.7.1
Edgeworth series 2.15.2—4, 19.9.5
EED 20.7.1
Elliptical compound bivariate normal 4.17.4, 7.2.12
Elliptically-contoured 7.2
Random variate generation 16.5.2
Exponential conditionals 8.2
Exponential marginals C9
Arnold T17.1, 17.3.4
Comparison of reliability properties, three exponential-marginals distributions 5.2.9
Cowan 9.14
Cross references 9.15
Downton. See **Bivariate distribution**—Downton's exponential
Gumbel-Hougaard 5.6.9.1
Gumbel's type I 5.10, 16.5.3, 17.2.2
Hawkes T17.1, 17.3.4, 17.3.6.2
Khintchine mixture 10.5.2
Limit of discrete distribution 3.8.2
Marshall and Olkin. See **Bivariate distribution**—Marshall and Olkin
Paulson T17.1, 17.3.4, 17.3.6.2
Sarkar 9.6, 9.7
Exponential power 7.4
Exponential-type, explanation fn4.1
Extreme-value 5.6.9.4, 5.10.9, 9.2.11, 9.17
Bivariate probit analysis 22.3.3
ϕ^2-bounded 4.6, 14.3
Farlie-Gumbel-Morgenstern 5.2
Bivariate probit analysis 22.2.1, 22.4.1
Compounding 20.6
Denominator in common 15.4.7
Iterated 5.2.13, 11.9, 13.2.2, 13.13
Khintchine mixture 10.5.3
Local dependence function fn13.5
Person years at risk estimator 9.15
Rank correlations 13.2.1
F-distribution 6.3. See also **Bivariate distribution**—Inverted beta
Adegboye and Gupta 6.3.9
Association property 12.3.3
Compounding, construction by 20.7.1, 20.7.2
Denominator in common, construction by 15.4.5
Jensen 6.3.9, 7.8, 15.4.6
Krishnaiah 6.3.9, 15.4.6
Liouville-Dirichlet 6.6
Noncentral 6.3.9
Pearson's system 3.9.1.2, 6.1
Regression and scedastic curves T3.1
Fifteen-constant 3.11.2
Form-free, sliding polynomials 8.12.5.7
Fox's H-function 10.6
Frank 3.2.5, 5.4
Rank correlations 13.4
Freund 9.3, 9.5.7, 9.7, 17.2.2, 17.4
Becker and Roux's generalisation 9.3.14
Bhattacharya and Holla's generalisations 9.3.15
Friday and Patil's generalisation 9.8
Hashino and Sugi's generalisation 9.4
Tosch and Holmes' generalisation 9.9
Gamma marginals
Association property 12.3.3
Cherian. See **Bivariate distribution**—Cherian's gamma
Diagonal expansion 14.6.2, T14.1
Dussauchoy and Berland 8.6.10
Farlie-Gumbel-Morgenstern 5.2.9
Interrelationships 7.7.12
Jensen 7.5.10, 7.7
Kibble. See **Bivariate distribution**—Kibble's gamma
Limit of discrete distribution 3.8.3
McKay. See **Bivariate distribution**—McKay's gamma
Schmeiser and Lal's methods 16.5.4.3
Transformed bivariate normal 19.7.3
Geometric compounding scheme 17.3
GGD 20.7.1, 20.7.3
GG1 20.7.1, 20.7.2
Gokhale 10.2.3
Gumbel-Barnett 5.11
Gumbel-Hougaard 5.6

Gunst and Webster 7.7.11
Houthakker effect 19.9.2
Hyperbolic 7.10
Inverse Gaussian 7.9
Univariate distribution used in construction by conditional method 8.10
Inverted beta 6.1, 6.5.15, 6.6, 15.4.7, 15.7.1, 20.7.1, 20.7.2. See also **Bivariate distribution**—F-distribution
Johnson and Tenenbein 15.7.2
Johnson's transformations of bivariate normal 4.14.2, fn4.8
Khintchine mixture 10.5
Kibble's gamma 3.4.2, 3.9.2, 7.5, 7.7.12, 17.3.5, 20.3.3, 20.7.1, 20.7.3
Kimeldorf and Sampson 10.7
Rank correlations 13.7
Konijn 10.9, 13.13
Laplace 7.4.5, 9.12.16
Lawrance and Lewis 9.10
Rank correlations 13.11
Lindgren 10.8
Liouville-Dirichlet 6.6
Logistic 5.6.9.3, 5.8.12, 5.9.5, 6.7, 15.7.1
Logistic normal 4.15, fn4.8
Lognormal 4.14.1, 4.14.5
Bivariate probit analysis 22.2.1, 22.2.3, 22.3.1
Longuet-Higgins 10.8
Malik and Trudel 7.5.10, 7.7.12
Marshall and Olkin 9.2
Bivariate probit analysis 22.4.1
Comparison of reliability properties with other exponential-marginals distributions 5.2.9
Comparison with Sarkar's and other distributions 9.6.1, 9.7
Cuadras and Augé 5.5, 9.2.9, 13.6
Extreme-values context 9.17.6.2
Friday and Patil's generalisation 9.8
Lawrance and Lewis' generalisation 9.10.8
Limit of discrete distribution 3.8.2
Rank correlations 13.6, 13.13
Reliability models T17.1, 17.2.1, 17.2.2, 17.3.4, 17.3.5, 17.3.6.2, 17.6.2
Tosch and Holmes' generalisation 9.9
Variables in common, construction using 15.5
McKay's gamma T3.1, 3.9.1.4, 6.1, 8.6, 8.12.1, 10.6
Meixner 15.3.2
Diagonal expansion T14.1, 14.6.3
Mihram and Hultquist. See **Bivariate distribution**—Beta-Stacy
Nachtsheim and Johnson 7.3
Noncentral chi-squared 7.6
Normal C4
Applications (miscellaneous) C19
Application to driver's injury severity 21.2
Bivariate probit analysis 22.2, 22.3.2, 22.4.1, 22.4.2.1, 22.4.2.4
Characterisations 4.5
Computer programs 4.12.8
Correlation, effect of transformations 4.14.1, 4.14.5, 11.6
Cumulative distribution, computation 4.12
Diagonal expansion T14.1, 14.6.1
Local dependence function 13.13, fn13.5
Moments and absolute moments 4.3.8
Pearson's system 3.9.1.3, 6.1
Random variate generation 16.5.1
Regression and scedastic curves T3.1
Tables 4.12.7
Tests for 4.13
Transformation of marginals fn4.6, 4.13.6, 4.14, 16.5.4.4
Normal conditionals 8.5
Normal marginals
Canonical sequence 14.6.1
Cross references 4.19
Diagonal expansion T14.1, 14.6.1
Distributions constructed for pedagogic purposes 4.18
Normal mixtures 4.17
Diagonal expansion 14.6.1
Normal, truncated 4.16, 17.6.4
Rank correlations 13.12
Normal variance-mean mixture 7.10.6
Normal variance mixture fn7.2
Obenchain 4.15
Pareto 5.8, 5.9, C20
Bivariate probit analysis 22.4.2.1
Denominator in common, construction by 15.4.7
Mardia's type 2 7.5.15
Marshall and Olkin 9.2.12
Pearson's system 3.9.1.2
Rank correlations 13.5
Regression and scedastic curves T3.1
Pareto conditionals 8.3
Pearson's fifteen-constant 1.6
Pearson's system 3.9.1, 3.9.2, 6.1
Evaluation of distribution function 6.4.10, 8.1.2
Pearson's type II 3.9.1.3, 6.4, 7.2.4
Diagonal expansion T14.1, 14.6.3
Pearson's type III T3.1
Phase-type 9.11.7, 9.13
Plackett 5.7
Product gamma 10.2.4
PSE (Proschan and Sullo) 17.2.2
Raftery 9.11
Rank correlations 13.11
Ratnaparkhi 8.6.11
Rayleigh 7.5.15
Rhodes 6.9
SGD 20.7.3
Sherwood and Taylor 13.11
Singularity along $Y = X$ 9.10.8, 10.9, 13.13, 21.6. See also **Bivariate distribution**—Marshall and Olkin
Stable 7.11, 8.10, 15.3.4
Takahasi-Burr. See **Bivariate distribution**—Burr
t-distribution 6.2
Association property 12.3.3
Denominator in common, construction by 15.4.4
Elliptical 7.2.4
Non-central 6.2.14
Pearson's system 3.9.1.3, 6.1
Poly 6.2.14
Random variate generation fn16.8
Regression and scedastic curves T3.1
Tosch and Holmes 9.9
Trivariate reduction. See **Trivariate reduction**
Triangular 15.3.5
Truncation
Bivariate normal 4.16, 13.12
L-shaped 24.12.3—4
Tukey lambda 6.5.14
Ulrich 10.2.4
Uniform within a convex polygon 19.3.3
van Uven (Pearson) 6.1
Wedge-shaped support 8.6
Weibull marginals 9.2.10, 9.2.12, 9.11.8, 9.12.13, 9.12.15, 9.15, 9.16, 19.8.3
Weighted linear combination. See **Weighted linear combination**
Z-distribution 15.6.3
Bivariate failure rate 18.3.2
Exponential conditionals distribution 8.2.7
Bivariate moments, existence fn2.1
Bivariate skewness and kurtosis 4.13.5.4, 4.13.5.8
Bivariate transforms (Laplace and Mellin) fn20.7
BMRL-t_0 (Better mean residual life at t_0) 18.2.7
Box-Cox transformation 2.4.4, 4.13.6
Brownian motion 7.10.6

Canonical sequence 14.6
Central limit theorem, bivariate 2.15.2—21, 4.4.4
Characteristic function 2.2.2.2, 2.15.2—2, fn17.7
Elliptical distributions 7.2.6
Infinite divisibility, condition for fn7.4
Random variate generation fn16.1
Chebyshev's inequality 11.5.4

Choleski decomposition fn4.4, 16.5.1
Cohen's kappa 10.9
Coherent systems, explanation fn18.5
Compatibility of conditional densities 8.1.3
Competing risks 9.2.7, 9.12.12, 9.15, 9.18
Compositions (Totalling 100%)
Bivariate beta distribution, applications of 6.5.12
Datasets 24.5
Logistic normal distribution 4.15
Neutrality 6.5.11, 6.6, 12.4.2, fn15.2
Compounding 3.4, 4.4.2, 5.6.5, 5.8.7, 5.8.13.2, 6.8, 7.10.4, C20
Denominator in common, relation to 15.4.2
Computer algebra 1.6, 3.1.4, 8.12.5.2, 10.6, 13.1
Conditional distribution method 3.5, C8
Random variate generation 16.4.2
Construction of bivariate distributions, methods for C3
Continuity/discontinuity of densities 8.11
Conventions, explanation of 1.5.2
Convergence in distribution, explanation fn11.1
Convex set, explanation fn11.16
Copula 3.2.3, C5
Cowan 9.14.10
Derivation, comparison of methods 1.6
Diagonal expansion T14.1, 14.7
Explanation 5.1.2
Lawrance and Lewis 9.10.6
Schweizer and Wolff's measure 11.9
Transformation 5.1.1
Cornish-Fisher expansion 2.15.2—4, 3.11
Correlation
Blomqvist quadrant measure fn5.2, 12.3.1
Computer program MONCOR 11.7.2
Cramér-Von Mises 11.8
Gideon and Hollister's outlier-resistant statistic 11.8.4
Grade correlation 11.8.3, 11.9, 12.3.1, C13
Particular distributions 5.7.5, 9.10.2, 9.10.3, 9.14.5, 9.17.8, 11.8.3, 13.1, 13.2, 13.3, 13.4, 13.5, 13.6, 13.7, 13.8, 13.9, 13.10, 13.11, 13.12
Kendall's tau 11.8.2, 12.3.1, C13
Data on sociopathy and hysteria 20.4.3
Particular distributions 9.17.8, 11.8.2, 13.1, 13.2, 13.3, 13.4, 13.5, 13.6, 13.7, 13.8, 13.9, 13.10, 13.11, 13.12
Kolmogorov-Smirnov 11.9
Maximal correlation 11.6
Matrix of correlations 11.10
Monotone correlation 11.7
Mean square contingency 11.9
Polychoric correlation. See Tetrachoric correlation
Product-moment 11.5
Appropriateness for skew distributions 9.1, 11.5.2
Chebyshev's inequality 11.5.4
Exponential marginals, lower limit 9.1
Particular distributions 4.3.6, 4.16.1, 5.2.5, 5.2.13, 5.3.5, 5.4.5, 5.6.4, 5.7.5, 5.8.5, 5.8.13.3, 5.10.5, 6.2.4, 6.3.5, 6.4.4, 6.5.4, 7.5.5, 7.7.4, 7.7.11, 7.9.4, 8.2.5, 8.6.5, 8.6.10, 8.12.1, 9.2.5, 9.3.5, 9.5.5, 9.6.5, 9.10.2, 9.10.3, 9.10.4, 9.11.1, 9.12.5, 9.14.5, 9.17.8, 10.3.3, 15.3.3, 20.7.4
Robustness of r 11.5.1
Sensitivity to marginal transformation 4.14.5, 20.4.3
Shortcomings 8.12.3, 11.5.2
Stable distribution, extension to 7.11.5
Trivariate reduction 15.3.3
Truncation, effect of fn4.9
Rank correlation. See **Correlation**—Kendall's tau; **Correlation**—Grade correlation
Regional dependence 11.15
Rényi's conditions 11.4
Schweizer and Wolff 11.9
Spearman's rho 11.8.3, 13.1. See also Correlation—Grade correlation
Spurious 12.4.2, fn15.2
Sup correlation 11.6
Tetrachoric correlation 11.11
Computer program 11.11
Ulam's metric fn11.9
Correlation ratio 11.5.3
Cumulant generating function 2.2.2.3

Decreasing 100α percentile residual life (DPRLα) 18.2.7. See also Ageing, concepts of
Decreasing failure rate (DFR). See Ageing, concepts of
Decreasing mean residual life (DMRL). See Ageing, concepts of
Deceasing variance of residual life (DVRL) 18.2.7. See also Ageing, concepts of
Dentistry. See also Biology; Medicine
Datasets 24.4.2—9, 24.9.1—1, 24.9.2—1, 24.9.2—2, 24.9.2—3, 24.9.3—3, 24.9.3—4
Dependence, concepts of 11.3, C12
Bounds for ρ_g in terms of τ 13.13
Extreme-value distributions, RTI property 9.17.8
Functional dependence 11.3.4
Global dependence function 11.14.1
Hoeffding's lemma 12.2.2
Implicit dependence 11.3.4
Interrelationships 12.3.6, 12.3.7
Negative 12.4.1
Kibble's distribution 7.5.12
Local dependence function 11.14, 13.13
Monotone dependence 11.3.3
Monotone quadrant dependence function 12.5
Monotone regression 5.9.5
Mutual complete dependence 11.3.2, 11.10
Negative 5.10.5, 12.3.2, 12.4
Neutrality. See **Compositions (Totalling 100%)**—Neutrality
Orderings 12.6
Particular distributions 5.3.9, 5.4.9, 9.2.13, 9.17.8
Quasi-independence 11.15.2
Random elements in common model 11.13
Regional dependence function 11.15
Rényi's conditions 11.4
Stringency of conditions 12.2.2
Total dependence 11.3
Total positivity, explanation 2.15.2—60, fn12.7
Dependence function 5.1.2, 9.17.3, 9.17.4
Dependent by total positivity (DTP). See Dependence, concepts of
DFR (Deceasing failure rate). See Ageing, concepts of
Diagonal expansion 3.10, 4.9, 6.4.9, 6.5.10, 7.7.2, 7.7.8, 10.3.8, 11.10, C14
Diagonal square dependent (DSD). See Dependence, concepts of
Differential equations 2.10, 3.9, 4.4.1, 5.3.6
Digression analysis 4.17.7
Dilogarithm fn9.4
Discrete datasets, bivariate 24.6
DMRL (Decreasing mean residual life). See Ageing, concepts of
DPRLα (Decreasing 100α percentile residual life) 18.2.7. See also Ageing, Concepts of
DSD (Diagonal square dependent). See Dependence, concepts of
DTP (Dependent by total positivity). See Dependence, concepts of
DVRL (Decreasing variance of residual life) 18.2.7. See also Ageing, concepts of

Edgeworth expansion 2.15.2—4, 3.11
Ellipse of concentration 7.2.10
Emptiness of reservoirs 7.5.9
Empirical characteristic function, tests based on bivariate 4.13.5.3
Empirical distribution function, tests based on
Bivariate 4.13.4.4
Univariate 4.13.5.2
Encyclopedia of Statistical Sciences 2.15
Exponential dispersion models 8.9
Extremes
Datasets 24.3

Extreme-value distributions, bivariate 9.17

ϕ^2-bounded, explanation 14.3
Factor analysis, transformation of variables 4.14.5
Fokker-Planck equation 3.9.2
Fréchet's bounds 3.1.2, 3.4.1
Illustration of contours (reference) 5.1.3
Test for independent action fn22.2
Functional dependence 11.3.4
Functions, explanation of various notations 1.5.4

Gamma function, definition fn2.2
GDSD (Generalised diagonal square dependent). See Dependence, concepts of
Generation of random variables 2.15.2—10
Generalised diagonal square dependent (GDSD). See Dependence, concepts of
Global cross ratio, constant 5.7.10
Gram-Charlier expansion 3.11
Graphical checks of bivariate normality 4.13.3

Haar distributions 4.13.3.6
Harmonically new better than used in expectation (HNBUE). See Ageing, concepts of
Hazard gradient 18.3.2
Hazard rate 2.15.2—12, 18.3.2
HNBUE (Harmonically new better than used in expectation). See Ageing, concepts of
Hoeffding's lemma 12.2.2
Human biology. See Medicine
Hypergeometric function, series representations of $_2F_1$ and $_3F_2$ fn13.3

IFR (Increasing failure rate). See Ageing, concepts of
IFRA (Increasing failure rate on average). See Ageing, concepts of
Illustrations of contours (references) 5.1.3
Implicit dependence 11.3.4
IMSL routines for random variate generation 16.3.6, 16.5.1
Increasing failure rate (IFR). See Ageing, concepts of
Increasing failure rate on average (IFRA). See Ageing, concepts of
Independence of maxima fn9.5
Indicator function, explanation 1.5.4
Inelegant methods of constructing distributions 3.3
Infinite divisibility 2.15.2—14, 7.5.6, 17.3.5
Explanation fn7.4
Information retrieval systems 23.4
Integer partition, explanation fn7.5
Integral transforms 2.15.2—15, fn20.7

Johnson's translation method 2.4.3, 3.2.2, fn4.8, 4.14.2,

Kendall's tau. See Correlation coefficient—Kendall's tau
Kernel method, bivariate density estimation 4.13.3.5
Khintchine's theorem 10.2.2
Krönecker delta, explanation 1.5.4
Kurtosis and skewness. See Skewness and kurtosis

Lack of memory property 2.6.1, 5.10.9, 9.2.12, 9.7, 17.2.2
Laguerre polynomials fn7.3, fn7.6
L, computation of 4.12
$\mathcal{L}$-distributions. See Ageing, concepts of
Left tail decreasing (LTD). See Dependence, concepts of
Limits of discrete distributions 3.8
Lognormal distribution. See Transformation of bivariate normal distribution; Bivariate distribution— Lognormal
L_p-distance, explanation fn11.10
LTD (Left tail decreasing). See Dependence, concepts of

MACSYMA. See Computer algebra
Markov process. See Physical and technological sciences— Markov chain
Mean square contingency fn14.4
Medicine. See also Biology; Dentistry
Bivariate life tables 5.9.6
Blood pressure 4.13.2, 19.4.2
Competing risks 9.2.7, 9.15, 9.18
Coronary occlusion, twins 19.5.3
Datasets 24.9
Disease concordance. See **Medicine**—Multifactorial model of disease transmission
Epidemiology 5.2.11, 19.4.1
Person years at risk estimator 9.15
Familial association of disease 5.9.6. See also **Medicine**—Multifactorial model of disease transmission
Fibres, airborne 19.4.6, 24.9.1—5, 24.9.2—4
Health and retirement 22.4.2.4
Impact tolerance 9.18.3
Injury severity, road accidents 20.3, C21
Leukemia survival times 8.2.8
Liability to disease. See **Medicine**—Multifactorial model of disease transmission
Lung disease
Bivariate probit analysis, coalminers 22.4.1
Fibres, airborne 19.4.6, 24.9.1—5, 24.9.2—4
Mongolism, twins 19.5.3
Multifactorial model of disease transmission 19.5, 20.4. See also **Medicine**—Familial association of disease
Multiple sclerosis, assessment 20.2
Neural tube defects, twins 19.5.3
Ordinal-scale assessments, multiple sclerosis 20.2
Pharmacokinetics 19.4.5
Probit analysis 22.4.1, 22.4.2.4
Psychiatry (inheritance of mental illness) 19.5.1, 19.5.3, 20.4
Pyloric stenosis, twins 19.5.3
Quantal response models 5.2.11, 5.6.8, 5.7.10, 5.9.6, 9.2.8
Radiology 23.3
Datasets 24.4.2—3
Remission times 9.5.9
Replicated observations 19.4.2
Schizophrenia, twins 19.5.3
Tactile stimulation, sensitivity of rats 5.6.8, 5.9.6
Threshold for disease transmission. See **Medicine**—Multifactorial model of disease transmission
Tolerance distribution. See **Medicine**—Quantal response models
Tumour incidence in mice or rats 5.2.11, 5.6.8, 9.2.8, 9.17.11
Twins. See **Medicine**—Multifactorial model of disease transmission
Veterinary science 19.5.5
X-ray images, perception of 23.3
Microcomputers and random variate generation 16.3.7
Mills' ratio, bivariate 4.12.2
Mixing 3.4
Mixtures, bivariate normal 4.17
Moment generating function 2.2.2.1, 2.15.2—9
Moments, bivariate 4.13.5.4, 4.13.5.8
Moments, conditional, subjectively specifying 8.12.3
Moment sequence 14.6
Monotone dependence 11.3.3
Monotone quadrant dependence function 12.5
Monotone regression dependence property 5.4.9, 10.4.4, fn12.8
Multinomial logit model 22.3.3
Mutual complete dependence 11.3.2, 11.10

NAG routines for random variate generation 16.3.6, 16.5.1
NBU (New better than used). See Ageing, concepts of
NBU-t_0 18.2.7
NBUE (New better than used in expectation). See Ageing, concepts of
NBUFR (New better than used in failure rate). See Ageing, concepts of
NBUFRA (New better than used in failure rate average). See Ageing, concepts of
Negative correlation, distributions oriented towards 5.10.5, 6.5.4, 8.2.5, 9.10.5. See also Dependence, concepts of—Negative
Neutrality. See Compositions (Totalling 100%)—Neutrality

New better than used (NBU). See Ageing, concepts of
NBU-t_0 18.2.7
New better than used in expectation (NBUE). See Ageing, concepts of
New better than used in failure rate (NBUFR). See Ageing, concepts of
New better than used in failure rate average (NBUFRA). See Ageing, concepts of
Normality, tests for 2.15.2—5
Numerical inversion methods, cumulative distribution fn16.3

Observer-agreement studies 20.2
Ordinal contingency tables
Compatibility with perfect rank ordering 11.12
Farlie-Gumbel-Morgenstern distribution 5.2.15
Tetrachoric correlation 11.11
Orthogonal polynomials T14.1
Laguerre fn7.3, fn7.6
Orthonormal functions, explanation fn14.1
Outlier-prone distribution 2.15.2—36, 4.13.2
Outliers, testing for 4.13.2, 5.8.10, 5.10.9

PDD (Positive definite dependence). See Dependence, concepts of
PDE (Positively dependent by expansion). See Dependence, concepts of
PDM (Positively dependent by mixture). See Dependence, concepts of
Peeling bivariate data 4.13.3.2
Perfect rank ordering, compatibility of ordinal contingency table with 11.12
Physical and technological sciences. See also Reliability
Astronomy 19.8.6
Datasets 24.13.1—7
Calibration 19.8.1
Datasets 24.13
Destructive testing 9.16, 9.18.3, 17.6.4, 19.8.3, 19.8.4, 24.2—4
Datasets 24.2—4, 24.13.1—1
Discharge of electricity between electrodes 8.6.10
Electrical signal processing 4.10, 7.2.1, 7.5.13, 14.8, 23.1.1
Electric counter system 7.5.13
Failure time 2.15.2—47, 2.15.2—48, 2.15.3—2, 5.2.9, 5.2.11, 5.6.8, 8.12.1, 9.2.8, 9.3.10, 9.11.8, 24.2. See also **Reliability; Ageing, concepts of**
Datasets 24.2
Fibres, airborne. See **Medicine**—Fibres, airborne
Instrument errors 19.8.1
Maintenance/production cycle 15.2
Markov chain 3.9.2, 5.2.14, 7.5.13, 9.12.13, 9.13, 14.8
Nuclear power industry, reliability models 9.2.8
Optics 19.8.8
Particle-size statistics 2.15.3—5
Phase problem of X-ray crystallography 2.15.3—6
Queueing systems 7.5.13, 9.12.13
Radar 19.8.7
Selection procedures 4.16.2
Signal detection, spacecraft command receiver 9.17.11
Dataset 24.3—4
Statistical physics 7.10.7
Statistical spectroscopy 3.11.4
Stress-strength models 17.6, 19.8.2
Structural reliability 19.8.2
Target shooting 19.8.9
Datasets 24.14.1—2, 24.14.1—4
Thermodynamics 19.8.5
Trace metal concentration, uranium exploration 5.2.11, 5.9.6, 20.6
Vehicle speeds 9.18.3
Wold's point process 9.12.13
Wood grinding forces 19.8.10
Wool, lightfastness, datasets 24.4.2—10
Planar random movement 4.10
Plotting positions, univariate distributions 4.13.3.1, 4.13.6
Polynomial biorthogonal property 14.5
Positive definite dependence (PDD). See Dependence, concepts of
Positive definite matrix, explanation fn7.1
Positive likelihood ratio dependence. See Dependence, concepts of
Positively dependent by expansion (PDE). See Dependence, concepts of
Positively dependent by mixture (PDM). See Dependence, concepts of
Positive quadrant dependence (PQD). See Dependence, concepts of
Positive quadrant dependence in expectation (PQDE). See Dependence, concepts of
Positive regression dependence (PRD). See Dependence, concepts of
Positive semi-definite matrix, explanation fn7.9
PQD (Positive quadrant dependence). See Dependence, concepts of
PQDE (Positive quadrant dependence in expectation). See Dependence, concepts of
PRD (Positive regression dependence). See Dependence, concepts of
Probability plotting 2.15.2—40
Probit analysis, bivariate C22
Biology 22.2.1
Driver's gap acceptance 22.2.3
Econometrics 22.2.2, 22.3.3, 22.4.2
Medicine 22.4.1, 22.4.2.4
Product-moment correlation. See Correlation—Product-moment
Projection pursuit 4.13.3.4
Psychology. See also Medicine—Psychiatry (inheritance of mental illness)
Bivariate normal distribution, applications of 19.6
Guessing in replicated paired comparison tests 20.5.1, fn20.5
Luce's choice axiom 2.15.2—25
Mental abilities 19.6.1
Datasets 24.4.2—15, 24.9.1—7, 24.10
Multiple choice tests, models for performance 23.5
Observer agreement
Datasets 24.4
Cohen's kappa 10.9
Psychotics and normals, scores on tests 24.9.1—7
Random utility models 2.15.2—45
Rats, tactile sensitivity, datasets 5.6.8, 5.9.6, 24.8.1—7
Replicated paired comparison tests 19.6.2, 20.5
Selection procedures 4.16.2, 19.6
Signal detection theory C23
Thurstone's theory of comparative judgment 2.15.3—10

Quantising of a two-dimensional random variable 4.10

Radii and angles, bivariate normality assessment 4.13.3.3
Random elements in common 11.13, 15.3.6
Random variate generation C16
Bivariate 16.4, 16.5
Cherian's gamma 10.3.8
Downton's exponential 9.12.11
Friday and Patil's model (also Marshall and Olkin's and Freund's) 9.8
Univariate 16.2, 16.3
RCSI (Right corner set increasing). See Dependence, concepts of
REDUCE. See Computer algebra
Relevation. See Reliability—Relevation
Reliability C9, C18. See also Ageing, concepts of; Physical and Technological Sciences—Failure time
Classes, univariate 18.2
Common environment 17.5
Common-mode/common-cause 9.2.8
Cumulative damage model 17.3.6
Datasets 24.2, 24.14.2
Dependence, concepts of 12.3.1, 12.3.3
Encyclopedia entries noted in univariate Chapter 2.15.2—46, 2.15.2—47, 2.15.2—48, 2.15.2—53, 2.15.3—4, 2.15.3—8, 2.15.3—9

Farlie-Gumbel-Morgenstern distribution 5.2.9, 5.2.11
Models leading to bivariate distributions C17
Shock models 2.15.2—53, 6.5.12, 9.2.6, 9.2.13, 9.3.8, 9.3.14, 9.3.15, 9.11.5, 9.12.9, 17.2, 17.3
Relevation 2.15.2—46, fn5.1
Stress-strength models 17.6, 19.8.2
Replicated observations
Datasets 24.4
Paired comparison tests 20.5
Reproductive exponential models 8.10
Reuse of published data 24.1.2
Reverse regular of order 2 (RR_2). See Dependence, concepts of—Negative
Right corner set increasing (RCSI). See Dependence, concepts of
Right tail increasing (RTI). See Dependence, concepts of
Robustness studies
Choice of distributions fn1.3
Elliptical distributions 7.2.11
ROC and similar curves 8.12.4, 23.1, 23.2
RR_2 (Reverse regular of order 2). See Dependence, concepts of—Negative
RTI (Right tail increasing). See Dependence, concepts of

Saran's triple hypergeometric function 20.7.3
Selection by a risk function 3.3, 5.2.11, 19.4.1
Semiparametric analysis 1.6, 2.15.2—52, 21.7.1
sgn, explanation 1.5.4
Sharpening of a scatterplot 4.13.3.2, 4.17.6
SIFR (Stochastically increasing failure rate) 18.2.7
Signal detection theory C23
ABX experiment 23.6
Applications 23.2, 23.3, 23.4, 23.5
Explanation 23.1
ROC curve 23.1, 23.2
Signal probability and slope of ROC 23.2
Skew bivariate distributions, appropriateness of r 9.1, 11.5.2
Skewness and kurtosis
Bivariate 4.13.5.4, 4.13.5.8, 4.13.5.9
Sample joint distribution 8.12.2
Univariate 2.3.2
Slepian's inequality 4.9, 7.2.10
SNBU (Stochastically new better than used) 18.2.7
Sociology, economics, and business
Actuarial statistics 2.15.3—1
Blackjack gambling 19.9.3
Business risk 8.12.3
Car ownership or use 9.11.8, 24.11.2—10
Commodities trading 19.9.2
Datasets 24.11
Econometrics
Bivariate probit analysis 22.2.2, 22.3.3, 22.4.2
Brand shares (consumer purchase behaviour) 6.5.12, 24.6—7
Competing risks 9.18.3
Duration variables 18.1
Income distribution models, including Pareto's 2.15.3—3
Injury severity, relevance of econometric methods 21.7.2
Investment price changes 7.11.7
Partial observability 22.4.2.5
Probit analysis, bivariate 22.2.2, 22.3.3, 22.4.2
Retirement and health 22.4.2.4
Sample selection equations 5.2.9, 5.9.5, 19.9.5, 22.4.2.3
Stable distribution (univariate or bivariate) 2.14.4, 7.11.7
Stochastic terms, thick tails 6.2.11
Supply and demand equations 19.9.5
Household size and income 19.9.1
Interview, survey, proxy responses, dataset 24.4.2—16
Ordinal scale, tetrachoric correlation 11.11
Policy analysis 19.9.4
Risk theory in insurance 2.15.3—7
Votes, proportions of 6.5.12
Whist data, Pearson's 6.2.11
Speeds of vehicles, competing risks analysis 9.18.3
Stable distributions, explanation 7.11.1
Stochastically increasing. See Dependence, concepts of
Stochastically increasing failure rate (SIFR) 18.2.7
Stochastically new better than used (SNBU) 18.2.7
Structure function of a system fn18.5, fn18.10
Support, explanation 5.1.2
Supremum, explanation fn11.6
Switched-source model 5.2.14
Switching regressions 4.17.7
Symbols, explanation of 1.5.3

Total positivity (TP). See Dependence, concepts of—Total positivity, explanation
Transformation, marginal
Construction of bivariate distribution 3.2
Effect on correlation 4.14.5
Transformation of bivariate normal distribution 3.2.2, 4.13.6, 4.14. See also Bivariate distribution—Lognormal; Johnson's translation method
Biology, size and shape of living things 19.3.1
Gamma marginals 19.7.3
Weibull marginals 19.8.3
Transformation of variates, effect on H and h 3.1.3
Trivariate reduction 3.6, C15
Additive 15.3
Beta distribution 6.5.7, 15.6.2
Cherian's bivariate gamma distribution 10.3.6, 15.6.1
Compounding and denominator in common 15.4.2
Denominator in common 15.4, 19.5.4
Drivers' injury severities 21.2
F-distribution 6.3.8, 6.3.9, 15.4.5, 15.4.6
Marshall and Olkin's distribution 15.5
Meixner classes 15.3.2
Multiplicative 10.2
Normal distribution 4.4.3
Pareto distribution 15.4.7
Phase-type distributions 9.13
Raftery's first special case 9.11.2
Random variate generation 16.5.4.2
Rank correlations 13.9
Symmetric stable distributions 7.11.4, 15.3.4
t-distribution 6.2.7, 15.4.4
Triangular distribution 15.3.5
Z-distribution 15.6.3
Two-dimensional sliding polynomials 8.12.5.7
Two-stage decision model 23.2, 23.3

Underwear, emperor's, possibility of holes in fn1.2
Uniform representation. See Copula
Unimodality of multivariate distributions 5.1.3
Univariate distribution C2. For various named distributions, see 2.15.1; For various theoretical issues, see 2.15.2
Beta 2.5
Random variate generation 16.3.3
Beta (shifted) 2.10
Burr 2.11, 2.15.1—2, 2.15.1—7
Cauchy 2.7.4, 2.14.4, 2.15.2—36, 2.15.2—42
Chi 2.6.3, 2.15.1—40
Chi-squared 2.6.3, 2.15.1—3a, 2.15.1—26, 2.15.1—55, 2.15.2—14
Circular, use by Nachtsheim and Johnson 7.3
Contaminated 2.4.7
Erlang 2.6.2
Exponential 2.6.1, 2.10, 2.15.1—4, 2.15.2—42, 2.15.2—48
Compound 2.6.7
Goodness-of-fit tests 2.15.2—62, fn18.1
Random variate generation 16.3.2
Exponential power 7.4.5
Random variate generation 16.3.5
Extreme-value 2.8, 2.15.1—6, 2.15.1—10, 2.15.2—42
F-distribution 2.10, 2.12.2, 2.15.1—27, 2.15.1—55, 2.15.2—14
Gamma 2.6.2, 2.10, 2.15.1—8, 2.15.1—39, 2.15.2—38, 2.15.2—48
Random variate generation 16.3.2
Gamma, reciprocal of 2.10
g and h family 2.4.5, 2.15.1—9

Generalised error 2.7.3
Generalised hyperbolic secant 2.15
Generalised Tukey lambda 2.7.1
Hyperbolic 2.14.3, 2.15.1—11
Hyperbolic secant 2.14.2
Inverse Gaussian 2.14.1, 2.15.1—14, 2.15.2—16
Random variate generation 16.3.5
Inverted beta 2.5.3, 2.10, 2.15.1—15, 20.7.2
Generalised 8.6.11
Johnson's system 2.4.3, 2.15.1—7, 2.15.1—16, 8.12.2
Laplace distribution as basis 2.7.2
Logistic distribution as basis 2.7.1
Kritsky-Menkel 2.6.5
Laplace 2.7.2, 2.15.1—17a, 2.15.1—20b, 2.15.1—30, 2.15.1—41, 2.15.2—38, 2.15.2—42
Location and scale 2.3.1
Logistic 2.7.1, 2.15.1—20a, 2.15.1—30, 2.15.1—34, 2.15.2—42
Lognormal 2.4.2, 2.15.1—21, 2.15.2—14, 2.15.2—42
Meixner hypergeometric 2.14.2
Noncentral (including doubly) beta, chi-squared, F, and t 2.13, 2.15.1—3b, 2.15.1—26, 2.15.1—27, 2.15.1—28, 2.15.1—59
Normal 2.4.1, 2.4.7, 2.10, 2.14.4, 2.15.1—29, 2.15.2—3, 2.15.2—4, 2.15.2—5, 2.15.2—14, 2.15.2—16, 2.15.2—18, 2.15.2—21, 2.15.2—30, 2.15.2—36, 2.15.2—38, 2.15.2—42, 2.15.2—45, 2.15.3—10
Compound 2.4.8, 2.15.1—24
Computation of d.f. 4.12.1
Random variate generation 16.3.1
Skew-normal 2.15.1—42
Slash 2.4.7, 2.15.1—43
Pareto 2.9, 2.10, 2.15.1—31, 2.15.2—42
Pearson system 2.10, 2.15.1—7, 2.15.1—33, 2.15.2—1, 2.15.2—19
Phase-type 2.15.1—35
Random variate generation 16.3.5
Rayleigh 2.6.3, 2.15.1—40, 19.7.1
Skewness and kurtosis 2.3.2
Slash 2.4.7, 2.15.1—43
Stable 2.7.4, 2.14.4, 2.15.1—44, 2.15.2—6
Random variate generation 16.3.5
Stacy 2.6.5
t-distribution 2.4.7, 2.10, 2.12.1, 2.15.1—28, 2.15.1—37, 2.15.1—47, 2.15.1—55, 2.15.2—14
Random variate generation 16.3.4
Tail behaviour 2.3.3
Uniform 2.5.2, 2.15.1—30, 2.15.1—49, 2.15.2—42
Wald 2.14.1
Weibull 2.6.4, 2.8.2, 2.15.1—51, 2.15.1—52, 2.15.2—38, 2.15.2—42, 2.15.2—48
Z-distribution 2.5.3
Univariate shape 2.3

Variables in common. See Trivariate reduction

Weighted linear combination 3.7, 9.11.3, 10.4
Random variate generation 16.5.4.2
Rank correlations 13.8